BIOLOGY OF GRASSHOPPERS

BIOLOGY OF GRASSHOPPERS

Edited by

R. F. CHAPMAN

Division of Neurobiology
University of Arizona
Tucson, Arizona

A. JOERN

School of Biological Sciences
University of Nebraska
Lincoln, Nebraska

A WILEY-INTERSCIENCE PUBLICATION

JOHN WILEY & SONS

New York/Chichester/Brisbane/Toronto/Singapore

Library of Congress Cataloging in Publication Data:

Biology of grasshoppers/edited by R.F. Chapman, A. Joern.
p. cm.

"A Wiley-Interscience publication."
Includes bibliographical references.
ISBN 0-471-60901-3
1. Locusts. I. Chapman, R. F. (Reginald Frederick) II. Joern,
A. (Anthony)

QL508.A2B55 1990
595.7'26—dc20 89-22666
CIP

Printed in the United States of America

10 9 8 7 6 5 4 3 2 1

Contributors

H. C. Bennet-Clark, Department of Zoology, University of Oxford, South Parks Road, Oxford OX1 3PS

E. A. Bernays, Department of Entomology, University of Arizona, Tucson, Arizona 85721

W. M. Blaney, Department of Biology, Birkbeck College, Malet Street, London WC1E 7HX

R. F. Chapman, Arizona Research Laboratories, Division of Neurobiology, University of Arizona, Tucson, Arizona 85721

M. A. Chappell, Department of Biology, University of California, Riverside, California 92521

J. F. Dale, Department of Zoology, University of Toronto, Toronto, Ontario M5S 1A1

J. M. Dearn, University of Canberra, P.O. Box 1, Belconnen, ACT 2616, Australia

R. A. Farrow, Division of Entomology, CSIRO, P.O. Box 1700, Canberra City, ACT 2601, Australia

S. B. Gaines, School of Biological Sciences, University of Nebraska, Lincoln, Nebraska 68588

G. J. Goldsworthy, Department of Biology, Birkbeck College, Malet Street, London WC1E 7HX

M. D. Greenfield, Department of Biology, University of California, Los Angeles, California 90024

A. Joern, School of Biological Sciences, University of Nebraska, 348 Manter Hall, Lincoln, Nebraska 68588

W. Loher, Department of Entomological Sciences, University of California, Berkeley, California 94720

M. R. McGuire, Plant Polymer Research, Northern Regional Research Center, USDA/ARS, 1815 N. University Street, Peoria, Illinois

T. E. Shelly, Hawaiian Evolutionary Biology Program, University of Hawaii, 3050 Maile Way, Honolulu, Hawaii 96822

M. S. J. Simmonds, Jodrell Laboratory, Royal Botanic Gardens, Kew, Surrey TW9 3AB

S. J. Simpson, Department of Zoology, University of Oxford, South Parks Road, Oxford OX1 3PS

D. A. Streett, Rangeland Insect Laboratory, USDA/ARS, Montana State University, Bozeman, Montana 59717

S. S. Tobe, Department of Zoology, University of Toronto, Toronto, Ontario M5S 1A1

D. W. Whitman, Department of Biological Sciences, Illinois State University, Normal, Illinois 61761

Preface

Migratory locust plagues have legendary status and the economic impact of grasshoppers and locusts has often been the driving force for research. While we fully recognize the importance of these aspects of the biology of grasshoppers and locusts (Acridoidea), we also believe that these insects are ideal organisms for studying problems of basic biology. It was in this belief that *Biology of Grasshoppers* was conceived and compiled. We have based the construction of the book on two underlying premises: (1) that acridoids are representative and interesting insects with key features that facilitate research on topical questions in a number of disciplines, and (2) that a multidimensional view of acridoids is necessary to appreciate fully their observable biological responses.

Our organization and choice of topics is focused on how individuals function. We unabashedly promote an organismal approach to the study of grasshoppers, a view that readily takes in studies from the molecular to the evolutionary scales. Our aim is to present a diversity of topics in a context that emphasizes the functioning of the whole organism. We have not attempted to be comprehensive and, due to a variety of constraints, have not been able to include chapters on a number of important topics. We think, however, that the selection of topics presented is a fair sample. Some of the missing topics have been adequately dealt with elsewhere in the literature, although not, of course, within the context of the grasshopper as an organism.

Uvarov summarized much of the known information on grasshoppers and locusts in his monumental works *Locusts and Grasshoppers* (1928) and *Grasshoppers and Locusts* (1966, 1977).* Their impact can be readily appreciated by the frequency with which they are cited in the chapters of this book, as well as elsewhere. They are still key sources of reference, but they were completed before the newer technologies of radar, electron microscopy, and molecular biology were generally applied to entomology and before new paradigms, especially in population biology, had been developed. As a result there have been great advances in some fields since Uvarov's books were completed. We are, in part, attempting to bring new ideas and results into focus, building on Uvarov's work as the base of reference.

Despite the ready availability of many grasshopper species worldwide, much of what we know is based on a limited number of species, especially the readily cultured locusts. Compare the entries for *Schistocerca gregaria* and *Locusta migratoria* in the species index with those for all the other species! And this is true not only of laboratory studies, but of population and other ecological work, too. We hope that the topics included in *Biology of Grasshoppers* will spark new interest and foster new studies. We believe that using grasshoppers within disciplines will provide key insights into general biological principles. It is also likely that important insights gained from such research can be exploited to develop more effective control programs. It is our hope that our emphasis on an organismal approach with the juxtaposition of chapters from scientists with very different backgrounds will provide links among disciplines that are too often ignored.

We are grateful to Mary Conway and Margery Carazzone at Wiley who have invariably been helpful and patient during the planning and compilation of the book. And we are indebted to our colleagues who, in writing their chapters for the book, have given freely of their time and expertise.

R. F. CHAPMAN
A. JOERN

Tucson, Arizona
Lincoln, Nebraska
May 1989

* Uvarov, B. 1928. *Locusts and Grasshoppers: A Handbook for Their Study and Control.* Imperial Bureau of Entomology, London.

Uvarov, B. P. 1966, 1977. *Grasshoppers and Locusts: A Handbook of Acridology*, Vols. 1 and 2. Cambridge Univ. Press, Cambridge, and Centre for Overseas Pest Research, London.

Contents

BIOLOGY OF GRASSHOPPERS

1

The Chemoreceptors

W. M. BLANEY and

M. S. J. SIMMONDS

1.1 INTRODUCTION

The structure of chemoreceptor sensilla in insects has been reviewed by Lewis (1970), Slifer (1970), Sturckow (1970), Altner and Prillinger (1980), and Zacharuk (1980, 1985), and many different types are recognized on the basis of differing structures and function. Early workers, limited to light microscopy, relied heavily on the form of the cuticular parts to classify sensilla. Therefore functions were ascribed to the different types by inference from the structure, from the position of the sensilla on the insect's body, and from the behavior of the insect when specific groups of sensilla were incapacitated. Subsequent studies, using scanning (SEM) and transmission (TEM) electron microscopy, have yielded much more detailed information, and in most cases electrophysiology of individual sensillum types has corroborated the earlier, inferred function. In the grasshoppers, studies of fine structure and electrophysiology of chemoreceptors are scarce, but work on other insects allows the earlier conclusions to be accepted with reasonable confidence.

In this chapter we review the types and distribution of chemoreceptors found in grasshoppers, relating the fine structure to the mode of functioning wherever possible. Finally we attempt to show how the chemoreceptors are used in some of the important aspects of the insect's life.

1.2 THE CHEMORECEPTORS: STRUCTURE

In order for insects to detect chemicals in the environment the dendrites of their chemosensory neurones must be exposed to the environment. The potential problem of desiccation is circumvented by enclosure of the cells within cuticular structures that are penetrated by one or more minute pores, through which access to the environment is achieved. This cuticular structure, together with the sensory and other cells associated with it, constitutes the sensillum.

1.2.1 Sensillum Types

Of the nine or so basic types of sensilla currently recognized (Altner, 1977; Zacharuk, 1980, 1985), only four have been described as chemoreceptors in grasshoppers (Fig. 1.1). Sensilla trichodea are slender and hairlike with an insertion on the cuticle that is movable by means of a membranous socket; sensilla chaetica are similar but more robust and, although generally tactile, may sometimes be chemoreceptors; sensilla basiconica are also hairlike but very short and may be reduced to pegs or cones; sensilla coeloconica are pegs set within depressions or pits in the cuticle.

A simple set of criteria may be used to relate the cuticular fine structure to its likely function (Altner, 1977). Namely, if there are no pores it cannot be a chemoreceptor (may be tactile or may monitor temperature or humidity), if there is a single terminal pore its major sensitivity will be to taste (sensitivity

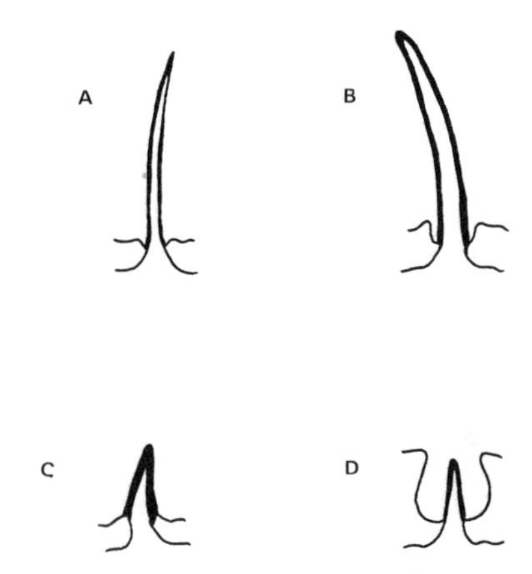

Fig. 1.1. The main cuticular structures of different types of chemoreceptors found in grasshoppers. (A) trichodea; (B) chaetica; (C) basiconica; (D) coeloconica.

to some vapors is not precluded), and if there are many pores it will mediate olfaction.

1.2.2 The Subepidermal Components

Despite these structural and functional differences, the complement of cells associated with the sensilla is very constant in arrangement and function. All the cells are epidermal in origin and when a new sensillum is being formed, say, prior to the molt of a nymph from one stadium to the next, a single epidermal "mother cell" undergoes division (Hansen, 1978). A second division yields four cells, three of which become concentrically arranged around the innermost cell. This inner cell develops into a bipolar neurone and sends an axon to the central nervous system (CNS) and a dendrite into the peripheral, cuticular part of the sensillum. A ciliary region separates the broad proximal part of the dendrite from the narrow, distal region in which the only subcellular organelles are longitudinal microtubules; all other structures such as Golgi apparatus and mitochondria are restricted to regions proximal to the ciliary bodies. The physical form and chemical sensitivity of the distal dendrite varies with the type of sensillum (see below). If the sensillum has more than one neurone, as is commonly the case with chemoreceptors, then further cell divisions will have occurred before differentiation took place. We have no knowledge of the factors governing cell division, differentiation, or the acquisition of particular chemical sensitivities. The neurones are primary sense cells and their axons reach the CNS without any intervening synapses (see, e.g., Blaney and Chapman, 1969; Steinbrecht, 1969).

The remaining cells, the so-called accessory cells, form a series of sheaths and sinuses around the neurones. Their numbers, and consequently their functions, vary but the general pattern is as follows. The innermost sheath cell, or thecogen cell, wraps round the neurone cell bodies and ensheathes the proximal region of each of the dendrites separately. It encloses the ciliary sinus (Fig. 1.2) and during the formation of the sensillum it secretes the dendritic sheath, or scolopale, which encloses all or part of the distal dendrites. The middle sheath cell, often called the trichogen cell, secretes the cuticular hairlike part of the sensillum, and the outer sheath cell, the tormagen cell, secretes the cuticular socket. All three sheath cells contribute to the boundary of a second, much larger sinus, the sensillar sinus, or receptor lymph cavity as it was formerly called (Fig. 1.2). Some authors refer to the ciliary sinus as receptor lymph 1 and the sensillar sinus as receptor lymph 2 (e.g., Wunderer and Smola, 1982). Where they border that sinus, the cell membranes of all three sheath cells have numerous microvilli and appear to be metabolically active, probably secreting fluid into the sinus. A band of specialized intercellular junctions in the membranes of all these cells holds the cells together and prevents any intercellular continuity between the sinuses and the hemolymph

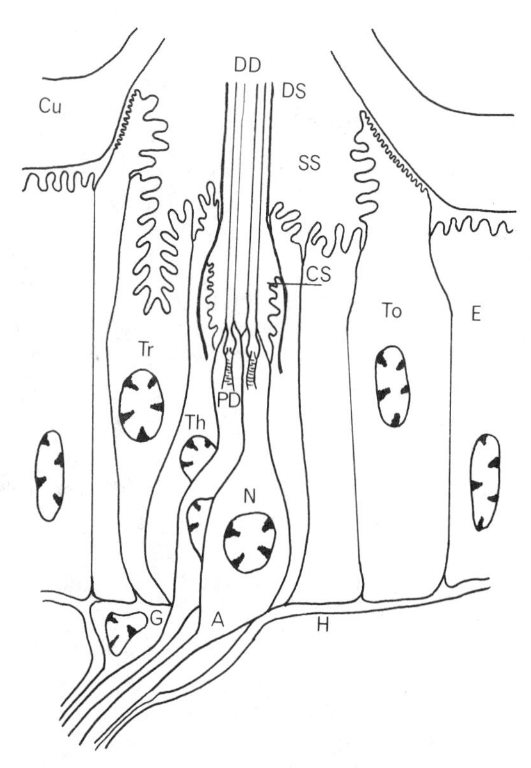

Fig. 1.2. The subepidermal components of a sensillum. N, neurone; A, axon; PD, proximal dendrite; DD, distal dendrite; Th, thecogen cell; CS, ciliary sinus; DS, dendritic sheath; Tr, trichogen cell; To, tormagen cell; SS, sensillar sinus; E, epidermal cell; Cu, cuticle; G, glial cell; H, hemolymph.

beneath. The axons, which run through the hemolymph, are ensheathed by glial cells.

The formative role of the enveloping cells during the development of a new sensillum has been studied, both at molts of nymphs of *Locusta migratoria* between stadia (Blaney et al., 1971) and in the embryo before emergence of the first instar (Altner and Ameismeier, 1986; Ameismeier, 1987). During development of the embryo a very thin first embryonic cuticle is formed; then, after apolysis, a thicker second embryonic cuticle develops. Apolysis occurs again and the first larval cuticle is secreted. The sensory and sheath cells are formed in the epidermis between the two apolyses and therefore the cuticular parts of the sensilla are developed only in the larval cuticle and not in the embryonic cuticle. However, there is evidence of some early or rudimentary development of the sensilla associated with the second embryonic cuticle. In an olfactory sensillum on the antenna of *Locusta migratoria* (Section 1.2.3) the cells of the developing sensillum can be indentified in the epidermis just after apolysis of the second embryonic cuticle (Ameismeier, 1987). At this stage these cells do not extend beyond the distal surface of the surrounding epidermal cells but a dendritic sheath does extend from the center of the putative sensillum to the apolysed cuticle where it penetrates the inner, lamellate layer and fuses with the outer, more electron-dense layer. Subsequently, the trichogen cell extends in a fingerlike process, secrete the cuticle of the hair, then withdraw to a basal position. Meanwhile the tormagen cell secretes the socket in a similar way and distal dendrites grow out into the dendritic sheath, which, as is typical of olfactory receptors, emerges into the ecdysial space through the base of the hair. At the next apolysis the sheath breaks off at this point and then, or earlier, the dendrites grow into the lumen of the hair. A similar development has been described (Altner and Ameismeier, 1986) for a contact chemoreceptor (Section 1.2.4) on the same antennae. Here the dendritic sheath extends from the terminal pore at the tip of the hair, across the ecdysial space to the second embryonic cuticle, where it is associated with an irregularly shaped modular structure in the outer layer of that cuticle. It is suggested by these authors that the modules may be rudimentary sensilla and that the second embryonic cuticle may therefore be considered as an ancestral larval cuticle, evolutionary pressure (Sehnal, 1985) having acted to reduce the number of molts.

Most studies on the development of insect sensilla, like the two just described, have been concerned with development from undifferentiated epidermal cells. In the grasshoppers, and other hemimetabolous insects, new sensilla are produced in this way at the molts between stadia but the majority of sensilla are carried over from one stadium to the next. The existing cells continue with their specialized functions and the major development is the production of new cuticular parts. This must be done, however, by cells that have already differentiated in previous stadia and, at the time of the approaching molt, are no longer secreting cuticle. This situation has been studied in an investigation of the contact chemoreceptors (Section 1.2.4) at the tips of the maxillary palps of fifth instars of *Locusta migratoria* as they molt to adults (Blaney et al., 1971).

Before apolysis, the trichogen cell enlarges and extends through the receptor lymph cavity to the base of the hair. As the cuticle separates from the epidermis, the tormagen cell remains attached to the inside of the socket, so maintaining the integrity of the receptor lymph cavity and keeping the scolopale, with the dendrites inside, in its normal bathing medium, rather than being exposed to the ecdysial fluid. Later, the tormagen cell pulls away and the trichogen cell extends as a fingerlike process around the existing scolopale to secrete the new hair on its outer surface, partly within the old hair. At this stage the old hair remains attached to the new one by the scolopale extending between the tips of each, and the dendrites look normal. When the new hair is fully developed, about one day before ecdysis, the dendrites in the distal part of the scolopale degenerate. However, at the base of the new hair normal dendrites are present and they grow out to the tip. Thus when ecdysis occurs the old scolopale breaks off at the tip of the new hair and is pulled away with the exuviae. The new sensillum is apparently fully functional.

In addition to forming the cuticular part of the sensillum, the sheath cells are believed to have a major physiological role in the functioning of the sensillum. The basal region of each of these cells is adjacent to or near the hemolymph, and there is evidence that the cells sequester materials from the hemolymph and secrete them into the sensillar sinus (Phillips and Vande Berg, 1976a,b). These materials include protein, mucopolysaccharides and, additionally, selected ions, especially potassium. The distribution of these ions maintains a transepithelium potential such that the sinus is 60–80 mV positive with respect to the hemolymph (Thurm, 1972). This contributes to the effective functioning of the sensillum (Kaissling and Thorson, 1980). In some olfactory sensilla the distal dendrites run through the fluid of the sensillar sinus; in others, and in taste receptors, they are separated from the fluid by the dendritic sheath, which may be permeable to some ions (Broyles et al., 1976) that provide an environment suitable for the processes of transduction and initiation of action potentials in the neurones (Zacharuk, 1985).

The above description forms the basis for an understanding of how chemoreceptors work. Details of the structures and functions associated with stimulus acquisition and transduction vary according to the modality of the stimulus and are best considered separately.

1.2.3 Olfactory Sensilla

The fine structure of three types of olfactory sensilla that occur in *Locusta migratoria* has been described in some detail: a thick-walled coeloconic sensillum on the antennae (Steinbrecht, 1969), a thin-walled trichoid sensillum on the maxillary palps (Blaney, 1977), and two similar thin-walled basiconic sensilla on the antennae (Ameismeier, 1987).

The coeloconic sensilla have a widely scattered distribution on the antennae (Thomas, 1966) and each consists of a short basiconic peg situated in the bottom of a more or less spherical pit in the cuticle (Fig. 1.1). The dendrites of one to three neurones project into the lumen of the peg. The structure of the peg is of the type described by Altner (1977) as a "double walled sensillum with spoke canals" (Fig. 1.3). The surface indentations take the form of longitudinal grooves that are readily visible in surface view with the SEM (Fig. 1.4). For this reason Zacharuk (1985) refers to this type of sensillum as "multiporous grooved." The pores, occurring along the inner surface of each groove, lead to the "spoke canals" of Altner. These canals connect the dendritic chamber to the exterior of the sensillum. Longitudinal, fluid-filled channels running in the thickness of the wall between adjacent spoke canals, from the sensillar sinus beneath the cuticle up to near the tip of the sensillum, give the sensillum its double-walled appearance. The dendritic sheath fuses with the cuticle of the peg and the dendrites within the peg are bathed in a fluid that contains an electron-dense material (Steinbrecht, 1969). This dense material fills the spoke canals and appears to occur in the grooves on the outer surface of the peg as well. Because of its position, it is assumed to have a role in the trapping of odorous molecules and their transfer inward to encounter the chemosensory membranes of the dendrites. Details of this process are not known. The dendrites lie, unbranched, within the inner lumen of the peg and have no evident structural modifications that could be implicated with the transduction process. As is commonly the case, the parts of the dendrites distal to the ciliary region contain no organelles other than longitudinal microtubules, although in other sensilla vesicular inclusions have been reported in this region (Zacharuk, 1985). In grasshoppers these sensilla have been shown to be involved in the detection of plant odors (see below).

On the flexible dome at the tip of each maxillary palp in *Locusta migratoria* there are, in nymphs of the fifth stadium, about 370 sensilla. Most are about 25 μm long and are contact chemoreceptors (Haskell and Schoonhoven, 1969; Blaney et al., 1971) but about 20 of them are less than 20 μm long and have an enlongate conical shape and a distinctively large socket. Le Berre et al. (1967) believed them to be mechanoreceptors but ultrastructural studies

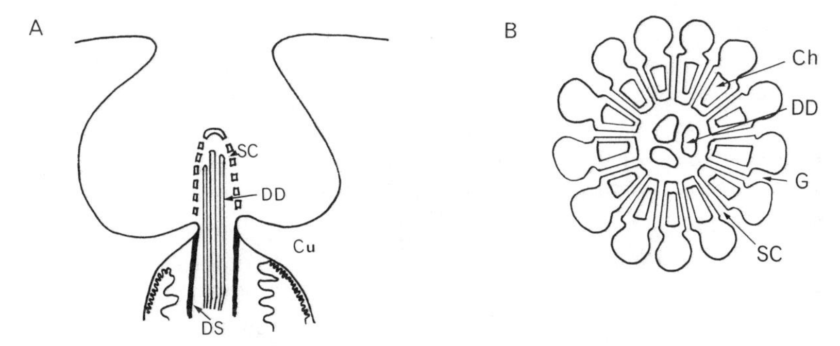

Fig. 1.3. Diagram of coeloconic olfactory sensillum in longitudinal section (A) and the peg only in transverse section (B). Cu, cuticle; DD, distal dendrite; DS, dendritic sheath; SC, spoke canal; Ch, fluid-filled channel; G, groove.

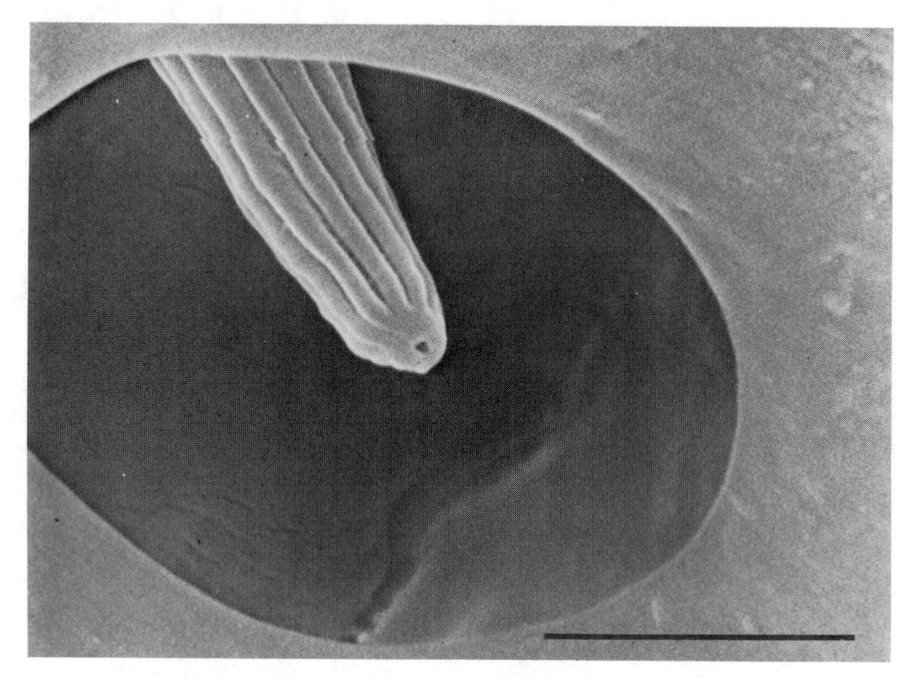

Fig. 1.4. Scanning electron micrograph of coeloconic olfactory sensillum on antenna of fifth-stadium nymph of *Locusta migratoria* (note longitudinal grooves and molting pore at tip). Scale line 2 μm.

(Blaney, 1977) have shown that they are olfactory sensilla. Most have 15 bipolar neurones and the usual subcuticular sheathing cells are present. There is a small ciliary sinus, associated with a complex labyrinth of sinuses within the inner sheath cell. The sensillar sinus is large and extends uninterrupted into the lumen of the peg where it bathes the dendrites. The peg is of the type described by Altner (1977) as "single walled" and, with a wall thickness of about 0.3 μm, it is categorized as thin walled. A significant feature of the wall is the presence of very numerous, evenly distributed pores, about 50 nm in diameter. This feature has led Zacharuk (1985) to propose the category "multiporous pitted" for sensilla such as this. Each pore is slightly bulbous within the thickness of the cuticle, and from this region dense pore tubules extend into the lumen of the peg (Fig. 1.5). These tubules, which are cuticular in origin, make contact with the dendrites and form a continuous link from the atmosphere to the dendrite membranes. Perhaps, therefore, they circumscribe the receptor sites with which stimulus molecules may interact (Blaney, 1977).

As is typical of this type of sensillum, the dendritic sheath ends at the base of the peg and the dendrites emerge through openings in the sheath and continue into the lumen of the peg (Fig. 1.5). Branching of the dendrites within the peg is also a common, but by no means obligatory, feature of such

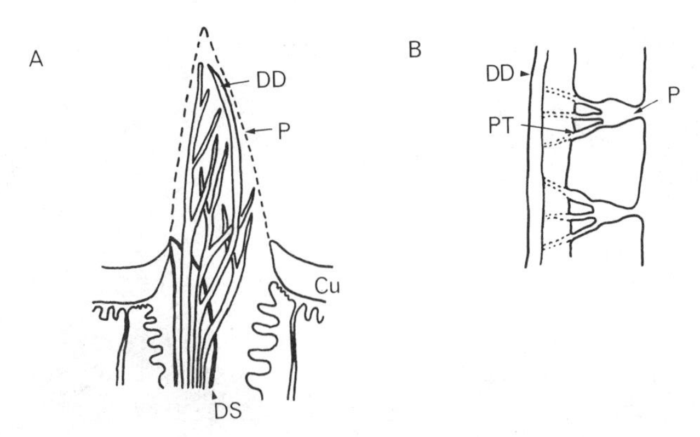

Fig. 1.5. Diagram of thin-walled olfactory sensillum in longitudinal section (A) and part of the peg enlarged (B). DD, distal dendrite; P, pore; DS, dendritic sheath; PT, pore tubule; Cu, cuticle.

sensilla. When branching occurs, the dendrites get smaller at successive branches and have fewer microtubules until, in the distal region of the peg, each branch contains only one microtubule. In the present case, however, the distal region of the peg contains some large-diameter dendrites with as many microtubules (20–50) as in the unbranched, proximal regions. In addition there are many small branches having only one microtubule, and a wide range of sizes in between. The diameter of a dendrite influences the efficiency with which stimuli of different intensities are received, and it is suggested that the mode of branching seen in this sensillum may enable it to respond efficiently to a wide range of stimuli differing widely in intensity (Blaney, 1977). These sensilla may be expected to have a role in the sampling of food odors, by virtue of their location, but unfortunately no electrophysiological evidence is available to confirm this.

A further two types of thin-walled, single-walled olfactory sensilla occur on the antennae of *Locusta migratoria* (Ameismeier, 1987). The two types are similar to each other and to the palp sensilla just described, but they do have some interesting differences. Ameismeier distinguishes them as type A and type B. Both are of similar length, about 16 μm, but type A has a greater diameter (5 μm) than type B (3 μm) and a much greater density and total number of surface pores (Fig. 1.6). Type A has 20–30 neurones with dendrites subdivided to give up to 120 branches, whereas type B has only three neurones with unbranched dendrites. A major difference is to be found in the sheathing cells: both have one thecogen cell; type A has four trichogen cells whereas type B has two; type A has two tormagen cells whereas type B has only one. On the basis of comparison with sensilla of hymenopterans, Ameismeier (1987) concludes that the occurrence of supernumerary enveloping cells is an advanced, rather than primitive characteristic. Ameismeier's study includes an investigation of the embryonic development of the sensilla, but it is not clear what advantages derive from the multicellular organization. In some cases bifurcate

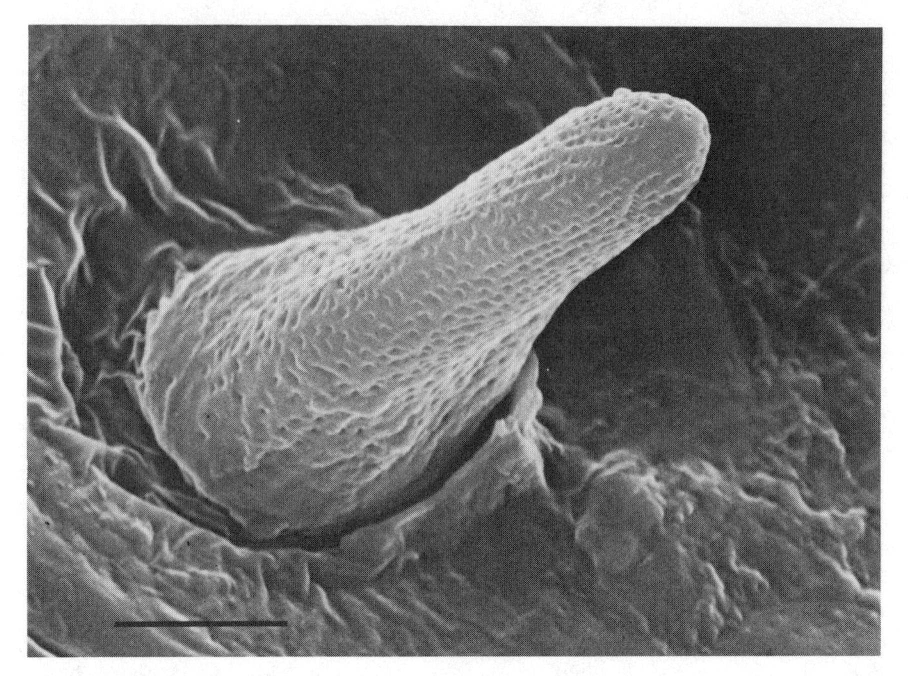

Fig. 1.6. Scanning electron micrograph of thin-walled olfactory sensillum on antenna of fifth-stadium nymph of *Locusta migratoria*. Scale line 2 μm.

versions of the type A sensilla were found, which could result from failure of the intercellular junctions to hold all four trichogen cells together while the peg is being secreted.

1.2.4 Gustatory Sensilla

Sensilla serving the sense of taste are variously referred to as gustatory sensilla or contact chemoreceptors. The latter term recognizes the fact that the sensillum comes into contact with the stimulus in solid or liquid form. That contact is made via a single, terminal pore, hence the further term uniporous sensilla (Zacharuk, 1985).

Most contact chemoreceptors of grasshoppers are trichoid or basiconic sensilla and occur in large numbers on many parts of the body, especially the antennae, mouthparts, and legs (Thomas, 1966; Chapman, 1982). Some account is given of the fine structure of basiconic sensilla on the tarsi of *Schistocerca gregaria* by Kendall (1970). However, the most complete studies are of trichoid sensilla on the maxillary palp tips of both *Schistocerca gregaria* (Blaney and Chapman, 1969) and *Locusta migratoria* (Blaney et al., 1971), and on the antennae of *Locusta migratoria* (Wunderer and Smola, 1982), and of truncated basiconic sensilla on the clypeo-labrum of *Locusta migratoria* (Cook, 1972,

1979). All these sensilla have the same basic arrangement of sensory and sheathing cells as already described (Section 1.2.2).

The palp tip sensilla are similar in the locust species *Schistocerca gregaria* and *Locusta migratoria*. The peg is about 25 μm long and is movably mounted in a flexible socket. The dendrites of the sensory neurones run inside the dendrite sheath, up the center of the peg, to end just below the tip (Fig. 1.7). At the tip, which is about 1 μm in diameter, the peg abruptly tapers to form a crest about 0.8 μm high, 1 μm long, and 0.5 μm wide (Blaney et al., 1971). The peg opens by a longitudinal slit along the top of the crest, through which the dendrites may be exposed to environmental stimuli (Fig. 1.8). Similar contact chemoreceptors with terminal pores occur on the antennae (Fig. 1.9).

The number of neurones in these sensilla varies but commonly there are six to nine. Studies of the fine structure (Blaney et al., 1971) revealed that, in some sensilla, a dendrite ends at the base of the peg and has a terminal tubular body of the type indentified in mechanoreceptors (Thurm, 1965). Electrophysiological studies of these sensilla in *Locusta migratoria* (Blaney, 1974) revealed a mechanoreceptor response in about 50% of them.

The final assessment of potential food material is made by chemoreceptors in the buccal cavity, including a large array on the posterior surface of the clypeo-labrum. These occur in discrete groups (Fig. 1.10) and have been

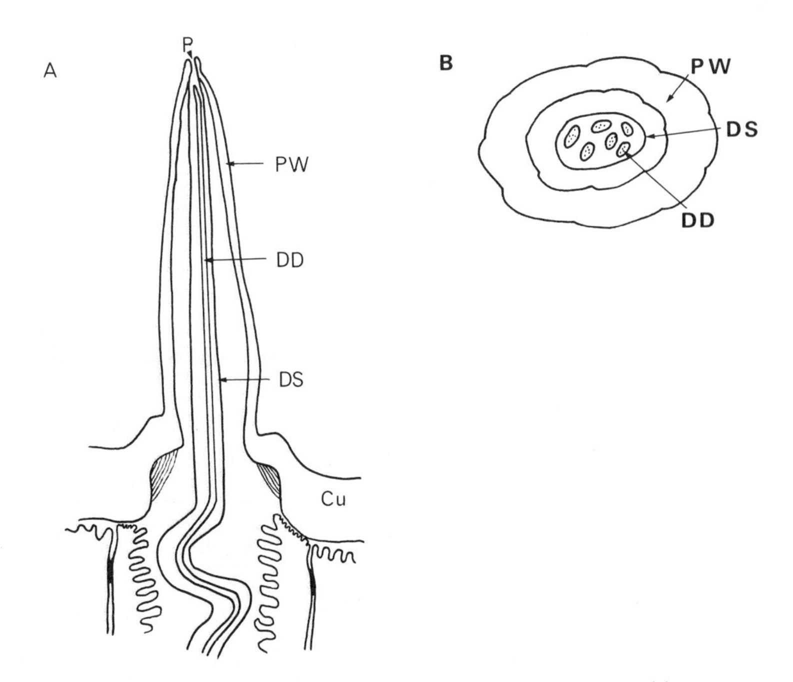

Fig. 1.7. Diagram of trichoid gustatory sensillum in longitudinal section (A) and the peg in transverse section (B). P, single terminal pore; PW, peg wall; DD, distal dendrite, DS, dendritic sheath; Cu, cuticle.

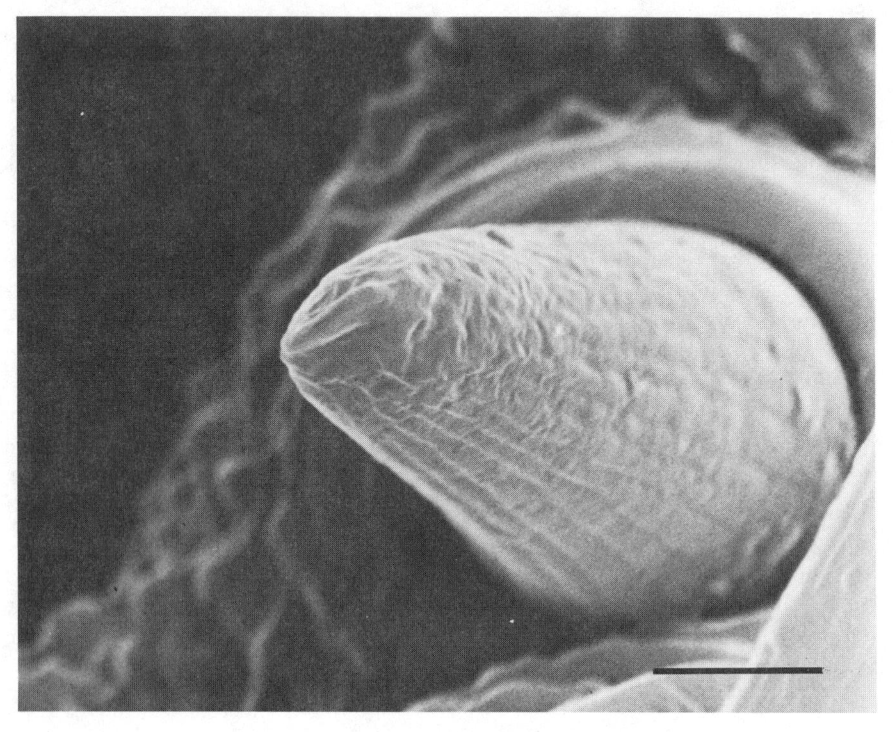

Fig. 1.8. Scanning electron micrograph of trichoid gustatory sensillum on the maxillary palp tip of fifth-stadium nymph of *Locusta migratoria*. Scale line 2 μm.

described in *Schistocerca gregaria* (Thomas, 1966), *Xenocheila zarudnyi* (Chapman, 1966), and *Locusta migratoria* (Cook, 1977). The groups identified as A1, A2, A3, and A10 (Thomas, 1966) are made up of a number of short basiconic pegs. In an electrophysiological study Haskell and Schoonhoven (1969) showed that the A1, A2, and A3 sensilla in *Schistocerca gregaria* and *Locusta migratoria* are contact chemoreceptors. Cook (1972) studied the ultrastructure of the A1 sensilla, the group nearest to the mouth aperture in *Locusta migratoria*. The squat, conical peg is 4.5 μm high and is sunk in a pit of similar depth (Fig. 1.11). The pore at the tip of the peg is very small and was not seen in TEM, but SEM photographs suggest its presence and electrophysiological recording with stimulating capillaries of 1 μm diameter confirm this (W. M. Blaney, unpublished). There are five neurones and their distal dendrites terminate just beneath the cuticular peg (Fig. 1.11). Haskell and Schoonhoven (1969) reported mechanoreceptor responses from these sensilla but no specializations in structure were seen in the distal dendrites by Cook (1972).

A type of contact chemoreceptor occurring among the wind-sensitive hairs on the head of *Locusta migratoria* is very similar to the palp tip sensilla and has been described by Wunderer and Smola (1982). They vary from 30 to 70 μm

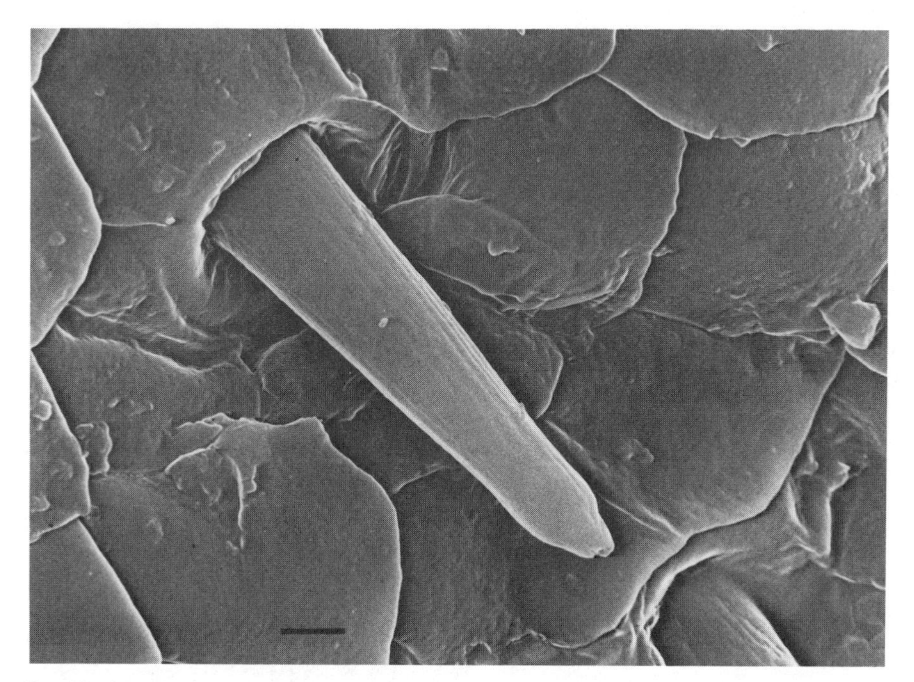

Fig. 1.9. Scanning electron micrograph of trichoid gustatory sensillum on the antenna of fifth-stadium nymph of *Locusta migratoria*. Scale line 2 μm.

in length, about half the length of the wind-sensitive hairs, which also differ in being curved. The chemoreceptors have five dendrites, one of which has the typical tubular body of a mechanoreceptor. There are five accessory cells; one thecogen, two trichogens, one tormagen, and a fifth cell that is an undifferentiated epidermal cell surrounding the complex. Wunderer and Smola speculate that the possession of more than the normal three sheath cells is an "ancestral characteristic" associated with phylogenetically older groups of insects such as the Orthoptera. The position of these sensilla, on the upper surface of the head, is an unusual one for contact chemoreceptors and it may be that they are involved in olfaction rather than taste.

In a further study, Cook (1979) found that the ultrastructure of sensilla in the A2 and A3 groups was very similar to that of the A1 sensilla, but some differences were found in sensilla of the A10 group. In the latter, the apical pore was readily visible with the SEM and took the form of a short transverse slit, situated to one side of the tip of the conical cuticular peg and not symmetrically at the apex as in the A1, A2, and A3 sensilla. Further, whereas the other A-type sensilla had five neurones, the A10 sensilla had only two. This is unusual, but not unknown, for contact chemoreceptors. Apart from this, the fine structure suggests that they are chemoreceptors, although electrophysiological confirmation is lacking.

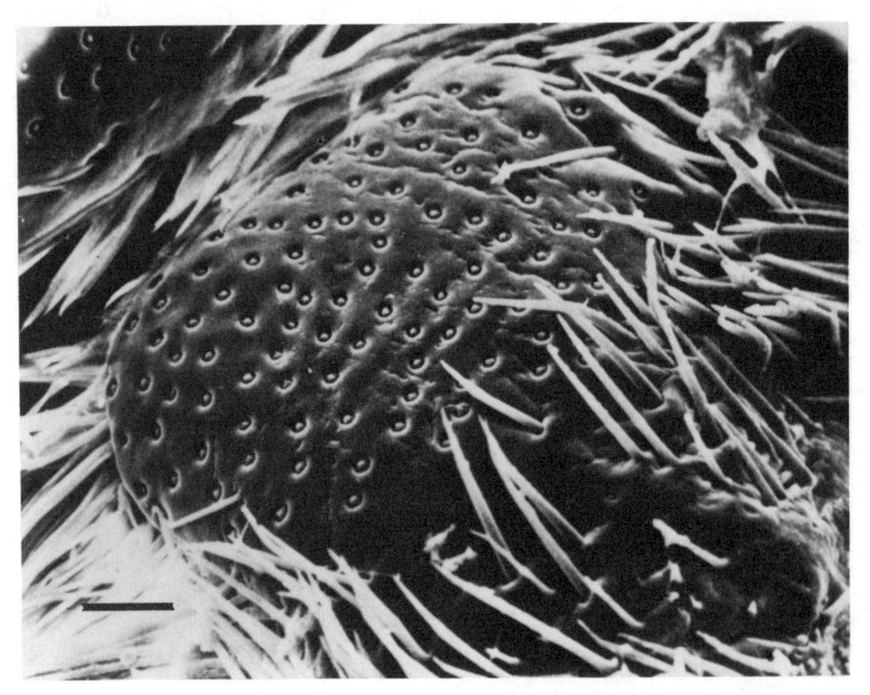

Fig. 1.10. Scanning electron micrograph of group of A1 basiconic gustatory sensilla on posterior face of clypeo-labrum of fifth-stadium nymph of *Locusta migratoria*. Scale line 50 μm (Courtesy A. G. Cook.)

1.2.5 Thermoreceptive and Hygroreceptive Sensilla

The fine structure of the coeloconic sensilla on the antenna of *Locusta migratoria* was investigated by Steinbrecht (1969) and has been described in Section 1.2.3. The structure is typical of olfactory receptors, and this role was confirmed by the electrophysiological investigations of Boeckh (1967) and Kafka (1971) described in Section 1.5.1. However, in a further electrophysiological study, Waldow (1970) reported that some of these sensilla showed no response to olfactory stimulation but instead had two antagonistic hygroreceptor neurones, one responding to moist air and one to dry air, and a temperature receptor neurone that responded to a decrease in temperature. This particular combination of cells has been found in other insects and has been termed a triad (Loftus, 1976). With the light microscope, no distinction could be made between these and the coeloconic olfactory sensilla, and consequently they are included in this chapter.

This anomaly of apparently similar sensilla having totally different functions was investigated in an elegant study by Altner et al. (1981). Electrophysiological recordings were made from individual sensilla; then the tip of a cactus spine

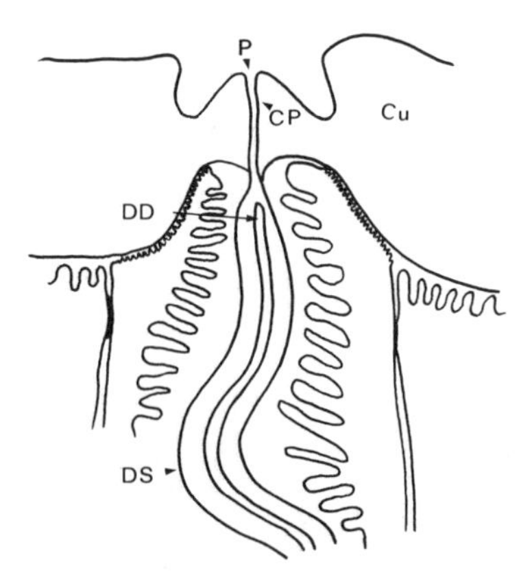

Fig. 1.11. Diagram of basiconic gustatory sensillum in longitudinal section. P, pore, CP, cuticular peg; DD, distal dendrite; DS, dendritic sheath; Cu, cuticle.

was inserted into the cuticle near the sensillum. The sensillum could then be identified with certainty and examined with the SEM. Further, because the cactus spine does not damage ultramicrotome knives, the sensillum could be sectioned for TEM study and its fine structure elucidated.

This TEM analysis revealed that the coeloconic sensilla fell into two main morphological types: "multiporous grooved" sensilla of the type already described (Section 1.2.3) having wall pores and two concentric cuticular walls (wp-dw sensilla, i.e., wall pore–double walled), and sensilla lacking wall pores (np sensilla, i.e., no pore). The correlated electrophysiology carried out by Altner et al. (1981) showed, as would be expected, that olfactory receptors only occurred in wp-dw types and never in np sensilla. Cold-sensitive receptors occurred in both types but hygroreceptors were observed only in np sensilla.

The np sensillum is similar in size to the wp-dw sensillum and has a peg about 5 μm long. Externally, apart from the absence of pores and grooves, the np sensillum differs in having a slightly dilated tip. Internally, the differences are more profound (Fig. 1.12). Only two of the three dendrites enter the peg and they fit tightly inside it, leaving no space for the sensillar sinus found in the wp-dw type. Similarly, there is no lymph space within the dendrite sheath. The membranes of the dendrite outer segments are attached to the sheath and the dendrites themselves are densely filled with microtubules. The third dendrite ends at the base of the peg, and all three dendrites are unbranched. The inner surface of the peg wall consists of two layers that look different in TEM and are presumably structurally different. The socket region at the base of the peg is penetrated by numerous clefts containing electron-dense material, which extends into the sensillar sinus (Fig. 1.12). At highest resolution, the electron-dense material is seen to consist of curled, threadlike structures set in a slightly more lucent matrix.

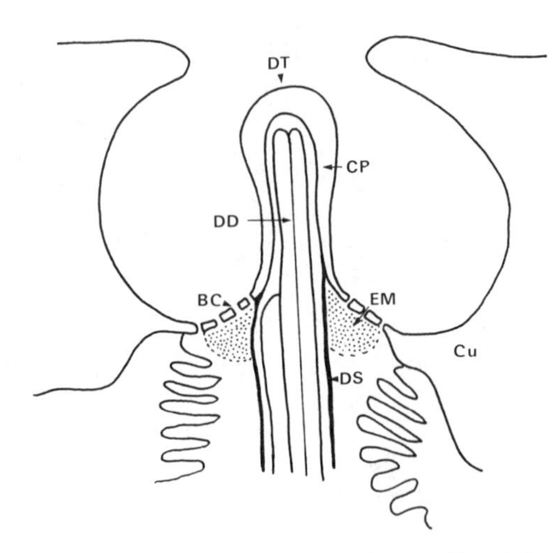

Fig. 1.12. Diagram of coeloconic thermo/hygroreceptor sensillum in longitudinal section. CP, cuticular peg; DT, dilated tip; DD, distal dendrite; BC, basal cleft; EM, electron-dense material; DS, dendritic sheath; Cu, cuticle.

1.3 DISTRIBUTION AND NUMBERS OF CHEMORECEPTORS

In common with many insects, probably most, grasshoppers have chemoreceptors on many parts of the body. The appendages, however, being manipulative or sensory, are particularly well endowed, as are the head and the ovipositor. On those regions bearing numerous chemoreceptors, the sensilla are seldom evenly spread but tend to be grouped together in "sensory areas," often with different types of sensilla in different groups. The disposition and density of chemoreceptors not only reflect the functioning of different parts of the body but relate to the life style of the insect and to its evolutionary antecedents.

1.3.1 The Distribution of Chemoreceptors

The principal aggregations of chemoreceptors occur on the antennae, the mouthparts, the tarsi, and the ovipositor. Studies of these aggregations have been carried out over the past quarter of a century, during which the available techniques have changed markedly. In consequence, the information available does not always reflect the complexity of the real situation.

The only study of the distribution of sense organs on the abdomen is that of Thomas (1965), using the light microscope to investigate adult females of *Schistocerca gregaria*. It is likely that the sensilla described by her as types F, I, and K are chemoreceptors (Chapman, 1982) and there are about 80, 300, and 200, respectively, of these types on the dorsal and ventral valves of each side of the ovipositor. It is probable that some of the ovipositor sensilla are humidity receptors for moisture content of the soil can affect choice of oviposition site.

Chemoreceptors on the legs of grasshoppers occur mainly on the tarsi; their distribution in adult *Schistocerea gregaria* has been studied by Kendall (1970). Basiconic sensilla with six neurones and "canal sensilla" with one neurone are taken to be chemoreceptors. In total they account for about 1000 chemoreceptor neurones on each of the prothoracic and mesothoracic legs and only about 500 on the metathoracic. There is little difference between males and females. A similar situation occurs in *Tettigonia viridissima* (Henning, 1974).

Contact chemoreceptors occur on the antennae but the majority of sensilla are olfactory receptors. Changes in distribution and numbers of antennal sensilla during growth of *Locusta migratoria* have been studied by Chapman and Greenwood (1986). The most distal annuli bear their full complement of sensilla at hatching but new annuli are added at each molt and additional sensilla are added to these annuli at successive molts. Sensillum differentiation proceeds proximally along the antenna and trichoid contact chemoreceptors are developed before olfactory sensilla. Most olfactory sensilla occur in sensory fields on the posterior and anterior faces of each annulus. The distribution of the two types of olfactory sensilla, basiconic and coeloconic, is consistent with separate but interacting chaetogens governing their differentiation.

Most studies on the distribution of chemoreceptors in grasshoppers have focused on, the mouthparts (Chapman, 1966; Thomas, 1966; Blaney and Chapman, 1969; Le Berre and Louveaux, 1969; Viscuso, 1971, 1974; Louveaux, 1972, 1973; Chapman and Thomas, 1978). The most detailed studies have been made of *Schistocera gregaria*, *Locusta migratoria*, *Zonocerus variegatus*, and *Xenocheila zarudnyi* but a number of other species have been studied less exhaustively (see Chapman, 1982, for details). The general distribution of sensilla is very similar across all the species. All of the mouthparts bear more or less densely scattered trichoid contact chemoreceptors but in addition there are discrete groups of different types of sensilla. On the epipharyngeal face of the labrum there are four paired groups of conical contact chemoreceptors, the A1, A2, A3, and A10 groups (Fig. 1.13). Two similar groups occur on the hypopharynx, close to the mouth. The tips of the maxillary and labial palps bear large numbers of trichoid contact chemoreceptors (Fig. 1.8), among which a few trichoid olfactory receptors are to be found (Blaney, 1977).

1.3.2 The Significance of Chemoreceptor Numbers

The total number of chemoreceptor neurones on the antennae is large: each antenna of an adult *Melanoplus differentialis* has about 4000 sensilla (Slifer et al., 1959), giving nearly 100,000 neurones. The corresponding figures for males of *Schistocerca gregaria* are 5000 sensilla and 106,000 neurones (Chapman, 1982), whereas with *Melanoplus bivittatus* there are only 1800 sensilla and about 40,000 neurones (Riegert, 1960). In general, the increase in numbers of sensilla during nymphal development is proportional to the increase in length of the antenna. Males of *Melanoplus differentialis*, *Truxalis nasuta*, and *Schistocerca gregaria*

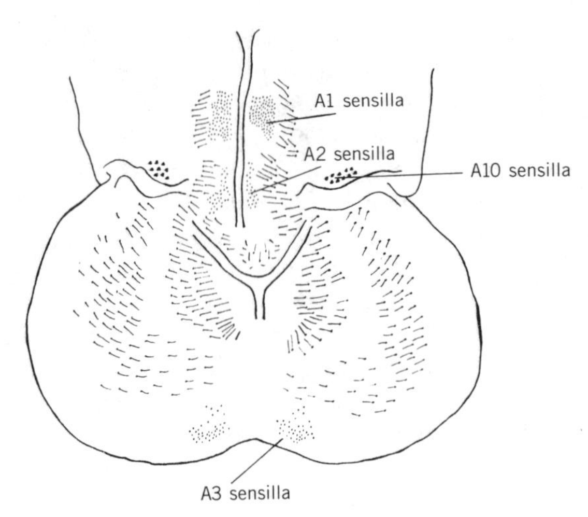

Fig. 1.13. Distribution of sensilla on the epipharyngeal face of the labrum.

have many more olfactory receptors on the antennae than females do, and these sensilla may be involved in sexual recognition (Chapman, 1982). Most antennal sensilla are involved in food recognition but in *Schistocerea gregaria* some probably perceive social pheromones (Gillett, 1975). Greenwood and Chapman (1984) have shown that solitarious *Locusta migratoria* have more olfactory sensilla on the antenna than insects of the gregarious phase. They suggest that social facilitation in the gregarious phase may allow a decrease in individual sensitivity with associated reduction in numbers of sensilla. The change can occur within a single generation and the mechanism is unknown.

When all the different types of chemoreceptors are included, the total number of chemosensitive neurones associated with the mouthparts of an adult *Locusta migratoria* is on the order of 15,000–20,000 (Chapman, 1982); the numbers increase through the stadia (Table 1.1). In general, within the Acridoidea the number of chemoreceptors is proportional to the size of the insect. For a given size of insect, however, there are differences in the numbers of some sensilla groups that can be correlated with the range of plant species on which the insect feeds (Chapman and Thomas, 1978). In a survey of the Acridoidea, Chapman and Thomas found that grass-feeding species had rather fewer sensilla that those feeding on broad-leaved plants and that species with a restricted and specialized diet had slightly fewer sensilla than closely related polyphagous species. An example of this is seen in Figure 1.14, which shows the range of sensillum numbers in the A1 group of the epipharynx in a range of species within the Pyrgomorphidae and the Gomphocerinae. *Poekilocerus* species feed mainly on Asclepiadaceae and have fewer A1 sensilla than other Pyrgomorphidae of comparable size, which are more catholic in their diet. Similarly, *Anablepia granulata*, which feeds only on *Brachyaria* spp., and *Xenocheila zarudnyi*, which specializes on *Ephedra*, have fewer sensilla than other species of

TABLE 1.1. Numbers of Neurones in Chemoreceptor Groups in Different Developmental Stages of *Locusta migratoria* (after Chapman, 1982)

Sensillum Group	Stadium 1	Stadium 5 (male)	Adult (male)
A1	320	1,430	2,105
A2	395	660	765
A3	425	475	395
A10	44	160	214
Hypopharynx	245	350	505
Galea	340	825	3,580
Maxillary palp tips	492	4,440	4,812
Paraglossa	296	736	1,192
Labial palp tips	600	3,360	3,260
Total	3,157	12,436	15,036

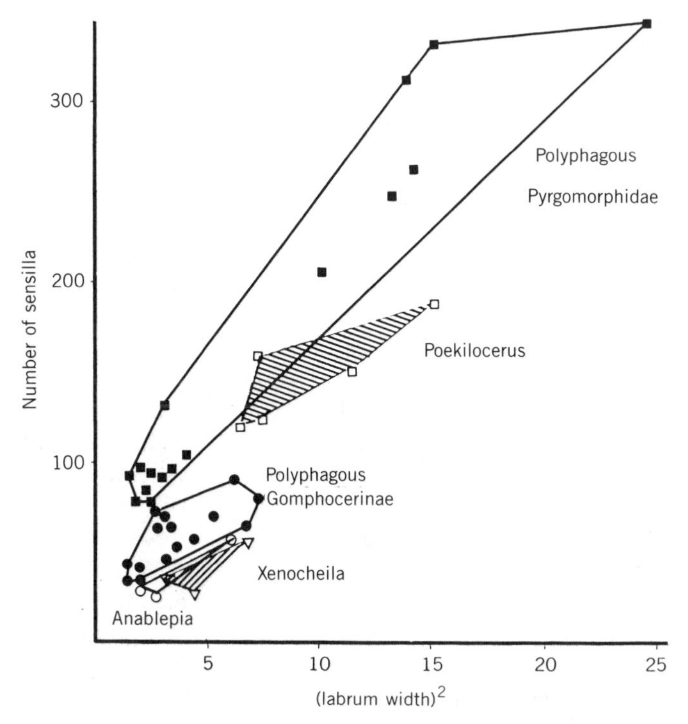

Fig. 1.14. Numbers of A1 sensilla in polyphagous members of the Pyrgomorphidae and Gomphocerinae compared with those in the specialist feeders *Poekilocerus* (Pyrgomorphidae) and *Xenocheila* and *Anablepia* (Gomphocerinae). (After Chapman, 1982.)

Gomphocerinae, which feed on a variety of grass species. These differences may reflect the differing emphasis placed on the balance between phagostimulants and antifeedants by oligophagous and polyphagous species (see also Section 1.5.2).

1.4 THE FUNCTIONING OF CHEMORECEPTORS

Despite the obvious differences, to us, between the senses of taste and smell, the perception of stimulating molecules of both modalities involves very similar events in the receptor cells. Much information about the functioning of sensilla may be gleaned from a study of their fine structure, as we have seen, but more critical information can be obtained from neurophysiological studies. Such studies are often technically very difficult and interpretation of the results is often susceptible to subjective bias. It is as well, therefore, to present here a brief account of the techniques and the problems to aid the reader in assessing the results obtained (see also Chapman and Blaney, 1979; Frazier and Hanson, 1986).

1.4.1 Neurophysiology

The essential feature of this technique is to detect and amplify, for permanent recording and analysis, bioelectric activity occurring in the insect between two metal electrodes placed onto or into the animal's body. Generally, one electrode, the recording electrode, is placed as near as possible to the organ being studied and the other, the indifferent electrode, enters the hemolymph some distance away. Precision of placement is enhanced by using a tungsten electrode etched to a point less than 1 μm in diameter, or by inserting a silver electrode into electrolyte in a glass capillary drawn out to a similar size.

Electrodes inserted at either end of an organ such as the antenna can record a summed response (an electroantennogram, or EAG) from all the stimulated receptors in the organ. More commonly, a single sensillum is the target. With olfactory sensilla, the recording electrode is inserted into the cuticle near the base of the sensillum, and air containing the test odor is passed over the organ (Kafka, 1971). Artifacts can occur when the odor is added to the airstream if the temperature or rate of flow is inadvertently changed during the stimulation.

Taste receptors are commonly stimulated by placing over the terminal pore of the sensillum a glass microcapillary containing the recording electrode, an electrolyte to ensure electrical continuity, and the test solution (Blaney, 1974). The principal limitations with this technique are the ubiquitous presence of electrolyte, which may stimulate the sensillum or interfere with the effect of another stimulant (Haskell and Schoonhoven, 1969), and the restriction of other stimulants to water-soluble compounds, unless they can be dispersed by some other means, such as sonication (Blaney, 1975).

When action potentials from a sensillum are recorded, the output from all the active cells appear in the recording. The sensory message is encoded in the train of action potentials and, because different neurones within the sensillum generally have different sensitivities and react differently to a given stimulus, it is highly desirable to be able to identify the output from individual cells. Ideally this identification may readily be achieved where there are few cells in the sensillum, each responding to different, narrow ranges of stimulants, and where fortuitously, their action potentials are of distinctly different amplitudes. In reality these criteria are very seldom met and the difficulty of associating action potentials with particular cells is a major handicap in analyzing the functioning of chemoreceptors and elucidating the sensory code.

1.4.2 Reception of the Stimulus

Olfactory receptors on the antennae or palps of grasshoppers are well positioned to monitor odors emanating from the environment generally, or from food materials in particular. The constant movement of these organs ensures that the sensilla continually receive fresh stimulation. It is suggested that the form and size of the insect sensilla cause airborne molecules to be attracted to them, and experiments on moth antennae using radio-labeled odorants show that the surface of the cuticle readily adsorbs such molecules and scarcely desorbs them at all (Kasang and Kaissling, 1972; Kanaujia and Kaissling, 1985). Measurements have shown that molecules need not hit the pores of olfactory receptors to be effective, and that they readily diffuse over the surface of the sensillum (Steinbrecht, 1974).

Gustatory receptors in the buccal cavity of grasshoppers normally encounter stimulants in aqueous solution, in the form of plant juices. However, palp tip receptors, and probably those on the tarsi too, typically respond to chemicals on dry surfaces (Bernays et al., 1975; Chapman, 1977). It has been suggested that the viscous material at the sensillum tip, indentified as a mucopolysaccharide, may have a role as a stimulus-conducting material (Bernays et al., 1975) but details of the mechanism are not available.

1.4.3 Transduction

In both gustatory and olfactory receptors, when stimulus molecules have reached the surface of the dendrite they interact with receptor sites in the membrane. This results in changes in the permeability of the membrane to certain ions, which in turn causes a change in the electrical potential across the membrane. The nature of this interaction is not fully understood: it may be analogous to an enzyme–substrate reaction or may involve other less specific processes (Blaney et al., 1988). It is at this stage, however, that specificity of response does occur and the molecular configuration of the stimulus molecule may have a very critical effect. Thus Kafka (1970) tested the coeloconic sensilla on the antenna of *Locusta migratoria* (Section 1.2.3) with 370 compounds and observed a high degree of sensitivity to only three of them. These were closely related hexanoic acids. One neurone was highly stimulated by 2-oxohexanoic acid but the structurally very similar 2-hydroxyhexanoic acid produced no response at all.

The biological effect of the interaction between the stimulus molecule and the receptor site on the neurone depends on the nature of the ion channels in the membrane that are activated. In some cases, the effect is to increase the electrical polarization across the cell membrane and thereby to prevent further transmission of electrical activity in the cell. Stimuli producing this effect are inhibitory stimuli. Conversely, excitatory stimuli result in permeability changes which depolarize the cell membrane. In both cases, the electrical change is proportional to the concentration and effectiveness of the stimulant and changes in magnitude vary with change in stimulus: the chemical "message" has been transduced into an electrical effect, with analogue characteristics. In the case of excitatory stimuli, the depolarization occurring at the receptor site spreads across the cell membrane to a region, near the cell body, where it generates a series of action potentials which are then propagated along the axon toward the CNS. The frequency of action potentials (firing rate) reflects the magnitude of the depolarization; hence the analogue code has been converted into the digital code of spike frequency and it is that code that is most commonly recorded in experimental electrophysiology.

1.4.4 Sensory Coding in Chemoreceptors

A high degree of specificity commonly occurs with pheromone receptor neurones in other insects (see, e.g., Kaissling, 1974) and the same may apply with grasshoppers, but data are not available on this. High specificity may also occur when, in food selection, a particular compound signals an acceptable plant (Chapter 2) or, alternatively, when the compound is a very effective antifeedant or repellant. In these cases, the output from a neurone responding in this specific manner signals to the CNS the presence of a particular biologically active compound. In terms of the total sensory input, such a neurone is said to constitute a "labeled line" to the CNS, and may be described as a specialist neurone.

In assessing its chemical environment, the grasshopper will encounter a very large number of different compounds and, even with the large numbers of receptor neurones that they have (Chapman and Thomas, 1978; Chapman, 1982) it would probably not be practical to have a different group of neurones tuned specifically to each compound likely to be encountered. Even if it were practical, such a system would leave analysis and decision making entirely to the CNS. This is not what happens. Instead, the majority of chemoreceptor neurones are "generalists," each responding to a range of compounds that are sometimes quite unrelated (Blaney, 1974), and each giving quite imprecise information to the CNS. How is this information interpreted?

The most detailed study of generalist receptors in the grasshoppers concerns the sense of taste, mediated by the palp tip sensilla on the maxillary palps of

Locusta migratoria (Blaney, 1981). Individual sensilla may respond to a range of compounds from a number of chemical classes greater than the number of neurones in the sensillum (Blaney, 1974). In different sensilla a variable number of neurones would respond to the same simple solution, such as 0.1 M sodium chloride. It is well established that in some insects, such as the blowfly *Phormia regina*, separate neurones in the sensillum respond to salt or to sugar (Dethier, 1976). However, in the palp tip sensilla of *Locusta migratoria*, cross-adaptation experiments revealed the same neurones responding, albeit with different firing rates, to stimulation with 0.1 M sodium chloride ("salt") on the one hand, and to 0.1 M sodium chloride containing 0.025 M fructose ("sugar") on the other (Blaney, 1975). A clear differentiation between the response to salt and that to sugar, based on firing rates, was more often seen when the output of the whole sensillum was considered, and Blaney (1975) suggested that the sensillum is the basic unit of response. Even so, most sensilla could not individually distinguish unambiguously between salt and sugar although the insects clearly made the distinction in behavioral experiments. This and similar experiments with a wider range of stimulants lead Winstanley and Blaney (1978) to conclude that there is no unique pattern of sensillum response associated with either acceptance or rejection. Interpretation of the electrophysiological data is further confounded by the variability in responses. The response of a given sensillum varied by as much as ± 10% between successive stimulations with the same solution (Blaney and Winstanley, 1980).

Experiments have shown that, during palpation of potential food, about 30 sensilla at the tip of each palp make contact with the substrate (Blaney and Chapman, 1970). In an attempt to model the sensory input available to a single locust, and to take account of the variability of response, Blaney (1975) stimulated 20 sensilla on the same palp four times with salt and four times with sugar in an alternating sequence. Six of the sensilla were more responsive to sugar, four to salt, and the remaining 10 did not show greater responsiveness to either stimulant, or were inconsistent in doing so; that is, they appeared not to distinguish reliably between the two. Even in those sensilla showing a "preference" there was sometimes overlap in the response range to the two stimulants because of the variability. However, the technique of "across-fiber" analysis (Pfaffmann, 1941) yields a solution. According to this hypothesis, discrimination is achieved by considering the output from a number of receptors simultaneously, provided that (1) the receptors have unique but overlapping action spectra and (2) each substance discriminated generates a unique total pattern of response. Analysis of the data from the above experiment (Blaney, 1975) revealed a highly significant variance between responses of individual sensilla to a single solution, indicating that sensilla differ from each other in their response — the first tenet of the across-fiber hypothesis. Variance between the responses to different solutions was highly significant, indicating that on the basis of the total information available the insect could differentiate between solutions, as indeed behavioral experiments confirmed. Finally, the variance due to interaction was highly significant, indicating that different sensilla respond in different ways to the two stimulants, or in other words, each solution generates a different total pattern of response — the second tenet of the across-fiber hypothesis. Thus, despite the variability, these sensilla produce a response that allows unequivocal discrimination between the two test solutions, provided the output from a population of sensilla is considered simultaneously.

It follows from this that the possession of generalist receptors need not be a handicap but may actually confer an advantage when broad-spectrum receptors allow responsiveness to a wide range of chemicals, and across-fiber analysis ensures that discrimination can be achieved. Thus the code for taste quality in this case can be read by considering the simultaneous output of many sensilla: this does not preclude the possibility that the insect actually reads the output of neurones rather than of sensilla, as the units of response. It also allows for the existence of sensilla or neurones that have evolved to respond to specific cues. We have no knowledge of higher-order neural networks that might indicate whether it is sensilla or neurones that act as the basic unit, and this experiment demonstrates what could be done, rather than proves what is done by the insect. Nor should we fail to accept that the categories specialist and generalist may merely represent the ends of a continuous spectrum and that the occurrence of intermediates does not invalidate the concept.

1.4.5 Thermoreceptors and Hygroreceptors

We have a very imperfect understanding of the mechanisms involved in reception of the stimulus in the no pore type of coeloconic sensilla (Section 1.2.5) that are found on the antennae of *Locusta migratoria*, and no doubt occur in other species as well. Two of the three neurones respond to changes in humidity and one to changes in temperature (Altner et al., 1981). One dendrite ends at the base of the peg and the other two enter it. It may be that the 2 + 1 morphological differentiation is associated with the 2 + 1 functional differentiation. The cuticle of the whole sensillum is sufficiently thin that changes in temperature could easily be conducted through it so that, on that basis, the basal dendrite could just as easily be the cold receptor as either of the other two. If it were, then the other two, being closer to the surface, might be more readily affected by variations in the water vapor in the air.

Although most neurones have some sensitivity to changes in temperature we do not know how transduction of the temperature stimulus operates in this sensillum. The same can be said of the hygroreceptors, but some aspects of the structure of the sensillum invite conjecture (Altner et al., 1981). There is evidence from other insects (Bernard, 1974; Becker, 1978; Yokohari, 1978) that mechanical stimulation of hygroreceptor sensilla modulates the response of the hygroreceptor neurones, and it is suggested that the normal mechanism of transduction may involve mechanical effects caused by changes in atmospheric water vapor. In the present case, such effects could be mediated by the thin, multilayered sensillum wall, which may be differentially porous. Alternatively, the clefts in the socket region of the peg may allow humidity-induced volume changes to occur in the electron-dense material protruding into the sensillar sinus (Fig. 1.12). Another possibility, which might operate instead of, or in addition to, any mechanical effects of volume change, derives from the fact that if the electron-dense material were sufficiently hygroscopic to withdraw water from the air in the sensillum pit, it would most likely exert a similar effect on the dendrites adjacent to it. Any tendency of the dendrites to lose water could be countered by the activity of ion pumps in the dendritic membrane that would adjust the osmotic potential of the cell, but at the same time alter the electrochemical gradient across the cell membrane (Altner et al., 1981). Thus the water vapor in the air would indirectly affect the degree of depolarisation of the dendritic membrane. Critical testing of these hypotheses awaits more detailed investigation of the sensillum.

1.5 CHEMORECEPTION AND THE LIFE OF THE INSECT

The detection of chemical stimuli is utilized by insects in locating food, in recognizing mates, in identifying suitable oviposition sites, and in a number of other activities. In the grasshoppers, investigations have centered almost exclusively on the involvement of chemoreceptors in food selection, to some extent at distant locations, but principally when the insect is already in contact with the plant.

1.5.1 Plant Odor Perception

Grasshoppers are known to detect plants by odor from a distance (see Section 3.3.1) but there are relatively few studies of this. In controlled laboratory conditions, Haskell et al. (1962) showed that fifth-stadium nymphs of *Schistocerca gregaria* could be induced to walk upwind in the laminar airflow of a wind tunnel when crushed grass was placed in an upwind section of the tunnel. They postulated that the locusts were detecting and orienting to the odor gradient which they believed to exist in the tunnel, the concentration decreasing with distance from the source. This interpretation was reassessed in further experiments by Kennedy and Moorhouse (1969) using the same apparatus. Their experiments showed that the upwind orientation involved anemotaxis, not tropotaxis as originally thought. The effect of the grass odor was to increase the state of excitation of the insects, so switching on the anemotactic response.

Another example of food odor enhancing a feeding-related behavior is reported by Mordue (1979), who found that fifth-stadium nymphs of *Schistocerca gregaria* with some mouthpart receptors removed bit an inert substrate more readily in the presence of food odor (see also Section 3.3.1 for other examples of the influence of host odors).

In most cases, electrophysiological studies of the host odor response in grasshoppers are lacking or equivocal (see Section 3.3.1), but the coeloconic sensilla (Section 1.2.3) on the antennae of *Locusta migratoria* have been thoroughly investigated. Boeckh (1967) found that cells in this sensillum are sensitive to a range of so-called green odors, such as the odor of fresh grass and of hexenal and hexanol. In all, he tested about 80 compounds, including short-chain fatty acids, aliphatic unsaturated aldehydes, and alcohols. Subsequently Kafka (1971) and Kafka et al. (1973) investigated the reaction spectra of a wide range of related molecules with this sensillum in an attempt to elucidate particularly the transduction mechanisms (see Section 1.4.3). There is no doubt that this sensillum is involved in the detection of food sources at a distance but there may be other sensilla, as yet undetected, that have a similar role. It is also likely that further investigations will reveal olfactory receptors sensitive to allelochemics that are signal stimuli for particular host plants. This is most likely to occur in insects with limited or specialist host ranges (see Section 3.3.1 and Staedler, 1984).

1.5.2 Assessment of Plants by Contact Chemoreceptors

We readily accept that the sense of taste is appropriate for the assessment of food, even though we mainly use the sense of smell ourselves, usually without realizing it. When we consider the insect tasting a plant not only with its mouthparts but with its antennae, feet, and possibly its ovipositor as well, the analogy is more tenuous. All these organs bear taste receptors, which may contribute to the assessment of a plant's suitability on the basis of its taste.

When a grasshopper stands on a potential host plant, as it commonly does, it is highly likely that it tests the quality of the plant surface with sensilla on its tarsi. Unfortunately it has not yet been shown conclusively that the tarsal chemoreceptors (Section 1.2.4) are capable, like palp tip sensilla (Blaney, 1974), of responding to plant surface waxes. However, internal, water-soluble constituents of leaves are known to leach out onto the surface, where they may stimulate tarsal sensilla. In nymphs of *Locusta migratoria*, tarsal contact has been shown to result in different responses to wet filter paper depending on the state of hydration of the animal (Kendall and Seddon, 1975). Dehydrated insects attempted to feed on the strips of paper encountered in an arena, whereas hydrated insects were repelled by them. Movements made by the antennae during this experiment suggest that the nymphs also sometimes responded to the humidity of the air close to the paper strips, presumably perceived by the coeloconic humidity receptors (Section 1.2.5).

Ablation experiments by Haskell and Mordue (1969) and electrophysiological investigations by Haskell and Schoonhoven (1969) have shed some light on the roles of various mouthpart receptors in food selection by adults and nymphs of *Schistocerca gregaria* and *Locusta migratoria*. In the experiments of Haskell and Mordue, insects were presented with filter paper disks impregnated with the phagostimulant, sucrose, or alternatively with a "deterrent factor," azadirachtin, and the effect of various ablations of mouthparts or groups of sensilla was determined by measuring the amount of impregnated paper consumed. It was clear that although all the receptor systems provided the CNS with information on both phagostimulants and deterrents, some specialization occurred between receptor groups. Thus the terminal sensilla on the palp tips responded particularly to the phagostimulant and had an important role in stimulating feeding; the A3 sensilla (Section 1.2.4) responded mainly to the deterrent.

The electrophysiological experiments of Haskell and Schoonhoven (1969) gave some clues as to the way in which some of these sensilla might be used by the insect in food selection. For example, they reported the occurrence of a mechanoreceptor response in each of the palp tip sensilla and suggested that information gained during palpation of the leaf surface could be used by the insect to gauge the hardness of the leaf. Blaney (1974) later disputed the ubiquitous presence of the mechanoreceptor response in these chemoreceptor sensilla (Section 1.2.4), but enough sensilla have this capability to afford the insect a very adequate sense of touch. This is used in guiding food while it is being manipulated by the mouthparts, even if an assessment of its hardness is not being made.

Haskell and Schoonhoven (1969) also investigated the chemoreceptor capability of the palp tip sensilla. They found that a single neurone responded with large-amplitude spikes to stimulation with sodium chloride in the range 0.1–1 M, although spike amplitude and frequency varied between preparations. Similarly, another neurone, firing with a smaller-amplitude spike, responded to sucrose solutions. Some interaction between these two responses was noted such that increasing salt concentration progressively diminished the response of the sugar cell. This phenomenon has been noted also by Blaney (1974) in *Locusta migratoria* and in other insects as well (W. M. Blaney, unpublished) and is probably an important aspect of sensory coding of taste quality (see below).

When Haskell and Schoonhoven (1969) stimulated the palp tip sensilla of *Locusta migratoria* with sap expressed mechanically from the grass *Holcus lanatus*, an acceptable food, the response was vigorous. It involved several neurones, and after a few seconds gave way to an intermittent "bursting" discharge. Such "bursting" discharges are normally associated with unpalatable materials (Ma, 1972; Blaney et al., 1984). The response which Haskell and Schoonhoven found with *Schistocerca gregaria* was different: with a much more dilute sap a single neurone response was obtained, with no bursting. An aqueous extract of seeds from the neem tree (identified as *Melia azadirachta* in their paper) also produced a response in only one neurone. These seeds are extremely deterrent to *Schistocerca*, so presumably a different neurone was involved from that responding to grass.

The work of Haskell and Schoonhoven (1969) was the first of its kind with grasshoppers and illustrated many of the difficulties encountered when attempting to understand how the acridoid chemosensory system works. Subsequent, more detailed studies by Blaney (1974, 1975, 1980, 1981), Blaney and Winstanley (1980), and Winstanley and Blaney (1978) revealed further pitfalls and suggested ways in which they might be circumvented. In a series of experiments they compared the electrophysiological responses of the palp tip sensilla with the behavioral responses to the same stimulants in fifth-stadium nymphs of *Locusta migratoria* and *Schistocerca gregaria*.

Both species appear to use the across-fiber method of analysis (Section 1.4.4) and their assessment of palatable, nutritious substances like sucrose is very similar. Allelochemics, on the other hand, are assessed differently by the two species. To the oligophagous *Locusta migratoria* most grasses contain no deterrent allelochemics, and are eaten, whereas most nongrasses do contain deterrents, and are rejected. By contrast, the response of the polyphagous *Schistocerca gregaria* to allelochemics is much more varied and depends not only on the nature of the allelochemic but also on its concentration (Bernays and Chapman, 1978). This would suggest that the sensory system of *Schistocerca gregaria* makes a more detailed and responsive assessment of allelochemics than does that of *Locusta migratoria*. This hypothesis was tested electrophysiologically and behaviorally by stimulating with sucrose as a phagostimulant and a range of allelochemics to which the two species reacted differently (Fig. 1.15). A control solution contained salt (as electrolyte) and sucrose, and each test solution differed from the control by the addition of one allelochemic, at a concentration occurring in plants. Thus the electrophysiological response to a test solution could be compared with that to the control solution, and the difference ascribed to the presence of the single allelochemic. Taking account of the variability that occurs (Section 1.4.4), three categories of response to the addition of each allelochemic were found, and "no change," "increase," and "decrease" sensilla identified (see legend to Fig. 1.15).

In an attempt to model the sensory information available to a given locust, recordings were made from a number of sensilla on one insect (Section 1.2.4). A "relative sensory input" was computed for each allelochemic, based on the mean rate of firing for the control solution, the mean magnitude of change with the test solution, and the number of sensilla in the three categories of change (Blaney, 1980). The validity of the model was tested by correlating the sensory response to each compound with the behavioral response, expressed as the percentage of insects failing to feed on glass-fiber discs, each soaked with one of the test solutions.

With *Schistocerca gregaria* the correlation was good (Fig. 1.15) except with azadirachtin (A), the most potent deterrent yet tested. It seems likely that, in the case of azadirachtin the sensory input is qualitatively distinct, perhaps associated with a "labeled line" to the CNS (Section 1.4.4). This is not to say that all the other inputs only differ quantitatively: the experiment merely shows that, on the basis of quantitative differences, the sensory code of this model system can be read tolerably well. There is no reason to suppose that the insect cannot do any better.

With *Locusta migratoria*, the model suggests that the insect discriminates well between different concentrations of sucrose but the capacity to make that discrimination is much reduced when an allelochemic is present as well (Fig.

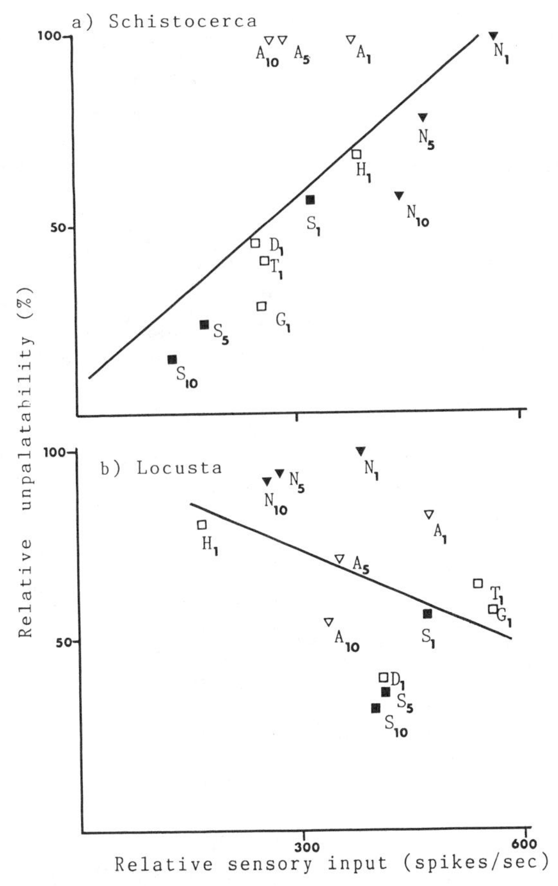

Fig. 1.15. Relationship between behavior and sensory input from the maxillary palp tip sensilla of *Schistocerca gregaria* and *Locusta migratoria.* All solutions tested contained the electrolyte sodium chloride (0.05 M). The compounds tested were S, sucrose; T, sodium hydrogen (+) tartrate (2×10^{-3} M); N, nicotine hydrogen (+) tartrate (2×10^{-3} M); D, sodium sulfate (7×10^{-3} M); H, hordenine hemisulfate (7×10^{-3} M); G, sinigrin (2×10^{-3} M); A, azadirachtin (1.8×10^{-8} M). Sucrose was tested on its own, or in combination with one of the other compounds. The subscript describes the concentration of sucrose used: 1 = 0.01 M, 5 = 0.05 M, and 10 = 0.1 M. Relative unpalatability is percent of insects failing to feed on glass fiber disks soaked in test solution. Relative sensory input is computed by comparing response to each solution with that to sucrose alone (the control solution) and identifying "no change," "increase," and "decrease" sensilla. The computation is: (av. spikes/sec for control × no. of "no change" sensilla) + (av. spikes/sec for increase × no. of "increase" sensilla) + (av. spikes/sec for decrease × no. of "decrease" sensilla) = relative sensory input (spikes/sec), where mean percentage increase or decrease is converted to spikes/sec and added to or subtracted from control level, respectively.

1.15). This may reflect the oligophagous habit, the insect feeding on grasses, between which it can discriminate, but rejecting, rather than discriminating between, nongrasses when deterrent allelochemics are present.

The reliance of this approach on quantitative differences between qualitatively similar inputs has already been mentioned and has prompted Chapman (1982, 1988) to question its validity. On the basis of Blaney's (1980) data he postulates the existence of a stimulatory input, an unspecialized inhibitory input, and possibly a specialized inhibitory input (e.g., tuned to azadirachtin). We are handicapped in resolving this discussion by our complete ignorance of neural processing at the level of second- and higher-order neurones in this system. We assume that, by a series of convergent synapses, qualitatively different inputs are resolved in a quantitative fashion to give a single decision: feed or don't feed. That could be achieved, say, by labeled lines for phagostimulants activating excitatory synapses and labeled lines for deterrents activating inhibitory synapses, the magnitude of synaptic response reflecting quantitatively the effectiveness of the phagostimulants and deterrents as stimulators of the peripheral receptors. But, as we have been, there is evidence that chemicals interfere with the response to other chemicals by the peripheral receptors. Thus in the present case, as Blaney (1981) has pointed out, interaction between inhibitory and stimulatory chemicals could occur at the receptor level and the input to the CNS be determined by that conflict. Alternatively, input from different "labeled lines" could signal stimulation and inhibition concurrently, leaving the conflict to be resolved by the CNS. It seems that both mechanisms operate, though to different extents, in the two locust species, and the good correlation of behavior and sensory input found by Blaney (1980) would argue for a good measure of peripheral resolution.

The disproportionate interest in the functioning of palp tip receptors, compared with that of other mouthpart receptors, stems from the essential role they play in food selection (Blaney and Chapman, 1970) in grasshoppers that are not deprived of food for abnormally long periods (Blaney et al., 1973). During testing of food before a meal, and more or less continuously throughout a meal, the palps make extensive, rapid vibrations so that the palp tip sensilla touch the substrate 10–15 times per second, each contact lasting about 20 ms (Blaney and Chapman, 1970). The effect of these intermittent contacts on the functioning of the palp-tip sensilla was investigated by Blaney and Duckett (1975), who showed that, by minimizing sensory adaptation, palpation increases the amount of sensory information reaching the CNS compared with that obtained from sustained contact. It also has the effect of increasing the surface area sampled by the sensilla. Bernays and Chapman (1974) have suggested that during a meal, feeding is driven by continued sensory input from the palps. Enclosing the palp tips with glass microcapillaries, Blaney and Duckett (1975) were able to manipulate the quality of stimulus reaching the receptors during a meal by leaving the tubes empty or by filling them with a deterrent or a phagostimulant solution. They found that when a locust was eating a favored grass, input from the palp tip sensilla was not necessary to sustain feeding. However, a deterrent input could reduce meal size, and when a less favored food was being eaten, phagostimulatory input from the palps could enhance feeding.

After completion of a meal, the locust enters a quiescent phase, lasting about an hour, during which the palp tip sensilla are largely unresponsive. Studies with SEM (Blaney and Chapman, 1969) showed that the pore at the tip of each sensillum was capable of opening and closing, and this was investigated by Bernays et al. (1972) using electrical and electrophysiological measurements. They found that the electrical resistance across the palp tips was high immediately after a meal but fell to a steady level during the next two hours. Likewise, few sensilla responded electrophysiologically to chemical stimulation immediately after feeding, but the number responding increased with food deprivation. The closing of the sensillum pores was found to be mediated by hormones from the storage lobes of the corpora cardiaca (Bernays and Mordue, 1973) and the release of the hormones into the hemolymph was found to be initiated by nerve impulses from the full crop (Bernays and Chapman, 1972). The significance of this opening and closing mechanism is by no means obvious. Bernays et al. (1972) have argued that it will minimize the sensory input from up to 10,000 neurones in the tips of the four palps when, after feeding, such input might be distracting. In *Locusta migratoria*, however, vibration of the palps virtually ceases for a time after feeding (Blaney and Chapman, 1970), so that stimulation of the sensilla is not likely to occur anyway.

During periods of food selection behavior the palp tip sensilla are generally open and fully functional. It does not follow, however, that the sensory input is constant at all such times or that the sensilla make an unvaried assessment of the chemical environment. Fifth-stadium nymphs of *Locusta migratoria* react to dilution of dietary protein by reducing intermeal intervals (Simpson and Abisgold, 1985) a reaction regulated by osmolality and free amino acid levels in the hemolymph (Abisgold and Simpson, 1987; and see Chapter 2). Several studies have shown that the sensitivity of chemoreceptors varies with factors such as developmental stage, time of day, and feeding history (see Blaney et al., 1986; Schoonhoven et al., 1987). It is possible, therefore, that the early initiation of feeding, resulting in reduced intermeal intervals, in protein-deprived locusts is mediated by a change in the functioning of the palp tip sensilla. This was investigated by Abisgold and Simpson (1988). The found that "low-protein" insects had significantly increased sensory sensitivity to the amino acid leucine and to sucrose, and that injecting amino acids into low-protein insects to bring the hemolymph amino acid profile up to that of normal insects markedly reduced the sensitivity to 8 of the 10 amino acids injected, but did not reduce the sensitivity to sucrose. This reduction was independent of the effect that injection had on blood osmolality and was sustained for 50 min after the injection. Such specific nutrient feedbacks, modulating the responses of taste receptors, may also underlie dietary selection in locusts (Simpson et al., 1988). It is not known whether the feedback is a direct effect of the dietary components, or whether it is indirect, perhaps involving a mechanism in which hormones have a role.

Similar feedback effects may also influence CNS neurones associated with the chemosensory system and thereby alter food selection behavior. Szentesi and Bernays (1984) have shown behavioral habituation following dietary exposure to a feeding deterrent, nicotine, in nymphs of *Schistocerca gregaria*. In addition, associative learning in nymphs of *Locusta migratoria* has been shown by Blaney and Winstanley (1982) and Blaney and Simmonds (1985) to influence food selection behavior involving the palp tip sensilla. In neither of these cases is there any information on the possibility of sensitivity changes occurring in peripheral receptors as well.

1.6 CONCLUSIONS

One of the most exciting and compelling aspects of chemoreception in grasshoppers is that so much remains to be discovered. The work described in this chapter and related work described elsewhere in this book form a firm basis for future research that will tell us more about the grasshoppers, but will also reveal basic information, applicable to other insects and to biological systems in general. The functioning of the chemoreceptor system seems certain to have a fundamental role in the determination of host range, a topic of profound behavioral and ecological importance, and in many insects a matter of major economic importance. We need a more detailed understanding of sensory coding, and to this end it will be necessary to pursue electrophysiological studies of the receptors of both generalist and specialist insects. Such studies will need to be complemented by electrophysiological studies of second- and higher-order neurones to assess whether the peripheral system is really reporting information in the way we think it is.

Another aspect of fundamental importance concerns the mechanisms that influence the sensitivity of peripheral receptors. Feedback mechanisms appear to influence feeding behavior by modulating peripheral sensitivity. These mechanisms could involve ingested nutrients, or deterrents, acting either directly on the receptors or via some intermediary, such as the hormonal system.

Many other topics that have attracted interest in recent years are incompletely explained: the role of olfaction in short-range host selection; the significance of the large number of chemoreceptors; the difference in numbers of sensilla between specialist and generalist feeders, and between solitarious and gregarious phases; the role and significance of multiple sheath cells in some sensilla; the control of differentiation of receptor sensitivity and the molecular mechanisms associated with modulation of that sensitivity; and the transduction mechanisms that enable the tasting of dry materials. The chemoreceptor system of grasshoppers offers scope for important advances to be made by ecologists, behavioralists, physiologists, and molecular biologists.

REFERENCES

Abisgold, J. D. and S. J. Simpson. 1987. The physiology of compensation by locusts for changes in dietary protein. *J. Exp. Biol.* **129**, 329–346.

Abisgold, J. D. and S. J. Simpson. 1988. The effect of dietary protein levels and haemolymph composition on the sensitivity of the maxillary palp chemoreceptors of locusts. *J. Exp. Biol.* **135**, 215–229.

Altner, H. 1977. Insect sensillum specificity and structure: An approach to a new typology. In J. Magnen and P. MacLeod, Eds., *Olfaction and Taste VI*. Information Retrieval Ltd., London, pp. 295–303.

Altner, H. and F. Ameismeir. 1986. Tubular bodies in dendritic outer segments projecting to second embryonic cuticle from *anlagen* of contact chemoreceptors in *Locusta migratoria* L. (Orthoptera: Acrididae) and *Periplaneta americana* (L.) (Dictyoptera: Blattidae). *Int. J. Insect Morphol. Embryol.* **15**, 253–262.

Altner, H. and L. Prillinger. 1980. Ultrastructure of invertebrate chemo-, thermo-, and hygroreceptors and its functional significance. *Int. Rev. Cytol.* **69**, 69–139.

Altner, H., C. Routh, and R. Loftus. 1981. The structure of bimodal chemo-, thermo-, and hygroreceptive sensilla on the antenna of *Locusta migratoria*. *Cell Tissue Res.* **215**, 289–308.

Ameismeier, F. 1987. Ultrastructure of the chemosensitive basiconic single-walled wall-pore sensilla on the antennae in adults and embryonic stages of *Locusta migratoria* L. (Insecta, Orthoptera). *Cell Tissue Res.* **247**, 605–612.

Becker, D. 1978. Elektophysiologische Untersuchungen zur Feuchterezeption durch die styloconischen Sensillen bei *Mamestra brassicae* L. (Lepidoptera, Noctuidae). Doctoral Dissertation, University of Regensburg.

Bernard, J. 1974. Etude électrophysiologique de recepteurs impliqués dans l'orientation vers l'hôte et dans l'acte hémophage chez un Hemiptere: *Triatoma infestans*. Thèse, Université de Rennes.

Bernays, E. A. and R. F. Chapman. 1972. The control of changes in the peripheral sensilla associated with feeding in *Locusta migratoria* (L.). *J. Exp. Biol.* **57**, 755–763.

Bernays, E. A. and R. F. Chapman. 1974. The regulation of food intake by acridids. In L. Barton Browne, Ed., *Experimental Analysis of Insect Behaviour*. Springer-Verlag, Berlin, pp. 48–59.

Bernays, E. A. and R. F. Chapman. 1978. Plant chemistry and acridoid feeding. In J. B. Harborne, Ed., *Biochemical Aspects of Plant and Animal Coevolution*. Academic Press, New York, pp. 99–141.

Bernays, E. A. and A. J. Mordue (Luntz). 1973. Changes in the palp tip sensilla of *Locusta migratoria* in relation to feeding: The effects of different levels of hormone. *Comp. Biochem. Physiol. A* **45A**, 451–454.

Bernays, E. A., W. M. Blaney, and R. F. Chapman. 1972. Changes in chemoreceptor sensilla on the maxillary palps of *Locusta migratoria* in relation to feeding. *J. Exp. Biol.* **57**, 745–753.

Bernays, E. A., W. M. Blaney, and R. F. Chapman. 1975. The problems of perception of leaf-surface chemicals by locust chemoreceptors. In D. A. Denton and J. P. Coghlan, Eds., *Olfaction and Taste V*. Academic Press, New York, pp. 227–230.

Blaney, W. M. 1974. Electrophysiological responses of the terminal sensilla on the maxillary palps of *Locusta migratoria* (L.) to some electrolytes and non-electrolytes. *J. Exp. Biol.* **60**, 275–293.

Blaney, W. M. 1975. Behavioural and electrophysiological studies of taste discrimination by the maxillary palps of larvae of *Locusta migratoria* (L.). *J. Exp. Biol.* **62**, 555–569.

Blaney, W. M. 1977. The ultrastructure of an olfactory sensillum on the maxillary palps of *Locusta migratoria*. *Cell Tissue Res.* **184**, 397–409.

Blaney, W. M. 1980. Chemoreception and food selection by locusts. In H. van der Starre, Ed., *Olfaction and Taste VII*. Information Retriveal Ltd., London, pp. 127–130.

Blaney, W. M. 1981. Chemoreception and food selection in locusts. *Trends Neurosci.* Feb., pp. 35–38.

Blaney, W. M. and R. F. Chapman. 1969. The anatomy and histology of the maxillary palp of *Schistocerca gregaria. J. Zool.* **157**, 509–535.

Blaney, W. M. and R. F. Chapman. 1970. The functions of the maxillary palps of Acrididae (Orthoptera). *Entomol. Exp. Appl.* **13**, 363–376.

Blaney, W. M. and A. M. Duckett. 1975. The significance of palpation by the maxillary palps of *Locusta migratoria* (L.): An electrophysiological and behavioural study. *J. Exp. Biol.* **63**, 701–712.

Blaney, W. M. and M. S. J. Simmonds. 1985. Food selection by locusts: The role of learning in rejection behaviour. *Entomol. Exp. Appl.* **39**, 273–278.

Blaney, W. M. and C. Winstanley. 1980. Chemosensory mechanisms in locusts in relation to feeding: The role of some secondary plant compounds. In *Insect Neurobiology and Pesticide Action (Neurotox 79)*. Chemical Industry, London, pp. 383–389.

Blaney, W. M. and C. Winstanley. 1982. Food selection behaviour in *Locusta migratoria.* In J. H. Visser and A. K. Minks, Eds., *Proceedings of the 5th International Symposium on Insect-Plant Relationships*. Pudoc, Wageningen, pp. 365–366.

Blaney, W. M., R. F. Chapman, and A. G. Cook. 1971. The structure of the terminal sensilla of the maxillary palps of *Locusta migratoria* (L.), and changes associated with moulting. *Z. Zellforsch. Mikrosk. Anat.* **121**, 48–68.

Blaney, W. M., R. F. Chapman, and A. Wilson. 1973. The pattern of feeding of *Locusta migratoria* (L.) (Orthoptera, Acrididae). *Acrida* **2**, 119–137.

Blaney, W. M., M. S. J. Simmonds, S. V. Evans, and L. E. Fellows. 1984. The role of the secondary plant compound 2,5-dihydroxymethyl 3,4-dihydroxylpyrrolidine as a feeding inhibitor for insects. *Entomol. Exp. Appl.* **36**, 209–216.

Blaney, W. M., L. M. Schoonhoven, and M. S. J. Simmonds. 1986. Sensitivity variations in insect chemoreceptors; a review. *Experientia* **42**, 13–19.

Blaney, W. M., M. S. J. Simmonds, S. V. Ley, and P. S. Jones. 1988. Insect antifeedants: A behavioural and electrophysiological investigation of natural and synthetically derived clerodane diterpenoids. *Entomol. Exp. Appl.* **46**, 267–274.

Boeckh, J. 1967. Reaktionsschwelle, Arbeitsbereich und Spezifitat eines Geruchsrezeptors auf der Heuschreckenantenne. *Z. Vergl. Physiol.* **55**, 378–406.

Broyles, J. L., F. E. Hanson, and A. M. Shapiro. 1976. Ion dependence of the tarsal sugar receptors of the blowfly, *Phormia regina. J. Insect Physiol.* **22**, 1587–1600.

Chapman, R. F. 1966. The mouthparts of *Xenocheila zarudnyi* (Orthoptera, Acrididae). *J. Zool.* **143**, 277–288.

Chapman, R. F. 1977. The role of the leaf surface in food selection by acridids and other insects. *Colloq. Int. C. N. R. S.* **265**, 133–149.

Chapman, R. F. 1982. Insect chemoreceptors. *Adv. Insect Physiol.* **16**, 247–356.

Chapman, R. F. 1988. Sensory aspects of host-plant recognition by acridoidea: Questions associated with the multiplicity of receptors and variability of response. *J. Insect Physiol.* **34**, 167–174.

Chapman, R. F. and W. M. Blaney. 1979. How animals perceive secondary compounds. In G. A. Rosenthal and D. J. Janzen, Eds., *Herbivores: Their Interaction with Secondary Plant Metabolites*. Academic Press, New York, pp. 161–198.

Chapman, R. F. and M. Greenwood. 1986. Changes in distribution and abundance of antennal sensilla during growth of *Locusta migratoria* L. (Orthoptera: Acrididae). *Int. J. Insect Morphol. Embryol.* **15**, 83–96.

Chapman, R. F. and J. G. Thomas. 1978. The numbers and distribution of sensilla on the mouthparts of Acridoidea. *Acrida* **7**, 115–148.

Cook, A. G. 1972. The ultrastructure of the A1 sensilla on the posterior surface of the clypeo-labrum of *Locusta migratoria migratorioides* (R & F). *Z. Zellforsch. Mikrosk. Anat.* **134**, 539–544.

Cook, A. G. 1977. The anatomy of the clypeo-labrum of *Locusta migratoria* (L.) (Orthoptera: Acrididae). *Acrida* **6**, 287–306.

Cook, A. G. 1979. The ultrastructure of some sensilla on the clypeo-labrum of *Locusta migratoria* (L.) (Orthoptera: Acrididae). *Acrida* **8**, 47–62.

Dethier, V. 1976. *The Hungry Fly*. Harvard Univ. Press, Cambridge, Massachusetts.

Frazier, J. L. and F. E. Hanson. 1986. Electrophysiological recordings and analysis of insect chemosensory responses. In J. R. Miller and T. A. Miller, Eds., *Insect-Plant Interactions*. Springer-Verlag, New York, pp. 285–330.

Gillett, S. D. 1975. The action of the gregarisation pheromone on five non-behavioural characters of phase polymorphism of the desert locust, *Schistocerca gregaria* (Forskål). *Acrida* **4**, 137–149.

Greenwood, M. and R. F. Chapman. 1984. Differences in number of sensilla on the antenna of solitarious and gregarious *Locusta migratoria* L. (Orthoptera: Acrididae). *Int. J. Insect Morphol. Embryol.* **13**, 295–301.

Hansen, K. 1978. Insect chemoreception. *Receptors and Recognition, Series B* **5**, 233–292.

Haskell, P. T. and A. J. Mordue (Luntz). 1969. The role of mouthpart receptors in the feeding behaviour of *Schistocerca gregaria. Entomol. Exp. Appl.* **12**, 591–610.

Haskell, P. T. and L. M. Schoonhoven. 1969. The function of certain mouthpart receptors in relation to feeding in *Schistocerca gregaria* and *Locusta migratoria migratorioides*. *Entomol. Exp. Appl.* **12**, 423–440.

Haskell, P. T., M. W. J. Paskin, and J. E. Moorhouse, 1962. Laboratory observations on factors affecting the movement of hoppers of the desert locust. *J. Insect Physiol.* **8**, 53–78.

Henning, B. 1974. Morphologie und Histologie des Tarsen von *Tettigonia viridissima* L. (Orthoptera, Ensifera). *Z. Morphol. Tiere* **79**, 323–342.

Kafka, W. A. 1970. Molekulare Wechselwirkungen bei der Erregung einzeiner Riechzellen. *Z. Vergl. Physiol.* **70**, 105–143.

Kafka, W. A. 1971. Specificity of odor-molecule interaction in single cells. In G. Ohloff and A. F. Thomas, Eds., *Gustation and Olfaction*. Academic Press, London, pp. 61–72.

Kafka, W. A., G. Ohloff, D. Schneider, and E. Vareschi. 1973. Olfactory discrimination of two enantiomers of 4-methyl-hexanoic acid by the migratory locust and the honeybee. *J. Comp. Physiol.* **87**, 277–284.

Kaissling, K.-E. 1974. Sensory transduction in insect olfactory receptors. In L. Jaenicke, Ed., *Biochemistry of Sensory Functions*. Springer-Verlag, Berlin and New York, pp. 243–273.

Kaissling, K.-E. and J. Thorson. 1980. Insect olfactory sensilla: structural, chemical and electrical aspects of the functional organisation. In D. B. Sattelle, L. M. Hall, and J. G. Hildebrand, Eds., *Receptors for Neurotransmitters, Hormones and Phermones in Insects*. Elsevier, Amsterdam, pp. 261–282.

Kanaujia, S. and K.-E. Kaissling, 1985. Interactions of phermones with moth antennae: adsorption, desorption and transport. *J. Insect Physiol.* **31**, 71–81.

Kasang, G. and K.-E. Kaissling. 1972. Specificity of primary and secondary olfactory processes in *Bombyx* antennae. In D. Schneider, Ed., *Olfaction and Taste IV*. Wiss. Verlagsges., Stuttgart, pp. 200–206.

Kendall, M. D. 1970. The anatomy of the tarsi of *Schistocerca gregaria* Forskal. *Z. Zellforsch. Mikrosk. Anat.* **109**, 112–137.

Kendall, M. D. and A. M. Seddon. 1975. The effect of previous access to water on the responses of locusts to wet surfaces. *Acrida* **4**, 1–9.

Kennedy, J. S. and J. E. Moorhouse. 1969. Laboratory observations on locust responses to wind-borne grass odour. *Entomol. Exp. Appl.* **12**, 487–503.

Le Berre, J. R. and A. Louveaux. 1969. Equipement sensoriel des mandibles de la larve du premier stade de *Locusta migratoria* L. *C.r. Hebd. Seances Acad. Sci.* **268**, 2907–2710.

Le Berre, J. R., Y. Sinoir, and C. Boulay. 1967. Etude de l'équipement sensoriel de l'article distal des palps chez la larvae de *Locusta migratoria migratorioides* (R. et F.). *C. R. Hebd. Seances Acad. Sci., Ser. D* **265**, 1717–1720.

Lewis, C. T. 1970. Structure and function in some external receptors. *Symp. R. Entomol. Soc. London* **5**, 59–76.

Loftus, R. 1976. Temperature-dependent dry receptor on antenna of *Periplaneta*. Tonic response. *J. Comp. Physiol.* **111**, 153–170.

Louveaux, A. 1972. Equipement sensoriel et système nerveux périphérique des pièces buccales de *Locusta migratoria* L. *Insectes Soc.* **19**, 359–368.

Louveaux, A. 1973. Etude de l'innervation sensorielle de l'hypopharynx de larve de *Locusta migratoria migratoriodes* R. & F. (Orthoptere, Acrididae). *Insectes Soc.* **22**, 3–12.

Ma, W. C. 1972. Dynamics of feeding responses in *Pieris brassicae* L. as a function of chemosensory input: A behavioural, ultrastructural and electrophysiological study. *Med. Landbouwhogesch. Wageningen* 72 (11), 1–162.

Mordue (Luntz), A. J. 1979. The role of the maxillary and labial palps in the feeding behaviour of *Schistocerca gregaria. Entomol. Exp. Appl.* **25**, 279–288.

Pfaffmann, C. 1941. Gustatory afferent impulses. *J. Cell. Comp. Physiol.* **17**, 243–258.

Phillips, C. E. and J. S. Vande Berg. 1976a. Mechanism for sensillum fluid flow in trichogen and tormogen cells of *Phormia regina* (Meigen) (Diptera: Calliphoridae). *Int. J. Insect Morphol. Embryol.* **5**, 423–431.

Phillips, C. E. and J. S. Vande Berg. 1976b. Directional flow of sensillum liquor in blowfly (*Phormia regina*) labellar chemoreceptors. *J. Insect Physiol.* **22**, 425–429.

Riegert, P. W. 1960. The humidity reactions of *Melanoplus bivittatus* (Say) (Orthoptera, Acrididae): Antennal sensilla and hygro-reception. *Can. Entomol.* **92**, 561–570.

Schoonhoven, L. M., W. M. Blaney, and M. S. J. Simmonds. 1987. Inconstancies of chemoreceptor sensitivities. In V. Labeyrie, G. Fabres, and D. Lachaise, Eds., *Insects-Plants.* Junk, Dordrecht, The Netherlands, pp. 141–145.

Sehnal, F. 1985. Growth and life cycles. In G. A. Kerkut and L. I. Gilbert, Eds., *Comprehensive Insect Physiology, Biochemistry and Pharmacology*, Vol. 2. Permagon, Oxford, pp. 1–86.

Simpson, S. J. and J. D. Abisgold. 1985. Compensation by locusts for changes in dietary nutrients: Behavioural mechanisms. *Physiol. Entomol.* **10**, 443–452.

Simpson, S. J., M. S. J. Simmonds, and W. M. Blaney. 1988. A comparison of dietary selection behaviour in larval *Locusta migratoria* and *Spodoptera littoralis. Physiol. Entomol.* **13**, 225–238.

Slifer, E. H. 1970. The structure of arthropod chemoreceptors. *Annu. Rev. Entomol.* **15**, 121–142.

Slifer, E. H., J. J. Prestage, and H. W. Beams. 1959. The chemoreceptors and other sense organs on the antennal flagellum of the grasshopper (Orthoptera; Acrididae). *J. Morphol.* **105**, 145–191.

Staedler, E. 1984. Contact chemoreceptors. In W. J. Bell and R. T. Carde, Eds., *Chemical Ecology of Insects.* Chapman & Hall, London, pp. 3–35.

Steinbrecht, R. A. 1969. Comparative morphology of olfactory receptors. In C. Pfaffmann, Ed., *Olfaction and Taste III.* Rockefeller Univ. Press, New York, pp. 3–21.

Steinbrecht, R. A. 1974. Odorant uptake and transport in insect sensilla. In T. M. Poyner, Ed., *Transduction Mechanisms in Chemoreception.* Information Retrieval Ltd., London, pp. 49–57.

Sturckow, B. 1970. Responses of olfactory and gustatory receptor cells in insects. *Adv. Chemorecept.* **13**, 107–159.

Szentesi, A. and E. A. Bernays. 1984. A study of behavioural habituation to a feeding deterrent in nymphs of *Schistocerca gregaria. Entomol. Exp. Appl.* **9**, 329–340.

Thomas, J. G. 1965. The abdomen of the female desert locust (*Schistocerca gregaria* Forskål) with special reference to the sense organs. *Anti-Locust Bull.* **42**, 1–22.

Thomas, J. G. 1966. The sense organs on the mouthparts of the desert locust *Schistocerca gregaria*. *J. Zool.* **148**, 420–448.

Thurm, U. 1965. An insect mechanoreceptor. Fine structure and adequate stimulus. *Cold Spring Harbor Symp. Quant. Biol.* **30**, 75–82.

Thurm, U. 1972. The generation of receptor potentials in epithelial receptors. In D. Schneider, Ed., *Olfaction and Taste IV.* Wiss. Verlagseges., Stuttgart, pp. 95–101.

Viscuso, R. 1971. Studio morphologico e istologico della parete epifaringea di *Eyprepocnemis plorans* (Charp.) (Orth. Acrid.). *Boll. Sedute Accad. Gioenia Sci. Nat. Catania* **10**, 633–655.

Viscuso, R. 1974. Studio comparato della lamina epifaringea di vari insetti ortotteri. *Redia* **55**, 129–141.

Waldow, U. 1970. Elektrophysiologische Untersuchungen an Feuchte-, Trocken- und Kalterezeptoren auf der Antenne der Wanderheuschschrecke *Locusta. Z. Vergl. Physiol.* **69**, 249–283.

Winstanley, C. and W. M. Blaney. 1978. Chemosensory mechanisms of locusts in relation to feeding. *Entomol. Exp. Appl.* **24**, 750–758.

Wunderer, H. and U. Smola. 1982. Contact chemoreceptors among wind-sensitive head hairs of *Locusta migratoria* L. (Orthoptera: Acrididae). *Int. J. Insect Morphol. Embryol.* **11**, 147–160.

Yokohari, F. 1978. Hygroreceptor mechanism in the antenna of the cockroach *Periplaneta. J. Comp. Physiol.* **124**, 53–60.

Zacharuk, R. Y. 1980. Ultrastructure and function of insect chemosensilla. *Annu. Rev. Entomol.* **25**, 27–47.

Zacharuk, R. Y. 1985. Antennae and Sensilla. In G. A. Kerkut and L. I. Gilbert, Eds., *Comprehensive Insect Physiology, Biochemistry and Pharmacology*, Vol. 6. Pergamon, Oxford, pp. 1–69.

2

Food Selection

R. F. CHAPMAN

2.1 INTRODUCTION

Food selection by grasshoppers was reviewed by Mulkern (1967) and Uvarov (1977). The latter review, however, was completed before 1970, when Uvarov died. Since that time extensive analyses of the food eaten by grasshoppers in the field, mainly in north America, have been carried out (e.g., Campbell et al., 1974; Joern, 1983) in parallel with in-depth studies on particular species (e.g., Bernays and Chapman, 1977; Parker, 1984; Sinoir, 1969). At the same time entomologists have become increasingly aware of the importance of plant chemistry as a basis for food selection in insects generally (Hsiao, 1985) and in grasshoppers in particular (Bernays and Chapman, 1978). Rowell (1978) has also drawn attention to the possible bias resulting from the extensive studies on species from grassland-dominated areas, notably in North America and Africa. He stresses the need to study species from the forests of Central America where a greater degree of host specificity may occur.

In this chapter I attempt to draw together these different approaches with the aims, first, of demonstrating the range of selectivity that occurs within the Acridoidea and then of describing the underlying proximate factors that determine whether an insect will consume or reject a particular plant.

2.2 FOOD EATEN IN THE FIELD

Probably all species of grasshoppers exhibit some degree of selectivity in the food they eat. Certainly this is true of all the species that have been studied. For example, Uvarov (1977) notes that although *Schistocerca gregaria* is known to feed on some 400 species of plants, there are others that are not eaten at all. Husain et al. (1946), for example, list 160 species from 54 families that were readily eaten in cages, 29 species eaten with great reluctance, and 9 plants from 7 families not eaten at all. This led Uvarov to doubt whether any species was really polyphagous, in the sense of feeding on "everything green." The more recent studies on the food of grasshoppers reinforce this view.

Most authors do not use the term polyphagous in the sense of Uvarov, and there are also differences in use of the terms monophagous and oligophagous. For example, Bernays and Chapman (1978) regard grass-feeding taxa as oligophagous because they feed almost exclusively on Poaceae, although within this family they may eat many species. Otte and Joern (1977), however, are prepared to call this polyphagy. Both views are tenable. However, semantic differences must not be allowed to conceal, or to create, real differences, so it is important to be as precise as possible about terminology. In this article monophagous means feeding on only one genus of plant; other genera are totally rejected. Oligophagous means feeding on a limited range of plants, usually within one family; other plants are usually totally rejected, but not invariably so. In disjunct oligophagy (Otte and Joern, 1977) the insect feeds on a limited number of unrelated plants usually in different families. Polyphagy means feeding on a range of plants from a number of different families. It grades into disjunct oligophagy. Polyphagy does not imply that all plants are equally well accepted; in other words, the insect exhibits preferences, totally rejecting some plants, eating some species in small amounts, and eating others in large amounts (see Fig. 2.4). In practice, as will be seen, precise definitions of monophagy and oligophagy are difficult to sustain and categorization contains a subjective element.

Another difficulty arises because, as Uvarov (1977) and Rowell (1985a) point out, the range of plants eaten, as determined by laboratory experiment, may not coincide with food plant use in the field. In this paper I use the terms monophagy, oligophagy, and polyphagy with reference to what the insect is known to eat in the field, generally as determined by analysis of foregut contents or feces. In most cases the terms are applied with reference to a species rather than a population. When a particular population is considered, this is apparent.

2.2.1 Monophagy

The only species of acridoid known to be strictly monophagous is *Bootettix argentatus*. It is known to feed only on *Larrea tridentata*. Analyses of crop contents reveal the presence of this plant only (Otte and Joern, 1977), and it will not eat other plant species even after 20 h without food (Chapman et al., 1988). Grasses are occasionally nibbled, that is, bitten repeatedly without ingestion, but the structure of the mandibles is such that the insect cannot bite through the veins of grass leaves.

No other well-documented cases of strict monophagy are known. *Hypochlora alba*, which is able to survive on only *Artemisia* (Knutson, 1982) and feeds almost exclusively on it (Table 2.1), nevertheless does occasionally eat other plants, especially grasses. *Melanoplus bowditchi* may also be specific to *Artemisia* (Mulkern et al., 1969), and *Melanoplus discolor* is almost entirely restricted to *Kuhnia*, another member of the Asteraceae (Mulkern et al., 1969). Greenfield and Shelly (Chapter 10) have demonstrated that *Ligurotettix planum* is monophagous on *Flourensia*. Other North American species that may be monophagous are listed by Otte and Joern (1977), but these are not based on critical analyses.

Anablepia granulata appears to be monophagous on the grass *Brachyaria brachylopha* (Hummelen and Gillon, 1968), although this is based on a single small sample. The only other species that is known to specialize on a particular grass is *Opeia abscura*. Although it eats a variety of grass species, its main food is *Bouteloua gracilis*, which normally makes up about 90% of its food (Table 2.2).

Sometimes a particular population may be monophagous although the species is known to feed on other plants elsewhere. This is reported for a population of *Rhachicreagra nothra* living on *Clibadium* (Rowell, 1985b), and populations of some grass-feeding species, for example, *Syrbula fuscovittata*, may approach this condition (Joern, 1979). *Ligurotettix coquilletti* is also in this category (see below).

TABLE 2.1. Food Eaten by *Hypochlora alba*, Based on Analysis of Crop Contents[a]

Asteraceae										
Artemisia	*Ambrosia*	*Antennaria*	*Aster*	*Achillea*	Fabaceae	Rosaceae	Brassicaceae	Solanaceae	Poaceae	Reference
97	1	T	T		T	T			2	Mulkern et al., 1964
91	4			1	3				1	Mulkern et al., 1969
88									16	Mulkern et al., 1969
97	1			1	1		2	2	T	Mulkern et al., 1969
94					1		2	2	T	Joern, 1985

[a] Figures are percentages of insects containing each category of food. Because some contained more than one type of food, the total percentage may exceed 100. T = trace.

2.2.2 Oligophagy

Increasing numbers of reports of oligophagy are accumulating. Rowell et al. (1983) record that *Hylopedetes nigrithorax* feeds on ferns, although within this taxon it feeds on a variety of genera. *Drymophilacris bimaculata* feeds only on a number of genera in the Solanaceae (Rowell, 1978). *Hesperotettix viridis* is largely restricted to genera within the tribe Astereae of Asteraceae (Parker, 1984), although individuals are sometimes found to have eaten small amounts of plants from other families (Table 2.3). *Hesperotettix speciosus* also feeds mainly on Asteraceae.

Oligophagy on grasses has been recorded for all or most species of Hemiacridinae, Tropidopilinae, Acridinae, Gomphocerinae, and Truxalinae (Table 2.4) and can be regarded as a general attribute of these subfamilies. Table 2.4 is based primarily on analyses of gut contents or feces. Insects are placed in a category provided that at least 90% of the records fall within it. Among the Oedipodinae grass feeding is a common phenomenon, although many species are polyphagous, whereas in other subfamilies grass feeding occurs in isolated genera. This is true of *Oxya hyla* among the Oxyinae examined (Chapman, 1964) and of occasional species of Catantopinae (Chapman, 1964). It will no doubt prove to be the case in numbers of other species in otherwise polyphagous taxa. The more extensive studies on North American species, however, show that all the grass feeding species are occasionally found to have eaten forbs.

TABLE 2.2. Food Eaten by *Opeia obscura*, Based on Analysis of Crop Contents[a]

Poaceae									
Bouteloua gracilis	Other *Bouteloua* spp.	*Calamovilfa*	*Stipa*	*Andropogon*	*Buchloe*	Other Grasses	Cyperaceae	Dicotyledons	Reference
94		1	1			1	2	2	Mulkern et al., 1964
92	5			5	2	7	2	6	Mulkern et al., 1969
88	1	2	3	1	5	5	1		Mulkern et al., 1969
85		1	11			3			Mulkern et al., 1969
95	1	1	1			1	3	3	Mulkern et al., 1969
66					19	14		1	Joern, 1979
94				4				3	Joern, 1979
94		T				4	T	1	Pfadt and Lavigne, 1982
75	5	4	5	8	3				Joern, 1985

[a] Figures are percentages of insects containing each category of food. Because some contain more than one type of food, the total percentage may exceed 100. T = trace.

TABLE 2.3. Food Eaten by *Hesperotettix* Species, Based on Analysis of Crop Contents[a]

	Asteraceae													
	Ambrosia	*Aster*	*Artemisia*	*Gutierrezia*	*Helianthus*	*Solidago*	Others	Boraginaceae	Chenopodiaceae	Fabaceae	Other Dicotyledons	Poaceae	Cyperaceae	Reference
H. speciosus	49					41						6		Mulkern et al., 1969
					89		4	1		1	1	3		Mulkern et al., 1969
	37				48		1			2	7	1		Joern, 1985
H. viridis	6	2	1		T	68	2	1		2	2	6	1	Mulkern et al., 1964
	42	2	11			17	1			2	1	2		Mulkern et al., 1969
	7	2	1			69	3	1		3	2	4	1	Mulkern et al., 1969
	1	13		62	3	1		1	3			1		Mulkern et al., 1969
	4	T			10	54	T		13	T		2		Joern, 1985

[a] Figures are percentages of insects containing each category of food. Because some contain more than one type of food, the total percentage may exceed 100. T = trace.

TABLE 2.4. Range of Food Types Eaten by Grasshoppers in the Field[a]

Taxon	Number of Species	Ferns	Number Feeding on: Dicots	Number Feeding on: Dicots + Monocots	Number Feeding on: Grasses	Diet Breadth: >3 Families	Diet Breadth: 2 or 3 Families	Diet Breadth: 1 Family[b]	Diet Breadth: Poaceae	Diet Breadth: One Genus	References[c]
Proctolabinae	1		1					1			1
Dericorythinae	1		1								2
Hemiacridinae	3				3				2		3,4
Tropidopolinae	5				5				3		3,4
Oxyinae	4			3	1						3,5
Coptacridinae	4		3	1							3,4
Calliptaminae	4		3	1			1				2,4
Eyprepocnemidinae	9		2	7		2					2–4
Cyrtacanthacridinae	7		5	1	1	3		1			2,4,6–10
Catantopinae	17		10	6	1	1					3–5,11,12,22
Melanoplinae	45		22	23		29		2		4	6,7,10,13–25
Leptysminae	7		2		5					1	23
Ommatolampinae	13		5	8		7	4	1		1	23,26,27
Rhytidochrotinae	1	1									28
Oedipodinae	55		7	27	21	14		1	10		2–7,10,13–16,18,20 21,23–25,29–36
Acridinae	25		1	2	22				12		2–4,11
Gomphocerinae	72		5	10	57	3	1	3	2	3	2–4,6,7,10,13–16, 18,20,21,23–25,31, 32,37,38
Truxalinae	2				2				1		2,4

[a] Mainly based on crop or fecal analysis. In many instances food was categorized only as "grass" or "forb"; data from these insects are not included in the diet breadth category.
[b] Other than Poaceae.
[c] References: 1, Rowell, 1978; 2, Gangwere and Agacino, 1973; 3, Chapman, 1964; 4, Hummelen and Gillon, 1968; 5, Monk, 1987; 6, Mulkern et al., 1964; 7, Mulkern et al., 1969; 8, Chapman, 1957; 9, Chapman, 1959; 10, Gangwere et al., 1976; 11, Chapman, 1962; 12, Anderson, 1964; 13, Mulkern et al., 1962; 14, Joern, 1983; 15, Banfill and Brusven, 1973; 16, Campbell et al., 1974; 17, Rogers and Uresk, 1974; 18, Pfadt & Lavigne, 1982; 19, Bailey & Mukerji, 1976; 20, Ueckert and Hansen, 1971; 21, Ueckert et al., 1972; 22, Bland, 1976; 23, Gangwere and Ronderos, 1975; 24, Otte and Joern, 1977; 25, Sheldon and Rogers, 1978; 26, Rowell, 1985a; 27, Braker, 1986; 28, Rowell et al., 1983; 29, Boutton et al., 1980; 30, Bailey and Riegert, 1971; 31, Boys, 1981; 32, Stebaev and Pshenitcina, 1978; 33, Bernays and Chapman, 1973; 34, Bernays et al., 1976b; 35, Launois, 1973; 36, Boys, 1978; 37, Bernays and Chapman, 1970a; 38, Capinera, 1985.

Disjunct oligophagy has been demonstrated in relatively few cases. Rowell (1985a) shows that the various species of *Rhachicreagra* exhibit disjunct oligophagy, feeding on a few species from up to six families, although certain genera of Asteraceae are important for a majority of species. Each species in captivity would accept any plant eaten by any congeneric species, but rejected most other plants that were offered.

Ligurotettix coquilletti is a more extreme example. Populations of this species are known to occur on *Larrea* (Zygophyllaceae), *Atriplex* (Chenopodiaceae), and *Lycium* (Solanaceae) (M. D. Greenfield, personal communication) and these are all eaten readily (Chapman et al., 1988). At any one locality, however, the population is entirely or almost entirely (Otte and Joern, 1977) restricted to feeding on one plant species. So the individual populations are monophagous, although the species exhibits disjunct oligophagy.

2.2.3 Polyphagy

Polyphagy has been demonstrated in numerous species, sometimes based on field observations, sometimes on the analysis of gut or fecal contents. In an extensive series of field observations, Braker (1986) observed *Microtylopteryx hebardi* to feed on species from nine families of monocotyledons and seven families of dicotyledons, although monocotyledons predominated. Most Melanoplinae and many Oedipodinae have been shown to be polyphagous in critical studies of the food eaten (Table 2.4) and this may also be true of Catantopinae, although most studies have failed to differentiate between plant species. Eyprepocnemidinae are also polyphagous (few critical studies, but all species eat both grasses and dicotyledonous plants). But among the taxa in which most species are grass feeders, polyphagy is an uncommon phenomenon.

2.3 FOOD SELECTION BEHAVIOR

Food selection behavior may involve attraction from a distance as well as selection when in contact with the plant.

2.3.1 Attraction from a Distance

Little is known about attraction from a distance, but olfactory and visual stimuli may be involved. Moorhouse (1971) showed that nymphs of *S. gregaria* were increasingly attracted by the odor of grass in a wind tunnel as their period without food became longer. Lee et al. (1987) also showed that nymphs of *Schistocerca americana* moved upwind in a wind tunnel to the odor of mint, on which they had been feeding previously. E. A. Bernays and R. F. Chapman (unpublished) observed that nymphs of *Chortoicetes terminifera* marching in the field in a hot, dry environment turned upwind and fed on tufts of fresh grass. It is not known if the insects were responding to host-plant odor or the local increase in humidity.

The experiments of Wallace (1958) with *S. gregaria* showed that this insect was attracted by vertical stripes and shapes around an otherwise featureless arena, and previously Kennedy (1939) had demonstrated in the field that the insect when on the ground often moved toward clumps of vegetation. Various workers later showed similar responses by other species (Mulkern, 1967).

None of the experiments on attraction by olfaction or vision suggest any degree of specificity to host plants, but this is not surprising in view of the fact that only polyphagous species have been tested. Bernays and Wrubel (1985), however, demonstrated that *Melanoplus sanguinipes*, another polyphagous species, learned to associate food with color and it found food more quickly when the food was in an appropriately colored area. At least some acridoids are known to move away from the immediate vicinity of their food after a meal, and the experiments of Bernays and Wrubel (1985) suggest that learning may enable them to find the food more readily before the next meal.

Parker (1984) showed that *Hesperotettix viridis* moved from its food plant more quickly when other plants of the same species were nearby than when they were several meters away, suggesting that it was responding to the visual stimuli presented by the close plants, although the possibility of olfaction is not excluded. The visual response need not have been specific because the host, *Gutierrezia*, was probably the most conspicuous feature in the insect's environment.

Evidence that specific olfactory responses may occur is provided by the experiments of Boppré et al. (1984). They showed that *Zonocerus elegans* is attracted by pyrrolizidine alkaloids. It is not known if this species is attracted by plants containing the chemicals, but *Zonocerus variegatus* is attracted, at least over short distances, by the odor of *Chromolaena odorata*, which is known to contain these alkaloids (Modder, 1984).

Other indirect, though still equivocal, evidence is provided by the electroantennogram responses of *Hypochlora alba* to the odors of its host plant. Blust and Hopkins (1987a) concluded that males, but not females, have developed a specific sensitivity to these odors. This conclusion is based on the reduced sensitivity of the males to geraniol, a nonhost odor, compared with the generalist feeder *Melanoplus sanguinipes*. However, the response to host odors relative to the response to geraniol is higher in both sexes of *H. alba* than in *M. sanguinipes*, so the results could be interpreted as showing that both sexes of the specialist have a specific sensitivity to the host odors. More work is needed to clarify this issue.

2.3.2 Selection When in Contact with a Plant

Although attraction to a plant from a distance may be important in the initial stages of food selection for some species, particularly if they are geophilous, many other species of grasshopper rest between meals on potential food plants. In either case, selection ultimately depends on behavior when the insect is in contact with a plant. The choice they are faced with is, consequently, "shall I eat *this* plant or not?"

An insect already on a plant is receiving two sets of chemical stimuli: the smell of the plant close to the surface probably as perceived by the antennal receptors, and the gustatory input from the chemicals on the leaf surface perceived by tarsal chemoreceptors. Usually, before selecting or rejecting a plant, the insect also drums the leaf surface with its maxillary and labial palps (palpation) and then it bites the leaf.

There is very little information about the roles of olfaction and tarsal contact in food selection. Mordue (1979) showed that nymphs of *S. gregaria* with the palps removed bit more frequently on elderberry pith (an inert substrate) in the presence of food odor than in its absence. Clearly the odor was acting as a stimulant and some evidence with *Bootettix argentatus* suggested a similar response in that insect (Chapman et al., 1988). This species was also shown to move off nonhost plants without palpation, but the effects of olfaction and tarsal contact cannot be separated in these experiments. Szentesi and Bernays (1984) record that *S. gregaria* rejected wheat coated with nicotine hydrogen tartrate on tarsal contact alone on the first few encounters. Other studies on food selection emphasize the roles of the palps and the receptors in the cibarial cavity (see Section 2.3.3).

The different sensory inputs resulting from these behaviors do not form a rigid hierarchical sequence because insects will feed on a leaf while standing on some quite different substrate. In an extreme example, *S. americana* has been observed feeding on a leaf coated with nicotine hydrogen tartrate while holding the forelegs high in the air to avoid contact with the surface. This followed previous rejection of the leaf on tarsal contact alone (White and Chapman, in press). Clearly, in this instance a strong negative signal from the tarsal chemoreceptors was overridden by positive signals from the cibarial receptors. The insect then adopted a stance that minimized the negative input. Grasshoppers will also eat following antennectomy and palpectomy, but insects without palps often failed to bite sucrose-impregnated filter paper although intact insects did so (Mordue, 1979), and palpation is probably a normal preliminary to biting.

2.3.3 What Makes a Grasshopper Bite?

Blaney and Chapman (1970) and Bernays et al. (1976a) demonstrated that the nymphs of *Locusta migratoria* were stimulated to bite by an extract of the surface wax of the normal host plant *Poa*. Biting followed contact with the dry surface of a test filter paper by the palps. Biting was not induced by contact with the wax from a nonhost plant. Nordihydroguaiaretic acid (NDGA) in the surface resin of *Larrea* also promoted biting by *B. argentatus*, whereas the extracts of nonhost plants had no effect (Chapman et al., 1988).

Despite this evidence, most work on the surface waxes has tended to emphasize the deterrent properties of nonhost waxes. Nonhosts are often seen to be rejected following palpation without biting and this behavior has now been recorded for a number of species with different feeding habits: *Schistocerca gregaria* (E. A. Bernays and R. F. Chapman, unpublished), and *S. americana* (R. F. Chapman, unpublished), both polyphagous species; *Chorthippus parallelus* (Bernays and Chapman, 1970b, 1975), *Chortoicetes terminifera* (Bernays and Chapman, 1973), *Euchorthippus pulvinalis* (Boys, 1981), and *Locusta migratoria* (Chapman, 1977), all essentially graminivorous; *Ligurotettix coquilletti* (R. F. Chapman and E. A. Bernays, unpublished), an oligophagous species; and *Bootettix argentatus* (R. F. Chapman and E. A. Bernays, unpublished), monophagous on *Larrea tridentata*. Removal of the surface waxes with a suitable solvent has been shown to increase the amount of biting on nonhosts by *L. migratoria* (Woodhead and Chapman, 1986) and *Euchorthippus pulvinalis* (Boys, 1981). The discriminating power of the palps is also demonstrated by the experiments of Mordue (1979). Insects stimulated by a host plant odor, but without palps, bit more frequently on an inert substrate than did intact insects.

Thus it seems likely that all acridoids have the ability to recognize nonhosts via the palps by virtue of the leaf surface properties.

Seedling sorghum is among the plants normally rejected by *L. migratoria* at palpation and analysis of the wax of these plants revealed that deterrence occurred in the fractions containing *n*-alkanes, esters and *p*-hydroxybenzaldehyde (Woodhead, 1983). Among the alkanes, those with chain lengths of 19, 21 and 23 carbon atoms were deterrent; others were not. Short-chain (C_{12}) acids esterified with tetracosanol (C_{24} alcohol) were also deterrent, whereas a long-chain (C_{24}) acid was not. The short-chain alkanes were only minor constituents of the wax of mature sorghum plants and these were readily eaten.

Blaney and his associates (Blaney and Simmonds, 1985; Blaney and Winstanley, 1982) demonstrated that the deterrent effect of the leaf surface chemicals of some nonhosts was learned by *L. migratoria* (see below), and this too may be a general phenomenon. This finding led Blaney et al. (1985) to suggest that biting is a normal response for a grasshopper following contact with any substrate if it does not receive information via the palps leading it to reject that substrate. Biting "does not depend on favourable input from the palps, but rather on lack of unfavourable input." Such an approach certainly helps to make sense of situations where the insects are seen to bite on inert substrates, but it does not account for those instances where palpation has been shown to stimulate feeding (see above), nor the fact that grasshoppers with normal access to food usually do not bite on inert substrates. More detailed studies with naive insects are necessary before we can answer this aspect of the general question, what makes a grasshopper bite a potential host plant?

Sinoir (1969) demonstrated that biting was induced by the stimulation of mechanoreceptors on the labrum and galea, and it is probably true that the effect of palpation is not to induce biting directly, but to cause the insect to lower its head. As a result, the mechanoreceptors on the labrum are stimulated and biting follows. It is possible that the apparently random biting that occurs on inert substrates is a consequence of such mechanical stimulation, although this is not the view taken by Blaney et al. (1985, see above), and it is clearly a positive activity, not the result of accidentally bringing the labrum into contact with the substrate.

2.4 PLANT CHEMISTRY

2.4.1 Internal Chemistry of the Leaf

When the insect bites into a leaf, it is assumed that the chemicals in solution within the cells spread over the inner surface of the epipharynx guided by hydrophilic cuticular setae (Cook, 1977a) to the groups of chemoreceptors just outside the mouth (see Chapter 1). The information received from these receptors provides the critical information on which acceptance or rejection of the food is finally based.

The chemicals occurring within the leaf can conveniently be considered in two categories: those of known or assumed nutritional value to the plant and insect, and plant secondary compounds, which are generally assumed to have no nutritional value but are assumed to have ecological value to the plant.

2.4.1.1 Nutrient Chemicals

Most of the Acridoidea so far examined are stimulated to feed by the common hexose and disaccharide sugars found in plants (Bernays and Chapman, 1978). The species that have now been shown to be stimulated by sucrose are *Bootettix argentatus* (R. F. Chapman, unpublished), *Camnula pellucida* (Thorsteinson, 1960), *Chorthippus curtipennis* (Thorsteinson, 1958), *Chortoicetes terminifera* (Browne et al., 1975), *Ligurotettix coquilletti* (Chapman et al., 1988), *Locusta migratoria* (Cook, 1977b), *Melanoplus sanguinipes* (Mulkern et al., 1978), *Poecilocerus bufonicus* (Abushama, 1968), *Schistocerca americana* (E. A. Bernays, unpublished observation), *S. gregaria* (Azzi, 1975), and *Zonocerus variegatus* (Bernays and Chapman, 1978). These insects represent a sufficiently wide range of feeding habits and taxa to conclude with some confidence that sucrose is a phagostimulant for Acridoidea. The same is probably true for glucose and fructose, although the data are less extensive. However, it is important to note that sucrose is not a phagostimulant for *Drymophilacris bimaculata*, an oligophagous species feeding on Solanaceae (Rowell, 1978), although no critical data are presented.

The amount of substrate eaten in experiments is proportional to the concentration of the sucrose up to an optimum. Sugars are usually presented dry on an inert substrate. The molar concentration refers to the concentration in the original solution so the dry weight available to the insect varies with the methodology used by different authors (compare Cook, 1977b, and Sinoir, 1969). The method of analysis also affects the results (Cook, 1976), but concentrations from 0.01 M upward are always stimulating. Consequently the levels of sucrose, glucose, and fructose commonly found in plants are normally sufficient to stimulate feeding unless their effect is offset by other factors. Other sugars are not usually present in high enough concentrations to play an effective role (Bernays and Chapman, 1978).

Starch and, in grasses, fructosans commonly occur as storage polysaccharides in leaves, sometimes exceeding 10% of the dry weight. Both compounds are slightly stimulating to *L. migratoria*, although these effects are probably normally outweighed by the soluble sugars. This is also true for most other nutrient components of the leaf. Some amino acids and plant phospholipids have been shown to be phagostimulatory for several species. However, their effects are so slight at concentrations normally found in leaves that they probably have little overall effect on food selection (Bernays and Chapman, 1978). It may be, however, that at least the amino acids are involved in enabling the insect to select between foods of different nutrient qualities (Simpson and Abisgold, 1985). The various phagostimulatory components (sugars, amino acids, phospholipids) tend to be additive in their effect on *L. migratoria* (Bernays and Chapman, 1978).

Unlike other nutrient chemicals in the leaf, inorganic salts do not stimulate feeding and are deterrent at higher concentrations, but there is some evidence that some salts in low concentrations slightly increase the quantity of sucrose consumed on test disks (Bernays and Chapman, 1978; Boys, 1981).

2.4.1.2 Responses to Plant Secondary Compounds

The responses of grasshoppers to plant secondary compounds fall into five classes, four of which have been observed with *S. gregaria*, the species most extensively studied (Bernays and Chapman, 1978). In class I, food intake is reduced as the concentration increases; no enhancement of feeding occurs in the range over which the chemicals normally occur (Fig. 2.1a). Many terpenoids are in this class for *S. gregaria* as well as some alkaloids and nonprotein amino acids (Table 2.5). Class II chemicals stimulate feeding at low concentrations, but become deterrent as the concentration increases (Fig. 2.1b). Some alkaloids are in this class. Class III contains chemicals that are phagostimulatory at intermediate levels, but have no effect at higher or lower concentrations (Fig. 2.1c). They might become deterrent at still higher concentrations, but such concentrations do not occur in plants. Most phenolic glycosides are in this category for *S. gregaria*. Chemicals in class IV have no effect on feeding in their normal concentration range (Fig. 2.1d; Table 2.5), whereas finally in class V the chemical is phagostimulatory at all naturally occurring concentrations. This last category is known only for the response of *Bootettix argentatus* to nordihydroguaiaretic acid (NDGA) (Fig. 2.2), but perhaps includes the response of *Zonocerus elegans* to pyrolizidine alkaloids (Boppré et al., 1984).

The response of different species to different compounds varies; for example, although *B. argentatus* exhibits a class V response to NDGA, *Ligurotettix coquilletti* and *Cibolacris parviceps* exhibit a class II response to the same compound over

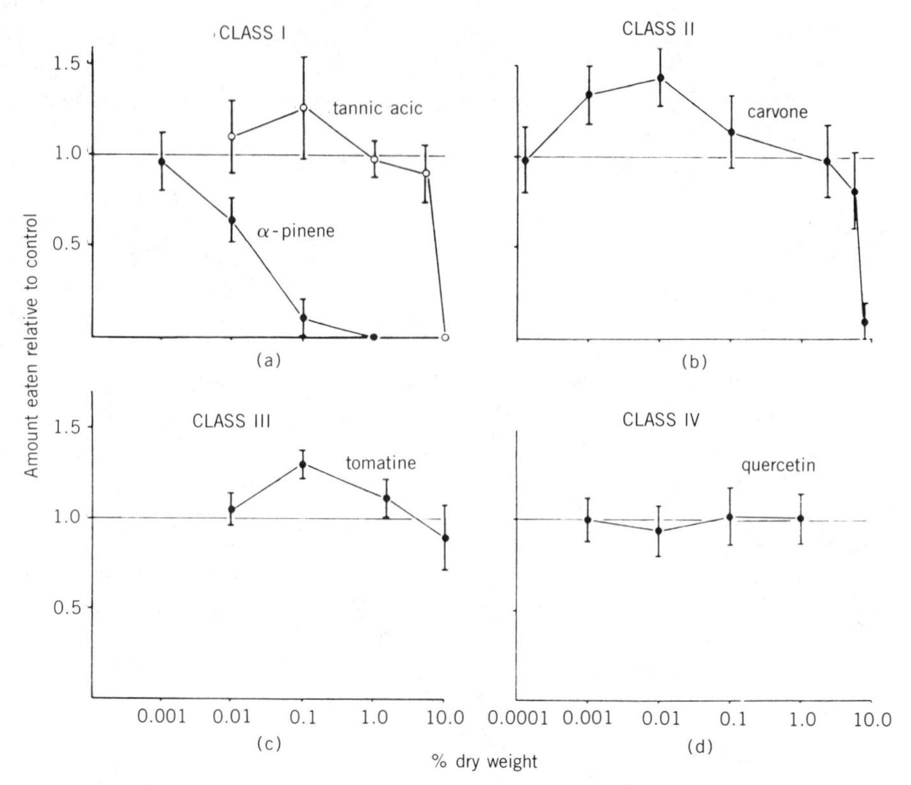

Fig. 2.1. Different types of effects produced by different concentrations of non-nutrient chemicals on the amounts eaten by *Schistocerca gregaria.* (After Bernays and Chapman, 1978.)

the same concentration range (Chapman et al., 1988; Fig. 2.2). Large numbers of compounds have been studied only in relation to *S. gregaria* and *L. migratoria* (Bernays and Chapman, 1978) so that useful comparisons are possible only between these two species. In this case 83 compounds were tested over ranges of concentrations at which they became deterrent. Sixty of these compounds were deterrent to *L. migratoria* at markedly lower concentrations than to *S. gregaria* (Fig. 2.3); only one, azadirachtin, was more deterrent to *S. gregaria* than to *L. migratoria*. No compounds were found that were phagostimulants for *L. migratoria* (class II, III, or V). These differences in sensitivity are probably associated with the fact that *L. migratoria* is oligophagous on grasses whereas *S. gregaria* is polyphagous.

Adams and Bernays (1978) demonstrated that the deterrent effects of secondary plant compounds to *L. migratoria* were additive. This was true when compounds from different chemical classes were added to test substrates and also when different compounds from the same class were added. Using a range of 14 phenolic compounds at concentrations of 0.01–0.4% dry weight they could detect no statistically significant reduction in the amounts eaten in

TABLE 2.5. Response by *Schistocerca gregaria* to Different Chemicals [a]

	Class of Response			
Chemical Group	I	II	III	IV
Nonprotein amino acids	3	1	3	9
Sulphur compounds	0	0	0	5
Monoterpenes	8	1	0	0
Sesquiterpenes	3	0	0	11
Triterpenoids	8	0	1	14
Carboxylic acids	0	2	2	0
Fatty acids	0	1	1	3
Phenolics	2	2	4	28
Phenolic glycosides	0	0	6	1
Tannins	3	0	0	0
Purines and miscellaneous N compounds	0	0	2	4
Cyanogenic glycosides	0	0	0	4
Alkaloids	5	4	2	6
Inorganic salts	0	0	0	7

[a] Responses divided into four classes corresponding with those illustrated in Fig. 2.1 and the numbers show the number of compounds in each chemical group that induced the particular class of response (after Bernays and Chapman, 1978).

short-term tests when each compound was tested individually, although at higher concentrations they were all known to be deterrent. However, when all 14 compounds were added together at the same concentrations, a marked deterrent effect was observed.

2.4.2 Plant Chemistry and Food Selection

There is ample evidence that phagostimulant and deterrent chemicals interact to determine meal size. Figures 2.1 and 2.2 show the effects of increasing concentrations of various chemicals added to phagostimulatory substrates. For many chemicals, meal size decreases as the concentration of deterrent is increased. The effects of such interaction in the selection of hosts was examined in an extensive series of experiments with *L. migratoria* and *S. gregaria* (Bernays and Chapman, 1977, 1978). Meal size was determined on a range of host and nonhost plants. Then each plant was extracted, separately, with chloroform, methanol, acetone, and water. Each extract was added to a phagostimulating wafer and its effect on the amount eaten determined.

Nymphs of *L. migratoria* that had been deprived of food for 5 h before determination of meal size on the plants in most cases ate a full meal or nothing at all. When the effects of extracts were examined it was found that plants that were totally rejected or eaten in small amounts invariably yielded

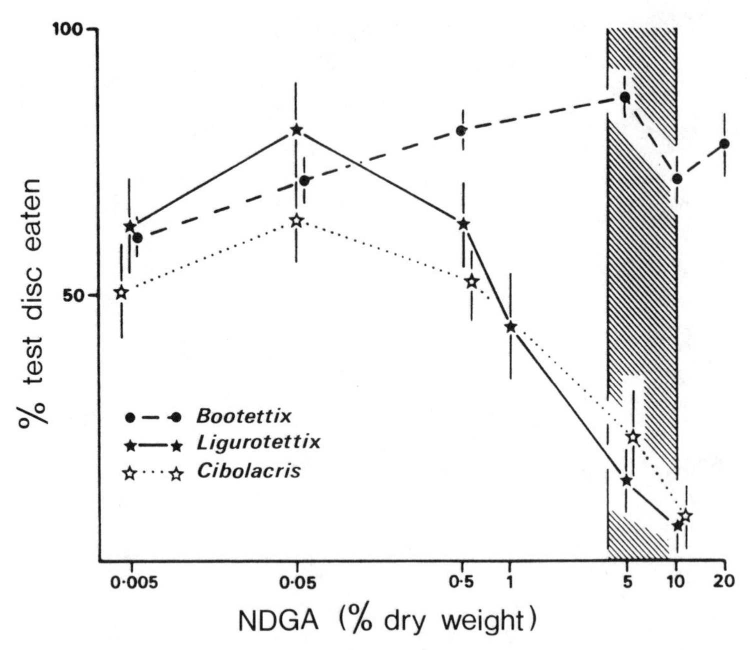

Fig. 2.2. Responses of three *Larrea*-feeding grasshoppers to different concentrations of NGDA on sucrose-impregnated glass-fiber disks. The hatched area shows the range of concentrations normally occurring on *Larrea* leaves. (After Chapman et al., 1988.)

at least one extract that was deterrent (Fig. 2.4). Plants that were eaten readily yielded no deterrent extracts but sometimes had an increased phagostimulatory effect. Essentially similar results were obtained in less extensive studies on *Chorthippus parallelus* (Bernays and Chapman, 1978) and *Truxalis grandis* (Abushama and Elkhider, 1976), both of which feed on grasses.

In contrast, the meal sizes of *S. gregaria* on a range of plants formed a continuum. Plants that were not eaten, or eaten only in small amounts, yielded at least two deterrent extracts, whereas those eaten in increasing amounts yielded progressively more extracts that were phagostimulatory and progressively fewer that were deterrent (Fig. 2.4).

These results demonstrate clearly that food intake in both species is governed by a balance of phagostimulatory and deterrent chemicals in the plants. The greater sensitivity of *L. migratoria* demonstrated with pure chemicals (Fig. 2.3) is apparently reflected in its all-or-nothing response to plants. Only those plants are normally eaten that contain no detectable deterrent in a behavioral assay of extracts, but the level of phagostimulation above a certain minimum seems to be unimportant. Meal size in this case is governed to a very large extent by volumetric considerations (see Chapter 3). *S. gregaria* is less sensitive to the deterrent effects of secondary compounds and is often stimulated by them (Figs. 2.1b, 2.1c). Its meal size is the result of a balance between

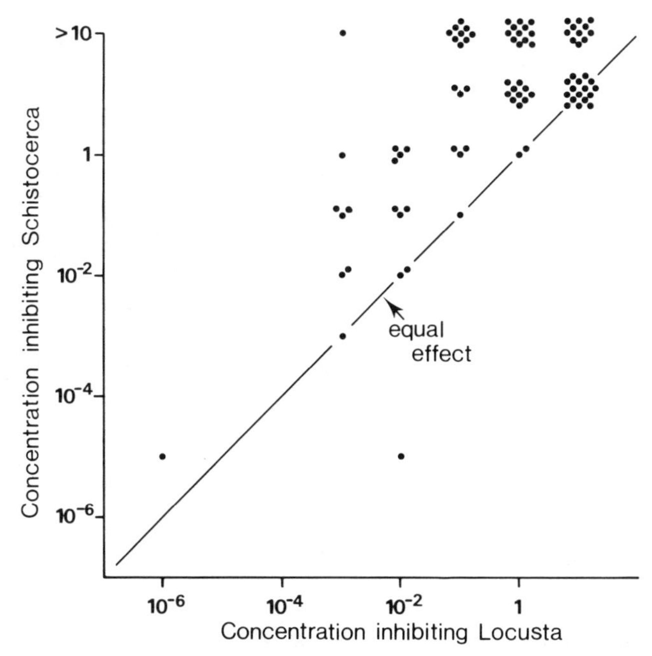

Fig. 2.3. Dry-weight concentrations of chemicals inducing a 50% reduction in the intake of wheat flour wafers by *Locusta migratoria* and *Schistocerca gregaria*. Each dot represents one compound. (Data from Bernays and Chapman, 1978.)

phagostimulatory and deterrent properties of the plant. Volumetric considerations may be relatively unimportant in governing meal size, although it is known that they are involved (Rowell, 1963).

Although it seems possible to account for the choice of foods and meal size in polyphagous acridids, such as *S. gregaria* and graminivorous species, such as *L. migratoria*, in terms of a balance between commonly occurring phagostimulants and deterrents without involving specific chemicals, there is evidence for specific phagostimulants in other species. *Bootettix argentatus*, which is monophagous on *Larrea*, is stimulated to feed by nordihydroguaiaretic acid, a chemical characteristic of the host (Chapman et al., 1988). *Hypochlora alba* is also stimulated to feed by an extract of its host plant, *Artemisia*, and activity has been found to reside in a fraction consisting primarily of monoterpene hydrocarbons (Blust and Hopkins, 1987b). Similarly a key chemical or group of chemicals is reported to govern host plant selection by *Drymophilacris bimaculata* by Rowell (1978). This species feeds only on Solanaceae. It will eat other plants if these are smeared with the juice from crushed *Solanum* leaves, but sucrose-impregnated filter paper is rejected. The identity of the chemical(s) involved is unknown. It is to be expected that such specific phagostimulants will be found only in relation to species feeding on a single plant species or group of species characterized by taxon specific chemicals.

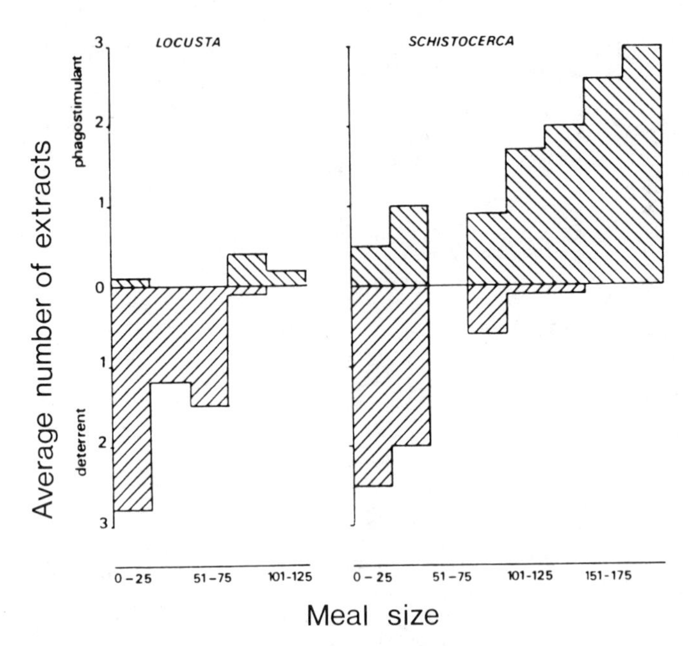

Fig. 2.4. Average numbers of extracts/plant (with a maximum of four) that stimulated or inhibited feeding by *Locusta migratoria* and *Schistocerca gregaria* in relation to the meal size on each plant species in milligrams. Based on extracts from 201 and 97 plant species, respectively. (After Bernays and Chapman, 1978.)

2.4.3 Pharmacophagy

Pharmacophagy, the ingestion of specific chemicals for purposes other than nutrition (Boppré et al., 1984), is known to occur only in one species of acridoid, *Zonocerus elegans*. Boppré et al. (1984) have demonstrated that this species is attracted to, and feeds on, filter paper impregnated with pure pyrrolizidine alkaloids. It is not known whether host plant selection is normally affected by this response, but *Z. variegatus* is attracted by the flowers of *Chromolaena odorata*, which contain pyrrolizidine alkaloids (Modder, 1984). It is suggested that the alkaloids are a source of a pheromone (Chapman et al., 1986).

2.5 PLANT QUALITY AND AVAILABILITY

2.5.1 Water Content

Food selection is affected both by the water content of the food and by the state of hydration of the insects. This is most comprehensively illustrated by the experiments of Lewis and Bernays (1985) with *S. gregaria* (see also Chapter 5). Three groups of nymphs were fed from the molt to fifth instar, one on fresh seedling wheat with a water content of 90%, one on lyophilized seedling wheat (2–5% water) supplemented each day by one meal of fresh wheat, and one on a choice of fresh and lyophilized wheat. After 3 days on these treatments the insects were offered a choice of fresh and lyophilized leaves. Insects previously fed on fresh wheat almost invariably started to feed on the lyophilized leaves within a few minutes of the start of the experiment, whereas those previously fed dry leaves fed on the fresh leaves. Insects previously given a choice of wet and dry foods showed no preference for either and started to feed after a longer time interval. Thus hydrated insects selected the dry food, dehydrated insects the wetter food. Similar results were obtained with *S. gregaria* by Roessingh et al. (1985) in no-choice experiments.

Experiments by various authors demonstrate a similar phenomenon in *L. migratoria*. Ben Halima et al. (1983) showed that the intake of lyophilized corn was proportional to the amount of water ingested and Sinoir (1966, 1968) showed that the consumption of sucrose-impregnated filter paper increased with the water content of the food to about 70%, and then declined. Loveridge (1974) found that adult *L. migratoria*, *S. gregaria*, and *Chortoicetes terminifera* ate very little when fed for prolonged periods on dry grass and bran, without water, but continued to ingest large amounts of these foods when water was also available. Conversely, Bernays and Chapman (1978) record that *L. migratoria*, which normally rejected *Rumex* and *Aethusa*, showed an increased tendency to feed on them if it was deprived of other sources of food and water; in the presence of water neither plant was eaten.

The mechanism by which these behavioral responses are mediated is not clear, but from the accounts of Lewis and Bernays (1985) it appears that the choice of whether to feed or not is made before the insect bites. Kendall and Seddon (1975) showed that well-hydrated nymphs of *L. migratoria* were repelled by damp strips of filter paper, but dehydrated insects attempted to feed on the same strips. The insects apparently responded to the humidity of the air immediately above the strip, which was perceived by the antennae, although tarsal contact with the damp paper also had a role in the response.

There is no doubt that the selection of food is affected by its water content and that of the insect, but it is not clear to what extent this influences food selection naturally because of the role of other factors. Lewis (1979, 1982) showed that *Melanoplus differentialis* fed preferentially on the wilted leaves of sunflower and that a reversal of preference between two host plants occurred when the water contents of the leaves were reversed. Bernays and Lewis (1986) also found that the palatability of several plant species for *S. gregaria* changed with wilting, and Chapman (1957) showed a change in preference by *Nomadacris septemfasciata* associated with a difference in the moisture contents of a sedge and a grass. However, as Lewis (1982) points out, these changes cannot be separated from possible chemical differences. Bernays et al. (1977) showed that an increase in the palatability of cassava leaves to *Zonocerus variegatus* associated with loss of turgor was due to a change in the rate at which those leaves produced hydrogen cyanide.

In synthesis, it may be presumed that acridoids adjust their intake of water, which is normally a constituent of their food, so as to maintain a body water content of about 70% (see Chapter 5). Consequently insects that are highly desiccated may be expected to reject relatively dry food and to favor food with a high water content. This may sometimes cause the insect to accept relatively unpalatable foods, as in the case of *L. migratoria* outlined above. Ben Halima et al. (1984) showed that under dry conditions in Morocco adult *Dociostaurus maroccanus* ate very large amounts of the composite *Scorzonera* even though the females failed to mature when fed exclusively on the plant. They concluded that the insect ate this plant primarily to maintain its water content. Schäller and Köhler (1981) showed that *Gomphocerus rufus* and *Chorthippus biguttulus*, but not *Chorthippus parallelus*, selected grasses that were high in water content, although whether this was due to the water content or some other feature was not determined. There are no field examples of insects that become water-loaded because of an excess of very wet food then selecting drier foods, though this may be expected to occur.

2.5.2 Differences in Plant Quality

Superimposed on the restrictions of host range due to plant chemotaxonomy are limitations due to chemical differences within a plant species. Such differences may occur temporally, between plant parts, or as a result of disease. They may also reflect genetic differences between individual plants.

The availability of nitrogen imposes a potential limitation on insect development and reproduction and might be expected to be involved in food selection by grasshoppers. Simpson et al. (1988) showed that nymphs of *L. migratoria* preconditioned on an artificial diet with and without protein adjusted their intake when subsequently given a choice. In the first hour of the choice test, the protein-fed insects selected mainly the carbohydrate (no-protein) diet, whereas those from a diet containing no protein fed mainly on the protein-only diet. It is not clear how the amounts of diet were adjusted, but the experiments leave no doubt that short-term regulation of protein intake can occur.

Evidence for the field is conflicting. Heidorn and Joern (1987) showed that *Ageneotettix deorum* ate larger meals on leaves of *Calamovilfa* that had higher nitrogen levels, but with *Phoetaliotes nebrascensis* (Joern and Alward, 1988) there was no consistent pattern. In only one of four grass species were meal sizes bigger on the nitrogen-rich leaves. This failure to find a consistent effect is perhaps not surprising because, apart from other possible differences between the host species, these insects were probably not deficient in nitrogen. Further, in their experiments with *L. migratoria*, Abisgold and Simpson (1987) showed that compensation for a dietary deficiency was achieved by a reduction in the interval between meals, not by changes in the meal length. If this is generally true, experiments that examine the size of only a single meal would not demonstrate compensation.

That grasshoppers do select different foods to regulate the intake of nitrogen is suggested by the observation that mature female *Oedaleus senegalensis* feed preferentially on the developing grains of millet (Boys, 1978; see Section 2.6.3).

Bernays et al. (1974) demonstrated that seedling grasses were unpalatable for a number of different acridid species, but mature plants of the same species were highly acceptable. In the case of *Sorghum* it was shown that the initial unpalatability was due to the combined effects of a cyanogenic glucoside, several phenolic compounds, and *p*-hydroxybenzaldehyde in the surface wax (Woodhead and Bernays, 1978; Woodhead, 1982). As the concentrations of these compounds declined, the leaves became more palatable. In other grasses unpalatability of seedlings is probably due to different compounds, such as alkaloids (Bernays and Chapman, 1976). Bernays et al. (1977) also demonstrated that yellow senescent leaves of cassava were readily eaten by *Zonocerus variegatus*, but green leaves were not, apparently because the senescent leaves lacked cyanogenic glucosides. When cassava plants are attacked by this species mature leaves with lower glucoside content are eaten before young leaves with high glucoside levels. Cooper-Driver et al. (1977) showed that there were seasonal variations in the acceptability of bracken (*Pteridium*) to *S. gregaria*. A fall in acceptability early in the season was attributed to rising levels of cyanogenic glucoside, whereas a late fall may have been due to increased tannins. Changes in palatability comparable with these examples may commonly occur for it is well known that leaf chemistry changes with plant age.

There is little evidence demonstrating differences in the selection of plant parts, apart from those due to aging. However, Boys (1978) showed that mature females of *Oedaleus senegalensis* fed preferentially on the milky heads of millet. Males normally ate only the leaves (see below). Modder (1984) records that *Zonocerus variegatus* was attracted to and fed actively on the flowers of *Chromolaena odorata*, although the leaves were not attractive.

Lewis (1979) showed that *Melanoplus differentialis* fed preferentially on the leaves of *Helianthus* with a visible fungus infection and Bernays et al. (1977) found that the early instars of *Z. variegatus*, which did not eat intact leaves of cassava, did feed on diseased spots. Smith (1939) also reported a correlation between grasshopper damage and the presence of rust in wheat. The reasons for this preference are unknown, although Bernays et al. (1977) showed that disease spots on cassava leaves did not produce cyanide when crushed. *Hesperotettix viridis* tends to move away from host plants that have been damaged previously by other grasshoppers (Parker, 1984). Presumably it is responding to changes induced in the chemistry of the plants.

2.5.3 Physical Properties of the Plant

Some plants are not eaten because of their physical properties, such as hairiness, hardness, or size.

Knutson (1982) showed that the hairs on the leaves of *Artemisia* largely prevented *Melanoplus bivittatus* and *M. sanguinipes* from feeding on it; when the hairs were removed the insects were able to feed more readily and survival was improved. Smith and Grodowitz (1983) observed similar, although much less extreme, effects of the trichomes of *Artemisia* on survival and development of *M. sanguinipes*. On most plants the trichomes are less dense than on *Artemisia*, but they may nevertheless influence feeding. Bernays and Chapman (1970b) found that the feeding of first-instar nymphs of *Chorthippus parallelus* on *Holcus* was delayed if the leaves were very hairy; adults were not affected. The adults were sometimes seen to be affected by the hairs on other plants.

For most grasshoppers leaf hardness is unlikely to be of major importance in the choice of food, but Bernays and Chapman (1970b) observed a greater tendency for both nymphs and adults to reject a grass after the first bite if it was very hard, although most plants in the field were not sufficiently hard for this to be important.

Selection may also be affected by the thickness of a leaf. First-instar *Chorthippus parallelus* were unable to eat most leaves of *Festuca* because the diameter of the rolled leaves was too great for the insects to bite (Bernays and Chapman, 1970b). As a result, *Festuca* was underrepresented, compared with its abundance, in the diets of early instars. Later instars rarely encountered this problem although very large leaves were still sometimes too thick for adult males to eat.

Although only a few critical studies have been made, it is clear that the physical properties of leaves may sometimes affect the choice of food made by acridoids, especially in the early instars. It is often true that the physical factors only delay feeding, but in a natural environment this would probably result in the insect moving to another plant.

2.5.4 Plant Availability and Foraging Behavior

There is ample evidence to show that the food eaten by a grasshopper is partially governed by the relative abundance of potential host plants. For example, Bernays and Chapman (1970a) showed that *Chorthippus parallelus* consumed the grasses in the habitat roughly in proportion to their abundance. Rowell (1985a) showed that the principal food plant consumed by *Rhachicreagra* spp. was always relatively common in the habitat. Ueckert et al. (1972) showed that there was a significant correlation between the percentage frequency with which plants occurred in the diets of grasshoppers and the percentage frequency with which the plants occurred in the habitat for 13 out of 14 species, but the correlation was less good in insects eating a wider range of plants. However, it is not to be expected that a perfect match will often occur between plant availability and food eaten because not all plants are equally palatable. Thus Bernays and Chapman (1970a,b) found that *Anthoxanthum* was eaten in smaller quantities than expected from its abundance, probably because of the relatively high levels of coumarin it contained, and *Festuca* was underrepresented in the diet of the early instars because they could not bite through the rolled leaf blade.

There is also laboratory evidence concerning the effects of availability on food choice. Chandra and Williams (1983) showed that when *S. gregaria* was given an array of two plants of equal palatability, the plants were eaten in proportion to their relative abundance. On the other hand, if one plant was less palatable it was eaten in relatively greater amounts as it became less abundant in the array. Conversely, Cottam (1985) showed that *Omocestus viridulus* often grazed disproportionately on the commoner of two grasses, especially when it was the more palatable. The behavioral basis for these differences in selection is not known, but in both experiments the insects were starved beforehand and the results may have been influenced by this. But whatever the behavioral basis, it is clear that the insect's response to a combination of plants varies with their relative frequencies in an unpredictable way and this is likely to affect food selection in the field as well as the laboratory.

In an interesting and important study of *Dociostaurus maroccanus*, a polyphagous species, Ben Halima et al. (1985) found that most individual adults confined in a large field cage consumed food from one or two different plant species each day and some ate as many as five plant species. Males tended to eat more different plants than females, probably because of their greater mobility. By the end of 4 days 1 out of 11 females had fed exclusively on one plant species, but the average number for females was 2.8 plant species and, for males, 5.5, with a maximum of 8. Over the 4-day period the range of plants eaten by individual insects was less than the range eaten by the population. The authors believe this to indicate a tendency for individuals to specialize and that the differences do not arise from restriction to specific microhabitats. They suggest that the insects differ genetically. An alternative explanation might be that they differed in their experience of different host plants both before and during the experiment.

Parker (1984) investigated foraging behavior of *Hesperotettix viridis* on its principal host, *Gutierrezia*, in the field. He found that movement to a new plant was influenced by its proximity, by the nutritional state of the insect, and by the quality of the present host. Despite the relative specificity of this species, continuous movement among individual hosts was a major factor in plant exploitation.

Gutierrezia is a relatively conspicuous shrub so that movement from plant to plant could probably be directed visually from a distance. This would not be true of species living on herbaceous plants in most habitats; nevertheless the importance of plant quality and nutritional experience of the insects may be widespread. If this is true, it is probable that active foraging for appropriate food is a common phenomenon in many grasshoppers although it has not been investigated.

Heidorn and Joern (1987) showed that graminivorous grasshoppers tended to accumulate in areas of the grass, *Calamovilfa*, that had received nitrogenous fertilizer. This took place over a 3–6 week period.

2.6 DIFFERENCES IN THE INSECT

2.6.1 Effects of Other Behaviors

Feeding behavior cannot be viewed in isolation: it is influenced by other aspects of behavior. Although it is obviously of overriding importance, feeding occupies less than 10% of the total nymphal period and even over more limited periods rarely takes more than 20% of the time (Chapman and Beerling, in press). Thus, in terms of time, feeding is not a dominant activity.

Temperature regulation is a major feature of the behavior of many acridoids and this may affect the availability of potential food plants. On a gross scale it may be primarily temperature that restricts the distribution of forest-dwelling species to light gaps (Rowell, 1978) and this, in turn, limits their feeding to plant species growing in those gaps. How important this may be is not known. Nor is there any field data on how the small-scale movements of acridoids associated with temperature regulation in more open habitats affects this food selection, but Kaufmann (1968) showed that early-instar nymphs of *M. differentialis* maintained in cages that were partially shaded ate plants that were in the sun even though they were not the most preferred.

Avoidance of predators is also a dominant aspect of the behavior of many species. Chapman (1957) showed that there was a diurnal change in the food of *Nomadacris septemfasciata* associated with the night roosting behavior of that species, which is assumed to have a predator-avoidance function. Many grass-feeding species are elongate, presumably an adaptation to avoid predation (Bernays and Barbehenn, 1987), and some are only found in tall grasses (*Acrida* spp.). This certainly restricts the foods available to them, and Joern (1983) has shown that among 31 species from Nebraska the diet breadth of ground-dwelling species is about twice that of species habitually found on vegetation.

Males of *Microtylopteryx hebardi* eat a greater range of plants than females, and Braker (1986) suggests that this may be a result of the greater activity of the males, whereas the food of the mature females may be affected by the suitability of plants for oviposition, because this species oviposits endophytically. The food of the early-instar nymphs is almost certainly affected by the plant on which they hatch.

2.6.2 Effects of Experience

There is clear laboratory evidence that food selection of acridids is affected by experience. Habituation, food aversion learning, and associative learning are known to occur.

Habituation is the waning of the response to a repeatedly presented stimulus. In terms of grasshopper feeding, this means that a deterrent compound may become increasingly acceptable over time. Jermy et al. (1982) and Szentesi and Bernays (1984) showed that when nymphs of *S. gregaria* were forced to feed on food treated with nicotine hydrogen tartrate for a limited time each day (so that they did not become short of food) they became progressively more ready to eat the food. This occurred only with relatively weak deterrents; concentrations of deterrents that totally inhibited feeding for 12 h or more did not elicit habituation (Jermy, 1987). *L. migratoria*, an oligophagous species, did not exhibit habituation to any compound.

Associative learning is demonstrated by the experiments of Blaney and Simmonds (1985) with *L. migratoria*. They showed that when a nymph was first presented with a nonhost, it usually bit it before rejecting. On subsequent encounters, however, the plant was rejected with increasing frequency at palpation, the insect having associated the distasteful internal features of the leaf with its surface properties. This was clearly due to the association of specific features (presumably chemical) of the plants, because an insect that had learned to reject *Senecio vulgaris* on palpation did not recognize *Brassica oleraceae* as a nonhost until it had bitten it. When transferred to *S. jacobeae*, however, some insects continued to reject at palpation. Perhaps the waxes of the two *Senecio* species have some feature in common. Memory did not persist if the insects were allowed a meal on a normal host plant. An association between food and color (Bernays and Wrubel, 1985) and food and odor (Lee et al., 1987) has been demonstrated in *Melanoplus sanguinipes* and *S. americana*, respectively.

Food aversion learning has been demonstrated in *S. americana* by Bernays and Lee (1988). In this case it has been shown that a food that is normally tolerated, spinach, may become totally unacceptable if feeding is associated with adverse effects to the insects induced, in this case, by an injection of nicotine. The insect continued to avoid the plant after a meal on wheat, and the memory was retained for 2 days, but was lost after 4 (Lee and Bernays, in press). There is also evidence that spinach alone has some adverse effects and the insect learns to avoid it (Lee and Bernays, 1988).

It is not known how much these phenomena influence food selection by grasshoppers in the field, but there are some references to "conditioning," by which is meant altered preference as a result of experience. For example, Bernays and Chapman (1970a) found that a population of *Chorthippus parallelus* from a habitat in which *Dactylus glomerata* was common readily accepted the plant, whereas insects from a population where the plant was uncommon did not do so. Similarly, Boys (1978) found that *Oedaleus senegalensis* from different habitats accepted the commonest grass from each habitat more quickly than other grasses. In neither of these instances can genetic variation between populations be ruled out. On the other hand, Rowell (1985b) obtained no evidence of conditioning in *Rhachicreagra nothra* and concluded that differences between populations were probably the result of evolutionary change.

2.6.3 Changes due to the Nutritional Status of the Insect

As grasshoppers are deprived of foods for longer periods they tend to eat plants that previously were unpalatable. This is demonstrated for *Chortoicetes terminifera* by Bernays and Chapman (1973). Second- and third-instar nymphs ate only 5 of 23 plant species after 1 h without food, but after 32 h 14 species were accepted and some were eaten in moderate amounts.

It is impossible in such cases to distinguish shortage of nutrients from shortage of water, but the experiments of Loveridge (1974) suggest that water is more critical (see Section 2.5.1). This may also account for records of unusual food plants in the diets of species that are otherwise monophagous or oligophagous. However, McGinnis and Kasting (1967) demonstrated that *Melanoplus bivittatus* adjusted its intake of a diet diluted with different amounts of cellulose so that the levels of nutrients consumed and growth of the insects remained relatively constant despite wide variation in the nutrient content of the food. More recently, Simpson and Abisgold (1985) have shown that nymphs of *L. migratoria* given artificial diets with high- or low-protein levels adjusted the amounts eaten so that over a 12-h period they partly compensated for the reduced protein in the low protein diet. This adjustment was achieved by shortening the interfeed length in response to lower levels of blood osmolality and lower concentrations of various free amino acids in the blood (Abisgold and Simpson, 1987; and see Chapter 4).

There is no direct evidence that food selection in the field is affected by the nutritional status of the insects, but the results of Boys (1978) and Parker (1984) are suggestive. Boys (1978) found that mature females of *Oedaleus senegalensis* selected the milky grains of millet in the field and, in laboratory tests, exhibited a preference for this stage of the plant. Immature females, or males at any stage, did not eat the grains. This can be interpreted as reflecting the increased protein demands of the female during oogenesis.

Parker (1984) maintained *Hesperotettix viridis* for 7 days either on high-quality plants of *Gutierrezia microcephala* or on plants that had been previously damaged and were deficient in nitrogen. When the insects from both sources were transferred to a second set of poor-quality plants those from the good plants rapidly moved away, but those from the poor plants stayed for longer. When transferred to good-quality plants, both sets of insects remained for an equally long time. The implication is that the nutrient status of the insects affected their responses to the poor plants.

2.7 CONCLUSIONS

The overall limits of diet breadth in the field appear to be set by the chemical taxonomy of plants, although these limits are themselves not rigidly fixed. Within this framework, the availability of plants in the habitat plays a dominant role. Availability is determined not only by the relative abundance of a particular plant species, but also by its phenological, nutritional, and pathological condition. Diet breadth may also be affected by the size of the insect, its nutritional status, its previous experience, and in particular, other aspects of its behavior. As a consequence of these constraints the diet breadth of a particular individual and even population may be only a very limited part of the range of plants that are eaten by a particular species under experimental conditions and that is, presumably, genetically determined. The probability that genetic variation between individuals also influences food selection has not yet been investigated.

Despite the wealth of information it is still not possible to claim a full understanding of the basis of food selection in any one species of grasshopper. Extensive studies on *L. migratoria* (Bernays and Chapman, 1977) and *C. parallelus* (Bernays and Chapman, 1975) have failed to demonstrate the occurrence of phagostimulants that characterize the host plants, and only in the instances of species with a limited range of hosts have such phagostimulants been found (Blust and Hopkins, 1987b; Chapman et al., 1988; Rowell, 1978). The critical factor, setting the outer limits of host plant choice, seems to be the distribution of deterrent compounds in nonhost plants.

Intuitively, this explanation seems unsatisfactory because it implies that each individual must sample every plant it encounters to determine its palatability and hosts are recognized only by a process of elimination of the nonhosts. Perhaps, however, this is not a bad strategy if we suppose that the Acridoidea as a group were originally polyphagous.

The only major change from this condition, on the basis of the evidence currently available, is in the evolution of graminivory. There is no evidence that the Poaceae are characterized as a taxon by the possession of characteristic secondary compounds and no evidence of such compounds has been found in the work on food selection by graminivorous grasshoppers. Chemically, the recognition of grasses appears to depend on recognizing what are *not* grasses; everything except grasses "tastes bad" and so only grasses are eaten. Do graminivorous species really recognize grasses (as a group) in some other way, perhaps leaf form? Or are grasses so abundant in the appropriate habitats that recognition is not a problem? Within the grasses most species appear to be relatively unselective, although still exhibiting preferences related to the age and condition of the plant. Where specificity does occur it may be related to the specific chemistry of the host, but there is no information on this.

Monophagy and oligophagy on plants other than grasses have arisen in several subfamilies and in three cases, *Drymophilacris bimaculata*, *Hypochlora alba*, and *Bootettix argentatus*, are associated with specific phagostimulants. Rowell (1985a) also discusses the possibility of chemical recognition of the host plants by *Rhachicreagra* spp. Perhaps it will prove that taxonomically relevant phagostimulants will be involved in many cases of oligophagy, yet the fact that most of these species commonly eat a wide range of plants, albeit in small quantities, argues against any rigid connection between characteristic chemicals and behavior.

Chapman (1982) has argued that the multiplicity of chemoreceptors on the mouthparts of grasshoppers may have had an inhibiting effect on the development of host plant specialization by tending to act against the development of labeled sensory lines. The validity of this argument will not be known until the chemosensory systems of monophagous and oligophagous species have been investigated neurophysiologically.

Environmental factors and behavior not immediately associated with feeding are seen to have considerable influences on food selection and perhaps similar factors have driven evolution toward more limited host plant ranges, as Rowell (1985b) implies with the *Rhachicreagra* species. This is in keeping with the arguments of Bernays and Graham (1988) that host plant specificity in phytophagous insects may often have arisen for reasons other than plant chemistry and the causative factors underlying host plant specificity should not necessarily be seen as primarily chemical, even though plant chemistry may now play an important part in host recognition. As Bernays and Chapman (1978) concluded, there is still little or no evidence to indicate that chemical coevolution of plants and grasshoppers occurred.

REFERENCES

Abisgold, J. D. and S. J. Simpson. 1987. The physiology of compensation by locusts for changes in dietary protein. *J. Exp. Biol.* **129**, 329–346.

Abushama, F. T. 1968. Food-plant selection by *Poecilocerus hieroglyphicus* (Klug) (Acrididae: Pyrgomorphinae) and some of the receptors involved. *Proc. R. Entomol. Soc. London, Ser. A* **43**, 96–104.

Abushama, F. T. and El T. M. Elkhider. 1976. Food preference of the acridid grasshopper *Truxalis grandis grandis* (Klug.). *Acrida* **5**, 245–255.

Adams, C. M. and E. A. Bernays. 1978. The effect of combinations of deterrents on the feeding behaviour of *Locusta migratoria*. *Entomol. Exp. Appl.* **23**, 101–109.

Anderson, N. L. 1964. Observations on some grasshoppers of the Rukwa Valley, Tanganyika. *Proc. Zool. Soc. London* **143**, 395–502.

Azzi, A.-E. A. 1975. A study of some of the factors affecting food consumption by the desert locust, *Schistocerca gregaria* (Forskål). Unpublished Thesis, University of Wales, Bangor.

Bailey, C. G. and M. K. Mukerji. 1976. Consumption and utilization of various host plants by *Melanoplus bivittatus* (Say) and *M. femurrubrum* (DeGeer) (Orthoptera: Acrididae). *Can. J. Zool.* **54**, 1044–1050.

Bailey, C. G. and P. W. Riegert. 1971. Food preferences of the dusky grasshopper, *Encoptolophus sadius costalis* (Scudder) (Orthoptera: Acrididae). *Can. J. Zool.* **49**, 1271–1274.

Banfill, J. C. and M. A. Brusven. 1973. Food habits and ecology of grasshoppers in the Seven Devils Mountains and Salmon River breaks of Idaho. *Melanderia* **12**, 1–33.

Ben Halima, T., A. Louveaux, and Y. Gillon. 1983. Rôle de l'eau de boisson sur la prise de nourriture sèche et le développement ovarien de *Locusta migratoria migratorioides*. *Entomol. Exp. Appl.* **33**, 329–335.

Ben Halima, T., Y. Gillon, and A. Louveaux. 1984. Utilisation des ressources trophiques par *Dociostaurus maroccanus* (Thunberg, 1815) (Orthopt: Acrididae). Choix des espèces consommées en fonction de leur valeur nutritive. *Acta Oecol.* [Ser.]: *Oecol. Gen.* **5**, 383–406.

Ben Halima, T., Y. Gillon, and A. Louveaux. 1985. Spécialisation trophique individuelle dans une population de *Dociostaurus maroccanus* (Orthopt.: Acridae). *Acta Oecol.* [Ser.]: *Oecol. Gen.* **6**, 17–24.

Bernays, E. A. and R. Barbehenn. 1987. Nutritional ecology of grass foliage-chewing insects. In F. Slansky and J. G. Rodriguez, Eds., *Nutritional Ecology of Insects, Mites & Spiders and Related Invertebrates*. Wiley, New York, pp. 147–175.

Bernays, E. A. and R. F. Chapman. 1970a. Food selection by *Chorthippus parallelus* (Zetterstedt) (Orthoptera: Acrididae) in the field. *J. Anim. Ecol.* **39**, 383–394.

Bernays, E. A. and R. F. Chapman. 1970b. Experiments to determine the basis of food selection by *Chorthippus parallelus* (Zetterstedt) (Orthoptera: Acrididae) in the field. *J. Anim. Ecol.* **39**, 761–776.

Bernays, E. A. and R. F. Chapman. 1973. The role of food plants in the survival and development of *Chortoicetes terminifera* (Walker) under drought conditions. *Aust. J. Zool.* **21**, 575–592.

Bernays, E. A. and R. F. Chapman. 1975. The importance of chemical inhibition of feeding in host-plant selection by *Chorthippus parallelus* (Zetterstedt). *Acrida* **4**, 83–93.

Bernays, E. A. and R. F. Chapman. 1976. Antifeedant properties of seedling grasses. *Symp. Biol. Hung.* **16**, 41–46.

Bernays, E. A. and R. F. Chapman. 1977. Deterrent chemicals as a basis of oligophagy in *Locusta migratoria* (L.). *Ecol. Entomol.* **2**, 1–18.

Bernays, E. A. and R. F. Chapman. 1978. Plant chemistry and acridoid feeding behaviour. In J. B. Harborne, Ed., *Biochemical Aspects of Plant and Animal Coevolution*. Academic Press, New York, pp. 99–141.

Bernays, E. A. and M. Graham. 1988. On the evolution of host range in phytophagous insects. *Ecology* **69**, 886–892.

Bernays, E. A. and J. C. Lee. 1988. Food aversion learning in the polyphagous grasshopper *Schistocerca americana*. *Physiol. Entomol.* **13**, 131–137.

Bernays, E. A. and A. C. Lewis. 1986. The effect of wilting on palatability of plants to *Schistocerca gregaria*, the desert locust. *Oecologia* **70**, 132–135.

Bernays, E. A. and R. Wrubel. 1985. Learning by grasshoppers: Association of colour/light intensity with food. *Physiol. Entomol.* **10**, 359–369.

Bernays, E. A., R. F. Chapman, J. Horsey, and E. M. Leather. 1974. The inhibitory effect of seedling grasses on feeding and survival of acridids (Orthoptera). *Bull. Entomol. Res.* **64**, 413–420.

Bernays, E. A., W. M. Blaney, R. F. Chapman, and A. G. Cook. 1976a. The ability of *Locusta migratoria* L. to perceive plant surface waxes. *Symp. Biol. Hung.* **16**, 36–40.

Bernays, E. A., R. F. Chapman, J. McDonald, and J. E. R. Salter. 1976b. The degree of oligophagy in *Locusta migratoria* (L.). *Ecol. Entomol.* **1**, 223–230.

Bernays, E. A., R. F. Chapman, E. M. Leather, A. R. McCaffery, and W. W. D. Modder. 1977. The relationship of *Zonocerus variegatus* (L.) (Acridoidea: Pyrgomorphidae) with cassava (*Manihot esculenta*). *Bull. Entomol. Res.* **67**, 391–404.

Bland, R. G. 1976. Effect of parasites and food plants on the stability of a population of *Melanoplus femurrubrum*. *Environ. Entomol.* **5**, 724–728.

Blaney, W. M. and R. F. Chapman. 1970. The functions of the maxillary palps of Acrididae (Orthoptera). *Entomol. Exp. Appl.* **13**, 363–376.

Blaney, W. M. and M. S. J. Simmonds. 1985. Food selection by locusts: The role of learning in rejection behaviour. *Entomol. Exp. Appl.* **39**, 273–278.

Blaney, W. M. and C. Winstanley. 1982. Food selection behaviour in *Locusta migratoria*. In J. H. Visser and A. K. Minks, Eds., *Proceedings of the 5th International Symposium on Insect-Plant Relations*. Pudoc, Wageningen, pp. 365–366.

Blaney, W. M., C. Winstanley, and M. S. J. Simmonds. 1985. Food selection by locusts: An analysis of rejection behaviour. *Entomol. Exp. Appl.* **38**, 35–40.

Blust, M. H. and T. L. Hopkins. 1987a. Olfactory responses of a specialist and a generalist grasshopper to volatiles of *Artemisia ludoviciana* Nutt. (Asteraceae). *J. Chem. Ecol.* **13**, 1893–1902.

Blust, M. H. and T. L. Hopkins. 1987b. Gustatory responses of a specialist and a generalist grasshopper to terpenoids of *Artemisia ludoviciana*. *Entomol. Exp. Appl.* **45**, 37–46.

Boppré, M., U. Seibt, and W. Wickler. 1984. Pharmacophagy in grasshoppers: *Zonocerus* being attracted to and ingesting pure pyrrolizidine alkaloids. *Entomol. Exp. Appl.* **35**, 115–117.

Boutton, T. W., B. N. Smith, and A. T. Harrison. 1980. Carbon isotope ratios and crop analysis of *Arphia* species in southeastern Wyoming grassland. *Oecologia* **45**, 299–306.

Boys, H. A. 1978. Food selection by *Oedaleus senegalensis* (Acrididae: Orthoptera) in grassland and millet fields. *Entomol. Exp. Appl.* **24**, 78–86.

Boys, H. A. 1981. Food selection by some graminivorous Acrididae. Unpublished Thesis, University of Oxford.

Braker, H. E. 1986. Host plant relationships of the neotropical grasshopper, *Microtylopteryx hebardi* Rehn (Acrididae: Ommatolampinae). Unpublished Thesis, University of California, Berkeley.

Browne, L. B., J. E. Moorhouse, and A. C. M. van Gerwen. 1975. Sensory adaptation and the regulation of meal size in the Australian plague locust, *Chortoicetes terminifera*. *J. Insect Physiol.* **21**, 1633–1639.

Campbell, J. B., W. H. Arnett, J. D. Lambley, O. K. Jantz, and H. Knutson. 1974. Grasshoppers (Acrididae) of the Flint Hills native tallgrass prairie in Kansas. *Agric. Exp. Sta., Kans. State Univ., Res. Pap.* No. 19, pp. 1–147.

Capinera, J. L. 1985. Determination of host plant preferences of *Hemileuca oliviae*, *Paropomala wyomingensis*, and *Diapheromera reli* by choice test and crop analysis. *J. Kans. Entomol. Soc.* **58**, 465–471.

Chandra, S. and G. Williams. 1983. Frequency-dependent selection in the grazing behaviour of the desert locust *Schistocerca gregaria*. *Ecol. Entomol.* **8**, 13–21.

Chapman, R. F. 1957. Observations on the feeding of adults of the red locust (*Nomadacris septemfasciata* (Serville)). *Br. J. Anim. Behav.* **5**, 60 75.

Chapman, R. F. 1959. Field observations on the behaviour of hoppers of the red locust (*Nomadacris septemfasciata* Serville). *Anti-Locust Bull.* **33**, 1–51.

Chapman, R. F. 1962. The ecology and distribution of grasshoppers in Ghana. *Proc. Zool. Soc. London* **139**, 1–66.

Chapman, R. F. 1964. The structure and wear of the mandibles in some African grasshoppers. *Proc. Zool. Soc. London* **142**, 107–121.

Chapman, R. F. 1977. The role of the leaf surface in food selection by acridids and other insects. *Coll. Inter. C.N.R.S.* **265**, 133–149.

Chapman, R. F. 1982. Chemoreception. The significance of sensillum numbers. *Adv. Insect Physiol.* **16**, 247–356.

Chapman, R. F. and E. A. M. Beerling. in press. The pattern of feeding of first instar nymphs of *Schistocerca americana*. *Physiol. Entomol.*

Chapman, R. F., W. W. Page, and A. R. McCaffery. 1986. Bionomics of the variegated grasshopper (*Zonocerus variegatus*) in west and central Africa. *Annu. Rev. Entomol.* **31**, 479–505.

Chapman, R. F., E. A. Bernays, and T. Wyatt. 1988. Chemical aspects of host-plant specificity in three *Larrea*-feeding grasshoppers. *J. Chem. Ecol.* **14**, 557–575.

Cook, A. G. 1976. A critical review of the methodology and interpretation of experiments designed to assay the phagostimulatory activity of chemicals to phytophagous insects. *Symp. Biol. Hung.* **16**, 47–54.

Cook, A. G. 1977a. The anatomy of the clypeo-labrum of *Locusta migratoria* (L.) (Orthoptera: Acrididae). *Acrida* **6**, 287–306.

Cook, A. G. 1977b. Nutrient chemicals as phagostimulants for *Locusta migratoria* (L.) *Ecol. Entomol.* **2**, 113–121.

Cooper-Driver, G., T. Swain, E. A. Bernays, and S. Finch. 1977. The seasonal changes in chemistry of *Pteridium aquilinum* and the related changes in palatability to *Schistocerca*. *Biochem Syst. Ecol.* **5**, 177–184.

Cottam, D. A. 1985. Frequency-dependent grazing by slugs and grasshoppers. *J. Ecol.* **73**, 925–933.

Gangwere, S. K. and E. M. Agacino. 1973. Food selection and feeding behaviour in Iberian Orthopteroidea. *Ann. Inst. Nacl. Inv. Agrar. Ser. Prot. Veg.* **3**, 251–337.

Gangwere, S. K. and R. A. Ronderos. 1975. A synopsis of food selection in Argentine Acridoidea. *Acrida* **4**, 173–194.

Gangwere, S. K., F. C. Evans, and M. L. Nelson. 1976. The food-habits and biology of Acrididae in an old-field community in southeastern Michigan. *Great Lakes Entomol.* **9**, 83–123.

Heidorn, T. J. and A. Joern. 1987. Feeding preference and spatial distribution of grasshoppers (Acrididae) in response to nitrogen fertilization of *Calamovilfa longifolia*. *Funct. Ecol.* **1**, 369–376.

Hsiao, T. H. 1985. Feeding behavior. In G. A. Kerkut and L. I. Gilbert, Eds., *Comprehensive Insect Physiology Biochemistry and Pharmacology*, Vol. 9. Pergamon, New York, pp. 471–512.

Hummelen, P. and Y. Gillon. 1968. Etude de la nourriture des Acridiens de la savane de Lamto en Côte d 'Ivoire. *Ann. Univ. Abidjan, Ser. E* **1**, 199–206.

Husain, M. A., C. B. Mathur, and M. L. Roonwal. 1946. Studies on *Schistocerca gregaria* (Forskål). XIII. Food and feeding habits of the desert locust. *Indian J. Entomol.* **8**, 141–163.

Jermy, T. 1987. The role of experience in the host selection of phytophagous insects. In R. F. Chapman, E. A. Bernays, and J. G. Stoffolano, Eds., *Perspectives in Chemoreception and Behavior*. Springer-Verlag, New York, pp. 143–157.

Jermy, T., E. A. Bernays, and A. Szentesi. 1982. The effect of repeated exposure to feeding deterrents on their acceptability to phytophagous insects. In J. H. Visser and A. R. Minks, Eds., *Proceedings of the 5th International Symposium on Insect-Plant Relationships*. Pudoc, Wageningen, pp. 25–32.

Joern, A. 1979. Feeding patterns in grasshoppers (Orthoptera: Acrididae): Factors influencing diet specialization. *Oecologia* **38**, 325–347.

Joern, A. 1983. Host plant utilization by grasshoppers (Orthoptera: Acrididea) from a sandhills prairie. *J. Range Manage.* **36**, 793–797.

Joern, A. 1985. Grasshopper dietary (Orthoptera: Acrididae) from a Nebraska sand hills prairie. *Trans. Nebr. Acad. Sci.* **13**, 21–32.

Joern, A. and R. Alward. 1988. Effect of nitrogen fertilization on choice among grasses by the grasshopper *Phoetaliotes nebrascensis* (Orthoptera: Acrididae). *Ann. Entomol. Soc. Am.* **81**, 240–244.

Kaufmann, T. 1968. A laboratory study of feeding habits of *Melanoplus differentialis* in Maryland (Orthoptera: Acrididae). *Ann. Entomol. Soc. Am.* **61** (1), 173–180.

Kendall, M. D. and A. H. Seddon. 1975. The effect of previous access to water on the responses of locusts to wet surfaces. *Acrida* **4**, 1–8.

Kennedy, J. S. 1939. The behaviour of the desert locust (*Schistocerca gregaria* (Forsk.)) (Orthopt.) in an outbreak centre. *Trans. R. Entomol. Soc. London* **89**, 385–542.

Knutson, H. 1982. Development and survival of the monophagous grasshopper *Hypochlora alba* (Dodge) and the polyphagous *Melanoplus bivittatus* (Say) and *Melanoplus sanguinipes* (F.) on Louisiana sagewort, *Artemisia ludoviciana* Nutt. *Environ. Entomol.* **11**, 777–782.

Launois, M. H. 1973. "L'alimentation du criquet migrateur dans le sud-ouest malgache," Programme des Nations Unies pour le Développement, Rome (FS) MML/BIO/4.

Lee, J. C. and E. A. Bernays. 1988. Declining acceptability of a food plant for the polyphagous grasshopper *Schistocerca americana* (Drury) (Orthoptera: Acrididae): the role of food aversion learning. *Physiol. Entomol.* **13**, 291–301.

Lee, J. C. and E. A. Bernays. (in press). Food tastes and toxic effects: associative learning by the polyphagous grasshopper *Schistocerca americana* (Drury) (Orthoptera: Acrididae). *Anim. Behav.*

Lee, J. C., E. A. Bernays, and R. P. Wrubel. 1987. Does learning play a role in host location and selection by grasshoppers. In V. Labeyrie, G. Fabres, and D. Lachaise, Eds., *Insects-Plants*. Junk, Dordrecht, Netherlands. pp. 125–127.

Lewis, A. C. 1979. Feeding preference for diseased and wilted sunflower in the grasshopper, *Melanoplus differentialis*. *Entomol Exp. Appl.* **26**, 202–207.

Lewis, A. C. 1982. Conditions of feeding preference for wilted sunflower in the grasshopper *Melanoplus differentialis*. In J. H. Visser and A. K. Minks, Eds., *5th International Symposium on Insect-Plant Relationships*. Pudoc, Wageningen, pp. 49–56.

Lewis, A. C. and E. A. Bernays. 1985. Feeding behavior: Selection of both wet and dry food for increased growth in *Schistocerca gregaria* nymphs. *Entomol. Exp. Appl.* **37**, 105–112.

Loveridge, J. P. 1974. Studies on the water relations of adult locusts. II. Water gain in the food and loss in the faeces. *Trans. Rhod. Sci. Assoc.* **56**, 1–30.

McGinnis, A. J. and R. Kasting. 1967. Dietary cellulose: effect on food consumption and growth of a grasshopper. *Can. J. Zool.* **45**, 365–367.

Modder, W. W. D. 1984. The attraction of *Zonocerus variegatus* (L.) (Orthoptera: Pyrgomorphidae) to the weed *Chromolaena odorata* and associated feeding behaviour. *Bull. Entomol. Res.* **74**, 239–247.

Monk, K. A. 1987. A comparison of the diets of some rainforest grasshoppers (Orthoptera: Acrididae) in Malaysia. *Malay. Nat. J.* **41**, 383–391.

Moorhouse, J. E. 1971. Experimental analysis of the locomotor behaviour of *Schistocerca gregaria* induced by odour. *J. Insect Physiol.* **17**, 913–920.

Mordue (Luntz), A. J. 1979. The role of the maxillary and labial palps in the feeding behaviour of *Schistocerca gregaria*. *Entomol. Exp. Appl.* **25**, 279–288.

Mulkern, G. B. 1967. Food selection by grasshoppers. *Annu. Rev. Entomol.* **12**, 59–78.

Mulkern, G. B., J. F. Anderson, and M. A. Brusven. 1962. Biology and ecology of North Dakota grasshoppers. I. Food habits and preferences of grasshoppers associated with alfalfa fields. *N. D. Res. Rep.* **7**, 1–26.

Mulkern, G. B., D. R. Toczek, and M. A. Brusven. 1964. Biology and ecology of North Dakota grasshoppers. II. Food habits and preference of grasshoppers associated with the Sand Hills prairie. *N. D. Res. Rep.* **11**, 1–59.

Mulkern, G. B., K. P. Pruess, H. Knutson, A. F. Hagen, J. B. Campbell, and J. D. Lambley. 1969. Food habits and preferences of grassland grasshoppers on the north central great plains. *N. D. Agric. Exp. Stn., North Cent. Reg. Publ.* **196**.

Mulkern, G. B., J. C. Records, and R. B. Carlson. 1978. Attractants and phagostimulants used for control and estimation of grasshopper populations. *Entomol. Exp. Appl.* **24**, 550–561.

Otte, D. and A. Joern. 1977. On feeding patterns in desert grasshoppers and the evolution of specialized diets. *Proc. Acad. Nat. Sci. Philadelphia,* **128**, 89–126.

Parker, M. A. 1984. Local food depletion and the foraging behavior of a specialist grasshopper, *Hesperotettix viridis*. *Ecology* **65**, 824–835.

Pfadt, R. E. and R. J. Lavigne. 1982. Food habits of grasshoppers inhabiting the Pawnee site. *Wyo. Agric. Exp. Stn. Sci. Monogr.*, 42.

Roessingh, P., E. A. Bernays, and A. C. Lewis. 1985. Physiological factors influencing preference for wet and dry food in *Schistocerca gregaria* nymphs. *Entomol. Exp. Appl.* **37**, 89–94.

Rogers, L. E. and D. W. Uresk. 1974. Food plant selection by the migratory grasshopper (*Melanoplus sanguinipes*) within a cheatgrass community. *Northwest Sci.* **48**, 230–234.

Rowell, C. H. F. 1963. A method of chronically implanting stimulating electrodes into the brains of locusts, and some results of stimulation. *J. Exp. Biol.* **40**, 271–284.

Rowell, H. F. 1978. Food plant specificity in neotropical rain-forest acridids. *Entomol. Exp. Appl.* **24**, 651–652.

Rowell, C. H. F. 1985a. The feeding biology of a species-rich genus of rainforest grasshoppers (*Rhachicreagra*: Orthoptera, Acrididae). I. Food Plant use and food plant acceptance. *Oecologia* **68**, 87–98.

Rowell, C. H. F. 1985b. The feeding biology of a species-rich genus of rainforest grasshoppers (*Rhachicreagra*, Orthoptera, Acrididae). II. Food plant preference and its relation to speciation. *Oecologia* **68**, 99–104.

Rowell, C. H. F., M. Rowell-Rahier, H. E. Braker, G. Cooper-Driver, and L. D. Gomez. 1983. The palatability of ferns and the ecology of the tropical forest grasshoppers. *Biotropica* **15**, 207–216.

Schäller, G. and Köhler, G. 1981. Untersuchungen zur Nahrungspraferenz und zur Abhängigkeit biologischer parameter von der Nahrungsqualität bei zentraleuropäischen Feldheuschrecken (Orthoptera: Acrididae). *Zool. Jahrb. Abt. Syst.* (*Oekol.*) *Geogr. Biol.* **108**, 94–116.

Sheldon, J. K. and L. E. Rogers. 1978. Grasshopper food habits within a shrub-steppe community. *Oecologia* **32**, 85–92.

Simpson, S. J. and J. D. Abisgold. 1985. Compensation by locusts for changes in dietary nutrients: behavioural mechanisms. *Physiol. Entomol.* **10**, 443–452.

Simpson, S. J., M. S. J. Simmonds, and W. M. Blaney. 1988. A comparison of dietary selection behaviour in larval *Locusta migratoria* and *Spodoptera littoralis*. *Physiol. Entomol.* **13**, 225–238.

Sinoir, Y. 1966. Interaction du déficit hydrique de l'insect et de la teneur en eau de l'aliment dans la prise de nourriture chez le criquet migrateur, *Locusta migratoria migratorioides* (F. et R.). *Hebd. C. R. Seances. Acad. Sci., Ser. D* **262**, 2480–2483.

Sinoir, Y. 1968. Etude de quelques facteurs conditionnant la prise de nourriture chez les larves du criquet migrateur, *Locusta migratoria migratorioides*. I. Facteurs externes. *Entomol. Exp. Appl.* **11**, 195–210.

Sinoir, Y. 1969. Le rôle des palpes et du labre dans le comportement de prise de nourriture chez la larve du criquet migrateur. *Ann. Nutr. Aliment.* **23**, 167–194.

Smith, R. W. 1939. Grasshopper injury in relation to stem rust in spring wheat varieties. *J. Am. Soc. Agron.* **31**, 818–821.

Smith, S. G. F. and G. V. Grodowitz. 1983. Effects on nonglandular trichomes of *Artemisia ludoviciana* Nutt. (Asteraceae) on ingestion, assimilation, and growth of the grasshoppers *Hypochlora alba* (Dodge) and *Melanoplus sanguinipes* (F.) (Orthoptera: Acrididae). *Environ. Entomol.* **12**, 1766–1772.

Stebaev, I. V. and L. B. Pshenitcina. 1978. Food selection of dominant Acridoidea species of Irtysh steppe and water meadows defined by the method of diagnostics of botanical composition of grasshopper's waste (in Russian). *Vopr. Ekol., Novosibirsk*, pp. 18–59.

Szentesi, A. and E. A. Bernays. 1984. A study of behavioural habituation to a feeding deterrent in nymphs of *Schistocerca gregaria*. *Physiol. Entomol.* **9**, 329–340.

Thorsteinson, A. J. 1958. The chemotactic influence of plant constituents on feeding by phytophagous insects. *Entomol. Exp. Appl.* **1**, 23–27.

Thorsteinson, A. J. 1960. Host selection in phytophagous insects. *Annu. Rev. Entomol.* **5**, 193–218.

Ueckert, D. N. and R. N. Hansen. 1971. Dietary overlap of grasshoppers on sandhill rangeland in northeastern Colorado. *Oecologia* **8**, 276–295.

Ueckert, D. N., R. M. Hansen, and K. C. Terwilliger. 1972. Influence of plant frequency and certain morphological variations on diets of rangeland grasshoppers. *J. Range Manage.* **25**, 61–65.

Uvarov, B. 1977. *Grasshoppers and Locusts. A Handbook of General Acridology*, Vol. 2. Centre for Overseas Pest Research, London.

Wallace, G. K. 1958. Some experiments on form perception in the nymphs of the desert locust, *Schistocerca gregaria* Forskål. *J. Exp. Biol.* **35**, 765–775.

White, P. R. and R. F. Chapman. in press. Tarsal chemoreception in the polyphagous grasshopper *Schistocerca americana*: Behavioral assays, sensilla distributions and electrophysiology. *Physiol Entomol.*

Woodhead, S. 1982. p-hydroxybenzaldehyde in the surface wax of sorghum: Its importance in seedling resistance to acridids. *Entomol. Exp. Appl.* **31**, 296–302.

Woodhead, S. 1983. Surface chemistry of *Sorghum bicolor* and its importance in feeding by *Locusta migratoria*. *Physiol. Entomol.* **8**, 345–352.

Woodhead, S. and E. A. Bernays. 1978. The chemical basis of resistance of *Sorghum bicolor* to attack by *Locusta migratoria*. *Entomol. Exp. Appl.* **24**, 123–144.

Woodhead, S. and R. F. Chapman. 1986. Insect behaviour and the chemistry of plant surface waxes. In B. Juniper and R. Southwood, Eds., *Insects and the Plant Surface*. Arnold, London, pp. 123–135.

3

The Pattern of Feeding

S. J. SIMPSON

3.1 INTRODUCTION

In order to make sense of the pattern of behavior exhibited by an animal it is necessary first of all to identify the relevant inputs and then to describe and quantify how they interrelate to produce behavior. These have been the aims of ethologists and experimental psychologists who have investigated the patterning of feeding in animals as diverse as zebra finches, rats, chickens, and horses (Slater, 1974a; Le Magnen and Devos, 1980; Clifton, 1979; Mayes and Duncan, 1986). The first insect to be studied was the African migratory locust, or more specifically, the fifth−instar nymph of the gregarious phase of that species (Blaney et al., 1973). We now understand the organization of feeding behavior in this graminivorous locust better than in any other animal. Detailed analyses of feeding patterns have not been published for another grasshopper, and only a few other insects have been investigated.

The aim of this chapter is to present and assess available information on the patterning of feeding in *Locusta migratoria.* To date nymphs have been studied under standardized and simplified conditions. During experiments, gregarious insects, reared in crowds up to the time of the experiment, have usually been kept alone in a relatively small container with ample, palatable, highly nutritious food (seedling wheat or artificial diet) under controlled temperature (30°C) and lighting regimes (L:D 12:12) (Fig. 3.1). The reason for using nymphs has been that their behavioral repertoire is less extensive than that of adults. Feeding behavior in nymphs is not complicated by the requirements of flight and reproduction, nor is it affected by senescence, although the changing demands of growth throughout a nymphal stadium add complications enough, as discussed in Chapter 4. By observing nymphs feeding ad libitum under standardized nutritional and environmental conditions it was hoped to have a clearer view of the mechanisms regulating ingestion. Once mechanisms are established it becomes possible to study the system under more complicated conditions with some reasonable chance of success. Hence an investigation of feeding patterns in nymphs given highly nutritious food has facilitated the study of the mechanisms underlying compensatory feeding for diets varying in their nutritional quality (see Chapter 4).

No animal spends its entire time feeding. Most eat intermittently, taking their food in bouts or meals. It follows that there are two questions that must be answered before the patterning of feeding can be explained: first, what determines when meals begin, and second, what determines when meals end? Before answers to these can be sought, however, the definition of a meal must be established.

3.2 DEFINING A MEAL

Choosing the correct definition of a meal is critical if an analysis of feeding patterns is to be valid. This is more difficult than it sounds and delimiting a meal requires careful consideration. There are gaps of variable duration in the feeding pattern of a locust (Fig. 3.2). At one extreme food intake stops very briefly as the mandibles open between chews, whereas at other times there are gaps of up to several hours between periods of ingestion. A range of possible interval lengths exists between these extremes. While eating grass a locust

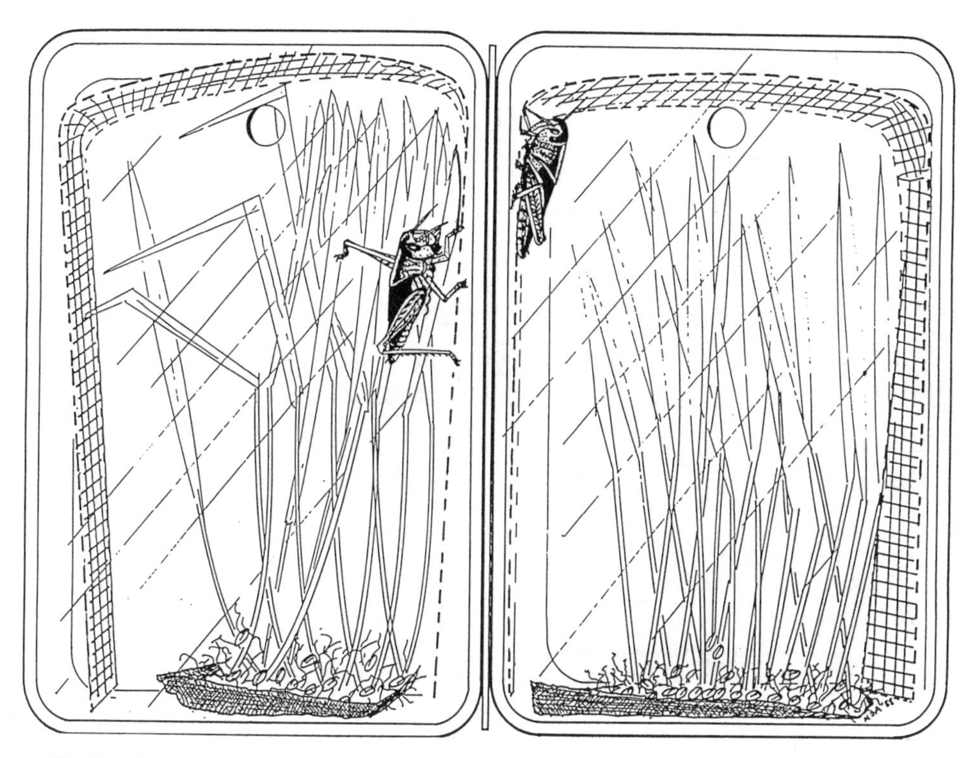

Fig. 3.1. The experimental boxes used for recording the feeding patterns of fifth-instar nymphs of *Locusta migratoria*. Each box contains ample seedling wheat, still growing on capillary matting. Note how one insect is feeding while the other is in the resting position, which is typical of intermeal intervals.

frequently stops pulling a blade into its mouthparts with its forelegs, lifts its face clear of the plant, and spends a few seconds chewing the food already in its preoral cavity. On other occasions it lifts its head and pauses without chewing. During longer gaps the locust stops chewing and begins walking. After a few seconds, or up to several minutes later, it either returns to the food and starts feeding again or else locates a perch and becomes quiescent. The problem of deciding what gaps represent pauses within a meal and which are intermeal intervals is a thorny one and has caused a deal of controversy, best seen in the literature on the white rat (e.g., Panksepp, 1978; Castonguay et al., 1986). The first study in which an attempt was made to define gaps in insect feeding patterns was that of Blaney et al. (1973), although results were not fully analyzed. A suitable method, involving the use of log-survivorship curves, had been used by Wiepkema (1968) on feeding patterns in mice, and Slater (1974a,b, 1975) went on to use the technique extensively in his studies on the temporal patterning of behavior in the zebra finch. The principle of the method is described below with respect to locust feeding. Further details are

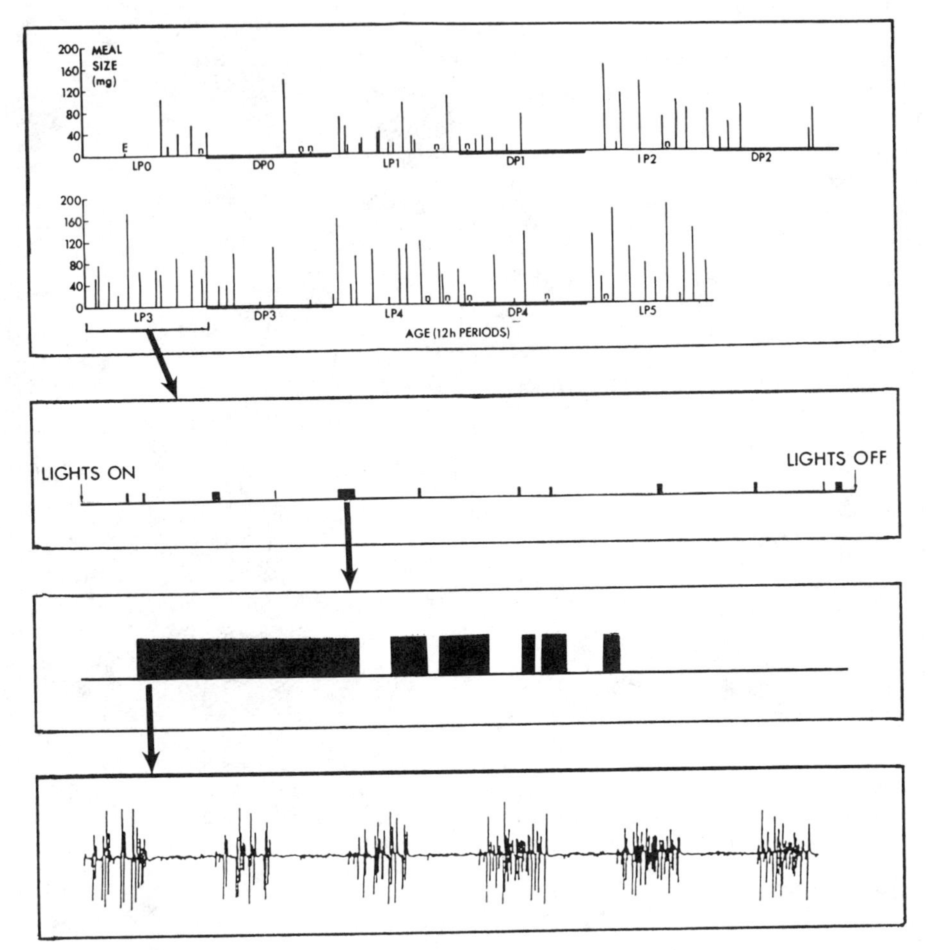

Fig. 3.2. Elements of a feeding pattern recorded from a locust nymph for the first 130 h of the fifth instar. The insect was kept under an L:D 12:12 photoregime at 30°C in a box as in Figure 3.1. Under these conditions the fifth instar lasts about 240 h. The top panel shows the timing of meals and the amounts eaten throughout the 130-h period. E, time of ecdysis; LP, DP, light and dark phases; n, "nibbles," that is, feeding periods of less than 1 min in duration. The second panel expands one of the 12-h light phases (LP3). Here meal duration is shown but not the amount eaten in each meal. The third panel expands the fifth meal eaten during LP3 to show intrameal gaps between periods of ingestion. The entire meal was 16 min long. The bottom panel shows 2.5 s of activity recorded from the mandibular adductor muscles during a period of ingestion in an unrestrained nymph feeding on wheat. (Upper two panels from Simpson, 1981.)

available in Slater's papers, in Fagen and Young (1978), and in Simpson (1982a).

If feeding occurred randomly with time during an observation period then interfeed gaps would follow an exponential distribution. In such a case feeding would not be concentrated into bouts or meals, the probability of commencing ingestion after a pause being independent of the length of that pause. Concentration of eating into bouts is suggested when the probability of starting feeding is greater after short than after long pauses. Whether or not this is the case can readily be seen by plotting the distribution of gaps as a log-survivorship function (Fig. 3.3). The slope of the curve at any point is proportional to the probability of feeding starting. If feeding is not "bouted" then the curve is a

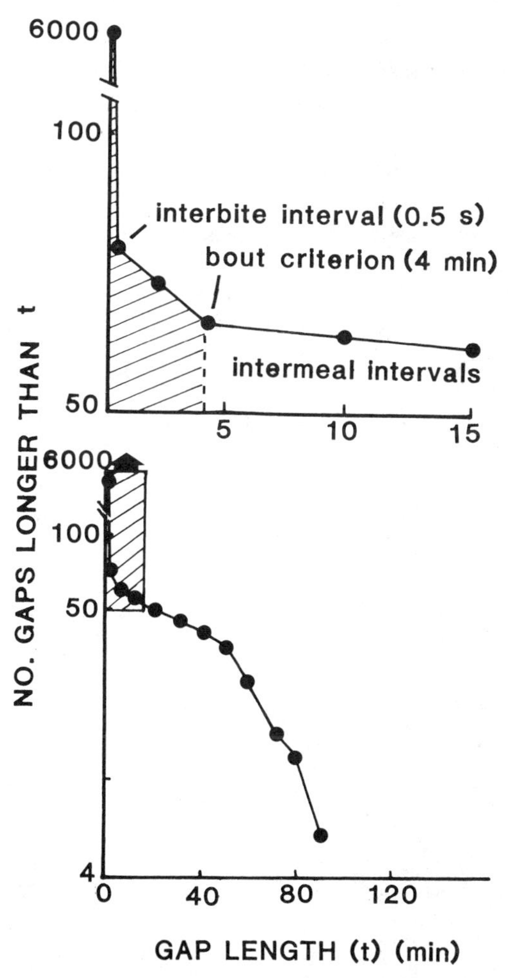

Fig. 3.3. Log-survivorship curves showing the distribution of gaps between periods of feeding. The data are from a single locust nymph for light phases 1–5 during the fifth instar (see Fig. 3.2). The upper graph expands the hatched portion in the lower and illustrates the technique for the selection of a bout criterion to distinguish intrameal pauses from intermeal gaps. Note that the time scale (*x* axis) is expanded close to the origin in order that the slope change in the survivor curve at 0.5 s, corresponding to mandibular movements, can be seen.

straight line, whereas "bouting" produces a concave curve. If it is assumed that a concave slope is made up of at least two behaviorally distinct populations of gap then it is possible to select objectively a criterion for gaps between rather than within meals. Figure 3.3 (top graph) shows a distribution that contains three distinct populations of gap, each distributed negative exponentially. The points at which the slope of the curve changes represent bout criteria. The first slope change occurs at about 0.5 s and corresponds to the frequency of mandibular movements during chewing. The second is at 4 min and indicates that bouts of chewing are themselves grouped into longer bouts. There is also a behavioral correlate for this slope change: in more than 90% of cases where a gap of 4 min or longer occurs after feeding the locust has left the food and entered a characteristic resting position, with hind femora held parallel to its abdomen and its hind tibiae reflexed. This 4-min criterion distinguishes intrameal pauses from intermeal gaps. The actual criterion varies slightly between individual nymphs but is almost invariably between 2 and 6 min in length. The shape of the log-survivorship curve for gaps longer than the bout criterion (Fig. 3.3, lower graph) tells another story and is discussed in the next section.

The lack of a behavioral correlate, for instance, as occurs in the feeding of *Manduca sexta* larvae (Reynolds et al., 1986), where the distribution of gaps is strongly concave but where the insect does not leave the food between pauses, necessitates that the criterion be chosen with care, for misallocation of gaps causes serious inaccuracies in analyses of feeding patterns. If a slope change is obviously localized, as it is in *Manduca*, then selection by eye is acceptable, but if the discontinuity is not sharply defined, more sophisticated techniques are required (see Slater and Lester, 1982). Another complication that can arise in analyzing log-survivorship functions is when feeding actually occurs in bouts but, because there are very few intrameal pauses, the survivorship curve is effectively straight. This is the case in male Australian sheep blowflies feeding on 1.0 M glucose solution (Simpson et al., 1989).

3.3 WHAT DETERMINES WHEN MEALS BEGIN?

The patterns of feeding exhibited by locust nymphs kept under controlled and simplified conditions are notable for being anything but simple. They are complex in structure (Fig. 3.2) and vary considerably between individuals (Simpson, 1981, 1982a). Analyses of such patterns have shown that a number of factors determine, or at least are associated with, the initiation of feeding. Until recently it has been impossible to quantify the relative contributions of these factors to the initiation of feeding and to describe how they interrelate to produce complex behavioral patterns. Recent advances in statistics, namely, the fitting of proportional hazards models to distributions of intervals using the statistical package GLIM, have made this possible. These techniques were developed by medical statisticians (see Whitehead, 1983) to quantify the efficacy of drug treatments and were modified for use on locust feeding data by A. R. Ludlow (Simpson and Ludlow, 1986). Procedural details can be found in that paper. The technique has since been used to analyze walking patterns in locusts on treadmills (Moorhouse et al., 1987) and promises to be an important tool in the study of causation in animal behavior.

3.3.1 The Influence of the Previous Meal

3.3.1.1 Time Since the Previous Meal

One of the most obvious influences on the likelihood that a recently fed locust will walk to the food and begin eating again is the effect of the previous meal. The log-survivorship curve for gaps longer than the bout criterion is convex in shape for most locusts. This change in slope is usually smooth (Fig. 3.3, lower graph), indicating that the probability of beginning a meal increases with time since the last meal ended. The way in which the probability varies is quantified in Figure 3.4. In this graph "feeding tendency," which is the "hazard" from the GLIM analysis and is directly proportional to the probability of a meal beginning (see Simpson and Ludlow, 1986), is plotted on a logarithmic scale against time since the last meal ended. The curve for an average meal eaten by an average insect is shown. Individuals do vary, with some locusts being consistently more likely to begin feeding than others. This is indicated by the fact that there was no significant difference between nymphs in the shape of the curve for feeding tendency but individuals did differ in the position of the curve's intercept along the y axis. Such variation in behavior between individual locusts was also noted by Blaney et al. (1973).

3.3.1.2 Size of the Previous Meal

Not surprisingly, the nature of the increase in tendency to feed with time since the last meal depends on the amount of food eaten during that meal. This is demonstrated in Figure 3.5. Large meals inhibit feeding more and for longer than do small ones. When the curve becomes parallel to the x axis it means that the likelihood of a meal beginning is independent of time since the last meal. This occurs soon after a small meal (about 20 min), about 60 min after an average sized meal and after more than 120 min for a large meal.

3.3.1.3 Nutritional Quality of the Previous Meal

In addition to its bulk, the nutritional quality of the previous meal influences the change in tendency to feed. Locust nymphs given one of four artificial diets varying in their protein and digestible carbohydrate content show differences

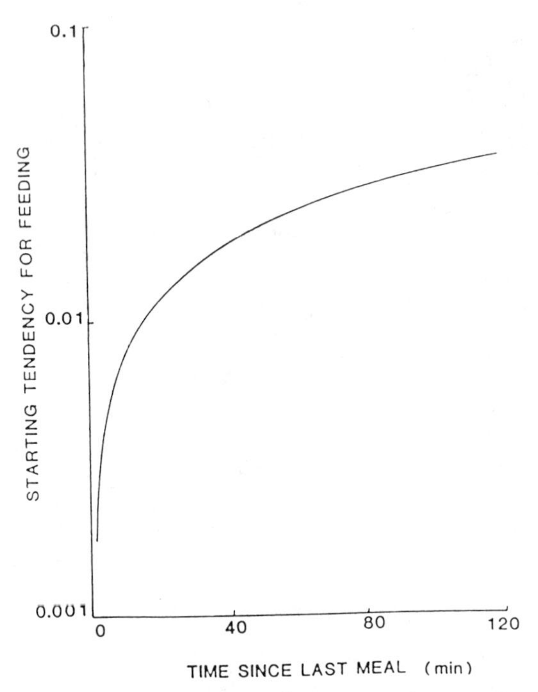

Fig. 3.4. Fitted line showing the starting tendency (which is directly proportional to the probability of initiating a meal during the next minute) plotted on a logarithmic scale against time since the last meal. The curve is based on the data from 10 different locust nymphs and represents the change in probability of commencing feeding with time since an average meal taken by an average insect. (From Simpson and Ludlow, 1986.)

in their feeding patterns. Nymphs given a diet containing 14% protein eat more than those given a diet with 28%, irrespective of whether the diet contains 14 or 28% digestible carbohydrate in addition to the protein. This compensatory increase in feeding on the low-protein diets is a result of insects eating the same-sized meals more often than nymphs given the high-protein diets (Simpson and Abisgold, 1985, and see Chapter 4). GLIM analyses show that the probability of feeding rises with time since the last meal but, although the shape of the curve is not influenced by the protein content of the diet, the intercept of the curve along the *y* axis is. Insects fed the low-protein diet are consistently more likely to commence feeding, irrespective of the time since they last fed.

3.3.1.4 Physiological Correlates

There are clear and quantifiable consequences of feeding on the probability that a new meal will begin. The most parsimonious explanation of the effects described is that feeding inhibits itself and that these inhibitory consequences decline with time after a meal. There are a variety of physiological correlates

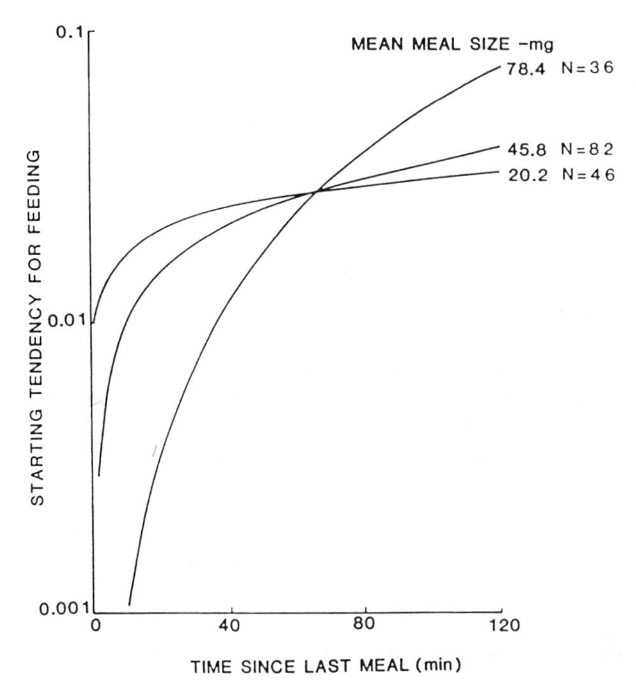

Fig. 3.5. Fitted lines showing the tendency of locust nymphs to start feeding plotted on a logarithmic scale against the time since the last meal ended. The curves for the change in tendency following large, average, and small meals are shown. Each curve is based on data from the same 10 nymphs. Intermeal intervals recorded for these locusts over a 24-h observation were classed according to the size of the meal that preceded each interval. The numbers beside each curve give the number of intermeal intervals contributing to the fitted line. (From Simpson and Ludlow, 1986.)

of satiation, some of which have been demonstrated experimentally to alter the probability of feeding. These have been discussed in recent reviews (e.g., Bernays and Simpson, 1982; Simpson and Bernays, 1983) and are only summarized here.

As food enters the gut it fills and distends the crop and, in the case of soft seedling grasses, also passes to the midgut and shunts food already there back to the hindgut. In so doing stretch receptors associated with both the crop and the ileum are stimulated (Bernays and Chapman, 1973; Simpson, 1983a; Section 3.3.3). Input from these is instrumental in terminating feeding, as is discussed in Section 3.4. In addition, stretching of the crop stimulates the release of hormones from the storage lobes of the corpora cardiaca. Such hormones have a variety of effects, including stimulating diuresis (Mordue, 1969) and crop emptying (Cazal, 1969), reducing locomotor activity (Bernays and Chapman, 1974; Bernays, 1980), and causing the pores at the tips of taste sensilla on the palps to close, thus possibly reducing sensory input on contacting food (Bernays et al., 1972; Bernays and Chapman, 1972a; Bernays and Mordue,

1973). Such effects decline with time after a meal taken by a previously deprived locust and have largely disappeared by about 2 h. Detailed studies of the time-courses of such effects after meals taken by locusts feeding ad libitum are needed, however, if the behavioral changes described above are to be fully explained.

Feeding also results in a change in the composition of the hemolymph, as a result of movement of water into the gut and absorption of digested nutrients from the gut into the blood (Bernays and Chapman, 1974). Work by Abisgold and Simpson (1987, 1988) has shown that specific nutrient feedbacks as well as blood osmolality control intermeal interval (these experiments are discussed in depth in Chapter 4). As outlined in Section 3.3.1.3, nymphs fed a low-protein diet were consistently more likely to start a meal than were those fed a high protein food. Two blood parameters, free amino acid concentration and osmolality, determine the lengthening of intermeal interval that occurs with increased dietary protein. Levels of these parameters do not vary markedly with time since a meal in locusts fed the low-protein diet, whereas there is a sudden increase in blood amino acids and osmolality during a meal of the high-protein diet. Elevated levels are then maintained in the blood of high-protein-fed insects until a steep decline occurs just before the length of an average intermeal interval. Because a relatively constant difference between the blood composition of locusts fed the two diets is maintained for much of the time, with rapid transitions between elevated and nonelevated levels in high-protein-fed nymphs, it is not surprising that the intercept on the *y* axis of the curve for feeding tendency, but not the curve's shape, differs between locusts eating the two diets. It is clear that neither the behavioral nor the physiological data support a simple set-point explanation for the effect of dietary protein on the initiation of feeding.

3.3.2 A Short-Term Endogenous Rhythm

A short-term endogenous rhythm influences the initiation of feeding in fifth-instar nymphs of *L. migratoria.* It runs with a period of about 15 min, with the exact duration of the cycle varying somewhat between individual locusts, and was discovered from analyses of feeding patterns (Simpson, 1981). When a cycle of the critical length is fitted to a locust's feeding pattern, meals tend to begin at a standard phase in the oscillation (Fig. 3.6). The same cycle fits the data from an insect on several consecutive days. Feeding does not begin on every cycle, however; sometimes a meal begins on the very next cycle after the last meal ended, whereas on other occasions eight cycles may elapse between feeds.

A number of other behaviors exhibited between meals also occur in the same rhythm as the initiation of feeding. These include locomotion without subsequent feeding and a variety of small movements exhibited by the insect as it rests between meals (Fig. 3.7). None of the latter behaviors has any obvious relationship with the initiation of ingestion and their thresholds are

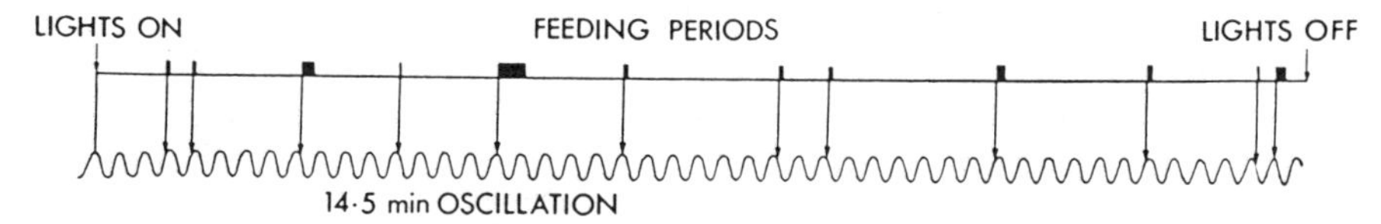

Fig. 3.6. A feeding record for the third 12-h light phase during the fifth stadium with a 14.5-min oscillation imposed from the time of lights on. The width of the dark horizontal bars represent meal durations. Feeding always started at a similar phase in the oscillation and the same period of oscillation independently fitted the feeding data from that insect on each of the first 5 days of the instar. (From Simpson, 1981.)

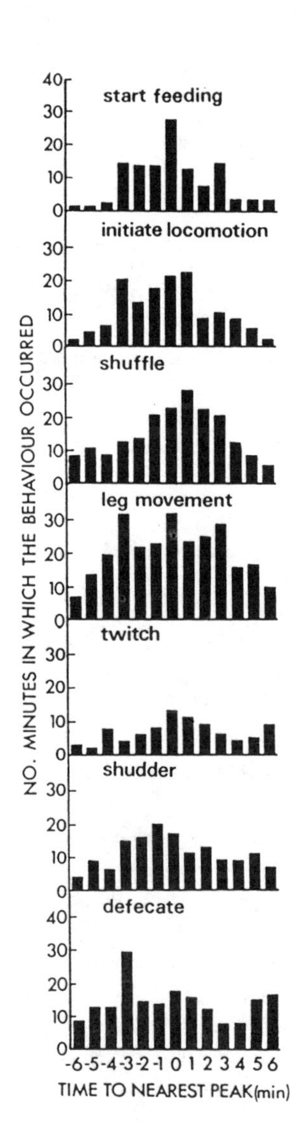

Fig. 3.7. Frequency distributions for the total number of times certain behaviors occurred during periods of resting in relation to the peaks of the short-term oscillation. Distributions are based on data from eight locust nymphs, which varied from 12 to 14.5 min in the period of their oscillation. All of the behaviors shown except defecation occurred significantly more often in the peak than in the trough half of the cycle. Rhythmicity was lost, however, during periods of feeding and locomotion. (From Simpson, 1981.)

dissociated from that for feeding. This is indicated by the fact that, although the tendency to feed rises with time since the last meal, the probability that these behaviors will occur remains constant (Simpson, 1981; Simpson and Ludlow, 1986). Recently a short-term cycle with a period of about 7–8 min has also been found in the walking patterns of locust nymphs mounted on a treadmill (J. E. Moorhouse, personal communication). Whether this is the same phenomenon as described above is not known, but it seems likely that the two are related.

Simpson and Ludlow (1986) analyzed the effect of the rhythm on the tendency to commence a meal. If the rhythm, for convenience sake, is assumed to be a square wave, the probability that feeding will start is four times greater in the peak than in the trough half of the cycle. Additionally, this increase in probability is independent of time since the last meal ended, so that an average locust is four times more likely to start feeding during the next minute in the peak half than in the trough half of the cycle, irrespective of whether the previous meal ended 15 or 105 min before.

So far the rhythm has only been characterized statistically. All that is known about its physiological basis is that it is endogenous and that it is phase-set by the lights in the controlled environment room coming on in the morning. Several insects kept in the same room at the same time exhibit different, asynchronous cycle periods (a range of 12–17 min has been found), indicating that there is no exogenous stimulus, such as the switching of a thermostat, causing the effect.

3.3.3 Defecation

Locust nymphs with ample food defecate, on average, every 30 min or so, although defecation does not occur at a regular phase in the short-term rhythm (Simpson, 1981). When a locust starts a meal, in 50% of cases it will have defecated less than 4 min previously. In fact, the production of a fecal pellet is followed for the next 4 min by a sevenfold increase in the tendency that feeding will start (Simpson and Ludlow, 1986). This enhanced tendency to feed is not affected by time since the last meal. Nor is there any interaction between the influence of defecation and the short-term oscillation, there being a sevenfold elevation in the probability that a meal will begin irrespective of whether the insect is near the peak or the trough in a cycle.

Either defecation stimulates feeding or else it just precedes it as part of the behavioral sequence associated with beginning a meal. Care must be exercised in attributing causality. There is, however, reasonable behavioral evidence that defecation actually results in feeding rather than simply accompanying the start of a meal. For example, defecation is followed by an increase not only in the tendency to feed but also in the probability that locomotion without feeding and a number of behaviors shown during resting periods will occur. As discussed above, these are also influenced by the short-term rhythm, even though the rhythm and defecation do not appear to interact with each other.

Also, about half of all meals are not preceded by defecation. Whether defecation stimulates feeding directly or indirectly, for instance, by promoting locomotion which then results in the insect contacting food and hence feeding, is a more difficult question and is discussed in a later section (3.3.6).

That defecation should stimulate feeding makes functional sense, because the production of a fecal pellet indicates that a significant portion of the gut is empty and it is time to fill the foregut with fresh food. Taking another meal helps shunt partly digested food in the midgut back to the hindgut where it can be processed ready for voiding (Simpson, 1983a). What, though, is the mechanism whereby defecation stimulates feeding and other behaviors? A study of the anatomy of the hindgut provided a possible answer, or at least a correlate for the temporary stimulating effect of defecation (Simpson, 1983a). The hindgut consists of the ileum and the rectum (Fig. 3.8). The ileum folds back on itself when empty and this fold straightens as food passes back from the midgut. During defecation the longitudinal muscle bands that surround the ileum contract, causing the ileum to fold, thus forcing digested food into the rectum and pushing the rectal contents out of the anus (Goodhue, 1963). Associated with the nerves that innervate the ileum are populations of stretch receptors. These are stimulated by changes in the shape of that part of the hindgut. Their response, like most stretch receptors, consists of an initial, powerful, phasic burst of activity, lasting about 30 s, followed by tonic firing, which lasts for as long as the change in shape of the ileum is maintained (Fig. 3.8). It is possible that the transitory increase in sensory input to the central nervous system, which occurs as the ileum folds during defecation, provides the excitation necessary to cause feeding. The tonic firing of stretch receptors that accompanies filling of the ileum as food moves back from the midgut plays another role, that of the regulation of meal size (Section 3.4) and, paradoxically, tonic input probably also causes inhibition of feeding between meals. This possibility of a dual excitatory and inhibitory role for the hindgut stretch receptors needs further investigation.

3.3.4 Environmental Stimuli

An insect's central nervous system is constantly bombarded with all manner of sensory stimuli from the environment. Two sources of stimuli that have been shown to influence the tendency to feed in locusts are food and the light regime. Although there are clearly many others, the effects have been quantified only for these two.

3.3.4.1 Stimuli Provided by the Food

If a locust is watched until it completes a meal during ad libitum feeding, carefully removed from its box, placed into a new one, and allowed to settle into the resting position, the time until it next commences locomotion depends upon whether the new box is empty or contains food. A locust is six times

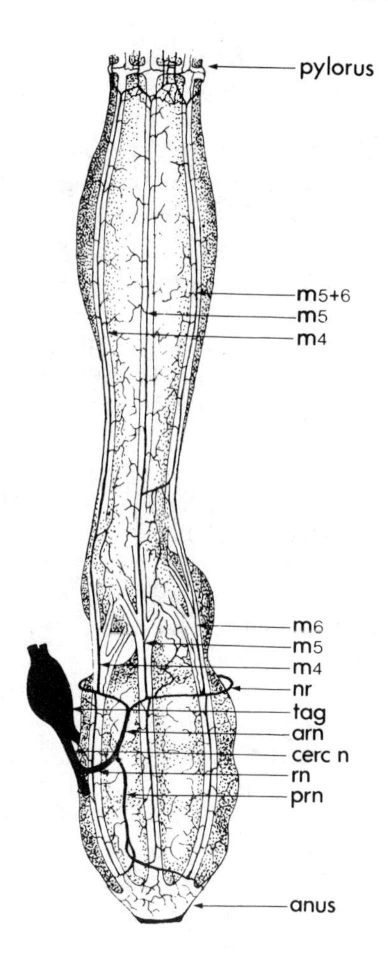
pylorus
m5+6
m5
m4
m6
m5
m4
nr
tag
arn
cerc n
rn
prn
anus

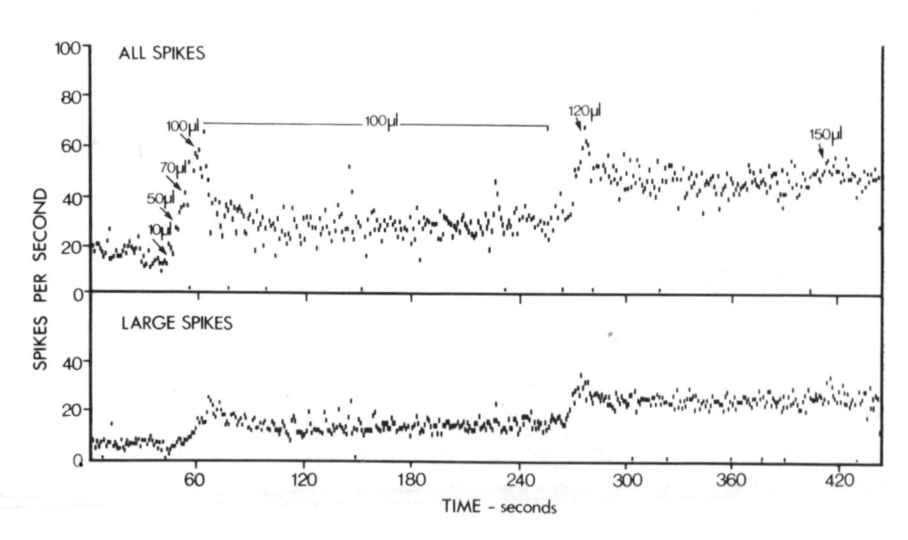
ALL SPIKES
LARGE SPIKES
SPIKES PER SECOND
TIME - seconds
10µl
50µl
70µl
100µl
100µl
120µl
150µl
0
20
40
60
80
100
60
120
180
240
300
360
420

more likely to break postprandial quiescence by locomoting during the next minute if food is present than if food is absent (Simpson and Ludlow, 1986). Those insects that have food available will feed during the minute following the initiation of locomotion in 90% of cases, so the tendency to locomote gives a good measure of tendency to feed under these conditions. The sixfold increase in the probability that locomotion will commence is independent of time since the last meal (Fig. 3.9).

Not only the probability of feeding is enhanced by the presence of food. As for the short-term rhythm and defecation, a number of behaviors exhibited during resting are also more likely to occur (Simpson, 1982b).

Food provides both visual and olfactory stimuli to a resting locust. Whether one or both of these is responsible for stimulating activity in a quiescent, recently fed locust is not known, although it seems that olfactory stimuli are the more important, for when locusts are kept in the dark, and so cannot see the food, the probability that feeding will commence is reduced by only a half, as compared with a sixfold reduction in the absence of all food stimuli. Grass odor is known to stimulate upwind locomotion in locusts (Kennedy and Moorhouse, 1969; Moorhouse, 1971; see Section 3.3.6) and will increase the responsiveness of nymphs to an otherwise relatively unstimulating food such as filter paper soaked in 0.1 M sucrose (Mordue, 1979).

3.3.4.2 The Light Regime

Locusts feed predominantly during light phases and analyses of feeding patterns indicate that average intermeal intervals are longer in the dark and less is eaten during each meal (Simpson, 1982a). The tendency to begin a meal during the next minute is twice as high during light than during dark phases. Like the effects of food stimuli, the rhythm, and defecation, this is not related to time since the last meal (Simpson and Ludlow, 1986).

As discussed by Simpson and Ludlow (1986), light could influence the tendency to feed in a number of ways. Light enables the insect to see the food as well as acting as an arousing stimulus per se (Cassier, 1965). Blaney et al. (1973) found that intermeal intervals could be shortened by switching a light

Fig. 3.8. (Upper) Anatomy of the hindgut of the locust showing the arrangement of nerves from the terminal abdominal ganglion (tag) and the position of the fold in the ileum. Each of the paired cercal nerves sends a branch posteriorly along the rectum (prn) and anteriorly along the ileum (arn). The arn form a ring (nr) around the rectum from which further branches extend anteriorly along the longitudinal muscle bands to join the midgut nerves at the pylorus (m nerves) and posteriorly to join the prn and encircle the anus. (Lower) Spike-frequency histograms showing neural activity recorded from the nerve ring (nr) of a locust that was cannulated with petroleum jelly into the ileum. Amounts cannulated are given on the record. Note the phasic response to changing fullness of the ileum and the tonic response to maintained changes. The response probably comes from multipolar nerves whose cell bodies are found in branches of the m nerves innervating the ileal fold. (From Simpson, 1983a.)

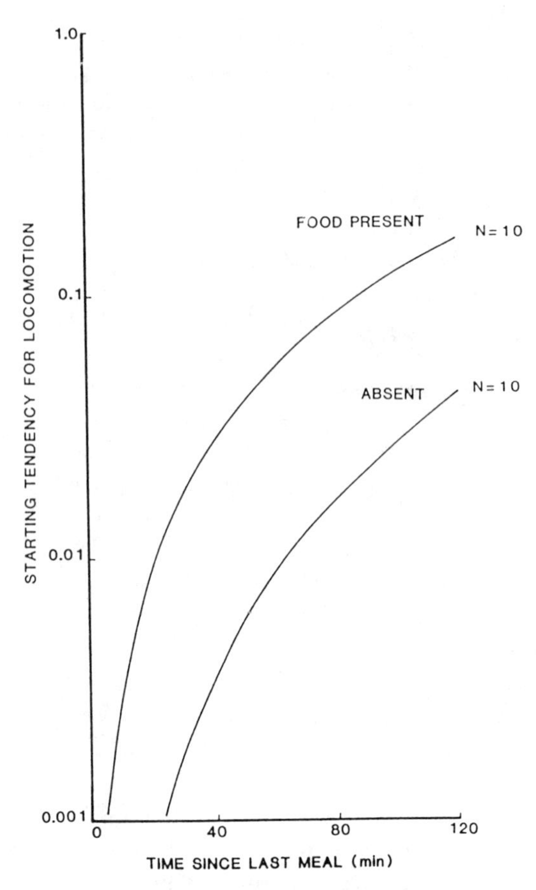

Fig. 3.9. Fitted lines showing the tendency to start locomoting plotted on a logarithmic scale against time since the last meal. The presence of food multiplies the probability of locomotion by about sixfold, independent of time since the last meal. A single intermeal interval from 20 nymphs was used, each nymph being removed from its container at the end of the first meal taken after lights on and then placed in a container either with or without food. The time to the initiation of locomotion following settling into the resting position was then recorded. (From Simpson and Ludlow, 1986.)

bulb on and off every 15 min. The difference in tendency to feed between light and dark phases could also be partly due to circadian changes in activity. The efficiency with which food is utilized and the rate at which the gut empties are lower during dark phases (Simpson, 1982c, 1983b), but it is not known whether these changes are controlled exogenously or by a circadian clock.

3.3.5 The Interrelationships between Factors

It is apparent from the preceding discussion that a number of factors influence the likelihood that a recently fed locust will start another meal. There are bound to be others, especially when the insects are under more complex and natural conditions. However, even taking the ones already characterized it was not possible to quantify their interrelationships until recently. The value of the statistical technique modified by Ludlow is that, because the contribution of each of the factors is measured on the same scale, the extent to which it multiplies the probability of feeding, they can be readily compared (Fig. 3.10). Also, it is possible to predict the probability that feeding will begin under a particular set of conditions by superimposing the various factors on a logarithmic scale (Fig. 3.11). It is possible to see from Figure 3.11 why it is that a locust will sometimes locomote and feed only 15 min after the last meal ended, whereas on other occasions it remains quiescent for 2 h: it is the coincidence of such factors as a peak in the oscillation and defecation that causes the difference.

Interactions can be described in the manner shown in Figure 3.11 only if the factors are independent of each other in their effect on the probability of feeding. For all but one of the factors tested (but not necessarily those that have yet to be characterized), perhaps surprisingly, this would appear to be the case, as evidenced by the lack of statistical significance in the interaction terms of analyses. The size of the previous meal has a more complex effect, however. Meal size influenced both the position and the shape of the curve for starting tendency versus time since the last meal. Hence the curves in Figures 3.10 and 3.11 represent the change in probability after a meal of average size. The rhythm, defecation, food stimuli, and the light regime acted independently of each other and of time since the last meal, simply shifting the curve for feeding tendency after an average-sized meal up and down its logarithmic scale.

As other factors are discovered and quantified they can be readily incorporated into the existing model for the initiation of feeding. For instance, what is the effect of other locusts nearby or of fluctuating temperature or of competition from other behavioral systems (thermoregulation, drinking, flight and reproductive behavior in adults, etc.)? Now that a conceptual framework and a suitable methodology exist it should be possible to address such questions.

3.3.6 The Relationship between Postprandial Quiescence and the Initiation of Feeding

The available data from the "behavioral dissection" of feeding patterns suggest a scheme for the control of the initiation of feeding, locomotion, and the behaviors performed while resting on the perch between meals. This is illustrated in Figure 3.12. So far in this discussion, however, the relationship between the initiation of locomotion after a period of resting following a meal and the actual initiation of ingestion has been sidestepped.

Once a locust starts to locomote after a period of resting, in 9 times out of 10 it feeds within the next minute; or at least this is the case when it is confined in a small container with ample, highly stimulating food such as seedling wheat. The time between starting to locomote and feeding need not

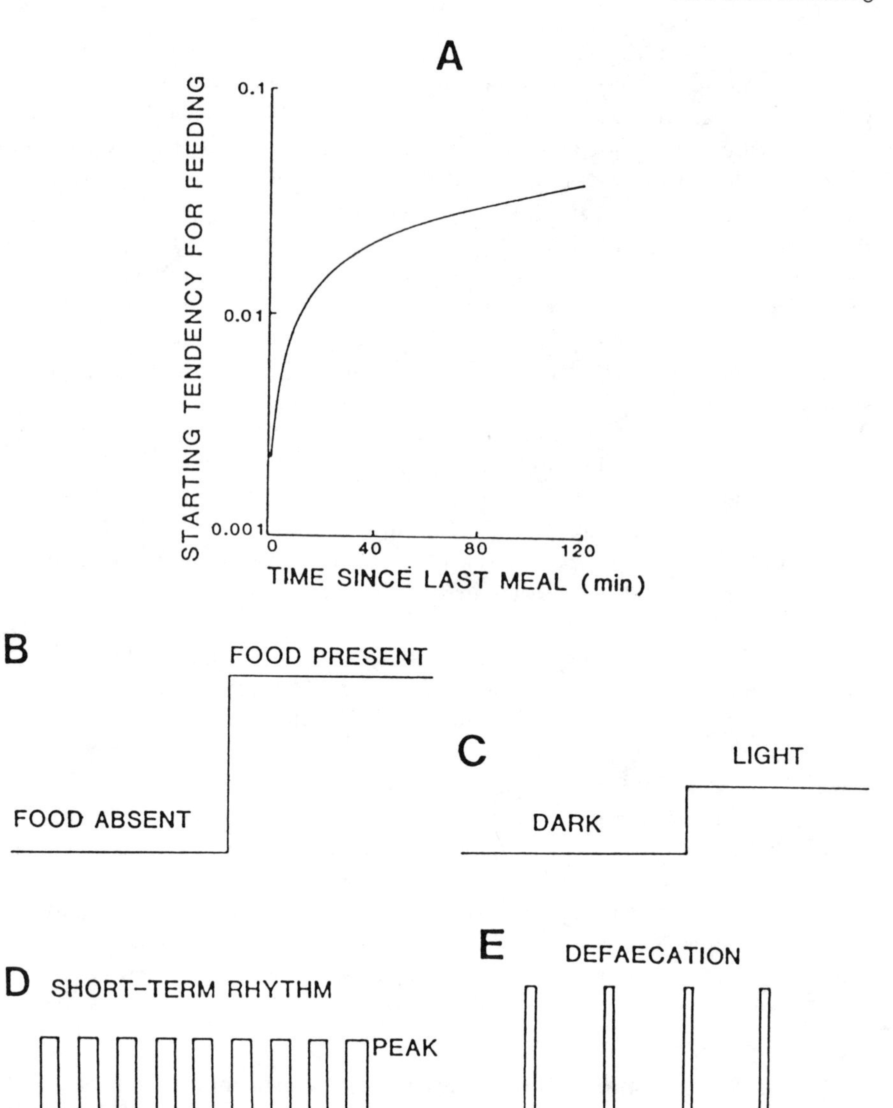

Fig. 3.10. Summary of the effects of various factors on the tendency to start feeding. (A) The starting tendency for feeding plotted on a logarithmic scale against time since a meal of average size. (B) The effect of food stimuli on the starting tendency for locomotion (locomotion leads to feeding within 1 min in 90% of cases under the conditions tested). (C–E) The effect of other factors on the starting tendency for feeding (also on a log scale). Each factor multiplied the probability of feeding independently of time since the last meal and of each other. This is equivalent to shifting curve A up or down by the amount indicated in B–E. (From Simpson and Ludlow, 1986.)

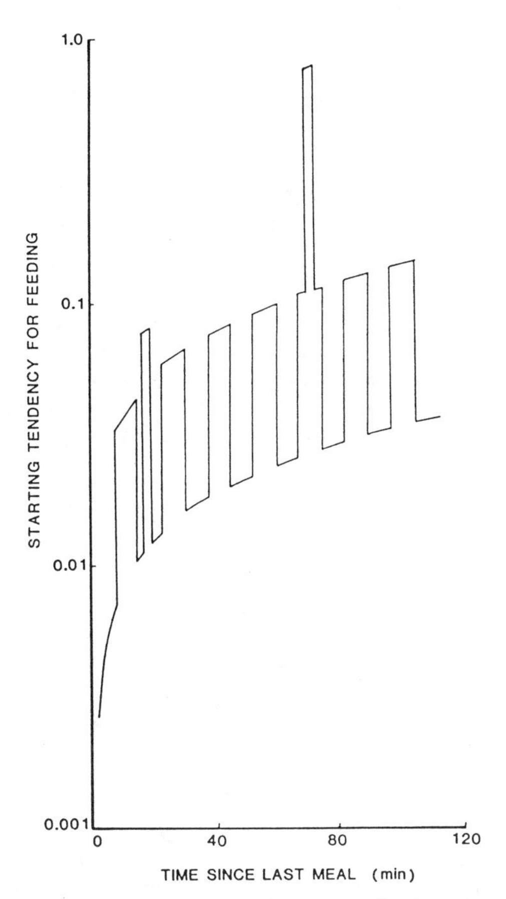

Fig. 3.11. The starting tendency for feeding plotted on a logarithmic scale against time since the last meal. An example showing how the probability of commencing a meal fluctuates as a result of defecation and the short-term rhythm. (From Simpson and Ludlow, 1986.)

be so small, however, as shown in locusts given artificial diets. Longer intermeal intervals are exhibited by nymphs fed a high rather than a lower-protein diet. This is both because insects on the high-protein diet remain quiescent for longer after a meal and also because they reject the food more often once it is contacted (Abisgold and Simpson, 1987, 1988). This indicates that the threshold for initiating locomotion and that for feeding are dissociated.

A closer investigation of the behavior of nymphs feeding on seedling wheat shows a similar distinction between postprandial quiescence and the initiation of feeding. Two basic types of locomotion precede feeding. Most commonly the insect leaves its perch and walks directly toward the food with its head lowered and its palps tapping the substrate. When the locust arrives at the wheat, a blade is grasped between its forelegs and ingestion begins immediately. On other occasions locomotion commences but rather than moving directly to the food and initiating ingesting, the insect meanders about the container and does not commence feeding until the wheat has been contacted several times. This type of locomotion and feeding is often preceded by defecation (Simpson and Ludlow, 1986). In instances when the first category of behavior is exhibited it seems that the particular combinations of causal factors present have stimulated directed locomotion (i.e., orientation to food cues) and excitation is such that contacting the wheat invariably stimulates the initiation of feeding. When the latter category of locomotion is exhibited just the threshold for nondirected locomotion is exceeded, and it is the repeated contact with the food, as well as the arousing effect of moving per se, that subsequently excites the insect to feed.

Evidence for a distinction between directed and nondirected locomotion in locusts comes from the work of Kennedy and Moorhouse (1969) and Moorhouse (1971). They showed that nymphs of the desert locust, *Schistocerca gregaria*, move upwind if exposed to grass odor in a wind tunnel. This response is inhibited for some time after feeding but a variety of stimuli, such as shaking the insect, can overcome this inhibition. The extent of feeding-induced inhibition depended both on the amount of food eaten in a previous test meal (given after a prior deprivation period of 17 h) and the time since that meal. The fact that mechanical, olfactory, and visual stimuli interact with the effect of the previous meal to influence the likelihood of upwind locomotion led Kennedy and Moorhouse (1969) to infer an "unspecified central nervous 'arousal' system to which several sensory inputs have access." The results from analyses of feeding patterns (Fig. 3.12) support and extend this suggestion to include feeding as well as those behaviors performed while resting.

Further support for the separate control of postprandial quiescence and the initiation of feeding can be gained by a comparative study of feeding in other species. Patterns of feeding are available in the literature for larval *Manduca sexta* (Reynolds et al., 1986) and for adult Australian sheep blowflies, *Lucilia cuprina* (Simpson et al., 1989). In both these species there is no correlation in most individuals between meal size and the length of the following intermeal interval, nor are the log-survivorship curves for the distribution of intermeal intervals convex. In other words, in marked contrast to the locust, there is no observable influence of the previous meal on the probability of feeding. Not that the previous meal is unimportant, however. Altering the nutritional quality of the diet of both the caterpillar and the blowfly results in altered intermeal intervals. Also, in the blowfly there is a strong correlation between the volume imbibed during a meal and the duration of postprandial quiescence: it is just that the inhibition of locomotor activity lasts, on average, for only 20% of an average intermeal interval, as compared with 98% for locusts fed seedling wheat. Although postprandial quiescence in the fly is observably affected by meal size, the time between first initiating locomotion and becoming responsive to tarsal contact with the food is not. Larvae of *Manduca*, unlike the

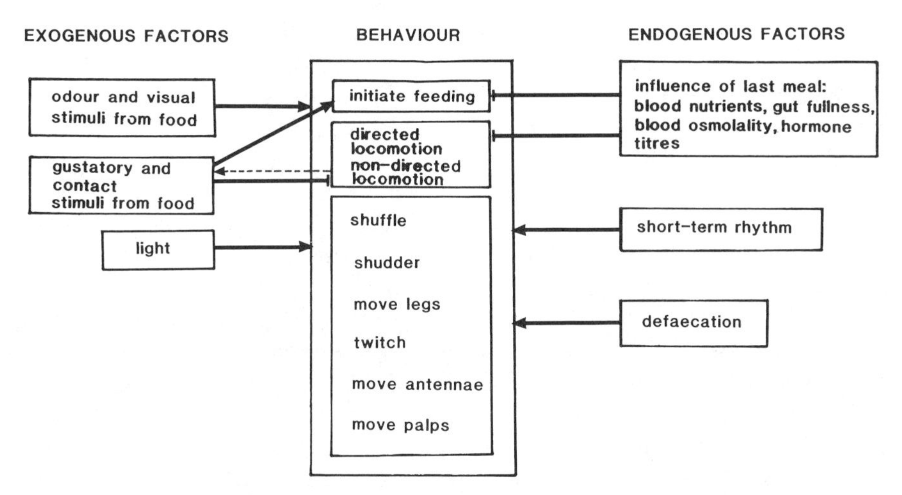

Fig. 3.12. A summary of the interrelationships between various causal factors and behaviors, including feeding, locomotion, and a number of behaviors exhibited during periods of resting between meals. Lines ending in arrows indicate an excitatory effect and T-bars represent inhibition (i.e., a reduction in the probability of occurrence of the behavior). Causal factors influence those behaviors within the box at which the arrow or T-bar terminates. Hence defecation has a less specific effect than does the influence of the last meal. The dotted line from locomotion to gustatory stimuli from the food indicates that locomotion leads to such stimulation.

locust and the fly, do not leave the food between meals and hence do not locomote before feeding.

3.4 WHAT DETERMINES WHEN MEALS END?

Bouts of feeding are usually short in comparison to intermeal intervals in locusts with constant access to food. An average meal on seedling wheat lasts about 5 min, whereas interfeeds are around 60 min. There are two parameters that delimit a meal and could provide clues as to the control of meal termination. These are the amount eaten during a meal (meal size) and meal duration (often termed meal length). Strictly speaking the latter is the period from the initiation of feeding until the beginning of the first gap longer than the bout criterion, not the total time during a meal spent ingesting, disregarding pauses. The reason for selecting a bout criterion is to distinguish between those gaps that are a part of a bout of feeding behavior and those that are between meals. Thus in terms of feeding behavior a pause within a meal is just as much part of a bout as is the time ingesting. Failure to recognize the distinction between meal duration and the time spent eating during a meal has led to a degree of confusion in the literature, and there are a number of studies that do not define what was meant by "meal length" in their analyses.

Analyses of feeding patterns recorded from locust nymphs given seedling wheat have provided an apparent paradox regarding the control of meal termination. When plotted as log-survivorship curves the distribution of meal sizes does not differ significantly from a negative exponential random model in most individuals. This suggests that the probability of ending a meal is not related to the amount of food already eaten during the meal, for if this were the case the curves would be convex in shape. Work on locusts previously deprived of food, however, shows quite clearly that the amount of food eaten is important, with feedback from stretch receptors on the crop and hindgut playing a major role in determining when a meal ends (Bernays and Chapman, 1973; Simpson, 1983a; Roessingh and Simpson, 1984). When the distribution of meal durations is plotted, survivorship curves are strongly convex. Hence, although meal termination is not affected by the amount of food eaten it would appear to be related to time since the meal started. This also contradicts data from other experiments. The only likely candidate for a time-dependent but volume-independent regulator of meal termination on a palatable food is adaptation of chemoreceptors (see Chapter 1) and, although this might be important when locusts are fed solutions of single chemicals (Barton Browne et al., 1975), other work suggests that adaptation is not important in terminating feeding on more natural foods (Bernays and Chapman, 1974).

3.4.1 The Relationship between Meal Size, Duration, and Ingestion Rate

The key to explaining the conflicting results from insects feeding ad libitum and those previously deprived of food was provided by Simpson and Bernays in a review article published in 1983. The parameter that had hitherto been virtually ignored was the rate of ingestion. Because meals are of a relatively standard duration but the amount eaten during a meal is highly variable, it must follow that, on average, bigger meals are eaten faster than smaller ones. This prompted a study of the relationship between meal size, meal duration, and ingestion rate (Simpson et al., 1988a). In this investigation a number of variables known from previous work (e.g., Bernays and Chapman, 1972b) to affect meal size and/or duration were combined in a factorial experiment. These included palatability of the food, the duration of prior food deprivation, age during the instar, and the presence of other locusts nearby.

There were a number of interactions between the various experimental treatments in their effect on the test meal. For instance, increasing the palatability of the wheat by dipping it in sugar solution resulted in the insects eating larger meals that were ingested at a faster rate than when the wheat was not sugar-coated. However, this effect was seen only in nymphs that had been deprived for the shortest of those times tested (2 h). After either 5- or 8-h deprivation the more extended deprivation overrode any effect of increasing phagostimulation (Fig. 3.13).

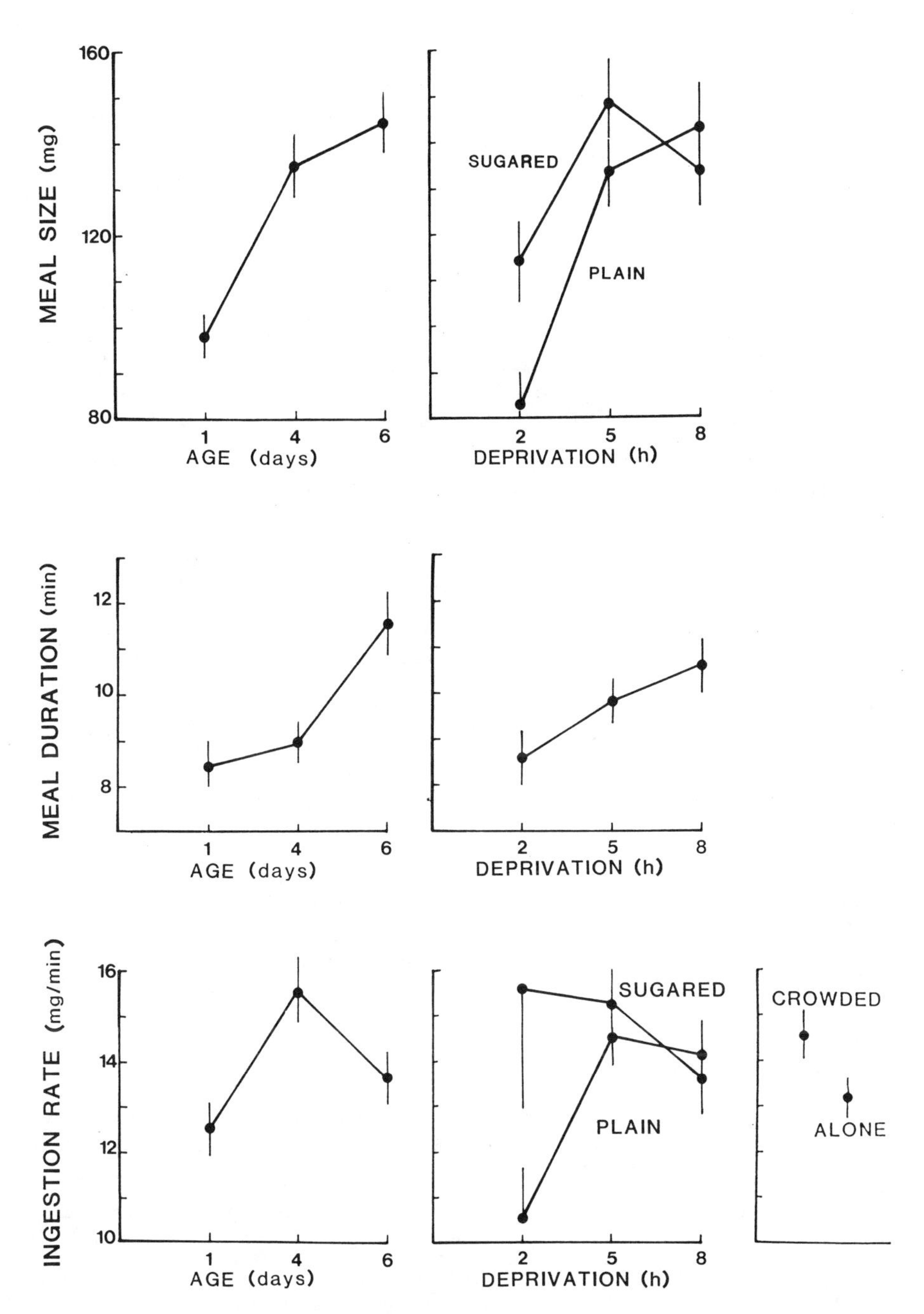

Fig. 3.13. The effects of differing periods of deprivation (2, 5, or 8 h), age during the fifth instar (1, 4, or 6 days), levels of phagostimulation (seedling wheat dipped in either 1 M sucrose solution or water and allowed to dry) and the presence of other locusts (alone or with two others in the box) on the amount of food eaten, the duration, and the mean rate of ingestion during a test meal. Treatments were tested in a randomized factorial design and statistically significant effects from the resulting ANOVA are graphed. (From Simpson et al., 1988a.)

Results showed that meal duration was less affected by the experimental variables tested than was meal size but that it did not remain constant (Fig. 3.13). When data from the whole experiment were pooled it was clear that mean ingestion rate during a meal increased proportionally with meal size up to meals of about 150 mg. Hence meals of less than 150 mg were relatively constant in duration. However, for meals greater than 150 mg, ingestion rate did not increase further and meal duration subsequently increased proportionally with meal size (Fig. 3.14). The change in ingestion rate that accompanied larger meals was due to both a decline in the time during a meal spent pausing and locomoting and an increase in the rate of ingestion during feeding periods (Fig. 3.14). Only rarely does a nymph feeding ad libitum take meals greater than 150 mg in size.

The most parsimonious explanation for the results obtained is that a variety of exogenous and endogenous factors, including chemosensory stimuli provided by the food, the presence of other locusts nearby, age, and deprivation-related factors, are integrated within the central nervous system and together contribute to the level of excitation present as a meal begins. The amount of excitation determines both the rate of ingestion and the amount of inhibition needed to end the meal. Because such inhibition comes largely from stretch receptors on the gut when the insect feeds on palatable food, meal size and ingestion rate are positively correlated and, as a result, meal duration remains constant. If excitation is above a certain level, however, ingestion becomes rate limited and so the insect must eat for longer for sufficient inhibitory feedback to accrue and terminate the meal. It is likely that other factors also contribute to the levels of excitation present as a meal begins. It is not yet known, for example, what are the effects of defecation and the short-term rhythm, both of which are known to influence the initiation of feeding. Such factors acting independently of time since the last meal would tend to produce a random distribution of meal sizes and also ensure that meal size correlates poorly with length of the preceding intermeal interval, as occurs in locusts feeding ad libitum (Simpson, 1982a).

3.4.2 Changes in Ingestion Rate throughout a Meal

The rate of ingestion does not remain constant during the course of a meal. In fact, at least in nymphs that have been deprived of food before a test meal for 2 h or more, it is, on average, fastest at the beginning and subsequently declines exponentially as the meal progresses. Such a decline during a meal will occur if it is assumed that ingestion rate is continually readjusted as the inhibitory consequences of eating reduce net levels of excitation. In other words, an exponential fall would occur if ingestion rate is limited in a negative feedback loop.

Unlike the situation in the rat [the other animal studied in this regard (Davis et al., 1978; McCleery, 1977)] the initial ingestion rate as a meal begins is relatively constant, not varying with meal size or the level of excitation

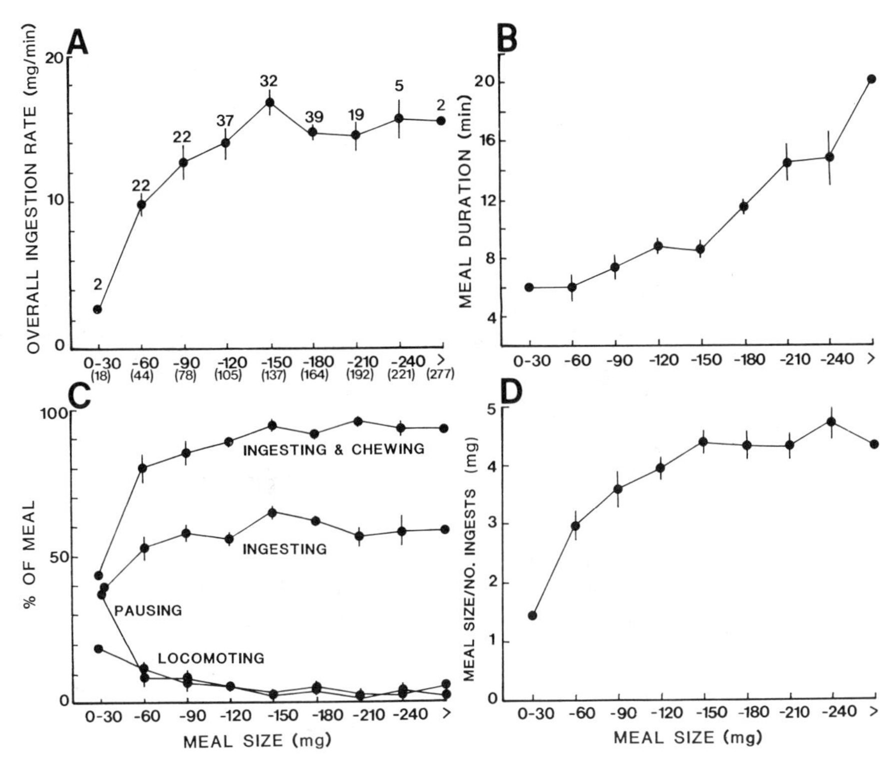

Fig. 3.14. Graphs showing various parameters of a test meal plotted against the size of the meal. Data are pooled from the experiment described in Figure 3.13, the meal size taken by each of 180 locust nymphs being placed into one of the 30-mg size categories 0–30, 31–60, 61–90, and so on. The number of locusts contributing to each mean is given above the points in (A). Values in parentheses below the meal size categories in (A) give the mean meal size for that category. Thus there were 22 insects that took a meal of 31–60 mg, the mean of these being 44 mg. "Meal size/no. ingests" in (D) refers to the amount of food eaten during the test meal divided by the number of 10-s time samplings during the meal when the insect was observed to be ingesting rather than pausing, locomoting, or chewing food already in the preoral cavity. It gives a measure of the rate of ingestion during feeding. (From Simpson et al., 1988a.)

as a meal begins. Also, the rate at which feeding declines in locusts is a function of excitation, being faster for small than big meals (Figs. 3.15, 3.16). This again differs from the rat where the exponential decay function seems to remain constant.

The model for the control of ingestion rate and meal size can be stated as follows:

$$\text{Excitation as a meal begins } (E) = f\,(\text{phagostimulation, time since last meal, age, crowding, etc.})$$

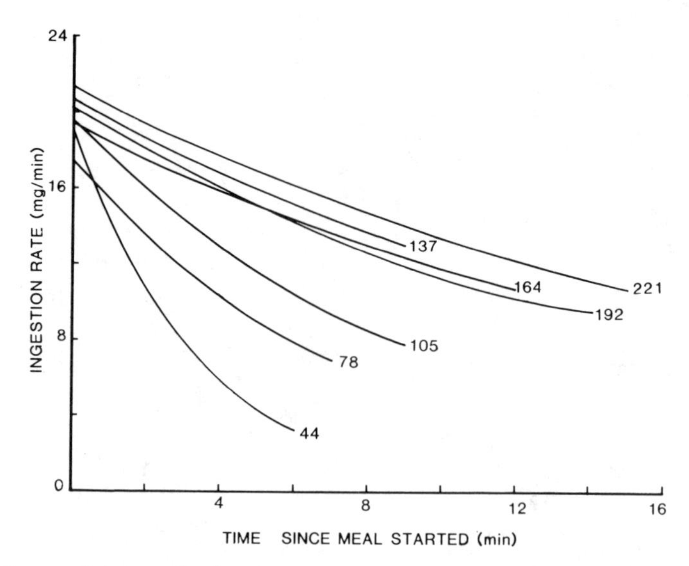

Fig. 3.15. Exponential curves for the change in ingestion rate throughout meals of various sizes (mean meal size is given in mg at the end of each curve). The data are based on 180 locusts, as in Figure 3.14. (From Simpson et al., 1988a.)

Initial ingestion rate (I_0) = k = 20 mg/min for seedling wheat but will vary with the texture, toughness, and chemical properties of the food

Exponential decay function ($-b$) = $f(E)$, with a minimum of -0.05 for seedling wheat

Ingestion rate at time t during the meal (I_t) = $I_0 \exp(-bt)$

Negative feedbacks (n) build up during the meal, such that $n = \int_0^T I_t \, dt$

Excitation declines with the increase in n, until $E - n = 0$, when the meal ends.

The limit on mean ingestion rate throughout a meal, seen for meals of greater than 150 mg in size (Fig. 3.14), is due to a minimum rate of decay rather than a maximum initial ingestion rate.

The model assumes that maximal levels of excitation are found at the start of a meal (which is likely; e.g., Bernays and Chapman, 1974) and that thereafter levels are whittled away by inhibitory feedbacks but do not decline of their own accord. This would mean that once feeding started it should continue in the absence of further phagostimulation, and that if negative feedbacks could be removed entirely, feeding would continue until terminated by muscular fatigue. The former is not the case, however: continued phagostimulation is needed for feeding to be maintained during a meal (see

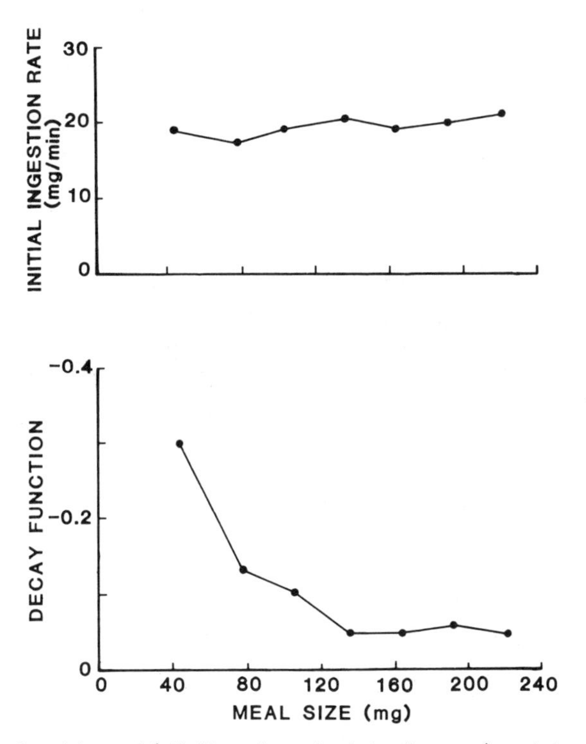

Fig. 3.16. Values for *y* intercept (initial ingestion rate during the meal) and decay function from the exponential curves shown in Figure 3.15 for changes in ingestion rate during meals of various sizes. Note how the initial ingestion rate does not vary, whereas the rate at which ingestion declines is fastest for small meals. (From Simpson et al., 1988a.)

Simpson and Bernays, 1983; Barton Browne, 1975). Both the immediate as well as the "perseverating" excitatory effects of continued phagostimulation [i.e., the generation of an enhanced "central excitatory state" (Barton Browne et al., 1975)] serve to maintain the level of excitation, as required not only by the mathematical model but also by the insect, because if excitation were not maintained feeding would terminate far too soon after being initiated.

Severing the nerves that carry the axons of gut stretch receptors results in a considerable prolongation of meal duration and meal size when locusts are fed a palatable food such as wheat (Bernays and Chapman, 1973; Roessingh and Simpson, 1984). In insects that have an empty gut because of prior food deprivation the major source of inhibitory stretch feedback comes from the crop. Locusts with constant access to food rarely fill their crop to the same fullness as those previously deprived of food at least partly because feedback from ileal stretch receptors contributes extra inhibition (Simpson, 1983a; Roessingh and Simpson, 1984). As well as being larger, meals eaten by insects previously deprived of food for 5 h or longer are much more consistent in size than those taken during ad libitum feeding. The increased variability does not imply that volumetric feedbacks are not important in terminating feeding in locusts feeding ad libitum. It results from a greater variability in levels of excitation as a meal begins. Prolonged deprivation standardizes excitation by overriding other excitatory factors, as seen in Fig. 3.13. It is important to note that there is no set point of gut fullness required to terminate feeding. A meal ends when there is enough negative feedback to cancel out the excitatory input. Just as excitation may vary, so does the amount of inhibition needed. Some other work suggests the presence of an extra volumetric feedback from elsewhere in the body, although its importance and precise nature are unknown (Barton Browne et al., 1976; Roessingh and Simpson, 1984).

An important source of inhibitory feedback that has not been mentioned so far is the presence of deterrent compounds in the food. If the concentration of deterrents relative to phagostimulants is too high, feeding is not initiated but if levels are lower, small meals are taken. Bernays and Chapman (1972b), for instance, showed that more palatable grasses are eaten in larger meals and at faster overall rates than are less palatable ones. Inhibition from deterrents in the food could be incorporated into the model described above but further studies will be needed to describe the time course of their action during a meal. Presumably the effect will also vary with the nature of the deterrent. For instance, compounds that act by blocking the response of taste receptors to phagostimulants, and thus reduce excitatory input at the periphery, might act in a different way to those that stimulate firing in other receptors.

3.5 CONCLUSION

There is now a considerable amount known about the pattern of feeding in nymphs of the African migratory locust. Studies of behavior under controlled and simplified conditions have enabled the formation of a theoretical framework that not only describes and quantifies the relationships between the physiological and environmental factors controlling feeding but also provides the opportunity to probe clearly defined hypotheses and predictions. Such a framework does not exist for any other animal, not even the white rat.

The techniques established for laboratory-reared nymphs of *L. migratoria* are enabling the mechanisms controlling dietary selection and other compensatory feeding behavior to be elucidated in this oligophagous species, as well as in the polyphagous acridid, *Schistocerca gregaria*, and in oligophagous and polyphagous species of *Spodoptera* (Simpson et al., 1988b; S. J. Simpson, M. S. J. Simmonds and W. M. Blaney, unpublished work). The interaction between nutrients and allelochemicals is also being investigated in specialist and generalist acridids and lepidopteran larvae, using feeding patterns as the starting point to a detailed physiological investigation. Fundamental differences in the way in which various species organize their feeding behavior with changes in the nutritional properties and secondary chemistry of their food may provide clues as to the evolution and functional significance of feeding strategies.

The whole area of the interaction between feeding and other behavioral systems is wide open to investigation. Perhaps now is the time to readdress the relationship between thermoregulation, water balance, feeding, locomotion, and reproduction, as first investigated in locusts by Chapman, Ellis, and others some 30 years ago.

However, without leaving the laboratory or *Locusta*, there is plenty still to be done. How do nymphs in the fifth stadium compare with their younger or older conspecifics? Do gregarious and solitarious forms differ? These and many other questions remain to be answered.

REFERENCES

Abisgold, J. D. and S. J. Simpson, 1987. The physiology of compensation by locusts for changes in dietary protein. *J. Exp. Biol.* **129**, 329–346.

Abisgold, J. D. and S. J. Simpson. 1988. The effect of dietary protein levels and haemolymph composition on the sensitivity of the maxillary palp chemoreceptors of locusts. *J. Exp. Biol.* **135**, 215–229.

Barton Browne, L. 1975. Regulatory mechanisms in insect feeding. *Adv. Insect Physiol.* **11**, 1–116.

Barton Browne, L., J. E. Moorhouse, and A. C. M. van Gerwen. 1975. An excitatory state generated during feeding in the locust, *Chortoicetes terminifera. J. Insect Physiol.* **21**, 1731–1735.

Barton Browne, L., J. E. Moorhouse, and A. C. M. van Gerwen. 1976. A relationship between weight loss during food deprivation and subsequent meal size in the locust *Chortoicetes terminifera. J. Insect Physiol.* **22**, 89–94.

Bernays, E. A. 1980. The post-prandial rest in *Locusta migratoria* nymphs and its hormonal regulation. *J. Insect Physiol.* **26**, 119–123.

Bernays, E. A. and R. F. Chapman. 1972a. The control of changes in peripheral sensilla associated with feeding in *Locusta migratoria. J. Exp. Biol.* **57**, 755–763.

Bernays, E. A. and R. F. Chapman. 1972b. Meal size in nymphs of *Locusta migratoria. Entomol. Exp. Appl.* **15**, 399–410.

Bernays, E. A. and R. F. Chapman. 1973. The regulation of feeding in *Locusta migratoria.* Internal inhibitory mechanisms. *Entomol. Exp. Appl.* **16**, 329–342.

Bernays E. A. and R. F. Chapman. 1974. The regulation of food intake by acridids. In L. Barton Browne, Ed., *Experimental Analysis of Insect Behaviour.* Springer-Verlag, Berlin, pp. 48–59.

Bernays, E. A. and A. J. Mordue (Luntz). 1973. Changes in palp tip sensilla of *Locusta migratoria* in relation to feeding: The effects of different levels of hormone. *Comp. Biochem. Physiol. A*, **45A**, 451–454.

Bernays E. A. and S. J. Simpson. 1982. Control of food intake. *Adv. Insect Physiol.* **16**, 59–118.

Bernays, E. A., W. M. Blaney, and R. F. Chapman. 1972. Changes in chemoreceptor sensilla on the maxillary palps of *Locusta migratoria* in relation to feeding. *J. Exp. Biol.* **57**, 745–753.

Blaney W. M., R. F. Chapman, and A. Wilson. 1973. The pattern of feeding of *Locusta migratoria* (Orthoptera, Acrididae). *Acrida* **2**, 119–137.

Cassier, P. 1965. Le comportement phototropique du criquet migrateur (*Locusta migratoria migratorioides* R. et F.): Bases sensorielles et en doctrines. *Ann. Sci. Nat., Zool. Biol. Anim.* **7**, 213–358.

Castonguay, T. W., L. L. Kaiser, and J. S. Stern. 1986. Meal pattern analysis: Artefacts, assumptions and implications. *Brain Res. Bull.* **17**, 439–443.

Cazal, M. 1969. Actions d'extraits de corpora cardiaca sur le péristaltisme intestinal de *Locusta migratoria. Arch. Zool. Exp. Gen.* **110**, 83–89.

Clifton, P. G. 1979. Temporal patterns of feeding in the domestic chick. I. *Ad libitum. Anim. Behav.* **27**, 811–820.

Davis, J. D., B. J. Collins, and M. W. Levine. 1978. The interaction between gustatory stimulation and gut feedback in the control of the ingestion of liquid diets. In D. A. Booth, Ed., *Hunger Models.* Academic Press, London, pp. 109–143.

Fagen, R. M. and D. Y. Young. 1978. Temporal patterns of behaviour; durations, intervals, latencies and sequences. In P. W. Colgan, Ed., *Quantitative Ethology.* Wiley, New York, pp. 79–114.

Goodhue, D. 1963. Some differences in the passage of food through the intestines of the desert and migratory locusts. *Nature (London)* **200**, 288–289.

Kennedy, J. S. and J. E. Moorhouse. 1969. Laboratory observations on locust responses to wind-borne grass odour. *Entomol. Exp. Appl.* **12**, 487–503.

Le Magnen, J. and M. Devos. 1980. Parameters of the meal pattern in rats: Their assessment and physiological significance. *Neurosci Biobehav. Rev.* **4, Suppl. 1**, 1–11.

Mayes, E. and P. Duncan. 1986. Temporal patterns of feeding behaviour in free ranging horses. *Behaviour* **96**, 105–129.

McCleery, R. H. 1977. On satiation curves. *Anim. Behav.* **25**, 1005–1015.

Moorhouse, J. E. 1971. Experimental analysis of the locomotor behaviour of *Schistocerca gregaria* induced by odour. *J. Insect Physiol.* **17**, 913–920.

Moorhouse, J. E., I. H. M. Fosbrooke, and A. R. Ludlow. 1987. Stopping a walking locust with sound: An analysis of variation in behavioural threshold. *Exp. Biol.* **46**, 193–201.

Mordue, W. 1969. Hormonal control of Malpighian tubule and rectal function in the desert locust, *Schistocerca gregaria. J. Insect Physiol.* **15**, 273–285.

Mordue (Luntz), A. J. 1979. The role of the maxillary and labial palps in the feeding behaviour of *Schistocerca gregaria. Entomol. Exp. Appl.* **25**, 279–288.

Panksepp, J. 1978. Analyses of feeding patterns: Data reduction and theoretical implications. In D. A. Booth, Ed., *Hunger Models.* Academic Press, London, pp. 143–166.

Reynolds, S. E., M. R. Yeomans, and W. A. Timmins. 1986. The feeding behaviour of caterpillars (*Manduca sexta*) on tobacco and artificial diet. *Physiol. Entomol.* **11**, 39–51.

Roessingh, P. and S. J. Simpson, 1984. Volumetric feedback and the control of meal size in *Schistocerca gregaria. Entomol. Exp. Appl.* **36**, 279–286.

Simpson, S. J. 1981. An oscillation underlying feeding and a number of other behaviours in fifth-instar *Locusta migratoria* nymphs. *Physiol. Entomol.* **6**, 315–324.

Simpson, S. J. 1982a. Patterns in feeding: A behavioural analysis using *Locusta migratoria* nymphs. *Physiol. Entomol.* **7**, 325–336.

Simpson, S. J. 1982b. The control of food intake in fifth-instar *Locusta migratoria L.* nymphs. Ph. D. Thesis, University of London.

Simpson, S. J. 1982c. Changes in the efficiency of utilisation of food throughout the fifth instar in nymphs of *Locusta migratoria. Entomol. Exp. Appl.* **31**, 265–275.

Simpson, S. J. 1983a. The role of volumetric feedback from the hindgut in the regulation of meal size in fifth-instar *Locusta migratoria L.* nymphs. *Physiol. Entomol.* **8**, 451–467.

Simpson, S. J. 1983b. Changes in the rate of crop emptying during the fifth-instar of *Locusta migratoria* and relationship to feeding and food utilisation. *Entomol. Exp. Appl.* **33**, 235–243.

Simpson, S. J. and J. D. Abisgold. 1985. Compensation by locusts for changes in dietary nutrients: Behavioural mechanisms. *Physiol. Entomol.* **10**, 443–452.

Simpson, S. J. and E. A. Bernays. 1983. The regulation of feeding: Locusts and blowflies are not so different from mammals. *Appetite* **4**, 313–346.

Simpson, S. J. and A. R. Ludlow. 1986. Why locusts start to feed: A comparison of causal factors. *Anim. Behav.* **34**, 480–496.

Simpson, S. J., M. S. J. Simmonds, A. R. Wheatley, and E. A. Bernays. 1988a. The control of meal termination in the locust. *Anim. Behav.* **36**, 1216–1227.

Simpson, S. J., M. S. J. Simmonds, and W. M. Blaney. 1988b. A comparison of dietary selection behaviour in larval *Locusta migratoria* and *Spodoptera littoralis*. *Physiol. Entomol.* **13**, 225–238.

Simpson, S. J., L. Barton Browne, and A. C. M. van Gerwen. 1989. The patterning of compensatory feeding behaviour in the Australian sheep blowfly. *Physiol. Entomol.* **14**, 91–105.

Slater, P. J. B. 1974a. The temporal pattern of feeding in the zebra finch. *Anim. Behav.* **122**, 506–515.

Slater, P. J. B. 1974b. Bouts and gaps in the behaviour of zebra finches, with special reference to preening. *Rev. Comp. Anim.* **8**, 47–61.

Slater, P. J. B. 1975. Temporal patterning and the causation of bird behaviour. In P. Wright, P. G. Caryl, and D. M. Vowles, Eds., *Neural and Endocrine Aspects of Behaviour of Birds*. Elsevier, Amsterdam, pp. 11–33.

Slater, P. J. B. and N. P. Lester. 1982. Minimising errors in splitting behaviour into bouts. *Behaviour* **32**, 179–210.

Whitehead, J. 1983. Fitting survival models using GLIM. In J. E. Gentle, Ed., *Computer Science and Statistics: The Interface*. North-Holland Publ., Amsterdam, pp. 97–103.

Wiepkema, P. R. 1968. Behaviour changes in CBA mice as a result of one gold thioglucose injection. *Behaviour* **32**, 179–210.

4

Nutrition

E. A. BERNAYS and

S. J. SIMPSON

4.1 THE BASIC NEEDS

The classic way to elucidate the detailed nutritional requirements of an animal is to develop a chemically defined artificial diet. The only attempt to do this for a grasshopper was made by R. H. Dadd in the early 1960s. Dadd did not succeed in developing a diet for *Locusta migratoria* or *Schistocerca gregaria* that could sustain growth and development as well as natural food plants but his studies provided much valuable information on grasshopper nutrition.

For nymphs to grow satisfactorily the diet had to contain protein, digestible carbohydrates, fatty acids, salts, and vitamins. In addition, cellulose was required, despite its indigestibility (Martin, 1983). With respect to protein needs, *Locusta migratoria* grew best when the diet contained a mix of casein, albumen, and peptone, whereas *Schistocerca gregaria* could grow with just casein. The specific amino acid requirements of the two species were not established, although most insect species seem to need the same essential amino acids as do members of other phyla (Dadd, 1985; see also Section 4.3.2). Hexoses, oligosaccharides, dextrins, and starches were all utilized but pentoses, sorbose, and galactose were not. The ability to digest and utilize starches contrasts with a total inability to do so in many species of Lepidoptera (Dadd, 1985). The only dietary lipids required were linoleic acid and cholesterol. Carotene, although not essential for growth, was needed for normal pigmentation in *Schistocerca*. Essential vitamins were found to be ascorbic acid (as for most insects that feed on fresh plant foliage), calcium pantothenate, choline, folic acid, nicotinic acid, pyridoxine, riboflavin, and thiamin. Inositol and biotin were also important both for growth and, in the case of inositol, for pigmentation. Another key nutrient is water, and the regulation of water balance is discussed in Chapter 5. For a fuller account of work on grasshopper nutrition the reader is referred to the original papers of Dadd and to his later accounts (Dadd 1960a–e, 1961a–c, 1963, 1970a, 1985).

That artificial diets were never as successful as fresh plant foliage in rearing locusts from egg hatch to reproductive status probably indicates a lack of some physical or chemical factor. There may also be imbalances of known nutrients in the diet. This has been stressed by House (1969) and by Dadd (1985) and probably applies to proportions of amino acids in different proteins as well as to proportions of major nutrients. Leaf proteins have never been employed in artificial diets but they are likely to be superior (Vanderzant, 1958). Another contributor to the inadequacy of artificial diet may simply be its chemical homogeneity. As is discussed in the next section, the nutritional requirements of an insect are constantly changing. The cost of meeting these needs by adjusting feeding behavior through eating more of a food, such as an artificial diet that contains a fixed ratio of nutrients, may be greater than that of the alternative strategy of feeding selectively on different parts of a chemically heterogeneous food, such as a plant (see Section 4.3.4).

4.2 SPECIFIC NEEDS AND CHANGES IN NUTRITIONAL REQUIREMENTS

Specific needs of acridids are little known but they may be important. It has been suggested that such well sclerotized hemimetabolous species at least have different needs from those of caterpillars (Bernays, 1982, 1986; Simpson et al., 1988). One of the reasons for this is clear in acridids: the cuticle makes up a large proportion of the dry body weight, and the investment in the cuticular proteins represents the greater part of all body protein. The cuticular proteins have their own particular amino acid profiles and so the need for different proportions of the various amino acids is different in insects having different proportions of the body protein in the cuticle. Among the particular requirements for cuticle are aromatic amino acids that are utilized eventually in sclerotization, and these can be limiting for acridids in some foliage (Bernays and Woodhead, 1984). The tree locust *Anacridium melanorhodon* is able to meet some of this need by using plant phenolics, which then become an important component of a low-protein diet (Bernays and Woodhead, 1982; Bernays et al., 1983).

Another particular need for some species (perhaps the majority) is relatively low water levels in the food. This is discussed in detail in the chapter on water regulation (see Chapter 5).

The quantitative and qualitative nutritional requirements of a grasshopper are not static. As in any animal they vary, for example, with growth and development, reproductive status, and the behavior of the insect. In *Locusta migratoria*, growth during the first half of a nymphal instar mainly comprises an increase in lipid and protein, largely in the form of fat body, muscle, and cuticle. Later in the instar lipid growth gives way to an increase in body carbohydrates, mainly glycogen in the fat body (Hill and Goldsworthy, 1968; Simpson, 1982b). Presumably the cyclic production of cuticle also has some impact on needs.

Examination of growth curves for acridids during an instar often reveal a short period of slowed growth about one-third of the way through the instar (e.g., Bernays, 1982, p. 12). It is possible that when it occurs this represents a limitation imposed by a suboptimal nutrient balance for the particular growth stage.

As shown for holometabolous phytophages, carbohydrate requirements become relatively greater during development through the nymphal instars (Dadd, 1960e, 1985). This probably relates to the slower growth rates of the larger instars and thus a diminished proportional need for protein (Peters, 1983). This has important implications for nutritional ecology. Ideal protein/carbohydrate ratios in the diet, which thus fall during nymphal life, imply changing needs in diet selection: the best food plant for first instars may not be the best for fifth instars. Differences in host plant use during development have been recorded but whether this is due to changes in nutrient requirements as opposed to physical constraints, plant availability, and plant phenology is not known. For example R. Lewis (unpublished) showed that first-instar *Chortoicetes terminifera*, the Australian plague locust, fed predominantly on leguminous species whereas later instars fed on grasses. The legumes may be expected to be higher in protein but are also softer and hence were readily eaten by these small insects.

The fat content of adult locusts increases during the first two weeks of somatic growth (Cheu, 1952; Loveridge, 1973). The increase is more marked in gregarious than in solitarious forms, which may relate to the fact that lipids are the main energy source for prolonged flight. Solitarious forms often make extended night migrations but gregarious forms are more generally active and probably fly long distances over more extended periods. Once somatic growth is completed in females, egg production requires extra lipid and protein as well as metabolic energy (Weed Pfeiffer, 1945; Hill et al., 1968; McCaffery, 1975) so that sexual differences in dietary requirements are expected. No detailed budgets have been worked out, but there are examples of large differences in food preference that are presumed to reflect differences in needs. For example, Pfadt and Lavigne (1982) record clear differences between foods eaten by males and females of several North American grasshoppers, and Boys (1978) showed that mature females of *Oedaleus senegalensis* ate relatively much more of the milky seed heads of millet whereas males ate relatively more leaf material.

4.3 MEETING NUTRITIONAL REQUIREMENTS

4.3.1 Plants as Food

Even at the best of times plant foliage has limitations as a source for building animal tissues, because it is often relatively low in protein. It is also sometimes low in digestible carbohydrate. Compounding this nutritional unsuitability is the considerable variation in general nutritional quality of plants, both spatially and temporally, and between and within individuals of the same plant species (Scriber and Slansky, 1981; Slansky and Rodriguez, 1987). An insect faced with a change in its own nutritional requirements or a change in the nutritional quality of its food plant can do one of three things: tolerate the change and pay the price of an extended development, reduced fecundity, and/or increased mortality; leave and attempt to find an alternative food supply; or compensate for the change.

Compensation can involve a number of responses. First, the insect can eat more or less of the same food until it meets the limiting nutritional requirement. Second, it may be able to adjust its digestive and assimilatory physiology so that more or less efficient use is made of ingested nutrients. Third, the insect can mobilize any reserves of a limiting nutrient that it has stored, for example, in the hemolymph or fat body. Finally, it can select between available sources of foods that are close at hand. These are not mutually exclusive, and examples of each are known among the grasshoppers (see Simpson and Abisgold, 1985; Simpson et al., 1988).

4.3.2 Tolerating Change or Leaving the Plant

Little is known in detail of behavior associated simply with tolerance of changes in physiological needs or of detrimental changes in plants, but clearly individuals confined to unsuitable food suffer fitness costs as measured by survival, growth, or fecundity in laboratory conditions. In the field this is likely to occur only where there are monospecific stands of plants, but in no case has nutritional inadequacy been separated from effects of specifically detrimental plant secondary compounds. In addition, conditions that lead to nutritionally poor food plants, such as an advancing dry summer, are often associated with conditions of food shortage. Studies with grass-feeding species may yield the most information on the severity of nutritional changes for grasshoppers, for grasses change rapidly during growth to become so low in protein that growth and fecundity are almost certainly affected (Bernays and Barbehenn, 1987).

Because acridids are so exploratory and mobile it may be expected that leaving unsatisfactory plants will be more common than tolerance. Parker (1984) examined the foraging behavior of the specialist grasshopper *Hesperotettix viridis* in New Mexico. He found that as the insects damaged the host plant *Gutierrezia microcephala*, the protein concentration fell to levels much lower than that of undamaged plants. Tenure times of marked insects distributed among the damaged and undamaged plants were dramatically greater on relatively undamaged plants, which also supported better growth. These results were interpreted to indicate that the insects could make decisions concerning food nutrient quality. There is also field evidence that populations (Mulkern, 1972) and individuals of polyphagous species (Ben Halima et al., 1985) do move from place to place eating a variety of different plants. This selection in turn may partly reflect an ability to reject and accept food on the basis of specific nutrient needs and is discussed further in Section 4.3.5.

4.3.3 Compensating for Changes in Growth and Reproductive Status

The increasing growth shown by nymphal grasshoppers up to mid-instar is sustained by increased consumption of food. This has been most studied in fifth-instar *Locusta migratoria*. When nymphs are reared at a constant 30°C under L:D 12:12, the instar lasts about 10 days. Nymphs eat progressively more during successive days up to mid-instar, with most being eaten during the light phases. The behavioral mechanism whereby more is eaten on successive days depends on the nature of the food plant. Nymphs fed on seedling wheat eat more primarily because they take larger meals, not because they feed any more often (Simpson, 1982a,b). However, if mature *Agropyron* is the food plant, the increased consumption is due to the same-sized meals being eaten more frequently (Blaney et al., 1973). The difference appears to relate at least partly to the physical properties of the two plants. The tougher mature grass is more restricted to the crop during a meal than is seedling wheat and hence stimulates stretch receptors on the crop wall that powerfully regulate meal size (Simpson, 1983a; see Chapter 3).

Associated with changes in amount eaten during the instar are variations in the rate of passage of food through the gut. Studies by Simpson (1983b) showed that the rate of crop emptying does indeed change, being faster in the middle than at the start of the instar. Because meals are more frequent in the middle of the instar, a locust has on average a similar amount of food in its gut when it begins a meal, irrespective of its age during the instar. This change in the rate of crop emptying is independent of the size of meals taken on the different days and seems likely to be due to changes in composition of the hemolymph. During mid-instar, blood osmolality is low due to removal of nutrients for growth and to the rapid increase in blood volume that occurs as a result of increased retention of water ingested in the food (Simpson, 1982b, 1983b). Hemolymph osmolality is known to alter crop motility and thus rate of emptying (Baines, 1979; Bernays and Simpson, 1982). There are also changes in the levels of digestive enzymes through the instar, with a peak at the time of greatest food intake (Chapman, 1985).

Superimposed upon the change in rate of crop emptying described above there is a slowing of rate of emptying on the fourth day of the instar. Correlated with these changes is variation in the efficiency at which food is digested and absorbed from the gut (Simpson, 1982b, 1983b). This is due mainly to a change in the efficiency with which protein is digested and absorbed, this being high soon after the molt and declining up to the middle of the instar with a secondary peak on day 4. During this day levels of protein in the blood are high, as is the rate of protein uptake by the growing tissues (Tobe and Loughton, 1967; Simpson, 1982b).

In addition to eating more and exhibiting altered digestive and assimilatory physiology, another way of meeting the special nutritional requirements of growth is for the nymph to select between available foods. Whether or not fifth-instar *L. migratoria* nymphs feed selectively as they grow was investigated by giving newly molted insects a choice of two artificial diets, one containing 28% protein but no digestible carbohydrate and the other containing 28% digestible carbohydrate but no protein. The ratio selected varied throughout the instar, being close to 50:50 at the start and end of the instar but being biased toward carbohydrate at mid-instar (Simpson et al., 1988). There is a temptation to suggest that such a choice represents the "optimal" ratio for the insect (see, for example, Cohen et al., 1987). The ratio selected could in fact be peculiar to the diets and the experimental procedure whereby highly mobile, opportunistic insects are placed in small containers with artificial diets (see Simpson et al., 1988, for further discussion; also see Section 4.3.4).

It is apparent that there is a complex interaction between properties of the food, feeding behavior, blood composition, digestive and assimilatory physiology, and the metabolic and specific nutritional requirements of the growing locust. The role of hormones in the control of these processes has yet to be elucidated, and which factors are cause and which effect are largely unknown.

Food intake and utilization is high during periods of somatic growth in adult acridids (Hill et al., 1968; McCaffery, 1975). Mordue and Hill (1970) showed that female *Schistocerca gregaria* prefer bran to lettuce during these periods and suggested that this is because bran provides a good source of digestible carbohydrates. It is probably also important for them to eat the dry food to prevent problems of water loading (see Chapter 5). Female acridids also increase consumption between production of successive egg batches (Hill et al., 1968; McCaffery, 1975). Unlike many other insects that have been studied, such as blowflies, they do not require a separate supply of protein.

Requirements are usually met by increased feeding, although it may be that mature female acridids actively select protein-rich plants or plant structures if given the opportunity (see Section 4.2).

4.3.4 Compensating for Low Levels of Dietary Nutrients by Increased Food Consumption

Acridids are remarkably good at compensating for dilution of the nutrients in an artificial diet. The most astonishing case is that of *Melanoplus sanguinipes*, which is able to compensate by increasing the amount it eats when its diet is diluted sevenfold by the addition of cellulose (McGinnis and Kasting, 1967). The only attempt to investigate the behavior and underlying physiological mechanisms of compensation has been on fifth-instar *Locusta migratoria*. Nymphs were given one of four artificial diets on the third day of the instar and their feeding behavior was recorded in detail over the next 12 h. The diets were variations on Dadd's artificial diet and contained either 28 or 14% protein, with either 28 or 14% digestible carbohydrate in an otherwise similar and nutritionally complete mix. Insects ate more of the lower-protein diets, irrespective of the amount of digestible carbohydrate present. They did so by eating the same-sized meals more frequently than insects fed the higher-protein diets (Simpson and Abisgold, 1985). Such behavioral compensation resulted in no loss of digestive efficiency for protein. Although locusts did not compensate for the changes in dietary digestible carbohydrates this does not mean that they are unable to do so. In fact they can, as is shown in Section 4.3.4. Rather, at the levels tested, it seems likely that protein was more limiting to the growing nymphs and, by regulating for protein, sufficient carbohydrate was ingested.

The changes in intermeal interval in relation to protein level in the food was partly influenced by osmolality of the hemolymph. This was found to be significantly higher after a meal of the high-protein diet than after a low-protein meal (Fig. 4.1). Also, blood osmolality was not affected by dietary carbohydrates, these apparently being removed from the blood to the fat body as soon as they were absorbed from the gut (Abisgold and Simpson, 1987). It might be expected that the higher blood osmolality incurred after eating the higher-protein diet would lead to a slower rate of gut emptying and hence to longer intermeal intervals. This was not the case, however: the passage of food did not differ in rate between insects fed the two diets (Fig. 4.2). Insects were able to increase consumption of the lower-protein diet without an increase in the rate of gut emptying because an amount of food equivalent to that eaten during a meal had cleared the gut by 40 min since the last meal, this being less than an average intermeal interval. The amount of food already in the gut as a meal began did not differ between locusts fed the two diets (Fig. 4.2). The effect of blood osmolality on gut emptying, if any, was clearly not a direct one.

An analysis of blood composition showed that free amino acids contributed 40% of the osmolality difference between locusts fed the high- and lower-

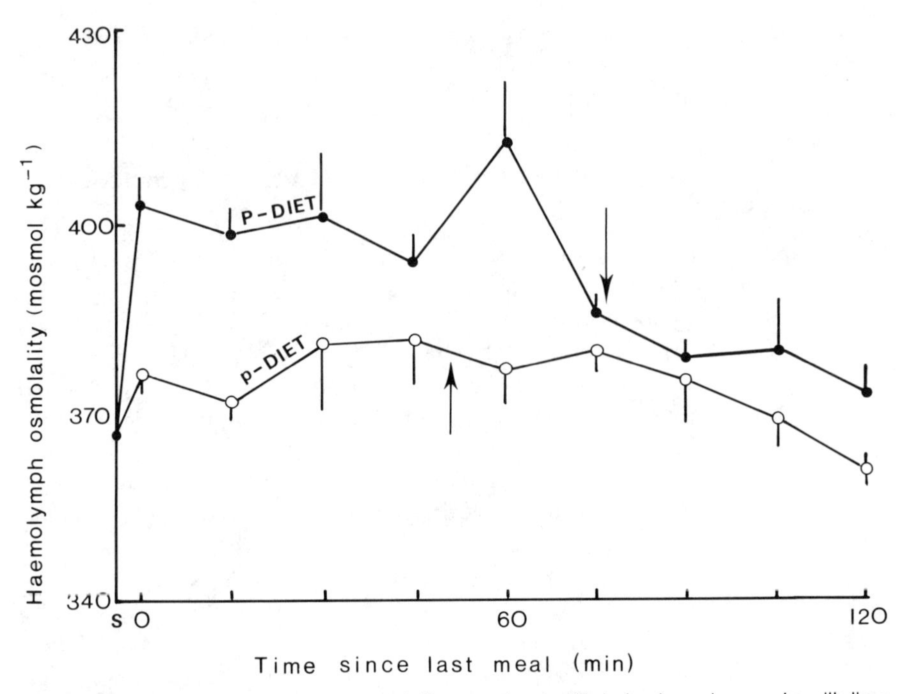

Fig. 4.1. The change in hemolymph osmolality occurring in fifth-instar *Locusta* nymphs with time since completing a meal during ad libitum feeding on either an artificial diet containing 28% protein (P-diet) or one with 14% protein (p-diet). Each point gives the mean ± S.E. for 12–15 insects. Arrows indicate the average time at which the next meal would have been taken. S, start of meal; O, end of meal. (From Abisgold and Simpson, 1987.)

protein diets. There were 11 amino acids that occurred in significantly different concentrations in the hemolymph of the two groups of nymphs. In order to test whether a general osmotic or a specific amino acid feedback was involved, locusts were taken as they completed a meal of the lower-protein diet, left for 40 min without food (10 min less than the average intermeal interval), and then injected with one of four solutions. These injections were designed to raise the levels of blood amino acids and/or osmolality to those that would have been found had the insect's last meal been of the high-protein diet. After the injection, insects were placed with food and the time to the next meal was recorded. Injections that increased either blood amino acids or osmolality caused a delay in feeding relative to the control, with the injection that increased both resulting in the longest intermeal intervals (Abisgold and Simpson, 1987; Fig. 4.3). Further experiments (J. D. Abisgold, unpublished) have shown that 8 of the 11 amino acids that occur in higher concentrations in the blood of high-protein-fed insects are critical in eliciting compensation (L-glutamine, serine, methionine, leucine, phenylalanine, lysine, valine, and

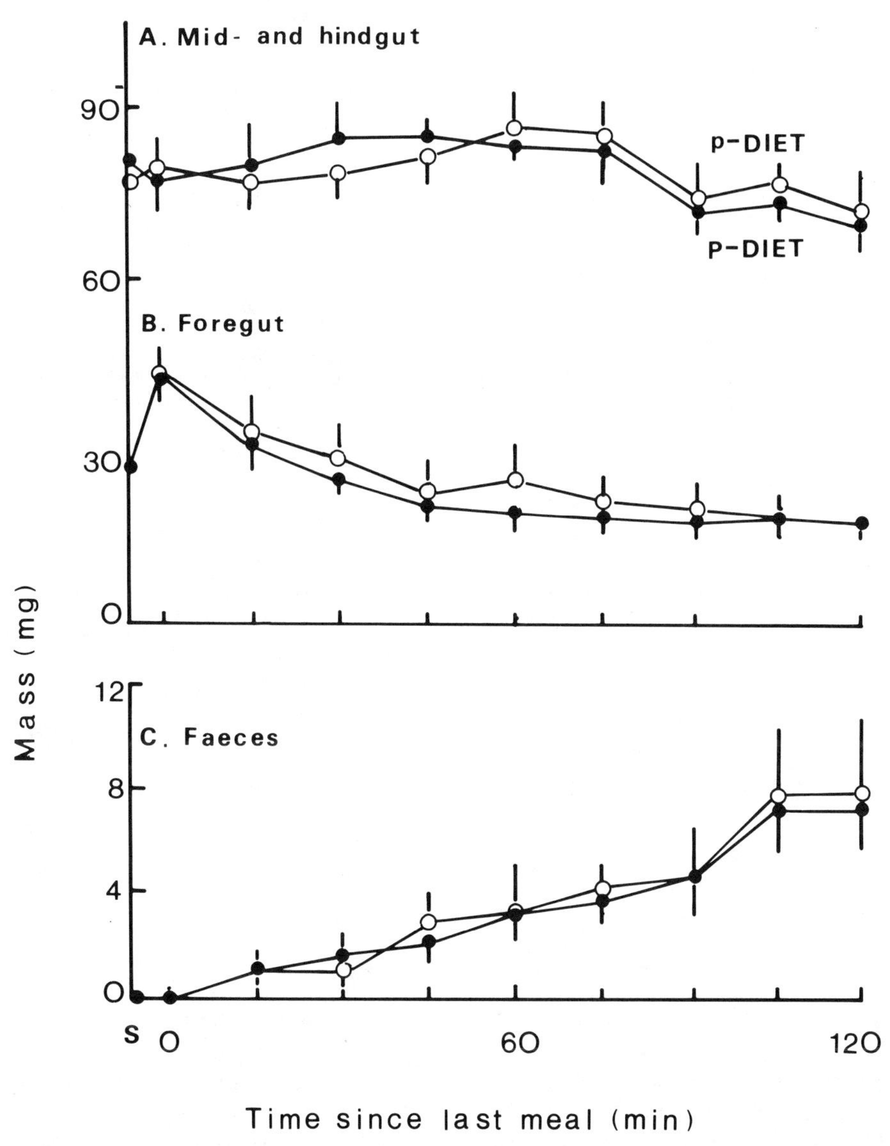

Fig. 4.2. The mass of the gut and cumulative fecal production at different times after a meal for *Locusta* nymphs fed either the P- or p-diet (see Fig. 4.1). (From Abisgold and Simpson, 1987.)

alanine). Unless all eight of these are present in elevated levels in the diet the locust exhibits a compensatory increase in feeding. The precise ratio in which they are present is not important.

Why these eight amino acids are the ones that regulate compensatory feeding is unclear. Five of them are among the 10 amino acids considered to

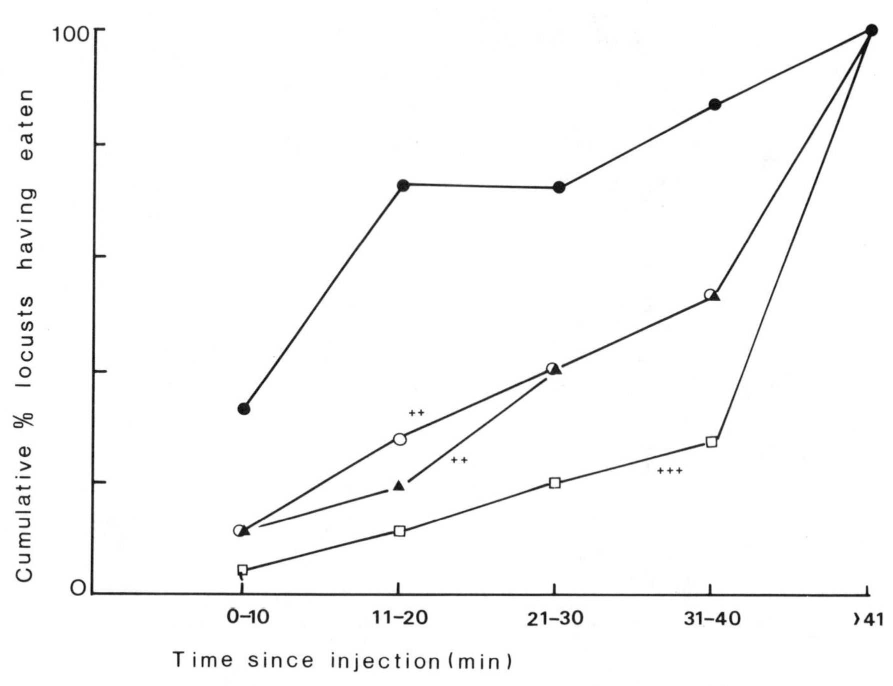

Fig. 4.3. The effect on subsequent feeding behavior of injecting *Locusta* nymphs with one of four solutions at 40 min (just less than an average intermeal interval) after completing a meal of the low-protein diet. Open squares, amino acids in saline solution (a mix 10 amino acids in the amounts and ratio required to raise blood levels up to those found at that time after a meal of high-protein diet); closed triangles, xylose in saline (an osmotic control for the amino acid in saline injection); open circles, the same amino acid mix but in distilled water rather than saline (this increased amino acid levels but not blood osmolality); closed circles, saline isosmotic with the blood (the control). ++, $P < .05$; +++ $P < .01$, relative to the control saline injection. (From Abisgold and Simpson, 1987.)

be essential to most animals (methionine, leucine, lysine, phenylalanine, and valine). The eight are a mix of acidic, basic, and neutral polar and nonpolar amino acids, and are representatives of all the major synthetic pathways. It may be that the locust has evolved to use these eight because they are a representative mix and by compensating for this mix the insect gains a balanced intake of all amino acids. In other words, the suite of eight amino acids could be, in effect, a sign stimulus. That the precise ratio of the eight is not critical is hardly surprising, considering the enormous variability in amino acid profile that occurs in foliage (Lyttleton, 1973). Because feeding behavior and the efficiency of utilization of protein varies throughout the fifth stadium in *L. migratoria* (Simpson, 1982b; J. D. Abisgold, unpublished; see Section 4.3.2), it would be interesting to see whether the same eight amino acids trigger compensatory feeding at all stages during the instar, as well as during other phases in the life of the insect.

Phenylalanine has been shown to be limiting to growth in *Schistocerca gregaria* nymphs when eating certain foliage (Bernays and Woodhead, 1984). Supplementing the diet with phenylalanine increased the efficiency of conversion of ingested food to body substance (see Section 4.4) and in one of the experimental replicates there was a suggestion that this was associated with reduced consumption of food. That consumption was not reduced in all experiments was apparently because intake was also limited in some cases by excess water. The phenylalanine apparently improved the overall balance of nutrients and improved growth rates.

An important question is how is the effect of blood amino acid profile and osmolality on behavior mediated? It is possible that blood composition is monitored by peripheral receptors associated with the gut or elsewhere in the body. Additionally or alternatively, receptors could be within the central nervous system. Neither mechanism is known in insects. Yet another possibility is that the sensitivity of mouthpart and tarsal gustatory receptors varies with blood composition. This could affect intermeal interval, and hence contribute to compensatory feeding for protein, by altering the nature of taste information passing to the central nervous system from taste receptors when food is contacted. It is well known that the initiation and maintenance of feeding when a locust contacts food depends on the presence of phagostimulants, and that phagostimulatory input to the central nervous system is coded in the firing of gustatory receptors (see Chapter 1). The most powerful phagostimulants for locusts are sugars, but amino acids are also known to excite feeding (Cook, 1977). Do levels of free amino acids and/or blood osmolality influence the responsiveness of taste receptors specifically to amino acids in the food? In order to answer this, locusts were taken as they completed a meal of the low-protein diet and the responses of taste sensilla on the maxillary palps were recorded to a mix of the eight amino acids, a sugar solution, and a salt solution. Responses were recorded before and at various times after injection with one of the same four solutions used in the behavioral studies described above.

There was a marked and rapid reduction in the total number of spikes elicited from hairs stimulated with the amino acid mix after injections that increased the free amino acid concentration of the blood up to levels expected in a high-protein-fed insect. This was independent of the osmotic effect of the injection. There was no significant difference between any of the injections when the same sensilla were stimulated with sugar. There was a statistically significant reduction in the response to salt stimulation, although the effect was not nearly so marked as that shown to amino acid stimulation. The salt response was decreased by injections that increased either blood amino acids or osmolality (Abisgold and Simpson, 1988; Fig. 4.4).

These experiments showed a clear and specific effect of levels of free amino acids in the blood on the responsiveness of taste sensilla to stimulation by those amino acids in the food. Because changing blood composition apparently did not influence the response of sensilla to the other major macronutrient and

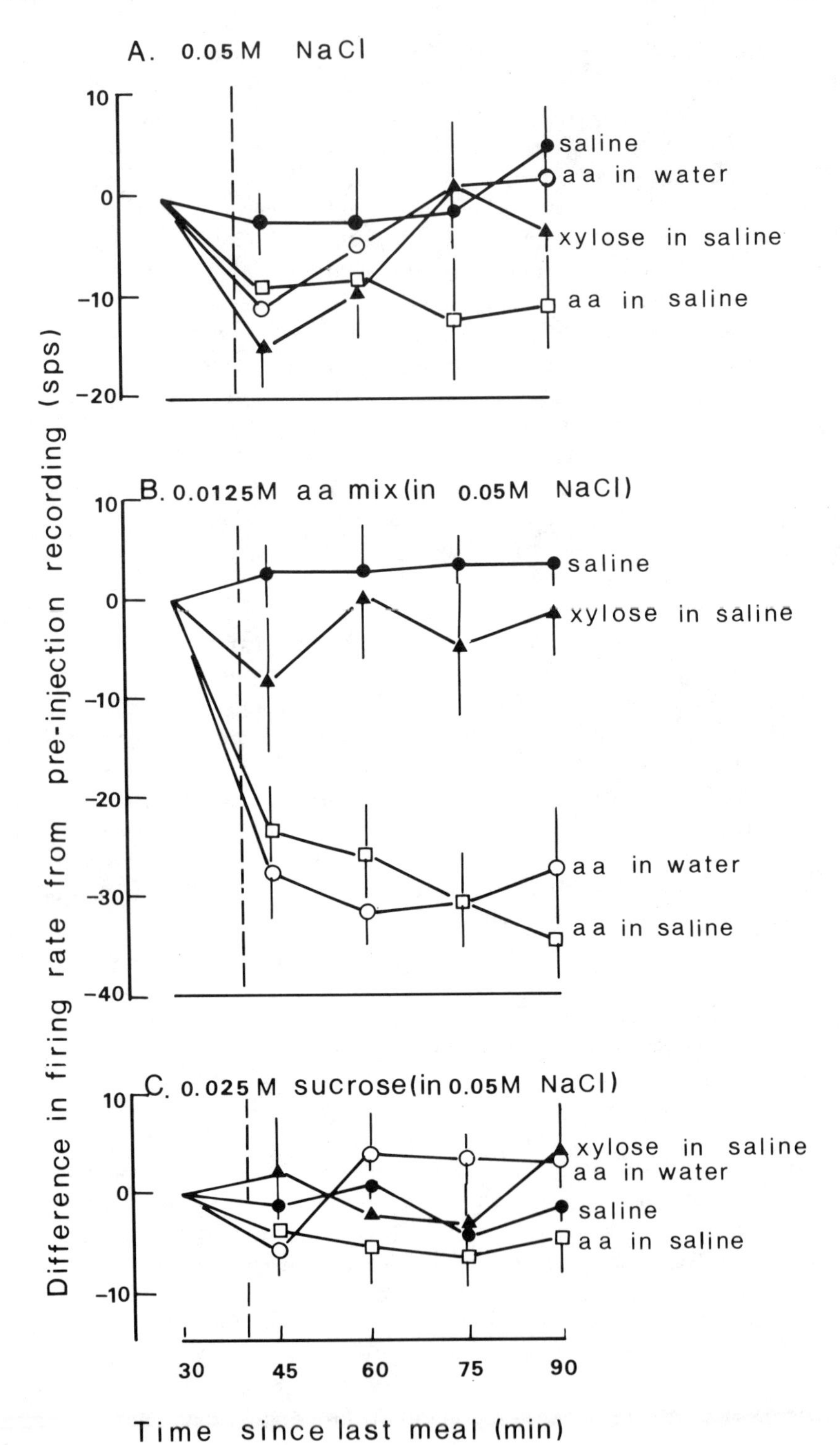
A. 0.05M NaCl
saline
aa in water
xylose in saline
aa in saline
B. 0.0125M aa mix(in 0.05M NaCl)
saline
xylose in saline
aa in water
aa in saline
C. 0.025M sucrose(in 0.05M NaCl)
xylose in saline
aa in water
saline
aa in saline
10
0
−10
−20
−30
−40
30
45
60
75
90
Difference in firing rate from pre-injection recording (sps)
Time since last meal (min)

phagostimulant, sugar, the central nervous system of a locust that is well fed on protein receives different levels of sensory input from a locust that has fed on a less-protein-rich food. This was the first time in any animal that a specific nutrient feedback influencing peripheral sensitivity and feeding behavior had been demonstrated. The mechanisms whereby blood composition alters taste sensitivity are yet to be elucidated and the possibility that blood levels of other nutrients also have specific peripheral effects has not been investigated. Also, the role of blood osmolality in extending intermeal intervals is still unclear, although it appears that it is important in determining the length of postprandial quiescence as compared with the specific effect of amino acids on the probability that feeding is initiated once food is contacted (Abisgold and Simpson, 1988).

4.3.5 Dietary Selection Behavior

Apart from altering the total amount of food eaten, the other compensatory behavioral response is to select between alternative sources of nutrients. Dietary self-selection or diet mixing, as this is termed, has received little attention in the insects, although it is generally considered to occur and is likely to be most important to mobile, polyphagous species. In the field there is evidence that individuals of polyphagous species do select a variety of plant species over time, although the spectrum of plants eaten may be different for different individuals (Ben Halima et al., 1985); local specialization seems less extreme than in some polyphagous Lepidoptera (Fox and Morrow, 1981). MacFarlane and Thorsteinson (1980) showed that *Melanoplus bivittatus* had a higher growth and survival rate when it was allowed to feed on several plant species rather than just one, and similar results have been obtained for other species (Kaufmann, 1965). Diet selection might involve more unusual choices of food: maybe the facultative carnivory (usually cannibalism) that many acridids show is an example.

A detailed study of selection behavior for nutrient mixtures has been carried out with fifth-instar *Locusta migratoria*. As in the experiments on dietary dilution described above, artificial diets have been used. This is necessary, at least initially, because it obviates the problem of the uncontrollable changes in plant chemistry that accompany gross variation in the levels of macronutrients.

Fig. 4.4. The difference in response of sensilla on the maxillary palps of *Locusta* nymphs stimulated 30 min after completing a meal of low-protein diet and the same sensilla stimulated at successive times after an injection given 40 min after the meal. The solutions injected, shown on the right of the figure, were the same as those in Figure 4.3. The three solutions used to stimulate the sensilla were (A) 0.05 M NaCl; (B) 0.0125 M amino acid mix; and (C) 0.025 M sucrose. Note the marked and specific decline in response recorded from sensilla to stimulation with an amino acid mix after injections that increased blood levels of amino acids up to those found in a high-protein-fed nymph. (From Abisgold and Simpson, 1988.)

Once mechanisms are established it may be possible to work with plants and to determine to what extent selection behavior optimizes nutrient balance in nature.

Locust nymphs were fed for 4, 8, or 12 h on one of four artificial diets during the third day of the fifth instar. One of the diets (PC) contained both protein (20%) and digestible carbohydrate (10%) in an otherwise nutritionally complete mix. The other three diets had the carbohydrate, the protein, or both carbohydrate and protein removed and replaced with indigestible cellulose. (These diets will be referred to as P, C, and O, respectively.) After the initial conditioning period on one of these diets, nymphs were given a choice of two diets, P and C, and hence a chance to select for the nutrients, if any, that were lacking in their previous food.

During the conditioning period those locusts given the P diet ingested it in amounts similar to those eaten by insects given PC, despite the fact that the P diet lacked digestible carbohydrate. Those locusts fed the C diet ate more than those fed PC for the first 8 h on the diet, after which intake fell sharply. A possible explanation is that the high levels of phagostimulation from sugar coupled with the lack of post-ingestive inhibition from blood amino acids stimulated the extra feeding, but by 8 h the lack of protein in the diet caused rejection of the food. Whether this was due to changes in taste sensitivity or aversion learning (see Chapter 2) is not yet known. Insects fed the O diet ate only small amounts during the conditioning period.

During the first hour of choice, locusts selected the P diet if they had previously been fed C and the C diet if previously fed P. Those insects fed the diet lacking both digestible carbohydrate and protein ate large quantities of both P and C in the choice test (Fig. 4.5). During the subsequent 8 h of choice this selection for nutrients that were missing in the previous diet continued but became masked by a tendency, shown by all nymphs, to select C over P.

It is striking that after feeding for only 4 h on a nutritionally inadequate diet a locust nymph is able to select for the missing nutrients when given a choice, and recent experiments show that there is even a tendency for such selection after feeding for only one meal on an inadequate diet. The mechanism of modulation of taste sensitivity described in Section 4.3.3 could account for selection of protein by insects prefed on the C diet. Another possibility is that aversion learning is involved in rejection of the incomplete diet as in vertebrates.

It has been shown experimentally that *Schistocerca americana* is capable of food aversion learning (Bernays and Lee, 1988) and that on certain foods this may be a mechanism by which even reasonable foods are rejected in favor of others that might provide a superior overall diet (Lee and Bernays, 1988). One example is with spinach. Initially individuals feed readily on this food, but in successive meals, the amount ingested becomes steadily reduced until finally rejection occurs. An acceptable food at this time is always eaten readily whereas spinach will be eaten again only after a long intermeal and several bouts of rejection behavior if no other food is available (Fig. 4.6). In these

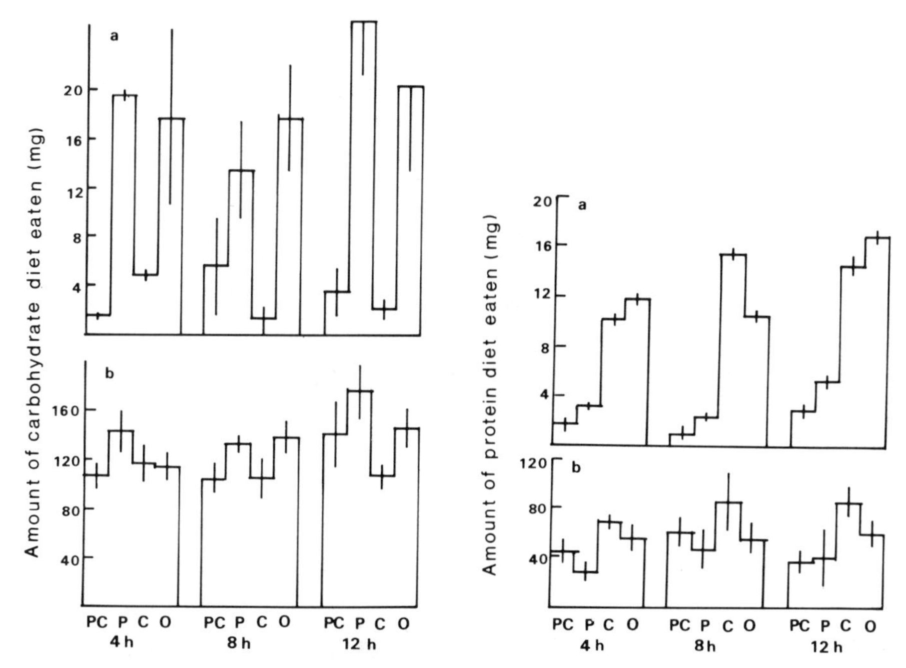

Fig. 4.5. Amounts eaten by *Locusta* nymphs in a choice test after having been conditioned for 4, 8, or 12 h on one of four diets: a nutritionally complete artificial diet (PC); or the same with the protein removed and replaced with indigestible cellulose (C); the digestible carbohydrate replaced (P); or both protein and digestible carbohydrate replaced (O). The left graph shows the amount eaten from a dish of C diet and the right graph shows the amount eaten by the same insects from a dish of P diet when both were presented in a choice test after the conditioning period. The upper graphs give amounts eaten during the first 60 min of choice, and the lower are for the following 8 h. Note how insects fed selectively for those nutrients that were missing in the conditioning diet. If both protein and carbohydrate were missing (O-conditioned nymphs) then consumption of both P and C was increased in the choice test. (From Simpson et al., 1988.)

examples with actual foliage it is difficult to separate nutrient factors from other effects. For example, concentrations of non-nutrients may be deleterious and dietary mixing could be an advantage if it results in dilution of individually noxious compounds in different leaves.

4.4 DIGESTIVE AND ASSIMILATORY PHYSIOLOGY

The wide variations in digestive efficiency found in the literature probably can be explained largely by the very wide variation in foliage protein levels (Bernays and Barbehenn, 1987). These may vary from as low as 5% dry weight to as high as 40%, but more commonly range from 10 to 20%. Within

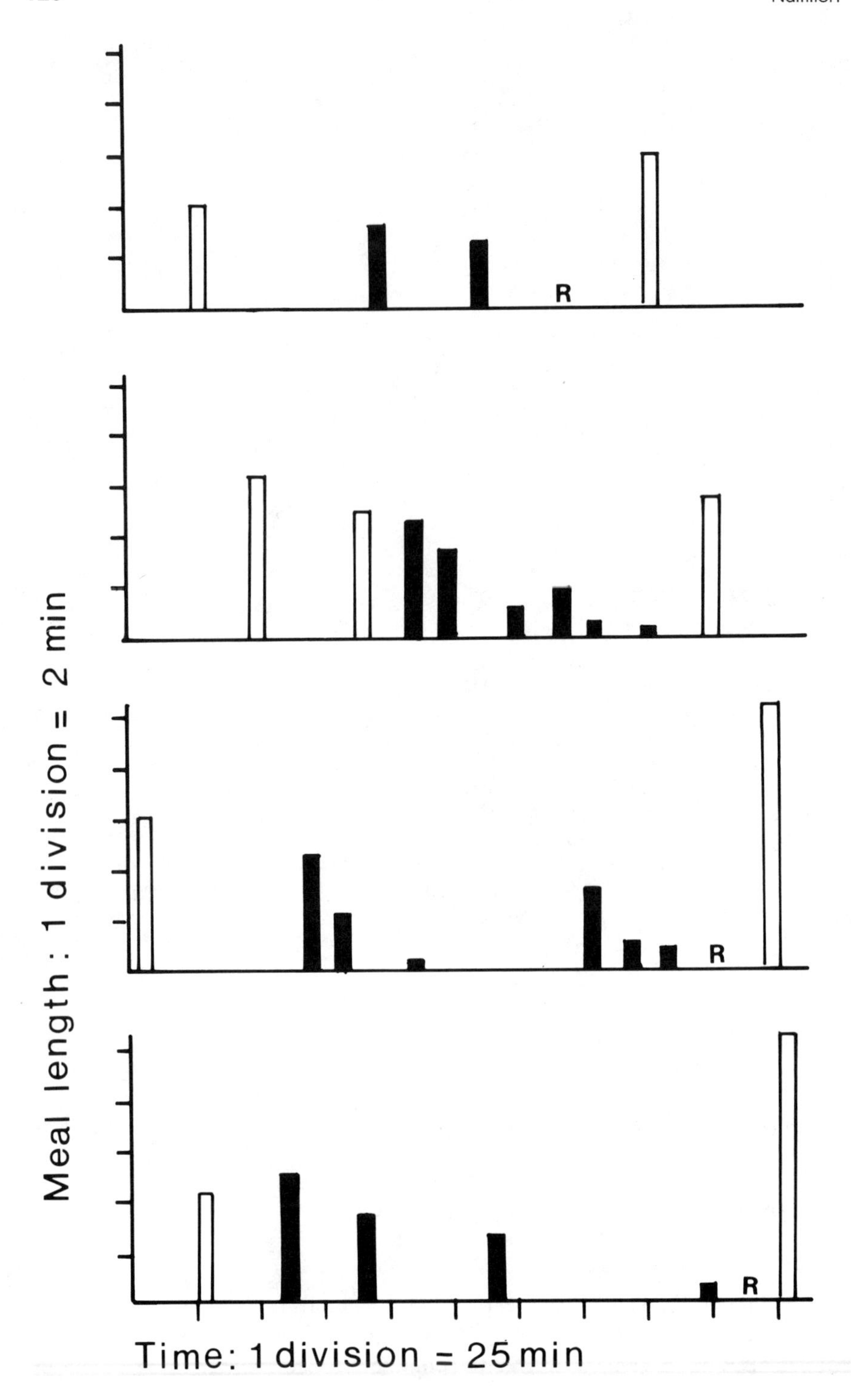

R
R
R
Meal length : 1 division = 2 min
Time: 1 division = 25 min

this range it is likely that protein is not in excess and that additional protein results in generally higher digestive efficiencies.

It has recently become clear that variation may also be due to experimental problems hitherto neglected. Small changes in actual or estimated food dry weight can lead to large errors of 50% or more in values for digestion and assimilation (Schmidt and Reese, 1986; Van Loon, 1988). In addition, fecal weights must be corrected for variable quantities of uric acid, and even when this is done there are problems. For example, acridids that regularly ingest food containing tannins and other phenolics tend to have more substantial peritrophic membranes (Table 4.1) and in the extreme case of the creosote bush grasshopper, *Bootettix argentatus*, protein investment in peritrophic membrane may amount to 10% of ingested protein (E. A. Bernays, unpublished). In estimating digestibility gravimetrically the protein of the membrane that surrounds the feces is wrongly considered as undigested food, potentially causing errors of 20–30%.

Another similar cause of error relates to cases where detrimental plant compounds cause breakdown and sloughing of midgut tissue into the lumen. This occurs in grass-feeding species when tannic acid is ingested (Bernays et al., 1980) and adds to the fecal weight, which is normally considered undigested food. Estimates of nitrogen uptake based on N measurements of food and feces may therefore be confounded by uric acid, peritrophic membrane, and midgut cellular material, and so give falsely low measures of digestive efficiency. Such effects may partly account for the apparent "antidigestive" effects of tannin, now known not to occur (Bernays, 1981; Martin et al., 1987).

Compensatory feeding to obtain the minimum quantities of an essential nutrient must result in poor utilization of others. It is only by careful studies, adding single additional nutrients to food plants in long experimental series, that naturally limiting nutrients will be identified. In experiments with *Schistocerca gregaria* feeding on lettuce for the length of the fifth instar, it was found that a few additional milligrams of phenylalanine resulted in a 25% increase in growth rate when both experimental and control insects ate similar amounts of foliage (about 1 g dry weight over the instar) (Table 4.2). This could be accounted for by increased efficiency of use of ingested food because the lettuce leaves contained relatively low levels of this amino acid (Bernays and Woodhead, 1984). No other amino acid had much effect with this foliage. In nature, such a polyphagous species may improve its nutrient balance by selecting a variety of plants.

Fig. 4.6. Declining acceptability of spinach in successive meals by *Schistocerca americana*, followed by a meal on wheat. Four representative examples. Open bars = meals on wheat; solid bars = meals on spinach; R = rejection of spinach. After a meal on wheat, the wheat was replaced by spinach; after a series of meals on spinach, the spinach was replaced again by wheat. (From Lee and Bernays, 1988.)

TABLE 4.1. Peritrophic Membrane Dry Weights in Relation to Dry Body Weight of Six Acridids Having Diets with Different Levels of Tannins and Related Phenols

	Peritrophic Membrane		
Species	Relative Amount of Tannin in Diet	Weight (mg)	% Body Wt.
Chorthippus parallelus	0	0.04	0.04
Locusta migratoria	0	0.1	0.04
Zonocerus variegatus	+	0.3	0.11
Schistocerca gregaria	++	1.1	0.36
Anacridium melanorhodon	+++	3.2	0.89
Bootettix argentatus	+++	0.4	1.11

TABLE 4.2. Feeding and Efficiency of Food Use in Fifth-instar Nymphs of *Schistocerca gregaria* When Fed on Lettuce With and Without Approximately 1 mg Added Phenylalanine per Day

Experiment	Consumption (mg ± S.E.)	Efficiency of Conversion of Ingested Food (ECI ± S.E.)	Dry Wt. Increase (% ± S.E.)	Mean Additional Phe Ingested (mg)
1				
Control	1080 ± 51	12.2 ± 0.9	82 ± 10	0
With Phe	1076 ± 46	14.3 ± 0.6*	110 ± 5*	7
2				
Control	1066 ± 33	12.6 ± 0.8	89 ± 9	0
With Phe	1041 ± 37	14.2 ± 0.6*	110 ± 4*	6

Source: From Bernays and Woodhead, 1984.

* Significantly different at $p < .01$, t-test.

Experimentally it has been demonstrated that when protein levels are very high, consumption is reduced, but in addition, digestion and utilization of the food overall is reduced (Bernays, 1985, p. 25). It is not known, however, whether this is because it is impossible to process so much protein or whether, with the higher levels, some other nutrient becomes limiting and this then results in inefficient use of the protein.

The efficiency of conversion of ingested food to body weight (ECI) can be influenced by the amount of water in the diet. In *Schistocerca gregaria* fresh wheat seedlings having 90% water gave significantly lower ECI values than a mixture of fresh and freeze-dried wheat seedlings when total food ingested was less than 50% water (Lewis and Bernays, 1985). Although this effect has not been examined in other species, there are various reports of acridids ingesting dry food when suitable lush food is available (see Chapter 2) and it may be a common phenomenon at least among savanna or desert species. It is possible that, in contrast to caterpillars, the gut physiology is best adapted for generally low water levels in the diet. The short midgut and anterior midgut ceca, together with the cycling of water into the posterior midgut and forward into the ceca, make very efficient use of limited amounts of water (Dow, 1981, 1986).

Work on digestive enzymes has been limited since the studies discussed by Uvarov (1966). Useful general reviews including acridid references are given by Dadd (1970b) and Applebaum (1985); identification and surveys of carbohydrates in acridids can be found in the work of Morgan (1975).

One of the major gaps in our knowledge is the changing need for different proportions of dietary components over the life history. This difficult topic needs clarification to understand the ecological role of nutrient selection behavior. It will also help in determining the importance of changing concentrations of nutrients in plants. This is an area rich in theory and controversy, but with little factual information.

4.5 SIGNIFICANCE OF PLANT SECONDARY COMPOUNDS

4.5.1 Negative Effects

Three types of compounds have been postulated to interfere with digestion or uptake of nutrients from the guts of insects. Tannins and other phenolics are best known in this regard but exhaustive studies with acridids indicate that in vivo they do not specifically interfere with these processes although they may be detrimental to unadapted species, particularly by causing damage to the midgut epithelium (Bernays, 1981; Bernays and Chamberlain, 1980; Bernays et al., 1981; Martin et al., 1987). On the other hand, in at least one species, *Anacridium melanorhodon*, hydrolyzable tannin can be hydrolyzed and the phenolic breakdown products utilized (see Section 4.2).

The second group of plant compounds that might be expected to influence digestion are the proteinase inhibitors. Weiel and Hapner (1976) found that grasshopper damage in a North American habitat was negatively correlated with the presence of proteinase inhibitor in the grasses but experimental work on any causal relationship was inconclusive. Work is needed to establish whether proteinase inhibitors reduce digestion, as appears to be the case for tomato proteinase inhibitor in certain caterpillars (Broadway and Duffey, 1986).

Less well known are the effects of plant saponins. Ishaaya (1986) has summarized the inhibitory effects of plant saponins on digestive enzymes. None of the studies involve acridids and most activities have been demonstrated in vitro, so that work is required in this area to establish whether or not there is any effect in vivo.

4.5.2 Positive Effects

Positive effects of plant secondary compounds have been little investigated, and as potential plant defenses such effects are not really expected. One established case is that of 3,4-dihydroxyphenols in the case of *Anacridium melanorhodon* (see Sections 4.2 and 4.5.1). Other reported significant improvements in growth include that caused by the cyanogenic glycoside amygdalin in the diet of *Locusta migratoria* (Bernays, 1983, p. 325) and the steroids tigogenin and hecogenin in the diet of *Melanoplus bivittatus* (Harley and Thorsteinson, 1967). The modes of action here are unknown but deserve more study. Finally, there are numerous examples of secondary compounds, including alkaloids, acting as phagostimulants to the highly polyphagous *Schistocerca gregaria* (Chapman and Bernays, 1977; and see Chapter 2), and probably also the very polyphagous *Romalea guttata* (Jones et al., 1987). It would be of interest to investigate whether these compounds have some kind of value in facilitating the utilization of the primary nutrients.

REFERENCES

Abisgold, J. D. and S. J. Simpson. 1987. The physiology of compensation by locusts for changes in dietary protein. *J. Exp. Biol.* **129**, 329–346.

Abisgold, J. D. and S. J. Simpson. 1988. The effect of dietary protein levels and haemolymph composition on the sensitivity of the maxillary palp chemoreceptors of locusts. *J. Exp. Biol.* **135**, 215–229.

Applebaum, S. W. 1985. Biochemistry of digestion. In G. A. Kerkut and L. I. Gilbert, Eds., *Comprehensive Insect Physiology, Biochemistry and Pharmacology*, Vol. 4. Pergamon, Oxford, pp. 279–312.

Baines, D. M. 1979. Studies of weight changes and movements of dyes in the midgut and caeca of fifth instar *Locusta migratoria migratorioides* (R & F), in relation to feeding and food deprivation. *Acrida* **2**, 319–332.

Ben Halima, T., Y. Gillon, and A. Louveaux. 1985. Spécialization trophique individuelle dans une population de *Dociostaurus maroccanus* (Orthopt.: Acrididae). *Acta Oecol. [Ser.]: Oecol. Gen.* **6**, 17–24.

Bernays, E. A. 1981. Plant tannins and insect herbivores — an appraisal. *Ecol. Entomol.* **6**, 353–360.

Bernays, E. A. 1982. The insect on the plant — a closer look. In J. H. Visser and A. K. Minks, Eds., *Proceedings of the 5th International Symposium on Insect-Plant Relationships*. Pudoc, Wageningen, pp. 3–17.

Bernays, E. A. 1983. Nitrogen in defence against insects. In J. A. Lee, S. McNeill, and I. H. Rorison, Eds., *Nitrogen as an Ecological Factor*. Blackwell, Oxford, pp. 321–344.

Bernays, E. A. 1985. Regulation of feeding behaviour. In G. A. Kerkut and L. I. Gilbert, Eds., *Comprehensive Insect Physiology, Biochemistry and Pharmacology*, Vol. 4. Pergamon, Oxford, pp. 1–32.

Bernays, E. A. 1986. Evolutionary contrasts in insects: Nutritional advantages of holometabolous development. *Physiol. Entomol.* **11**, 377–382.

Bernays, E. A. and R. Barbehenn. 1987. Nutritional ecology of grass foliage-chewing insects. In F. Slansky and J. G. Rodriguez, Eds., *Nutritional Ecology of Insects, Mites, Spiders and Related Invertebrates*. Wiley, New York, pp. 147–176.

Bernays, E. A. and D. J. Chamberlain. 1980. A study of tolerance of ingested tannin in *Schistocerca gregaria. J. Insect Physiol.* **26**, 415–420.

Bernays, E. A. and J. Lee. 1988. Food aversion learning in the polyphagous grasshopper *Schistocerca americana. Physiol. Entomol.* **13**, 131–137.

Bernays, E. A. and S. J. Simpson. 1982. Control of food intake. *Adv. Insect Physiol.* **16**, 59–118.

Bernays, E. A. and S. Woodhead. 1982. Incorporation of dietary phenols into the cuticle in the tree locust *Anacridium melanorhodon. J. Insect Physiol.* **28**, 601–606.

Bernays, E. A. and S. Woodhead. 1984. The need for high levels of phenylalanine in the diet of *Schistocerca gregaria* nymphs. *J. Insect Physiol.* **30**, 489–493.

Bernays, E. A., D. J. Chamberlain, and P. McCarthy. 1980. The differential effects of ingested tannic acid on different species of Acridoidea. *Entomol. Exp. Appl.* **28**, 158–166.

Bernays, E. A., D. J. Chamberlain, and E. M. Leather. 1981. Tolerance of acridids to ingested condensed tannin. *J. Chem. Ecol.* **17**, 247–256.

Bernays, E. A., D. J. Chamberlain, and S. Woodhead. 1983. Phenols as nutrients for a phytophagous insect *Anacridium melanorhodon. J. Insect Physiol.* **29**, 535–539.

Blaney, W. M., R. F. Chapman, and A. Wilson. 1973. The pattern of feeding of *Locusta migratoria* (Orthoptera, Acrididae). *Acrida* **2**, 119–137.

Boys, H. 1978. Food selection by *Oedaleus senegalensis* (Acrididae: Orthoptera) in grassland and millet fields. *Entomol. Exp. Appl.* **24**, 278–286.

Broadway, R. M. and S. S. Duffey. 1986. Plant proteinase inhibitors: Mechanism of action and effect on the growth and digestive physiology of larval *Heliothis zea* and *Spodoptera exigua. J. Insect Physiol.* **32**, 827–834.

Chapman, R. F. 1985. Coordination of Digestion. In G. A. Kerkut and L.I. Gilbert, Eds., *Comprehensive Insect Physiology, Biochemistry and Pharmacology*, Vol. 4. Pergamon, Oxford, pp. 213–240.

Chapman, R. F. and E. A. Bernays. 1977. The chemical resistance of plants to insect attack. *Pontif. Acad. Sci. Scr. Varia* **41**, 603–633.

Cheu, S. P. 1952. Changes in the fat and protein content of the African migratory locust, *Locusta migratoria migratorioides* (R. & F.). *Bull. Entomol. Res.* **43**, 101–109.

Cohen, R. W., S. L. Heydon, G. P. Waldbauer, and S. Friedman. 1987. Nutrient self-selection by the omnivorous cockroach *Supella longipalpa. J. Insect Physiol.* **33**, 77–82.

Cook, A. G. 1977. Nutrient chemicals and oligophagy in *Locusta migratoria. Ecol. Entomol.* **2**, 113–121.

Dadd, R. H. 1960a. Observations on the palatability and utilization of food by locusts, with particular reference to the interpretation of performances in growth trials using synthetic diets. *Entomol. Exp. Appl.* **3**, 283–304.

Dadd, R. H. 1960b. Some effects of dietary ascorbic acid on locusts. *Proc. R. Soc. London, Ser. B* **153**, 128–143.

Dadd, R. H. 1960c. The nutritional requirements of locusts. I. Development of synthetic diets and lipid requirements. *J. Insect Physiol.* **4**, 319–347.

Dadd, R. H. 1960d. The nutritional requirements of locusts II. Utilization of sterols. *J. Insect Physiol.* **5**, 161–168.

Dadd, R. H. 1960e. The nutritional requirements of locusts. III. Carbohydrate requirements and utilization. *J. Insect Physiol.* **5**, 301–306.

Dadd, R. H. 1961a. The nutritional requirements of locusts. IV. Requirements for vitamins of the B complex. *J. Insect Physiol.* **6**, 1–12.

Dadd, R. H. 1961b. The nutritional requirements of locusts. V. Observations on essential fatty acids, chlorophyll, nutritional salt mixtures, and the protein or amino acid components of synthetic diets. *J. Insect Physiol.* **6**, 126–145.

Dadd, R. H. 1961c. Observations on the effects of carotene on the growth and pigmentation of locusts. *Bull. Entomol. Res.* **52**, 63–81.

Dadd, R. H. 1963. Feeding behaviour and nutrition in grasshoppers and locusts. *Adv. Insect Physiol.* **1**, 47–109.

Dadd, R. H. 1970a. Arthropod nutrition. In M. Florkin and B. Scheer, Eds., *Chemical Zoology*, Vol. 5A. Academic Press, New York, pp. 35–95.

Dadd, R. H. 1970b. Digestion in Insects. In M. Florkin and B. Scheer, Eds., *Chemical Zoology*, Vol. 5A. Academic Press, New York, pp. 117–146.

Dadd, R. H. 1985. Nutrition: Organisms. In G. A. Kerkut and L. I. Gilbert, Eds., *Comprehensive Insect Physiology, Biochemistry and Pharmacology*, Vol. 4, Pergamon, Oxford, pp. 313–390.

Dow, J. A. T. 1981. Countercurrent flows, water movements and nutrient absorption in the locust midgut. *J. Insect Physiol.* **27**, 579–585.

Dow, J. A. T. 1986. Insect midgut function. *Adv. Insect Physiol.* **19**, 187–328.

Fox, L. R. and P. A. Morrow. 1981. Specialization: Species property or local phenomenon? *Science* **211**, 887–893.

Harley, K. L. S. and A. J. Thorsteinson. 1967. The influence of plant chemicals on the feeding behavior, development, and survival of the two-striped grasshopper, *Melanoplus bivittatus* (Say), Acrididae: Orthoptera. *Can. J. Zool.* **45**, 305–319.

Hill, L. and G. Goldsworthy. 1968. Growth, feeding activity and the utilization of reserves in larvae of *Locusta*. *J. Insect Physiol.* **14**, 1085–1098.

Hill, L., A. J. Luntz, and P. A. Steele. 1968. The relationships between somatic growth and feeding activity in the adult desert locust. *J. Insect Physiol.* **14**, 1–20.

House, H. L. 1969. Effects of different proportions of nutrients on insects. *Entomol. Exp. Appl.* **12**, 651–669.

Ishaaya, I. 1986. Nutritional and allemochemic insect-plant interactions relating to digestion and food intake. In J. R. Miller and T. A. Miller, Eds., *Insect-Plant Interactions*. Springer-Verlag, New York, pp. 191–224.

Jones, C. G., T. A. Hess, D. W. Whitman, P. J. Silk, and M. S. Blum. 1987. Effects of diet breadth on autogenous chemical defense of a generalist grasshopper. *J. Chem. Ecol.* **13**, 283–297.

Kaufmann, T. 1965. Biological studies on some Bavarian Acridoidea (Orthoptera) with special reference to their feeding habits. *Ann. Entomol. Soc. Am.* **37**, 47–67.

Lee, J. and E. A. Bernays. 1988. Declining acceptability of a food plant for a polyphagous grasshopper *Schistocerca americana*: The role of food aversion learning. *Physiol. Entomol.* **13**, 291–301.

Lewis, A. C. and E. A. Bernays. 1985. Feeding behavior: Selection of both wet and dry food for increased growth in *Schistocerca gregaria* nymphs. *Entomol. Exp. Appl.* **37**, 105–112.

Loveridge, J. P. 1973. Age and the changes in water and fat content of adult laboratory-reared *Locusta migratoria migratorioides*. R. & F. *Rhod. J. Agric. Res.* **11**, 131–143.

Lyttleton, J. W. 1973. Amino acids, peptides and ureides. *In* G. Butler and R. Bailey, Eds., *Chemistry and Biochemistry of Herbage*. Academic Press, New York, pp. 63–103.

MacFarlane, J. H, and A. J. Thorsteinson. 1980. Development and survival of the two striped grasshopper, *Melanoplus bivittatus* (Say) (Orthoptera: Acrididae), on various single and multiple plant diets. *Acrida* **9**, 63–76.

Martin, M. M. 1983. Cellulose digestion in insects. *Comp. Biochem. Physiol. A.* **75A**, 313–324.

Martin, M. M., J. Martin, and E. A. Bernays. 1987. Failure of tannic acid to inhibit digestion or reduce digestibility of plant protein in gut fluids of insects: Implications for theories of plant defense. *J. Chem. Ecol.* **13**, 605–621.

McCaffery, A. R. 1975. Food quality and quantity in relation to egg production in *Locusta migratoria migratorioides*. *J. Insect Physiol.* **21**, 1551–1558.

McGinnis, A. J. and R. Kasting. 1967. Dietary cellulose: Effect on food consumption and growth of a grasshopper. *Can. J. Zool.* **45**, 365–367.

Mordue (Luntz), A. J. and L. Hill. (1970). The utilization of food by the adult female desert locust *Schistocerca gregaria*. *Entomol. Exp. Appl.* **13**, 352–358.

Morgan, M. R. J. 1975. A qualitative survey of the carbohydrases of the alimentary tract of the migratory locust, *Locusta migratoria migratorioides*. *J. Insect Physiol.* **21**, 1045–1054.

Mulkern, G. B. 1972. The effects of preferred food plants on distribution and numbers of grasshoppers. In C. F. Hemming and T. H. C. Taylor, Eds., *Proceedings of the International Study Conference on the Current and Future Problems of Acridology*. Centre for Overseas Pest Research, London, pp. 215–218.

Parker, M. A. 1984. Local food depletion and the foraging behavior of a specialist grasshopper, *Hesperotettix viridis*. *Ecology* **65**, 824–835.

Peters, R. H. 1983. *The Ecological Implications of Body Size*. Cambridge Univ. Press, London and New York.

Pfadt, R. E. and R. J. Lavigne. 1982. Food habits of grasshoppers inhabiting the Pawnee site. *Wyo., Agric. Exp. Stn., Sci. Monog.* **42**, 1–72.

Schmidt, D. J. and J. C. Reese. 1986. Sources of error in nutritional index studies of insects on artificial diet. *J. Insect Physiol.* **32**, 193–198.

Scriber, J. M. and F. Slansky. 1981. The nutritional ecology of immature insects. *Annu. Rev. Entomol.* **26**, 183–211.

Simpson, S. J. 1982a. Patterns in feeding. A behavioural analysis using *Locusta migratoria* nymphs. *Physiol. Entomol.* **7**, 325–336.

Simpson, S. J. 1982b. Changes in the efficiency of utilisation of food throughout the fifth-instar of *Locusta migratoria* nymphs. *Entomol. Exp. Appl.* **31**, 265–275.

Simpson, S. J. 1983a. The role of volumetric feedback from the hind gut in the regulation of meal size in the fifth instar *Locusta migratoria*. *Physiol. Entomol.* **8**, 451–467.

Simpson, S. J. 1983b. Changes in the rate of crop emptying during the fifth instar of *Locusta migratoria* and their relationship to feeding and utilization. *Entomol. Exp. Appl.* **33**, 235–243.

Simpson, S. J. and J. D. Abisgold. 1985. Compensation by locusts for changes in dietary nutrients: Behavioural mechanisms. *Physiol. Entomol.* **10**, 443–452.

Simpson, S. J., M. S. J. Simmonds, and W. M. Blaney. 1988. A comparison of dietary selection behaviour in larval *Locusta migratoria* and *Spodoptera littoralis*. *Physiol. Entomol.* **13**, 225–238.

Slansky, F. and J. G. Rodriguez (Eds.) 1987. *Nutritional Ecology of Insects, Mites, Spiders and Related Invertebrates*. Wiley, New York.

Tobe, S. S. and B. G. Loughton. 1967. The development of blood proteins in the African migratory locust. *Can. J. Zool.* **45**, 975–985.

Uvarov, B. P. 1966. *Grasshoppers and Locusts*. Cambridge Univ. Press for the Anti-Locust Research Centre, London.

Vanderzant, E. S. 1958. The amino acid requirements of the pink bollworm. *J. Econ. Entomol.* **51**, 309–311.

Van Loon, J. 1988. Sensory and nutritional effects of amino acids and phenolic plant compounds on the caterpillars of two *Pieris* species. Doctoral Thesis, Agricultural University, Wageningen.

Weed Pfeiffer, I. 1945. Effect of the corpora allata on the metabolism of adult female grasshoppers. *J. Exp. Zool.* **99**, 183–233.

Weiel, J. and K. D. Hapner. 1976. Barley proteinase inhibitors: A possible role in grasshopper control. *Phytochemistry* **15**, 1885–1887.

5

Water Regulation

E. A. BERNAYS

5.1 INTRODUCTION

Acridids have a rather generalized terrestrial insectan morphology, and accounts of the normal adaptations of land arthropods to prevent water loss may be found in good general texts (Chapman, 1982; Edney, 1977). Edney's (1977) book on water balance in land arthropods is the most comprehensive recent coverage of the general topic. Wharton (1985) has reviewed the physiological processes involved in water balance in insects, and Phillips et al. (1986) have reviewed the cellular processes involved in water reabsorption in insect hindguts. Information on particular species of grasshoppers with respect to water content at different times during development, water loss, and fatal limits of desiccation may be found in Uvarov (1966). The most thorough study of the control of spiracular and cuticular water loss was done by Loveridge (1968a,b, 1974), and a very good general study of water balance is given by Loveridge (1975).

Although grasshoppers occur in temperate zones the Acridoidea are primarily tropical, being abundant in most habitats of the tropics and subtropics, from very wet to very dry. For example, species richness in a Costa Rican wet rainforest area (Finca La Selva) is similar to that of an equivalent-sized arid region of the Sonoran desert and adjacent mountains in southern California (Boyd Deep Canyon Research Station) (Rowell, 1978; R. F. Chapman, unpublished). High-density populations and outbreak species are, however, characteristically found in geographic areas with pronounced dry seasons, especially in grasslands, and compared with other major groups of insect herbivores, a relatively large proportion of the known species of Acridoidea is associated with grassland (Bernays and Barbehenn, 1987).

There is no work to date that suggests that there may be specialized structures or modifications of the excretory system to deal with wet extremes such as those encountered by rain forest species or by *Paulinia*, which feeds on water hyacinth. Neither are there obvious specialized structures in species of dry extremes such as the desert locust *Schistocerca gregaria*, or the desert species of the southwestern United States that have been examined (Bradley, 1985). In one comparative study Abushama (1970) found that the mesic species *Anacridium melanorhodon* lost water more rapidly than the xeric *Poecilocerus hieroglyphicus*, and presumably even greater differences will be found between species of very wet habitats and those of xeric habitats. Massion (1983) suggested that species at higher altitudes in dry tropics might tend to lose water at a higher rate but the threat of desiccation was relatively low because of more succulent vegetation. Some species of acridids may spend part of their lives in very wet conditions, and part in extremes of drought. Thus in the Niger River floodplain of West Africa, large populations of the migratory locust, *Locusta migratoria*, develop in flooded areas that then become very arid. The ability of grasshoppers to deal with these different extremes resides in a combination of behavioral and physiological adaptability.

5.2 BEHAVIORAL ASPECTS OF WATER REGULATION

5.2.1 Food Selection

The most important behavioral element in regulation of water balance is the ingestion of foods having appropriate water content. When reared in the laboratory under dry conditions, grasshoppers reject drier foliage in preference for wet, lush foliage (Loveridge, 1974). Rearing under wet conditions leads to a preference for drier foods, which may then also be eaten in larger than average quantity (Fig. 5.1) (Bernays and Chapman, 1978; Chapman and Bernays, 1977). Given mixtures of foods having high and low water content, nymphs of both *Schistocerca gregaria* and *Locusta migratoria* were found to ingest quantities of each such that overall water intake was approximately one-third water and two-thirds dry matter (Lewis and Bernays, 1985; Roessingh et al.,

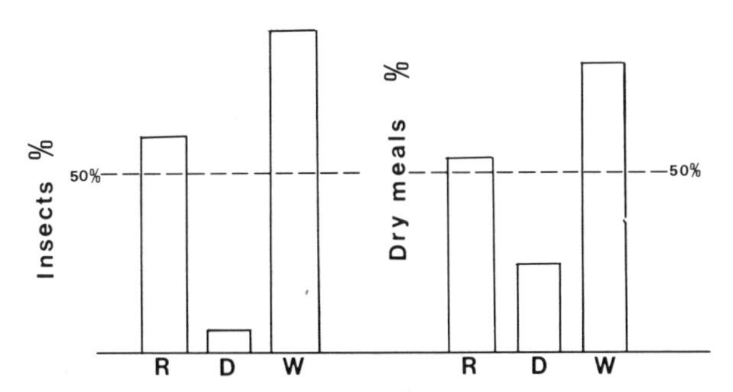

Fig. 5.1. Feeding by fifth-instar nymphs of *Schistocerca gregaria* on lyophilized seedling wheat, after three different pretreatments. Pretreatments were: R, regulated insects with both lush and lyophilized food available; D, dry insects with only lyophilized food; W, wet insects with only lush seedling wheat. On the left is shown the percent of insects choosing dry food for the first meal of over 4 min ($n = 19$ for R, 20 for D, 21 for W). On the right is shown the percent of total meals by the same insects on dry food over a 2-h experiment. In both cases the differences are significant with a sign test at $p < .01$. (After Lewis and Bernays, 1985.)

1985; Sinoir, 1966). When offered foods consisting of fresh wheat seedlings (90% water) and lyophilized wheat seedlings (2–5% water), insects ate alternately from the wet and dry food, and showed improved growth rates over insects having only wet food (Table 5.1).

There is also evidence that in the field grasshoppers select a mixture of wet and dry foods (Lewis, 1984; McKinlay, 1981), although there is always the possibility that factors other than water are important. For example, the grasshopper *Melanoplus differentialis* was found to feed selectively in the field on wilted and partly desiccated leaf material of several plant species (Lewis, 1984), but this could also have been due to favorable changes in plant chemistry that sometimes occur during wilting (Bernays and Lewis, 1986).

The preference of locusts with low hemolymph volume for wet food is not surprising: a decrease in body water must be compensated for, and such consumption is a universal animal response. In the absence of wet food, feeding on dry food (i.e., water content of less than 5%) is halted after about 24 hr. This is an essential water conservation behavior because feeding and feces production in such circumstances inevitably cause water loss (Bernays, 1977; Loveridge, 1975). The exact amounts of water required for an insect to eat maximally on dry food have been demonstrated for one species, *Locusta migratoria*. When individual adults were kept entirely on lyophilized wheat leaves in the laboratory, the amount of this dry food eaten was directly proportional to the amount of water each insect was allowed to drink (Ben Halima et al., 1983). Thus, when given 100 μl water per day, insects ate approximately 120 mg dry food per day; when given 185 μl/day, insects ate approximately 250 mg of dry food per day. The maximum intake of dry food was 400–500 mg and this occurred when water was provided ad libitum. These data are consistent with those quoted above where approximately one

TABLE 5.1. Growth Rate and Indices of Consumption and Utilization by Fifth-instar Male Nymphs of *Schistocerca gregaria* Raised on Either Lush Seedling Wheat or a Mixture of Lush Seedling Wheat and Lyophilized Seedling Wheat.

	Wet Food	Wet + Dry Food	*P*
Weight increase (%)	117 ± 4	137 ± 3	<.001
Relative growth rate (mg/mg/day)	0.063 ± 0.02	0.076 ± 0.001	<.001
Relative consumption rate (mg/mg/day)	0.464 ± 0.007	0.496 ± 0.007	<.01
Efficiency of conversion of ingested food	13.7 ± 0.4	15.3 ± 0.2	<.01

Source: From Lewis and Bernays, 1985.

part water to two parts dry matter was found to be the chosen balance when wet and dry foods were both available.

The preference of fully hydrated locusts for dry food probably results less from a positive stimulation of dry food than from a deterrence of water to such locusts (Kendall and Seddon, 1975; Roessingh et al., 1985). *Schistocerca gregaria* nymphs previously fed only very wet food and then given a choice of wet and dry food tended to reject the wet food until long meals on dry food had been taken (Lewis and Bernays, 1985).

Subtle but perhaps important changes may occur in relation to molting. Lewis and Bernays (1985) noted that *Schistocerca gregaria* having the choice of wet and dry food ate more of the dry than the wet during most of the fifth instar, but on the day prior to ecdysis, only wet food was ingested. It is also known that in *Schistocerca gregaria* hemolymph volume is greatest at ecdysis, after which it falls rapidly (Lee, 1961). It may be speculated that physiological processes tend to be more efficient when hemolymph volumes are relatively low, but that ecdysis is most efficient with a higher volume, which would enhance the ability of the insects to create high pressures and thus maximally expand the folded cuticle.

Plants that are not normally hosts may be eaten by rather dehydrated insects when the host plant is particularly dry (Bernays and Chapman, 1973, 1978). Water regulation in these circumstances is usually hard to separate from nutritional factors.

5.2.2 Drinking

Drinking can be important in maintaining water balance when the water content of the food is below 50% (Loveridge, 1974). Detailed studies have been carried out on the regulatory mechanisms with the migratory locust, *Locusta migratoria* (Bernays, 1977). The readiness to drink, measured as a positive response when freely moving insects make contact with water, was associated with a 10% reduction in body volume. This could be induced by removal of hemolymph alone, and in fifth-instar nymphs, previously deprived of food for 10 h to allow emptying of the gut, removal of 50 μl caused positive responses (Fig. 5.2). The monitoring of body volume changes presents no problem to the insect, because there are known to be body wall stretch receptors, but the effect may not be direct for it took about an hour after removal of hemolymph for full responsiveness to develop. It is possible that antidiuretic factors (see below) may contribute to behavioral changes, just as angiotensin does in mammals.

It is to be expected that osmotic stress should play some part in causing drinking, but massive injections of various salts and sugars, causing a rise in osmotic pressure of the hemolymph, had no effect on responsiveness of *Locusta migratoria* to water. On the other hand, the amount of water ingested after drinking had been initiated was correlated with an increase of the osmotic pressure of the hemolymph (Fig. 5.3). The drop in hemolymph osmotic pressure during drinking appears to be the direct cause of termination of the drink. It is assumed that osmoreceptors of some sort are involved. Results of Barton Browne and van Gerwen (1976) on the Australian plague locust, *Chortoicetes terminifera*, can also be interpreted as indicating an osmotic control over the amount of water ingested. The work so far indicates that initiation

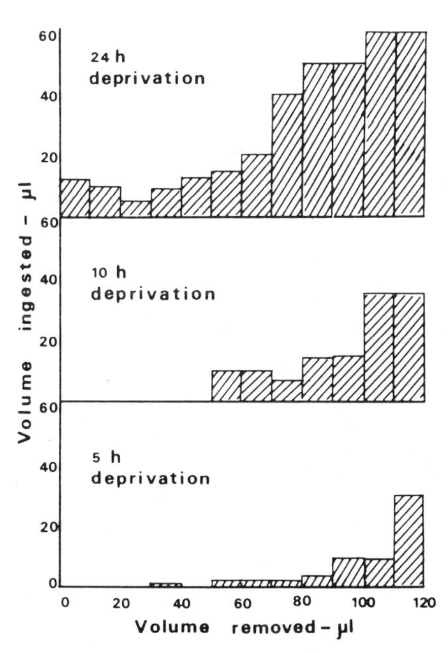

Fig. 5.2. Effect of removal of hemolymph on the volume of water ingested by fifth-instar nymphs of *Locusta migratoria*, in three different situations of water deprivation. (From Bernays, 1977.)

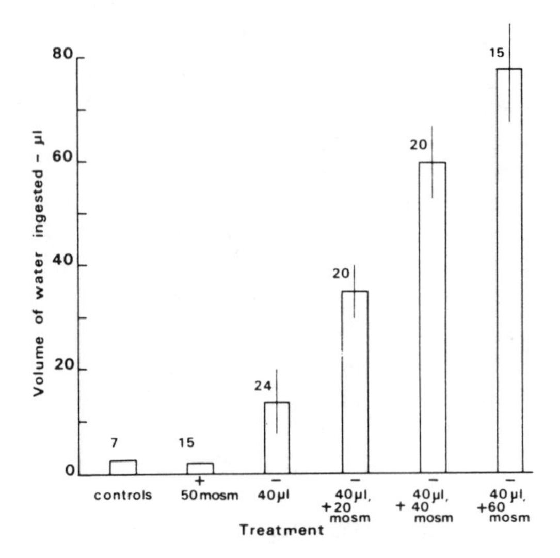

Fig. 5.3. The amounts of water ingested by fifth-instar nymphs of *Locusta migratoria* with hemolymph removed (−40 µl) and/or hemolymph osmotic pressure increased (+mosm). All animals were totally deprived beforehand for 12 h at 30°C. Numbers above each column = numbers tested; bars = S.E. (From Bernays, 1977.)

and termination of drinking have different controlling mechanisms (Fig. 5.4) (Bernays, 1977, 1985; Bernays and Simpson, 1982).

There are a number of reports (Gangwere, 1960; Edney, 1977) of drinking at wet places in field conditions but how common drinking is in the field is not known. The prevalence of dew in many arid regions makes it conceivable that drinking could be important in such areas.

5.2.3 Interaction with Other Behaviors

A number of behaviors influence water balance indirectly, though they are probably not normally part of water regulation. Thus increased activity, especially flight, causes greatly increased spiracular water loss particularly as a result of ventilatory movements (Loveridge, 1968b). Also as air temperature increased from 30 to 40°C and more, ventilation in *Locusta migratoria* was found to increase sharply, with a concomitant increase in water loss. This increase in ventilatory movement would then be accompanied by evaporative cooling, and was therefore assumed by Loveridge to result in depression of body temperature although this was not measured. Stower and Griffiths (1966) estimated that only 5% of heat loss in the desert locust, *Schistocerca gregaria*, was from evaporative cooling under average conditions, but as Edney (1977) points out, evaporative cooling may be important for short-term emergencies, at least in relatively large insects such as acridids. The interaction of temperature regulation and water regulation is an area that has been little studied.

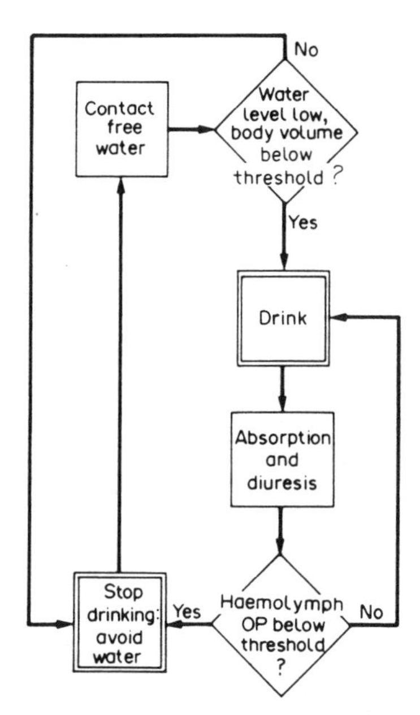

Fig. 5.4. A model to illustrate the regulation of drinking in nymphs of *Locusta migratoria*. (From Bernays, 1985; Bernays and Simpson, 1982.)

5.3 PHYSIOLOGICAL ASPECTS OF WATER REGULATION

In acridids, as in other terrestrial arthropods, the excretory system normally keeps the insect in positive water balance. In a typical locust, urine passes into the rectum at a rate of 10–15 μl/h, and this is less than the 17–20 μl/h basal rate of water uptake from the rectal lumen (Mordue et al., 1980). However, some water will still be lost in the feces as well as through the cuticle and spiracles. As described above, deficits are relieved by eating and drinking. To promote water loss, the insect releases a diuretic hormone and the volume of urine entering the gut via the Malpighian tubules is then higher than the absorption capacity of the rectum. In times of water deprivation, the lowest possible quantities of fluid should enter the Malphigian tubules and gut, and much of this is then apparently recovered by the tissues of the rectum.

5.3.1 Hemolymph

The tolerance of variation in water content is probably not very different from that of other insects of similar size. Examples of variation with age, sex, food intake, and egg production for various different insects are given by Edney (1977). In these cases, it is the hemolymph that appears to act as the reservoir, with increases of over 100% in its volume being associated with molting, decreases or increases of 50% with dietary levels of water, and a decrease of 25% with 1 day of food deprivation. In hot dry field conditions in east Africa, Chapman (1958) was unable to collect hemolymph at all from some individuals of *Nomadacris septemfasciata* in the heat of the day, although all the insects contained ample hemolymph at night, suggesting very large daily changes. Loveridge (1975) considers that in normally hydrated locusts 20% of the total water content is accounted for by the hemolymph, and that most of this is available as a water reserve. This is possible because of the ability of insects to regulate the hemolymph osmotic pressure by sequestration of osmoeffectors in the fat body (Tucker, 1977).

5.3.2 Cuticle and Spiracles

Overall rates of water loss from cuticle and spiracles are given for eight species in different conditions by Farlow and MacMahon (1988). These vary from about 0.58% body weight per hour to 0.82% body weight independent of body size at a relative humidity of $<5\%$. Although the cuticle and its associated wax are obvious means of conserving water, some loss from the body occurs here. In the case of *Locusta migratoria* adults weighing approximately 1 g, water loss through the cuticle at low humidities (high saturation deficits) may be as high as 5 mg/h per insect (0.5% of body weight per hour), but at high humidities water loss through the cuticle is minimal (Loveridge, 1968a). These differences do not just reflect an effect of the environment on a simple physical system. For example, insects from a moist environment placed at high saturation deficits initially lose water rather quickly from the cuticle itself. Thereafter in insects deprived of food, perhaps due to the consequent reduction in intermicellar pore dimensions of the cuticle, water loss falls from up to 10 mg/h to less than 4 mg/h for an adult (Loveridge, 1968a, 1975). Decreases in rate of water loss have also been shown in several North American species by Anderson et al. (1979) and Farlow and MacMahon (1988).

At low humidities or when food is dry, spiracular water loss alone is reduced to less than 1 mg/h in adult *L. migratoria*, but at high humidities or when food is wet the same insect may lose 4–5 mg/h through the spiracles. The change is brought about by changes in ventilatory rates of the locusts (Loveridge, 1968b), and is the same type of change as the hyperventilation noted for *S. gregaria* in wet conditions (Roessingh et al., 1985), where it was suggested that this activity was a means of losing some excess water.

5.3.3 Hormonal Regulation

The physiological regulation of water content involving hormones is primarily brought about by changes in secretion of water by the Malpighian tubules — diuresis — and by changes in the extent of water reabsorption from the rectum.

A large grasshopper weighing about 1 g may eat its own weight of fresh foliage each day. Thus, if the foliage is assumed to be 80% water, the insect may take in over 800 μl water per day, much of which moves into the hemolymph, which must be maintained at a volume of only about 400 μl (Uvarov, 1966). To assist in the removal of the excess water, the intake of a large lush meal stimulates the release of diuretic hormone, which will increase the volume of fluid secreted into the Malpighian tubules from around 5–10 μl/h to 25 μl/h (Mordue, 1972). The diuretic hormone is a peptide synthesized by the cerebral neurosecretory cells and released from the storage lobes of the corpora cardiaca. Release probably occurs as a result of neural inputs originating in the wall of the foregut distended by newly ingested food (Mordue et al., 1980). Smaller meals are taken on dry food, and little diuretic hormone is released as a result of feeding. With no feeding, no diuretic hormone is produced and fluid secretion by the tubules remains at a low level.

The rectum is the most important organ in water conservation. Water reabsorption from the rectum may balance the water input into the hindgut from the Malpighian tubules. In a large grasshopper the resting level of 17–20 μl/h is against an osmotic gradient of up to about 500 mosmol. An antidiuretic hormone (Cazal and Girardie, 1968), produced by the glandular lobes of the corpora cardiaca, increases the water uptake ability of the rectum, which in a dehydrated locust can reabsorb water against an osmotic gradient of 1000 mosmol (Phillips, 1964, 1981). Such an ability is highly adaptive for those species inhabiting arid regions.

It is a common observation (e.g., Norris, 1961) that under dry conditions dry feces are produced by acridids, and there are records of water content varying from 10% water to over 80% water when the food is lush (Uvarov, 1966). Individuals can obviously regulate their water loss through changes brought about by the hormonal milieu. There are species differences, however, which presumably reflect their adaptations to different environmental conditions. The two locust species, *Locusta migratoria* and *Schistocerca gregaria*, have been extensively studied in the laboratory. The latter inhabits extremely arid regions and perhaps typifies species adapted to dry extremes. When fed on lush wheat seedlings in high humidities individuals become water loaded (Roessingh et al., 1985). Performance is improved only by providing dry food (Lewis and Bernays, 1985). *L. migratoria*, on the other hand, is a species inhabiting areas subject to extensive flooding. When fed on lush food at high humidity it grows maximally and is able to produce fluid feces. This fluid spreads over the abdomen, leaving a characteristic white residue as it dries (Simpson, 1982).

5.3.4 Interactions with Other Physiological Processes

Metabolism itself produces water. Oxidation of fats yields over 100 g water for every 100 g fat, while oxidation of a similar weight of glucose yields 56 g water. Metabolic water is therefore a component of the total water income, and in dry conditions may be a large component. The general consensus is that insects are not in a position to regulate the production of metabolic water although starving or flying insects switch to fat metabolism (Edney, 1977). Weis-Fogh (1967) showed that in flight the desert locust could produce 8 mg water/g/h and that under certain conditions insects could maintain a positive water balance when flying over very long distances. However, at very high temperatures and low humidities, the thoracic ventilation occasions a large respiratory loss, and there is a net water loss (Loveridge, 1968b). Loveridge and Bursell (1975) demonstrated that, although starving locusts shifted to fat metabolism, this was counterproductive in terms of maintenance of water balance, partly because the production of ATP requires a smaller weight of fat and that the fat is slightly more efficiently used than carbohydrate.

Xenobiotics may interact with systems controlling water balance. Ingested toxins may require additional water for their rapid removal into the Malpighian tubules (Maddrell, 1981). Neurotoxins may cause release of diuretic hormone into the hemolymph and thereby cause desiccation (Maddrell, 1980). A specific effect of a nonprotein amino acid in the legume *Vicia sativa* has been described for *Locusta migratoria*. β-Cyano-L-alanine given in the diet in physiological amounts resulted in production of very wet feces apparently due to an effect on the rectum, reducing the normal reabsorption and eventually causing desiccation (Schlesinger et al., 1976). It is likely that other interactions will be found in the future.

5.4. CONCLUSIONS

The combination of behavioral and physiological processes provides a suite of mechanisms that clearly serves to allow survival of acridids in a range of habitats and environmental extremes and on a range of foods. A water balance sheet for a typical acridid is illustrated in Figure 5.5, based on the extensive studies of Loveridge on the migratory locust *Locusta migratoria*. The most important factors in maintaining water balance in normally feeding insects are water gain in the food and loss in the feces. Temperature and humidity extremes, however, can have profound effects on water loss through cuticle and spiracles. The finer controls of needs occasioned by such environmental changes, together with needs for different physiological functions, are largely wrought by hormonal factors acting on the Malpighian tubules and rectum and possibly on water intake behavior.

The acridids studied to date indicate that they are highly adapted for xeric environments. Unlike Lepidoptera, where digestion and growth are maximized with increasing leaf water content, acridid physiology seems to be most efficient with relatively low dietary water. A number of interesting questions arise from this: (1) What is the relative importance of basking for temperature regulation and water loss? (2) To what extent are foods selected for their low water content, and does this need override chemical deterrence? (3) How do species in wet habitats get rid of excess water? Can they produce free urine? (4) Are

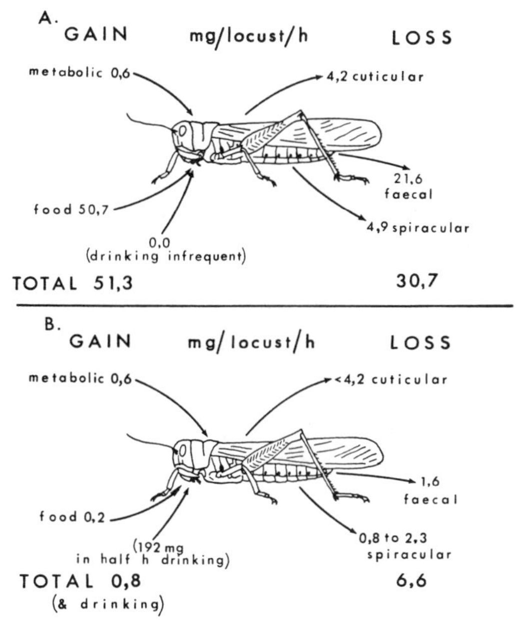

Fig. 5.5. The water balance of "standard" acridids of mass 1.6 g fresh weight at 50% relative humidity and 30°C. (A) Insect feeding on fresh, green grass. (B) Insect feeding on dry food. Components of water gain and weight gain are listed on the left, and components of water loss and weight loss are listed on the right in each case. (From Loveridge, 1975.)

our studies biased by the use of more common grassland species, including locusts?

Additional comparative studies are desirable to understand the adaptations and overall patterns of water regulation in acridids.

REFERENCES

Abushama, F. T. 1970. Loss of water from the grasshopper *Poecilocerus hieroglyphicus* (Klug), compared with the tree locust *Anacridium melanorhodon melanorhodon* (Walker). *Z. Angew. Entomol.* **66**, 160–167.

Anderson, R. V., C. R. Tracy, and Z. Abramsky. 1979. Habitat selection in two species of short horned grasshoppers: The role of thermic and hydric stress. *Oecologia* **38**, 359–374.

Barton Browne, L. and A. C. M. van Gerwen. 1976. Regulation of water ingestion by the locust, *Chortoicetes terminifera*: The effect of injections into the haemolymph. *Physiol. Entomol.* **1**, 159–167.

Ben Halima, T., A. Louveaux, and Y. Gillon. 1983. Rôle de l'eau des boisson sur la prise de nourriture sèche et le développement ovarien de *Locusta migratoria migratorioides*. *Entomol. Exp. Appl.* **33**, 329–335.

Bernays, E. A. 1977. The physiological control of drinking behaviour in nymphs of *Locusta migratoria*. *Physiol. Entomol.* **2**, 261–273.

Bernays, E. A. 1985. Regulation of feeding behaviour. In G. A. Kerkut and L. I. Gilbert, Eds., *Comprehensive Insect Physiology, Biochemistry and Pharmacology*, Vol. 4. Pergamon, Oxford, pp. 1–32.

Bernays, E. A. and R. Barbehenn. 1987. Nutritional ecology of grass foliage-chewing insects. In F. Slansky and J. G. Rodriguez, Eds., *Nutritional Ecology of Insects, Mites, Spiders and Related Invertebrates*. Wiley, New York, pp. 147–175.

Bernays, E. A. and R. F. Chapman. 1973. The role of food plants in survival and development of *Chortoicetes terminifera* (Walker) under drought conditions. *Aust. J. Zool.* **21**, 575–592.

Bernays, E. A. and R. F. Chapman. 1978. Plant chemistry and acridoid feeding behaviour. In J. B. Harborne, Ed., *Biochemical Aspects of Plant and Animal Coevolution*. Academic Press, New York, pp. 99–141.

Bernays, E. A. and A. C. Lewis. 1986. The effect of wilting on palatability of plants to *Schistocerca gregaria*, the desert locust. *Oecologia* **70**, 132–135.

Bernays, E. A. and S. J. Simpson. 1982. Control of food intake. *Adv. Insect Physiol.* **16**, 59–118.

Bradley, T. J. 1985. The excretory system: Structure and physiology. In G. A. Kerkut, and L. I. Gilbert, Eds., *Comprehensive Insect Physiology, Biochemistry and Pharmacology*, Vol. 4. Pergamon, Oxford, pp. 421–465.

Cazal, M. and A. Girardie. 1968. Contrôle humoral de l'équilibre hydrique chez *Locusta migratoria migratorioides*. *J. Insect Physiol.* **14**, 655–668.

Chapman, R. F. 1958. A field study of the potassium concentration in the blood of the red locust *Nomadacris septemfasciata* (Serv.), in relation to its activity. *Anim. Behav.* **6**, 60–67.

Chapman, R. F. 1982. *The Insects: Structure and Function*. Hodder & Stoughton, London.

Chapman, R. F. and E. A. Bernays. 1977. Chemical resistance of plants to insect attack. *Pontif. Acad. Sci. Scr. Varia* **41**, 603–633.

Edney, E. B. 1977. *Water Balance in Land Arthropods*. Springer-Verlag, New York.

Farlow, L. J. and J. A. MacMahon. 1988. A seasonal comparison of metabolic and water loss rates of three species of grasshoppers. *Comp. Biochem. Physiol. A.* **89A**, 51–60.

Gangwere, S. K. 1960. Notes on drinking and the need for water in Orthoptera. *Can. Entomol.* **92**, 911–915.

Kendall, M. D. and A. Seddon. 1975. The effect of previous access to water on the responses of locusts to wet surfaces. *Acrida* **4**, 1–7.

Lee, R. M. 1961. The variation of blood volume with age in the Desert Locust (*Schistocerca gregaria* Forsk.). *J. Insect Physiol.* **6**, 36–51.

Lewis, A. C. 1984. Plant quality and grasshopper feeding: Effects of sunflower condition on preference and performance in *Melanoplus differentialis*. *Ecology* **65**, 836–843.

Lewis, A. C. and E. A. Bernays. 1985. Feeding behavior: Selection of both wet and dry food for optimal growth by *Schistocerca gregaria* nymphs. *Entomol. Exp. Appl.* **37**, 105–112.

Loveridge, J. P. 1968a. The control of water loss in *Locusta migratoria migratorioides* R. & F. I. Cuticular water loss. *J. Exp. Biol.* **49**, 1–13.

Loveridge, J. P. 1968b. The control of water loss in *Locusta migratoria migratorioides* R. & F. II. Water loss through the spiracles. *J. Exp. Biol.* **49**, 15–29.

Loveridge, J. P. 1974. Studies on water regulation of adult locusts. II. Water gain in the food and loss in the faeces. *Trans. Rhod. Sci. Assoc.* **56**, 1–30.

Loveridge, J. P. 1975. Studies on the water relations of adult locusts. III. The water balance of non-flying locusts. *Zool. Afr.* **10**, 1–28.

Loveridge, J. P. and E. Bursell. 1975. Studies on the water relations of adult locusts (Orthoptera, Acrididae). I. Respiration and the production of metabolic water. *Bull. Entomol. Res.* **65**, 13–20.

Maddrell, S. H. P. 1980. The insect neuroendocrine system as a target for insecticides. *Insect Neurobiology and Pesticide Action*. Society for Chemical Industry, London, pp. 329–334.

Maddrell, S. H. P. 1981. The functional design of the insect excretory system. *J. Exp. Biol.* **90**, 1–15.

Massion, D. D. 1983. An altitudinal comparison of water and metabolic relations in two acridid grasshoppers (Orthoptera). *Comp. Biochem. Physiol. A* **74A**, 101–105.

McKinlay, K. S. 1981. The importance of dry plant material in the diet of the grasshopper *Melanoplus sanguinipes*. *Can. Entomol.* **113**, 5–8.

Mordue, W. 1972. Hormones and excretion in locusts. *Gen. Comp. Endocrinol., Suppl.* **3**, 289–298.

Mordue, W., G. J. Goldsworthy, J. Brady, and W. M. Blaney. 1980. *Insect Physiology*. Blackwell, Oxford.

Norris, M. J. 1961. Group effects on feeding in adult males of the Desert Locust *Schistocerca gregaria* (Forsk.) in relation to sexual maturation. *Bull. Entomol. Res.* **51**, 731–753.

Phillips, J. E. 1964. Rectal absorption in the desert locust, *Schistocerca gregaria* Forskål. I. Water. *J. Exp. Biol.* **41**, 15–38.

Phillips, J. E. 1981. Comparative physiology of insect renal function. *Am. J. Physiol.* **241**, R241–R257.

Phillips, J. E., J. Hanrahan, M. Chamberlin, and B. Thomson. 1986. Mechanisms and control of reabsorption in insect hindgut. *Adv. Insect Physiol.* **18**, 329–422.

Roessingh, P., E. A. Bernays, and A. C. Lewis. 1985. Physiological factors influencing preference for wet and dry food in *Schistocerca gregaria* nymphs. *Entomol. Exp. Appl.* **37**, 89–94.

Rowell, H. F. 1978. The grasshoppers of La Selva. Unpublished lists and photographs at Finca La Selva, Costa Rica.

Schlesinger, H. M., S. W. Applebaum, and Y. Birk. 1976. Effect of beta-cyano-L-alanine on the water balance of *Locusta migratoria*. *J. Insect Physiol.* **22**, 1421–1426.

Simpson, S. J. 1982. Changes in the efficiency of utilisation of food throughout the fifth instar nymphs of *Locusta migratoria*. *Entomol. Exp. Appl.* **31**, 265–275.

Sinoir, Y. 1966. Interactions du déficit hydrique de l'insecte et de la teneur en eau de l'aliment dans la prise de nourriture chez le criquet migrateur *Locusta migratoria migratorioides* (R. & F.). *C.R. Hebd. Séances Acad. Sci.* **262**, 2480–2483.

Stower, W. J. and J. F. Griffiths. 1966. The body temperature of the desert locust (*Schistocerca gregaria*). *Entomol. Exp. Appl.* **9**, 127–178.

Tucker, L. E. 1977. Regulation of ions in the haemolymph of the cockroach *Periplaneta americana* during dehydration and rehydration. *J. Exp. Biol.* **71**, 95–110.

Uvarov, B. P. 1966. *Grasshoppers and Locusts*, Vol. 1. Cambridge Univ. Press, London and New York.

Weis-Fogh, T. 1967. Respiration and tracheal ventilation in locusts and other flying insects. *J. Exp. Biol.* **47**, 561–587.

Wharton, G. W. 1985. Water balance of insects. In G. A. Kerkut, and L. I. Gilbert, Eds., *Comprehensive Insect Physiology, Biochemistry and Pharmacology*, Vol. 4. Pergamon, Oxford, pp. 565–601.

6

Grasshopper Thermoregulation

M. A. CHAPPELL and

D. W. WHITMAN

6.1 INTRODUCTION AND BASIC PRINCIPLES

Temperature is of profound importance to grasshoppers, as it is to all organisms. At the most fundamental level, grasshoppers can tolerate only a relatively narrow range of temperatures; outside this domain survival is impossible. More subtly (and probably more importantly in terms of day-to-day existence and evolutionary fitness), temperature affects virtually all biological activities. An incomplete list of aspects of grasshopper biology known to be influenced by the thermal regime includes metabolic rate, locomotion, rate of water loss, food consumption, digestibility and assimilation, fecundity and egg survival, growth, maturation, and survival rates of juveniles, life-span of adults, predator avoidance, courtship and mating behavior, communication, and habitat selection (Uvarov, 1966, 1977). It is safe to assume that there is no component of grasshopper biology that is not influenced by temperature, either directly or indirectly.

Given the near-universal importance of thermal factors, it is not surprising that many species of grasshoppers have evolved effective mechanisms for regulating their body temperatures. For the purposes of this review, *thermoregulation* is defined as the maintenance of body temperature (T_b) which is relatively independent of a generalized "operative" or "environmental" temperature (T_e; Bakken, 1976) by means of active physiological and/or behavioral responses. The "environmental" temperature represents the summed thermal effects of air and substrate temperatures, convective heat exchange, and radiative heat exchange. Because these parameters are different for different species, T_e is a property of *both* animal and environment. Thermoregulation does not require or imply that the animal maintain a constant T_b, but merely that for a given period, and over a range of environmental temperatures, T_b differs from what would be expected if the animal responded passively or randomly to a changing external thermal milieu (Fig. 6.1). In most cases, thermoregulatory responses act to stabilize T_b relative to T_e, and the effectiveness of thermoregulation is measured as the slope of a regression relating T_b to T_e: a slope of 1 indicates no thermoregulation and a slope of 0 indicates perfect thermoregulation (Fig. 6.2). In natural environments, grasshoppers may be able to thermoregulate only within part of the temperature range they encounter, for example, only at moderate temperatures when sunlight is available. When the full range of ambient conditions is considered, the overall relationship between T_b and T_e is best described by logistic or other nonlinear functions (Fig. 6.3; Kemp, 1986).

In grasshoppers, as in all thermoregulating organisms, regulation of body temperature is accomplished by a variety of mechanisms that adjust rates of heat gain and heat loss. Before discussing these mechanisms in detail, we present a brief review of the fundamental concepts and units of temperature and thermoregulation.

6.1.1 Temperature and Heat

A basic law of physics states that if two bodies are not in thermal equilibrium (i.e., if their temperatures are not equivalent), heat energy will flow from the region of higher temperature to the region of lower temperature. The quantity of heat transferred is determined by the temperature difference between the bodies and by their respective heat capacities (i.e., the temperature change

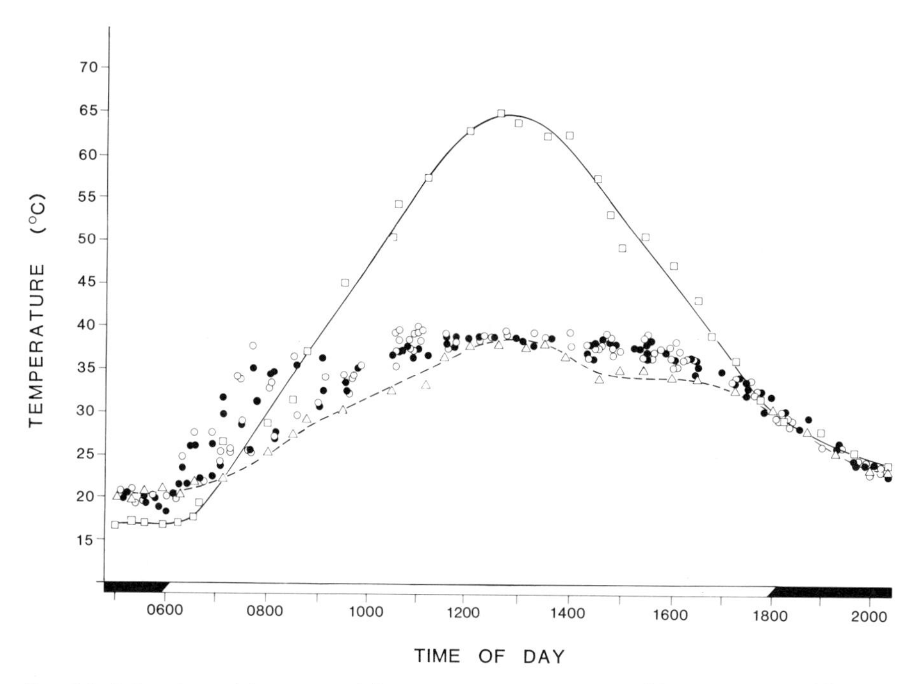

Fig. 6.1. Soil surface (□) and air (△) temperatures, and male (○) and female (●) body temperatures for a population of *Taeniopoda eques* monitored throughout a typical hot day in Arizona. (From Whitman, 1987.)

resulting from a specified heat flux). Within the temperature range experienced by living organisms (roughly 0–50°C) the specific heat of water remains close to 4.18 J/g/°C whereas that of animal tissue is approximately 3.43 J/g/°C.

For a grasshopper, the two thermal systems of concern are the animal's body and the environment. Although a considerable oversimplification compared with most natural habitats, it is convenient for the purposes of this analysis to consider both animal and environment to be thermally uniform at temperatures T_b and T_e, respectively. In the vast majority of natural situations, the heat capacity of the environment is much greater than that of the animal, and hence the environment functions as an essentially infinite heat source or sink for the animal. Accordingly, given enough time T_b will equilibrate at T_e. Exceptions to this general rule occur when the animal produces substantial quantities of heat through metabolic activity (thereby keeping T_b elevated above T_e), or when the animal loses heat continually through the evaporation of body water (thereby keeping T_b below T_e).

6.1.2 Pathways for Heat Transfer

Heat transfer between grasshoppers and the environment normally occurs through three major pathways: (1) conduction to or from the solid substrate,

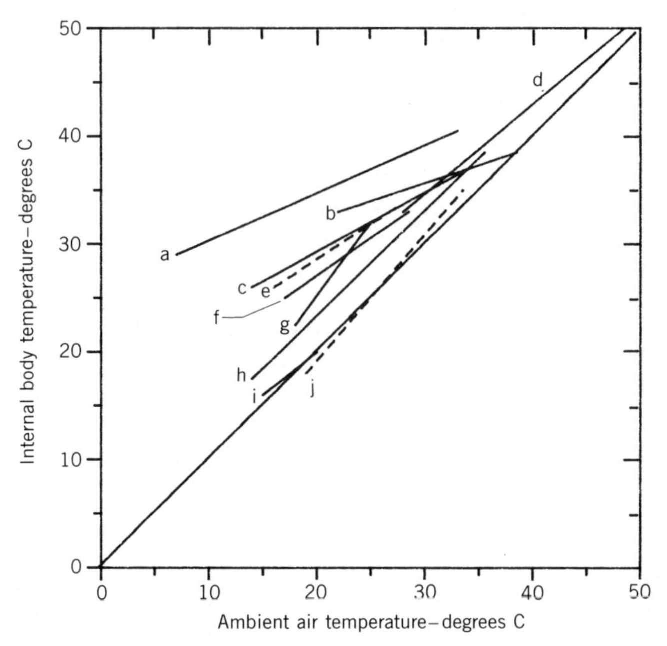

Fig. 6.2. Regression slopes of body temperatures on air temperatures under natural conditions for various grasshopper species. (a) *Melanoplus sanguinipes* (Chappell, 1982); (b) *Taeniopoda eques* females, during the day (Whitman, 1987); (c) *Psoloessa delicatula* (Anderson et al., 1979); (d) *Trimerotropis pallidipennis* (Chappell, 1982); (e) *Trimerotropis suffusa* (Gillis and Possai, 1983); (f) *Circotettix rabula*, montane population (Gillis and Smeigh, 1987); (g) *Chorthippus brunneus*, in sunlight (Begon, 1983); (h) *Eritettix simplex* (Anderson et al., 1979); (i) *Taeniopoda eques* females, dawn and dusk (Whitman, 1987); (j) *Chorthippus brunneus*, in shade (Begon, 1983).

(2) convective exchange with the air surrounding the animal, and (3) the absorption and emission of radiant energy (Fig. 6.4). For these pathways, heat flows across the body surface at rates determined by the difference between the body surface temperature (T_s) and the environmental temperature (Porter and Gates, 1969; Monteith, 1973). Owing to their small size, hemolymph circulation, and other factors, T_s is approximately equal to T_b for most medium- and small-sized grasshoppers (Bakken, 1976; Uvarov, 1966). For large grasshoppers, however, temperature differences as high as 7°C can occur in different parts of the body during radiative heating, and 3°C differences are not uncommon under equilibrium conditions (Whitman, 1987). Factors influencing heat transfer rates include surface properties such as texture, color, and area, as well as the insect's morphology, posture, and orientation, and environmental parameters such as the radiant intensity, substrate conductivity, and wind speed and turbulence. These exchange pathways are complex and interdependent, making their quantification a formidable undertaking even in controlled laboratory situations (Bakken, 1976).

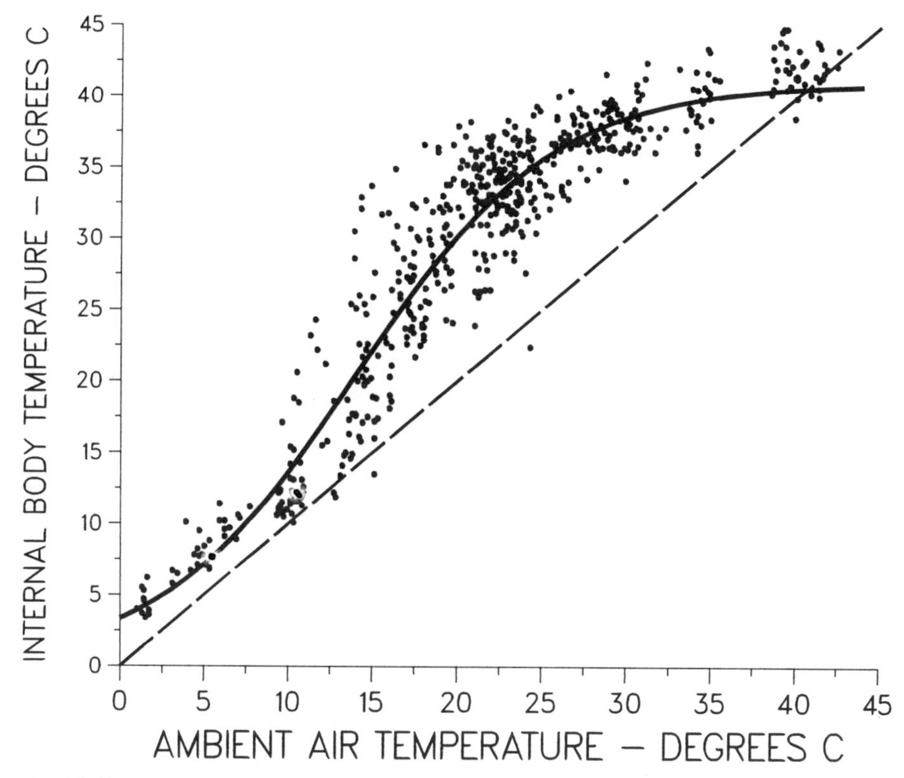

Fig. 6.3. The relationship between internal body temperature and ambient air temperature for *Aulocara elliotti* females over a broad range of ambient temperatures. Solid line represents temperatures predicted from a logistic equation. (From Kemp, 1986.)

Results of most studies of insect thermal relations indicate that heat loss is primarily by means of convective exchange (Digby, 1955; Church, 1960a,b; Casey, 1981; Chappell, 1982, 1983). For a given body size and morphology, convection increases approximately as a linear function of the temperature excess (T_{ex}, the difference between T_s and the air temperature T_a). In addition to T_{ex}, convective exchange depends on the convection coefficient (h_c), which varies approximately with the square root of wind speed (V). Convection is enhanced by turbulent airflow, which predominates in natural environments. In the dense vegetation and near-ground habitats favored by most grasshoppers wind speed is low, but turbulent flow increases h_c by up to 1.7-fold compared with laminar (nonturbulent) flow (Kowalski and Mitchell, 1976). Convection decreases in importance as body size increases, but in all cases the overall effect of convection is to "drive" body temperature toward T_a.

Heat gain for most grasshoppers is through the absorption of radiant energy. Short-wave solar radiation in the ultraviolet, visible, and near-infrared spectra (wavelength 0.3–2.6 μm) is usually the most important source

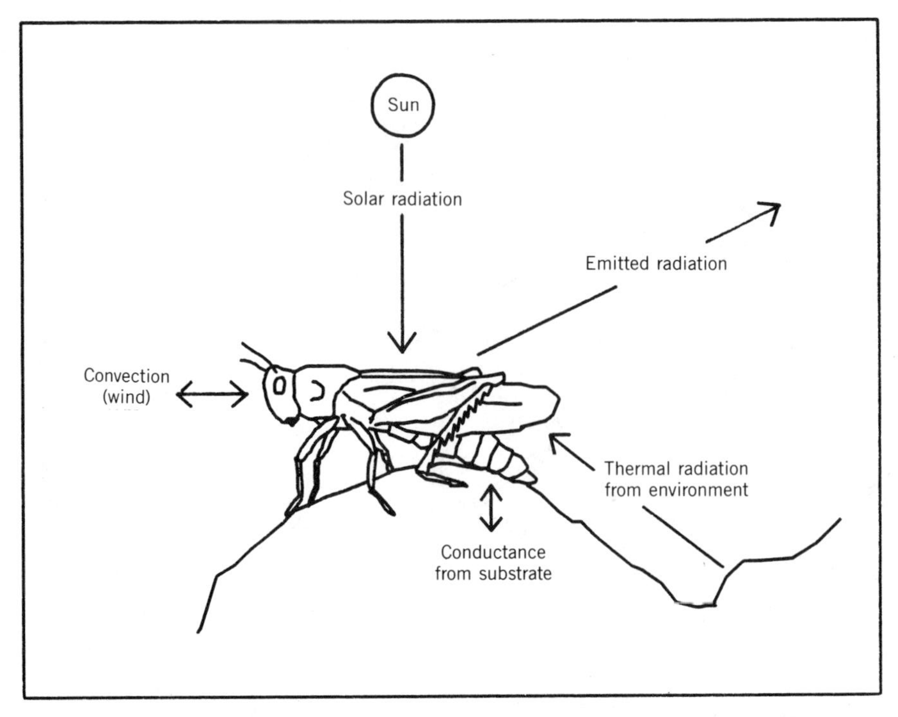

Fig. 6.4. Heat-exchange pathways for a grasshopper on the ground.

(Monteith, 1973). The quantity of solar radiation absorbed (Q_{abs}) is relatively independent of T_b or T_s, but is strongly linked to absorptivity (i.e., color), size, shape, and orientation. The absorptivity of solar radiation by animal surfaces varies between 50% and 90%, but is normally 70–75% (Porter and Gates, 1969; Anderson et al., 1979; Willmer and Unwin, 1981). Rates of radiation absorption are also influenced by atmospheric conditions, altitude, and the position of the sun in the sky. The intensity of sunlight at sea level is roughly 1000 W/m^2 at noon on a clear day, but drops sharply if the air is dusty or humid or if the sun is near the horizon.

Grasshoppers also exchange heat through the emission and absorption of long-wave "thermal" infrared radiation (wavelengths 10–20 μm). Net exchange of thermal radiation varies approximately as $E_a(T_s^4 - T_e^4)$, where E is emissivity (=absorptivity at the same wavelength), a is the Stephan–Boltzmann constant, and temperatures are in degrees Kelvin. A grasshopper experiences a net gain of radiant heat if it is cooler than the environment and a net loss of radiant heat if it is warmer than the environment. For the most part there is relatively little net gain or loss of heat through this pathway compared with convective exchange, because T_s usually does not differ markedly from T_e. Occasionally a grasshopper may absorb significant quantities of thermal radiation from hot objects in the environment, such as sun-heated rocks.

Thermal emission and absorption are affected by morphology and orientation, but color is relatively unimportant (the emissivity and absorptivity of biological materials are approximately 95–99% at these longer wavelengths; Porter and Gates, 1969).

Heat transfer by conductance occurs when a grasshopper's body is in direct contact with a solid substrate. Ordinarily the only body parts in contact with the substrate are the tarsi; because of the small area of contact, conductive heat transfer is of minor importance. However, when the body is pressed against the substrate, as in a crouching animal, conductance may significantly affect T_b. The total conductive heat flux depends on the gradient between body and substrate temperature, the area of contact, and the heat capacity and thermal conductivity of the substrate.

6.1.3 Equilibrium Heat Balance

Although complex in the quantitative sense, the main features of equilibrium heat balance in grasshoppers can be distilled to a few simple qualitative rules:

A. In the absence of a strong source of radiant energy, T_b will closely approximate T_a (unless the animal produces significant amounts of metabolic heat or evaporates significant quantities of water). Wind speed and turbulence levels are unimportant in determining T_b in this situation: because there is no gradient between T_b and T_a, there is no net convective or net radiative heat flux.

B. If a significant radiant source such as sunlight is present, a temperature excess is attained as absorbed heat causes T_b to rise above T_a until the rate of radiative heat gain is balanced by increasing convective and radiative heat loss. When heat loss equals heat gain, T_b stabilizes.

C. The magnitude of the temperature excess ($T_{ex} = T_b - T_a$) increases with the intensity of radiation, with absorptivity, and with increasing body size. Postural adjustments that maximize the surface area exposed to radiant heating also increase T_{ex}. Conversely, T_{ex} decreases with increasing wind speed, with increased turbulent flow, and with postural adjustments that enhance convection.

D. The overall environmental temperature T_e can be operationally defined as equivalent to T_b, if metabolic heat production and evaporative water loss are small. This logic simplifies the analysis of thermal environments that often include highly asymmetric radiant sources, temperature distributions, and wind orientations.

6.1.4 Nonequilibrium Conditions

The above approximations apply only to grasshoppers in thermal equilibrium with their environment, that is, when heat gain precisely balances heat loss and T_b is stable. In nature, however, this is seldom the case. The thermal environment is a constantly changing temporal and spatial mosaic. Because grasshoppers are small and highly mobile, they can move rapidly and take advantage of small and often transient favorable microclimates. Accordingly, transient states are often more significant to T_b than equilibrium situations for most species.

In response to transient conditions, T_b changes because there is a net flow of heat into or out of the body. The rates of heat transfer can be approximated with a simple "Newtonian model" incorporating body and air temperatures. This model assumes that net heat influx is proportional to temperature excess, which engenders an exponential approach over time to the equilibrium situation of $T_b = T_a$:

$$\frac{dH}{dt} = C(T_b - T_a) \tag{1}$$

where dH/dt is the rate of heat transfer (in joules/min or watts) and C is thermal conductance (in joules/min/°C or watts/°C), an index of the overall ability of heat to flow between animal and environment. Thermal conductance is a function of both the radiative and convective milieu (Monteith, 1973; Robinson et al., 1976) and therefore is affected by wind speed, turbulence, body size, orientation, color, and postural changes. Thermal conductance can also be influenced by internal physiological adjustments such as circulatory control (Heinrich, 1974; Kammer, 1981; Wasserthal, 1980), although these mechanisms have not been reported for grasshoppers.

The standard Newtonian model is difficult to apply to natural habitats, because it assumes a uniform thermal environment surrounding the animal and an equilibration temperature equal to T_a. In nature the thermal environment is usually nonuniform (e.g., sunlight is essentially a point source) and thermal equilibrium is attained when $T_b = T_e$. The problem can be overcome by substituting environmental temperature for T_a in Eq. 1:

$$\frac{dH}{dt} = C(T_b - T_e) \tag{2}$$

Unfortunately, it is often difficult to determine T_e in the field (see Section 6.1.6). An additional challenge is establishing the proper value of C for the complex convective regimes in natural habitats. Thermal conductance is usually measured in the laboratory from passive cooling rates in the absence of forced convection or at controlled wind speeds. It is difficult to determine or simulate natural turbulence levels, and approximations derived from meteorology or heat-transfer engineering are often employed (Wathen et al., 1974; Mitchell, 1976; Kowalski and Mitchell, 1976; Anderson et al., 1979).

6.1.5 Endothermy and Evaporative Cooling

For all practical purposes, most grasshoppers are exclusively ectothermic; that is, body temperatures depend almost entirely on T_e. However, many species are endothermic (i.e., they produce substantial quantities of metabolic heat) during flight, and possibly during preflight warm-up. During endothermy, T_b increases above T_e until a steady state is achieved when metabolic heat production (MHP) is balanced by convective and radiative heat loss. The T_b at thermal steady state is a function of both MHP and C:

$$T_b = T_e + \frac{\text{MHP}}{C} \tag{3}$$

Except for the source of heat, this condition is similar to that described for heat input from solar radiation.

An analogous equation can be used if substantial quantities of heat are lost by means of evaporative cooling. Maximal rates of evaporative heat loss (EHL) are limited by the temperature and humidity of the air surrounding the animal. However, in most circumstances, grasshoppers restrict water loss to rates much lower than the maximum possible. The overall effect of EHL is to reduce T_b below T_e:

$$T_b = T_e - \frac{\text{EHL}}{C} \tag{4}$$

6.1.6 Measurement Techniques

Measurements of thermal parameters for animals as small as grasshoppers are technically difficult and potentially misleading. Direct measurements of T_b are usually accomplished with small thermistors or fine-gauge thermocouples. For rapid single-point determinations of T_b, a sensor is often incorporated into the tip of a fine-gauge hypodermic needle probe to facilitate insertion into the animal. These measurements must be made quickly (within a few seconds), because the small size and concomitant low heat capacity of grasshoppers lead to rapid changes of T_b when the animals are removed from thermal equilibrium or steady state. Moreover, care must be taken to ensure that T_b measurements are not distorted by heat flow between the animal and the needle probe (or the hands of the operator), or by EHL arising from discharged hemolymph. For continuous monitoring of T_b, chronically implanted thermocouples are the method of choice. It is necessary to use extremely thin insulated leads in order to minimize both heat transfer along the highly conductive wires and interference with the animal's normal behavior patterns.

Grasshoppers are relatively small, and as a result most standard thermal and meteorological instruments are of limited use for studying their temperature relations. Many sensors lack the ability to track adequately the small-scale spatial or temporal variations to which grasshoppers are subjected. This problem is particularly acute for studies that attempt to determine T_e, or for other examinations of microclimates experienced by grasshoppers in natural habitats. Variation in convection and radiation regimes large enough to be highly significant for the heat balance of grasshoppers may occur within seconds, or over distances that are much smaller than the dimensions of typical anemometers or radiation meters. Even relatively small temperature sensors such as thermistors or thermocouples often produce inaccurate readings in air, owing to the very low heat capacity of air and the conductance of heat along the connection wires.

Because of these and other considerations, a number of investigators have employed mathematical models to analyze temperature regulation in natural habitats. Typical approaches use a combination of empirical measurements and theoretical treatments of metabolic heat production, evaporation, and radiative and convective heat transfer to generate an energy budget equation describing the total heat balance in terms of body temperature, for example,

$$M - E + Q_{abs} = Ea(T_b^4 - T_e^4) + h_c(T_b - T_a) \tag{5}$$

By incorporating suitable values into the various terms, this equation can be solved for T_b or other variables of interest (Porter and Gates, 1969). Alternative models use electrical circuit analogues to simulate the movement of heat (Robinson et al., 1976; Bakken, 1976).

Application of the appropriate mathematical model allows calculation of T_b to be extrapolated to any conceivable thermal regime. However, the accuracy of mathematical models is often contingent on error-prone approximations of important parameters (e.g., turbulent enhancement, surface area exposed to sunlight), and other essential data may be difficult or tedious to obtain. The empirical verification of a model may also be challenging. Nevertheless, several studies have developed models that demonstrated good predictive power in field environments (Stower and Griffiths, 1966; Henwood, 1975; Anderson et al., 1979; Chappell, 1983).

A simpler and more direct method for obtaining environmental temperatures makes use of the fact that most of the heat fluxes important to grasshoppers (e.g., radiation and convection) occur across the integument and are essentially independent of internal physiological parameters. An inert object with morphology and surface characteristics similar to those of a living grasshopper will respond similarly to the thermal environment. Accordingly, a thermocouple or thermistor-equipped physical model of a grasshopper, or a dried mounted specimen, can serve as a "thermometer" to measure T_e directly (Chappell, 1982). Because they respond in the same way to wind, radiation, and air temperatures, models and live animals reach thermal equilibrium at the same T_b. However, the time course of changes in T_b during transient conditions will differ, unless the physical model has the same heat capacity as a live animal (Bakken, 1976).

6.2 TEMPERATURE TOLERANCES AND THERMAL PREFERENCES

6.2.1 Temperature Tolerances

To survive, grasshoppers must maintain their body temperatures within a range bounded by the maximum and minimum tolerable T_b. Typically, grasshoppers become more and more sluggish as T_b decreases below normal levels. Movements become progressively slower and less coordinated as T_b is further reduced, and the animals ultimately enter a "cold torpor" characterized by complete immobility. At the opposite extreme, as T_b is raised above normal levels, movements initially become increasingly rapid; this stage is followed by reduced coordination at higher T_b, and finally by paralysis or "heat torpor." Further increases in T_b are rapidly lethal (Table 6.1; Hussein, 1937; Whitman, 1987).

Although it is not difficult to obtain data on the overall effects of temperature, establishing precise and ecologically relevant values for maximum and minimum tolerable temperatures is problematic. First, temperature effects depend on numerous parameters, including age, nutritional status, hydration state, ambient light intensity, humidity, and overall thermal acclimatization history (Uvarov, 1966, 1977). No single study of thermal tolerances can hope to incorporate all these factors. Second, behavior and physiology seldom change abruptly with changing T_b, so it is impossible to specify an exact correspondence between temperature, physiology, and behavior. Instead, tolerances are statistical means, often with wide variances. Researchers have attempted to resolve

TABLE 6.1. Minimum Body Temperatures (°C) for Various Activities for Three Species of Adult Grasshoppers[a]

Behaviour	*M. mexicanus*	*S. gregaria*	*T. eques*
Movement	−5	2	10
Walking	12	7	11
Defecation	—	—	12
Feeding	12	17	14
Righting self	12	—	15
Jumping	−3	—	15
Molting	22	26	22
Flight	—	20	25
Stridulation	—	—	36
Preference	—	41	36
Maximum voluntarily tolerated	46	46	42
Critical thermal maximum	53	—	45
Heat torpor	57	49	46
Instantaneous death point	58	54	47

Melanoplus mexicanus (Parker, 1930), *Schistocerca gregaria* (Gunn et al., 1948; Hussein, 1937; Uvarov, 1966, 1977; Volkonsky, 1939), and *Taeniopoda eques* (Whitman, 1988).

some of these interpretive problems by defining a hierarchical series of behavioral responses to increasing temperature. These include minimal temperatures for movement or for coordinated locomotion, maximum voluntarily tolerated temperature (at which animals actively attempt to escape), the critical thermal maximum (at which animals are unable to right themselves), and the heat torpor temperature (at which movement stops). These limits are usually quantified by slowly heating the animal under conditions of 100% relative humidity. Avoidance responses may comprise negative thermotaxis, jumping, flight, or postural reorientation (May, 1979; Casey, 1981).

Lower temperature limits vary greatly for different behaviors and among different grasshopper species (Table 6.1; Fraenkel, 1929, 1930; Uvarov, 1966, 1977). Complete cold-induced torpor has been reported for T_b between −5 and 17°C, slight movements begin a few degrees higher, and walking is possible at T_b of 5−20°C. There may be some correlation between cold tolerances and temperatures in natural habitats; lower temperature thresholds for ground locomotion and higher temperature-specific metabolic rates are observed in species or populations native to cold alpine regions, as compared to species or populations from warmer environments (Uvarov, 1966, 1977; Massion, 1983; Hadley and Massion, 1985). However, such comparisons must be viewed with caution in light of the reasons discussed above, and because of differences in measurement techniques.

Upper temperature limits are often more clearly defined than lower temperature limits, because the progression from normal activity through loss of coordination to heat torpor occurs within a fairly narrow range of T_b. Detrimental effects on locomotion usually begin at T_b of 44−46°C and paralysis is severe at T_b of 47−51°C, although higher temperature tolerances have been recorded (Table 6.1; Uvarov, 1966, 1977). There are surprisingly few data on the maximum voluntarily tolerated temperature (MVT) for grasshoppers. However, field measurements of T_b indicate that MVT is usually at least 2−3°C less than the range of temperatures that degrade locomotion. For example, heat torpor in *Trimerotropis pallidipennis* from southern California deserts begins at 49−51°C and MVT is approximately 45−48°C (Chappell, 1983). Observations such as these must be regarded cautiously, because grasshoppers can recover from laboratory exposures to thermal conditions they would be extremely unlikely to survive in nature. For example, normal behavior can often be restored in animals suffering from moderate heat torpor merely by reducing T_b to normal levels. In nature, heat-torpid grasshoppers would be unable to escape from predators or dangerously hot microclimates. Even a modest reduction in locomotor coordination due to unusually low or high T_b may significantly increase both visibility and vulnerability to predators. Thus the "ecological" survival limits for T_b in nature may differ from physiological tolerances established in the laboratory.

6.2.2 Thermal Preferences

The most common laboratory procedure for determining preferred temperatures employs a thermal gradient within which the animals can move freely. The preferred temperature is operationally defined from the distribution of animals in the gradient, or as the statistical mean of their T_bs (e.g., Chapman, 1965). In the field, it is often possible to deduce preferred temperature from the range of T_bs observed in free-living grasshoppers. However, for field studies it is crucial to demonstrate that the animals have the option to control T_b by selecting their position within a range of available microclimates.

Results from a number of field and lab studies suggest that the preferred temperatures of grasshoppers range from roughly 30 to 44°C, with most species showing preferences between 35 and 42°C (Table 6.1; Uvarov, 1977). However, there is often considerable variation between individuals. Interspecific variation in preferred temperature may be correlated with habitat differences, with species from cool environments preferring lower T_b than species from hot environments (Uvarov, 1977). Often in these comparisons the differences in preferred temperature for two grasshoppers from different habitats are considerably smaller than the difference in environmental temperature between the habitats. For example, in California the T_es of desert-dwelling *Trimerotropis pallidipennis* and alpine *Melanoplus sanquinipes* differ by 15–20°C but the mean T_bs of the two species differ by only 4–7°C (Chappell, 1982). In some species (*Locusta migratoria, Schistocerca gregaria*) the preferred T_b is a function of developmental stage, with immature stages preferring lower temperatures than later stages and adults (Chapman, 1955, 1965). The thermal acclimatization regime may influence thermal preferences in some species of grasshoppers, but has no effect in others (Chapman, 1965).

6.3 BEHAVIORAL THERMOREGULATION

In most grasshoppers, behavioral thermoregulation is the only available mechanism for adjusting body temperature. These behaviors are well known — indeed, some of the earliest work on behavioral thermoregulation in insects was conducted on the desert locust *Schistocerca gregaria* (Fraenkel, 1929, 1930). Because grasshoppers produce little metabolic heat, except during sustained flight, and generally lack the water reserves necessary for extensive and sustained evaporative cooling, these methods of heat exchange are usually unimportant.

6.3.1 Postural and Orientation Control

Grasshoppers can dramatically alter their rates of heat gain or heat loss in a particular microhabitat by simply changing their posture and/or orientation. These changes produce their effects by altering the way the insect is affected by the radiative, conductive, and convective properties of the environment.

The most important and commonly described thermoregulatory postures in grasshoppers are those that function to increase or decrease the absorption of solar radiation (Uvarov, 1977). By exposing as much of their body as possible to the rays of the sun, grasshoppers can quickly elevate their body temperatures considerably above ambient air temperatures (Table 6.2). Species such as *Brachystola magna*, which are broad and flat, "dorsal bask" by aligning their dorsum perpendicular to the solar beam. Most grasshoppers are taller than broad, and "flank" by presenting their side to the sun. The behavior of *Taeniopoda eques* (Whitman, 1987) is typical of flanking grasshoppers (Fig. 6.5a). This species orients the long body axis perpendicular to the incident solar beam and tilts the entire body so that insolation is incident upon the broad flat flank (instead of the narrower, peaked dorsal surface). The hind leg on the sunlit side is slightly lowered, fully exposing it to the sun, while the hind leg on the opposite side is raised out of the shadow of the body. If the sun is high in the sky, the animal may actually lie on its side in order to maximize its exposure (Fig. 6.5b). Flanking *T. eques* expose four times the body surface to the sun than when facing the sun, and thus heat rapidly (Fig. 6.6a). In some flanking grasshoppers, the effective absorption area may be equivalent to as much as 20–25% of the total body surface area (Anderson et al., 1979).

The efficacy of basking in elevating T_b above air temperature depends on a number of factors, notably radiation intensity, size, color, and convective conditions. Except when the sun is low in the sky or on hazy or overcast days, the intensity of sunlight is relatively constant between 700 and 1100 W/m^2. Other factors being equal, large grasshoppers attain higher temperature excess than small grasshoppers, and T_{ex} decreases as wind speed increases (Fig. 6.6b; Digby, 1955; Stower and Griffiths, 1966; Polcyn and Chappell, 1986; Whitman, 1987). The thermal significance of color in grasshoppers is surprisingly controversial; its theoretical importance is obvious but empirical obser-

TABLE 6.2. Greatest Temperature Excess ($T_b - T_a$) Recorded for Free-living Grasshoppers in Natural Habitats

Species	Temperature Excess (°C)	Source
Melanoplus sanguinipes	18	Chappell (1982); Kemp (1986)
Aulocara elliotti	18	Kemp (1986)
Melanoplus packardii	16	Kemp (1986)
Taeniopoda eques	16	Whitman (1986)
Dactylotum bicolor	12	Kemp (1986)
Trimerotropis suffusa	11	Gillis and Possai (1983)
Chorthippus brunneus	10	Begon (1983)
Psoloessa delicatula	10	Anderson et al. (1979)
Schistocerca gregaria	10	Waloff (1963)
Eritettix simplex	9	Anderson et al. (1979)
Trimerotropis pallidipennis	7	Chappell (1982)

Fig. 6.5. Various postures of *Taeniopoda eques* important for thermoregulation: (a) flanking, (b) ground flanking, (c) stilting, (d) shading. (From Whitman, 1987.)

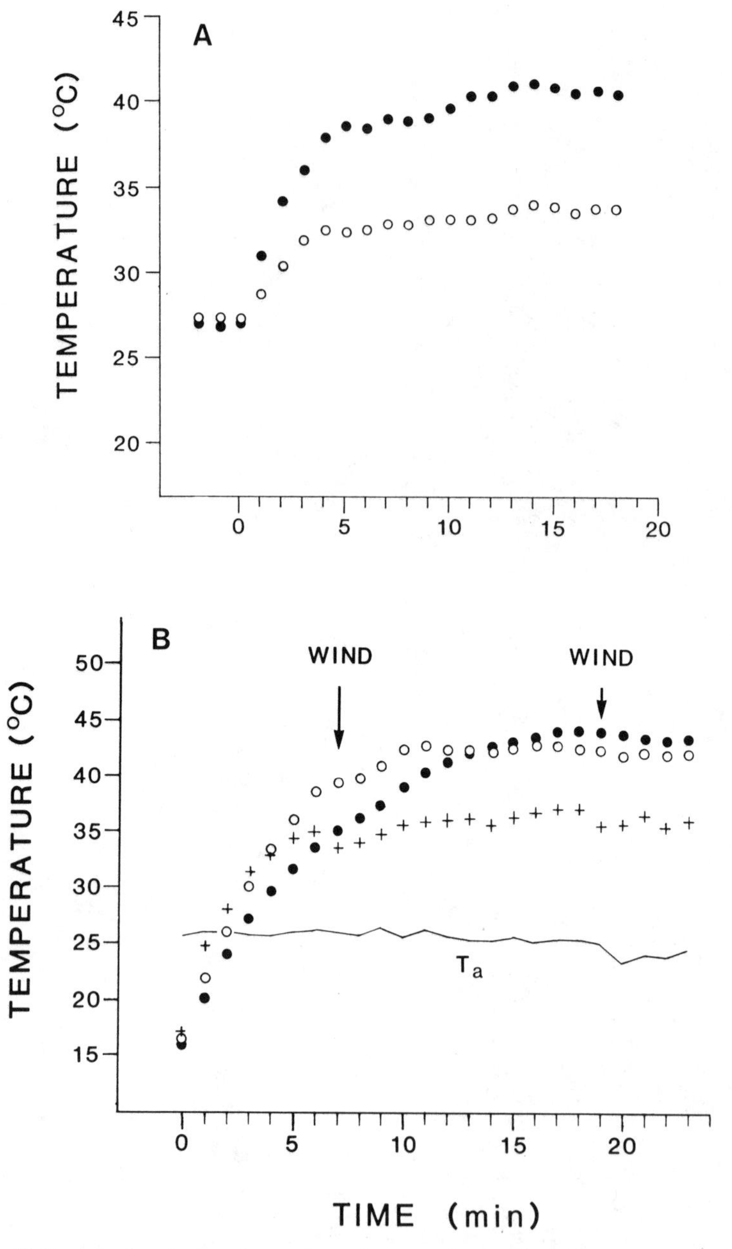

Fig. 6.6. (A) The effect of flanking behavior during radiative heating. Thoracic temperature for an adult male *Taeniopoda eques* restrained in the flanking posture (•), and an adult male restrained in a parallel position with respect to the rays of the sun (○). Insects were moved from shade to sun at time zero. (From Whitman, 1987.) (B) The effect of size on radiative heating. Thoracic temperatures of a 0.3-g second instar (+), a 1.4-g adult male (○), and a 4.4-g adult female (•) *T. eques*, restrained in the flanking position and exposed to sunlight. Wind gusts are indicated by vertical arrows. (From Whitman, 1987.)

vations have yielded inconsistent results. Artificially blackened grasshoppers attain higher T_{ex} than controls (Hill and Taylor, 1933; Digby, 1955; Joern, 1982) and similar data have been reported for other insects (Watt, 1968; Hamilton, 1975). However, Pepper and Hastings (1952) found insignificant effects of color on the T_{ex} of the grasshopper *Melanoplus differentialis*, and Stower and Griffiths (1966) found no difference in the T_{ex} of red/black and green morphs of *Schistocerca gregaria* in a series of paired comparisons. It is likely that the lack of color effects in these studies was due to the small size of the individuals studied and to minimal differences between the overall solar absorptivities of the various color morphs (Willmer and Unwin, 1981). A few grasshoppers are apparently able to enhance absorption by changing color. The Australian species *Kosciuscola tristis* is blackish at low T_a and when basking, but turns bright blue at high temperatures (Key and Day, 1954). Body color in many species is a function of temperature during ontogeny (Parker, 1930; Shotwell, 1941; Uvarov, 1966). Larvae reared at low temperatures become melanic. The thermoregulatory significance, if any, of these long-term color changes is unknown.

When overheated, grasshoppers exposed to sunlight assume postures that reduce solar heat loads by minimizing the effective absorption area (Volkonsky, 1939; Rainey et al., 1957; Chapman, 1959; Waloff, 1963; Chappell, 1982; Joern, 1982; Gillis and Smeigh, 1987). By facing directly toward or away from the sun with the long axis parallel with the solar beam, grasshoppers can reduce absorbed solar radiation by a factor of four to five or more compared with the full basking posture (Uvarov, 1977; Anderson et al., 1979).

Postural adjustments may also influence body temperature through conductive heat transfer. Several authors have described "crouching" behavior (Waloff, 1963), which entails close contact of the ventral thorax and abdomen with the substrate (usually the ground surface; Fig. 6.7). The large area of contiguity greatly facilitates direct conduction of heat between grasshopper and substrate, while reducing the surface area exposed to convective exchange. In most circumstances, crouching is a mechanism for heat gain; it is normally observed when overall environmental temperature is low, but ground temperature is higher than near-ground T_a. In locusts (*Schistocerca gregaria*, Waloff, 1963), crouching can bring T_b to within 1°C of ground temperature and 5°C above T_a. Crouching related to thermoregulation has also been reported for *Psoloessa delicatula* (Anderson et al., 1979), *Trimerotropis pallidipennis* and

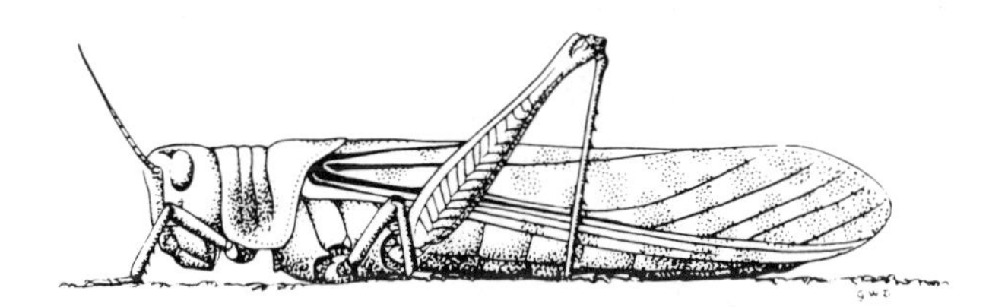

Fig. 6.7. Crouching posture of *Schistocerca gregaria*. (From Waloff, 1963.)

Melanoplus sanguinipes (Chappell, 1982, 1983), *Cercotettix rabula* (Gillis and Smeigh, 1987), and *Taeniopoda eques* (Whitman, 1987). The behavior is probably widespread among acridids (Uvarov, 1977).

A contrasting posture, "stilting" (Waloff, 1963; Anderson et al., 1979; Gillis and Smeigh, 1987; Whitman, 1987), is observed when T_b and ground temperatures are high. Stilting grasshoppers extend their legs and elevate the body as high as possible above the substrate (Fig. 6.5c). Occasionally one or more legs may be lifted and held off the ground (Whitman, 1987). This posture is effective in reducing the heat load for several reasons. Conductance from the hot ground is minimized because only the tips of the legs are in contact with the substrate. Convective heat exchange is enhanced because the body surface area exposed to wind is maximized, and wind speeds are higher above the unstirred boundary layer of air near ground level. Finally, because there is normally a steep gradient of air temperature in the few centimeters above ground level, stilting lifts the grasshopper's body into cooler air than near the ground surface (Chappell, 1983; Whitman, 1987). Waloff (1963) reported that stilting allowed *Schistocerca gregaria* to maintain a T_b of 43°C when T_a was 40°C and ground temperature exceeded 56°C. In *Trimerotropis pallidipennis* the T_e for a stilted animal was more than 5°C below that of a crouched individual in an open desert habitat, and for individuals sitting on small pebbles instead of directly on the ground the difference was almost 10°C (Chappell, 1983).

Grasshoppers display a variety of other postures at high temperatures. For example, *Psoloessa delicatula* assume "dropped" or "straddle" positions, which expose the maximum surface area to convective cooling (Anderson et al., 1979). Other postures intermediate between stilting and crouching presumably yield correspondingly intermediate heat exchange characteristics (Waloff, 1963; Uvarov, 1977; Chappell, 1982, 1983).

Owing to the complex body forms of insects, the effects on convective heat transfer of orientation changes that shift the apparent wind direction are difficult to predict (Polcyn and Chappell, 1986) and little studied. According to heat transfer theory, convective heat flux from an inanimate cylindrical body is minimized when wind direction is parallel with the long axis of the body (yaw angle 0°) and maximized when it is perpendicular to the long axis of the body (yaw angle 90°). For a standard cylinder the difference in convection between the two orientations is about twofold (Digby, 1955; Wathen et al., 1974; Mitchell, 1976). Results from wind-tunnel studies of freshly killed insects or model insects are generally in accordance with predictions, although the magnitude of change is often considerably less than for theoretical cylinders. For example, Digby (1955) reported 12–15% increases in convection when *Schistocerca gregaria* were perpendicular to the wind as opposed to parallel with it. Similarly, Anderson et al. (1979) reported 30–33% increases for *Psoloessa delicatula* and *Eritettix simplex*, and Chappell (1983) observed 20–25% increases for *Trimerotropis pallidipennis*.

In natural habitats, orientation with respect to wind and sun are not independent, because at a given time and place there is a fixed angular relationship between sun position and wind direction. Therefore a basking insect usually cannot select an orientation that is simultaneously optimal for both maximizing solar heat gain and minimizing convective heat loss; that is, orientations yielding maximum T_b may differ from those actually maximizing radiative heating (Polcyn and Chappell, 1986). These unavoidable interactions compromise the effectiveness of solar basking. Nevertheless, under favorable conditions the T_{ex} of basking grasshoppers may be impressively large (Table 6.2).

Postural adjustments may also affect convective heat transfer. Grasshoppers may position their legs to increase the surface area exposed to convection (Anderson et al., 1979); unfortunately the thermal significance (if any) of this behavior remains unquantified.

6.3.2 Microhabitat Selection

Differential use of microhabitats with disparate thermal attributes is perhaps the most common mechanism for body temperature control in grasshoppers. Small body size and high mobility allow grasshoppers to exploit numerous and diverse microclimates within relatively small areas (Fig. 6.5d). Microhabitat selection usually involves moving between sunlit and shaded regions. This can be as simple as moving from one side of a plant stem to the other (e.g., Rainey et al., 1957; Chapman, 1959). Grasshoppers often perform daily cyclical movements between vegetation and the soil, which facilitate temperature regulation (Fig. 6.8). At night, these grasshoppers roost in various plants. Following sunrise, they bask on the plants, warm, and descend to the ground to feed. During midday, when ground temperatures become intolerably high, the grasshoppers move back into the relatively cooler vegetation, where they seek shade and shelter from the sun. When air and ground temperatures fall in the late afternoon, the animals once again move to the ground, before ascending vegetation for the night (Waloff, 1963; Chapman, 1955; Joern, 1982; Whitman; 1987). Some species such as *Dactylotum bicolor* remain in vegetation throughout the day, basking during the morning and afternoon, and moving to shade during midday (Parker, 1982). Other species remain on or near the ground during basking (e.g., *Psoloessa delicatula*, Anderson et al., 1979; and *Melanoplus sanguinipes*, Chappell, 1982).

In gregarious locusts, sun-seeking and basking behaviors occasionally involve aggregations of hundreds or thousands of individuals. Basking aggregations are relatively quiescent and tightly packed, with each insect in contact with several other individuals (Chapman, 1955). Animals in aggregations actively avoid artificially shaded areas (Ellis and Ashall, 1957; Fig. 44 in Uvarov, 1977) but remain quiescent in sunlit areas presumably until T_b exceeds MVT. The close proximity of individuals during group basking probably facilitates heat gain by reducing surface–volume ratios and convective heat exchange (Casey, 1981).

The movement into vegetation during the hot midday not only allows

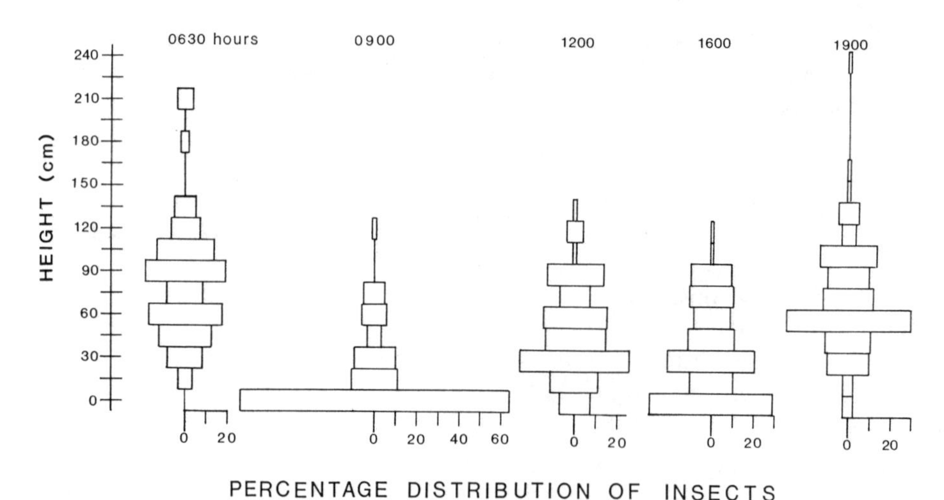

Fig. 6.8. Change in vertical distribution of *Taeniopoda eques* at various times during a typical Chihuahuan Desert day. Grasshoppers utilize the ground only during the relatively mild morning and late afternoon. (From Whitman, 1987.)

grasshoppers to avoid the hot ground surface and boundary layer: it also exposes them to increased convective heat loss, further reducing heat loads (Stower and Griffiths, 1966; Whitman, 1987). Even the use of minimal shade patches can have significant effects. *Zonocerus variegatus* reduces T_b from 50 to 40°C by moving from the top to the bottom of large leaves (Vuillaume, 1954, cited in Uvarov, 1977), and *Nomadacris septemfasciata* reduces T_b by 1–7°C by moving from the sunlit to the shaded side of plant stems (Rainey et al., 1957; Chapman, 1959). Shade is often scarce in desert habitats, which may result in clustering. Dense aggregations of *Schistocerca gregaria* form in the small patches of shade under scattered shrubs when T_e is high (Ellis and Ashall, 1957). The importance of refuges from solar radiation is illustrated by *Trimerotropis pallidipennis* from the hot deserts of southern California. This species spends most of the day under scattered shrubs, the only surface habitat cool enough for survival (T_e in open areas may exceed 70°C; Fig. 6.9). Even under shrubs, the animals must continually reposition themselves, "tracking" the shadows of individual branches as the sun moves across the sky. If it is forced into open areas by predators, *T. pallidipennis* will overheat rapidly, reaching lethal T_b within 1–1.5 min unless it returns to the shade (Chappell, 1982, 1983).

A few grasshoppers utilize subsurface refugia to avoid overheating. *Calliptamus turanicus* and several other species enter crevices in the soil to exploit the cool subsurface temperatures. *Eremogryllus hammadae*, from sandy areas of the Sahara, immediately digs itself below the surface upon landing when temperatures are high. At a depth of 5 cm, sand temperatures are 8–16°C cooler than at the surface (Uvarov, 1977).

Microhabitat selection is also employed by many grasshoppers (particularly

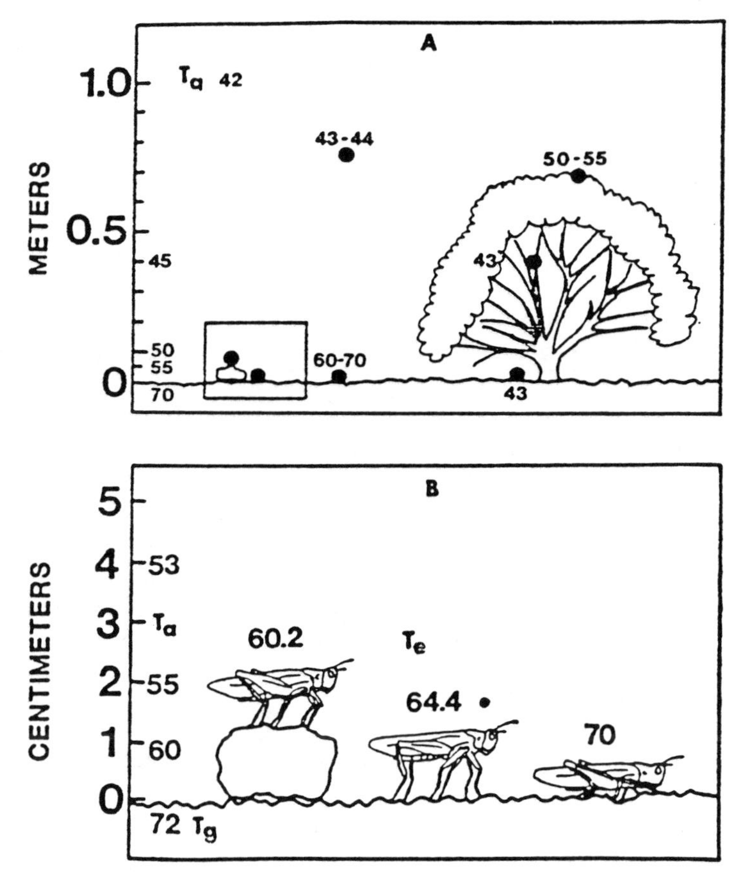

Fig. 6.9. (A) Predicted equilibrium temperatures for *Trimerotropis pallidipennis* in representative microhabitats on a typical day in its southern California desert environment. (From Chappell, 1983.) (B) Enlargement of the enclosed area in part A showing details of the temperature gradient near the ground, and the effects of stilting and rock climbing on grasshopper body temperatures. (From Chappell, 1983.)

species from temperate regions) to minimize their exposure to cold conditions at night. *Locustana pardalina* digs a small pit in the soil, in which it shelters, taking advantage of soil temperatures that (at least during the early evening) are warmer than air temperature (Uvarov, 1977). In many areas the open ground surface often becomes colder than more sheltered habitats after the sun sets, because the ground rapidly radiates heat to the night sky (Porter and Gates, 1969; Monteith, 1973). Some grasshoppers completely bury themselves in the evening and emerge in the morning (*Acrotylus junodi*, *A. insubricus*, *Chrotogonus concavus*, *Tmethis pulchripennis*, Uvarov, 1977). This behavior allows the animals to avoid the cold surface and take advantage of the warmer soil temperatures in subsurface layers. Other species shelter overnight in dense vegetation (*Schistocerca gregaria, Melanoplus sanguinipes*), soil or rock crevices (*S. gregaria, Aiolopus simulatrix, Locusta migratoria*), or rodent holes (*Anacridium*, Uvarov, 1977; Chappell, 1982).

A crucial factor for any grasshopper species utilizing behavioral thermoregulation is the availability of a suitable range of microhabitats (Chappell, 1983). The animals must be able to select among microhabitats that are thermally tolerable, and that, in addition, contain other necessities such as food supplies, shelter from predators, and so forth. This point is clearly illustrated by the careful study of Anderson et al. (1979) of two species from Colorado grasslands. *Eritettix simplex* occurred in dense vegetation on artificially fertilized and irrigated plots, whereas *Psoloessa delicatula* was found primarily on control plots that contained extensive open areas and considerably less vegetation. Utilizing an energy budget approach, the authors analyzed the thermal characteristics of the various microhabitats and the heat transfer characteristics and thermoregulatory behavior of the animals. They then computed the range of T_bs achievable by the two species and compared these predicted values to actual field measurements of T_b.

The predicted range of T_b for *E. simplex* was small (differing from T_a by only 1–3°C) throughout the day for two reasons: first, the dense vegetation minimized opportunities for solar basking, and second, plant transpiration produced relatively uniform air temperatures throughout the habitat. In contrast, *P. delicatula* could attain a wide range of T_b for most of the day because its open habitat was much more thermally heterogenous than that of *E. simplex*. For both species, there was a close correspondence between behavior and microhabitat characteristics (Fig. 6.10). *P. delicatula* exhibited an elaborate suite of thermoregulatory behaviors (postural adjustment) and maintained a T_b significantly more stable than T_a. In contrast, *E. simplex* apparently lacks the behavioral repertoire necessary for thermoregulation, and its T_b was always similar to T_a.

6.4 ENDOTHERMY

In grasshoppers, endothermy has been reported only during flight, and data are limited to a few large, highly migratory species. The physiology, behavior, energetics, and aerodynamics of the flight of desert locusts (*Schistocerca gregaria*) were painstakingly examined by T. Weis-Fogh and his colleagues during the 1950s and 1960s, and most of our knowledge of endothermy and its temperature relations in acridids comes from these studies.

Schistocerca gregaria is a powerful flier, able to continually maintain airspeeds of 3–6 m/s for many hours. It flies both during the day (if temperatures and radiation loads are moderate) and at night (Gunn et al., 1948; Waloff and Rainey, 1951; Uvarov, 1977). In this species, as in other insects, the contractile frequency and power output of muscle tissue depend greatly on temperature (Neville and Weis-Fogh, 1963; Josephson, 1981). Muscle performance is low

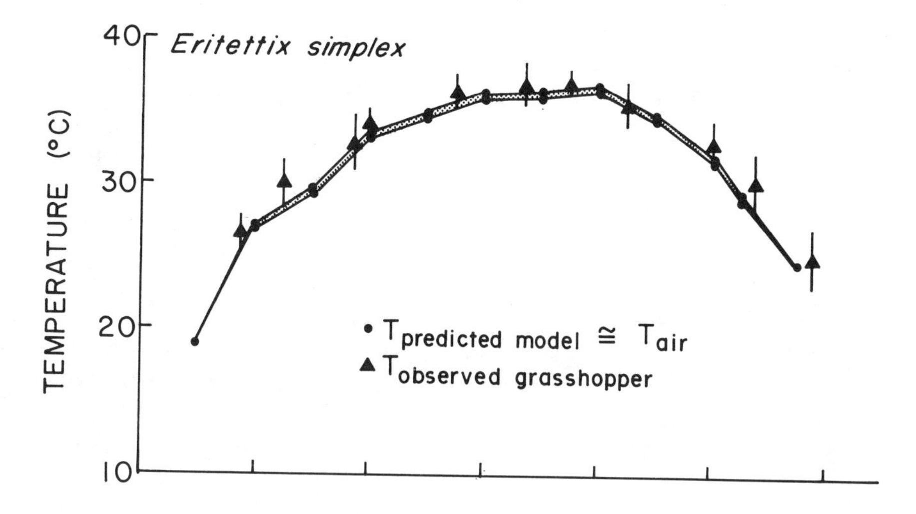

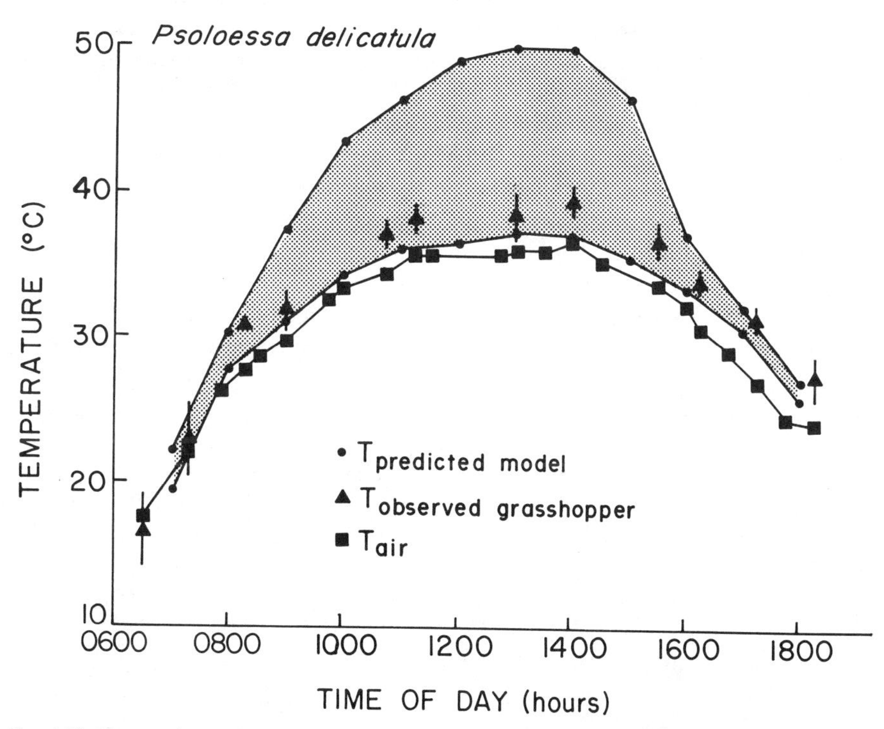

Fig. 6.10. Comparison of actual body temperatures with those predicted by estimations of heat-exchange parameters in two species of grasshoppers from Colorado grasslands. (From Anderson et al., 1979.)

at low T_b, increasing rapidly with increasing T_b until maximum tolerable temperatures are approached. Accordingly, the temperature of the thorax (T_{th}) must exceed a lower limit (the minimum flight temperature, or MFT) before the wing muscles can produce adequate power for sustained flight. In *S. gregaria*, MFT is approximately 20°C. At lower T_a (15–17°C), locusts attempt to fly when appropriately stimulated, but wingbeat frequency is only 10–12.5 hz and the animals cannot lift off. Normal flight performance (wingbeat frequencies of 17–20 hz) is attained at T_as of 22–24°C and higher (Waloff and Rainey, 1951; Weis-Fogh, 1956), but locusts can achieve the necessary T_{th} by basking even at T_a as low as 10–11°C (Rungs, 1946; cited in Uvarov, 1977). Other species apparently have similar thermal limits for flight (*Locusta migratoria, Nomadacris septemfasciata*; Barnish and Ferris, 1966; Rainey et al., 1957). Although preflight wing fanning has been recorded for a number of locust species (Rainey, 1974), it is unclear how this activity relates to body temperature. Preflight endothermic warm-up as observed in moths, bees, dragonflies, and beetles (Kammer, 1981) has not been described in detail in desert locusts or other acridids.

After the onset of flight, a locust's thoracic temperature increases rapidly because of the intense metabolic activity of the thoracic muscles — most (roughly 80–85%) of the metabolic power produced during flight appears as heat because of mechanical inefficiency (Jensen, 1956). If flight is sustained, T_{th} stabilizes after approximately 5 min at 3–16°C above T_a (Weis-Fogh, 1956). The actual thoracic temperature excess is apparently determined by the balance between convective heat loss and power output, being maximal at low wind speed and high metabolic rate (Weis-Fogh, 1964).

The metabolic rate of flying locusts has been determined independently by several unrelated techniques: measurements of rates of oxygen consumption, rates of fat consumption, and the sum of aerodynamic power requirements and heat loss rates. In general there is good agreement among the different methodologies (Weis-Fogh, 1952, 1956; Jensen, 1956; Uvarov, 1966). Under normal flying conditions (level flight at air speeds of 3–3.5 m/s), oxygen consumption is about 0.25 ml/g/min, or about 40- to 50-fold greater than that of resting animals (Krogh and Weis-Fogh, 1951; Neville, 1965). Flight metabolism is equivalent to a total metabolic power output of 0.087 W/g, and the resulting temperature excess is 6–8°C (Weis-Fogh, 1964). This gradient remains relatively constant over a wide range of T_a, decreasing only when T_a approaches the upper limit for sustained flight (35–37°C). At higher T_a (40–42°C) locusts can be induced to fly, but only for brief bursts that are followed by prolonged episodes of high ventilation rates ("panting"; Weis-Fogh, 1956). These laboratory data are consistent with field measurements of the T_{th} of flying locusts, which never exceed 41°C (Gunn et al., 1948).

Schistocerca gregaria may continue to fly at T_a slightly exceeding its maximum flight temperature by combining powered (flapping) flight with gliding (Roffey, 1963b; Neville, 1965; Uvarov, 1977). Roffey (1963a) reported that in locust swarms flying at high T_a, most of the individuals were gliding at any one time, with glide durations of up to 45 s. Gliding shuts off metabolic heat production and allows the animal to cool rapidly by convection, thereby reducing the $T_{th} - T_a$ gradient.

The constancy of the $T_{th} - T_a$ gradient over a T_a range of more than 10°C indicates that flying *Schistocerca* do not regulate their temperature; instead, the temperature excess is determined simply by the power required for flight, which is relatively independent of temperature. Unlike other endothermic fliers such as sphinx moths and bumblebees (May, 1985), desert locusts apparently lack the circulatory mechanisms for using the abdomen as a "radiator" to dissipate excess metabolic heat. In the absence of thermoregulatory control of T_{th}, *Schistocerca* have an impressive ability to keep wingbeat frequency and metabolic power output constant over a wide range of T_{th}, despite the well-known temperature sensitivity of important physiological parameters such as muscle contractility and nerve conduction velocity (Josephson, 1981).

The presence and extent of endothermy and thermoregulation during flight in other grasshopper species is uncertain, for none has been studied as thoroughly as the desert locust. Scattered field data from *Nomadacris septemfasciata* suggest that regulation is minimal (Rainey et al., 1957). However, in flying *Locusta migratoria*, T_b increased only 2–4°C between T_as of 17–20 to 28–29°C, indicating some regulation of T_b (Yurgenson, 1950, cited in Uvarov, 1977). The regulatory mechanism, if any, is undescribed.

6.5 EVAPORATIVE COOLING

Temperature is an important factor in determining rates of water loss from insects. However, most investigators have concluded that EHL is usually unimportant for insect thermoregulation, largely because insects such as grasshoppers lack adequate reserves of body water to sustain significant rates of evaporative cooling for other than short intervals (Digby, 1955; Church, 1960a; Shaw and Stobbart, 1972). The use of evaporative water loss as a thermoregulatory mechanism has not been clearly demonstrated in grasshoppers. Nevertheless, there are numerous reports of the effects of temperature on rates of cutaneous and spiracular (respiratory) water loss (Uvarov, 1966, 1977; Shaw and Stobbart, 1972). Water loss increases rapidly with increasing T_a, but in most cases rates of evaporation are too small to have significant effects on heat balance or T_b, even at very high T_a. For example, at $T_a = 30$°C and 20% relative humidity, the total evaporative water loss from *Schistocerca gregaria* is about 9 mg/h (Krogh and Weis-Fogh, 1951; A. Hamilton, 1964). This engenders a heat loss of 0.006 W, approximately equal to the resting metabolic heat production but nevertheless insufficient to influence T_b by more than a fraction of a degree. In *Locusta migratoria*, total transpiration at 50% relative humidity at $T_a = 30$°C is also approximately 9 mg/h. Water loss in *Locusta* increases abruptly to about 20–25 mg/h at 41–42°C, primarily because of greatly increased ventilatory movements (Loveridge, 1968a,b). Despite the large increase in spiracular transpiration, total evaporative heat loss is only 0.017 W at T_a = 42°C — enough to reduce T_b by perhaps 1°C in typical microclimate conditions. Interestingly, a recent report suggests that localized evaporation from the abdominal Slifer's patches of locusts increases survival time at high T_a (55°C) in dry air — a possible emergency cooling mechanism (Makings, 1987).

Water loss increases considerably during flight in *Schistocerca gregaria*, reaching 21 mg/h at 40% relative humidity at 30°C (Church, 1960a). As with hyperthermic *Locusta migratoria*, the increase results primarily from greatly enhanced tracheal ventilation (in this case, to support high aerobic metabolism; Weis-Fogh, 1967). This transpiration rate removes heat at a rate of about 0.014 W, representing only 10–12% of metabolic heat production. Even in completely dry air, evaporation would dissipate only 17% of the metabolic heat production at T_a = 30°C and (assuming the animals were willing to fly at this T_a) 35% at T_a = 40°C (Church, 1960a). The actual effect on thoracic temperature excess is probably smaller than calculated here, because while almost all MHP is localized in the thorax, only the fraction of total evaporation that occurs from the thorax will affect T_{th}. Church (1960a) estimated that about two-thirds of total evaporative heat transfer affected thoracic temperature. The net reduction in temperature excess due to evaporation would then be only 0.4°C under normal flight conditions (T_a = 30°C, 40% relative humidity).

6.6 ECOLOGY AND ADAPTIVE VALUE OF THERMOREGULATION

Most investigations of grasshopper thermoregulation document its occurrence, investigate the physical mechanisms of heat gain or loss, or (in some cases) examine temperature effects on particular behaviors or on various indices of physiological performance. Although these studies often speculate on the adaptive value of thermoregulation, quantitative data relating the effects of thermoregulation to evolutionary fitness (i.e., reproductive success) are extremely scarce. When effective thermoregulation makes the difference between survival and death (e.g., in hot desert habitats), its adaptive value is obvious. However, for most grasshoppers, the adaptive effects of thermoregulation are more subtle. Fitness is affected indirectly through increased growth rates, better physiological function, more effective predator avoidance, increased competitive ability, lengthened activity periods, exploitation of thermally marginal microhabitats, and so on. All of these effects enhance fecundity, but in ways that are difficult to measure.

One study that investigated the adaptive value of grasshopper thermoregulation in terms of reproductive success was performed by Whitman (1988) on *Taeniopoda eques* from southern Arizona desert habitats. In this species, T_b directly or indirectly influenced the ability to feed, digest, molt, disperse, escape from predators, and (most importantly) develop and mature. Developmental temperatures influenced body size (Whitman, 1986), a particularly important factor because increased body size is thought to provide greater water retention, more effective thermoregulation, deeper ovoposition, and easier escape from small invertebrate predators. Perhaps the most significant finding was that *T. eques* could not successfully reproduce without the ability to thermoregulate. *T. eques* requires 850 degree-days (DD) to complete its life cycle (hatch to ovoposition), but the short seasonal environment in southern Arizona supplies, on average, only 692 DD when measured as air temperature. By thermoregulating, *T. eques* makes up the 158 DD deficit, increasing its body temperature an average of 5°C per daylight hour over ambient temperatures.

REFERENCES

Anderson, R. V., C. R. Tracy, and Z. Abramsky. 1979. Habitat selection in two species of shorthorned grasshoppers. The role of thermal and hydric stresses. *Oecologia* **38**, 359–374.

Bakken, G. S. 1976. A heat transfer analysis of animals: Unifying concepts and the application of metabolism chamber data to field ecology. *J. Theor. Biol.* **60**, 337–384.

Barnish, G. and H. Ferris. 1966. The effect of temperature on the twitch duration of the dorsolongitudinal flight muscles of *Locusta migratoria migratorioides* (Reiche and Fairm.). *Proc. Trans. Rhod. Sci. Assoc.* **51**, 121–130.

Begon, M. 1983. Grasshopper populations and weather: The effects of insolation on *Chorthippus brunneus*. *Ecol. Entomol.* **8**, 361–370.

Casey, T. M. 1981. Behavioral mechanisms of thermoregulation. In B. Heinrich, Ed., *Insect Thermoregulation*. Wiley, New York. pp. 79–114.

Chapman, R. F. 1955. Some temperature responses of nymphs of *Locusta migratoria migratorioides* (R. and F.), with special reference to aggregation. *J. Exp. Biol.* **32**, 126–139.

Chapman, R. F. 1959. Observations on the flight activity of the Red Locust (*Nomadacris septemfasciata* Serville). *Anti-Locust Bull.* **33**, 1–51.

Chapman, R. F. 1965. The behaviour of nymphs of *Schistocerca gregaria* (Forskål)(Orthoptera, Acrididae) in a temperature gradient, with special reference to temperature preference. *Behaviour* **24**, 283–317.

Chappell, M. A. 1982. Metabolism and thermoregulation in desert and montane grasshoppers. *Oecologia* **56**, 126–131.

Chappell, M. A. 1983. Thermal limitations to escape responses in desert grasshoppers. *Anim. Behav.* **31**, 1088–1093.

Church, N. S. 1960a. Heat loss and the body temperatures of flying insects. I. Heat loss by evaporation of water from the body. *J. Exp. Biol.* **37**, 171–185.

Church, N. S. 1960b. Heat loss and the body temperatures of flying insects. II. Heat conduction within the body and its loss by radiation and convection. *J. Exp. Biol.* **37**, 186–213.

Digby, P. S. 1955. Factors affecting the temperature excess of insects in sunshine. *J. Exp. Biol.* **32**, 279–289.

Ellis, P. E. and Ashall, C. 1957. Field studies on diurnal behavior, movement, and aggregation in the desert locust (*Schistocerca gregaria* Forskål). *Anti-Locust Bull.* **25**, 1–94.

Fraenkel, G. 1929. Untersuchungen über Lebensgewohnheiten, Sinnesphysiologie und Sozialpsychologie der Wandernden Larven der Afrikanischen Wanderheuschrecke *Schistocerca gregaria* (Forsk.). *Biol. Zentralbl.* **49**, 657–680.

Fraenkel, G. 1930. Die Orientierung von *Schistocerca gregaria* zu strahlender Wärme. *Z. Vergl. Physiol.* **13**, 300–313.

Gillis, J. E. and K. W. Possai. 1983. Thermal niche partitioning in the grasshoppers *Arphia conspersa* and *Trimerotropis suffusa* from a montane habitat in central Colorado. *Ecol. Entomol.* **8**, 155–161.

Gillis, J. E. and P. A. Smeigh. 1987. Altitudinal variation in thermal behavior of the grasshopper *Circotettix rabula* (Rehn & Hebard) from central Colorado. *Southwest. Nat.* **32**, 203–211.

Gunn, D. L., F. C. Perry, W. G. Seymour, T. M. Telford, E. N. Wright, and D. Yeo. 1948. Behaviour of the desert locust (*Schistocerca gregaria* Forskål) in Kenya in relation to aircraft spraying. *Anti-Locust Bull.* **3**, 1–70.

Hadley, N. F. and D. D. Massion. 1985. Oxygen consumption, water loss and cuticular lipids of high and low elevation populations of the grasshopper *Aeropedellus clavatus* (Orthoptera: Acrididae). *Comp. Biochem. Physiol. A* **80A**, 307–311.

Hamilton, A. G. 1964. Occurrence of periodic or continuous discharge of carbon dioxide by male desert locusts (*Schistocerca gregaria* Forskål). *Proc. R. Soc. London, Ser. B* **160**, 373–395.

Hamilton, W. J. 1975. Coloration and its thermal consequences for diurnal insects. In N. F. Hadley, Ed., *Environmental Physiology of Desert Organisms.* Dowden, Hutchinson & Ross, Stroudsberg, Pennsylvania, pp. 67–89.

Heinrich, B. 1974. Thermoregulation in endothermic insects. *Science* **185**, 747–756.

Henwood, K. 1975. A field-tested thermoregulation model for two diurnal Namib Desert tenebrionid beetles. *Ecology* **56**, 1329–1342.

Hill, L. and H. J. Taylor. 1933. Locusts in sunlight. *Nature (London)* **132**, 276.

Hussein, M. 1937. The effect of temperature on locust activity. *Bull. Minist. Agric. Egypt. Tech. Sci. Serv.* **184**, 1–55.

Jensen, M. 1956. Biology and physics of locust flight. III. The aerodynamics of locust flight. *Philos. Trans. R. Soc. London, Ser. B* **245**, 137–169.

Joern, A. 1982. Importance of behavior and coloration in the control of body temperature by *Brachystola magna* Girard (Orthoptera: Acrididae). *Acrida* **10**, 117–130.

Josephson, R. K. 1981. Temperature and the mechanical performance of insect muscle. In B. Heinrich, Ed., *Insect Thermoregulation.* Wiley, New York, pp. 19–44.

Kammer, A. E. 1981. Physiological mechanisms of thermoregulation. In B. Heinrich, Ed., *Insect Thermoregulation.* Wiley, New York, pp. 115–158.

Kemp, W. P. 1986. Thermoregulation in three rangeland grasshopper species. *Can. Entomol.* **118**, 335–343.

Key, K. H. and M. F. Day. 1954. A temperature controlled physiological colour change in the grasshopper *Kosciuscola tristis* Sjost. (Orthoptera: Acrididae). *Aust. J. Zool* **2**, 340–363.

Kowalski, G. J. and J. W. Mitchell. 1976. Heat transfer from spheres in the naturally turbulent outdoor environment. *J. Heat Transfer* **96**, 649–653.

Krogh, A. and T. Weis-Fogh. 1951. The respiratory exchange of the desert locust (*Schistocerca gregaria*) before, during, and after flight. *J. Exp. Biol.* **28**, 344–357.

Loveridge, J. P. 1968a. The control of water loss in *Locusta migratoria migratorioides* R. & F. I. Cuticular water loss. *J. Exp. Biol.* **49**, 1–14.

Loveridge, J. P. 1968b. The control of water loss in *Locusta migratoria migratorioides* R. & F. II. Water loss through the spiracles. *J. Exp. Biol.* **49**, 15–30.

Makings, P. 1987. Survival value of Slifer's patches for locusts at high temperature. *J. Insect Physiol.* **33**, 815–822.

Massion, D. D. 1983. An altitudinal comparison of water and metabolic relations in two acridid grasshoppers (Orthoptera). *Comp. Biochem. Physiol. A* **74A**, 101–105.

May, M. L. 1979. Insect thermoregulation. *Annu. Rev. Entomol.* **24**, 313–349.

May, M. L. 1985. Thermoregulation. In G. A. Kerkut and L. I. Gilbert, Eds., *Comprehensive Insect Physiology, Biochemistry and Pharmacology*, Vol. 4. Pergamon, Oxford, pp. 507–552.

Mitchell, J. W. 1976. Heat transfer from spheres and other animal forms. *Biophys. J.* **16**, 501–509.

Monteith, J. L. 1973. *Principles of Environmental Physics*. Elsevier, New York.

Neville, A. C. 1965. Energy and economy in insect flight. *Sci. Prog. (Oxford)* **53**, 203–219.

Neville, A. C. and T. Weis-Fogh. 1963. The effect of temperature on locust flight muscle. *J. Exp. Biol.* **40**, 111–121.

Parker, J. R. 1982. Some effects of temperature and moisture on *Melanoplus mexicanus mexicanus* Saussure and *Camnula pellucida* Scudder (Orthoptera). *Bull. Mont., Agric. Exp. Stn.* **223**, 1–132.

Parker, M. A. 1930. Thermoregulation by diurnal movement in the barberpole grasshopper (*Dactylotum bicolor*). *Am. Midl. Nat.* **107**, 228–237.

Pepper, J. H. and E. Hastings. 1952. The effects of solar radiation on grasshopper temperatures and activities. *Ecology* **33**, 96–103.

Polcyn, D. M. and M. A. Chappell. 1986. Analysis of heat transfer in *Vanessa* butterflies: Effects of wing position and orientation to wind and light. *Physiol. Zool.* **59**, 706–716.

Porter, W. P. and D. M. Gates. 1969. Thermodynamic equilibria of animals with environment. *Ecol. Monogr.* **39**, 245–270.

Rainey, R. C. 1974. Biometeorology and insect flight. Some aspects of energy exchange. *Annu. Rev. Entomol.* **19**, 407–439.

Rainey, R. C., Z. Waloff, and G. F. Burnett. 1957. The behaviour of the red locust (*Nomadacris septemfasciata* Serville) in relation to the topography, meteorology and vegetation of the Rukwa Rift Valley, Tanganyika. *Anti-Locust Bull.* **26**, 1–96.

Robinson, D. E., G. S. Campbell, and J. R. King. 1976. A re-evaluation of heat exchange in small birds. *J. Comp. Physiol.* **105**, 153–166.

Roffey, J. 1963a. Observations on night flight in the desert locust (*Schistocerca gregaria* Forskål). *Anti-Locust Bull.* **39**, 1–32.

Roffey, J. 1963b. Observations on gliding in the desert locust. *Anim. Behav.* **11**, 359–366.

Rungs, C. 1946. Observations sur la biologie de *Schistocerca gregaria* Forsk. effectuées au cours des essais de lutte de 1945. *Bull. Off. Natl. Anti-Acrid. Algér.* **2**, 76–87.

Shaw, J. and R. H. Stobbart. 1972. The water balance and osmoregulatory physiology of the desert locust (*Schistocerca gregaria*) and other desert and xeric arthropods. *Symp. Zool. Soc. London* **31**, 15–38.

Shotwell, R. L. 1941. Life histories and habits of some grasshoppers of economic importance on the Great Plains. *U.S., Dep. Agric., Tech. Bull.* **774**, 1–47.

Stower, W. J. and J. F. Griffiths. 1966. The body temperature of the desert locust. *Entomol. Exp. Appl.* **9**, 127–178.

Uvarov, B. P. 1966. *Grasshoppers and Locusts: A Handbook of General Acridology*, Vol. 1. Centre for Overseas Pest Research, London.

Uvarov, B. P. 1977. *Grasshoppers and Locusts: A Handbook of General Acridology*, Vol. 2. Centre for Overseas Pest Research, London.

Volkonsky, M. A. 1939. Sur la photo-akinèse des acridiens. *Arch. Inst. Pasteur Algér.* **17**, 194–220.

Vuillaume, M. 1954. Biologie et comportement, en A.O.F., de *Zonocerus variegatus* L. (Orth. Acrididae) avec essais de comparaison entre Acridiens grands et petits migrateurs. *Rev. Pathol. Veg. Entomol. Agric. Fr.* **33**, 121–198.

Waloff, Z. 1963. Field studies on solitary and *transiens* desert locusts in the Red Sea area. *Anti-Locust Bull.* **40**, 1–93.

Waloff, Z. and R. C. Rainey. 1951. Field studies on factors affecting the displacements of desert locust swarms in eastern Africa. *Anti-Locust Bull.* **9**, 1–50.

Wasserthal, L. T. 1980. Oscillating hemolymph "circulation" in the butterfly *Papilio polyxenes* L. revealed by contact thermography and photocell measurements. *J. Comp. Physiol.* **139**, 145–163.

Wathen, P. J., J. W. Mitchell, and W. P. Porter. 1974. Heat transfer from animal appendage shapes — cylinders, arcs, and cones. *Trans. ASME* **97**, 536–554.

Watt, W. B. 1968. Adaptive significance of polymorphism in *Colias* butterflies. I. Variation of melanin pigment in relation to thermoregulation. *Evolution (Lawrence, Kans.)* **22**, 437–458.

Weis-Fogh, T. 1952. Fat combustion and metabolic rate of flying locusts (*Schistocerca gregaria* Forskål). *Philos. Trans. R. Soc. London, Ser. B* **237**, 1–36.

Weis-Fogh, T. 1956. Biology and physics of locust flight. II. Flight performance of the desert locust (*Schistocerca gregaria*). *Philos. Trans. R. Soc. London, Ser. B* **239**, 459–510.

Weis-Fogh, T. 1964. Functional design of the tracheal system of flying insects as compared with the avian lung. *J. Exp. Biol.* **41**, 207–227.

Weis-Fogh, T. 1967. Respiration and tracheal function in locusts and other flying insects. *J. Exp. Biol.* **47**, 561–587.

Whitman, D. W. 1986. Developmental thermal requirements for the grasshopper *Taeniopoda eques* (Orthoptera: Acrididae). *Ann. Entomol. Soc. Am.* **79**, 711–714.

Whitman, D. W. 1987. Thermoregulation and daily activity patterns in a black desert grasshopper, *Taeniopoda eques*. *Anim. Behav.* **35**, 1814–1826.

Whitman, D. W. 1988. Function and evolution of thermoregulation in the desert grasshopper *Taeniopoda eques*. *Anim. Ecol.* **57**, 369–383.

Willmer, P. G. and D. M. Unwin. 1981. Field analysis of insect heat budgets: Reflectance, size, and heating rates. *Oecologia* **50**, 250–255.

Yurgenson, I. A. 1950. Behaviour of the Asiatic locust (*Locusta migratoria*) in connection with its thermal regime. *Mater. Ekol. Konf., 2nd, Kiev* **1**, 248–251 (in Russian).

7

Jumping in Orthoptera

H. C. BENNET-CLARK

7.1 INTRODUCTION

Jumping is an escape from the physical limitations that walking imposes. In walking, the maximum speed is determined by the leg length and the stride frequency: the leg length is subject to various constraints, such as the problems of making a long thin structure of low weight and high strength, and the stride frequency is ultimately limited by the time that is required for the activation of the discrete antagonistic muscle contractions of the forward and backward swing of the leg. Thus a small walking animal will be quite slow, potentially easy to catch, and with a small feeding range.

By jumping, a small animal is able, by a single movement, to leap many times its body length, and to use this feat to escape predation, to catch prey, or in the case of larval Orthoptera, to achieve an average speed of movement that would otherwise be unattainable. Jumping has evolved many times in pterygote insects, so it must be an advantageous trick to achieve — it is found, for example, throughout both suborders of Orthoptera, among Hemiptera, Coleoptera, and Mecoptera, and is the notable feat of the Siphonaptera. Many jumping insects have become so specialized that they have, to all intents and purposes, lost some alternative means of locomotion: loss of the wings is the rule in fleas and is common in Orthoptera and jumping Hemiptera. Unlike flight, which is the realm of the adult insect, in certain cases jumping occurs throughout the instars so the advantage conferred may be exploited through the life cycle; such is the case with Orthoptera.

Because of the physical problems of jumping, the body design of a jumping animal may be very different from that of one that moves in a more conventional manner. It is to be expected that there will be skeletal, muscular, and nervous specializations linked to the jumping habit. To understand the significance of these specializations, it is necessary to understand the mechanics of the problem.

7.2 MECHANICS OF JUMPING

7.2.1 The Energetics of Jumping

A jump can be divided into three phases: a takeoff, where the animal pushes against the ground, a trajectory through the air, and a landing. This can be restated in terms of a projectile that is initially given potential energy, causing it to takeoff, which then causes it to rise from the ground, so that the kinetic energy is partly converted to potential energy, and then return to the ground, dissipating its remaining kinetic energy as heat (Fig. 7.1).

Because we are dealing with energies, the relations between height or range can be expressed simply in terms of the energy put into the system. The simplest case is a simple vertical jump, for which the height achieved is related to the energy by

$$E \text{ (energy)} = \frac{mv^2}{2} \text{ (the kinetic energy)} = mg_n h \text{ (the potential energy)} \quad (1)$$

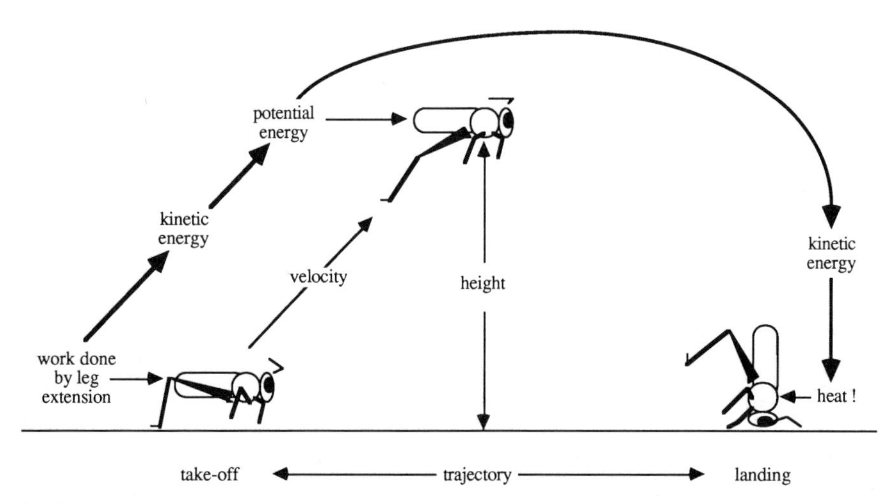

Fig. 7.1. Diagram of the energy relations within a jump. The insect accelerates by doing work with its legs to give itself kinetic energy, which is partly converted into potential energy at the top of the trajectory and then to heat as it returns to the ground.

where m is the mass, v is the initial velocity at takeoff, h is the height attained, and g_n is the acceleration due to gravity. This can be re-arranged to obtain the height:

$$\text{height} = \frac{v^2}{2g_n} \tag{2}$$

Similarly, the range of a long jump, which depends on the initial angle of the trajectory, ∅, can be obtained by

$$\text{range} = \frac{v^2 \sin 2\varnothing}{g_n} \tag{3}$$

This can be re-written in terms of the energy requirements for a given height: because from Eq. 1,

$$v^2 = \frac{2E}{m} \tag{4}$$

$$\text{height} = \frac{E}{mg_n} \tag{5}$$

This has the rather unexpected consequence that the height attained depends on the energy to mass ratio in the jumping animal. This can be expressed in another way: if animals have similar proportions of muscle and if the muscle can produce the same specific energy per unit weight, then the height (or range) achievable should be the same. In fact, small animals jump less far than large animals, partly because the basic mechanics outlined above is a gross oversimplification and partly because of the force–time requirements of the takeoff phase of the jump.

In order to take off, the jumping animal must produce a force that will cause it to accelerate to an appropriate speed. The force required can be obtained from simple Newtonian mechanics:

$$v^2 = \frac{f2d}{m} \tag{6}$$

where d is the distance of acceleration and is related to the length of the animal's legs. Because height $\propto$ velocity^2 it follows that with smaller animals, with shorter legs, the force that must be produced for a given jump will rise in inverse proportion to the decrease in leg length. It also follows that for a given animal to double its jump range it must double v^2 and, from Eq. 6, double the force that it produces during leg extension. So, in general, smaller jumpers must produced relatively larger forces than big ones.

The time relations do not favor the smaller jumper either. The time (t) between the start of leg extension and takeoff, assuming constant force production, is given by

$$t = \frac{vm}{f} = \frac{2d}{v} \tag{7}$$

So if the animal is smaller, the time available for takeoff is less, and if a given animal is to increase the range of its jump (range is proportional to v^2), it must accelerate to a higher speed in a shorter time: here the time available varies as range$^{-0.5}$.

The scaling of the force and time relations at the takeoff have important effects on the power (p) requirements for jumping:

$$p = fv \tag{8}$$

or, transposing from Eq., 7,

$$p = \frac{v^2m}{t} = \frac{v^3m}{2d} \tag{9}$$

So for a given animal, the range of its jump (proportional to v^2) is proportional to the power it can produce but, with smaller animals, where the distance through which they can accelerate, d, is small, the power they must produce for a given range (proportional to v^2) must increase as the distance of acceleration$^{-1.5}$. Thus tiny animals must produce proportionately far more power than large animals; examples of this are my calculations that a leopard jumping to a height of 2.5 m must produce 110 W/kg during its 1.5-m acceleration but that a rat flea jumping to only 0.1 m is required to produce 2.75 kg/kW (Fig. 7.2) during its 0.5-mm acceleration! The rat flea, too, must produce this power within a period of only 0.75 ms (Bennet-Clark, 1977) and produce forces that accelerate it at up to 2000 m/s^2 or about 200 g_n.

Both the remarkably high power involved and the extreme rapidity of the takeoff led to a comparison of the known mechanical properties of muscle with the power requirements for the jumps observed for various insects. Even in active endotherms, muscle does not appear to produce much above 0.86 kW/kg (reviewed by Bennet-Clark, 1977) and the best insect muscle known, the metathoracic *extensor tibiae* of the locust *Schistocerca gregaria*, only produces 0.45 kW/kg (Bennet-Clark, 1975) which, assuming that the animal has some 30% of its weight as jumping muscle (a very high proportion), gives an available power output from the animal's muscles of about 150 W/kg. This would allow the leopard to jump as it does but is far less than is required for most jumping insects (Fig. 7.2).

When, instead, one examines the energy requirements for jumping (Eqs. 1–5) it can be calculated that to jump to a height of 1 m requires about 10 J/kg. Because it is known that such muscles as are found in frog or crayfish produce

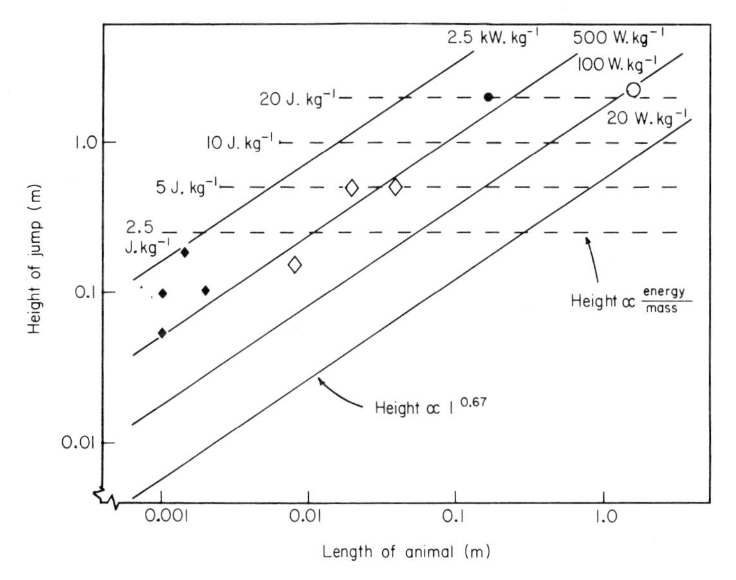

Fig. 7.2. Graphs showing the specific power and specific energy requirements for vertical jumps of animals of different sizes: the points show the estimated requirements for the known jumping performance of various animals, assuming that they can accelerate through a distance equal to their body length. For a given height of jump, the specific energy required is independent of size (broken lines, Eq. 4) but the height attained by an animal of a particular size scales as specific power$^{1.5}$ (from Eq. 8), so for given height of jump, an animal must produce specific power that scales as its body (and leg) length$^{0.67}$ (solid lines), so small jumping insects must produce very much higher specific powers than larger animals. Key: ○ leopard; ● lesser galago; ◇ locusts; ◆ fleas. (From Bennet-Clark, 1977.)

total energies of from 220 to 420 J/kg, it follows that even with under 5% of the body weight as muscle used for jumping, the required energy ought to be available to most animals — the problem is one of amplifying its power.

Expressed in another way, if the energy can be produced at low speed and hence low power, stored, and then released at a higher velocity and power, both the energy and power requirements can be met. Because insects of various taxa do indeed jump, despite the remarkable power requirements of the task, study of the jumping mechanism reduces to description and analysis of the energy stores and of the mechanism for providing, storing, and releasing the required energy.

7.2.2 Locust Jumping Performance

The jumping performance of locusts can be estimated from the mass of the animal and its range (Eq. 3). Typical male locusts weigh 1.5–2 g, and can jump over 0.8 m at 30°C whereas the heavier females, at 2.5–3.5 g, jump between 0.5 and 0.7 m. One-legged insects jump about half as far as intact ones (Bennet-Clark, 1975). These jumps require the animal to accelerate to between 2.2 and 2.8 m/s, which requires kinetic energies of between 6 mJ (for the smaller males) and 10 mJ (for the larger females): each hind leg, of course, produces half of this energy. In the case of females, the *extensor tibiae* muscle weighs about 70 mg per leg so this muscle must be producing about 5 mJ, which is a specific energy production of 71 J/kg, which is by no means an exceptional requirement for the muscle (see above).

Given the length of the hind tibia and hind femur (about 25 mm each) and knowing that the tibia extends through about 120° during the takeoff (Brown, 1963), giving an extension movement of about 43 mm, one may calculate that the time taken to accelerate to 2.5 m/s (using Eq. 7) is 34 ms. This requires a constant acceleration (from Eq. 7) of 73 m/s^2 (or 7.3g_n) and peak power output, at the moment of takeoff (From Eq. 9), of 0.45 W for a 2.5-g locust. When one now takes into account that this is achieved by about 140 mg of leg extensor muscle, one sees than an apparent specific power output is required from the muscle of 3.25 kW/kg, which is about a factor of 10 higher than the levels of power output typical for muscle (see above).

The approximate validity of these rough calculations is borne out from early studies of the course of jumping that showed that locusts extend the hind legs to accelerate the body from the ground (Brown, 1963, 1967) in a scissors-like opening movement lasting some 20–30 ms. However, it was also known that the extensor muscle was quite slow acting, only reaching peak force after about 0.5 s (Hoyle, 1955). It should be noted that neither of these descriptions attempted to describe the performance in terms of the energetics of the animal or its muscle, and that there was an incompatibility of a factor of around 20 in these early studies between the time relations of the jump impulse (20–30 ms) and of the production of full force by the *extensor tibiae* muscle (500 ms) which is in the same order as the incompatibility between the required power output of the muscle and the typical power outputs of known muscles.

Understanding of the mechanics of the jump requires a redescription of the leg anatomy and a description of the relevant features of the muscle physiology.

7.3 THE HIND LEGS OF ORTHOPTERA

7.3.1 Locust Hind Leg Anatomy

Jumping orthopterans, the majority of the order, have long and highly specialized hind legs. Early descriptions and illustrations, such as that by Snodgrass (1929, 1935), show the small coxa and trochanter, the swollen high femur, and a long tubular tibia. The major muscles of the hind leg reside in the femur and, in particular, some three-fourths of the femur volume is a vast pennate *extensor tibiae* muscle that originates over the inner surface of the top and sides of the femoral wall and inserts on a large, flat, bladelike extensor apodeme. The ventral muscle of the femur is a flexor muscle that, though also pennate, has longer fibers than those of the extensor. There are also various smaller femoral muscles, such as the retractor unguis that causes the tarsi to curl around stems but these are relatively unimportant in jumping.

The exoskeleton of the hind femur is roughly hexagonal in cross section, with cuticular ridges at the corners of the tube. It is thus well adapted to accept high internal stresses. At the distal end of the femur, the tibia articulates by a pair of specialized suspensory ligaments of unsclerotized cuticle that originate from a pair of heavily sclerotized crescent-shaped semilunar processes (Albrecht, 1953), externally flattened but forming a pair of banana shaped ridges internally (Fig. 7.3). The proximal ends of the semilunar processes fuse with a sclerotized arc of cuticle around the bottom of the distal end of the femur; this arc doubtless serves to distribute the stress of the extensor muscle onto the femoral cuticle. The lateral and dorsal walls of the femoral cuticle have chevron-shaped marks showing the separate origins of the many blocks of the extensor tibiae muscle.

Ventrally, the distal end of the femur has a small depression that internally forms a small lump, Heitler's lump, over which the flexor apodeme slides. The distal end of the flexor apodeme forms a pocket that slides over this lump and, when the flexor muscle is activated, the flexor apodeme locks over this lump and effectively prevents extension of the leg: when the flexor muscle relaxes, the catch is released (Heitler, 1974).

The extensor apodeme forms a blade of cuticle with preferred orientation of the chitin fibrils along its length. The apodeme becomes wider and thinner in the wider proximal part of the femur but its cross-sectional area is more or less constant over most of its length (Bennet-Clark, 1975). It twists from a horizontal strap at its insertion on the tibia to become a vertical blade in the femur (Fig. 7.3).

The hind tibia in locusts is a tubular structure, largely filled with a trachea.

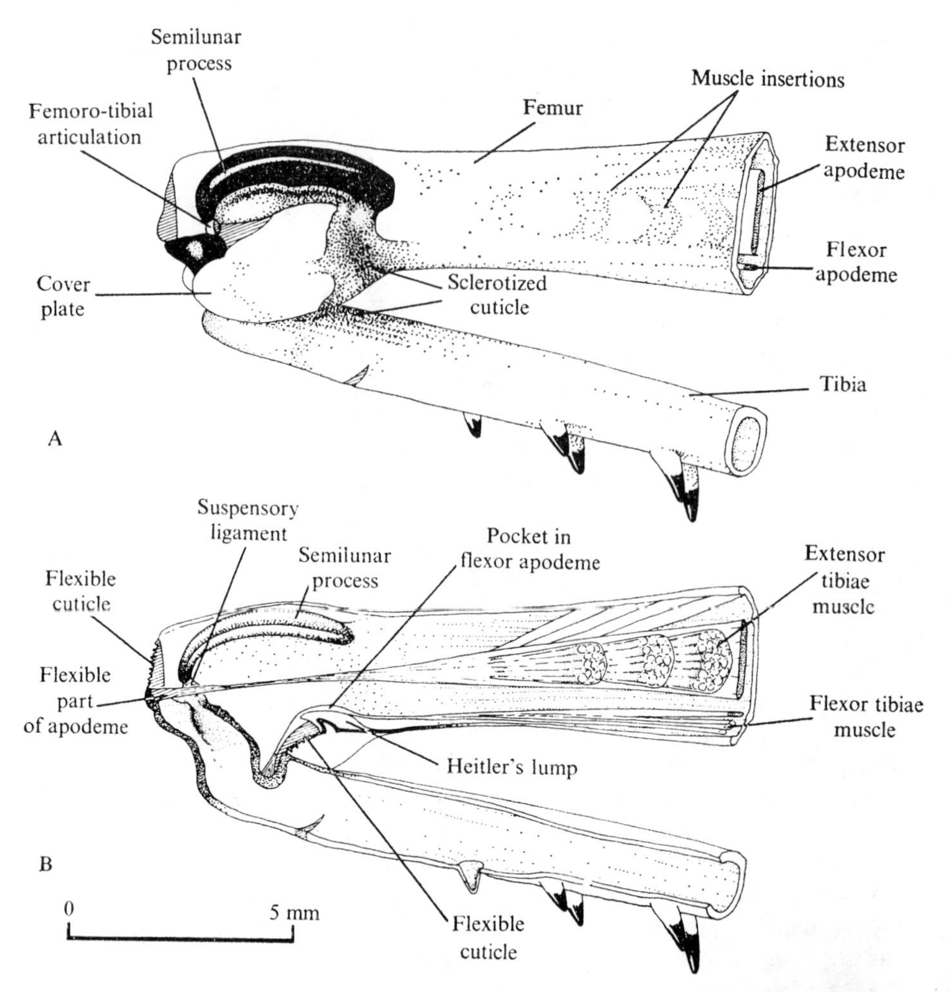

Fig. 7.3. Diagram of the femoro-tibial articulation of an adult locust *Schistocerca gregaria* to show (A) the external view and (B) the internal view. The tibia is suspended from flexible ligaments at the distal ends of the semilunar processes, which internally form banana-shaped thickenings: the semilunar processes are major energy stores. The pennate extensor tibiae muscle inserts on the straplike extensor apodeme, which is another important energy store. The pocket in the flexor apodeme slips over Heitler's lump to form a catch that locks the leg in the flexed position. The hollow trachea-filled tibia is a light, stiff beam. (From Bennet-Clark, 1975.)

It appears well adapted to withstand bending stresses: there is postecdysial deposition of cuticle, with a longitudinal preferred orientation, along the anterior, compression, side of the tibia (Neville, 1965, 1975) and the heavy spines along the posterior side appear to increase rigidity (Alia and Crovetti, 1961). The distal spines on the tibia are articulated and dig into the substrate as the animal accelerates.

7.3.2 Energy Storage in the Locust Hind Leg

Before a locust jumps, the hind legs are fully flexed and the animal remains stationary for a brief period before jumping. If the extensor muscle of the hind leg is stimulated, the leg will normally extend — but in a relatively slow kick. If, instead, the leg is held so that it cannot extend, stimulation of the extensor muscle leads to a contraction that causes a small movement of the femoro–tibial articulation owing to strain of the semilunar process along the line of the extensor apodeme (Fig. 7.4) (Bennet-Clark, 1975). Release of the tibia at this stage causes an explosive kick.

Though it is quite difficult to devise a setup to measure the stress–strain properties of the semilunar processes, this can be calculated from measurements made with intact legs. The extensor tibiae muscle of a typical leg produces a force of at least 14 N and up to 17 N (this is equivalent to over 1.4 kg inside 24-mm-long 120-mg femur!) which strains the paired semilunar processes through only 0.40–0.45 mm (Fig. 7.4) (Bennet-Clark, 1975). Although the stress–strain curve is not absolutely linear, a close approximation to the energy stored is given by

$$\text{energy} = \frac{\text{peak stress} \times \text{peak strain}}{2} \tag{10}$$

from which one can calculate that the two semilunar processes of one leg can store between 2.8 and 4 mJ: they are potentially major energy stores but cannot account for all the 5 mJ per leg required for the jump.

The extensor apodeme is highly stressed by the extensor muscle: there are reports in the literature that the extensor muscle is able to break its apodeme (Hoyle, 1958; Bennet-Clark, 1975). Its area of cross section, 25×10^{-9} m^2, can

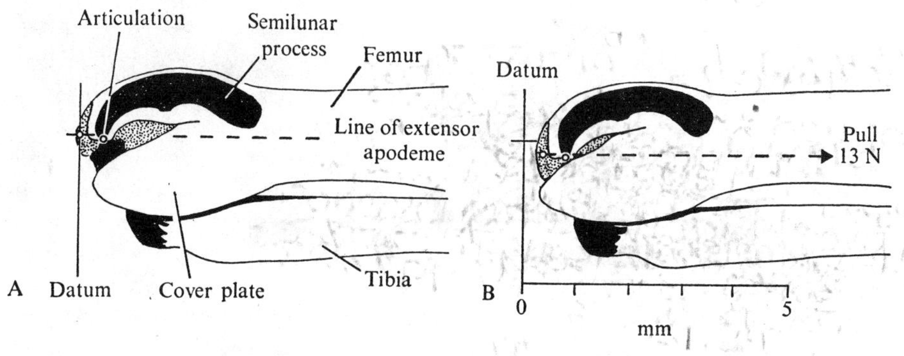

Fig. 7.4. Diagrams drawn from photographs to show the effect of stimulation of the *extensor tibiae* muscle on the flexed leg of a locust. Compared with the unstimulated leg, (A), the semilunar processes strain after muscle contraction, (B), through about 0.4 mm along the line of the extensor apodeme. (From Bennet-Clark, 1975.)

withstand a stress of about 15 N and gives the apodeme material an ultimate tensile stress of around 600 MN/m^2 which is similar to that of silk (Wainwright et al., 1976). The Young's Modulus of the apodeme has been measured both indirectly in bending (Bennet-Clark, 1975) and directly in tension (Ker, 1977), giving similar results of 14–19 GN/m^3. Using values in this range, one can calculate that, with a stress of 15 N, the apodeme strains between 0.45 and 0.55 mm and thus stores (from Eq. 10) a similar amount of energy to the semilunar processes (the situation here is more complicated because the apodeme is not stressed evenly along its length) (Bennet-Clark, 1975).

There is also likely to be a small amount of energy stored by the axial compression of the cuticle of the femur as a whole but this compression has proved too small to measure and so is probably insignificant.

The conclusion is that there are cuticular energy stores capable of storing a total of over 6 mJ per leg, which is appreciably more than is required to explain the observed jumping performance of locusts. The weight penalty of these energy stores is remarkably small. The two semilunar processes of one leg weigh about 2.6 mg and the extensor apodeme weighs about 0.4 mg: these structures can store the energy produced by 70 mg of muscles (Bennet-Clark, 1975). It is worth noting that the specific energy storage of the locust extensor tibiae apodeme is about five times that of steel and about three times that of resilin, a rubbery protein found in insect wing mechanisms (Weis-Fogh, 1960).

7.3.3 The Jumping Muscles of the Locust Hind Leg

The extensor tibiae muscle has rather long A bands; typical locust flight muscle has A bands that are 3.1 μm long (Weis-Fogh, 1956) compared with the 5.5 μm reported for those of the extensor tibiae (Cochrane et al., 1972). This can be expected to correlate with the production of high forces by the muscle rather than with fast velocity of contraction (Huxley and Niedergerke, 1954), which is borne out by measurements of tetanic force from the extensor tibiae muscle of 750–800 kN/m^2 (Bennet-Clark, 1975) compared with 350 kN/m^2 for the insect's flight muscle with its shorter sarcomeres (Weis-Fogh, 1956). Although it is hard to get direct measurements of the force–velocity relationships of the locust extensor tibiae muscle, these can be calculated. The basic measurement from which this is derived is a force versus time plot for a leg in tetanic contraction. From the force that is produced, the known geometry of the leg and muscle, and the properties of the elastic skeletal elements that are strained by its contraction, graphs of muscle performance against time can be obtained; these are shown in Figs 7.5 and 7.6.

As already indicated, the muscle shortens quite slowly: the peak force of 16 N is attained only after 300 ms at 30°C after a shortening of the fibers of about 20%, and the maximum velocity of shortening is only 7 mm/s, or around 2 lengths per second, far slower than is typical for fast twitch muscles. The power production of the muscle, however, is impressive, reaching a peak

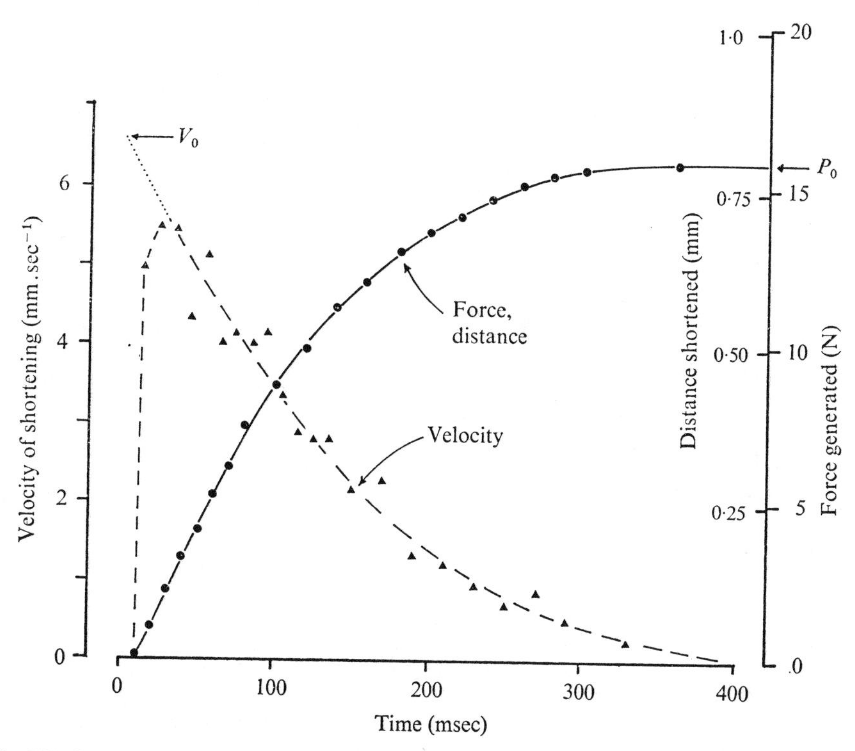

Fig. 7.5. Graphs of force, distance, and velocity against time for the *extensor tibiae* muscle of an adult locust *Schistocerca gregaria* in tetanus at 30°C. The force was calculated from measurements made at the distal end of the tibia of an intact leg and the distance shortened was derived from the force, assuming that the muscle strained a linear spring. The velocity of shortening was obtained by differentiating the force versus time plot. V_0 is the maximum velocity obtained by extrapolation to zero force and P_0 is the maximum force at zero velocity. (From Bennet-Clark, 1975.)

of about 450 W/kg, which is close to the highest value that has been recorded for any muscle (Bennet-Clark, 1975, 1977). One feature of the energy storage mechanism that is noteworthy is that it appears that the geometry of the muscle fibers is such that their peak power output is produced over the first half of their contraction, so the initial rate of energy storage is high, but that the muscle can then continue contracting, ever more slowly, producing more energy until maximum force is produced and contraction ceases. This suggests that the muscle is able to operate at close to its optimal rate of work production and hence energy storage over most of its rather slow contraction.

The *flexor tibiae* muscle is not only far smaller than the extensor but also has very different force–velocity relations: essentially, it is far faster. The tetanic force builds up to 3/4 of peak force in less than 50 ms and, at the cessation of stimulation, the force decays from 3/4 of peak to 1/10 in about 40 ms. At peak level, the catch mechanism described by Heitler (1974) is fully

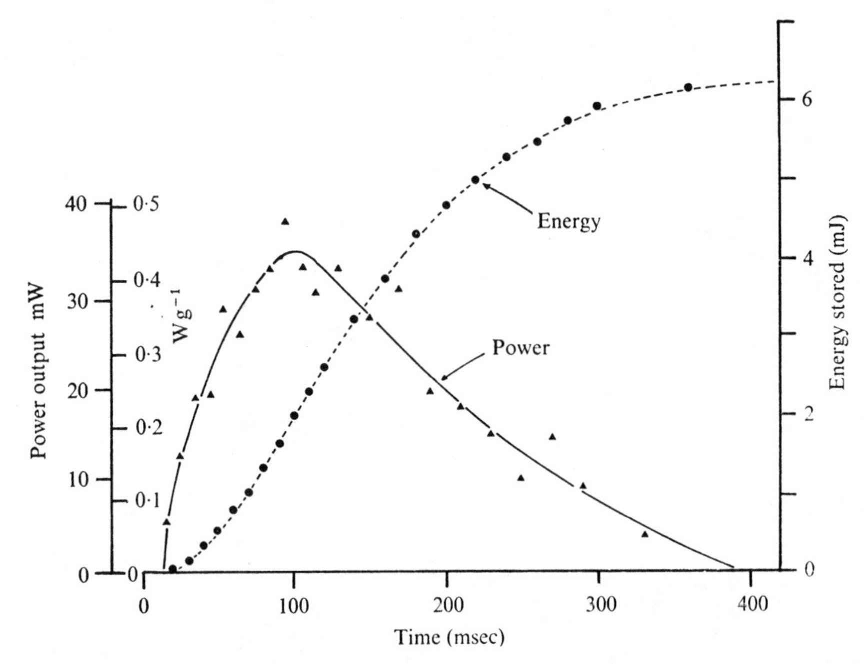

Fig. 7.6. Graphs of power and energy production against time for the *extensor tibiae* muscle of an adult locust in tetanus at 30°C. The power was calculated as the instantaneous product of the force and velocity shown in Fig. 7.5 and the energy production by integrating the curve for power output. (From Bennet-Clark, 1975.)

engaged, so the leg may be held locked in the flexed position while energy is stored in the skeletal springs by the extensor muscle. When subsequently the flexor relaxes, leg extension occurs by release of the stored energy via the still-contracted extensor muscle; this causes rapid extension of the now-relaxed flexor muscle.

The neural events that surround jumping involve an initial activation of the *flexor tibiae* muscle followed by co-activation of both flexor and extensor; this last stores the bulk of the energy required for the jump (see Section 7.6). The jump is released by cessation of stimulation to the *flexor tibiae* while stimulation of the extensor continues (Heitler and Burrows, 1977a).

7.3.4 Safety Factors in the Locust Hind Leg

One striking feature of the leg is how easily it breaks: if uncontrolled, the *extensor tibiae* muscle can break the extensor apodeme in a simple tetanic contraction, or if it is scratched, the cuticle of the semilunar processes can break as the *extensor tibiae* muscle contracts; the suspensory ligaments of the tibio–femoral articulation and the cuticle of the head of the tibia and the semilunar processes are also easily broken by forces only slightly in excess of those produced by the *extensor tibiae* muscle. In the design of the leg, it appears that the factor of safety of some regions is less than 1.2 and nowhere exceeds 2 (Bennet-Clark, 1975). There are two explanations for these very low factors of safety.

The first is that the structure of the leg is loaded by internally generated forces due to the muscles, which are under close proprioceptor control and thus, although highly stressed, remain safe. The second explanation concerns the cost effectiveness of the elastic mechanisms stressed by the muscles: unless these are highly stressed, they will not store as high a specific energy as is possible, so the efficiency of storage will be suboptimal and also, possibly, the loss between storage and release of energy would be higher (for a discussion, see Alexander, 1981).

7.3.5 The Hind Legs of Other Grasshoppers

Though there is little published information on their jumping performance, I have examined the anatomy of British and European field acridids such as *Oedipoda germanica* and *Chorthippus brunneus*, and of the apterous *Tetrix undulata* (Tetrigoidea). I found that all had a similar hind leg anatomy to that of the desert locust, with a large pennate *extensor tibiae* muscle and well-developed semilunar processes, and that there was the same type of internal lump onto which the extensor apodeme engaged to hold the hind tibia in the flexed position. From this and the way in which these insects prepare to jump, I conclude that the jumping mechanism is essentially similar to that described for locusts (Bennet-Clark, 1975). Some of the brachypterous and apterous acridioids have relatively large back legs and appear to jump, for their size, far further than do locusts: Gabriel (1984) states that *Tetrix undulata* has very well developed semilunar processes and that, although the insect weighs only 50 mg, it can jump 0.7 m.

7.4 KINEMATICS AND ENERGETICS OF JUMPING BY LOCUSTS

7.4.1 Distance–Time Relationships at Takeoff

By analysis of high-speed films of locust jumps, it has been possible to establish the time course of the acceleration that causes the jump (Fig. 7.7). By differentiation of the distance versus time graph, one can obtain the insect's velocity versus time and, by differentiation of velocity versus time, one obtains the insect's acceleration versus time. If the weight of the insect is known, the instantaneous force the insect produces can then be obtained, using the relationship

$$\text{force} = \text{mass} \times \text{acceleration} \qquad (11)$$

which shows that the force rises steadily, reaching a broad peak during the

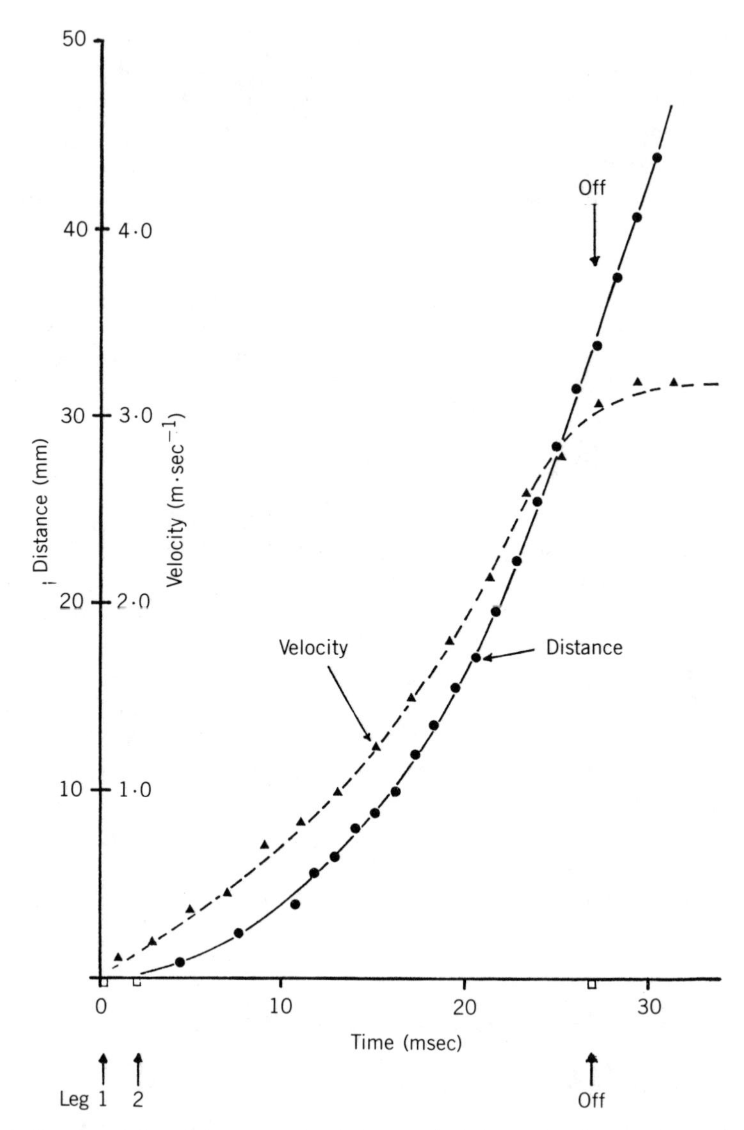

Fig. 7.7. Graphs of the distance and velocity against time of a jump of an adult locust *Schistocerca gregaria*, plotted from a film made at 930 frames/s. Distance was measured directly from the film and velocity was derived by differentiating the distance versus time curve. Arrows on the *x* axis show when each leg started to extend and when the insect left the ground. (From Bennet-Clark, 1975.)

first 15–30 mm distance or 15–25 ms duration of leg extension, and then falls to zero when the animal leaves the ground (Fig. 7.8); the peak force of 0.3 N causes an acceleration of 170 m/s^2 or 17 g_n, which can be compared with a far lower value calculated in Section 7.2.2. The discrepancies arise because the leg

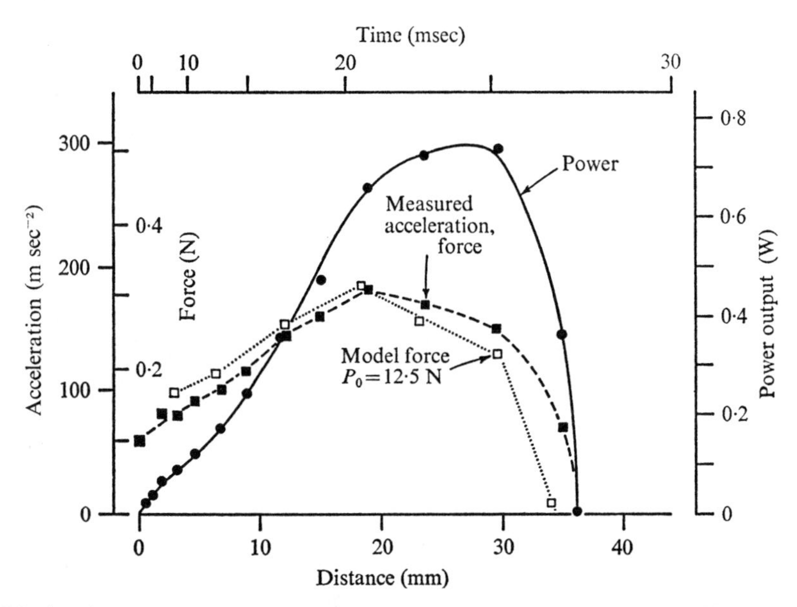

Fig. 7.8. Graphs of the force, acceleration, and power calculated from the data plotted in Fig. 7.7. Acceleration is obtained by differentiating the curve for velocity and force is obtained by multiplying acceleration by the animal's mass. Power is the instantaneous product of force and velocity (Eq. 7). The curve for model force is obtained by relaxing a linear spring stressed by 12.5 N and strained 0.75 mm into the lever system of the extending hind leg of a locust (shown in Fig. 7.9). (From Bennet-Clark, 1975.)

extension in the real insect is through a shorter distance than in the calculated case, and in the real insect, the force and hence the acceleration are not constant throughout the impulse. The instantaneous power delivered (from Eq. 8) is the product of the instantaneous velocity (Fig. 7.7) and the instantaneous force (shown in Fig. 7.8), which shows that the insect produces a peak power (for both legs) of about 0.75 W or a specific power of about 250 W/kg. These measured values, again, are rather greater than those calculated earlier as an illustration of the method, but are instructive because they illustrate that the real performance is not adequately explained by simple linear Newtonian mechanics.

It is worth noting that the *extensor tibiae* muscle of one leg produces a peak power of 35 mW (Fig. 7.6) but, that during the jump acceleration, each leg (producing half the total power shown in Fig. 7.8) produces 375 mW; the storage-release mechanism produces over tenfold power amplification. It also follows that there can be only a minor contribution by the extensor tibiae muscle to the total power produced during the period of acceleration.

7.4.2 A Model of the Jump Mechanism

The postulated mechanism for the jump that emerges from these observations is that the bulk of the energy is stored before the impulse and discharged through the extending legs: in this case, it should be possible to model the force–distance relationships of the impulse.

The energy store can be modeled as a spring that strains 0.75 mm with a stress of 12.5 N, storing 4.7 mJ. This spring, when applied to the extensor apodeme of the tibia, extends the leg through 34 mm (Fig. 7.9): the force–distance relationships can be obtained by dividing the instantaneous force produced by the relaxing spring by the instantaneous mechanical advantage of the leg. The mechanical advantage and distance moved at the apodeme versus distance moved by the jumping locust are shown in Fig. 7.9. Mechanical advantage falls from an initial value of over 160:1 over the first 50 μm of movement of the apodeme to about 30:1 after a movement of 0.75 mm at the apodeme, at the same time as the force produced by the semilunar processes and extensor tibiae apodeme in series falls from 12.5 N to zero. This model results in a plot of force versus distance (Fig. 7.8) which closely resembles that measured from plots of the performance of the jumping insect.

One attractive feature of this type of storage–release catapult mechanism is that maximum power may be produced at close to the maximum velocity attained by the insect and is independent of the force–velocity properties of the muscle. Were the power to be produced directly by muscle, it is likely that the maximum power would be produced at between one-third and one-half the maximum velocity *and* force attainable by the muscle: for jumping, this would require a very fast muscle, which would intrinsically produce less force per unit area and less work per unit mass than one with longer sarcomeres that was slower and, furthermore, because it must contract as fast as possible, could not achieve its maximum tetanic force while shortening and so would produce far less work per contraction than a muscle that contracts ever more slowly, as does the locust *extensor tibiae* muscle (Fig. 7.5), into an energy store (Bennet-Clark, 1976).

It should be noted in passing that this type of operation of the leg muscles requires a very different type of design of the muscular system to that where the muscle is required to produce long periods of oscillation at high power, such as in flight or swimming, where the muscle is tailored to its load so that it is able to contract at the force and velocity at which it is able to produce maximum power per cycle (Alexander, 1973; Weis-Fogh and Alexander, 1977) rather than maximum work per cycle, as in the locust hind leg (Bennet-Clark, 1976.)

7.4.3 Energy Losses in Leg Extension

In a mechanism of this type, there are potential losses due to the energy involved in accelerating and rotating the legs. This energy depends on the

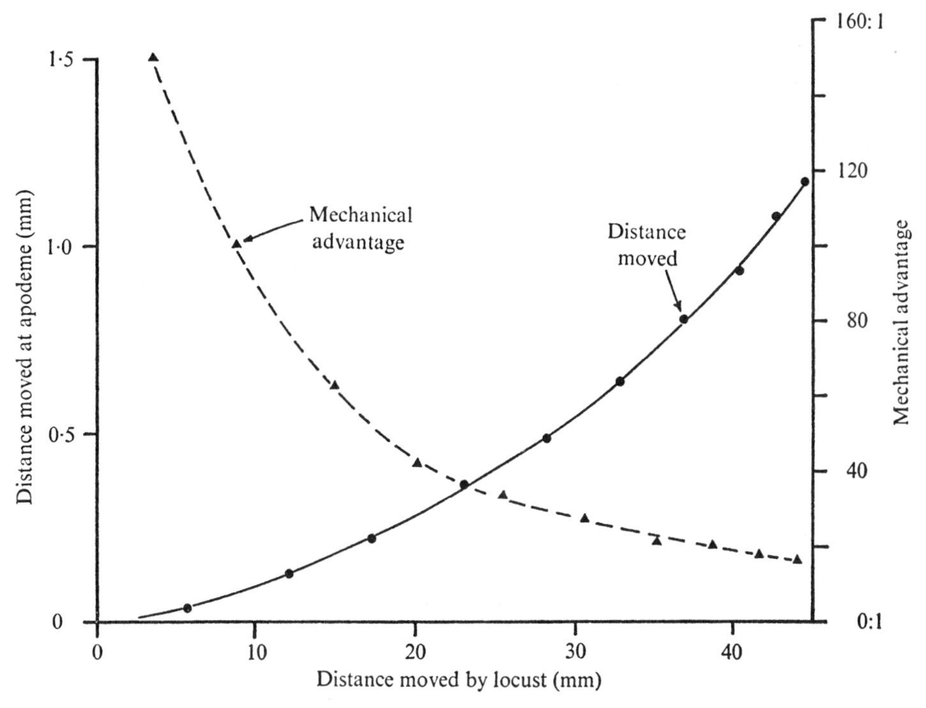

Fig. 7.9. Graphs of the mechanical advantage and distance moved at the extensor apodeme against the distance the tip of the hind tibia extends from the base of the hind femur of the locust *Schistocerca gregaria* (this is equivalent to the distance that the animal moves as the leg extends). The data so obtained were used to calculate the force that would be produced by a relaxing linear spring stressed by 12.5 N and strained 0.75 mm into the lever system of the extending hind leg to give the model force curve shown in Fig. 7.8. (From Bennet-Clark, 1975.)

square of the angular velocity of the leg podomeres and on the mass of the tibia; the implications are that it pays to minimize the angular velocity and mass of leg. Considering the whole leg, it will pay to have a relatively long slow impulse for this will reduce the angular velocity of the extending podomeres, with the additional advantage that the bending forces on the limb will also be reduced. Because the bulk of the energy-producing muscle is in the hind femora, the optimum will be for the muscle to produce as high a specific energy as possible, so as to minimize muscle weight, and for the angular velocity to be low: in practice the femur of a 2.5-g locust weighs only 120 mg, with 70 mg of extensor muscle, and is 24 mm long so that it rotates through less than 90° for 35 mm of leg extension. In the case of the tibia, the optimum will be a long thin light beam: the 30-mm-long hind tibia and tarsi of a locust are largely air-filled and weigh only 24 mg. It can be calculated that about 0.02 mJ of energy is wasted in rotation of the tibia and tarsi and a further 0.07 mJ is wasted in femoral rotation and cannot be converted into the kinetic energy of the insect (Bennet-Clark, 1975): this is only about 2% of the elastic energy delivered to each leg.

A further source of energy loss is that due to the viscosity of the spring and leg muscles. Here, from a plot of the course of the kick of an unloaded locust leg which made a damped oscillation at 130 Hz, I calculated the velocity-dependent loss factor of leg extension and from this the energy loss due to viscosity. This calculation gives energy losses of between 3 and 12% dissipated as heat within the leg in a normal jump (Bennet-Clark, 1975). Here the important point is that, again, the loss depends on the angular velocity, so it will pay to have longer legs that can extend over a longer period of time, to minimize the energy losses that their extension incurs.

7.4.4 Energy Losses due to Air Resistance

With small animals such as grasshoppers, the aerodynamic drag on the body may have a substantial effect on the range of the animal because the drag force of the passage of the body through the air absorbs energy. The initial velocity of the insect, from which the height that it should attain in vacuo (h_v), as calculated from Eq. 2, gives an oversimplified view of its behavior in the air. To examine the situation more exactly, the air resistance must be taken into account.

The drag force on a moving body is proportional to its frontal area, its drag coefficient, and the density of the fluid through which it is passing. It is also proportional to the square of velocity and, because velocity varies through the jump, the effect of air resistance on the height of a jump in air (h_a) can be obtained (Bennet-Clark and Alder, 1979) by application of a rather awkward equation:

$$h_a = \frac{m}{C_D \rho A} \text{n} \left(\frac{C_D \rho A V^2}{2 m g_n} + 1 \right) \tag{12}$$

where m is the mass, C_D is the drag coefficient of the body, ρ is the air density, A is the frontal area of the animal, V is the initial velocity at takeoff, and g_n is the acceleration due to gravity. This can be combined with Eq. 2 to give the ratio of the height attained in air, h_a, to that attainable in vacuo, h_v, by

$$\frac{h_a}{h_v} = \frac{1}{C_D} \left(\frac{m}{\rho h_v A} \right)_{\text{n}} \left(\frac{C_D \rho h_v A}{m} + 1 \right) \tag{13}$$

This reduces the performance to three dimensionless parameters: $\frac{h_a}{h_v}$ which may be termed the jump efficiency; C_D, which is a function of body shape, and the term $\frac{\rho h_v A}{m}$ which relates the kinetic energy, via h_v, to the frontal area to mass ratio and via ρ to the properties of the fluid. It can be seen by inspection of Eq. 13 that high values of jump efficiency will be achieved with low values of C_D, with a high mass to frontal area ratio and with low values of h_v.

This has been tested experimentally by observing the height attained when insect bodies of various species including locusts were propelled at a known initial velocity (Bennet-Clark and Alder, 1979). The experimental method was to propel the insect vertically from a calibrated spring gun to give an initial velocity appropriate for a known h_v and to observe the height h_a its body attained. The values of h_a measured are shown for various insect species in Fig. 7.10. Data were obtained from first-instar larval locusts weighing about 10^{-5} kg through to adults weighing 10^{-3} kg, as well as for fleas weighing less than 10^{-6} kg (1 mg). It wll be seen that there is a substantial loss of energy, amounting to about 35% loss of height for h_v of 1 m (resulting from dissipation of 35% of the energy) for the smallest locusts (Fig. 7.10) and that the energy loss is even greater for this h_v for the even smaller fleas.

This can be explained by reference to the jump efficiency, h_a/h_v, measured for these insects when plotted against their frontal area/mass (Fig. 7.11). It will be seen that insects with a frontal area per unit mass of about 1 mm^2/mg should achieve jump efficiencies of about 65% when attempting a 1-m high jump but will achieve efficiencies of over 90% when attempting jumps of less than 0.25 m. With fleas, where the frontal area per unit mass is far greater, lower efficiencies, as indicated by a lower h_a/h_v, are achieved at all heights. It appears that small insects would not find it economical to make attempts to jump spectacularly far or high (Bennet-Clark and Alder, 1979).

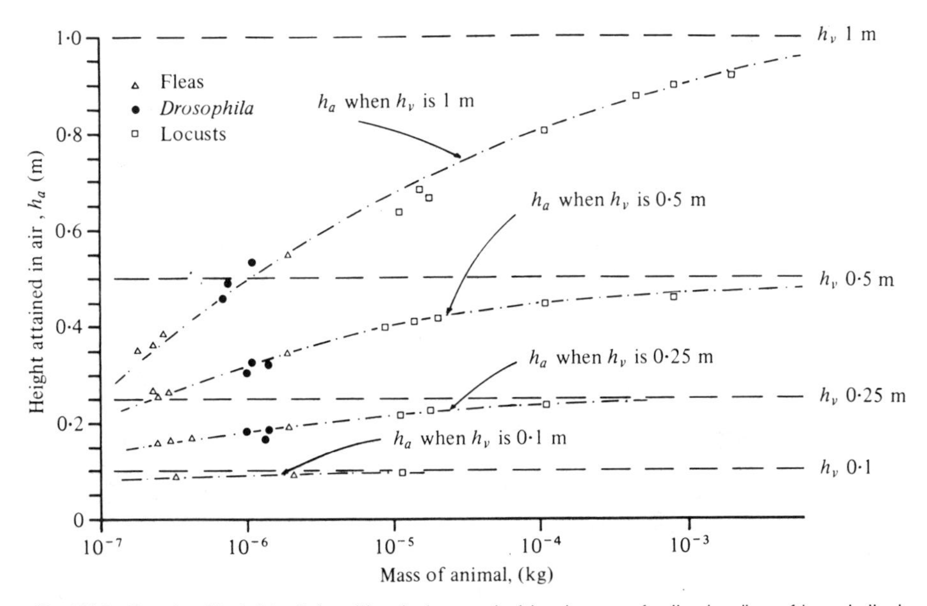

Fig. 7.10. Graph of height attained in air, h_a, against body mass for the bodies of insects that were propelled with a spring gun at initial velocities that would have projected them to the heights in vacuo shown by the horizontal dashed lines. The curved lines (– · – · –) are fitted by eye to the observed heights attained. Key: □ locusts of different instars; ● *Drosophila*; △ fleas. (From Bennet-Clark and Alder, 1979.)

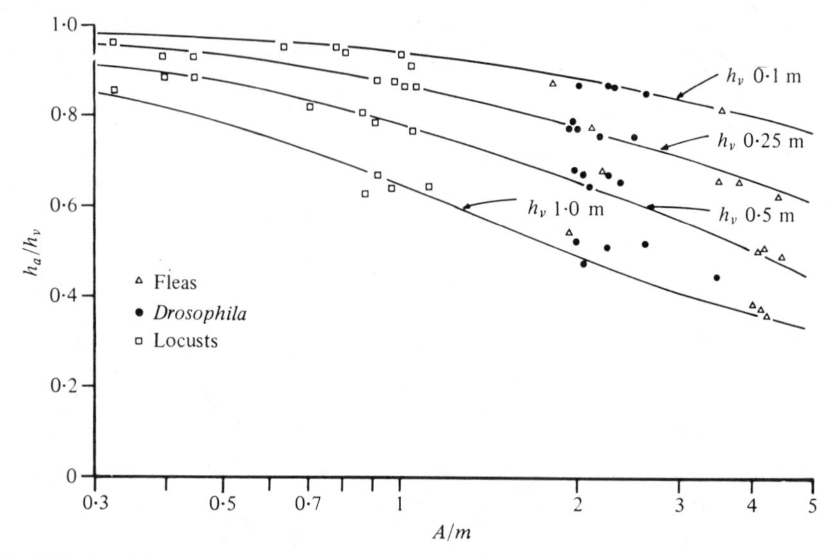

Fig. 7.11. Graph of jump efficiency or the ratio of height attained in air to height attainable in vacuo, h_a/h_v, against the frontal area to body mass ratio for various insects that were propelled with a spring gun at initial velocities that would have projected them to the heights in vacuo shown. The lines show the theoretical values for h_a/h_v calculated from Eq. 13 for bodies with a C_D of 1.0. Key: □ locusts of different instars; • *Drosophila*; △ fleas. (From Bennet-Clark and Alder, 1979.)

It follows from this that a well-adapted jumping insect should have a low C_D and a low frontal area to mass ratio. This was examined experimentally by comparing the energy loss due to drag for winged flies versus that for wingless locust larvae. The results are shown in Table 7.1, which expresses the performance in terms of C_D as calculated from Eq. 13. The presence of wings adds very substantially to the drag coefficient but their removal lowers the drag coefficient to close to that of an intact larval locust of similar size. Rather surprisingly, although removal of the legs further lowers the drag coefficient of the locusts and flies, this is accompanied by a 25% loss of weight for the locusts but a far smaller weight loss for the flies.

When the bodies were propelled with the spring gun to a standard initial velocity, it was found that removal of the wings caused the insects to reach substantially higher jump efficiencies, rising from 45 to 55% for winged versus wingless *Drosophila* at h_v = 1 m, but that leg removal had only a minor effect. With *Calliphora*, essentially similar effects were observed but the jump efficiency of the body at the same h_v was about 70%, as was that of the far lighter first-instar locusts (Bennet-Clark and Alder, 1979). With bodies of first-instar locusts, leg removal had a far smaller effect on jump efficiency, presumably because the reduction in drag coefficient that resulted was balanced by a similar reduction in the total mass of the moving insect and hence its kinetic energy.

TABLE 7.1. Some Aerodynamic Parameters for Insect Bodies of Different Species

Insect Species and Stage	Weight of body (mg)	C_D with Wings	C_D with Legs but no Wings	C_D without Legs or Wings	Frontal Area / Mass (mm^2/mg)
Schistocerca first-instar larvae	15	—	1.08	0.73	1.03
Drosophila adults	1.0	1.58	1.18	0.89	2.25
Calliphora adults	55	1.34	0.95	0.84	0.95

Source: From Bennet-Clark and Alder, 1979.

The frontal area to mass ratio of the 55-mg *Calliphora* was very similar to that of a 15-mg first-instar locust. This is because the interior of most flying insects is filled with substantial air sacs so that the body density is about a third that of water, whereas that of the compact bodies of larval locusts is similar to that of water. The inference, from this and further experiments with model insect bodies, is that where an insect is unable to control the attitude of its body through the air — as in jumping — it is advantageous to have as compact and dense a body as possible and to minimize the area of any protuberances from the body: this is in distinct contrast to the desiderata for body design of a flying insect, where strategically sited bristles and air sacs can reduce and optimise the drag coefficient along the line of flight but cause it substantially to worsen along other axes (Vogel, 1965; Bennet-Clark, 1980). It is noteworthy that the drag coefficient of the smooth body of first-instar locusts is less than that of the body of either fly species examined, both of which had long bristles on all parts of the body (Table 7.1).

7.5 JUMPING BY IMMATURE LOCUSTS

Immature acridids jump by essentially the same mechanism as do the adults. There are various problems to overcome, such as the detachment of a highly stressed cuticle from its muscles at apolysis, the growth and development of new energy stores, and the concomitant growth of muscles that will allow efficient jumping without undue risk of limb fracture.

7.5.1 Jumping Performance

In most of the larval instars, locusts do not jump nearly as well as when adult. From the first to fourth instars, the jump range is only about a third that of the adults and even in the fifth instar it is only about half that of the adults

(Fig. 7.12) (Gabriel, 1985a). Shortly after ecdysis, the range is less than one-half of the peak performance of the instar, and peak performance is not attained for several days (Gabriel, 1985b).

Over a period of time, the major difference between instars is that the adults make a few rather long jumps but that the larvae make many more shorter jumps. Typically, the jumping muscles produce about 3.2 kJ/kg per 10-min period of sustained jumping in all instars (Gabriel, 1985b). During this 10-min period, the total range of the jumps is 22 m for first instars, 27 m for fourths, and 40 m for adults or between 0.13 and 0.24 km/h. It is small wonder that locusts are such formidable pests (Gabriel, 1983, 1985a).

Within each instar, there is, in the first few hours, an initial increase in the stiffness of the semilunar processes, probably brought about by postecdysial tanning. This is followed by a further increase of stiffness over several days as postecdysial cuticle is deposited. This is accompanied by postecdysial deposition of cuticle onto the extensor apodeme, which becomes both wider and thicker: at the same time the extensor tibiae muscle grows and the femur as a whole becomes rounded in cross section as it becomes filled with muscle (Gabriel, 1983). This, in part, explains the steady increase in jumping performance through the instar. The increase of range from the fourth to fifth instar to adult suggests that the design of the jumping leg changes from instar to instar.

7.5.2 Allometry of the Jump Mechanism

In isometric growth, the linear dimensions vary as body mass$^{0.33}$; any departure from this basic scaling law is an indication of allometry. Because the jump range is greater in the later instars, their hind legs should become relatively larger and, in particular, their energy stores should become larger. This problem has been examined by Gabriel (1985a,b). She expressed leg dimensions as length/mass$^{0.33}$; any significant departures from isometry are given as relative measurements of length/mass$^{0.33}$. It was found that there was little change in the length and height of the femur from instars 1 to 4 but that there was a 1.25× increase in relative femur length from fourth instar to adult. At the same time, at the broadest part near the trochanter, the relative femur height increased 1.3× and the width increased 1.15×, so the total femur volume relative to that of the animal increased 1.9× (Gabriel, 1985a). The relative dimensions of the semilunar processes likewise changed little from instar 1 to instar 4 but from fourth instar to adult the relative length increased 1.4×, the height 1.5×, and the internal thickness of the process 1.4× (Gabriel, 1985a,b), so its relative total volume increased some 3×, and assuming that its specific energy storage capacity increased commensurately with its mass, could account for the threefold increase in jump range from the early instars to the adult: this was largely confirmed by measurements that Gabriel (1985b) made of the deformation under load of the semilunar processes, by which she showed that the processes of the adults were stiffer than those of the fourth and fifth instar larvae. The changes in the properties and relative dimensions

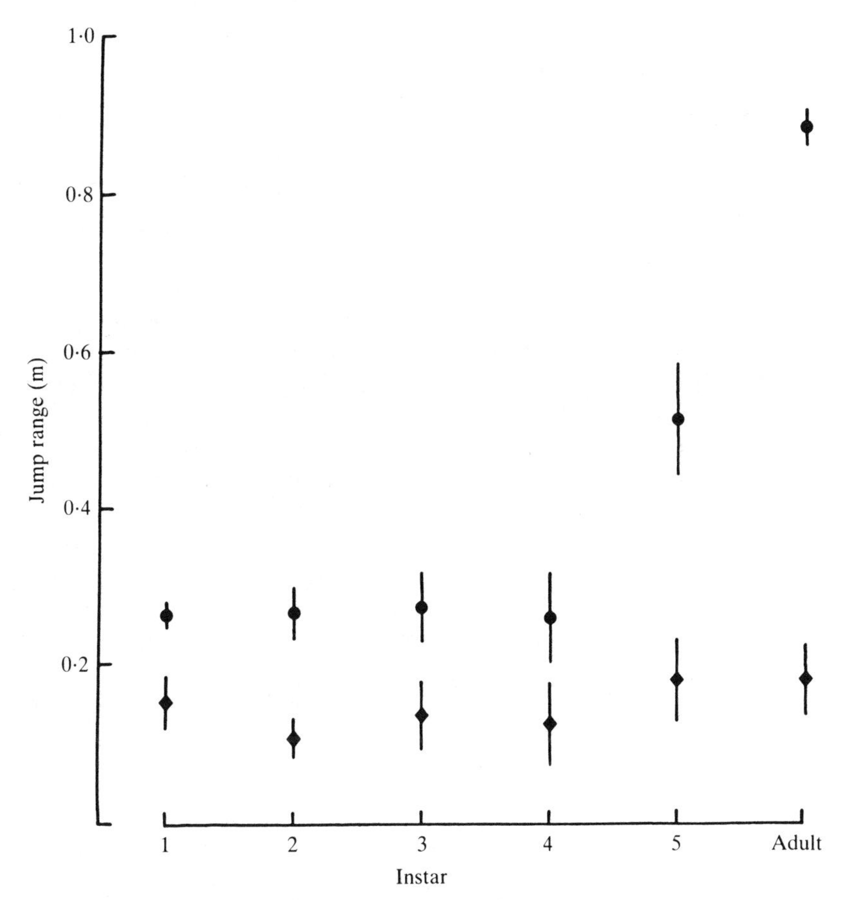

Fig. 7.12. Graph of range of the escape jump for different instars of the locust *Schistocerca gregaria.* Mean range and the standard deviation are shown for locusts on the first day after ecdysis (◆) and later, when peak performance was attained (•). It will be seen that well-developed insects in instars 1–4 have similar ranges but that fifth instars can jump twice as far and adults three times as far as the younger locusts. (From Gabriel, 1985a; courtesy of Company of Biologists Ltd.)

of the semilunar processes were accompanied by a 2.5× increase in the relative area of cross section of the extensor apodeme, which, with the 1.25× increase in the relative femur length, corresponds to a three-fold increase in the apodeme mass between fourth instar and adult. Concurrently, there are general increases in the relative thickness of the cuticle of the femur and tibia (Gabriel, 1985b): overall, the leg is designed to operate over a relatively higher range of forces in the adult than in the larvae.

Turning now to the mechanics of leg extension, the basic geometry of the lever that extends the tibia via the extensor apodeme and the lever ratio between the flexor apodeme and the extensor apodeme do not change from fourth instar to adult (Gabriel, 1985a).

There are also significant changes in the geometry of the *extensor tibiae* muscle: in the larvae, the fibers are relatively longer and have a lower angle of pennation (Fig. 7.13) than in the adult (Gabriel, 1986b): this correlates in part with the relatively less stiff semilunar processes and smaller relative area of cross section of the extensor apodeme of larvae compared with the adult. It is calculated that the muscle fibers of the fourth instars must shorten about 21% and produce a force of 0.33 N/mm^2 but that those of the adults must shorten 25% and produce a force of 0.71 N/mm^2 (Gabriel, 1985b). Rather surprisingly, there are no significant ultrastructural differences between such mechanically important parameters of the *extensor tibiae* muscles as A-band length and the ratio of numbers of actin filaments to myosin filaments between fourth instar larvae and adults (Gabriel, 1983).

The *extensor tibiae* muscles of the adults are probably working at close to their maximum isometric force whereas those of the larvae are working up to a maximum safe force in the energy stores that is far below the maximum isometric force of the muscle. When stimulating the *extensor tibiae* muscle, Gabriel (1983) noted that that of the larvae was capable of breaking the cuticle of the leg but that this rarely happened with the legs of adults.

It can be noted, too, that the 10-min total range does not differ as much as the range of the escape jump of locust larvae and adults (Section 7.5.1), further suggesting that their muscles are working under different load

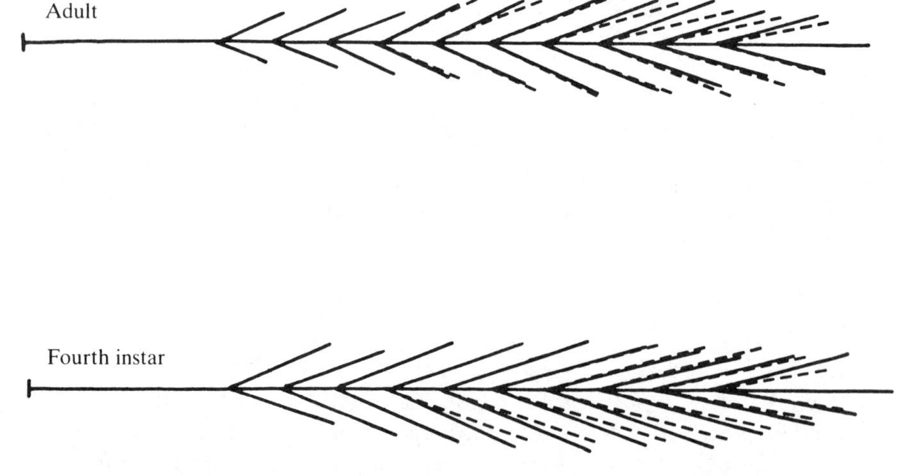

Fig. 7.13. Diagram of the relative fiber lengths and angles of pennation for the *extensor tibiae* muscle of the locust *Schistocerca gregaria.* The apodeme (central line) is 17.6 mm long in adults and 8.6 mm long in fourth instars. The solid lines represent ventral fibers and the dotted lines fibers within the body of the muscle. The fibers in the fourth instar are relatively longer and their angle of pennation is more oblique than in the adult. (From Gabriel, 1985b; courtesy of Company of Biologists Ltd.)

conditions. In larvae, it appears that the muscle geometry will be more appropriate for powering a relatively low force with rapid muscle contraction and energy storage, compared with that in adults where the high angle of pennation of the *extensor tibiae* muscle will produce a far higher force but where, to exploit the energy stores to the full, it will be necessary for the muscle fibers to shorten through a relatively greater distance, so the pre-jump phase of energy storage is likely to last longer and be metabolically relatively more costly to adults that to larvae. This may explain why, over a period, the larvae make many more, but shorter, jumps than do the adults.

7.6 NEURAL EVENTS SURROUNDING JUMPING

7.6.1 The Motoneurons

The *extensor tibiae* muscle receives quite simple excitatory innervation from fast (fast *extensor tibiae*, FETi), slow (SETi), and inhibitory motoneurons, but the *flexor tibiae* is innervated by seven excitatory motoneurons (FlTi), to different groups of fibers and of different speeds, and by two inhibitory motoneurons (Hoyle and Burrows, 1973; Heitler and Burrows, 1977a). The cell bodies of these various motoneurons in the metathoracic ganglion have been identified and their central projections are well known.

7.6.2 The Motor Program

The basic pattern of nervous activity that leads to a fast tibial extension is largely known from studies of the unilateral kicking of tethered locusts (Fig. 7.14; Heitler and Burrows, 1977a). At the start of a kick, there is an initial flection of the tibia, brought about by a 300-ms burst of stimuli in both the fast and slow *flexor tibiae* motoneurons (FFlTi, SFlTi; flexor excitors in Fig. 7.14), which is followed, after a pause, by a coactivation of motoneurons, which spike together for 300–600 ms: as the FETi starts to spike, there is an increase in the rate of firing of the FFlTi. This period of co-contraction ends with the inhibition of the excitatory FlTi motoneurons and a burst of spikes in the inhibitory FlTi which starts some 60 ms before the start of the leg extension even though spiking in the excitatory extensor motoneurons continues through until after the kick. Following the kick, there is a burst of activity in the FlTi neurons that re-flexes the tibia (Heitler and Burrows, 1977a).

This type of activity is entirely compatible with the requirements of an energy storage system of the type described in Section 7.3.2: there are flextion and maintenance of flextion to engage the tibial lock (Heitler, 1974); then there is a long contraction of the *extensor tibiae* muscle that stores the energy, while the leg is held flexed; then there is a release, by relaxation of the flexor leading to release of the catch, of the stored energy, the force of which is transmitted through the still-active and contracted extensor muscle.

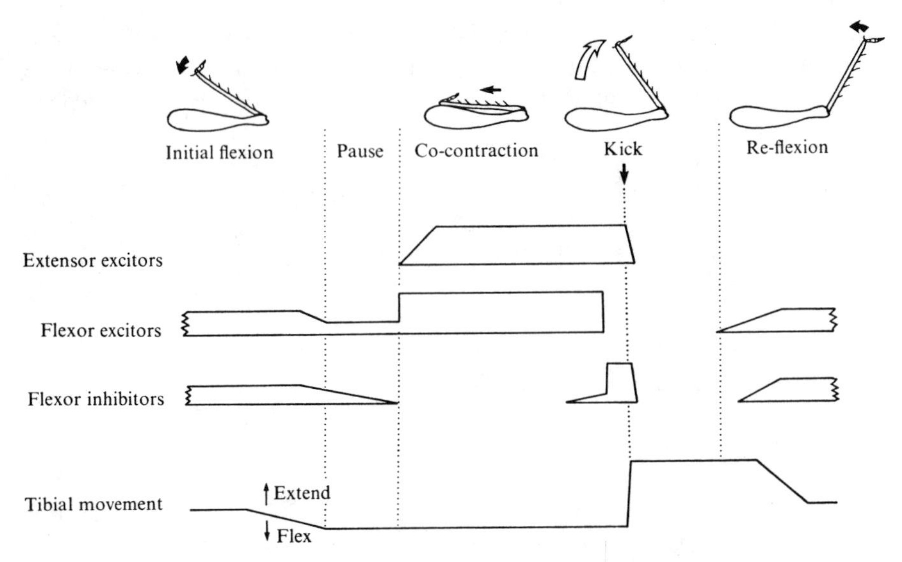

Fig. 7.14. A summary diagram showing the pattern of activity in various motoneurons in a defensive kick by a locust. The frequency of firing of the neurons is indicated by the height of the activity block. A similar motor program is shown during jumping. (From Heitler and Burrows, 1977*a*.)

7.6.3 Sensory Inputs and Neural Circuits

Both sensory inputs and neural circuits act to ensure that the various phases of the jump motoneuron program are appropriate, coordinated, and safe. Of the various types of sense organs that are involved, some fire in response to the position of associated structures and others respond to strain of the cuticle (Heitler and Burrows, 1977b); their properties are summarized below.

When the leg is fully flexed, the tibia fits into a groove on the ventral side of the femur: into this groove protrudes Brunner's organ, a soft cuticular tubercle associated with three sensory hairs and two campaniform sensilla (Fig. 7.15). When Brunner's organ is stimulated mechanically, a reflex burst of excitatory postsynaptic potentials is seen in the ipsilateral FETi motoneuron. Though these potentials are not large enough to cause spiking in the FETi, they are capable of adding to other inputs and so are possibly significant in elevating the rate of spiking of the FETi that occurs after leg flexion preceding the jump (Fig. 7.14). Locusts can, however, jump if Brunner's organ is removed (Heitler and Burrows, 1977b). Brunner's organ is also implicated in detecting leg folding after ecdysis but there is evidence of a backup reflex controlled by tibial receptors (Hughes, 1978).

The receptor beside Heitler's lump (Figs 7.2, 7.15; Section 7.3.1) is capable of monitoring active leg flexion. It appears to be stimulated only by the flexor apodeme when in the extreme flexed position and with the flexor tibiae muscle contracted. The apodeme, near its origin on the tibia, bifurcates and the

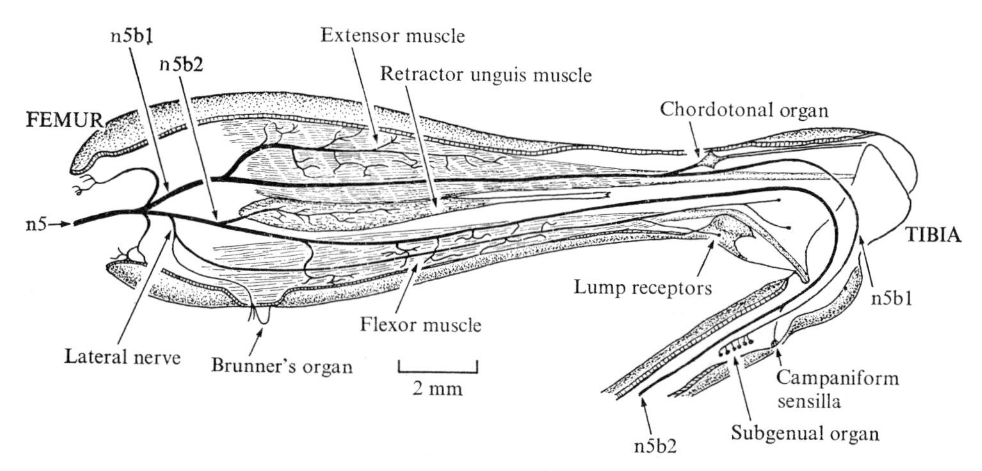

Fig. 7.15. Diagram of the principal muscles, nerves, and sense organ of the hind femur of the locust *Schistocerca gregaria.* The role of the sense organs in control of the motor program is discussed in the text. n5 is the fifth nerve, numbered from the front, arising from the metathoracic ganglion. n5b1 and n5b2 are major branches of this nerve. (From Heitler and Burrows, 1977b.)

posterior arm of the fork, in the fully flexed position, lies directly on the receptor. The effects of stimulation of this sense organ on activity in the flexor tibiae muscle are hard to interpret; electrical stimulation of the lateral nerve leading from the lump and joint receptors usually leads to excitatory post-synaptic potentials in the FlTi motoneurons but sometimes to inhibitory postsynaptic potentials. Whether the postsynaptic potentials are excitatory or inhibitory appears to depend on the state of excitation of the locust. If the lateral nerve (Fig. 7.15) is cut, the locust will not then jump but it will flex its leg and proceed to the co-activation of the FETi and FlTi motoneurons. Heitler and Burrows, (1977b) suggest tentatively that the stimulation of the lump receptor leads to excitation of interneurons in turn leading to inhibition of the *flexor tibiae* muscle, and thus to the release of the flexor catch (Heitler, 1974) (Section 7.3.1) and a successful jump.

The activity of the motoneurons of the *extensor tibiae* is partly controlled by activity of the subgenual organ in the tibia (Fig. 7.15). When the proximal tibia is pushed toward the femur, this stresses the subgenual organ in a manner similar to that occurring prior to a jump and leads to excitation of the FETi. Reexcitation of the FETi also occurs when the tibia is fixed at 90° to the femur and this can lead to a train of FETi spikes and a strong push by the tibia; these appears to result from positive feedback from the subgenual organ to the FETi and SETi via a reexcitation reflex (Heitler and Burrows, 1977b).

A rather similar — if curious — reflex results in excitatory postsynaptic potentials in the FETi but when the tibia is fully extended forcibly. This reflex is not mediated via the apodemes or tibial receptors for it can be elicited by

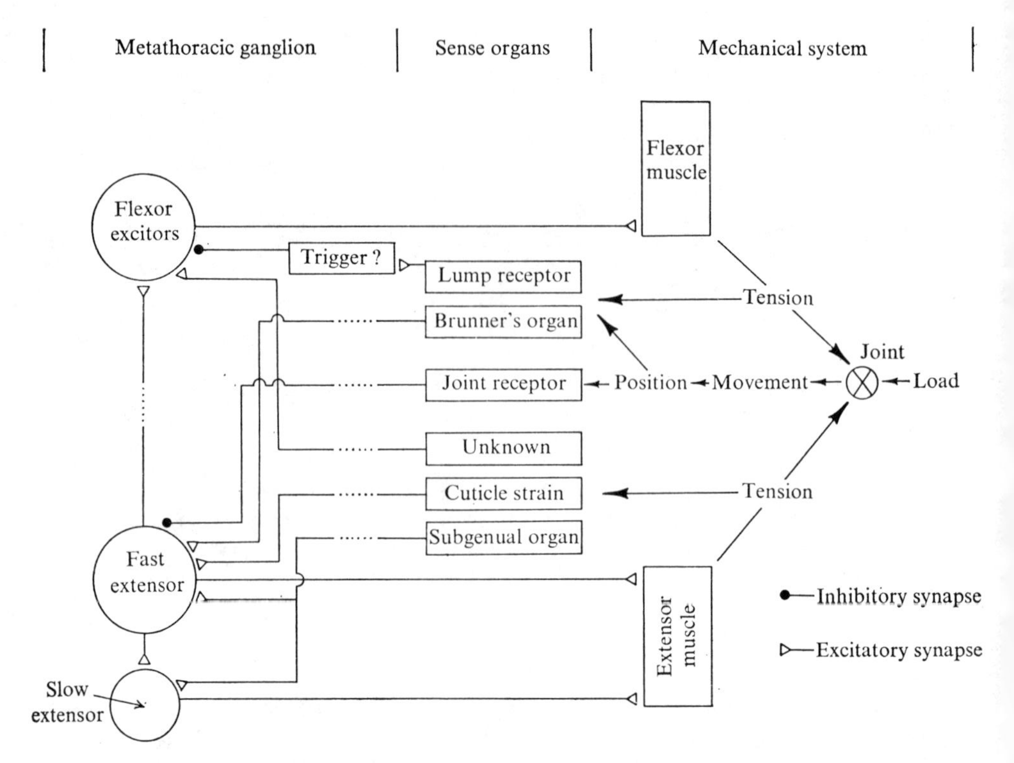

Fig. 7.16. Summary diagram showing the principal sense organs, neural circuits, and effector organs involved in the jump of locusts. Details of the more important pathways are given in the text. Dotted lines indicate the probable existence of interneurons. (From Heitler and Burrows, 1977b.)

distorting the distal end of the femur after removal of the entire tibia. Heitler and Burrows (1977b) suggest that forcible extension of the tibia provides a similar and mechanical stress in the femoral cuticle to that occurring when the extensor tibiae is actively contracting prior to the jump.

In addition to the sensory inputs, there are important central connections between groups of excitatory motoneurons. One such between the FETi and SETi leads to excitatory postsynaptic potentials in the SETi following spikes in the FETi motoneuron (Heitler and Burrows, 1977b). This reflex recruits the SETi in the early stages of the contraction of the extensor tibiae muscle, which has "catch" properties that maintain the tension after the cessation of stimulation (Wilson and Larimer, 1968; Bennet-Clark, 1975) and so may serve both to increase the muscle tension and to maintain this tension after the trigger activity that releases Heitler's catch (Heitler, 1974; Sections 7.3.1; 7.6.2)

Probably the most important central connection, however, is the excitatory connection whereby spikes in the FETi cause excitatory postsynaptic potentials in various flexor motoneurons. These postsynaptic potentials appear to be mediated by chemical synapses and have the important effect of causing the FlTi motoneurons to spike faster when the extensor tibiae muscle is stimulated (Hoyle and Burrows, 1973; Heitler and Burrows, 1977b), thereby ensuring that the leg does not extend but, instead, that energy is stored.

By and large, the pattern of sensory inputs and the central interconnections are understood and a model that describes their roles can be made (Fig. 7.16).

7.7 CONCLUSIONS

Jumping is a means of locomotion that confers great advantages on its possessors. Jumping is not without its penalties, however. In the case of grasshoppers, the hind limbs must be specialized to the point where they can only be used rather clumsily and slowly in walking.

Because the energy for the jump must be stored before the jump, the leg must be not only designed to store this energy in an efficient manner but also specialized so that the energy can be released effectively, which is accompanied by safety factors that are so low as to be positively alarming to human engineers. On the credit side, the weight penalty due to the energy stores is trivial: those of the two hind legs of a locust weigh less than 0.5% of the body weight (Section 7.3.2) and enable the animal to achieve speeds up to 10X those typical of running insects.

The skeletal specializations of the jumping mechanism are associated with extreme specialization of the neuromuscular system to ensure that the required energy is stored as fast as possible and then released without damage to either skeleton or muscles.

The process of jumping in grasshoppers is, among locomotor tasks, relatively well understood and forms a useful model for the mechanical and neuromuscular requirements of jumping in other animals.

REFERENCES

Albrecht, F. O. 1953. *The Anatomy of the Migratory Locust.* Univ. of London Press, London.

Alexander, R. McN. 1973. Muscle performance in locomotion and other strenuous activities. In L. Bolis, K. Schmidt-Nielsen, and S. H. P. Maddrell, Eds., *Comparative Physiology: Locomotion, Respiration, Transport and Blood.* North-Holland Publ., Amsterdam, pp. 1–21.

Alexander, R. McN. 1981. Factors of safety in the structure of animals. *Sci. Prog. (Oxford)* **67**, 109–130.

Alia, E. E. and A. Crovetti. 1961. Alcune indagini fotoelastografiche sul femore e la tibia delle zampe metatorariche di *Anacridium aegytium* (L.) (Orthoptera, Cantanopidae). *Studi sassar., Sez. 3* **9**, 517–543.

Bennet-Clark, H. C. 1975. The energetics of the jump of the locust *Schistocerca gregaria*. *J. Exp. Biol.* **63**, 53–83.

Bennet-Clark, H. C. 1976. Energy storage in jumping insects. In H. R. Hepburn, Ed., *The Insect Integument.* Elsevier, Amsterdam, pp. 421–443.

Bennet-Clark, H. C. 1977. Scale effects in jumping animals. In T. J. Pedley, Ed., *Scale Effects in Animal Locomotion*. Academic Press, London, pp. 185–201.

Bennet-Clark, H. C. 1980. Aerodynamics of insect jumping. In H. Y. Elder and E. R. Trueman, Eds., *Aspects of Animal Movement*. Cambridge Univ. Press, London and New York, pp. 151–167.

Bennet-Clark, H. C. and G. M. Alder. 1979. The effect of air resistance on the jumping performance of insects *J. Exp. Biol.* **82**, 105–121.

Brown, R. H. J. 1963. Jumping arthropods. *Times Sci. Rev., Summer 1963*, pp. 6–7.

Brown, R. H. J. 1967. Mechanism of locust jumping. *Nature (London)* **214**, 939.

Cochrane, D. G., H. Y. Elder, and P. N. R. Usherwood. 1972. Physiology and ultrastructure of phasic and tonic skeletal muscle fibres in the locust *Schistocerca gregaria*. *J. Cell Sci.* **10**, 419–441.

Gabriel, J. M. 1983. The development of the locust jumping mechanism. D.Phil. Thesis, Oxford University, England.

Gabriel, J. M. 1984. The effect of animal design on jumping performance. *J. Zool.* **204**, 533–539.

Gabriel, J. M. 1985a. The development of the locust jump mechanism. I. Allometric growth and its effect on jumping performance. *J. Exp. Biol.* **118**, 313–326.

Gabriel, J. M. 1985b. The development of the locust jump mechanism. II. Energy storage and muscle mechanics. *J. Exp. Biol.* **118**, 327–340.

Heitler, W. J. 1974. The locust jump: Specialisation of the metathoracic femoral-tibial joint. *J. Comp. Physiol.* **89**, 93–104.

Heitler, W. J. and M. Burrows. 1977a. The locust jump. I. The motor programme. *J. Exp. Biol.* **66**, 203–219.

Heitler, W. J. and M. Burrows. 1977b. The locust jump. II. Neural circuits of the motor programme. *J. Exp. Biol.* **66**, 221–241.

Hoyle, G. 1955. Neuromuscular mechanisms of a locust skeletal muscle. *Proc. R. Soc. London, Ser. B* **143**, 343–367.

Hoyle, G. 1958. The leap of the grasshopper. *Sci. Am.* **198**(1), 30–35.

Hoyle, G. and M. Burrows. 1973. Neural mechanisms underlying behavior in the locust *Schistocerca gregaria*. I. Physiology of identified motorneurones in the metathoracic ganglion. *J. Neurobiol.* **4**, 3–41.

Hughes, T. D. 1978. Physiological studies of ecdysis in locusts. D.Phil. Thesis, Oxford University, England.

Huxley, A. F. and R. Niedergerke. 1954. Structural changes in muscle during contraction. *Nature (London)* **173**, 971–973.

Ker, R. F. 1977. Some mechanical and structural properties of locust and beetle cuticle. D.Phil. Thesis, Oxford University, England.

Neville, A. C. 1965. Chitin lamellogenesis in locust cuticle. *Q. J. Microsc. Sci.* **106**, 269–286.

Neville, A. C. 1975. *Biology of the Arthropod Cuticle*. Springer-Verlag, Berlin.

Snodgrass, R. E. 1929. The thoracic mechanisms of a grasshopper and its antecedents. *Smithson. Misc. Collect.* **82**(2), 1–111.

Snodgrass, R. E. 1935. *Principles of Insect Morphology*. McGraw-Hill, New York.

Vogel, S. 1965. Studies on the flight performance and aerodynamics of *Drosophila*. Ph.D. Thesis, Harvard University, Cambridge, Massachusetts.

Wainwright, S. A., W. D. Biggs, J. D. Currey, and J. M. Gosline. 1976. *Mechanical Design in Organisms*. Edward Arnold, London.

Weis-Fogh, T. 1956. Tetanic force and shortening in locust flight muscle. *J. Exp. Biol.* **33**, 668–664.

Weis-Fogh, T. 1960. A rubber-like protein in insect cuticle. *J. Exp. Biol.* **37**, 889–907.

Weis-Fogh, T. and R. McN. Alexander. 1977. The sustained power output available from striated muscle. In T. J. Pedley, Ed., *Scale Effects in Animal Locomotion.* Academic Press, London, pp. 511–525.

Wilson, D. M. and J. L. Larimer. 1968. The catch properties of ordinary muscle. *Proc. Natl. Acad. Sci. U.S.A.* **61**, 909–916.

8

Hormonal Control of Flight Metabolism in Locusts

G. J. GOLDSWORTHY

8.1 INTRODUCTION

Those insects that can fly are often rather good at it, not only in that they may be particularly acrobatic, but also because they may fly for extraordinarily long distances. Locusts are good examples of the latter. Here we concentrate on the ability of migratory locusts to supply their muscles with sufficient quantities of fuel to support flights of some duration.

Migratory stages of many animals adopt similar strategies in terms of fueling long-term locomotor activities; thus birds, fish, and insects use fats as the major fuel during their migrations (see Goldsworthy and Cheeseman, 1978; Goldsworthy, 1983; Goldsworthy and Wheeler, 1984). The initial fuel is usually carbohydrate because it is readily available, both in the form of muscle glycogen deposits and as blood sugar. However, there are clear advantages in using fats in terms of weight of stored fuel and energy yield per stored weight. It appears therefore to be a common strategy to switch subsequently to the use of fats as quickly as possible. In running for a bus our leg muscles use glycogen and blood glucose as fuels, but athletes switch quickly to fat as a fuel during marathon running. One effect of training is to improve the speed of this change in fuel, and this may be true also for laboratory "trained" locusts (Jutsum et al., 1982). During the first 3–4 min of sustained flight in locusts, both muscle glycogen and carbohydrate in the hemolymph are used at a very high rate. After this there is gradual decline in the use of sugars (matched by an increase in the use of fats) until by 30 min or so their use is minimal and that of fat is maximal (see Goldsworthy, 1983).

Although locusts conform to a common strategy seen in many migratory vertebrates, perhaps as one might expect, there are important differences in detail relating to the nature of the fuels made available to the flight muscles in the hemolymph (Bailey, 1975). The major blood "sugar" of most insects is trehalose, a disaccharide of glucose, and fats are mobilized usually in the form of diacylglycerides from stores of triacylglycerides in the fat body. These features appear to be functionally related to the insects' possession of an "open" circulatory system — hemolymph is not directed to the tissues in blood vessels to any large extent, but circulates about the body almost in a "tidal flow" manner caused by the muscular movements of the insect itself. Consequently entry of metabolites into tissues depends on diffusion over relatively long distances compared with those in the "closed circulation" of vertebrates. To facilitate rapid diffusion, the concentrations of metabolites in insect hemolymph are thus characteristically higher than those in vertebrate blood: total carbohydrate may be 30 times that of glucose in human blood — a glucose level that would be fatal! The use of trehalose, a nonreducing sugar to which the tissues are not freely permeable, is an adaptation toward the requirements for high concentrations of carbohydrate: tissues take up trehalose only after hydrolyzing it to glucose (its rate of entry being determined by the activity of a membrane-bound enzyme, trehalose, which hydrolyzes the disaccharide to two molecules of glucose), and its contribution to the total osmotic pressure of the hemolymph is only half that of an equivalent number of free glucose molecules. Similarly, the high concentrations of diacylglycerides supplying the flight muscles could not be tolerated as an equivalent amount of fatty acid as carried in vertebrate blood. The metabolic "cost" to insects for transporting such neutral lipids to support energy metabolism is the need to employ special carrier lipoproteins for these water-insoluble molecules (see Section 8.5.2).

Fortuitously, the pattern of fuel utilization in locusts described above is reflected by changes in the concentrations of total carbohydrate and fat in the hemolymph during flight (see Goldsworthy, 1983). Progress in understanding the involvement of hormones in locust flight metabolism has been speeded enormously by the ease with which such changes in hemolymph metabolites can be measured and interpreted. This, together with the discovery in the late 1960s that a decapeptide neurohormone, adipokinetic hormone (AKH), is involved in the mobilization of fuels during flight has meant that we now have a clearer picture of the hormonal control of flight metabolism in the locust than in any other insect.

8.2 WHY ARE HORMONES INVOLVED IN THE REGULATION OF FLIGHT METABOLISM?

Increased energy utilization by muscles requires the availability of suitable metabolic fuels. If these are not readily available, they must be made so by mobilization from elsewhere (see Bailey, 1975; Candy, 1985). The fat body is not innervated by neurons from the central nervous system, so it is by the use of hormones, carrying the chemical message that mobilization of stored energy reserves is required, that glycogen or triacylglycerol breakdown in the fat body is stepped up. The stimulus for hormone release does not originate from the muscles, but from the central nervous system (see Orchard, 1987). We know only a little of these aspects of the endocrinology of flight, and the major part of this chapter is concerned with the roles of hormones in regulating fuel supply to the flight muscles.

Hormones may also serve to direct some aspects of muscle metabolism. Not only do hormones make energy substrates available to the muscles, but they play a part in regulating substrate utilization in the flight muscle, and perhaps in influencing the power output of the muscles (Candy, 1978; Whim and Evans, 1988). Consideration of some of these features of hormone action forms a later section of this account (see Section 8.4.2).

8.3 WHICH HORMONES ARE INVOLVED IN FLIGHT METABOLISM?

8.3.1 Peptides

At least two adipokinetic hormones (see Fig. 8.1) are synthesized in the glandular lobes of the corpora cardiaca of locusts (see Hekimi and O'Shea, 1987). Both the decapeptide, AKH-I, and either one (depending upon locust species) of two octapeptides, AKH-IIL or AKH-IIS (Goldsworthy and Wheeler, 1988), are thought to be released by a mechanism involving the catecholamine octopamine during flight in locusts (Orchard, 1987; but see Konings et al., 1988). The decapeptide and the octapeptides each release diacylglycerols (Goldsworthy et al., 1986a,b) from the fat body (see Section 8.4.1) to provide fuel for the flight muscles during long-term flight, but AKH-I is more potent than AKH-II (Goldsworthy et al., 1986a,b). At the flight muscles, AKH-I has a second direct action in stimulating the oxidation of diacylglycerols (see Section 8.4.2). These neuropeptides also bring about a reorganization of circulating hemolymph lipoproteins (see Section 8.5.2), and activate fat body glycogen phosphorylase. Again, AKH-I is more potent than AKH-II in these latter activities (Goldsworthy et al., 1986a,b). However, although both peptides are thought to act on the fat body by stimulating adenylate cyclase activity (see Section 8.4.1), AKH-II increases fat body cAMP levels to a greater extent than AKH-I (Goldsworthy et al., 1986a), suggesting that the two hormones act at different receptors and that AKH-II has other, as yet unknown, functions.

8.3.1.1 Adipokinetic Structure–Activity Requirements

In early pioneering studies, Stone et al. (1978) used a series of synthetic peptide analogues of AKH-I to show that for biological activity peptides must be blocked by L-pyroglutamic acid (cyclized L-glutamic acid) at the *N*- terminal, and by an amidated threonine residue at the *C*-terminal, and they must be between 8 and 10 amino acids long; substitutions in the area of residues 6 to 8, which were assumed to interfere with a β turn at position 6 (proline), reduced biological activity. It is now clear that AKH-I and AKH-II are members of a large family of neuropeptides (peptides produced in the nervous system) of which more than 10 are known currently (see Fig. 8.1). Recently, we used all the naturally occurring analogues of AKH found in different insects that were available to us to reinvestigate structure–activity relationships in this hormone family (Goldsworthy et al., 1986a,b; Wheeler et al., 1988). Our results are summarized in Figure 8.2 and have confirmed and extended the earlier findings of Stone et al. (1978), but most importantly have shown that although certain peptides elicit "full" or maximum possible responses, others elicit only attenuated adipokinetic (lipid-mobilizing, see Goldsworthy and Wheeler, 1986b) or hypertrehalosemic (trehalose-mobilizing, see Gade, 1986; Wheeler et al., 1988) responses.

Figure 8.1 shows the structures of naturally occurring members of the AKH family of insect peptides. We can conclude that in natural peptides of more than eight residues, position 10 is threonine and position 9 is glycine. Position 8 is tryptophan in all known peptides of this family. Position 7 is either asparagine or glycine, except in *Manduca* AKH, where it is serine.

Position 6 is proline in most of the natural peptides, the exceptions being the AKH-IIs of locusts and of a grasshopper (*Romalea*), where it is alanine in *Locusta* and threonine in *Schistocerca* and *Romalea*, and the AKH of *Manduca*, in which it is serine. Position 5 is either threonine or serine. In all the AKH peptides, position 4 is phenylalanine. Position 3 is either asparagine

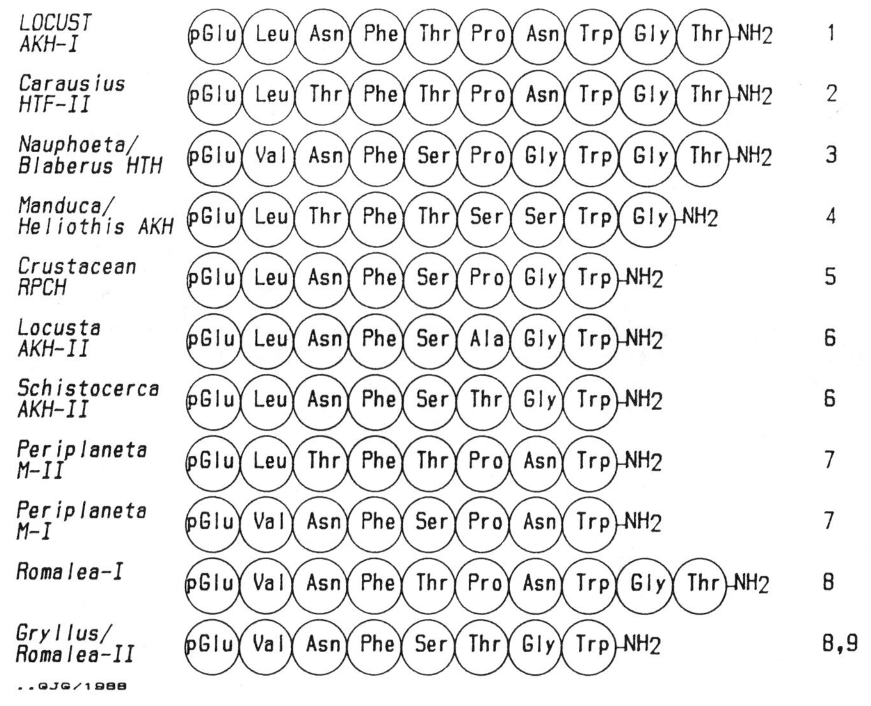

Fig. 8.1. The amino acid sequences of some members of the AKH/RPCH family of arthropod neuropeptides. Numbers at right show source of information: 1, Stone et al. (1978); 2, Gäde and Rinehart (1987a); 3, Hayes and Keeley (1986), Gade (1987); 4, Ziegler et al. (1985), Jaffe et al. (1986); 5, Fernlund and Josefsson (1972); 6, Gäde at al. (1984, 1986), Siegert et al. (1985); 7, O'Shea et al. (1984); 8, G. Gäde (personal communication); 9, Gäde and Rinehart (1987b), G. Gäde (personal communication).

or threonine. Position 2 is either leucine or valine. Position 1 is always pyroglutamate.

The potencies of drugs and hormones can be expressed in terms of their ED_{50} values — the amount needed to achieve a half-maximal response (see Table 8.1). When lipid-mobilizing activity is measured in the locust, AKH-I is the most potent peptide so far discovered and has an ED_{50} of 1 pmol. At doses above 3 pmol it gives what we regard as the maximum possible or "full" adipokinetic activity (ED_{max} = 3 pmol). This maximum rate of lipid mobilization possible has been used as a standard against which the effects of other peptides are measured. *Carausius* peptide (C-II) has only a single residue change of threonine instead of asparagine at residue 3. This doubles the ED_{50} from 1 to 2 pmol, and the ED_{max} increases so that doses in excess of 8 pmol are required for a "full" response.

Among the RPCH, M-I, and M-II octapeptides (Table 8.1), various substitutions have little effect overall. These octapeptides are all relatively "weak"

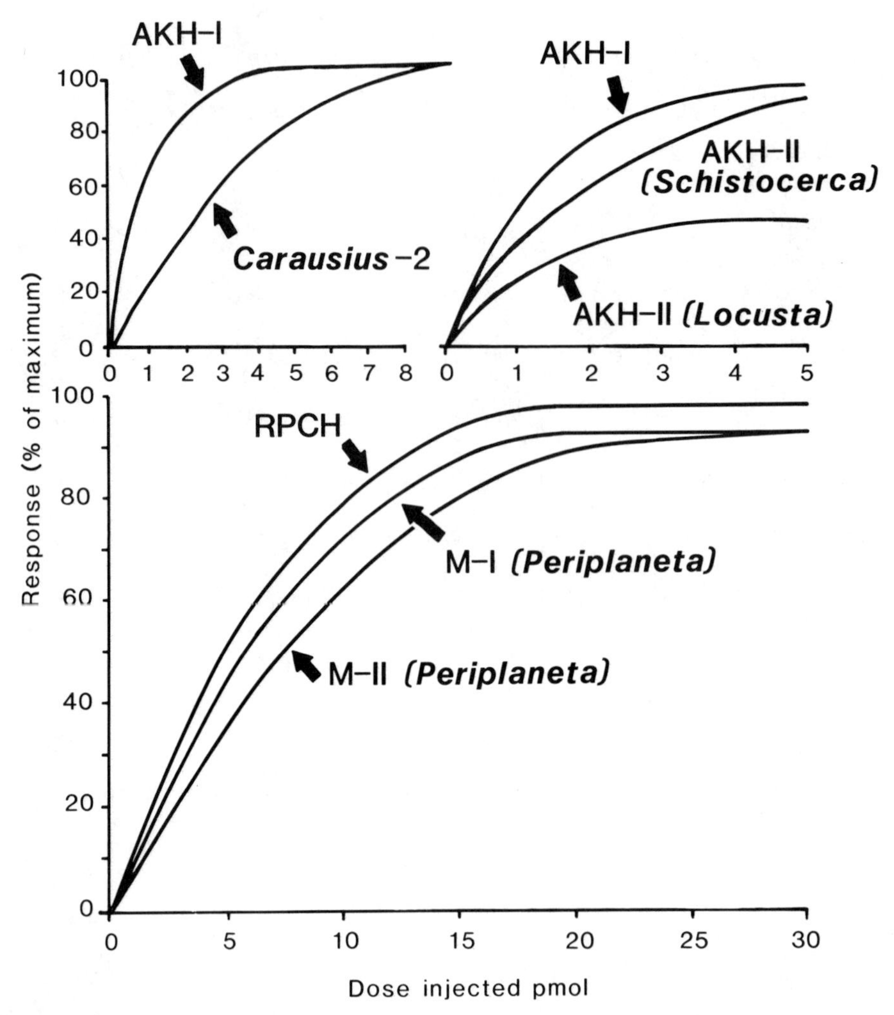

Fig. 8.2. Representative dose–response curves for the lipid-mobilizing activities of various naturally occurring neuropeptides. The responses were calculated as a percentage of the maximum response obtained to a crude extract of *Locusta* corpora cardiaca (0.02 pair equivalents/locust) tested on the same batch of locusts on the same day. Note the truncated response to *Locusta* AKH-II; this is seen even at doses up to 50 pmol/locust.

compared with AKH-I with ED_{50} values around 5 pmol; they do show "full" adipokinetic activity, but their ED_{max}'s are between 20 and 30 pmol.

Octapeptides lacking proline are of particular biochemical interest because of the prediction of a β turn at proline in AKH-I. This prediction is effectively an assessment of the probability of a turn calculated from the particular sequence of amino acids present, and is based on experience gained from studies of secondary structures in proteins (Chou and Fasman, 1974). It is

TABLE 8.1. The Relative Potencies of Members of the AKH/RPCH Family of Arthropod Neuropeptides in Causing Hyperlipemia in *Locusta* and, Where Indicated, in *Schistocerca*[a].

Neuropeptide	ED_{50} (pmol)	ED_{max} (pmol)	Maximum Response (%)
Locust AKH–I	1	3	100
Locust AKH–I	6	20	100[b]
Carausius HTF–II	2	8	100
Manduca/Heliothis AKH	10	40	45
Crustacean RPCH	4.6	20	100
Crustacean RPCH	15	—	—[b]
Locusta AKH–II	2	3	60
Locusta AKH–II	20	30	55[b]
Schistocerca AKH–II	2	5	95
Schistocerca AKH–II	12	25	60[b]
Periplaneta M–I	5	30	95
Periplaneta M–II	5	20	95

[a] The responses are calculated as a percentage of the maximum response obtained to a crude extract of *Locusta* corpora cardiaca (0.02 pair equivalents/locust) tested on the same batch of locusts on the same day (from Goldsworthy et al., 1986b, and unpublished observations). In our hands, with our strains of locusts, *Schistocerca* appears very much less sensitive than *Locusta* to those peptides tested. The dashes indicate that there are no data.
[b] Assayed in *Schistocerca*.

problematic whether such analyses are valid for short peptides but, when they are applied to the known peptides of the AKH family, all predict a high probability of a turn around position 6, except in those peptides lacking proline (G. J. Goldworthy, unpublished observations). Using circular dichroism (CD) spectroscopy, we have shown (A. Drake, G. J. Goldsworthy, C. H. Wheeler, and G. Gäde, unpublished observations) that in aqueous solution at room temperature none of the naturally occurring neuropeptides of the AKH family has a CD spectrum characteristic of a β turn. However, on addition of SDS (sodium dodecyl sulfate) micelles or liposomes (both of which act as mimics of biological membranes in this system) the CD spectra of AKH-I, RPCH, MI, and MII change to ones characteristic of a type-I β turn. Thus we anticipate that the peptides take up a type-I β turn configuration when binding to their receptors. None of the naturally occurring locust or grasshopper octapeptides, AKH-IIL, AKH-IIS, Ro-II, or *Manduca* AKH, show such changes in their CD spectra; they have no identifiable CD spectrum with or without liposomes or SDS micelles. Those members of the AKH family that readily take up a β turn configuration can show "full" hyperlipemic and hypertrehalosemic activity when tested at sufficiently high doses. When tested intraspecifically, the octapeptides for which the CD analysis shows no evidence of a β turn all elicit attenuated hyperlipemic and hypertrehalosemic responses. Unfortunately, the situation is not completely clear-cut: although AKH-IIL gives an attenuated response, even at doses greater than 50 pmol, whether tested in *Locusta* or in *Schistocerca*, AKH-IIS gives an attenuated response when tested intraspecifically in *Schistocerca*, but gives a full adipokinetic response (see Fig. 8.2) in *Locusta* (Goldsworthy and Wheeler, 1986b). Apparently, the presence of proline or of a β turn is not as essential for *full* adipokinetic activity (comparable with that seen in response to AKH-I) as we had suggested previously (Goldsworthy et al., 1986b).

The data indicate that AKH receptors on the fat body of *Locusta* can discriminate between the locust AKH-II octapeptides, whereas those of *Schistocerca* cannot. When injected intraspecifically, the two locust AKH-II octapeptides do not show "full" activity but give truncated responses. *Locusta* AKH-II is potent in comparison with AKH-I ($ED_{50} \simeq 2$ pmol), but only gives about 60% of the maximum response in *Locusta* even at doses up to 50 pmol. *Schistocerca* AKH-II is less potent in *Schistocerca* (ED_{50} = 12 pmol) and gives a truncated response. On the other hand, *Schistocerca* AKH-II is very potent in *Locusta* (ED_{50} = 2 pmol) and gives a "full" adipokinetic response. *Manduca* (and *Heliothis*) AKH is relatively inactive in *Locusta*, having an ED_{50} of 10 pmol and giving a truncated response even at doses up to 75 pmol.

Attempts to saturate the adipokinetic receptors in the fat body of *Locusta* with a large excess of AKH-IIL, and thus prevent any further response to AKH-I, have not been successful (Goldsworthy et al., 1986a). Nevertheless, when members of the AKH family are tested for their ability to mobilize lipids in *Manduca*, many peptides, although being relatively potent in themselves, give an attenuated response compared with that to *Manduca* AKH itself. In this case, however, some of the peptides, but not all, do act as antagonists for *Manduca* AKH (S. E. Reynolds and A. M. Fox, personal communication).

8.3.1.2 Evolutionary Considerations

The high degree of conservation of amino acid residues in the AKH peptides is a major reason for assuming that they belong to a single peptide family. But is the evidence for this totally convincing? The concept of peptide families assumes that a primary ancestral peptide can be identified from which all members of the family could have evolved. The precise codons used for each amino acid in the various insects are not yet known, but theoretical routes for single point mutations to change from one amino acid residue to another can be worked out. For example, the variations at residue 2 (valine or leucine) can be explained by three possible routes, and single point mutations can explain the variability at residues 5 (threonine or serine) and 6 (proline, threonine, or alanine). However, the variation at residue 7 (asparagine, glycine, or serine) is intriguing, because there is no route between asparagine and glycine, except via the sequence seen in *Manduca* AKH (which has serine at residue 7). Although this appears an acceptable situation on molecular biological grounds, few zoologists would view the highly specialized lepidopterans as likely candidates to provide evolutionary links between crustaceans and insects, or between locusts and cockroaches! No firm conclusions can be made about the evolutionary relationships of these peptides at present. We urgently need more information concerning the structures of the precursor molecules for these peptides, and for the structures of AKH peptides from a wider range of insect groups.

8.3.2 Octopamine

In *Schistocerca*, titers of hemolymph octopamine increase during flight: only the D enantiomer of octopamine is present, increasing fourfold in concentration within the first 10 min of flight, but then decreasing rapidly toward resting levels (Goosey and Candy, 1980). Octopamine is thus effectively present as a pulse between 5 and 15 min of flight initiation, and its overall role during flight must therefore be related to the very early stages of flight, rather than long-term events (see Section 8.6). Although octopamine can stimulate the release of diacylglycerols from the fat body during flight (see Orchard, 1987), Goldsworthy (1983) has argued that its major role may be in influencing metabolism in the flight muscle (see also Candy, 1989; Whim and Evans, 1988). Octopamine is present in high concentrations in various parts of the insect nervous system and the thoracic nerves supplying the flight muscles are likely candidates as sites of release, although this is not fully established (see Goldsworthy, 1983; Orchard, 1987; Wheeler, 1988).

Octopamine is a catecholamine related to noradrenalin of vertebrates, and may mediate a "fight or flight" response in locusts (see Orchard, 1987). The stress of handling certainly increases octopamine titers in the hemolymph, and this can cause a small but prolonged increase in the concentration of hemolymph lipid. However, the amounts of lipid mobilized in response to octopamine are very small and appear to be detectable only in nonstressed locusts that have been suitably "rested" and left undisturbed beforehand. Octopamine-mediated lipid mobilization is very slight in comparison with the maximum rates of mobilization possible in response to adipokinetic peptides (see Orchard, 1987). This mobilization does not involve the reorganization of hemolymph lipoproteins (Van Heusden et al., 1984) so characteristic of peptidic stimulation (see Section 8.5.2), perhaps not surprisingly in view of the small amounts of diacylglycerol released.

8.4 HOW DO THE HORMONES ACT?

8.4.1 Actions on the Fat Body

Both octopamine and the adipokinetic peptides stimulate adenylate cyclase activity in the fat body, and it is assumed that the pathways are similar to those involved in the activation of triacylglycerol lipase in vertebrates (see Goldsworthy, 1983; Orchard, 1987). The general pathways are illustrated in

Figure 8.3. Release of diacylglycerols from the fat body is assumed to work via an intracellular second messenger such as cAMP, whether stimulated by peptides or amines. This may activate a triacylglycerol lipase via a conventional protein kinase cascade (see Goldsworthy, 1983; Beenakkers et al., 1985; Orchard, 1987). It is possible that Ca^{2+}, acting as another second messenger, may also be involved because the lipid-mobilizing action of AKH on the fat body depends on Ca^{2+} in vitro (see Goldsworthy, 1983). Activation of the triacylglycerol lipase is thought to hydrolyze triacylglycerols to monoacylglycerols which are subsequently reacylated by a monoacylglycerol transferase to form stereospecific *sn*-1,2 diacylglycerols, and released from the fat body (see Beenakkers et al., 1985) as part of a carrier lipoprotein, which in AKH-injected or flying locusts is lipoprotein A^+ (see Section 8.5.2).

8.4.2 Actions on the Flight Muscles

8.4.2.1 Adipokinetic Hormones

The rates of fuel utilization (of locusts flying in the laboratory on flight mills) can be calculated from changes in the concentrations of hemolymph metabolites.

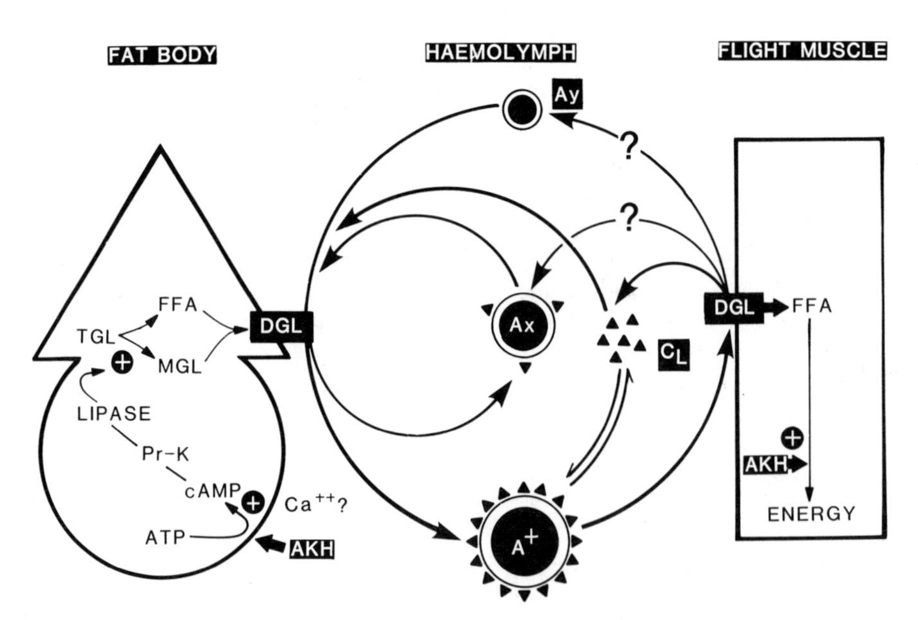

Fig. 8.3. Diagrammatic scheme of the major actions of AKH in locusts. Diacylglycerols (DGL), made available by the action of AKH in increasing triacylglycerol lipase in the fat body, are loaded onto lipoprotein A_{yellow} (Ay), which forms lipoprotein A^+, which reversibly binds C_L proteins. At the flight muscle, diacylglycerols associated with lipoprotein A^+ are hydrolyzed, and the fatty acids (FFA) are used to support flight muscle metabolism. Lipid unloading leads to the liberation of free C_L proteins. The lipoproteins thus form a reusable "shuttle," carrying energy to the flight muscles.

Such experiments have yielded a substantial body of circumstantial evidence that trehalose and diacylglycerols compete as fuels for the flight muscles at least during the first moments of flight. This evidence has been discussed previously at some length (Goldsworthy, 1983), and is not dealt with here.

Robinson and Goldsworthy (1977a) studied flight muscle metabolism directly by using half-thorax preparations of *Locusta.* These can be stimulated to contract in vitro, and the oxidation of radio-labeled substrates in such preparations used to determine substrate choice by the flight muscles: glucose or trehalose are good substrates for contraction, but carbohydrate utilization is decreased to about 50% by the addition of substrates containing hemolymph lipoproteins. This substrate competition is reversed by increasing the concentration of trehalose above that found in hemolymph. Addition of extracts of the corpora cardiaca containing AKH has no effect when only trehalose or trehalose plus lipoproteins from resting hemolymph are provided as substrate. However, if tissue extracts containing AKH (or purified natural AKH-1) are added to perfusions containing both carbohydrates and lipoproteins from AKH-injected locusts (i.e., lipoprotein A^+-containing substrates), then carbohydrate utilization is inhibited by about 90%, and this inhibition is noncompetitive because it cannot be relieved by even a doubling of the amount of trehalose present. Addition of AKH increases the rate of oxidation of lipids contained in lipoproteins prepared from AKH-injected locusts. This effect of AKH in stimulating the oxidation of lipids carried by lipoprotein A^+ may be due to a direct action on the mitochondria. This is possible because AKH-containing extracts not only increase the flux of lipids into the flight muscles but also stimulate the oxidation of palmitate and palmitoyl carnitine by isolated flight muscle mitochondria. The effect of AKH does not appear to be simply due to stimulation of lipid uptake, because prevention of lipid oxidation by 2-bromostearic acid also prevents increased lipid uptake (Robinson and Goldsworthy, 1977b).

How is the inhibition of glycolysis during AKH-stimulated oxidation of lipoprotein-bound diacylglycerols brought about? We know very little about this at the molecular level in terms of the direct actions of the hormone on the flight muscle, but we do have some clues as to the mechanism by which glycolysis may be regulated.

Fatty acids are transported into the mitochondria by a mechanism involving carnitine acyl transferase enzymes on both the inner and outer walls of the mitochondria. Bromostearic acid is an inhibitor of the outer mitochondrial transferase in vertebrates (Chase and Tubbs, 1972), and it relieves the noncompetitive inhibition of glycolysis observed in half-thorax preparations provided with trehalose and lipoprotein A^+ as fuels in the presence of AKH (Robinson and Goldsworthy, 1977a). This suggests that it is the formation of acyl carnitine and/or the increased movement of acyl groups into the mitochondria and their subsequent oxidation that is responsible for the inhibition of glycolysis. During increased oxidation of fatty acids in mammals, the rate of glycolysis is reduced by increased levels of citrate inhibiting phosphofructokinase, but in insects this enzyme is insensitive to citrate (see Goldsworthy, 1983). Instead, aldolase appears to be a regulatory enzyme of glycolysis in locusts (Ford and Candy, 1972), and Storey (1980) has demonstrated that this enzyme is sensitive to regulation by both citrate and palmitoyl carnitine. Certainly, acyl carnitine in the flight muscles increases fivefold in concentration during locust flight (Worm and Beenakkers, 1980; Worm et al., 1980). In mammals, 2-bromo derivatives of long-chain fatty acids inhibit the outer mitochondrial carnitine acyl tranferase, thus preventing further formation of acyl carnitine. Although existing acyl carnitine can pass into the mitochondrial matrix, further movement of acyl groups into the mitochondria is inhibited. Such inhibitors therefore lead to a reduction in intramitochondrial acyl carnitine, and thereafter the supply of acyl groups for the mitochondria declines. It seems that in both locusts and mammals, bromostearic acid can relieve the inhibition of glycolysis by intermediates of fatty acid metabolism.

The quality and quantity of lipid supplied to the flight muscles in vitro is of prime importance (see Goldsworthy, 1983). Substrates containing lipoprotein A^+ at concentrations equivalent to those found in the hemolymph inhibit glycolysis, but dilution of these substrates, even in the presence of AKH, relieves the inhibition proportionally. Furthermore, up to fourfold concentration of the resting substrate, lipoprotein A_{yellow}, does not increase the inhibition of glycolysis, even if AKH is added. The effect of lipoproteins on the inhibition of carbohydrate utilization in flight muscle thus depends not simply on the total amount of lipid, but also on the type of diacylglycerol-carrying lipoprotein available to the muscle. Candy (1978) was unable to show an effect of AKH on lipid utilization in flight muscle preparations, although he did show that octopamine and other compounds affected flight muscle metabolism. This problem may be related to the difficulties in producing preparations of hemolymph lipoproteins of suitable quality (see Goldsworthy, 1983).

It should be emphasized that the experimental conditions employed in these studies using half-thoraces in vitro are not truly representative of the situation in vivo. In the preparations in vitro the muscles are contracting only at 2 Hz, rather than about 20 Hz in flight, and in vivo the flight muscles never encounter energy substrates in the hemolymph in isolation. Indeed, hemolymph from resting, AKH-injected, and flown locusts always contains some carbohydrate (trehalose and free glucose) and mixtures of lipoproteins including intermediates in the transition from lipoprotein A_{yellow} to lipoprotein A^+) in different degrees of lipid-loading (Goldsworthy et al., 1985; G. J. Goldsworthy, unpublished observations). The conclusion from these studies in vitro that substrate competition at the flight muscles exists and is influenced by AKH is nevertheless valuable direct confirmation of earlier studies in which fuel utilization had been assessed in vivo in locusts that had been injected with carbohydrate or dipalmitin.

Although convenient to the investigator, dipalmitin is not an ideal substrate to use in vivo in such studies from a physiological standpoint. It is a relatively minor component of the natural diacylglycerol hemolymph pool and, because it is water-insoluble, it has to be injected as an emulsion. For these reasons, we attempted to provide further corroborative evidence in vivo for the effects of AKH on substrate utilization (Goldsworthy et al., 1979). A different approach was taken in that the concentrations of hemolymph metabolites were manipulated in vivo in a more indirect manner. Locusts were made to fly under their own propulsion on roundabouts for 30 min and then rested for 2 h before being flown for a further 30 min. When compared with locusts injected with saline, injection of a crude preparation of AKH just before flight caused a slight but significant reduction in the rate of disappearance of trehalose from the hemolymph during the first flight. But hormone injection before the second flight prevented completely any further decrease in trehalose concentration beyond that seen during the first flight. How can this be explained? In the first flight the locusts started with high resting levels of trehalose and low resting levels of diacylglycerol (carried mainly on A_{yellow}, but with some more highly lipid-loaded lipoproteins present: see Goldsworthy et al., 1985) — a situation very similar to a normal flight except that AKH was introduced (by injection) a few minutes earlier than normal. In the second flight, the locusts started with low levels of hemolymph trehalose, and high levels of hemolymph diacylglycerol (carried mainly as lipoprotein A^+). Associated with these changes in the rates of utilization of hemolymph trehalose were changes in the speed with which the locusts drove the roundabouts. These results were consistent with the hypothesis that when the hormone was administered prior to the first flight it induced an earlier utilization of diacylglycerol than normal. Thus fatty acid oxidation in the flight muscles was stimulated at a time when insufficient diacylglycerol was available in the hemolymph to support flight muscle energy expenditure fully. Glycolysis was nevertheless inhibited, and the locusts flew more slowly than normal whereas the mobilization of diacylglycerol gathered strength. When hormone was injected prior to the second flight, when levels of hemolymph diacylglycerol were still high (after the first flight, levels of hemolymph lipids continued to increase during rest because AKH persists in the hemolymph for some time), lipid oxidation powers the flight muscles, very little trehalose is oxidized, and flight speed is relatively constant. Overall, these studies support the view that the conclusions from the studies in vitro (using half thorax preparations) can be applied to the situation in vivo.

8.4.2.2 Octopamine and Other Factors

Circulating octopamine may play a role in determining substrate oxidation in the flight muscles. At concentrations comparable with those measured in the hemolymph during flight, octopamine stimulates the oxidation of a variety of substrates in working preparations of locust half thoraces (Candy, 1978): it increases also the twitch tension of the dorsal longitudinal muscle (Whim and Evans, 1988). It seems likely that octopamine works on the flight muscles via an activation of adenylate cyclase (Candy, 1978), but the exact mechanism is uncertain. The full physiological significance of these effects of octopamine on metabolism in the flight muscles is yet to be established. But Goldsworthy (1983) suggested that octopamine could play a vital role in the first 10–15 min of flight in maintaining the oxidation of glucose when at the same time increasing concentrations of diacylglycerol in the hemolymph are competing with declining levels of hemolymph trehalose. The increase in the oxidation of diacylglycerols in the flight muscles would be gradual, and would allow a smooth transition from the use of carbohydrate to the use of fats. In effect the process delays a "complete" switch to fats until amounts of diacylglycerol sufficient to support flight activity are made available.

8.4.3 Receptors

No information is available concerning the characterization of receptors for adipokinetic peptides. Progress in this area has been prevented by a lack of biologically active radio-labeled hormone analogues. Moreover, the pharmacology of AKH receptors is hardly explored; many agonists are known, but we know of no antagonist and none of the vertebrate α or β blockers that we have tested appear to be effective (G. J. Goldsworthy, unpublished observations). If progress is to be made in this area of insect molecular biology, there must be a substantial effort to identify and characterize the receptors for those hormones. The pharmacology of octopamine receptors in insects has been studied extensively, however (see Evans, 1985), and it appears that they are heterogeneous but resemble to varying extents the α receptors described in mammals.

8.5 HOW ARE FUELS TRANSPORTED TO THE FLIGHT MUSCLES?

8.5.1 Water-Soluble Fuels

Hemolymph trehalose is freely available to the flight muscles in solution. As such, it presents no immediate problems in terms of availability at the start of flight, nor in transport to the flight muscles. Other quantitatively minor fuels, such as amino acids, glycerol, and ketone bodies (see Goldsworthy, 1983), will of course be in the same category. It is intriguing that the trehalose-hydrolyzing enzyme, trehalase, in the flight muscles appears to be activated at the onset of flight (Candy, 1974); the mechanism remains unknown.

8.5.2 Water-Insoluble Fuels

Neutral lipids are essentially insoluble in water. Their transport through the hemolymph requires a special mechanism. As in vertebrates, this involves the utilization of special macromolecular particulate complexes called lipoproteins. The lipoproteins in locust hemolymph change in composition during flight (Mayer and Candy, 1967), and have been studied extensively by several groups (see Goldsworthy, 1983; Beenakkers et al., 1985; Chino, 1985; Wheeler, 1988).

Adipokinetic hormones bring about a major regrouping of hemolymph proteins and lipoproteins associated with an increase in the lipid-carrying capacity of the hemolymph (see Fig. 8.3). Essentially, a new lipoprotein particulate complex, A^+, forms during the action of adipokinetic peptides; A^+ particles, in comparison with those of A_{yellow} lipoprotein in resting blood, are of larger diameter (Wheeler et al., 1984a); they bind large quantities of other hemolymph proteins called C-apoproteins (Wheeler and Goldsworthy, 1983 a, b) to them; and they carry up to 18× more neutral lipid (see Goldsworthy, 1983). The formation of A^+ lipoprotein does not require protein synthesis de novo, and it is a reversible process, the lipoproteins forming a reusable shuttle, carrying diacylglycerols between the fat body and the flight muscles (see Fig. 8.3). In our laboratory we recognize two C-apoproteins in the hemolymph of *Locusta* (Goldsworthy et al., 1985), each of which binds to A^+ (Wheeler and Goldsworthy, 1986). In other laboratories, only a single (see Beenakkers et al., 1985; Chino, 1985) apoprotein has been described, but this appears broadly similar in chemical nature to the two that we have studied.

8.5.2.1 Lipid Unloading at the Flight Muscles

The mechanism by which the diacylglycerol is unloaded from the A^+ lipoprotein particles at the flight muscle appears to be specific and regulated indirectly by AKH. Locust flight muscles possess a membrane-bound lipoprotein lipase that hydrolyzes lipids carried as part of lipoprotein A^+ at over 4× the rate of those carried by A_{yellow}, and hydrolysis of lipids associated with A_{yellow}, is reduced by about 90% in the presence of lipoprotein A^+ (Wheeler et al., 1984b,1986; Van Heusden et al., 1986).

Vertebrate chylomicrons and VLDL (very low density lipoprotein) particles bind apoproteins, which play a role in the activation of membrane-bound lipoprotein lipase in the peripheral tissues (Nilsson-Ehle et al., 1980). By analogy, we envisaged that locust C-apoproteins could *activate* lipoprotein lipases, and were somewhat surprised, therefore, to find that locust C-apoproteins *inhibit* the membrane-bound flight muscle lipase. Nevertheless, this inhibition of lipoprotein lipase by C-apoproteins forms the basis of a mechanism, indirectly controlled by AKH, for the regulation of lipoprotein lipase during rest and flight (Wheeler and Goldsworthy, 1985). When the free (i.e., nonlipoprotein-bound) C-apoprotein concentration in the hemolymph is high, as in resting locusts, lipase activity is inhibited. During flight, or when AKH is injected into resting locusts, C-apoprotein inhibition of the lipase is relieved by the decrease in the concentration of free C-apoprotein consequent on the formation of lipoprotein A^+. Furthermore, if an excess of C-apoproteins is injected into locusts treated with AKH (90 min previously) before removal of the muscle tissue for assay, then the enzyme remains inhibited (Wheeler and Goldsworthy, 1985; Wheeler et al., 1986). Presumably, C-apoproteins are extracted bound to the muscle membrane preparations, and when the free concentration of these C-proteins in the hemolymph of test locusts is high, these bound proteins are responsible for inhibiting the lipase activity in vitro.

This mechanism of lipase regulation by the free C-apoprotein concentration ensures that rapid uptake of lipid by the flight muscles, and subsequent inhibition of trehalose oxidation by products of lipid oxidation, takes place only when lipoprotein A^+ (the preferred substrate) is present in sufficient quantities to sustain flight. The inhibition of flight muscle lipoprotein lipase by C-apoproteins is competitive at low concentrations but becomes mixed at higher concentrations of the apoproteins; both C-I and C-II apoproteins are effective (Wheeler et al., 1986; Wheeler and Goldsworthy, 1986).

The presence of lipoprotein lipase in locust flight muscle can be demonstrated via electron microscopy by a histochemical staining method (Fig. 8.4). This technique provides confirmation of our previous findings that the enzyme is membrane-bound, but also shows that the enzyme activity is restricted to the T-tubule system (see Fig. 8.5), and is therefore freely accessible to hemolymph-borne lipoproteins. Inhibition of enzyme activity in vitro in the presence of C-apoproteins, and activation by injection of AKH into donor locusts, can be confirmed by histochemical staining of membrane preparations obtained from flight muscles.

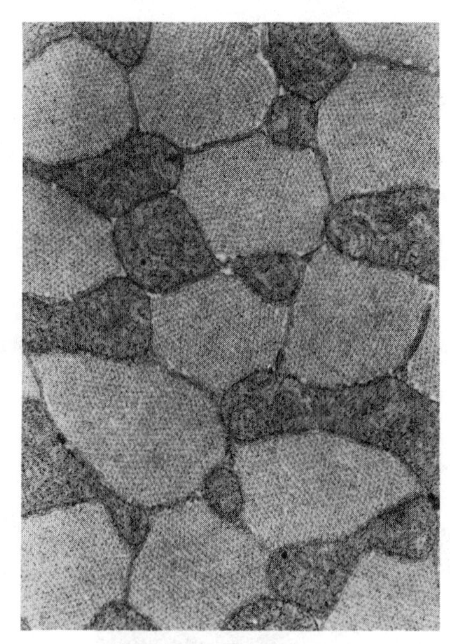

Fig. 8.4. Lipase activity in the flight muscles of locusts. Note the localization of the staining to the T-tubules (G. J. Goldsworthy, J. Mundy, and C. H. Wheeler, unpublished observations). Lipase staining is completely inhibited by addition of NaF to incubations (see also legend to Fig. 8.5). Staining methods were modifications of the techniques of Murata *et al.* (1968) and Murata and Nagata (1972).

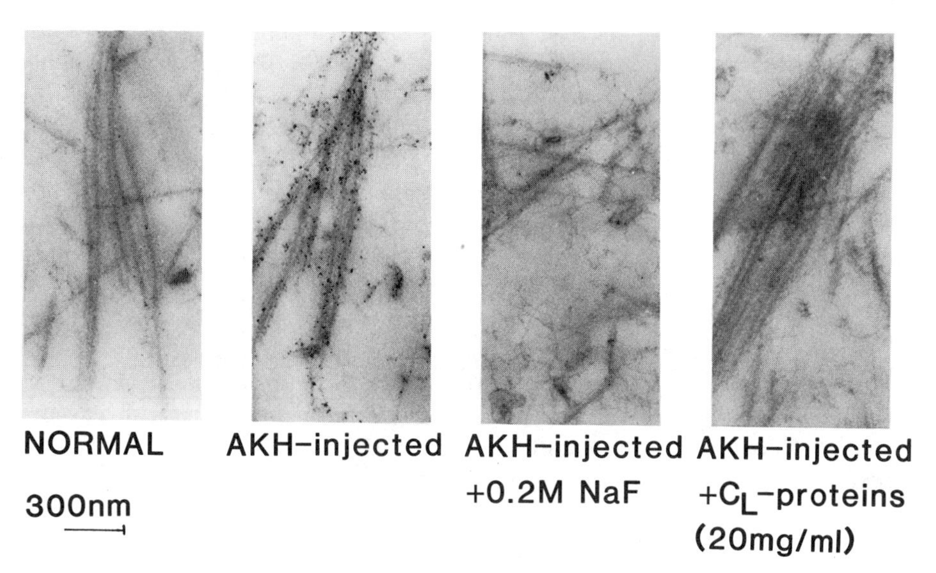

Fig. 8.5. Lipase activity in isolated membrane preparations from locust flight muscle. Note the increased lipase staining in preparations extracted from locusts injected with 2 pmol AKH-I 90 min before sacrifice; this enhanced staining is not seen when C_L proteins are injected just before sacrifice, or when an inhibitor of the enzyme such as NaF is included in the incubations (G. J. Goldsworthy, J. Mundy, and C. H. Wheeler, unpublished observations).

8.6 CONCLUSIONS

Parallels exist between hormonal and metabolic events that occur during flight in locusts and those in some exercising vertebrates (Goldsworthy and Cheeseman, 1978; Goldsworthy, 1983; Goldsworthy and Wheeler, 1984, 1986a). Thus increased muscular activity is fueled initially by fuels intrinsic to the muscles, and then by those from the "blood." But if activity is prolonged, the mobilization of fat stores is initiated and muscle energy metabolism is increasingly fueled by the oxidation of fatty acids. For these reasons, changes in the pools of tissue energy metabolites of a variety of animals can be superficially very similar during prolonged activity, or even during fasting. Both starvation and exercise impose a heavy metabolic stress, and the requirements of animals in these situations are essentially similar: they must avoid too rapid a depletion of their limited stores of immediately available fuels, and to do this they switch to the utilization of an alternative, more ergonomic, fuel such as lipid. There is thus a common metabolic strategy in migratory animals, from marathon runners, through birds and fish, to insects.

A major part of this strategy is that the muscles must reduce their utilization of carbohydrates *before these actually run out.* In effect, this is achieved by an inhibition of glycolysis in the "skeletal" muscles, an effect brought about indirectly by increased oxidation of fatty acids. Thus the provision and use of an alternative fuel provides the signal for conservation of carbohydrate. But why should animals need to conserve carbohydrate during chronic metabolic stress? We have insufficient information to be sure of the answer to this question for insects, but certainly in mammals some tissues have a continuous obligatory requirement for glucose. Nevertheless, the evidence shows that locusts reduce their utilization of hemolymph trehalose at a time when they have oxidized only about half that which was present before flight.

The broad operational strategy by which glycolysis is inhibited at times when the use of fatty acids is increased is essentially similar in mammals and locusts but, clearly, the details differ (see Section 8.4.2). Nevertheless, it is clear that in locusts the flight muscles are possible targets for a number of hormones which may act together to regulate the utilization of fuels during flight. These act both directly and indirectly in such a way as to effect a smooth changeover from the utilization of carbohydrate to fat as the major fuel, and prevent the possibility of "stalling". The strategy seems clear; the molecular details await further elucidation.

REFERENCES

Bailey, E. 1975. Biochemistry of insect flight. Part two. Fuel supply. In D. J. Candy, Ed., *Insect Biochemistry and Function*. Chapman & Hall, London, pp. 89–176.

Beenakkers, A. M. T., D. J. Van Der Horst, and W. J. A. Van Marrewijk. 1985. Biochemical processes directed to flight muscle metabolism. In G. A. Kerkut and L. I. Gilbert, Eds., *Comprehensive Insect Physiology, Biochemistry and Pharmacology*, Vol. 10. Pergamon, Oxford, pp. 451–486.

Candy, D. J. 1974. The control of muscle trehalase activity during locust flight. *Biochem. Soc. Trans.* **2**, 1107–1109.

Candy, D. J. 1978. The regulation of substrate transport and metabolism by octopamine and other compounds. *Insect Biochem* **8**, 77–181.

Candy, D. J. 1985. Intermediary metabolism. In G. A. Kerkut and L. I. Gilbert, Eds., *Comprehensive Insect Physiology, Biochemistry and Pharmacology*, Vol. 10. Pergamon, Oxford, pp. 1–41.

Candy, D. J. 1989. Utilization of fuels by the flight muscles. In G. J. Goldsworthy and C. H. Wheeler, Eds., *Insect Flight*. CRC Press, Boca Raton, Florida, pp. 305–319.

Chase, J. F. A. and P. K. Tubbs. 1972. Specific inhibition of mitochondrial fatty acid oxidation by 2-bromopalmitate and its coenzyme A and carnitine esters. *Biochem. J.* **129**, 55–65.

Chino, H. 1985. Lipid transport. In G. A. Kerkut and L. I. Gilbert, Eds., *Comprehensive Insect Physiology, Biochemistry and Pharmacology*, Vol. 10. Pergamon, Oxford, pp. 115–135.

Chou, P. Y. and G. D. Fasman. 1974. Prediction of protein conformation. *Biochemistry* **13**, 222–245.

Evans, P. D. (1985). Octopamine. In G. A. Kerkut and L. I. Gilbert, Eds., *Comprehensive Insect Physiology, Biochemistry and Pharmacology*, Vol. 11. Pergamon, Oxford, pp. 499–530.

Fernlund, P. and L. Josefsson. 1972. Crustacean color-change hormone: Amino acid sequence and chemical synthesis. *Science* **177**, 173–175.

Ford, W. C. L. and D. J. Candy. 1972. The regulation of glycolysis in perfused locust flight muscle. *Biochem. J.* **130**, 1101–1112.

Gäde, G. 1986. Relative hypertrehalosaemic activities of naturally occurring neuropeptides from the AKH/RPCH family. *Z. Naturforsch.* **41**, 315–320.

Gäde, G. 1987. Amino acid sequence and primary structure of the hypertrehalosaemic hormone in *Nauphoeta cinerea*. *Z. Naturforsch., C: Biosci.* **42C**, 225–230.

Gäde, G. and K. L. Rinehart. 1987a. Primary structure of the hypertrehalosaemic factor II from the corpus cardiacum of the Indian stick insect, *Carausius morosus*, determined by fast atom bombardment mass spectrometry. *Biol. Chem. Hoppe-Seyler* **368**, 67–75.

Gäde, G. and K. L. Rinehart. 1987b. Primary sequence analysis by fast atom bombardment mass spectrometry of a peptide with adipokinetic activity from the corpora cardiaca of the cricket *Gryllus bimaculatus*. *Biochem. Biophys. Res. Commun.* **149**, 908–914.

Gäde, G., G. J. Goldsworthy, G. Kegel, and R. Keller. 1984. Single step purification of locust adipokinetic hormones I and II by reversed phase high-performance liquid chromatography, and amino acid composition of the hormone II. *Hoppe-Seyler's Z. Physiol. Chem.* **365**, 391–398.

Gäde, G., G. J. Goldsworthy, M. H. Schaffer, J. C. Cook, and K. L. Rinehart. 1986. Sequence analyses of adipokinetic hormones II from corpora cardiaca of *Schistocerca nitans*, *Schistocerca gregaria*, and *Locusta migratoria* by fast atom bombardment mass spectrometry. *Biochem. Biophys. Res. Commun.* **134**, 723–730.

Goldsworthy, G. J. 1983. The endocrine control of flight metabolism in locusts. Insect Physiol. **17**, 149–204.

Goldsworthy, G. J. and P. Cheeseman. 1978. Comparative aspects of the endocrine control of energy metabolism. In P. J. Gaillard and H. H. Boer, Eds., *Comparative Endocrinology*. Elsevier, Amsterdam, pp. 423–426.

Goldsworthy, G. J. and C. H. Wheeler, 1984. Comparative aspects of the hormonal control of energy metabolism. *Nova Acta Leopold.* **255**, 21–37.

Goldsworthy, G. J. and C. H. Wheeler. 1986a. The endocrine control of flight metabolism in locusts. In W. Danthanarayana, Ed., *Insect Flight Dispersal and Migration.* Springer-Verlag, Heidelberg, pp. 49–59.

Goldsworthy, G. J. and C. H. Wheeler. 1986b. Structure/activity relationships in the adipokinetic hormone/red pigment concentrating hormone family. In A. B. Borkovec and D. B. Gelman, Eds., *Insect Neurochemistry and Neurophysiology 1986.* Humana Press, Clifton, New Jersey, pp. 183–186.

Goldsworthy, G. J. and C. H. Wheeler. 1988. Physiological and structural aspects of adipokinetic hormone function in locusts. *Pestic. Sci.* **25**, 85–95.

Goldsworthy, G. J., A. R. Jutsum, and N. L. Robinson. 1979. Substrate utilization and flight speed during tethered flight in the locust. *J. Insect Physiol.* **25**, 183–185.

Goldsworthy, G. J., C. M. Miles, and C. H. Wheeler. 1985. Lipoprotein transformations during adipokinetic hormone action. *Physiol. Entomol.* **10**, 151–164.

Goldsworthy, G. J., K. Mallison, and C. H. Wheeler. 1986a. The relative potencies of two known locust adipokinetic hormones. *J. Insect Physiol.* **32**, 95–101.

Goldsworthy, G. J., K. Mallison, C. H. Wheeler, and G. Gäde. 1986b. Relative adipokinetic activities of members of the AKH/RPCH family. *J. Insect Physiol.* **32**, 433–438.

Goosey, M. W. and D. J. Candy. 1980. The D-octopamine content of the haemolymph of the locust *Schistocerca americana gregaria* and its elevation during flight. *Insect Biochem.* **10**, 393–397.

Hayes, T. K. and L. L. Keeley. 1986. Isolation and structure of the hypertrehalosaemic hormone from *Blaberus discoidalis* cockroaches. In A. B. Borkovec and D. B. Gelman, Eds., *Insect Neurochemistry and Neurophysiology 1986.* Humana Press, Clifton, New Jersey, pp. 195–198.

Hekimi, S. and M. O'Shea. 1987. Identification and purification of two precursors of the insect neuropeptide adipokinetic hormone. *J. Neurosci.* **7**, 2773–2784.

Jaffe, H., A. K. Raina, C. T. Riley, B. A. Fraser, G. M. Holman, R. M. Wagner, R. L. Ridgway, and T. K. Hayes. 1986. Isolation and primary structure of a peptide from the corpora cardiaca of *Heliothis* zea with adipokinetic activity. *Biochem. Biophys. Res. Commun.* **135**, 622–628.

Jutsum, A. R., N. L. Robinson, and G. J. Goldsworthy. 1982. Effects of flight training on flight speed and substrate utilization in locusts. *Physiol. Entomol.* **7**, 291–296.

Konings, P. N. M., H. G. B. Vullings, M. Geffard, R. M. Buijs, J. H. B. Diedern, and W. F. Jansen. 1988. Immunocytochemical demonstration of octopamine-immunoreactive cells in the nervous system of *Locusta migratoria* and *Schistocerca gregaria. Cell Tissue Res.* **251**, 371–379.

Mayer, R. J. and D. J. Candy. 1967. Changes in haemolymph lipoproteins during locust flight. *Nature (London)* **215**, 987.

Murata, F. and T. Nagata. 1972. Supplemental studies on the method for electron microscopic demonstration of lipase in the pancreatic acinar cells of mice and rats. *Histochemie* **29**, 8–15.

Murata, F., S. Yokata, and T. Nagata. 1968. Electron microscopic demonstration of lipase in the pancreatic acinar cells of mice. *Histochemie* **13**, 215–222.

Nilsson-Ehle, P., A. S. Garfinkel, and M. C. Schotz. 1980. Lipolytic enzymes and plasma lipoprotein metabolism. *Annu. Rev. Biochem.* **49**, 667–693.

Orchard, I. 1987. Adipokinetic hormones — an update. *J. Insect Physiol.* **33**, 451–463.

O'Shea, M., J. Witten, and M. Schaffer. 1984. Isolation and characterization of two myoactive neuropeptides: Further evidence of an invertebrate peptide family. *Neuroscience* **4**, 521–529.

Robinson, N. L. and G. J. Goldsworthy. 1977a. Adipokinetic hormone and the regulation of carbohydrate and lipid metabolism in a working flight muscle preparation. *J. Insect Physiol.* **23**, 9–16.

Robinson, N. L. and G. J. Goldsworthy. 1977b. A possible site of action of adipokinetic hormone on the flight muscle of locusts. *J. Insect Physiol.* **23**, 153–158.

Siegert, K., P. J. Morgan, and W. Mordue. 1985. Primary structures of locust adipokinetic hormones II. *Biol. Chem. Hoppe-Seyler* **366**, 723–727.

Stone, J. V., W. Mordue, C. E. Broomfield, and P. M. Hardy. 1978. Structure-activity relationships for the lipid-mobilizing action of adipokinetic hormone. Synthesis and activity of a series of hormone analogues. *Eur. J. Biochem.* **89**, 195–202.

Storey, K. 1980. Kinetic properties of purified aldolase from flight muscle of *Schistocerca americana gregaria.* Role of the enzyme in the transition from carbohydrate to lipid-fuelled flight. *Insect Biochem.* **10**, 647–655.

Van Heusden, M. C., D. J. Van Der Horst, and A. M. T. Beenakkers. 1984. In vitro studies on hormone-stimulated lipid mobilization from fat body and interconversion of haemolymph lipoproteins of *Locusta migratoria. J. Insect Physiol.* **30**, 685–963.

Van Heusden, M. C., D. J. Van Der Horst, J. M. Van Doorn, J. Wes, and A. M. T. Beenakkers. 1986. Lipoprotein lipase activity in the flight muscle of *Locusta migratoria* and its specificity for haemolymph lipoproteins. *Insect Biochem.* **16**, 517–523.

Wheeler, C. H. 1988. Transport of fuels to the flight muscles. In G. J. Goldsworthy and C. H. Wheeler, Eds., *Insect Flight.* CRC Press, Boca Raton, Florida, pp. 326–334.

Wheeler, C. H. and G. J. Goldsworthy. 1983a. Qualitative and quantitative changes in *Locusta* haemolymph proteins and lipoproteins during ageing and adipokinetic hormone action. *J. Insect Physiol.* **29**, 339–347.

Wheeler, C. H. and G. J. Goldsworthy. 1983b. Protein–lipoprotein interactions in the haemolymph of *Locusta* during the action of adipokinetic hormone: The role of C-proteins. *J. Insect Physiol.* **29**, 349–354.

Wheeler, C. H. and G. J. Goldsworthy. 1985. Lipid transport to the flight muscles in *Locusta.* In M. Gewecke, Ed., *Insect Locomotion.* Parey, Berlin, pp. 126–135.

Wheeler, C. H. and G. J. Goldsworthy. 1986. Lipoprotein/apoprotein interactions during adipokinetic hormone action in *Locusta.* In A. B. Borkovec and D. B. Gelman, eds., *Insect Neurochemistry and Neurophysiology 1986.* Humana Press, Clifton, New Jersey, pp. 187–190.

Wheeler, C. H., J. E. Mundy, and G. J. Goldsworthy. 1984a. Locust haemolymph lipoproteins visualized in the electron microscope. *J. Comp. Physiol.* **154**, 281–286.

Wheeler, C. H., D. J. Van Der Horst, and A. M. T. Beenakkers. 1984b. Lipolytic activity in the flight muscles of *Locusta migratoria* measured with haemolymph lipoproteins as substrates. *Insect Biochem.* **14**, 261–266.

Wheeler, C. H., K. M. Boothby, and G. J. Goldsworthy. 1986. CL-proteins and the regulation of lipoprotein lipase activity in locust flight muscle. *Biol. Chem. Hoppe-Seyler* **367**, 1127–1133.

Wheeler, C. H., G. Gäde, and G. J. Goldsworthy. 1988. Humoral functions of insect neuropeptides. In M. C. Thorndyke and G. J. Goldsworthy, Eds., *Neurohormones of Invertebrates*. Cambridge Univ. Press, London and New York, pp. 141–157.

Whim, M. D. and P. D. Evans. 1988. Octopaminergic modulation of flight muscle in the locust. *J. Exp. Biol.* **134**, 247–266.

Worm, R. A. A. and A. M. T. Beenakkers. 1980. Regulation of substrate utilization in the flight muscle of the locust, *Locusta migratoria*, during flight. *Insect Biochem.* **10**, 53–59.

Worm, R. A. A., W. Luytjes, and A. M. T. Beenakkers. 1980. Regulatory properties of changes in the contents of coenzyme A, carnitine and their acyl derivatives in flight muscle metabolism of *Locusta migratoria. Insect Biochem.* **10**, 403–408.

Ziegler, R., K. Eckart, H. Schwarz, and R. Keller. 1985. Amino acid sequence of *Manduca sexta* adipokinetic hormone elucidated by combined fast atom bombardment (FAB)/tandem mass spectrometry. *Biochem. Biophys. Res. Commun.* **133**, 337–342.

9

Flight and Migration in Acridoids

R. A. FARROW

9.1 INTRODUCTION

Our understanding of the role of migration in insect life histories has undergone a radical shift over the past three decades. Although originally regarded as a passive reaction to adversity, resulting from habitat deterioration or competition for resources, migration is now seen as a positive mechanism that maximizes reproductive success or fitness in environments of spatially and temporally varying resources and survival values. In many species, migration plays an integral role in life-history strategies and spatial population dynamics (Dingle, 1972, 1985; Solbreck, 1985), which are in turn intimately linked to the environmental "template" of Southwood (1977).

In acridoid insects, migration was originally considered to be confined to the swarming phase of locusts, where it presented rather special features not manifested by other insects. It was not until the emphasis of research shifted to the study of movements by nonswarming individuals that migration could be interpreted within the more general, conceptual framework that was being developed for other migratory insects (Dingle, 1982; Johnson, 1969), and in the following section I focus on migration in relation to life histories and its importance in the evolutionary biology of acridoids.

Flight by insects in the upper air was once considered to be largely accidental, resulting in a passive transport by wind to destinations over which the migrants had no control and where most failed to reproduce, but it is now widely recognized as an active process by which species achieve significant spatial redistributions. In order to capitalize on the potential reproductive gains resulting from the population displacements brought about by migration, a range of morphological, behavioral, physiological, and genetic adaptations have evolved to regulate the timing of movements and their direction and distance in space (Dingle, 1985; Gatehouse, 1987; Johnson, 1969). Various cues may be used to restrict migration to periods when the chances of finding suitable habitats for reproduction are high, for example, insects that migrate at the onset of rain (Clark, 1969). Migrants that move downwind between geographically fixed breeding areas, influenced by varying wind patterns, may migrate only when winds are in appropriate direction (Hughes, 1979; Kennedy and Way, 1979; Taylor and Reling, 1986) and for durations that correspond to the distances that need to be traveled (Rogers, 1983). Baker (1978) goes further in proposing that permanent migratory pathways have evolved with respect to the fixed geographical relationships of seasonal breeding areas. These preferences are manifested by *navigation*, which can be independent of wind direction.

Environmental heterogeneity operates over a range of spatial and temporal scales. Most habitats are unfavorable to insect development and survival over large geographic areas at the same time each year because of seasonal climatic effects. In temperate latitudes these are represented by cold winters and, in certain regions, by dry summers, whereas in the seasonal tropics there are dry, but often cool, seasons contrasting with humid, overcast, rainy seasons that may be equally unfavorable to insects like acridoids, because of lack of insolation (Begon, 1983) and increased susceptibility to fungal pathogens (Chapman et al., 1986). The duration and timing of these changes vary with latitude and often correspond with seasonal shifts in wind systems associated with movements of the intertropical convergence (ITC) and polar fronts.

On a lesser scale, individual habitats are often patchily distributed owing to a range of site effects, some the result of human activities including deforestation and shifting agriculture. Habitat patches vary over time through the effects of weather (especially temporal and spatial variations in rainfall), natural succession of vegetation, and temporal differences in the growth and senescence of individual components of the vegetation. This ephemerality and unpredictability increases as climates become progressively more arid, and plant growth comes to depend entirely on rain and runoff. The survival values of different patches also vary according to the impact of natural enemies whose numbers are in turn affected by the same physical factors that affect their hosts, as well as by the numbers and availability of hosts. The complex spatial and temporal interactions of such systems are well represented by the Taylors' space–time reticulum (Taylor and Taylor, 1977).

Insects cope with physical changes to habitats in two main ways: (1) by *tolerating* extremes, by means of cold- or drought-resistant resting stages (often the egg), which are usually associated with a diapause; or (2) by *moving* to a less hostile environment for the duration of the unfavorable season. Reproduction may occur in such refuges although quiescence is also frequent. Insects also have to cope with the impact of biotic factors and have evolved mechanisms to minimize mortality from parasites, predators, or pathogens. Although camouflage, the sequestering of distasteful compounds, threatening reactions, and escape movements are well-known strategies to protect prey from attack by predators and parasites, movement into newly created habitats may reduce the impact of natural enemies on the colonizing generation of hosts, particularly by host-specific parasitoids. The specific strategy adopted in any particular environment will be the one by which a species maximizes the numbers of its descendants in the long term (Dingle, 1972; Southwood, 1977). For example, the benefits of migration to a new habitat could include increases in survival and in the number of generations per annum, and the costs could be associated with the failure to find a new habitat, failure to produce offspring because of insufficient time in a short-lived habitat patch, and possible declines in reproductive rate due to the demand on the reserves required for oogenesis by flight.

Where habitats deteriorate at the same time over large areas, staying put may be the preferred strategy if suitable refuges are too far away and are not connected by suitable wind systems for migration to be effective. In patchy habitats where the availability of suitable conditions for breeding is less predictable or synchronized, movement tends to be of higher reproductive value. Populations may spend the unfavorable season as reproductively quiescent adults rather than as immobile immature stages because the former can search out suitable habitat patches for reproduction as soon they become available.

The more permanent habitat patches are not necessarily more favorable if reproductive success is greater in ephemeral than in longer lived habitats. Freshly growing ephemeral grasses may provide better nutrition for insect growth and survival and for subsequent egg production than mature perennial grasses of more permanent habitats (McCaffery, 1975). Natural enemies generally show a delayed response to host numbers and would have a limited affect in a newly colonized habitat but can build up to high numbers when successive generations breed in the same place (Farrow, 1977a; Stower and Greathead, 1969). At any particular latitude of the tropics, the combination of insolation and moisture may also be more favorable for acridoids at the start and end of the rainy season than at its height, owing to enhanced insolation and low impact of fungal pathogens, and populations may move to maintain themselves in an ecoclimatic zone that is optimal for survival (Launois, 1978). Such zones would advance with the seasonal spread and intensification of the monsoon and then retreat. In general, movement on the scale of the ephemeral habitat patches ensures the long-term survival of the species in such an environment (Southwood, 1962). This strategy also spreads reproduction across a range of available habitats which is an important element in the stabilization of populations (Den Boer, 1968).

Emigration is often divided into *obligatory* or *facultative* movements. In situations where emigrants leave habitats when conditions start to deteriorate but stay if conditions remain favorable for further reproduction, emigration is said to be facultative. However, the distribution area of a given species may include mosaics of seasonally available habitats and areas where the habitat is more or less permanently favorable. In the former, regular intergenerational movements by entire regional populations give an impression of obligatory migrations (Farrow, 1975b), whereas in the latter area, populations appear relatively sedentary, suggesting that the species is, overall, a facultative rather than an obligatory migrant. If emigration occurs from habitats that are still favorable for reproduction, emigration is usually considered to be obligatory. In this situation the habitat does not necessarily remain vacant because it can be colonized by migrants dispersing from other locations. There is, however, no advantage in leaving a deteriorating habitat if a better one is unlikely to be found. Some species do not leave deteriorating habitats until they receive a cue that better habitats are to be found elsewhere, for example, from rain-bearing weather systems (Clark, 1969). If such systems occur frequently this strategy gives the appearance of being obligatory (Farrow, 1979a) rather than facultative.

The success of the migratory strategy has been measured in terms of the *reproductive value* (Dingle, 1972), which is the contribution that an individual makes to future population growth. As grasshopper populations generally have discrete generations and synchronized age structures, the *reproductive ratio*, R_0, is probably a better measure of population change (Farrow, 1979a). *Reproductive success* is an outcome of the interaction between reproductive rate and the likelihood of colonizing a new habitat, which have been succinctly expressed in Southwood's matrices (1977). If a species is an efficient colonizer, reproductive success can be equated directly with R_0. It has often been assumed that high reproductive success is related to high fecundity and rapid maturation (Dingle, 1972). This presents a possible paradox because competition for resources between reproductive needs and migration effort could be expected to reduce fecundity and delay maturation. However, it has been shown that variations in immature survival must also be taken into account, because this factor has the greatest impact on reproductive rate and small changes in its value can easily compensate for any reductions in fecundity (Farrow and Longstaff, 1986).

9.1.1 The Movement Continuum

Adult flight results in movements both *within generations*, which influence survival of the parent generation and its accumulation of resources for reproduction, and *between generations*, which affect survival of the offspring (Gatehouse, 1987).

Movements within generations, associated with the daily activities of locating food and oviposition sites, mating, shelter, and the avoidance of enemies, may achieve a local redistribution of populations between generations that is enough to optimize reproductive success [e.g., in the temperate grasshopper populations studied by Richards and Waloff (1954)]. These movements are termed *appetitive* or *trivial* and are the outcome of *foraging behavior* (Kennedy, 1986). Movements between generations often occur as a single event in an insect's life, usually at the *postteneral/prereproductive* stage of adult development, described by Johnson's *oogenesis/flight syndrome* (1969), in which migration and reproduction are supposed to be mutually exclusive activities. Such movements are typically more sustained (e.g., measured in hours rather than minutes) and transport individuals between habitat patches or geographic regions and are defined as *migratory* movements. However, like most biological attributes these categories are not well differentiated in practice. Extended foraging movements may result in displacements that are essentially migratory in nature (Jones et al., 1980); furthermore, the activities of feeding and reproduction are closely integrated into the movements of aggregations of acridoids and persist throughout the life of the aggregation. In solitarious acridoids the trivial movements involved in feeding and reproduction and dispersal are largely accomplished by walking and hopping, and flight is restricted to escape reactions and to reproductive activities by the male.

The term "migration" has been hard to define to the satisfaction of both ecologists and behavioralists. Ecological definitions in terms of the extent of the physical displacement of individuals and populations (Baker, 1978; Southwood, 1981; Taylor and Taylor, 1983; Taylor, 1986; Uvarov, 1977) have tended to overshadow the underlying behavioral processes (Kennedy, 1951, 1986). Most biologists, including myself, would support Gatehouse's proposal (1987) that migration is primarily a *behavioral process that has ecological consequences*.

9.1.2 The Flight Environment

Wind speed typically increases with altitude and for every airborne insect there is at any instant a level above which it will be displaced downwind, regardless of its orientation or air speed. Below this level an insect can fly in any direction by adjusting its heading to compensate for the speed and direction of the wind (Kennedy, 1951). This layer is defined by Taylor (1974) as the *flight boundary layer* (FBL), and its upper level corresponds to the altitude at which the air speed of the insect equals the wind speed.

There are important differences in the structure and motions of the air over land between day and night that influence the height of the FBL and the behavior of day- or night-flying insects (Drake and Farrow, 1988; Pedgley, 1982). By day, surface heating results in thermally driven vertical mixing, which causes wind velocity to increase rapidly with height from the surface, forming a steep wind shear, above which wind velocity increases more slowly. By night, atmospheric stability is reestablished as surface air cools by radiation and, under clear skies, a thermal inversion typically develops that causes the air near the ground to become highly stratified. Wind velocity gradually increases from often still conditions near the ground to a maximum at 100–300 m above the surface, that is, a weak but large wind shear. The wind at this altitude is approximately geostrophic (i.e., is not affected by frictional drag from the surface) and is typically stronger than at equivalent heights by day. This shear zone may contract as the layer of calm air deepens during the course of the night so that a steep, strong shear develops near the top of the inversion (Farrow, 1986). This effect may be further accentuated by the development of a wind velocity maximum at this altitude which represents one type of *low-level jet* (Browning and Pardoe, 1973). Although well documented for temperature and subtropical regions, low-level jets have not yet been widely identified overland from monsoon flows within the tropics (Grossman and Friehe, 1986). The air speeds of acridoids vary between 10 and 20 km/h and their flight boundary layers can vary in depth from a few tens of metres on windy days to a few hundred at night in a weak wind shear and in calms.

Within the boundary layer, migration can occur in fixed directions near the ground, either through a compensated, *light-compass* response, which is independent of wind direction (Williams, 1958), or as an *optomotor* response to the preferred speed and direction of surface patterns as they pass across the ommatidia (Baker et al., 1984; Kennedy, 1939, 1951; Pedgley, 1982) (Fig. 9.1). The latter hypothesis explains the frequency of upwind movements in light winds (Mikkola, 1986) as flying insects adjust their orientation so that surface images pass from front to back of the eye in a symmetrical, longitudinal direction. The theory also explains why the height of flight increases in response to a strengthening of the wind as fliers attempt to maintain this optimal retinal velocity. These movements contrast with the downwind displacements above the boundary layer where the optomotor response predicts a turning downwind to maintain the optimal front-to-back movement of the background across the retina, although with increasing height this may become undetectable (Kennedy's *maximum compensatory height*) (Fig. 9.1). The light-compass and optomotor responses are widely used by day-flying insects, but night-flying insects are also distinctly oriented (Riley, 1975), particularly when flying downwind above the FBL (Drake, 1983; Schaefer, 1976), and, if using the optomotor response, they would need to be able to detect angular velocities at very low levels of ground illumination (Riley and Reynolds, 1986).

Downwind flight above the FBL by insects is essentially migratory in nature (Farrow and Daly, 1987). In behavioral terms it depends on an extended ascent and a persistent horizontal flight, involving orientation and perhaps navigation, followed by a landing that may either be random or involve selection of a suitable habitat; in ecological terms it can lead to significant population displacements. Flight below the FBL can be both trivial and migratory depending largely on the degree of persistence of cross-country flight and the extent of its interruption by various stimuli (Kennedy, 1986).

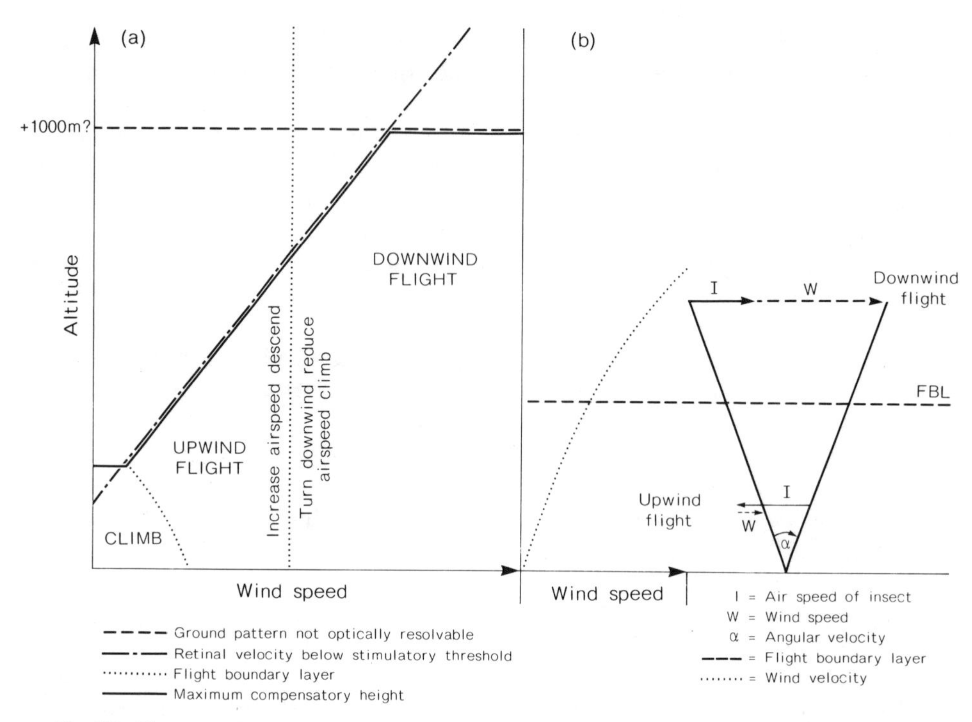

Fig. 9.1. Diagram of optomotor response showing (a) zones of upwind and downwind flight and climb and descent responses in relation to wind speed (after Kennedy, 1951) and (b) relation of upwind and downwind flight to the angular rate of apparent movement of the ground (modified from Pedgley , 1982).

9.1.3 The Scope of Migration Research on Acridoids

Locusts and, to a lesser extent, grasshoppers have figured prominently in advances in migration research, which have included the theory of downwind displacement of swarms of desert locust by day (Rainey, 1951), the discovery of nocturnal movements of solitarious locusts at night (Davey, 1953; Rao, 1960), and the detection by radar of individual locusts and grasshoppers flying at night (Roffey, 1969; Schaefer, 1969). Research has continued in four principal but not exclusive directions: (1) biogeographical studies of population displacements (Bennet, 1976; Pedgley, 1981; Rainey, 1963; Uvarov, 1977; Waloff, 1966); (2) studies of the dynamics of migratory movements with radar (Drake and Farrow, 1983; Riley and Reynolds, 1983; Reynolds and Riley, 1988; Schaefer, 1976), and, to a limited extent, with vertical photography Baker et al., 1984; Waloff, 1972); (3) investigations of atmospheric processes that influence the direction and distance of movements (Pedgley, 1981, 1982, Rainey, 1963) and the behavior of migrants (Drake, 1983; Rainey, 1976; Riley and Reynolds, 1986; Schaefer, 1976; Waloff, 1963, 1972); and (4) studies of the role of migration in population dynamics (Farrow, 1982a; Farrow and

Longstaff, 1986) and spatial dynamics (Taylor and Taylor, 1983). Less progress has been made in understanding (5) the physiological and genetic processes involved in the initiation, persistence, and termination of migratory flight (Gatehouse, 1988; McAnelly, 1985; Pener, 1985; Rankin, 1978, 1985), from which a rather confused picture of the role of the hormonal systems emerges; (6) the phenomenon of alary polymorphism (Chapman et al., 1978, 1986; Ritchie et al., 1987); and (7) the evolutionary and adaptive significance of migratory strategies (Baker, 1978; Reynolds and Riley, 1988), in which it has proved difficult to reconcile movements along preferred lifetime tracks with purely downwind displacements between seasonal breeding areas.

This review covers recent developments in our understanding of migratory movements in acridoids, concentrating more on the nature of such movements and their adaptive significance, than on the underlying physiological processes (Chapter 8). The effects of weather are central to this review because of their major influence on flight behavior and population displacement and their dominant role in the control of the numbers and distribution of acridoids (Dempster, 1963).

9.1.4 Grasshoppers or Locusts

Taxa in the superfamily Acridoidea (Orthoptera) are commonly called either grasshoppers or locusts. In principle this division separates those (the locusts) which readily aggregate in bands or swarms in response to density, a process termed *gregarization* (Uvarov, 1969), from those that do not (the grasshoppers). In most acridoids, however, increases in population density affect flight behavior. The range of these responses from both adults and nymphs of different acridoid species is so broad and continuous that no attempt is made in this chapter to separate "locusts" from "grasshoppers." Furthermore, the behavior of locusts and grasshoppers living at low density is similar and both groups undertake movements as individuals by day as well as by night. Accordingly this chapter distinguishes between the flight and movements of solitarious acridoids (Section 9.2) from those living in aggregations (Section 9.3).

Field and laboratory research on migration has tended to concentrate on species of economic importance, namely, various species of locust, and much of this review of flight behavior and movement must inevitably be derived from this group and extrapolated where necessary to other acridoids. The desert locust, *Schistocerca gregaria*, has been the main focus of studies on acridoid migration for a long time. It regularly exhibits spectacular daytime movements that are relatively easy to follow and study (Kennedy, 1951; Rainey, 1951, 1963; Waloff, 1966). Although this research led to discoveries such as the theories of optomotor response (Kennedy, 1951) and downwind displacement (Rainey, 1963), it has also tended to divert research activity from studies of flight and migration in other locusts and grasshoppers. More importantly, what are now known to be specialized features of the flight behavior of swarms of *S. gregaria* have often been incorrectly used as a general model of locust behavior (e.g., in Taylor and Taylor, 1983). Studies on recent plagues of other locusts and grasshoppers — for example, tropical migratory locust, *Locusta migratoria migratorioides*; spur-throated locust, *Nomadacris* (*Austracris*) *guttulosa*; Australian plague locust, *Chortoicetes terminifera*; and Senegalese grasshopper, *Oedaleus senegalensis*, among others — have provided opportunities for reassessing the dynamics and functional significance of flight and movement among a range of acridoids.

9.2 FLIGHT AND MOVEMENT IN SOLITARIOUS ACRIDOIDS

9.2.1 Daytime Flight and Movement

It has long been claimed that individual grasshoppers and solitarious locusts do not fly spontaneously by day from low-density populations (Davey, 1959; Uvarov, 1977; Waloff, 1963), other than in territorial displays or when males are seeking mates.

Other observations suggest, however, that diurnal movements by flight are often overlooked in the field, depending on conditions and species present. The evidence comes from visual observations (Chapman, 1959; Chapman et al., 1978), sometimes involving scanning with binoculars (Farrow et al., 1982; McAnelly and Rankin, 1986b), radar observations (Drake, 1982; Roffey, 1969), the detection of acridoids outside their normal range under conditions which preclude night flight (Alexander, 1964; Gurney, 1953), and a readiness of acridoids to undertake protracted flights in daylight in a flight mill (McAnelly and Rankin, 1986a).

Daytime flights have been detected both at low level, where at air speeds of 10–20 km/h, acridoids could cover 1–3 km in short flights of about 10 min (Farrow et al., 1982), and at higher altitudes (110–300 m), where they fly downwind for longer periods at speeds of 20 km/h and more (Drake, 1982), covering much longer distances. Short flights largely result in trivial movements, involving a redistribution of populations between habitats, whereas longer ones may result in migratory movements between habitats, depending on the duration of individual flights and their frequency.

More reports of daytime movements originate from temperate than tropical areas, perhaps because temperatures further from the tropics are often too cool at dusk (<20°C) to facilitate night flight. In the northern temperate zone, macroptery is commoner in years with warm summers (Ritchie et al., 1987), when conditions would be inherently more favorable for flight. Clearly more direct observations are needed in both temperate and tropical environments to determine the extent of diurnal flights.

Flight and movement by day are influenced by the same convective processes which affect aggregations, described in the section on gregarious populations, and by the same synoptic and subsynoptic scale events described for night-flying acridoids below. Heritable variation in the ability to undertake long (= migratory) flights has been discovered in a predominately day-flying species, *Melanoplus sanguinipes*, of North America (McAnelly, 1985; McAnelly and Rankin, 1986a) and intra- and interpopulation variation has been related to differences in habitat permanence.

9.2.2 Nighttime Flight and Movement

The discovery of long-range nocturnal movements in acridoids has revolutionized our understanding of the population dynamics of mobile insects in the seasonal grasslands of the semiarid zone and of the functional significance of such movements. The evidence has been slow to accumulate and hard to acquire but the processes of takeoff and ascent, of horizontal flight, and of the extrinsic and intrinsic factors governing these processes are being documented for an increasing number of acridoid species. Less is known about the duration of flight and the number of times that individual nocturnal flights are undertaken during an insect's lifetime, and almost nothing is known about descent and landing and subsequent site selection.

9.2.2.1 Evidence for Nocturnal Flights

Observations of dusk takeoff (Clark, 1971), attraction to light (Botha and Jansen, 1969; Farrow, 1977c; for other references, see Uvarov, 1977) and records of large-scale population displacements (Davey, 1959; Descamps, 1965; Duranton et al., 1979a; Farrow, 1977a; Golding, 1948; Joyce, 1952; Lecoq, 1978a,b) have provided circumstantial evidence for migration by acridoids at night. More direct evidence comes from marking (Davey, 1953), radar observations (Drake and Farrow, 1983; Rainey, 1976; Riley, 1974; Riley and Reynolds, 1979, 1983; Reynolds and Riley, 1988; Schaefer, 1976), aerial sampling (Rainey, 1976; Drake and Farrow, 1983), and observations with searchlights (Botha and Jansen, 1969; Roffey, 1963; Waloff, 1963).

Lights probably attract only acridoids that are flying close to the ground. Samples include individuals that are dispersing locally at low levels (Jago, 1979) as well as immigrants that have descended from higher levels of the atmosphere following a longer flight (Farrow, 1979a; Reid et al., 1979). Emigrants leaving a source area are not usually trapped by light at the site of departure (Farrow, 1979a).

Radar technology has not yet overcome the problem of target identity, although it has been possible to separate acridoids from moths (Lepidoptera) and larger from smaller species of acridoid, on the basis of differences in wingbeat frequency (Riley and Reynolds, 1983; Schaefer, 1976). Aerial sampling is the only unequivocal method of identifying migrants, although aerial densities are often too low to catch significant numbers of target insects (Drake and Farrow, 1983). In most radar studies the identity of airborne taxa of acridoids has been derived indirectly from ground survey and light trapping.

Evidence for nocturnal movements has been gathered from acridoid populations of the semiarid grasslands of west, east, and south Africa, the Middle

East, India, and Australia, but not from the temperate prairies and steppe of North America and Eurasia. The following species have been positively identified from movements that have been monitored by radar: *Aiolopus simulatrix* and *Diabolocatantops axillaris* in the Sudan (Schaefer, 1976; Rainey, 1976), *Oedaleus senegalensis*, *D. axillaris*, and *Schistocerca gregaria* in Mali and Niger (Riley and Reynolds, 1979, 1983; Reynolds and Riley, 1988; Schaefer, 1976), and *Chortoicetes terminifera* and *Aiolopus tamulus* in southeast Australia (Drake and Farrow, 1983). Many additional species of acridoid were almost certainly involved in the movements observed in Africa (Riley and Reynolds, 1979). Other species are known from population studies and light-trap observations to migrate at night in tropical and subtropical Africa, the Middle East, India, and Australia, including the locusts *Anacridium* spp., *L. m. migratorioides*, *Locustana pardalina*, and *N. guttulosa* and the grasshoppers *Gastrimargus* spp., among others. However, it should not be assumed that migration is an integral part of the life history of all tropical acridoids (Jago, 1983), and as the genetic studies of Shaw and co-workers (1980) suggest, narrow hybrid zones of a few hundred meters can be maintained between chromosome races of what must be a relatively sedentary species of acridoid in the Australian subtropics.

Radar studies indicate a common behavioral pattern to nocturnal flight behavior in all the acridoids studied, suggesting that the composite picture presented here is probably valid for all suspected night-flying acridoids whether or not a species has been specifically studied by radar.

9.2.2.2 Takeoff and Ascent

Takeoff starts at dusk in rapidly fading light and peaks just after dark (Fig. 9.2). Increases in aerial density of two to three orders of magnitude are observed in acridoids over periods of 30–60 min at dusk (Drake and Farrow, 1983; Riley and Reynolds, 1979). The numbers taking off then decline rapidly and no further takeoff or ascent has ever been detected over the rest of the night. Acridoids climb at a steep angle, initially upwind and then turning downwind when wind velocity exceeds about 2 m/s (Waloff, 1963). They ascend at a rate of about 0.3–0.5 m/s, reaching 150–500 m in 5–20 min after takeoff (Riley and Reynolds, 1979), and have been reported to climb as high as 1800 m in the Sudan, taking about an hour (Schaefer, 1976).

Many ascents are not sustained and the fliers return to ground after short displacements of less than 1 km (R. A. Farrow and V. A. Drake, unpublished), resulting in local turnover and redistribution of populations (Farrow, 1979a; Jago, 1983). Ascents that lead to persistent horizontal flight are considered to be migratory in nature, whereas the short flights are probably associated with local movements to more favorable habitat patches, possibly involving some form of site evaluation and selection.

Each individual undertakes a discrete movement on the night that takeoff and ascent occur; consequently the number of nights on which an individual migrates during adult life exerts a net cumulative impact on the spatial changes of acridoid populations at each generation.

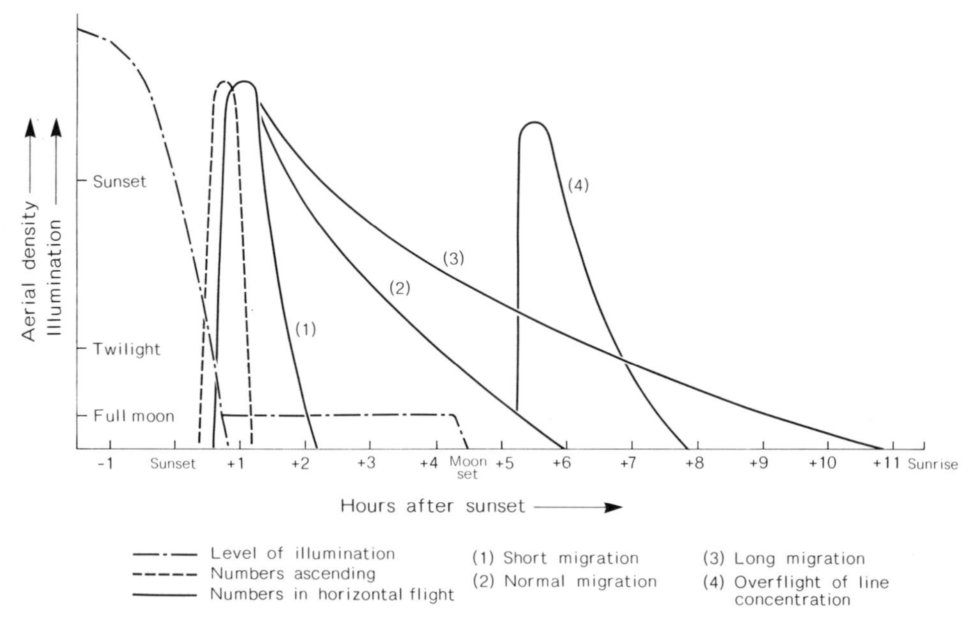

Fig. 9.2. Diagram of relative densities of airborne migrants during dusk takeoff, ascent, and horizontal flight during the night over an observation site, in relation to level of illumination.

9.2.2.3 Factors Governing the Incidence of Emigration

The readiness of individual acridoids to take off at dusk is influenced by a suite of interacting internal and external factors. The internal factors (see Adult Development, below) comprise muscle development and cuticle hardening, level of energy reserves, and stage of reproductive development (Hunter, 1982). There is little evidence that macropterous species vary genetically in their propensity to take off in migratory flight at night, as suggested for a day-flying species, *M. sanguinipes*, by McAnelly and Rankin (1986a), although more experimental work is needed in this area. The external factors (see Meteorological Conditions, below) include temperature, relative humidity, wind speed, rainfall, and atmospheric pressure, stability, and electromagnetic activity (Bergh, 1979; Clark, 1969; Johnson, 1969), as well as habitat conditions including food quality (Hunter, 1982). External factors such as temperature may also affect migratory flights indirectly through their effect on the timing of fledging and the duration of the teneral period.

The time of takeoff is remarkably constant from night to night and is determined by light intensity, being slightly earlier on overcast nights (Roffey, 1963; Schaefer, 1976). Riley and Reynolds (1979) showed that takeoff started when light intensity diminished to 10^{-5}–10^{-6} W/m^2/nm in the 300–500 nm band. Although nocturnal movements are easier to detect at high population densities than at low, there is little evidence that the level of takeoff in populations of solitarious individuals is influenced by density, as proposed for *L. pardalina* by Lea (1968), in which density-sensitive individuals leave centers of increasing density whereas density-insensitive individuals remain. The level of postteneral departures is as great in low- as in high-density populations of *C. terminifera*, provided no other factors are limiting (Farrow, 1982a).

Adult Development. Most acridoids caught at lights are at least 4–5 days old with fully hardened cuticles and well-developed fat bodies, and the females are generally immature with undeveloped terminal oocytes. This suggests that most nocturnal flight is restricted to the postteneral/prereproductive stage of development, which accords with Johnson's "oogenesis flight syndrome" (1969). However, it does not exclude the possibility that mature adults also undertake significant interovipositional movements at night (Farrow, 1975b), particularly if this stage is less attracted to light, as suggested by Jago (1983) for *O. senegalensis*, and traps are biased toward pre- and interreproductive females, including those that may be moving locally.

There are three main temporal patterns of emigration according to differences in seasonal patterns of reproductive development (Fishpool and Popov, 1984):

1. Species that enter reproductive diapause for the duration of the dry season, for instance, *N. guttulosa* and *D. axillaris*. In this group, flight activity by immature individuals peaks at the end of the rains following fledging, declines during the dry season, and rises (in the same individuals) at the start of the following rainy season and then declines as maturation advances and oviposition occurs (Baker and Casimir, 1986; Reynolds and Riley, 1988). This activity corresponds first to movements to suitable dry season diapause sites and later to reproductive sites. This second period of migration is associated with the rise in temperatures and humidities that precede the arrival of the rains and is accompanied by female maturation (Farrow, 1977b). Several nights of prereproductive migratory flight may take place in each individual.
2. Species in which reproductive development occurs without delay in all conditions, for example, *L. m. migratorioides*. Where habitats are short-lived, emigration of maturing adults with large fat bodies to complementary breeding areas is the norm (Farrow, 1975b), suggesting that migration and oogenesis are closely associated processes. Possibly only one night of prereproductive flight occurs, *between* fledging and breeding areas, but is supplemented by several nights of less extensive movements by maturing and ovipositing individuals *within* breeding areas.
3. Species in which there is facultative maturation, dependent on the presence of green food during nymphal development and at fledging, which enables lipid to be accumulated, as in *C. terminifera* (Symmons and McCulloch, 1980). Several options occur in this species: if environmental conditions are favorable during hopper development and at fledging, because of an abundance

of fresh green vegetation, lipid is accumulated and adults emigrate and mature. When conditions are dry at fledging, adults accumulate little lipid and neither mature nor emigrate, although there may be brief flights at low altitude resulting in local turnover and dispersal (Clark et al., 1969). If rain occurs after fledging and green vegetation becomes available, lipid accumulation and egg production proceed rapidly. In some situations, no emigration occurs and the adults all reproduce in the vicinity of the fledging area (Hunter et al., 1981; Hunter, 1982). In other situations, intense flight activity occurs at the onset of drought-breaking rains (see Meteorological Conditions) and is accompanied by maturation (Clark, 1969, 1972), suggesting that sufficient lipid has been accumulated prior to the rains and that emigration and reproduction had been controlled by extrinsic factors. Finally, emigration has also been observed after several cycles of egg production in this species (Farrow, 1977a), possibly because external factors inhibited emigration at an earlier stage. A similar situation may occur in *S. gregaria*, where maturation and reproduction without emigration may occur in fledging areas if conditions are locally suitable for further reproduction (Bennet, 1976; Roffey and Popov, 1968), whereas in dry conditions individuals probably disperse locally by short or low-level flights and do not migrate until suitable rain-bearing weather systems develop.

Egg production and migration are not well-defined alternative states in female acridoids, as suggested by the oogenesis/flight syndrome. Both processes depend on the accumulation of lipid and other reserves after fledging, in association with the development of the fat body, and are influenced by a range of external factors. There are obvious reasons why migratory behavior should decline once a female has reached a habitat that remains favorable for the duration of the reproductive period. In many circumstances, however, interreproductive movements may persist, particularly where a deterioration in habitats occurs. Male reproductive development appears less limited by the rate of lipid accumulation but it has not been shown that males necessarily emigrate before females. Most males are ready to copulate by the end of the teneral period and many females are mated before emigration, ensuring that if densities are excessively diluted during migration, most females can reproduce without further insemination (but see Gatehouse, 1988).

Meteorological Conditions. Evening takeoff is rarely observed if ambient temperatures at sunset are below 20°C, and large-scale takeoff in acridoids is limited to temperatures of 25°C or more (Clark, 1969; Hunter, 1982; Uvarov, 1977; Riley and Reynolds, 1979). At low temperatures, potential emigrants are sometimes observed "fanning" on vegetation at dusk in order to raise their body temperature by metabolic processes (Farrow, 1975b). The fall in migratory activity in continental areas of the subtropics during the dry season is at least partly associated with the decline in dusk temperatures as surface air cools rapidly by radiation at this season.

In temperate zones, takeoff by acridoids is often limited by low evening temperatures even during summer. Increases in temperature, suitable for takeoff, are associated with often short-lived, prefrontal airflows of poleward-moving tropical air (Farrow, 1975a). Opportunities for takeoff in cooler postfrontal flows and in anticyclonic conditions are generally more limited (Drake and Farrow, 1983). The number of nights on which prereproductive flights can occur is also closely regulated by synoptic weather. In a fast-moving frontal system, the zone of weather favorable for takeoff and emigration may pass across a site containing acridoid populations on a single night. In *C. terminifera*, light-trap records and surveys of population displacements confirm that migrations are often restricted to a single night (Clark, 1972; Farrow, 1975a, 1977a; Symmons and Wright, 1981). When prefrontal airflows have been slow-moving there have been occasional reports of substantial displacements over several successive nights, probably by the same individuals, in both *C. terminifera* (Symmons and Wright, 1981) and *N. guttulosa* (Baker and Casimir, 1986). Cumulative displacements of as much as 1000 km have been reported in the former species over four consecutive nights of favorable weather (e.g., 21–24 March 1979, Symmons and Wright, 1981).

In the tropics and subtropics, dusk temperatures rarely limit flight activity except in the middle of the dry season or winter; nevertheless, migratory activity is often enhanced by the conditions associated with atmospheric instability, which are usually referred to as *disturbed weather* (Clark, 1969, 1971, 1972; Farrow, 1979a; Pedgley, 1982; Riley and Reynolds, 1983). These periods are characterized by afternoon and evening storms and their associated outflow squalls. They occur within the monsoon flow, where they are often associated with westward-moving disturbance lines; in cyclonic centers along the Intertropical Convergence (ITC) (Fig. 9.10); in tropical troughs and depressions; ahead of temperate zone cold fronts (Fig. 9.12); and in association with upper atmospheric troughs. Although flight activity is high whether or not the storms produce rain, it is particularly enhanced after rain, especially following long dry periods during which migratory activity has been inhibited (Clark, 1969; Farrow, 1979a).

In tropical Africa, nightly migrations are a regular feature of the period at the end of the rains when large numbers of adult grasshoppers are recruited (Farrow, 1975b; Riley and Reynolds, 1983). Nevertheless, large-scale movements have coincided with weather disturbances. Pedgley (1982) found that a mass southward migration of solitary *L. m. migratorioides*, reported by Davey (1959), was related to the passage of a dry cold front, which had penetrated unusually far south for this time of year. A night of massive grasshopper migration across part of Niger coincided with the passage of a weak frontal system (Fig. 9.10) (Riley and Reynolds, 1983). The relatively settled weather of the dry season may also contribute to the reduced level of flight activity reported at that time of the year.

The mechanism responsible for the increased levels of flight activity in disturbed weather has not been elucidated, although an activity response to passing frontal systems has been demonstrated in the laboratory for *S. gregaria* by Bergh (1979, 1988). He attempted to find a correlation with changes in atmospheric electrical activity but failed to do so. Wellington (1954) suggested that falling atmospheric pressure could provide such a cue (for migratory insects in general), and Clark (1969) proposed that increases in humidity could be responsible in acridoids.

These observations suggest that in those parts of the subtropics where rainfall is unpredictable, such responses prevent adults migrating extensively when the probability of conditions becoming favorable for breeding is low; that is, movement is restricted when conditions are generally poor, and is immediately enhanced as soon as they improve (e.g., in *C. terminifera* and possibly *L. pardalina*, *S. gregaria*, and *Schistocerca cancellata*), provided the populations are not at the fledging stage (see earlier options).

On the other hand, in the seasonal tropics, movements often appear to be obligatory and generally coincide with a widespread deterioration of breeding conditions at the end of the rains, e.g., in *L. m. migratorioides* and some acridoids of west Africa (Farrow, 1975b; Riley and Reynolds, 1983). However in other parts of these species' ranges, in habitats where conditions are favorable throughout the year, migratory movements are not so obvious, suggesting that migration is either a facultative process or that there are genetic differences in migratory propensity between different subpopulations, as postulated for *M. sanguinipes* by McAnelly and Rankin (1986a).

Ascent at night is typically an active process on the part of emigrants because there are no known convective processes operating at dusk after the cessation of thermal convection in the late afternoon, although it could be occasionally assisted by a local ascent of warm air at the head of density currents, for example, those associated with lake breeze systems.

9.2.2.4 Horizontal Flight and Migration

Nocturnal migration of acridoids is characterized by remarkably uniform and persistent cross-country flights, typically at several hundred meters altitude. Analysis of these movements is often complex because of the need to incorporate (a) numerical changes over time with variations in numbers of migrating targets in both vertical and horizontal fields and (b) track and speed of individuals with direction and distance of displacements.

Aerial Densities. In uniform wind fields the numbers of migrating acridoids tend to vary over the range $<1 \times 10^{-6}$ to $1 \times 10^{-4}/m^3$; these values are equivalent to area densities over the ground of 10–600/ha (Drake and Farrow, 1983). The magnitude of the migration or "flux" is a function of density and speed of the migrants and typically ranges between <1 and $100/m^2/h$. The cumulative number migrating overhead during the course of a night can be very large: for example, over six successive nights of observation in southeast Australia in November 1979, an average of nearly two million *C. terminifera* migrated northward each night across a 1-km line, perpendicular to the prevailing direction of migration (Drake and Farrow, 1983). There is no reason to suppose that this movement was localized at the radar site, for suitable locust-containing habitats extended several hundred kilometers from the site, indicating that as many as 3×10^9 locusts migrated over a week in this part of southeast Australia. Reynolds and Riley (1988) estimated that 1.9 million acridoids flew across a 1-km line over their radar site in west Africa in 2.5 h on a single night. As with most migrations the numbers airborne in such movements depend primarily on the size of the contributory source populations, rather than on immediate weather conditions (Johnson, 1969).

Temporal Changes. Aerial densities over any site invariably start to decline after the peak of takeoff and generally fall to low levels 4–6 h after dusk (Fig. 9.2). This is primarily due to progressive descent and landing by migrants. Changes over time are also influenced by the numbers of insects leaving upwind source areas at different distances from the radar observation site. For example, if there are no insect sources further than 3 h flying time from the radar site, the numbers overflying will abruptly fall at 3 h after takeoff. Conversely, if dense populations ready to emigrate exist at some distance upwind of the radar, aerial densities may suddenly increase late in the night (Fig. 9.2, curve 4) (Drake and Farrow, 1983).

Layer Concentrations. During the course of the evening the vertical distribution of migrants generally becomes stratified with the progressive formation of one or more *layer concentrations* (Fig. 9.3). The lowest layer is usually the most important and its lower boundary develops as a sharp discontinuity at 100–300 m above the ground. It is typically 50–200 m thick, although its upper boundary is less well defined (Drake and Farrow, 1983; Riley and Reynolds, 1979). This layer usually intensifies during the night and persists until migration ceases. Higher layers are also occasionally reported (Riley and Reynolds, 1979), although it is not known if they contain acridoids.

Line Concentrations. The relatively uniform pattern of nocturnal movement in acridoids is frequently interrupted by the passage of intense *line concentrations*, in which densities temporarily increase by several orders of magnitude (Drake, 1983; Reid et al., 1979; Riley and Reynolds, 1983; Schaefer, 1976) (Fig. 9.4a). Acridoid densities as high as $1.5 \times 10^{-3}/m^3$ have been reported from such concentrations (Riley and Reynolds, 1983). The length of lines is unknown, because they generally extend beyond the limits of radar detection (10–100+ km), but the distance between the sharply defined advancing edge, moving at 10–50 km/h, and the less well-defined trailing edge is often only a few hundreds of meters. Multiple line concentrations are also reported in which successive advancing edges are about 1 km apart (Reid et al., 1979). Line concentrations are observed almost nightly in Australia during radar observations and slightly less frequently in Africa, where they are largely confined to the rainy season (D. Reynolds, personal communication).

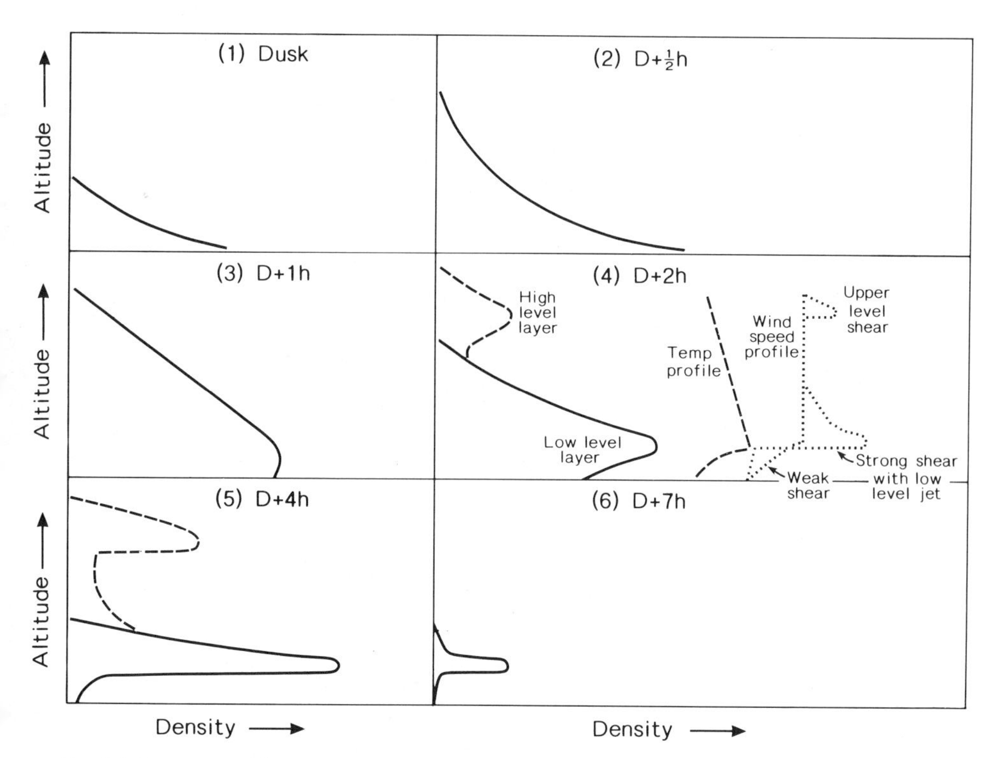

Fig. 9.3. Diagrams of the progressive stratification of migrants during the course of the night in relation to temperature and wind profiles.

Orientation. One of the most remarkable features of night-flying acridoids (and other insects) is that they often exhibit varying degrees of mutual alignment or collective orientation (Drake, 1983; Riley, 1975; Riley and Reynolds, 1979, 1983, 1986; Schaefer, 1976), suggesting that when flying at night at high altitude, acridoids attempt to influence their track, relative to the ground, in the same way as individuals flying within their flight boundary layer; that is, they *navigate*.

Orientation is quantified in terms of the orientation angle and the degree of collective orientation, which tends to be highest where migrants are concentrated into layers (Drake, 1983). Orientation patterns are usually unimodal, which occurs when all individual migrants are similarly oriented. Occasional bimodal orientation patterns occur in which two groups of insects are flying in different directions, usually centered at different altitudes. They are usually considered to be different species (Riley and Reynolds, 1986), oriented to different cues. At high wind speeds there is a tendency for migrants to orient within a few degrees of the downwind direction, although there is no evidence that acridoids ever attempt to fly upwind in a wind stronger than their flying speed. At lower wind speeds, which may approach an acridoid's flying speed of 3–6 m/s, the

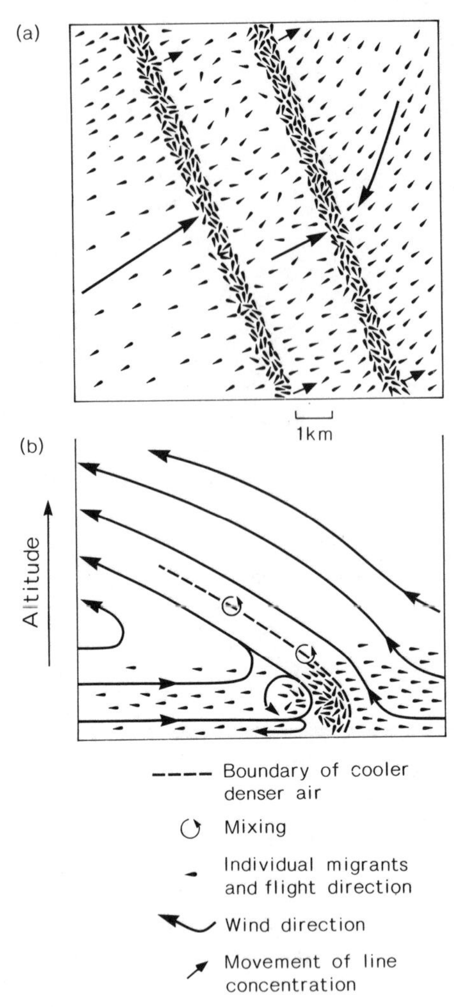

Fig. 9.4. Diagrams of (a) plan view of a double line concentration of migrants as seen on radar and (b) cross section of a single line concentration caused by a density current. (Modified from Pedgley, 1982.)

degree of collective orientation often declines and the direction of orientation may vary up to 90° from the wind direction, resulting in significant off-wind displacements (Riley and Reynolds, 1979, 1986) (Fig. 9.5). In light winds, migrants thought to be acridoids have been observed orienting on fixed compass bearings at high altitude in a range of wind directions (Riley and Reynolds, 1986; Schaefer, 1976) and were effectively migrating within their flight boundary layer. Summarizing their studies in Mali on migrants that were predominantly acridoids, Riley and Reynolds (1986) found that only a quarter of orientations were within 10° of a downwind direction and many were oriented toward

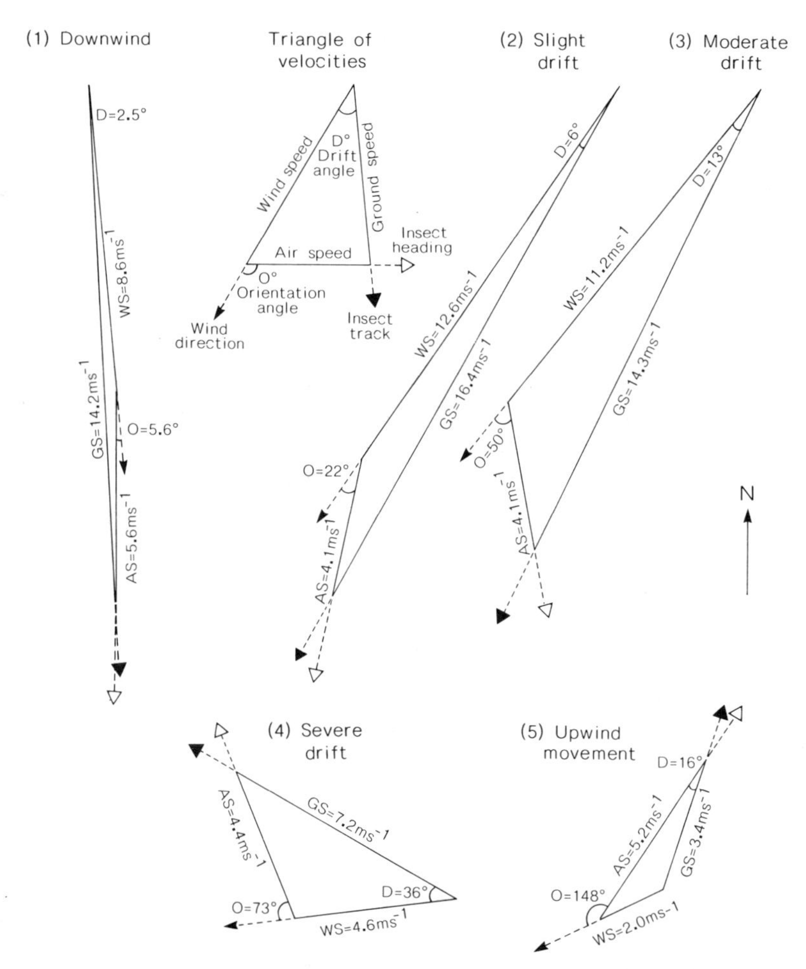

Fig. 9.5. Diagrams of the relationships between migrant orientation, track, and heading showing different degrees of drift and upwind displacement. (Modified from Riley and Reynolds, 1979.)

25–50°, in a range of wind directions. This may be a site-specific response, as discussed below.

At the end of the rainy season in Africa, north of the equator, the off-wind headings of acridoids usually produce a southerly bias to displacement direction (Riley and Reynolds, 1979; Reynolds and Riley, 1988; Schaefer, 1976), but in Australia no consistent orientation direction has been found (Drake, 1983). It is not known if the preferred off-wind heading is reversed (to the north) at the start of the rainy season in Africa north of the equator, when populations tend to move northward. The northward movements observed over the Central Niger Delta in Mali by Riley and Reynolds (1979) occur at the time when acridoids flying outside the delta are migrating south. They correspond to a unique local event, the northward retreat of the floods that progressively exposes suitable habitat from south to north at a time when habitats outside the floodplain are deteriorating from north to south (Farrow, 1975b). If the same species of acridoids migrate north when over the delta and south when to the east or west, then the migrants concerned are most likely reacting to specific environmental cue(s) over the delta. These are as yet unknown, although responses to topographical features seem unlikely (Riley and Reynolds, 1979).

Migratory Distance. The distance covered by migrants is a function of flight duration and ground speed. Ground speed is related to air speed and wind velocity by a triangle of velocities (e.g., Riley and Reynolds, 1979) (Fig. 9.5). Ground speeds of 20–50 km/h commonly occur in migrating acridoids and are equivalent to maximum displacements of 60–250 km in 3–5-h flights. However, many individuals cover much shorter distances (<50 km) because of progressive descent and landing during the course of the night starting immediately after the initial takeoff. In temperate areas occasional migrations of acridoids have occurred at ground speeds of 100+ km/h resulting in movements of 500+ km in a few hours of a single night (Farrow, 1975a).

Landing. Of all the processes involved in a nocturnal migration of acridoids, landing is the least understood. It is remarkable that species such as *C. terminifera* and *L. m. migratorioides* are rarely found in unsuitable habitats after substantial night flights (Clark, 1978; Farrow, 1975b), suggesting a high degree of selectivity immediately prior to and after landing, possibly assisted by repeated short flights to locate suitable conditions. Little is known about possible visual or olfactory responses involved in site selection, although the latter are known to be important in other migrant species (Solbreck, 1985).

9.2.2.5 Factors Governing Nocturnal Flight and Displacement

Flight duration at night is generally much shorter than limiting extrinsic factors such as temperature or the onset of daylight would indicate, suggesting that cessation of migratory flight is determined intrinsically. Many flights are not sustained following takeoff but little is known of the cause of the switch between short and sustained flight, although several of the factors discussed in the following sections probably have an important influence. The duration of migratory flight is unlikely to be influenced by depletion of energy reserves, for most migrant acridoids contain large fat bodies even after a substantial flight (Bond and Blackith, 1987). Flight may be extended if the migrants find themselves over water; Bond and Blackith (1987) have reported the example of a solitarious *S. gregaria* that crossed 3000 km of ocean over the course of 4–5 days and nights.

Intrinsic Variation in Flight Capacity. There is growing evidence from a wide range of migratory insects (Gatehouse, 1988; McAnelly and Rankin,

1986a; Rankin et al., 1986) that the duration of flight or *flight capacity* is genetically determined and that some of the observed variation is heritable.

Flight capacity is usually measured in flight mills. These tests generally show that most individuals undertake short flights and progressively fewer undertake flights of increasing duration (Gatehouse, 1988). The values provide relative rather than absolute measures of the flight potential for individuals migrating in the field.

It has also been suggested that the mean flight duration may reflect the distance that populations need to move to maximize reproductive success (Rogers, 1983). This programming could be important in species whose populations migrate over relatively short distances of 50–200 km between a fixed pattern of breeding areas (Farrow, 1975b), or in seasonally shifting zones of ecoclimatic optima (Launois, 1978), and would prevent migrants from "overshooting" suitable breeding areas in uniform wind fields. Such limitations would be less important in species that have adopted a strategy of downwind movement into areas of atmospheric convergence (described below). In these species it may be advantageous to migrate for as long as feasible to bring as large a proportion of the population as possible into the rain-affected area.

Flight Duration and Meteorological Conditions. Although the temperature threshold for sustained flight is lower than that for takeoff, it is usually difficult to show that flight has been terminated by falling air temperature. This is rarely known at the migration altitude and estimation by extrapolation from surface data is unreliable because of the frequent presence of an inversion (Farrow, 1986). However, where cold air abruptly replaces warm air during the passage of a storm outflow or a cold front and there is considerable vertical mixing of air, flight may suddenly cease. Washing out by rain may also terminate flight in such mesoscale systems (Drake and Farrow, 1988).

Flights by solitarious locusts such as *C. terminifera* appear to be of much longer duration in disturbed weather than in settled conditions, and these have been attributed to lower saturation deficits (Clark, 1969, 1972). Similar responses may occur in solitarious *S. gregaria* and other acridoids, leading to the extensive movement of populations in convergent wind systems and to large-scale translocations in prefrontal airflows.

Vertical Distribution and Atmospheric Stratification. Low-level layer concentrations of acridoids are typically associated with the top of the nocturnal temperature inversion (Fig. 9.3). Temperatures at this height are equivalent to late afternoon surface temperatures, at the cessation of thermal convection, and often remain so for most of the night. This provides an environment suitable for sustained flight long after the surface air has cooled below the flight threshold (Farrow, 1986). The top of the nocturnal wind shear also occurs just below the top of the temperature inversion and boundary layer wind maxima and low-level jets are often detected at this altitude (Drake and Farrow, 1988) (Fig. 9.3). As the layer of calm air deepens below these wind maxima, the optomotor response also predicts that insects will climb into zones of higher wind velocity and accumulate there (Pedgley, 1982). It is probable that layering is a response to variations in both temperature and wind gradients, although it has been suggested that some layers suspected to contain acridoids also occur at the interface between the cold and warm air flows of density currents, in gravity waves (Simpson, 1987), and along the sloping interface between the dry and warm air of the ITC (Schaefer, 1976).

The origins of high level layers are less certain. They are generally associated with wind shears (Drake and Farrow, 1988), although temperature ceilings may also be important (Riley and Reynolds, 1979).

Concentrations and Wind Convergence. Line concentrations of migrants are associated with wind shifts in which an advancing flow of cooler denser air undercuts and displaces a less dense air mass (Drake and Farrow, 1988) (Fig. 9.4b). These weather features occur at the subsynoptic scale as density currents and at the synoptic scale at the interfaces of the opposing winds in the ITC and polar fronts.

A density current is typically only a few hundred meters deep and at its leading edge there is a zone of strong horizontal convergence. Acridoids flying ahead of such a current are carried into it and, if they avoid being carried upward, concentrate at the interface of the convergence and are displaced in the direction of the flow (Fig. 9.4b) (Drake, 1982; Pedgley, 1982; Schaefer, 1976; Symmons and Luard, 1981). Sea breezes, storm outflows, and katabatic winds are forms of density current that concentrate migrant insects over comparatively featureless plains as well as over undulating country (Drake and Farrow, 1983; Riley and Reynolds, 1983; Reynolds and Riley, 1988; Schaefer, 1976).

Density currents derived from a sea breeze have been observed progressing westward as far as 400 km inland in eastern Australia. They appear as wind shifts to a more easterly direction and arrive at around midnight at distances of about 200 km from the coast and have been associated with sudden changes in direction and density of acridoid migrations (Drake and Farrow, 1983). Line concentrations including acridoids have been followed for over 20 km in storm outflows in east Africa (Schaefer, 1976; Pedgley et al., 1982) and for shorter distances in Australia (R. A. Farrow and V. A. Drake, unpublished). The downhill flow of katabatic winds at night is probably responsible for the line concentrations of acridoids observed in the Sahara in very shallow drainage basins. Multiple line concentrations occasionally form ahead of wind shifts and appear to be caused by the development of atmospheric solitary waves in which there is no change to an existing migration unless a circulation develops in the wave (Drake, 1985).

The localized concentrations of acridoids that appear on the ground after a night of migration in disturbed weather (Farrow, 1977a) may result from mass depositions of aerial concentrations in density currents, resulting from the mechanisms described earlier. Directed short-range movements and active concentration into suitable habitat patches by specific behavioral responses may also cause these concentrations.

At the synoptic scale, line concentrations rarely develop along polar fronts, possibly because the insect-transporting winds blow parallel with the advancing front. In the vicinity of the front, warm air rises over the advancing wedge of cooler air and unless migrants resist being transported upwards, densities do not increase. Lines containing acridoids have been observed ahead of a cold front (Reid et al., 1979), possibly in association with solitary waves. In this instance the front was dry but in other systems the migrants may be washed out by rain, for example, in the offshore movements of *C. terminifera*, described by Farrow (1975a).

There are few reports of insect concentrations in the interface of the ITC and only one instance that possibly involved acridoids (Schaefer, 1976), which occurred when the ITC crossed a radar observation site. The potential concentration of airborne migrants by the ITC has been calculated by Rainey (1976) to be equivalent to relative increases in aerial density of fourfold per hour. Although major airflows meet at the ITC, winds are very light in its vicinity and recent interpretations of its structure (Barry and Chorley, 1987) suggest that actual areas of convergence are often ephemeral or form part of moving circulations that would not be conducive to concentrating migrants into lines. The mechanisms involved in the large-scale concentration of migrating acridoids by cyclonic circulations (Clark, 1971, 1972) have not been studied and are largely beyond the range of current entomological radar techniques.

Causes of High-Level (100 m+) Orientation. Although acridoids fly at varying angles to the wind, wind shifts generally cause a corresponding change in orientation so that the same angle to the wind is maintained (Drake, 1983; Riley and Reynolds, 1986; Schaefer, 1976), suggesting that orientation angle is primarily determined by wind direction, possibly through the optomotor response. The locust eye is known to be very sensitive to angular movement at high light levels (Thorsen, 1966) but Riley and co-workers (1988) have shown that at night this response depends on the level of illumination. Their conclusions for *L. m. migratorioides* (in a laboratory study) suggest that acridoids can detect small amounts of cross-wind drift under moonlight, which are adequate to explain some cases of orientation observed in low-level layers but possibly not the close-to-downwind headings, which are more frequent. Furthermore, orientation patterns vary neither with altitude, as predicted by the optomotor response, nor with the level of nocturnal illumination (Riley and Reynolds, 1986).

A second mechanism of orientation to wind direction involves possible responses to directional accelerations/decelerations (anisotropic gusts) in the wind, perhaps caused by Kelvin–Helmholtz waves generated by wind shear (Riley and Reynolds, 1986), and to sudden accelerations and decelerations caused by vertical movements across the wind shear (Gillett, 1979). Acridoids are well endowed with sensory organs that could respond to such stimuli, but there is as yet no evidence that such cues are being used.

Orientation on fixed compass bearings in preferred geographic directions at night is restricted to light winds. However, the basis of this response has not been satisfactorily explained in acridoids. It is maintained when both the moon and the sun are well below the horizon, and when the former is obscured by cloud. Orientation to star patterns appears to be beyond the sensitivity of the apposition eye of acridoids (Wehner, 1984), although such responses are claimed for a noctuid moth with a superposition eye (Sotthibandu and Baker, 1979). The polarization pattern at night is unlikely to be used because it is too faint to be detected by the insect eye (Riley and Reynolds, 1986). It has been suggested by Gould (1980) that all organisms have a general sensitivity to the earth's magnetic field, but there is little evidence for responses to magnetism in selected migratory species of insect (Jungreis, 1987). The possibility that migrants follow an environmental gradient, such as an increase in humidity, is discounted because most such gradients are thought to be beyond the known limits of insect sensitivity.

9.2.2.6 Downwind Displacement and Wind Systems

Measurements of the direction and speed of displacement of acridoids migrating over an observation site can be used to construct *back trajectories* to the time of takeoff to determine source areas (Drake and Farrow, 1983) as well as *forward trajectories* to estimate potential destinations. These estimates assume that any changes in direction and speed over time at the observation site, because of changing wind systems, also occurred in areas upwind or downwind of the site. Error will increase with distance of displacement but is small in uniform wind fields over distances of 100–200 km (Reynolds and Riley, 1988). The origins of migrants passing over an observation site at successive intervals can be plotted to provide *source lines* (Drake and Farrow, 1983).

In most situations, however, radar is not available and the track of migrants is unknown. Where insects are flying in relatively strong winds it can be assumed that their track does not deviate significantly from the wind vector. Consequently, *wind trajectories* or *streamlines*, derived from real-time patterns of atmospheric circulation, can be used to represent regional displacements of acridoids (Pedgley, 1981, 1982; Rainey, 1951). At night it is assumed that acridoids are transported downwind in the nocturnal geostrophic flow at a speed exceeding that of the streamline by 15–25 km/h. In some situations, however, supergeostrophic flows develop that transport migrants faster than expected along "conveyor-belt" systems (Rose et al., 1975), which may not always be detected by standard meteorological observations.

Every biogeographical region is subject to a particular pattern of circulation derived from the general pattern of global circulation and modified by local climatic and geographic factors. These patterns are illustrated in Figure 9.6 to indicate the potential effect of regional wind trajectories on acridoids migrating downwind over the major grassland habitats around the world for the months of January and July. It should be emphasized that real-time streamline charts should always be used to simulate actual migrations.

The regional systems shown here occupy three well-defined climatic zones:

1. The *tropical* zone is a zone of large-scale convergence between the northeast or southeast trade winds and southwest or northwest monsoons (the direction depending on hemisphere, Fig. 9.6). The seasonal limits of the ITC mark the boundaries of this zone, although the only region where transequatorial trajectories occur over grassland habitat is in east Africa. Within this zone there are two subsidiary convergences that influence migration (Rainey, 1976), the Red Sea convergence and the equatorial or rift convergence (Fig. 9.6). As stated earlier, the ITC is no longer recognized as a zone of permanent convergence, as indicated by the generalized streamline charts in Figure 9.6. Daily weather charts (e.g., Fig. 9.10) show areas of localized convergence and subsidence that grow and decay in situ or are associated with westward-moving cyclonic circulations (Barry and Chorley, 1987).

2. The *subtropical* zone is a zone of large-scale atmospheric subsidence, which is typically formed of eastward-moving high-pressure cells in the southern hemisphere and semipermanent cells in the northern hemisphere. Winds tend to be light and variable at the center of this zone, inhibiting long-range movement between tropical and temperate zones in many regions and tending to favor movement away from this zone.

3. The *temperate* zone is dominated by a westerly flow that is highly disturbed by eastward-moving depressions. The cold fronts embedded in these depressions separate warm subtropical air, which is directed poleward and is relatively favorable for migration, from cooler polar air, which is directed equatorward. Winds tend to rotate, often through a full 360°, during the passage of successive systems that cross specific regions at frequent intervals and provide opportunities for migration in a range of directions when temperatures are high enough for flight to occur. Winds are generally stronger in temperate than tropical regions and trajectories are correspondingly longer.

9.2.3 Migration Systems and Population Movements

The movements of acridoid populations that occur in response to the spatial and temporal separation of habitats in particular regions and the wind systems that operate over these regions represent a *migration system* that is generally unique to that region (Drake and Farrow, 1988). Downwind movements produce three main kinds of spatial redistribution of acridoid populations: (1) *unidirectional* interregional displacement on the trade winds, monsoons, and prefrontal flows (Figs 9.7, 9.8, 9.10, 9.12c); (2) *centrifugal concentration* in

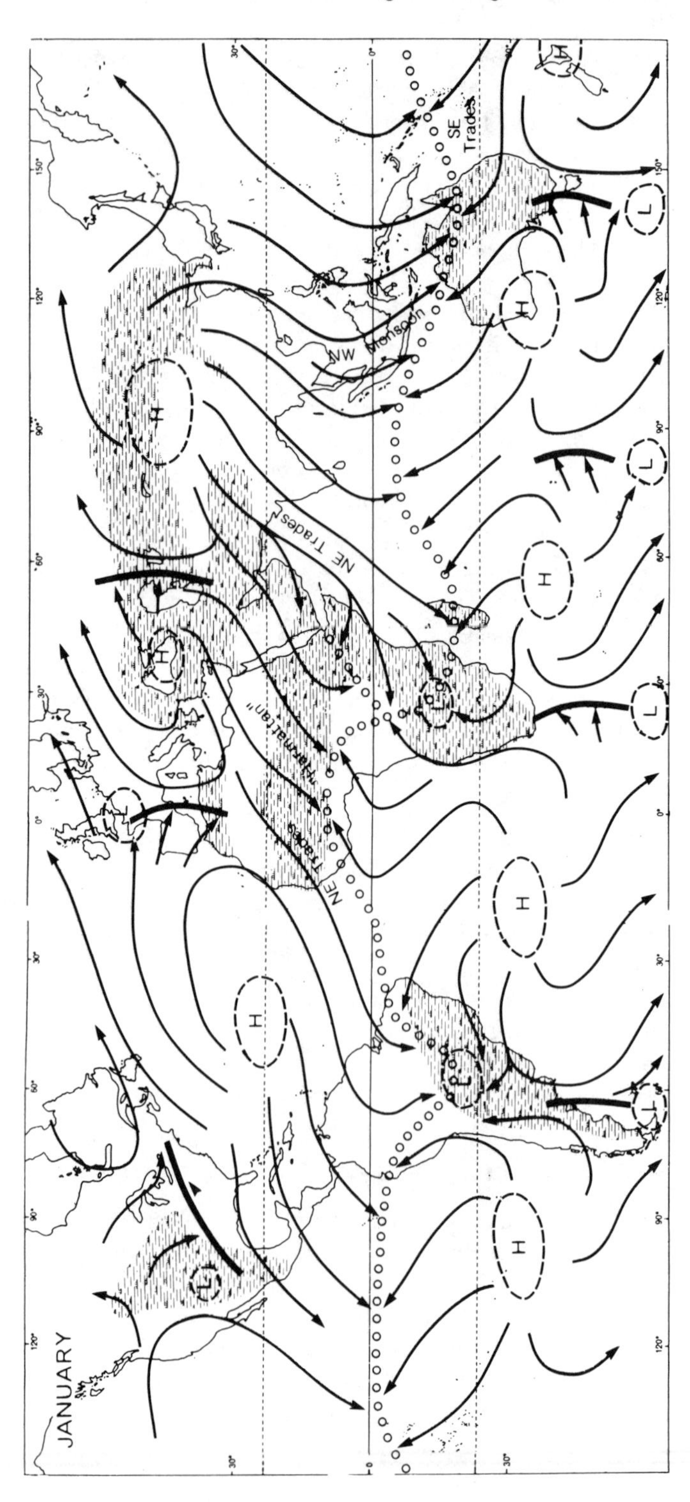
JANUARY
NE Trades
"Harmattan"
NE Trades
NW Monsoon
SE Trades
H
L

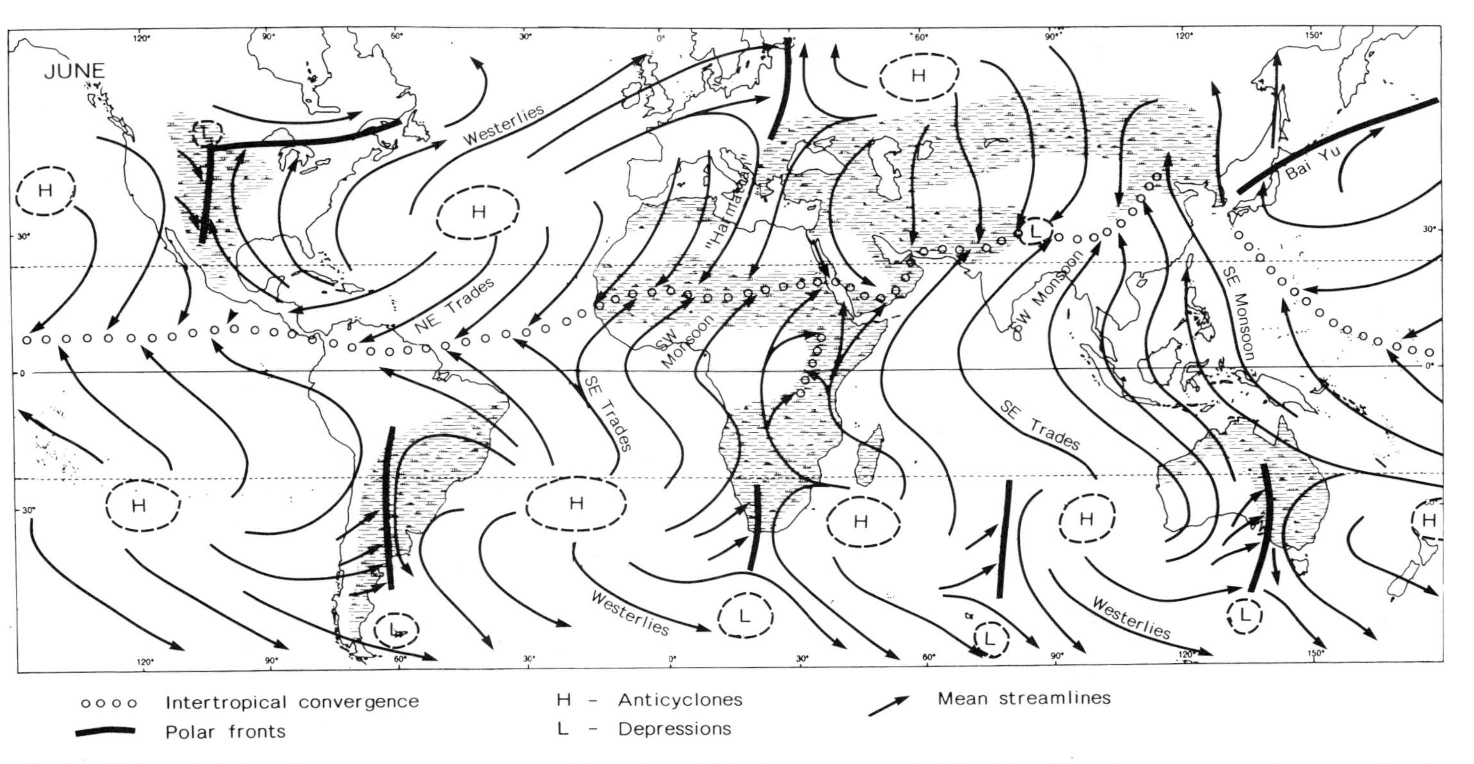

Fig. 9.6. Global wind circulation over major grassland habitats of acridoids, showing seasonal shifts of the ITC and associated monsoons and trade winds and the polar fronts and their associated prefrontal airflows.

cyclonic circulations and convergences of the tropics (Figs 9.9, 9.12c); and (3) *multidirectional dispersal* in anticyclonic zones, dominated by light and variable winds (Fig. 9.12a). The migration systems of selected key species or assemblages of species in five biogeographical regions, tropical and subtropical Africa, subtropical and temperate Australia, and temperate North America/Eurasia, are now examined to determine the extent to which population displacements between complementary breeding areas are explained by downwind movement in the local synoptic circulation.

9.2.3.1 Tropical Africa North of the Equator (10–20°N)

Favorable breeding habitats are associated with (1) the belt of rain that follows the poleward movements of the ITC, bringing a single rainy season to higher latitudinal zones of the tropics and a double season to equatorial zones and (2) the flood regimes of major river systems, most of which have northward components to their flow for at least part of their length (Rivers Senegal, Niger, Oubangui, and Chari).

The northern edge of the rain belt is situated several hundred kilometers to the south of the surface wind convergence in a deepening southwest airflow (Figs 9.7, 9.9). The rain belt progressively extends over the whole of the grassland zone of west Africa from May to August and retreats more rapidly in October (Fig. 9.7). Major floodplain grasslands tend to be progressively exposed from south to north at the end of the rains (Fig. 9.11a).

Large-scale southward displacements of acridoid populations occur at the end of the rains in the northeast airflow as it advances rapidly southward in the wake of the retreating ITC (Popov, 1988; Riley and Reynolds, 1983; Schaefer, 1976). Equivalent northward displacements have been less well documented and take the form of progressive intergenerational movements, in the wake of the advancing ITC, during the first half of the rainy season (Duranton et al., 1979a; Lecoq, 1978a,b; Popov, 1988) (Figs 9.7–9.9). Most migratory acridoids do not move to the furthest seasonal positions of the ITC (Fig. 9.7) but migrate over a range of 100–500 km within the zone 10–16°N, where rainfall conditions are probably optimal for reproduction (Launois, 1978). Movements over a similar range probably occur in the same latitudinal zones of east Africa. Variations in the timing of the advance and retreat of the ITC have a major impact on the dynamics of this migration system (Popov, 1988). If the ITC is late in retreating and is to the north of fledging populations at the end of the rains in October, emigrants are taken further north (if the winds are strong) at a time when populations would normally move south (Riley and Reynolds, 1983) (Fig. 9.10), although in light winds they preferentially move southward (Schaefer, 1976). Once the southward shift of the ITC is sustained, those acridoids that had initially moved north participate in the general southward displacement of populations.

In floodplain habitats, the most significant movements are the concentration of populations in the floodplains and the northward displacement in the wake

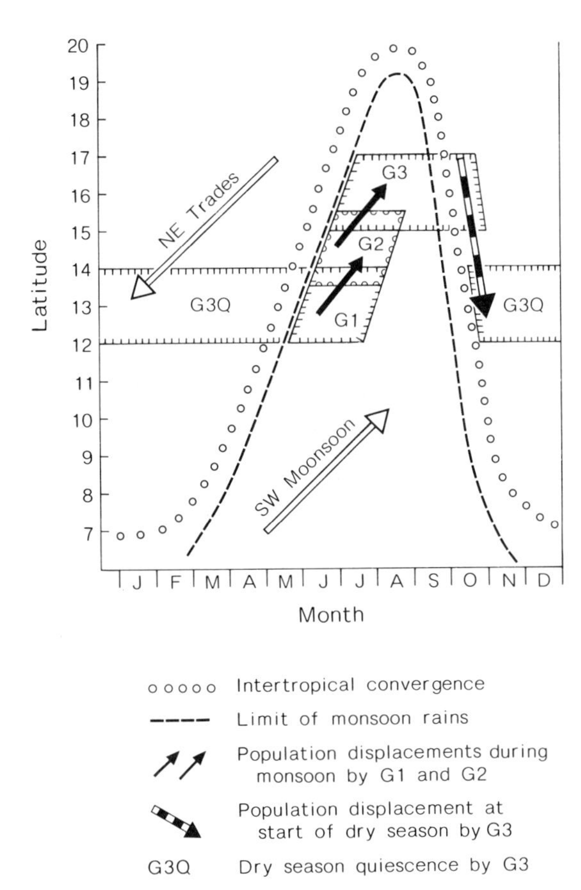

Fig. 9.7. Diagram of the seasonal location and population movements of a trivoltine acridoid in relation to the seasonal displacement of the west African monsoon and trade wind system. (Modified from Launois, 1978.)

of the retreating floods, as exemplified by *L. m. migratorioides* in the Central Niger Delta region (Davey, 1959; Farrow, 1975b) (Fig. 9.11). This species has a complex life cycle in this region: two generations are produced on the habitats exposed by the receding floods during the dry cool season (RF1 and RF2) and two to three generations are produced during the rainy season, first in the floodplain itself and later in the surrounding Sahelian habitats (R1–R3). The seasonal advance and retreat of the floods causes successive breeding habitats to be distinctly separated for each generation. Like the acridoids discussed previously, southward migrations occur in the northeast airflow established to the north of the retreating ITC in the period October–November, at the end of the rains (R_3, Fig. 9.11a). The earliest arrivals on the Niger floodplain occur at the time the ITC is situated at this latitude. However, as the ITC continues its rapid retreat south immigrants still arrive at the *same*

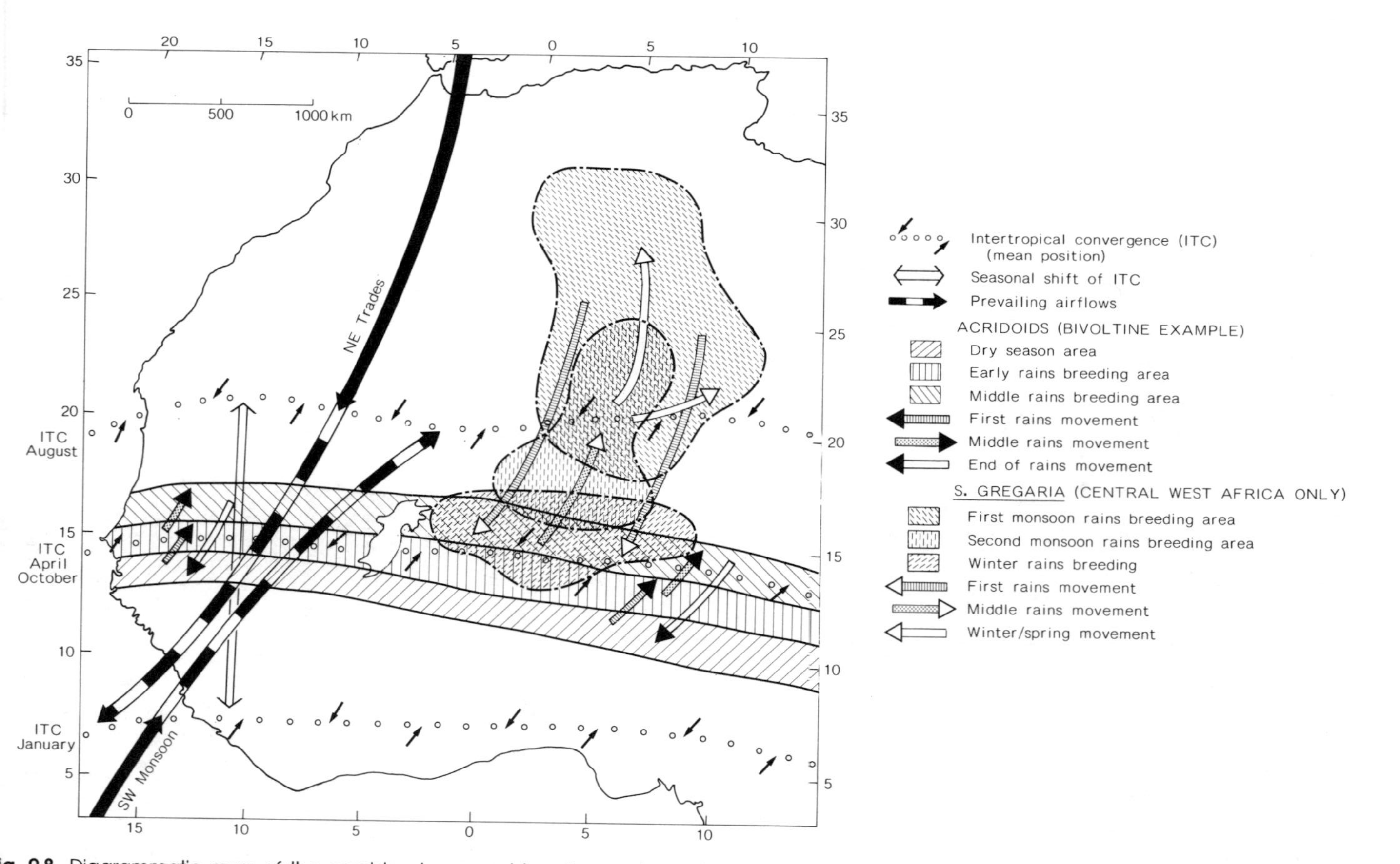

Fig. 9.8. Diagrammatic map of the combined seasonal location and population movements of a trivoltine acridoid and *S. gregaria* (central west African populations only) in relation to the seasonal displacement of the west African monsoon and trade wind system. (Partly based on Launois, 1978, and Popov, 1965.)

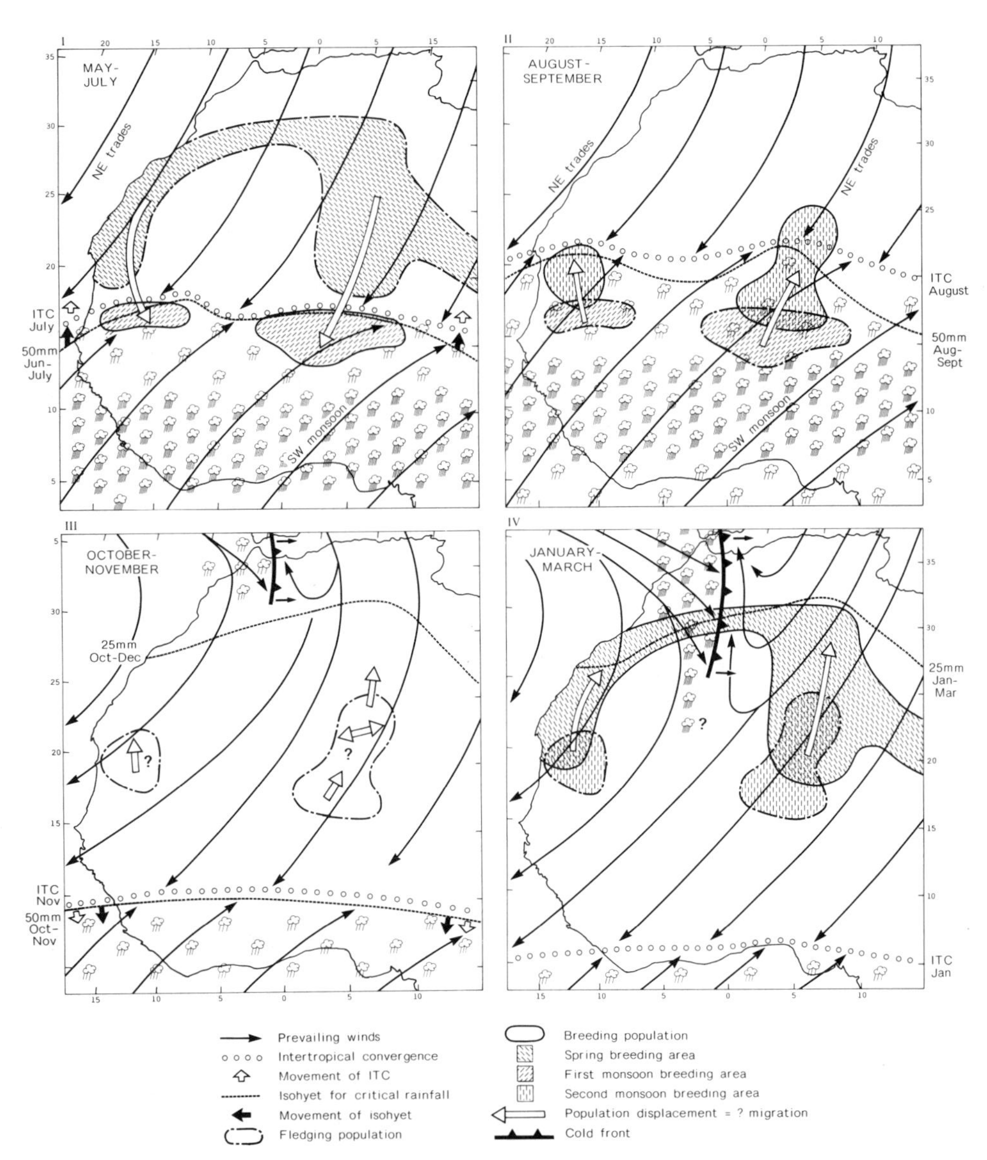

Fig. 9.9. Diagrammatic maps of the seasonal movements of each generation of *S. gregaria* in relation to the seasonal displacement of the ITC and polar fronts and the distribution of monsoon and winter rains in west Africa. (Partly based on Popov, 1965).

latitudes in the Niger and Bani floodplains. Populations also converge from the east and west of the delta, possibly assisted by the development of local offshore lake breezes at night, around the periphery of flooded areas. The same individuals then migrate northeast in the wake of the receding floods

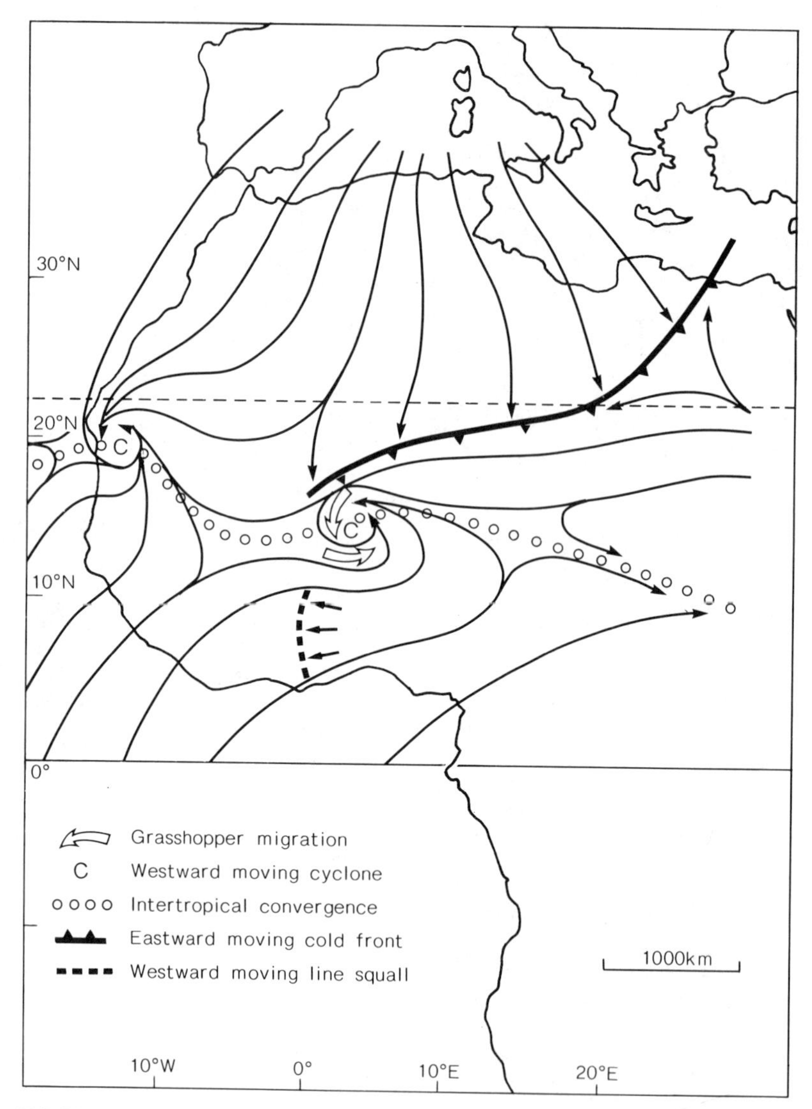

Fig. 9.10. Migrations of acridoids in west Africa in relation to the position of the ITC showing convergence in a cyclone embedded in the ITC and the effect of the location of the ITC on direction of population movement. (After Riley and Reynolds, 1986.)

(November–December). Their offspring (RF_1, Fig. 9.11a) also migrate north to the line of the receding floods (December–January). These latter movements take place *against* the prevailing wind and have been supported by radar observations (Riley and Reynolds, 1979). The next generation, produced in March–April toward the end of the dry season (RF_2, Fig. 9.11b), disperses over a short distance southward (with prevailing northeast winds), within the

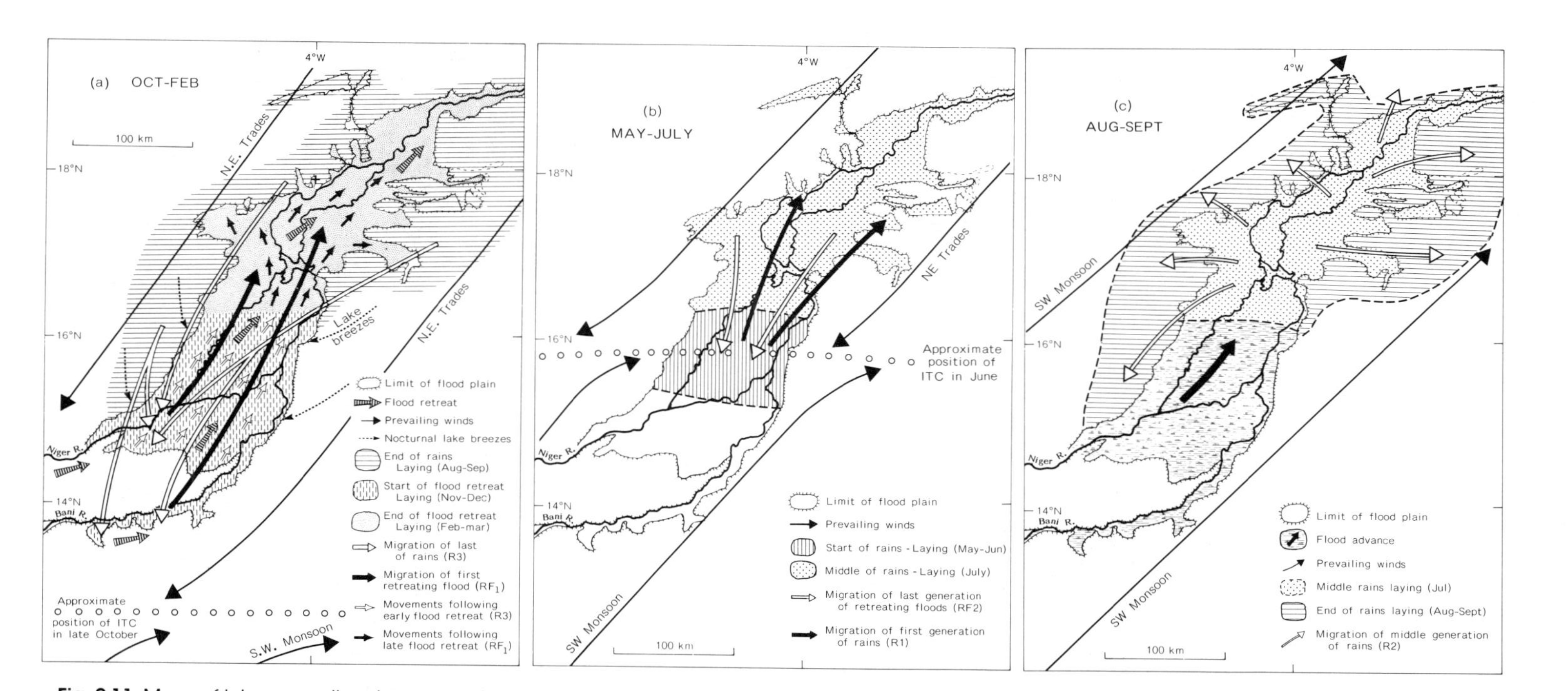

Fig. 9.11 Maps of intergenerational movements of *L. m. migratorioides* in the middle Niger area of Mali in relation to wind systems and flood regimes for each period of breeding. (After Farrow, 1972.)

floodplain, and starts laying *before* the arrival of the southwest monsoon. The progeny of this generation (R_1, Fig. 9.11b) are produced in the wake of the advancing ITC and migrate downwind to the northeast, to the northern floodplain in July. The generation of the middle rains (R_2, Fig. 9.11c) leaves the northern floodplain for the Sahelian zone, and movements are directed to habitats to the east and west of the floodplain. These also appear unrelated to local wind systems or to the position of the ITC. Similar complex seasonal displacements that are independent of wind direction at certain times of the year have also been established for *L. m. migratorioides* in Madagascar (Duranton et al., 1979b; Farrow, 1974; Têtefort et al., 1966).

Relatively few other acridoids exploit the habitats exposed by retreating floods, possibly because the retreat occurs in the cool season. *Paracinema tricolor*, *Aiolopus thalassinus*, and *Oxya* sp. are the commonest of these species and probably follow the northward-moving flood retreat against the direction of the prevailing winds.

The relatively short distance and directional bias of the intergenerational movements of most migrant acridoids in the tropics, compared with the potential extent and direction of movements suggested from actual wind fields, indicates that acridoid migration systems are strongly adapted to seasonal changes in the spatial pattern of the favorability of regional habitats and use the wind systems when movements in the desired directions can be achieved. There is little evidence that any migrant acridoids are concentrated at the interface in the ITC, except possibly in regions of cyclonic circulations; rather, there is a latitudinal shift of populations in the wake of the advance or retreat of the convergence. When the ITC moves north most migrants are found in the southwest airstream to the south of the ITC; conversely, when it retreats south, most are found in the northeast airflow to its north.

9.2.3.2 Subtropical Africa North of the Equator (20–30°N)

The migration system of this zone has been documented for only a single species, the desert locust *S. gregaria*, although the movements of its solitarious populations are considerably less studied than those of its swarms. Several species of migratory acridid are, however, found in both Mediterranean and tropical Africa and in Saharan oases. The recent upsurge of acridoids (including *L. m. migratorioides*) in newly developed center-pivot irrigation systems in what were previously extreme desert conditions in Libya (Duranton et al., 1983) suggests that acridoid migration across subtropical desert areas of Africa may be more significant than generally supposed.

Solitarious *S. gregaria* present a seemingly unique migration system in which populations persist in the region of the subtropical subsidence and depend on summer rainfall at the northern edge of the tropical zone (15–20°N, 100–500 mm rain/year) and winter rainfall extending from the southern edge of the temperate zone (25–35°N <50–100 mm rain/year) (Popov, 1965; Waloff, 1966). Populations appear to move north or at least disperse locally at the time the ITC is retreating south, the northeast trades are intensifying, and populations of other acridoids are moving south (Figs 9.8, 9.9). Mixed populations of immature *L. m. migratorioides* and *S. gregaria* have been observed departing over the same time period from a site in southern Mali in October during a persistent northerly airflow (Farrow, 1975b), presumably displacing in opposite directions. Southerly winds rarely occur before December at these sites and are caused by the approach of cold front systems to the north. The spring breeding areas of solitarious populations are poorly known from North Africa, although immigration is invariably associated with temporary, warm, prefrontal airflows from the south (Bennet, 1976; Pedgley, 1982). This suggests that the initial movements from the subtropics are localized and dispersive in nature, with major displacements being delayed until the passage of potentially rain-bearing disturbances. These may also result in the concentration of previously scattered populations. As Popov (1965) has shown, the monsoon and winter rains breeding areas overlap considerably (Fig. 9.8), suggesting that for some individuals and local populations intergenerational movements may be of limited extent. Because the probability of winter populations receiving adequate rain for breeding is so low (Bennet, 1976), it is possible that much of the population fails to breed in winter. It is not known whether these individuals persist in a relatively sedentary state in the highlands of the central Sahara until the following monsoon season when they disperse and breed, or whether there is heavy adult mortality. Similar poleward movements against the direction of the prevailing winds occur during winter throughout the *S. gregaria* distribution area between west Africa and India.

Individual *S. gregaria* reappear in the Sahelian zone at the start of the rains, presumably as a result of downwind movement on the northeast trades (Figs 9.8, 9.9). There is no evidence that this generation is influenced in the same way by frontal systems as its predecessor. This would cause populations to move further north to the Mediterranean region in late spring, a movement analogous to that observed in *C. terminifera*. The following generation, produced in the Sahelian zone, is overtaken by the northward-moving ITC and downwind movements can carry populations back into the central Sahara where populations appear to concentrate in areas of relief (Popov, 1965), where there is sufficient runoff for suitable breeding habitats to form. Similar movements between complementary breeding areas occur in east Africa, where the main difference is the concentration of populations along the Red Sea coast in early winter in response to the intensification of the Red Sea convergence (Fig. 9.6), which may produce sufficient rain for successful breeding to occur. This is followed by movement to spring breeding areas under the influence of prefrontal airflows.

9.2.3.3 Subtropical Australia (20–30°S)

Migration systems in the subtropical zone of Australia are exemplified by the movements of *C. terminifera*, which have been particularly well studied (Clark,

1969, 1971, 1972, 1978; Farrow, 1975a, 1977a). Nocturnal movements occur within the bounds of the subtropical grasslands (20–28°S) and comprise short-range dispersal in anticyclonic conditions and longer-range concentration during the development of low-pressure circulations (Fig. 9.12). There are no marked seasonal shifts in wind patterns that would favor large-scale seasonal movements within this zone in Australia, although between autumn and spring (through winter), prefrontal, poleward flows are more strongly developed (favoring unidirectional southward migrations out of this zone). In summer, tropical disturbances (causing concentration) tend to prevail but there is no evidence of downwind movement on the prevailing southeast trades into more equatorial latitudes, as suggested by Symmons (1986). Although most population movements can be interpreted in terms of downwind migration, activity is nevertheless restricted to particular patterns of circulation.

Similar population movements involving dispersal, concentration, and unidirectional displacement probably occur among acridoids of southern Africa and central South America.

9.2.3.4 Temperate North America and Eurasia (30–60°N)

Although poleward migrations of a great range of insects occur at night across the prairie belt of North America (Rabb and Kennedy, 1979), none of the numerous acridoids of this region has been identified participating in such movements. Trivial movements within habitats are attributed to walking (Uvarov, 1977; Joern, 1983) and to the short daytime flights discussed earlier. Local populations do not recover numerically from the effects of chemical control for several years, suggesting that migration rates are low. Little is known about the population movements of solitarious acridoids at night in Eurasia.

9.2.3.5 Temperate Australia (30–38°S)

Adults of several species of acridoids often appear in early summer on warm northerly airflows that penetrate this zone before the local populations of those species have fledged (Farrow, 1977a). These immigrants originate from populations that had fledged earlier under the warmer conditions experienced in the more northerly latitudes. Southward movements are not balanced by equivalent northward returns. Movements in these two directions, as well as in a range of other directions (as winds rotate during the eastward passage of anticyclonic cells), are as frequent in autumn as in spring and do not support the existence of a circuit migration, proposed for *C. terminifera* by Roffey (1972). Absence of a bias to orientation during migration also suggests that there are no preferred directions to nocturnal movements in this species.

During periods of above-average rainfall when suitable breeding conditions occur outside the normal range of tropical and subtropical species such as *L. m. migratorioides* and *N. guttulosa*, breeding populations are temporarily established by immigration over an expanded distribution area extending into

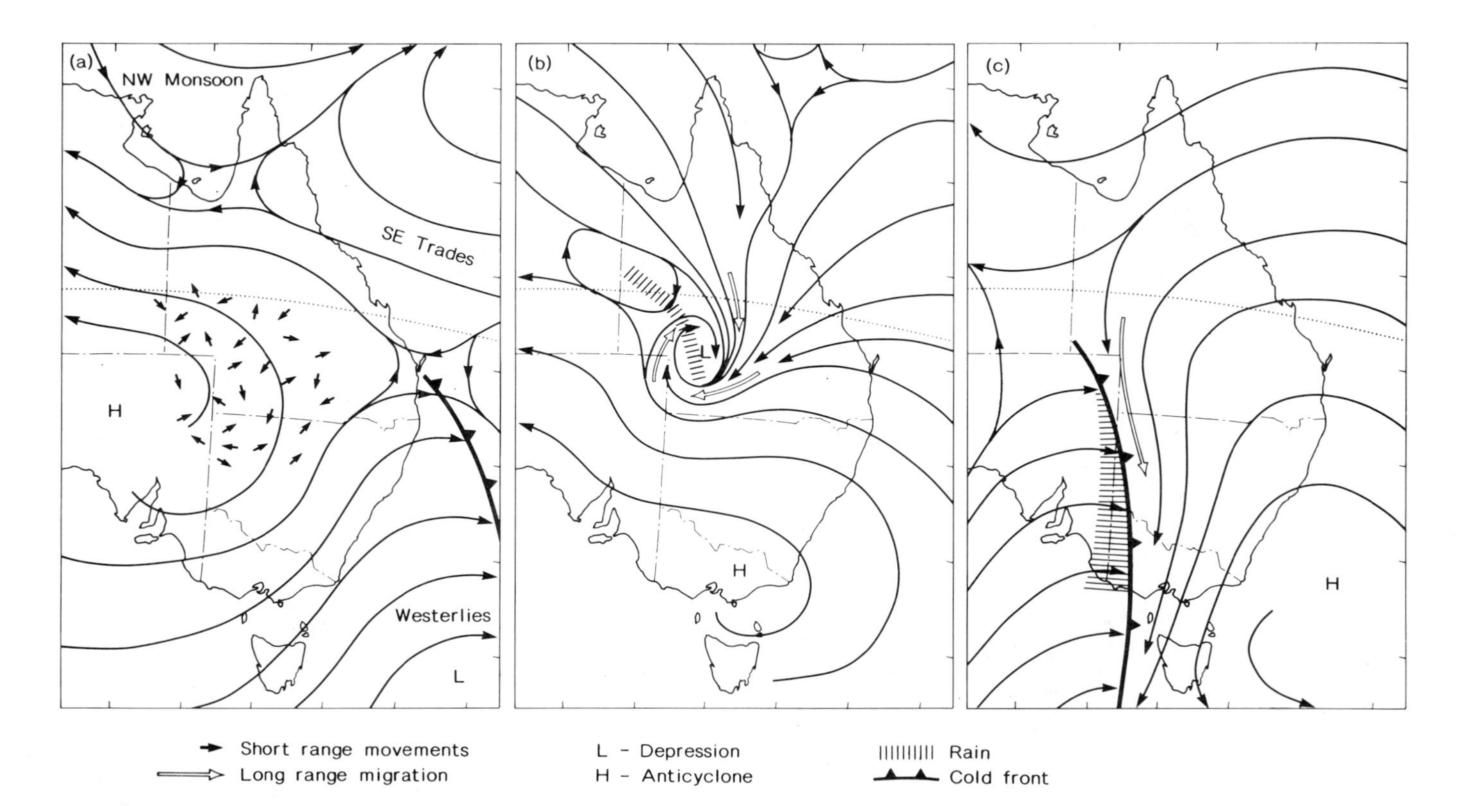

Fig. 9.12. Diagrammatic maps of (a) dispersal in anticyclonic conditions, (b) concentration in cyclones, and (c) unidirectional displacement in prefrontal airflows by *C. terminifera* in eastern Australia.

temperate latitudes of southern Australia (Farrow, 1977b, 1979b; Nicolas and Farrow, 1984) (Fig. 9.19).

9.2.4 Migration and Life Histories

Although acridoids are a diverse and often dominant insect group of grasslands throughout the world, their development and survival depend on two potentially conflicting requirements that are inversely related through the effects of cloud cover, namely, rainfall and insolation. The effects of rainfall may be indirect, through runoff and flooding and control of vegetation growth. Eggs require sufficient soil water to develop and hatch; nymphs and adults require fresh green food to survive and develop, adequate insolation to maintain a high metabolic rate, and shelter from extremes of weather and from natural enemies. Development and reproduction are adversely affected by lack of insolation (Begon, 1983).

The complementary requirements of adequate moisture and high insolation are the main reasons why acridoids reach their greatest abundance and diversity in grasslands of the semiarid tropics. The climate is characterized by long spells of sunny weather, interspersed with brief and often heavy falls of rain; the habitats are dominated by mosaics of grass tussocks and bare ground. Habitat favorability declines toward the equator as cloudiness and rainfall increase, probably because of the susceptibility of acridoids to pathogens in moist environments (e.g., entomopathogenic fungi). In the temperate steppe, prairie, and Mediterranean regions, summers are dry and warm and much of the moisture is made available by more generalized winter rain or snow when populations are dormant. In the more maritime climates, increasing summer rainfall and cloudiness cause a decline in environmental favorability and in the extent of natural grassland habitats.

Migration and diapause are usually seen as two alternative strategies to cope with gross seasonal changes in environmental favorability, habitat patchiness, and impact of natural enemies. *Suspended development* is generally a more appropriate term than diapause, because there is often uncertainty as to whether an organism is in true diapause or is simply quiescent. Furthermore, diapause per se does not necessarily confer tolerance to extremes but simply prevents organisms from emerging at a time when any activity would be difficult and when reproduction would be unsuccessful. Migration is often associated with direct development and continuous reproduction in a network of complementary breeding habitats whereas suspended development is typically associated with a sedentary existence in a single habitat in which extremes are simply tolerated a variety of ways. In acridoids, however, migration and suspended development (with or without diapause) are essentially complementary strategies, closely adapted to the scale of the local temporal and spatial discontinuities of the regional environments previously described. In seasonal environments, life-history strategies can be conveniently divided into those that represent adaptations to maximize multiplication during the reproductive season and those which minimize mortality during the unfavorable season.

9.2.4.1 Tropics (20 N–20°S)

Most tropical grasslands are restricted to zones in which there is a marked advance and retreat of an intense dry season. Acridoids have adopted a range of strategies to survive this period and to maximize reproduction during the short rainy season. The majority of species are fully winged, suggesting that flight at least within and between habitats and possibly between regions is ecologically essential in this environment (Reynolds and Riley, 1988). Of the 70 common acridoids of the west African Sahel, included by Fishpool and Popov (1984) in their survey of life-history strategies, about 55% spend the dry season as reproductively quiescent adults and comprise both multi- and univoltine species. Of the remainder, half (22% of the total) spend the dry season as quiescent eggs whereas half show no arrest but reproduce continuously. It is difficult to extrapolate results from well-documented studies of migration in species such as *D. axillaris* (Reynolds and Riley, 1988) and *O. senegalensis* (Riley and Reynolds, 1983) to allied species because of the different life-history strategies adopted by members of the same genus (Fishpool and Popov, 1984). Nevertheless, there is intensive migratory activity at the start of the dry season that results in a general population shift toward the equator. A less well documented migratory phase occurs at the start of the rains and at intervals during the rains, resulting in poleward population shifts.

In those species that spend the dry season in the egg stage, migratory movements have been described mainly from multivoltine species such as *O. senegalensis*, in which the post-rains migration is essentially prereproductive and enables adults to lay in a region where the eggs are less stressed by drought. The eggs can hatch earlier than in populations that lay at higher latitudes, owing to the earlier onset of the rains, a time difference that may be the equivalent of a generation. Of greater significance to survival are the subsequent intergenerational movements to higher latitudes during the rainy season, which presumably improve reproductive success (Launois, 1978). The movements of univoltine species that survive the dry season as eggs must necessarily be more limited if they are to lay eggs in similar locations each year, and in general, the latitudinal spread of these species appears to be relatively small (Fishpool and Popov, 1984).

In those species that spend the dry season as reproductively quiescent adults, the extent of movements by newly fledged adults at the start of the dry season and by maturing adults at the start of the rains varies considerably between species. At one extreme are species like *D. axillaris*, which undergo extensive population shifts toward the equator at the end of the rains into a higher rainfall zone. Here the vegetation is in a better condition to support and shelter adults for the duration of the dry season (i.e., more perennial grasses and shrubs). Although the complementary poleward movements at the start of the rains are less well documented (Joyce, 1952), they enable populations to colonize habitats in which optimal combinations of insolation and rainfall, and survival from natural enemies, maximize reproductive success. These displacements appear to be initiated by a rapid downwind migration between regions followed by more localized short-range movements in a range of directions both between and within habitats. As Reynolds and Riley (1988) point out, this life history can evolve only where the risks associated with the move to different regions outweigh those that may occur if individuals remain in the same locality (Dingle, 1980; Southwood, 1977). The strategy of delaying reproduction and moving to a more favorable aestivation site belongs to Johnson's class III migration (1969), and it has been suggested by Solbreck (1978) that this occurs where there is a simultaneous and widespread deterioration of habitats.

Not all acridoids in this group (with a reproductively quiescent adult) appear so migratory: the locust *N. septemfasciata* fails to show significant regional displacements between breeding and aestivating sites, although local movements between habitats are important as conditions deteriorate in the dry season and adults concentrate in more favorable refuges (Chapman, 1959).

In the fully winged, bivoltine species *Caledia captiva*, maintenance of a narrow hybrid zone between two chromosomal races in tropical Australia (Moran et al., 1980) suggests that individual displacements are quite limited and appear restricted to walking. This species is limited to higher rainfall zones in eastern Australia, where variations in seasonal conditions are widespread and where the chances of locating a more favorable patch elsewhere by migration are perhaps low.

Those species that reproduce continuously, such as *L. m. migratorioides*, require suitable refuges for breeding during the dry season. These are usually represented by marshes or riverine habitats, of which the most important in west Africa are the floodplains of the Central Niger Delta and the Lake Chad system. Substantial populations of species such as *L. m. migratorioides* are confined to the vicinity of such refuges. This species can be alternatively viewed as a primary marsh-inhabiting species (c.f., temperate races of this species) that temporarily colonizes rainy season habitats only when the marshes become seasonally inundated. This conclusion is supported by the observation that populations do not colonize the Sahelian region of Mali until midway through the rainy season as the floods start to rise (Farrow, 1975b). The subsequent concentration of populations in floodplains is primarily a result of directed movements into limited areas of habitat rather than of convergence by atmospheric processes. Although the prereproductive migration of species like *L. m. migratorioides* appear to be obligatory in these environments, in those regions in which conditions are favorable for breeding all year round (e.g., tropical grasslands where a moderate rainfall is evenly distributed year round), there is little evidence for synchronized population displacements (i.e., in Australia and Papua New Guinea). These responses are typical of a facultative class I migrant of Johnson (1969). Where the same seasonal habitats are occupied each year, following the circuits described earlier for *L. m. migratorioides* in Mali, populations are susceptible to high levels of mortality from a wide range of natural enemies (Farrow, 1975b), which appear to be less important for acridoids inhabiting the more arid areas to the north. The options presented here suggest that in a continuously breeding species with three to five generations annually and a high probability of locating suitable breeding sites, the average annual reproductive rate may be no greater than in species that breed over one or two generations in more arid habitats, where mortality from natural enemies is lower and where suitable habitat patches are more widely distributed.

The acridoid fauna of forested areas is relatively poor (Uvarov, 1977) and contains a preponderance of brachypterous species, suggesting that a sedentary habit has secondarily evolved in this habitat. The predominantly brachypterous pyrgomophid *Zonocerus variegatus* has become abundant in areas cleared for cropping in which varying proportions of macropterous forms are produced by a combination of high densities and food shortages that occur in the dry season (Chapman et al., 1986). It is suggested that this grasshopper was primarily adapted to temporary clearings in the original forest, formed by falling trees, and that macropterous forms fill the role of colonizers.

9.2.4.2 Subtropics (20°N/S–30°N/S)

The subtropics present different problems for acridoids: rains are typically unpredictable and localized and falls are highly variable. Rain is associated with cyclonic circulations and active cold fronts rather than with linear flows typical of the tropics and the principal strategy is one of *opportunism*.

Adults, rather than desiccation-resistant eggs, persist through extended unfavorable conditions in an immature, sedentary state. When rain-producing weather conditions develop, populations are quickly able to locate rain-affected areas where they reproduce and multiply with only limited checks by natural enemies. If conditions remain favorable at the initial breeding site as a result of further rain, follow-up breeding may occur without any intergenerational displacements, although mortality from natural enemies may then increase. A low level of migratory activity in settled weather prevents populations from (a) migrating at a time when reproduction cannot occur, (b) dispersing to produce densities so low that mating would not occur (Gatehouse, 1988), and (c) dispersing to less favorable tropical environments under the influence of the trade winds. However, migration in disturbed weather can present problems to emigrants when the fledging areas are covered by poleward airflows associated with passing cold fronts, for it can induce population movements to higher latitudes from which there may be no return. This problem appears to have been solved by *S. gregaria* by directed movements on preferred wind systems, but not by *C. terminifera*, where one-way migrations into temperate latitudes appear to be an inevitable consequence of a migratory behavior adapted for the subtropics.

9.2.4.3 *Temperate Zone (30°N/S–60°N/S)*

In the prairies and steppes of North America and Eurasia, the season for development and reproduction is short. Most acridoids are univoltine and the extremes of winter are spent as a cold-resistant, diapause egg. The apparent absence of long-range population movements from both continents appears to be related to a complex of factors: (1) Rain is typically deposited in eastward-moving belts aligned north/south (c.f., tropical regions) in the wake of active cold fronts, as well as in warm fronts in the northern hemisphere, and is regularly deposited over large areas. As a result, growth and senescence of grasses is relatively synchronized over large areas and varies little from year to year; consequently, substantial interhabitat movements are unnecessary to maximize reproduction. (2) Suitable winter refuges and grassland source areas in the subtropics are of limited extent in the northern hemisphere. (3) Dusk temperatures are often too low for flight, particularly at higher altitudes. (The flight threshold is higher in acridoids than in other nocturnal migrants such as noctuid moths). (4) Airflows suitable for transporting migrants southward in autumn are generally too cool to permit return flights. Not all acridoids of the temperate zone are sedentary; populations of at least one species in Australia, *C. terminifera* are, as shown below, largely sustained by immigration from lower latitudes.

Ritchie and co-workers (1987) have claimed that expression of macropterism in brachypterous grasshoppers of the northern temperate zone acridoids is not a dispersal mechanism, although its occurrences was accentuated in hot summers when conditions for flight were optimal. In contrast, expression of macroptery in the predominantly brachypterous acridoid, *Phaulacridium vittatum* is density-related in temperate Australia and flight results in the colonization of new habitats as well as dispersal from high-density populations (Farrow et al., 1982). As in *Z. variegatus*, macroptery appears to have been essential for colonizing naturally occurring woodland clearings, which probably formed the original habitat of this species.

9.2.5 Interactions between Migration, Fecundity, and Mortality

Migration has often been assumed to be a "costly strategy" because of (1) the potential diversion of resources from egg production into energy for flight and (2) the possible failure of migrants to locate habitats in which to reproduce (Angelo and Slansky, 1984; Dingle, 1972; Harrison, 1980; Jago, 1983; Roff, 1977). Such potential costs cannot be simply evaluated in terms of the differences in fecundity and adult mortality between migrants and nonmigrants, because these measurements usually fail to take account of the overall advantages of migrating or of staying, in terms of the numbers of viable offspring produced over time. On the other hand, high reproductive capacity is also regarded as one of a set of attributes of a successful colonizer and is required to exploit new environments (Parsons, 1982; Simberloff, 1981) and involves rapid maturation and high fecundity.

9.2.5.1 Fecundity and Maturation

Fecundity in acridoids is a function of the number of eggs per pod and the number of pods. The first is limited by the ovariole number, which is fixed in the embryo. Not all ovarioles function during maturation and there is a progressive resorption of oocytes, the extent of which is influenced by a range of environmental factors, including food quality and temperature (Farrow, 1975b, 1979a; McCaffery, 1975). The number of pods produced is similarly affected by environmental factors but tends to vary between two and three for the larger species, including most locusts. Partitioning of reserves between the eggs during oogenesis causes some variation in egg size, which affects the size and food reserve of the hatchling and may also influence reproductive success.

If migration were to affect the number and size of eggs in a pod, it would be expected that the first pod laid following immigration would contain fewer eggs or lighter eggs than those laid later. There is no such evidence of smaller clutches in two acridoids (Farrow, 1975b, 1979a; Farrow and Longstaff, 1986) or of smaller eggs (R. A. Farrow, unpublished). It is interesting to note that solitary populations of *L. m. migratorioides* lay fewer eggs per pod on average in the main outbreak area in Mali than in the outbreak area in Australia, where populations appear to be less mobile. However, this appears to be caused by differences in environmental conditions (Farrow and Longstaff, 1986).

In species polymorphic for wing length, flightless brachypterous females generally lay more eggs per pod than their macropterous counterparts (Ritchie et al., 1987). However, this may arise from the greater space available for all ovarioles to produce eggs when the wing muscles are not developed, rather than from any increase in resources.

Maturation in both *C. terminifera* and *L. m. migratorioides* is extremely rapid during migration and there is no evidence from field populations that production of the first pod is slower than the later ones (Farrow, 1975b). In a laboratory-based study on a day-flying species (*M. sanguinipes*), McAnelly and Rankin (1986b) could not detect any delays in oviposition between females that were flown for varying periods on flight mills and unflown controls. Instead flight was shown to speed up oogenesis, relative to controls, in this species. There is little evidence that this response is a general one among acridoids and would not, in any case, apply to the many species that migrate when in reproductive diapause.

Flight and migration are typically brief episodes in the life cycles of solitarious acridoids, and a brief diversion of resources into migration appears to have little long-term impact on fecundity. Compared with many other insects, the fecundity of both migratory and sedentary acridoids is relatively low, amounting to 50–300 eggs per female [c.f. females of migratory noctuids (Lepidoptera), which lay 2000–3000 eggs]. Nevertheless these provide a potential rate of population increase per generation of 25–150-fold. Unlike many other phytophagous insects, the eggs of acridoids are not laid on the host plant and individual eggs must be large enough to produce an active nymph with sufficient reserves to locate a suitable host. Consequently there is little flexibility for maximizing reproductive success by improving fecundity, rather, this is achieved by colonizing environments which improve the subsequent survival of eggs and nymphs. The high rates of multiplication observed in ephemeral habitats ($R_0 = 5-20$) appear to arise from relatively low rates of predation, parasitism, and disease, although perinatal mortality usually remains high (Farrow, 1982a).

9.2.5.2 Migration Losses

Total numbers of an insect species are determined by the net balance between births and deaths, where the latter includes the mortality of individuals that emigrate and fail to find habitats suitable for reproduction, either outside the species distribution area or within this area in a patchy environment. Most population studies on acridoids are carried out in discrete locations and produce estimates of the changes between parent and offspring generations (*net reproductive ratio* R_0) and the movement of individuals between fledging and reproduction, into and out of the study area (*net migration* index $M\pm$) (Farrow, 1979a). Neither of these parameters provides a value for the proportion of migrants that fail to reproduce following emigration. It has long been assumed that this value is high (Harrison, 1980), although Taylor and Taylor (1983) and Solbrek (1985) have provided evidence that challenges this conclusion.

In acridoids, complementary breeding areas are generally discrete and the cumulative values of the reproductive ratio over a number of generations will provide an indicator of the level of adult losses within generations. Long-term studies of *L. m. migratorioides* in the Middle Niger Area showed that although intra-annual fluctuations in numbers between generations are very large, interannual variations are small, equivalent to a long-term R_0 of about 1 (Farrow, 1975b). Although populations declined during the study period, this low productivity could not have sustained adult losses during migration between the complementary fledging and laying areas, shown in Figure 9.11, of more than 10% per generation. No immigration from outside areas was detected.

In *C. terminifera*, on the other hand, long-term annual productivity in central-western New South Wales, over a 3-year period, was much higher, equivalent to an average R_0 of about 6. Nevertheless, populations declined over the 3-year period due to net emigration. If this net reproductive increase is typical of the entire breeding area of this species in eastern Australia, it must be balanced by a net loss of adults at some stage. In view of the net southward migration of this species into less favorable habitats it can be concluded that there are substantial peripheral losses of adults as a result of emigration out of suitable breeding areas (Farrow, 1975a).

These two examples possibly represent the opposite ends of a continuum, comprising at one extreme those species that can afford to expend high adult losses in the location of ephemeral habitat patches, in which high reproductive rates are achieved, and at the other, those that occupy seasonally varying but temporarily and spatially predictable habitats, where reproductive success is lower (due largely to high rates of predation and parasitism) but the chances of locating complementary breeding areas are high.

9.2.6 Conclusions — Migration Strategies of Grasshoppers versus Solitarious Locusts

It is difficult to distinguish between solitarious locusts and grasshoppers in terms of both flight behavior, discussed here, and biological attributes such as reproductive capacity (Farrow and Longstaff, 1986). Locusts in their solitary state comprise both opportunistic but active migrants, like *S. gregaria* and *C. terminifera*, facultative migrants like *L. m. migratorioides*, and relatively sedentary species, like *N. septemfasciata*. A similar range of migratory activities occurs in the grasshoppers varying from highly migratory species such as *O. senegalensis* and *D. axillaris* to a wider range of more sedentary species. In terms of the exploitation of particular habitats, however, locusts appear more capable of locating habitats in which they achieve high rates of multiplication, attributes that contribute to their unique capacity to form outbreaks and plagues. Of those acridoids adapted to concentrate in, and migrate between, complementary regional habitats by means of directed movements, *L. m. migratorioides* appears to be more efficient at maintaining year-round productivity in such environments than other acridoids. Of those that attempt to concentrate in areas where rain has recently fallen, by means of downwind displacement, *S. gregaria* and *C. terminifera* appear to exploit such ephemeral habitats far more effectively than other acridoids.

9.3 FLIGHT AND MOVEMENT IN GREGARIOUS POPULATIONS

9.3.1 The Gregarious Response

All locust and many grasshopper species respond to increases in population density by aggregating in marching bands as nymphs and in airborne streams and swarms as adults, both of which are generally active by day only. Underlying this increase in locomotor activity is adaptation to the intraspecific encounters that occur as densities increase, so that individuals become attracted to one another instead of being indifferent or possibly even repelled. Acridologists have incorrectly termed this process habituation or conditioning (Uvarov, 1977). Those acridoids termed locusts are generally more sensitive to the effects of crowding, particularly in the juvenile stage, than those referred to as grasshoppers, although in some locusts (e.g, *Anacridium* spp. and *N. guttulosa*) only the adults aggregate in response to crowding. The range of physiological and morphological attributes associated with crowding in locusts (Uvarov, 1969) also occur in many grasshoppers when crowded (Jago, 1983). The change of state from the solitary to the gregarious condition is generally called *gregarization*.

In most acridoid species, gregarious behavior is never observed in the field. It cannot be assumed that this is necessarily due to the lack of a gregarious response as the population densities of many species are regulated at low levels by environmental factors and may never attain the levels required to induce gregarization. Where changes in environmental conditions have caused high densities to develop, gregarious responses have been observed in species in which this phenomenon had never been previously recorded (Ahluwalia et al., 1977; Anonymous, 1982; Farrow, 1982b). The number of species of "grasshopper" observed in some form of gregarious aggregation appears to be increasing in many parts of the world. This is probably a consequence of the loss of vegetation cover in grassland habitats as a result of overgrazing and other man-made changes (Farrow, 1984, 1988; Uvarov, 1957, 1977).

The intensity of the gregarious response in adults depends on differences in the length of time for which individuals, and their parents, have been aggregated as well as intrinsic differences in the level of response to density. Gregarious adaptation may be brief when previously scattered adults are brought together by meteorological processes or by the effects of a patchy habitat, or it may extend over longer periods if the hoppers have been crowded and band formation has occurred and may be particularly protracted if members of preceding generations have been aggregated. These last sequences are more typical of locusts than grasshoppers and lead to the initiation of mass daytime flights in swarms soon after fledging; if the adults aggregate after fledging, there is a more protracted period of recruitment and aggregation of flying individuals.

Adult aggregations, when airborne, may vary from streams, in which the fliers do not spread far from the direction of flight, to swarms where fliers advance over a broad front that is generally wider than the depth of the aggregation. Streams are typically diffuse, low-flying, and ephemeral. Swarms are more cohesive and persistent and vary from low-flying, slow-moving aggregations to those that are high-flying and generally faster-moving. In most species there is a tendency for gregarious behavior to intensify over time as streams consolidate into swarms and aerial densities increase. The pattern of daily behavior in a gregarious population may also involve a progression from streaming to swarming activity (Uvarov, 1977). The behavior associated with the low-flying, diffuse and often ephemeral aggregations of acridoids has been termed *facultative gregarization* (Uvarov, 1977), to distinguish it, in terms of displacement potential, from the more intense gregarious behavior observed in mobile swarms of *S. gregaria* and a few other locust species. Although the desert locust, *S. gregaria*, is often used as the model for gregarious behavior in acridoids (Uvarov, 1977), few locust and no grasshopper species ever come close to exhibiting a similar level of swarm structure and behavior, suggesting that the gregarious response of this species is atypical of the response in most acridoids. In many airborne aggregations of acridoids, however, there is often little to distinguish the gregarious behavior observed in those species, nominally termed grasshoppers, such as the Senegalese grasshopper (*O. senegalensis*), from the gregarious behavior of those species called locusts.

9.3.2 Initiation of Daytime Gregarious Flight

Gregarious acridoids typically spend the night "roosting" on vegetation in dense aggregations. Morning takeoff starts with "spontaneous" flights by a few individuals, which then stimulate further flights so that the whole aggregation rapidly becomes airborne and leaves on a sustained flight. Takeoff occurs only once temperatures have risen to a suitable level and other weather conditions are favourable.

Precursors of this kind of response have been observed in an outbreak of *L. m. migratorioides* in Australia. Where concentrations formed by the aggregation of adults after fledging (Farrow, 1979b), mass takeoff was observed only after individuals were provoked into flight by artificial disturbance. Escape flights by flushed individuals then provoked a ripple of short flights to spread throughout the concentration (R. A. Farrow, unpublished). Because many individuals took off before others flew overhead, it is probable that sound (e.g., rustling of beating wings) was as important a cue as sight, and sudden noises also stimulated takeoff, as discussed by Haskell (1957).

In aggregations of this species in the same area, where gregavization had been of limited duration during part of the preceding juvenile stage, takeoff was induced by a few individuals, but the duration of flight was short, causing simultaneous milling lasting only a few minutes (Fig. 9.14a) (Farrow, 1979b; Symmons, 1979). In the populations studied above, there was no evidence that such behavior ever developed into sustained flight in that generation. Similar kinds of flight activity, which do not lead to significant displacements of aggregations, have been reported from a range of species of grasshopper and locust (for references, see Uvarov, 1977) and have been termed *group flights* (Uvarov, 1977).

9.3.3 Components of Gregarious Flight Behavior

Gregarious flight behavior is manifested at three levels and involves: (1) movement of individuals within aggregations; (2) movements of individual aggregations; and (3) movements of gregarious populations (constituting a variable number of swarming units) during the life of plagues. Gregarious flight behavior has been studied in detail in only two species; *S. gregaria* (Kennedy, 1951; Waloff, 1972; Waloff and Rainey, 1951; and summarized by Uvarov, 1977) and *L. m. migratorioides* (Baker et al., 1984; Farrow, 1979b, and unpublished). The most important advances in this area have resulted from the use of vertical photographs through swarms flying over the observer (Waloff, 1972; Baker et al., 1984), so that orientation and track of individual locusts could be measured in relation to wind direction.

9.3.4 Individual Flight Movements within Aggregations

The most important feature of behavior in aggregations is the intermittent nature of individual flights, which are typically much shorter in duration than the periods spent feeding and resting on the ground. Occasionally all individuals in an aggregation may be simultaneously airborne for short periods, as in the milling flights described previously, but more usually, there is a continuous interchange between the aerial and terrestrial components. A second feature of gregarious flight is the common directionality adopted by fliers, causing individuals in close proximity to fly parallel to one another. Finally, there are well-defined regions of landing and takeoff in aggregations, probably in response to the loss of visual contact with other fliers and settled individuals. Takeoff appears to be a response to the decline of overflights and produces the characteristic trailing edge to aggregations. Descent and landing occur in response to the turning back of fliers at the edge of aggregations, which also maintains swarm cohesion (Waloff, 1972), and to the forward limit of settled individuals.

This behavior imparts a characteristic *rolling motion* (Fig. 9.13) to both low- and high-flying swarms, and typifies the movement of streams and swarms of all species of locust and grasshopper; it has been described by many observers (for references, see Uvarov, 1977). Such behavior is extremely efficient in maintaining the cohesion of aggregations, and stragglers are rarely found. Individuals that are left behind are usually parasitized or senile or are concealed from one another by dense vegetation (Waloff, 1972). At the same time, swarms will overrun and recruit scattered individuals in their path, and these are rapidly induced to aggregate. The direction taken by the overfliers largely determines the direction by which the aggregation advances (Waloff, 1972).

Periods of feeding and resting by individuals are closely integrated into the intermittent pattern of flight activity. Most gregarious acridoids start feeding, often voraciously, once they alight at the front of aggregations, whereas toward the rear they appear to be resting. Little is known of the time spent at these activities, but laboratory observations indicate that bouts of feeding in hoppers are separated by postprandial rests during which the foregut is emptied and the food is digested (Ellis, 1951). If a full meal is taken on alighting, feeding could last 10–15 min and would be followed by the postprandial rest, possibly lasting 15–45 min, by which time the individual would find itself at the rear of the aggregation from where it would be stimulated to takeoff. In some species there is more time spent resting than feeding by day and most food is consumed at night in the roosting site (Anonymous, 1982).

The relative durations of these aerial and terrestrial activities in a rolling aggregation are influenced by the time taken to overfly the settled part of the aggregation, which in turn is determined by the speed of fliers, the height to which they ascend, and the size of aggregations. The rate of advance of aggregations is in turn a function of the rate of settlement and spacing of individuals at the leading edge of aggregations, and the speed of overflight.

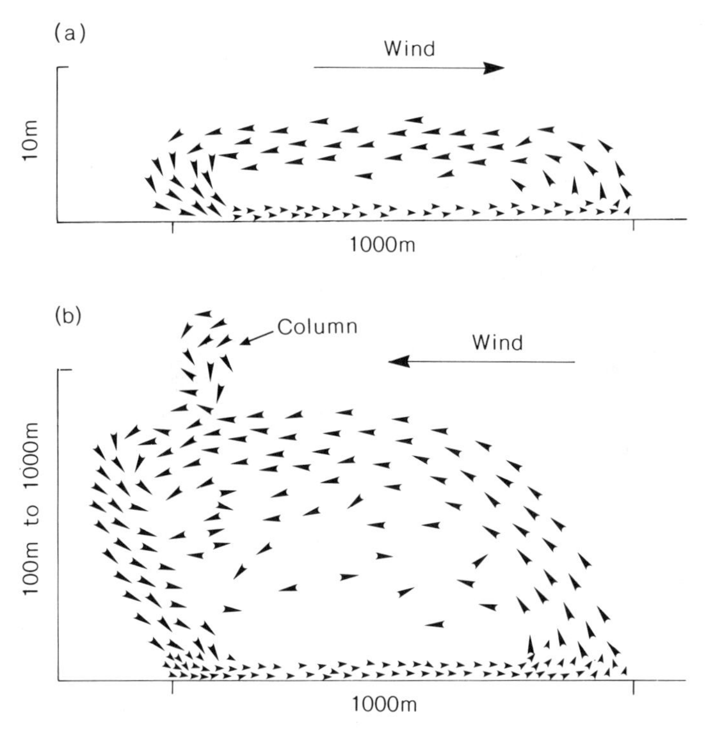

Fig. 9.13. Diagram of the structure of a rolling swarm, (a) low-flying and (b) high-flying, in relation to wind direction. (Modified from Uvarov, 1977.)

For example, in an advancing aggregation that measures 1 km from front to rear, an individual with a ground speed of 15 km/h (Baker et al., 1984), would fly from the back to the front of the aggregation in about 4 min. If individuals settle at high density, the aggregation may advance less than 10 km in a day and an individual will spend at least 60 min feeding and resting between flights. In this example, an individual locust or grasshopper would spend over 90% of its time on the ground and would undertake up to 10 short flights during the day, depending on the duration of swarm activity.

This simplified analysis suggests that the intervals between meals in rolling aggregations in favorable environments may be little different from those in solitarious individuals, which hardly move by day. The temporary suppression of appetitive responses to stimuli, such as those associated with feeding and reproduction, is a key element in the behavioral definition of migration, proposed by Kennedy (1961, 1986), suggesting that movement of such aggregations is by this definition not migratory, although the cumulative spatial displacements achieved in the long term clearly accord with the ecological concepts of migration of Rainey (1951), Taylor and Taylor (1983), and Uvarov (1977).

The direction taken by individual fliers is extremely variable but there is an overwhelming tendency for individuals to fly on parallel courses with their neighbors (Waloff, 1972; Baker et al., 1984) to form streaming groups. At low altitudes, most individuals adopt an upwind course (Baker et al. 1984; Chapman, 1959; Kennedy, 1951; Waloff, 1972), although there is invariably a large and varying difference between the insect track and the wind direction (see Fig. 9.5, 5). Periodic shifts in direction, similar to a yacht tacking, have been observed and appear to keep the fliers on an upwind course despite large offwind headings (Baker et al., 1984). Waloff (1972) found that the orientation of lower fliers in the interior of swarms was largely random but at their edges was predominantly into the swarm. At higher altitudes, probably where the wind speed is higher than the locust's air speed, most individuals adopt a downwind course, although again there is often a large and varying difference between insect track and wind direction. The maintenance of offwind courses appears to be the mechanism by which individual fliers or streams of fliers keep an upwind course in light winds, through a modification of the optomotor response. This leads to long-period oscillations in direction (Baker et al., 1984), similar to zigzagging by male moths flying upwind in a pheromone plume (Marsh et al., 1978). No evidence for orientation to compass direction or sun was found by Baker and co-workers (1984) in the swarms of *L. m. migratorioides* studied.

9.3.5 Movement of Aggregations

The structure and motion of acridid aggregations is influenced by the interactions of four principal variables: (1) *level of recruitment*, varying from local, producing *streams*, to widespread, producing *swarms*; (2) *degree of cohesion*, which affects density and size of aggregations; (3) *height of overfliers*, which separates *low-flying* aggregations, which are moving in winds whose velocity is less than that of the fliers so that they can advance in a direction that is independent of wind, from *high-flying* aggregations, in which the fliers are generally displaced downwind because wind velocity is higher than that of the fliers; and (4) *directionality* of individual and group motions producing *unidirectional*, *to-and-fro*, and *milling* movements. There are continuous shifts in flight behavior between these different categories as fliers respond to changing environmental conditions and levels of mutual interactions.

9.3.5.1 Streams

When adult grasshoppers or locusts reach high-enough densities, streaming is often the first manifestation of gregarious daytime movement, particularly if aggregation has occurred after fledging. Individuals fly within a few meters of the ground in thin streams in which there is a continuous interchange of acridoids between ground and air as individuals take off and alight after short flights that are probably on the order of 100 m. A stream may be on the order of a few tens of meters wide and several hundreds of meters long, although this is often difficult to determine. At the head of streams fliers tend to fan out when they alight and feed. By their very nature, streams contain similarly oriented individuals and straight courses may be followed over distances of several kilometers. However, within a given area streams are often oriented in different directions and fusion frequently occurs when streams meet. Streams may be split by topographical features as individuals attempt to fly around rather than over obstacles. The overall rate of progress of streams is slow and rarely amounts to more than a few kilometers a day in any one direction.

At least three variants of streaming movement have been observed in acridoids: (1) unidirectional cross-country movements with each stream following the same general direction all day; (2) to-and-fro movements involving a reversal of direction between morning and afternoon, although one overall direction usually prevails. These have been observed along a northwest–southeast axis, with southeast dominant, in *C. terminifera* (Farrow, 1977a, and unpublished); and (3) streaming from and to roosting sites in the morning and evening, respectively, typical of *C. terminifera* aggregations in cool autumn weather (Farrow, 1977a), and in overwintering *N. guttulosa* aggregations (R. A. Farrow, unpublished). Similar kinds of movements have been observed in aggregations of North American and Eurasian grasshoppers and locusts (for references, see Uvarov, 1977) and daily movements between shelter habitats and oviposition sites have been reported in *N. septemfasciata* (Woodrow, 1965). Streaming does not generally lead to substantial population displacements and falls into Uvarov's (1977) category of *group wandering*.

Flight usually stops during wind gusts or when the wind is consistently strong, even if it is in the same direction as the stream, and when the sun is obscured, even if air temperatures are favorable for flight (Casimir and Bament, 1974). Intermittent flights are characteristic of days with scattered fairweather cumulus and convection cells, the effects of which are described in more detail in the next section. In hot weather, flight may also cease around midday for several hours or more.

The direction taken by streams is affected by a complex of responses: (a) the tendency for individuals to fly parallel to one another, which maintains streams on a straight course (the *gregarious inertia* of Kennedy, 1951); (b) a visual response to relief features such as high ground and belts of trees which cause streams to be deflected; (c) the visual response to the passage of surface features as they pass by the ommatidia (the optomotor hypothesis of Kennedy, 1951, described in an earlier section, and the modifications to it proposed by Baker et al., 1984), which accounts for off-wind headings in light winds; (c) an as yet unproved sun-compass orientation which could account for those preferred flight directions that are maintained independent of wind direction and topographical features (Baker, 1978; Williams, 1958) and could control a change in direction between morning and afternoon, although most insects that use a sun-compass orientation, for example, butterflies, fly over rather than around obstacles.

9.3.5.2 Low-Flying Swarms

In some species of grasshopper and in most locusts, streams often consolidate into low-flying rolling swarms that may continue to grow by further recruitment of individuals and by fusion. Swarms also form directly from hopper bands without further consolidation. Concentration of fliers landing at the leading edge of aggregations leads to a lateral expansion at the front of swarms, so they tend to become much wider than long. Flight occurs within Taylor's boundary layer, defined in Section 9, and fliers rarely exceed 30 m in altitude; most are within 10 m of the ground (Balten, 1969; Reid et al., 1979). A layer concentration is often observed about 1–2 m above the top of the vegetation canopy (Fig. 9.14b). These characteristics have given rise to the term *stratiform swarm* (Waloff and Rainey, 1951), and this type of movement has again been included in the category of *swarm wandering* by Uvarov (1977) to distinguish it from large-scale *swarm migration* of high-flying swarms, described below.

Low-flying swarms advance as little as 3–5 km a day (Casimir and Bament, 1974; Farrow, 1979b; Uvarov, 1977). This slow progression is related to the short time that individuals spend airborne, the frequent cessation of flight altogether, the slow ground speed of fliers because of the lack of wind assistance, and the high density of settling individuals at the head of swarms. Waloff (1972) has estimated that only 13–30% of locusts were airborne at any one time in low-flying swarms of *S. gregaria.* Takeoff is often inhibited by strong winds. In gusty conditions, related to the passage of convective cells, flight is usually intermittent. Locusts take off and fly during lulls associated with updrafts and rising temperatures and land during gusts caused by downdrafts of cooler air (Casimir and Bament, 1974; Kennedy, 1951; Rainey and Waloff, 1951; R. A. Farrow and V. A. Drake, unpublished); radar has shown that fliers may be confined to the walls of the associated convective cells (Reid et al., 1979). Because the air speed of the fliers is higher than the wind velocity, low-flying swarms appear to progress in a series of surges in which waves of takeoff or landing pass through the swarm, corresponding to the rate of cell displacement and to the distance between successive downdrafts (R. A. Farrow, unpublished). These are on the order of 1 km apart, that is, about the same size as small swarms. Flight may cease during the hottest part of the day when temperatures of 40°C+ are recorded (Ahluwalia et al., 1976).

The direction taken by low-flying swarms appears to be controlled by the same complex of factors that affect streams, combining innate directionality, wind, and topography. In some studies the direction of swarm advance has been primarily correlated with wind direction and may be upwind, as reported on a few occasions in *S. gregaria* and *L. m. migratoria* by Waloff (1972), Zakharov (in Uvarov, 1977), and Baker et al. (1984), or downwind, as found occasionally in *C. terminifera* (Reid et al., 1979). As winds increased in strength the locusts would settle or fly cross-wind, probably in order to maintain the preferred retinal velocity. In other studies, however, including some on

Fig. 9.14. Photographs of (a) milling aggregation of *Locusta migratoria migratorioides* in northern New South Wales (Photo: R. C. Lewis CSIRO); (b) low-flying swarm of same species from the Central Highlands of Queensland, Australia, during the 1973–1976 plague. (Photo: J. Dowse CSIRO).

L. m. migratorioides, swarms have been observed advancing in relatively fixed directions, irrespective of wind direction (Ahluwalia et al., 1976; Casimir and Bament, 1974; Farrow, 1977a, 1979b), although they are often deflected from their general direction of advance by topographical features. This may produce well-defined routes in *C. terminifera* (Casimir, 1958; Magor, 1970) by which plagues regularly invade the same geographic regions.

The direction of swarm advance may remain the same throughout a season or it may show seasonal reversals in direction. In Australia, swarms of *C. terminifera* invariably advance southeast (Casimir and Bament, 1974), whereas swarms of *N. guttulosa* and *L. m. migratorioides* tended to move west in autumn and east in the following spring (Casimir and Edge, 1979; Farrow, 1979b), although the direction taken by individual swarms at any instant was rather variable. These movements occurred in a region influenced by the southeast trade wind with no marked seasonal shifts in prevailing winds. In Africa and South America, seasonal changes in the direction of swarm advance of several species have been reported (Lean, 1931; Waloff, 1966; Waloff and Pedgley, 1986; Symmons, in Uvarov, 1977), but because there is some uncertainty about the height of swarms and extent of downwind movement, this topic is discussed in the section on populations. Finally, there are instances where swarms have moved in different directions from their respective source areas at the same time (Brown, 1980; Parker et al., 1955).

In mountainous areas the direction of swarm movements may be entirely dictated by topography, and movements along valleys in a range of directions have been observed in places like the Markham Valley of Papua New Guinea during a plague of *L. m. migratorioides* (Baker, 1975; Young et al., 1984).

Low-flying swarms of most species move 20–100 km during a life-span of 3–6 weeks (Casimir and Bament, 1974; Uvarov, 1977) depending on weather and topography, although if there is an extended immature period, longer distances of 100–200 km may be achieved. In Australia there is a well-documented example of a swarm of *L. m. migratorioides* that covered 250 km in 11 weeks between fledging and oviposition (Farrow, 1979b), a movement quite atypical of the majority of the swarming population (Fig. 9.19).

9.3.5.3 Milling Swarms

When dense aggregations of fledglings are produced from hopper bands, the first flights produce milling aggregations in which individuals, groups of individuals, or streams move in a range of directions within the swarm. When fliers reach the edge of the swarm they may turn back into its interior, turn along its edge, or alight, producing a complex but very characteristic swirling motion. In swarms of *S. gregaria* this behavior is quickly superseded by unidirectional displacements, but it is particularly persistent in swarms of *L. m. migratorioides* observed in Australia and Papua New Guinea (Baker and Casimir, 1986; Farrow, 1977c, and unpublished) (Fig. 9.14a). Milling sustains very tight aggregations in this species and accounts for its particularly slow rate of swarm displacement of sometimes as little as 3–4 km a day. Milling behavior has been observed in a wide range of locusts and some grasshoppers (Uvarov, 1977), and is particularly common at oviposition and mating, when normal swarm displacement may temporarily cease (Casimir and Bament, 1974).

9.3.5.4 High-Flying Swarms

Under hot conditions, fliers from aggregations may ascend beyond their flight boundary layer into winds that are faster than their flying speed so that they are transported downwind regardless of their heading. Most high fliers are oriented downwind, although at lower levels in swarms, where many fliers are ascending or descending through their flight boundary layer and where turning movements are occurring at the edges of the swarm, orientation is more random (Waloff, 1972). Downwind orientation above the flight boundary layer is predicted by the optomotor response discussed earlier, and experimental evidence suggests that fliers may never reach a "maximum compensatory height" by day, at which they cease being able to detect the angular velocity of surface features (Riley and Reynolds, 1986).

Most high-flying swarms still proceed with the characteristic rolling motion although fliers often overshoot the front of the swarm and turn upwind as they descend to settle on the ground at its leading edge (Fig. 9.13b). At any one instant the greater part of the swarm is settled on the ground (Waloff, 1972) although the proportion of time that individuals remain airborne is probably longer than in low-flying swarms. Swarms rarely advance at more than half the upper wind speed (Rainey, 1963) owing to the high proportion of settled individuals. There is little information on swarm structure and flight behavior in situations where settling is inhibited, for example, over water or possibly in desert areas without food sources.

Downwind movements fall in two main categories, those that take place in fast-moving prefrontal airflows at up to about 100 m altitude and those that occur in strong convective conditions in which winds are generally light and fliers may attain altitudes of 1000 m or more.

The first category is typified by the swarm structure observed in *C. terminifera* under the influence of high temperatures, associated with prefrontal airflows that originate from the subtropics. Winds in these airflows may exceed 50 km/h at less than 5 m altitude, which in most circumstances would inhibit take off, but flights of up to 100 m altitude and sometimes higher have been observed in such conditions, resulting in extensive downwind swarm displacement by this species of 20–50 km a day (Casimir and Bament, 1974; Farrow, 1977a). It is unlikely that thermal convection has much influence on the vertical movements of fliers in this situation, although there may be some concentration in so-called "streets" of convective updrafts that occur in these airflows. Because such weather conditions rarely persist for more than a day or two at a time, owing to the eastward displacement of cold fronts, these southward swarm surges are short-lived. Similar types of movement may occur on prefrontal systems in the northern hemisphere during plagues of *S. gregaria* and *M. sanguinipes*, but details are lacking.

The second category is typified by the high-altitude swarms of *S. gregaria*, in which fliers often ascend to more than 1000 m, under the influence of intense thermal convection (Waloff and Rainey, 1951) and generally light winds. Daily displacement overland has been observed to vary from 5 to 100 km in the vicinity of the ITC in east Africa and has been related to wind speed (Rainey, 1963), although as already suggested, some trans-Saharan movements on prefrontal airflows are probably more rapid because of higher wind speeds that produce daily displacements of several hundred kilometers. Recent swarm crossings of the Atlantic Ocean (Ritchie and Pedgley, 1989) suggest that the displacement capacity of *S. gregaria* swarms is much greater than previously supposed, even taking gliding into account. In this instance swarms must have covered about 1000 km a day over 5–6 days, in winds that would have averaged about 40 km/h.

The airborne component of high-flying swarms, as exemplified by *S. gregaria*, usually exhibits a strongly reticulate structure (Uvarov, 1977), reflecting convective turbulence, and columns of fliers are often seen rising through the swarm giving rise to the term *cumuliform* or *towering* (Waloff and Rainey, 1951; Uvarov, 1977). These columns are associated with vigorous updrafts in walls of convective cells and were observed at spacings of 1–3 km, a distance similar to that described earlier for low-flying swarms. Towering swarms have been observed in few other locust or grasshopper species and then only rarely, including *Calliptamus italicus*, *L. m. migratoria*, and *M. sanguinipes* (for references, see Uvarov, 1977) and *C. terminifera* (Clark, 1972). In these species they appear to persist for relatively short periods in spells of intense convection during the course of major plagues. They have not been convincingly reported in plagues of *L. m. migratorioides* or *N. septemfasciata*, where the airborne component of swarms is generally stratiform in structure. Although common in *S. gregaria* such behavior is largely confined to immature swarms, where it may persist for long periods prior to reproduction. Modifications to the towering structure of *S. gregaria* swarms at the interface of the opposing wind systems of the ITC have been described by Sayer (1962), who observed "sheets" of locusts rising up the sloping interface of the convergence.

High-flying swarms invariably move with the wind because of the downwind displacement of the uppermost fliers as they advance to the swarm's leading edge (Rainey, 1951; Uvarov, 1977); Baker's suggestion (1978) that such swarms adopted a preferred compass direction or headed towards rain in sight or high ground has not been supported for the swarms in question (Draper, 1980).

9.3.6 Movements of Gregarious Populations

Although the process of gregarization is sometimes sufficiently localized over the distribution area of a species of grasshopper or solitarious locust that only a single aggregation is produced, conditions usually favor the simultaneous production of a varying number of aggregations, often concentrated in well-defined *outbreak areas*, which may vary in size from tens to hundreds of square kilometers. In some species, all known outbreaks originate from the same localized area or areas where conditions intermittently favor multiplication and gregarization, and from nowhere else (e.g., *L. m. migratorioides*, Farrow, 1974). In other species the location of outbreaks is more variable and often follows the varying and unpredictable patterns of rainfall distribution (e.g., *S. gregaria* and *C. terminifera* and most "grasshopper" species). The mobile aggregations progressively recruit solitarious individuals from intervening scattered populations so that entire local populations tend to be comprised of a number of gregarious units. It is probably at this stage or when the first gregarious progeny are produced from the initial aggregations that a *plague* is said to have started, which may persist through many gregarious generations over several years and spread over an enormous area as a result of the accumulated movements and multiplication of the component swarms. The area covered by gregarious populations in a plague is known as the *invasion area* and may be larger or smaller than the distribution area of solitary populations. Plagues are sustained in two ways, by reproduction of the swarming populations (typical of most locusts) and by repeated gregarization of solitarious populations (commoner in grasshoppers where swarms rarely last for more than a generation). Plagues terminate when breeding is unsuccessful and the offspring fail to survive, the inevitable outcome of all plagues.

The movements of gregarious populations of most major locust and grasshopper pests have been described in detail by Uvarov (1977), and two additional species are covered by Waloff and Pedgley (1986). Accordingly, this chapter restricts itself to summarizing the principal features of movements by gregarious populations and analyzing the factors controlling such movements.

For any species, each plague must be considered as an individual event, molded by the unique environmental processes operating at that period over the area invaded. For some species these processes have been documented only from single events (e.g., the 1928–1941 plague of *L. m. migratorioides* in Africa or the 1874–1877 plague of *Melanoplus spretus* in the western United States), whereas in others (e.g., *S. gregaria* in Africa and the Middle East and *C. terminifera* in Australia) plagues have occurred at more frequent intervals and the patterns of different plagues have been documented and compared (Bennet, 1976; Hemming et al., 1979; Waloff, 1966; Wright, 1987) so that the course of future invasions can be predicted in relation to synoptic weather with some confidence.

The patterns of population movements in plagues have been divided into two principal kinds, *expansive* and *circuit* (Uvarov, 1977). This is a convenient although somewhat arbitrary system, for as Chapman (1972) has pointed out, movements depend on the environmental processes operating during the plague and one system may evolve into another depending on weather and other factors. Examples of the kinds of population movements observed in plagues are summarized in Figure 9.15.

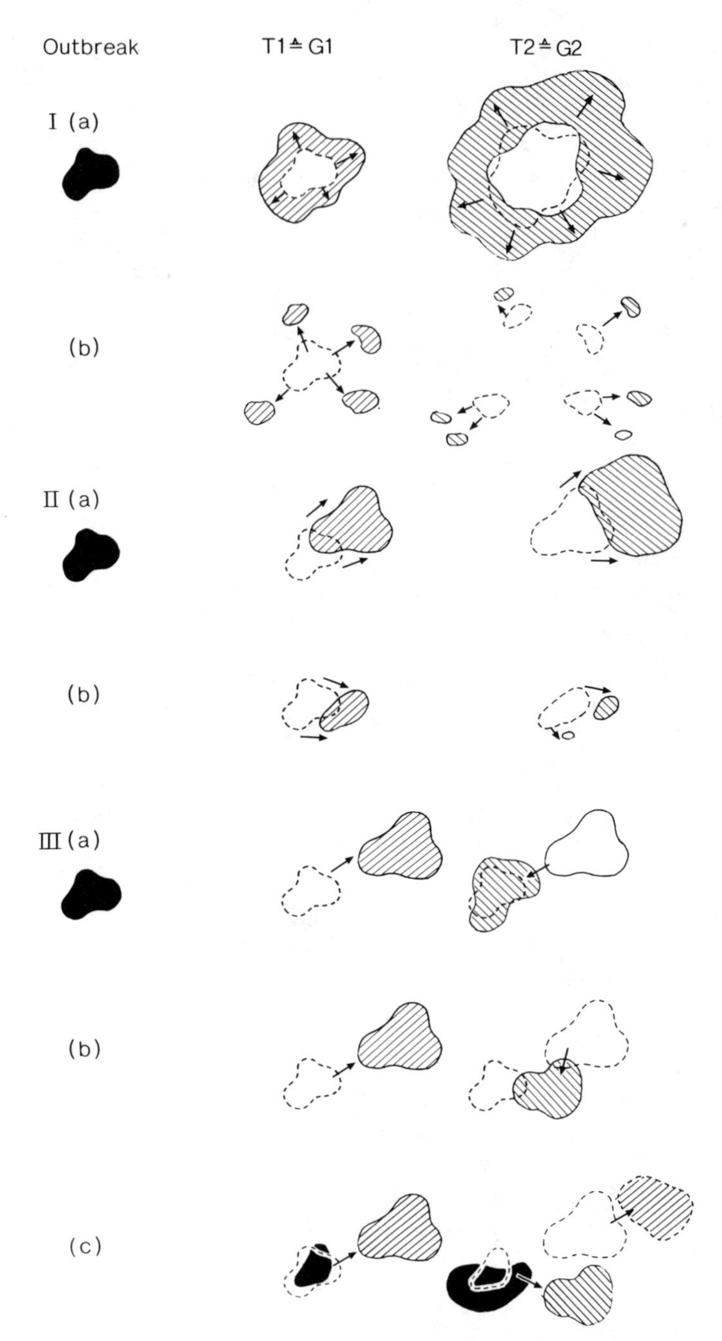

Fig. 9.15. Diagrams of patterns of locust invasions, showing in column 1 the original outbreak area and in columns 2 and 3, the spread of a generation between fledging and reproduction at Times T1 and T2, respectively, over two successive generations G1 and G2. (I) Expansive, multidirectional: (a) increasing in area; (b) in decline. (II) Expansive, unidirectional: (a) increasing in area; (b) in decline. (III) Circuit: (a) closed; (b) open = expansive; (c) open with continuously active outbreak area.

9.3.6.1 Expansive Invasions

In expansive movements, swarms move away from the original outbreak area(s) and although directions may vary, they do not, in subsequent generations, return to the original outbreak center. Many grasshopper outbreaks lead to short-lived expansive movements in streams and loose swarms. Displacements rarely cover more than a few kilometers prior to oviposition, and outbreaks typically last for just a single generation. Such ephemeral movements are generally poorly documented. Expansive plagues cannot persist indefinitely because swarms eventually spread to progressively less favorable habitats, although they may reach the edge of the species' range before they collapse.

Movements may be *multidirectional*, in which swarms move outward in a range of directions from a central source area (Fig. 9.15, Ia) as shown by *Dociostaurus maroccanus* in Western Turkey (1909–1918) (Uvarov, 1932), *C. italicus* in Kazakhstan (1948–1955) (Vasil'ev, in Uvarov, 1977), *L. m. migratorioides* in the Markham Valley region of New Guinea (1974–1976) (Baker and Casimir, 1986), and *L. m. manilensis* in eastern China (Tsou, in Uvarov, 1977) among others; or *unidirectional*, in which constituent swarms move in a single prevailing direction over the lifetime of the plague (Fig. 9.15 II). These are characteristic of plagues in temperate/subtropical latitudes of Asia, North America, and Australia and usually involve invasions of higher latitudes, as shown by *M. sanguinipes* in north America (1938–1940) (Parker et al., 1955) (Fig. 9.16), *L. m. migratoria* in central Asia in 1925–1926, 1947, and 1967 (for reference, see Uvarov, 1977), *N. guttulosa* in Australia (1974–1976) (Casimir and Edge, 1979), and *C. terminifera* in Australia (23 plagues 1934–1987) (Wright, 1987) (Fig. 9.17). However, two species have spread from high to low latitudes, *M. spretus* in the western United States (1874–1877) (Riley et al., in Uvarov, 1977) and *L. pardalina* in southern Africa (Botha, in Anonymous, 1982; Smit, 1939; Anonymous 1982); the plague of *M. sanguinipes* also showed a significant change in direction in 1940 toward low latitudes. Sequential changes to the common direction(s) of these invasions are frequent, as shown by the plague of *M. sanguinipes* (Fig. 9.16), and are usually seasonal or intergenerational in nature, but are insufficient to return swarming populations to their original source areas. The proposal that swarm movements of *C. terminifera* form part of an open circuit system (Roffey, 1972) (combining eruptive southeast movements by gregarious populations and northwest movements of solitarious individuals), rather than a one-way expansive system, is not supported because of the lack of evidence for a seasonal circuit by solitary populations as described earlier.

9.3.6.2 Circuit Invasions

In circuits, the directional shifts in population movements between complementary breeding areas are sufficient to return swarming populations to their original source areas. Provided swarm breeding remains successful, plagues could persist indefinitely in such a system, as suggested by Rainey (1963).

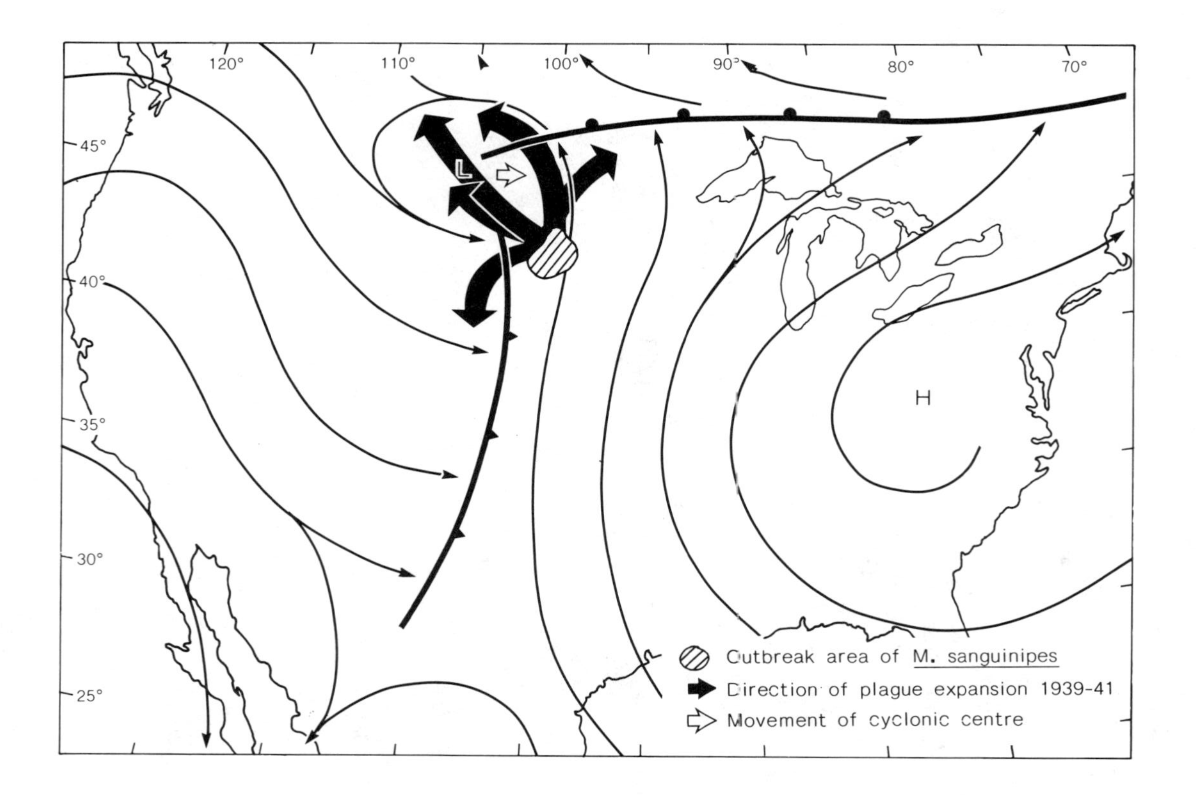

Fig. 9.16. Diagram of movements of gregarious populations of *M. sanguinipes* in the 1938–1941 plague in relation to a generalized weather pattern for summer. (Insect data modified from Parker et al., 1955.)

When the bulk of the population moves between complementary breeding areas, a *closed circuit* is established (Fig. 9.15, IIIa), but in many instances only part of the population does so, while the rest adopts an expansive mode, moving into progressively less favorable conditions; this has been termed *open circuit* (Uvarov, 1977) (Fig. 9.15, IIIb). Plagues may change from being dominated by a closed-circuit movement in one season to a more expansive or open-circuit movement in the next.

Circuits have been documented for locusts only in the genus *Schistocerca* and in several different geographic populations of *L. m. migratorioides*. In the genus *Schistocerca*, swarms shift seasonally between summer breeding areas in low latitudes of the subtropics and spring breeding areas in higher latitudes of the temperate zone. In *S. gregaria*, three subsidiary interconnecting circuits occur across Africa, north of the equator, and the Middle East (Waloff, 1966) (Fig. 9.18). The importance of these different circuits varies between and during individual plagues, depending, among other factors, on the original source of the plague and regional differences in swarm multiplication (Bennet, 1976; Hemming et al., 1979). The circuits are interconnected by extensive east–west displacements of swarms. *S. gregaria* is one of the few gregarious acridoids in which the seasonal displacements of swarming populations are analogous to those of the solitary populations described earlier. In addition, occasional swarms also leave the well-defined boundaries of these breeding areas in one-way movements into equatorial and temperate latitudes from which they or their progeny usually fail to return (i.e., open circuit). Similar seasonal movements are reported from the southern hemisphere for *S. cancellata* in South America during plagues in 1946–1947 and 1950–1951 (Daguerre, in Uvarov, 1977; Waloff and Pedgley, 1986) and for *S. g. flaviventris* in southern Africa during plagues in 1934–1935 and 1948–1949 (Waloff and Pedgley, 1986) (Fig. 9.18). In the former species there is also a major east–west component to movements between the Andean foothills and the plains to the east.

The circuits of *L. m. migratorioides* are seasonal north–south displacements across the tropical savannas of west Africa (1928–1941), Madagascar, and the Philippines (Batten, 1967; Farrow, 1974). In the latter two regions, where plagues are relatively short-lived, expansive movements from coastal outbreak areas are more conspicuous than return circuits. In west Africa the extent of the seasonal displacements in these circuits is much larger than that resulting from the migratory movements of solitary populations, described earlier. The 1928–1941 plague of *L. m. migratorioides* also spread eastward to east Africa and southward across the equator to southern Africa (Batten, 1966). These movements were not balanced by return movements of the kind seen in west Africa and are expansive in nature. The 1974–1976 plague of this species in Australia also featured a circuit between easterly summer breeding areas and the westerly overwintering areas, although expansive movements in other directions also influenced the spread of this plague (Farrow 1979b).

Although *N. septemfasciata* has only one generation annually, two distinct seasonal movements occurred in each swarming generation during the

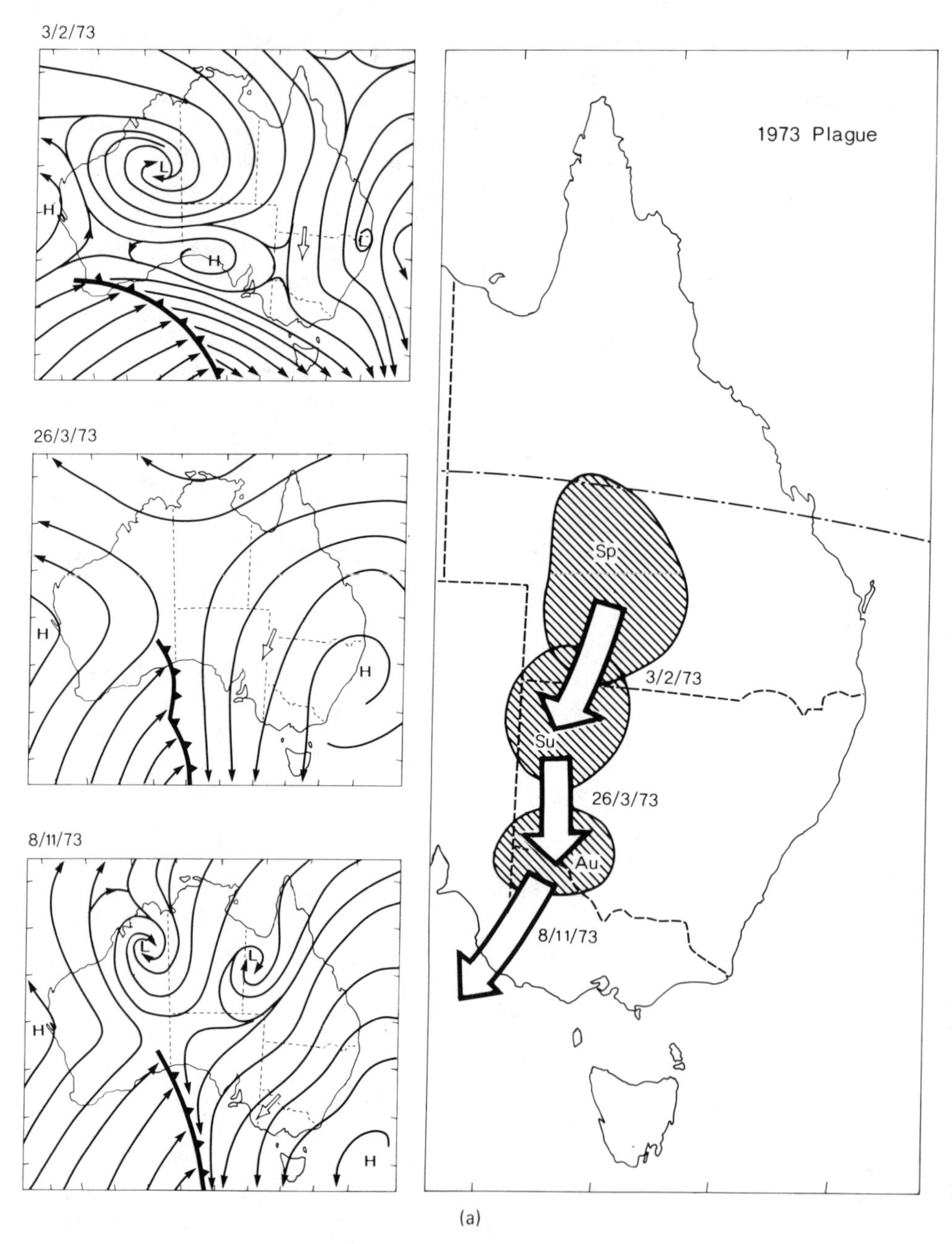

Fig. 9.17. Diagram of movements of gregarious populations of *C. terminifera* in the plagues of (a) 1973 and (b) 1979, in relation to the synoptic weather on the nights of migration. (Invasion modified from Farrow, 1975a, 1977a and Wright, 1988.)

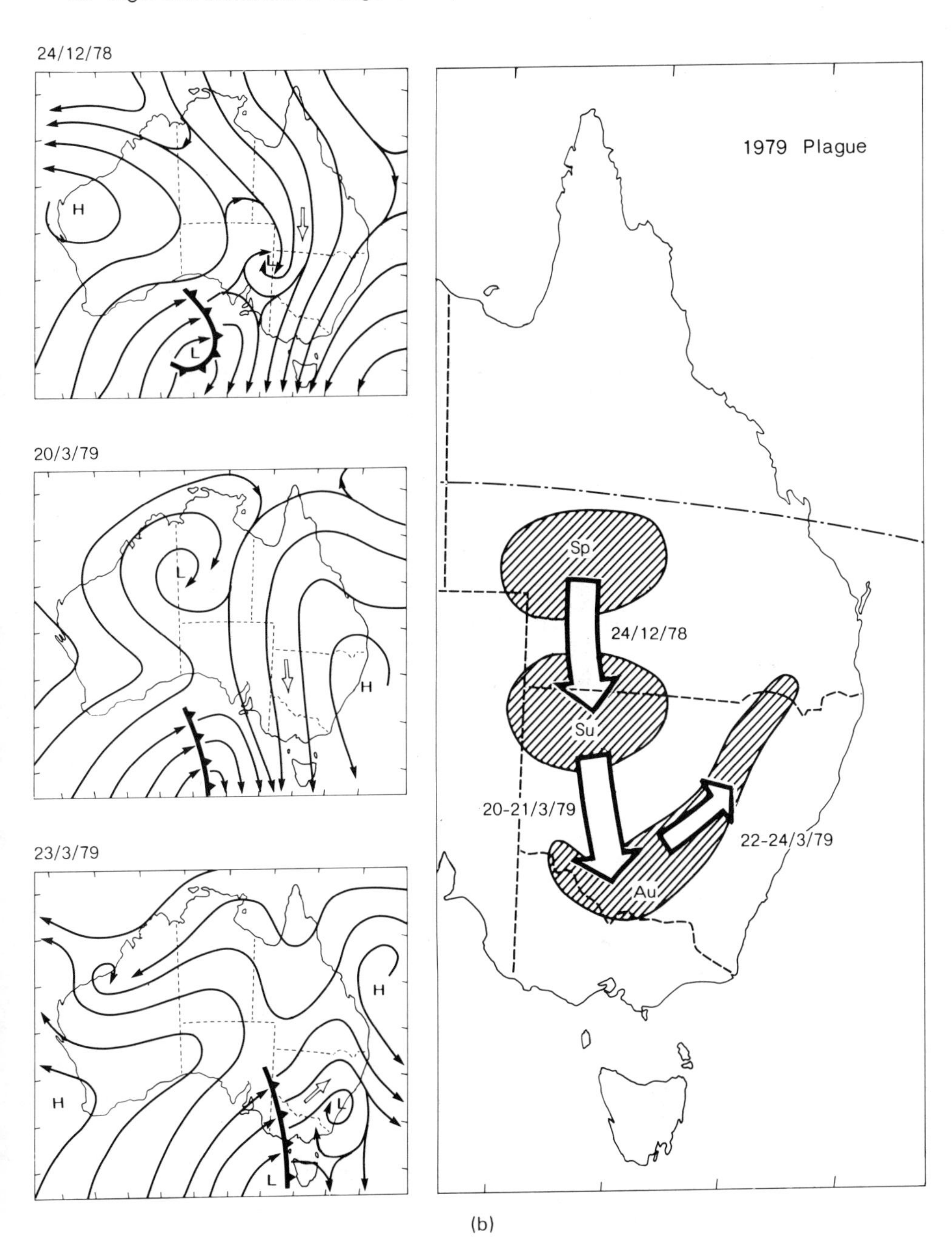

Fig. 9.17. (*Continued*).

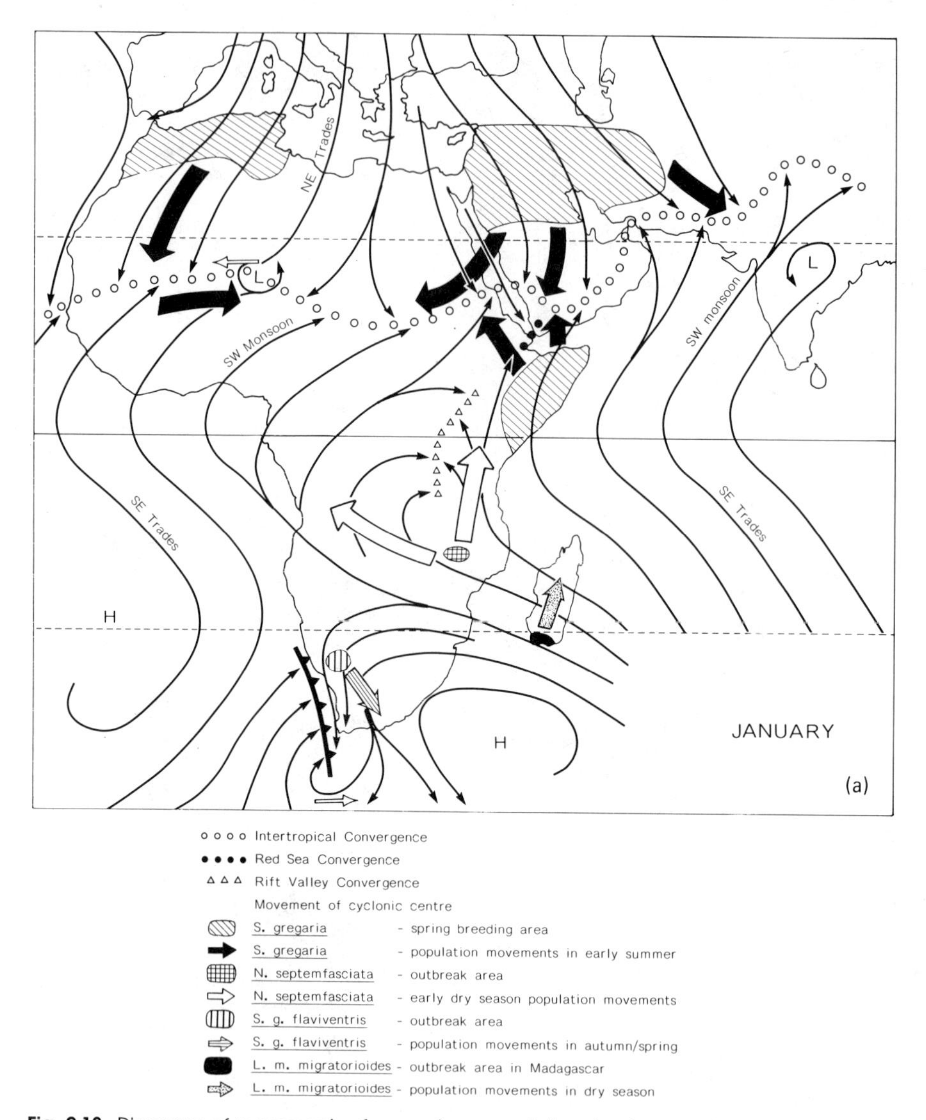

Fig. 9.18. Diagrams of movements of gregarious populations in plagues of selected locusts in Africa in relation to generalized synoptic weather for January and July. (Locust data modified from Uvarov, 1977, and Waloff and Pedgley, 1986).

1930–1945 plague in southern Africa (Symmons, in Uvarov, 1977): in sexually immature swarms at the start of the dry season (April–June) movements occurred to the north and west; in maturing swarms at the end of the dry season (September–November) movements were to the south and east (Fig. 9.18). Although a partial circuit may be established, most swarms or their progeny never returned to previous breeding areas and the plague was

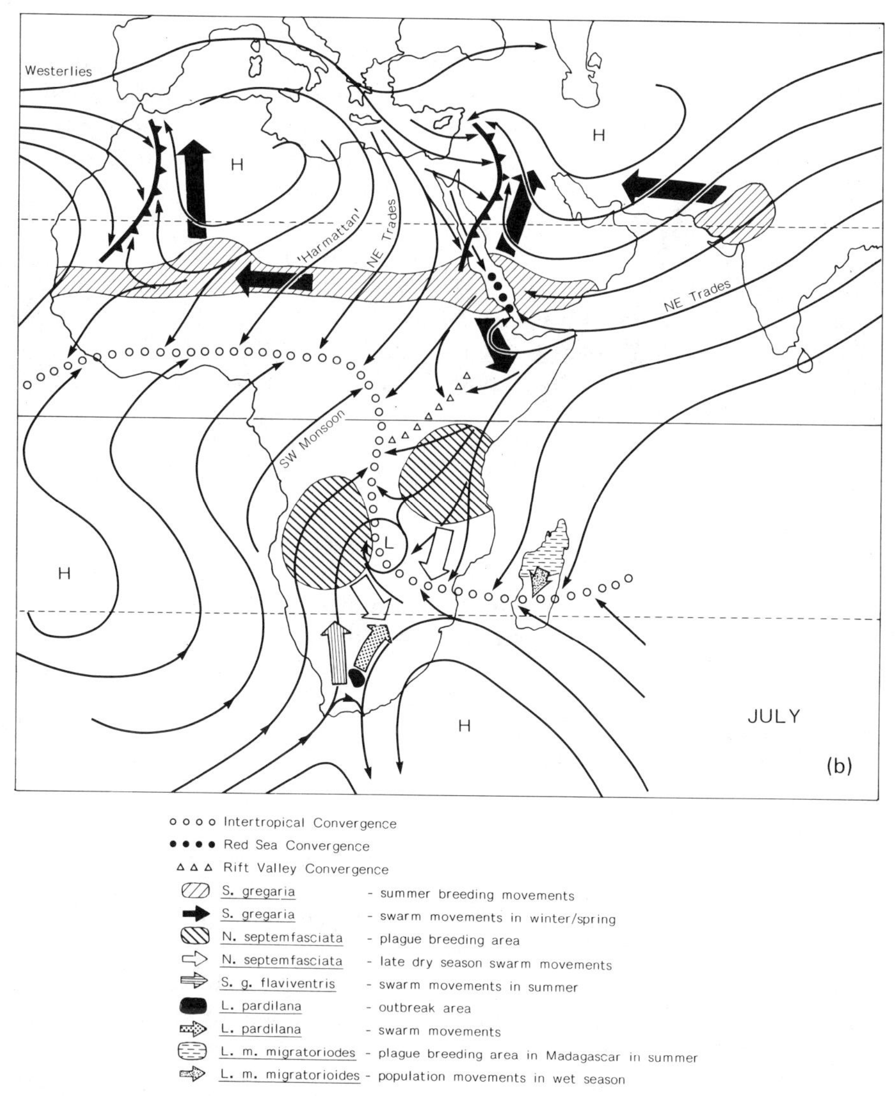

Fig. 9.18. (*Continued*).

essentially expansive in nature, dependent on the continued recruitment of swarming populations from outbreak areas (Fig. 9.15, IIIc).

9.3.7 Factors Influencing the Spread of Plagues

Although the spread of swarms during the course of a plague is controlled by varying numbers of external and internal factors, their relative significance is often poorly understood because of the lack of documentation of these factors at the time of displacement. This problem may reduce to such basic questions as whether or not flight occurred on a given day, under the weather conditions of that day, and if so, how much took place at high level with wind assistance, compared to the amount at low level in either a preferred direction (independent of wind direction) or in one that was determined by wind direction. The invasion pattern is also influenced by a range of interactions involving such factors as the incidence of individual migrations at night; the influence of maturation and oviposition on swarm behavior; the size of aggregations, with large swarms generally moving further than small ones; the state of the vegetation with swarms advancing faster in sparse and/or dry vegetation; the breeding success of successive gregarious generations; and the extent of continuing recruitment from solitarious populations. As Baker and co-workers (1984) have emphasized, the wide diversity of observed swarm behavior reflects the broad range of conditions to which gregarious locusts have to react and is in part responsible for our failure to develop a general theory of swarm behavior.

This review will limit itself to a discussion of the influence of synoptic winds on the movement of swarming populations because of the central role of the wind in the theory of downwind displacement of swarms (Rainey, 1951, 1963; Uvarov, 1977). Although many authors have shown that downwind migration of swarms occurs during plagues of most locust species, others have observed situations in which there was no correlation between wind direction and swarm displacement (for references, see Uvarov, 1977). De Lepiney (1928) was probably the first to show that locust swarms were subject to transport by wind, *provided* wind velocity exceeded the air speed of the locust. The failure to find such correlations is, as suggested earlier, most likely due to movement at low level in preferred directions, where factors other than wind direction control the directions of locust orientation and swarm displacement, rather than, as Uvarov (1977) has proposed, the difficulty of making accurate observations. Analysis of this relationship nevertheless presents considerable difficulties in practice because of the general inadequacy of simultaneous measurements of the wind field at different altitudes and of the movement of gregarious units or individual swarms and of the gregarious population or plague that they make up.

If downwind movement on prevailing wind systems is the dominant force controlling the direction and spread of plagues, the following regional systems can be predicted. In tropical areas, seasonal reversals in wind direction, associated with the advance and retreat of the tradewinds and monsoon (NE/SW and SE/NW), favor development of circuit movements between the furthest seasonal limits of the ITC. In subtropical latitudes the trade winds (NE and SE) should favor unidirectional movements toward the tropics. In temperate latitudes, the eastward movements of cyclones and their associated fronts cause systematic rotations in wind direction that would favor multi-directional expansive plagues. However, differences in temperature between airflows of different origins will encourage expansive movement to higher latitudes in the warm air sectors of eastward-moving depressions. The opportunity for expansive movement is accentuated in continental areas of north America and eastern Asia by the development of sustained southerly airflows in summer under the influence of maritime subtropical anti-cyclones.

9.3.7 *Tropical Systems*

In *L. m. migratorioides*, the seasonal north–south shift of swarms in west Africa during the 1928–1941 plague (Batten, 1967) can be explained by downwind displacement in the northeast and southwest airflows that follow the displacement of the ITC (Batten, 1967). However, observations by Lean (1931) in Nigeria show that the direction of swarm advance changed from northeastward to southwestward *before* the onset of the northeast trades, when the ITC started to retreat. Furthermore, most swarms in this plague were low-flying, and observations on swarms of this subspecies elsewhere show a predominance of movements in preferred directions, suggesting a possible innate tendency for gregarious populations in West Africa to move poleward in early summer and equatorward in late summer. Eastward extensions of the plague to east Africa during 1930 probably occurred in the westerly flow of the monsoon. Once the plague had extended to east Africa, substantial southward downwind displacements were able to take place as the northeast trades extend as far as 20°S to the east of the Rift Valley convergence. However, the plague continued to advance southward well outside the limits of the ITC into southwest and southern Africa as far as 32°S, *against* the prevailing flow of the southeast trades, probably by means of low-altitude directed flight.

In Madagascar, plagues regularly spread from the southwest of the island to the northeast across the prevailing direction of the southeast trades (Farrow, 1974). In recent outbreaks of this locust in different parts of the tropics including Australia and Papua New Guinea only low-flying swarms have appeared and outbreaks have spread in a range of directions that have sometimes been upwind (G. L. Baker and Casimir, 1986; P. S. Baker et al., 1984) but more frequently appear to have been independent of wind direction (Farrow, 1977c).

The seasonal movements of gregarious populations of *N. septemfasciata* during the 1930–1945 plague (Symmons, in Uvarov, 1977) can be satisfactorily explained only in terms of a combination of downwind movements and movements in preferred directions that may be related to differences in flight behavior between sexually immature and mature swarms. Immature swarming populations, produced at the start of the dry season, move north and west under the influence of the southeast trades (Fig. 9.18). At the end of the dry season the movement is reversed and although part of the invasion area is influenced by the northeast monsoon, most of the movements take place to the south of the ITC under the influence of the opposing southeast trades. There is again little evidence that gregarious populations were concentrated in the vicinity of the ITC or that their location was related to seasonal positions of the convergence. Movements to the south and east took swarms into higher rainfall regions where breeding success was increased (Uvarov, 1977).

9.3.7.2 Subtropical Systems

In the subtropics, the migration systems in plagues tend to be region (and species) specific, and show strong adaptations to regional habitats and to the colonization of habitats where reproduction has the highest chances of success. These movements are achieved by a combination of downwind movements on preferred wind systems and movements in preferred directions.

In *C. terminifera* there is no evidence that plagues ever spread downwind toward the equator in northern Australia under the influence of the prevailing southeast trades as discussed by Symmons (1986). On most occasions swarms remain low-flying and plagues displace toward the southeast as a result of compass-directed and upwind movement. Downwind movements are restricted to the northerly airflows that develop in disturbed weather. This type of weather appears to promote increased flight activity particularly after rain (Casimir and Bament, 1974) as in solitarious individuals. However, these systems transport populations only limited distances by day and are *not* responsible for the spread of plagues in this species. This is due to invasive night flights by gregarious locusts flying as individuals at the postteneral/prereproductive stage of development. These movements are identical with those of solitarious individuals described earlier and are initiated only in periods of disturbed weather, when northerly airflows develop either ahead of eastward-moving cold fronts (active from autumn to spring) or in tropical troughs (active in summer) (Fig. 9.17). The directions of these nocturnal migrations do not depart significantly from the direction of the geostrophic wind trajectory.

The nocturnal flights in gregarious populations of *C. terminifera* result in eruptive intergenerational movements of 200–500 km over one to three nights at each generation and deposit gregarious populations in new areas where swarms re-form and fly mostly at low level toward the southeast, except on the occasional days of southward surges in prefrontal airflows. Invasion patterns, in terms of the distances and directions of intergenerational movements, vary according to the structure of the synoptic system at the time of migration and are shown for three recent plagues (Fig. 9.17). Changes in sizes of infestations are a reflection of multiplication between generations and concentration and dispersal by atmospheric processes during migration.

The prevailing southward direction of *C. terminifera* invasions maintains gregarious populations in the preferred steppe grassland habitats of this species in the 150–400-mm rainfall zone, and movements into summer rainfall savanna grasslands of the north and east and into the more arid tussock grasslands of the west appear to be avoided. Nevertheless, like all expansive movements, these are ultimately suicidal as populations move into progressively less favorable breeding habitats with the final intergenerational movements sometimes taking populations offshore (Farrow, 1975a).

In southern Africa, the northward spread of plagues in the spring generation of *L. pardalina* (Smit, 1939) does not accord with the expectation of a downwind movement to the south on warm prefrontal airflows (Fig. 9.18), observed at this time of year in *S. g. flaviventris* (in southern Africa) and *C. terminifera* (in Australia). Movements could occur on a southerly anticyclonic flow, although this would be relatively cool at this time of year. According to Smit (1939), easterly movements predominate in the summer generation. These are decidedly against observed airflows (Fig. 9.18). Because swarms are generally low-flying in this species, a preferred direction is probably maintained, independent of wind direction, in both spring and summer. These movements would take swarms into areas of higher summer rainfall where the chances of survival are likely to be higher.

Seasonal displacements of swarming populations of *S. gregaria* (Waloff, 1966) follow changes in distribution of the seasonal rains of northern Africa and the Middle East more closely than the shifts in seasonal wind patterns and are no more influenced by convergent airflows at the interface of the ITC than by other synoptic scale systems. The directions of displacement can be summarized as follows. (1) At the end of the northern tropical rainy season, in October–November, a new generation of gregarious populations has, like solitarious populations, been produced between west Africa and west India in a zone corresponding to the northernmost position of the ITC (in July–August). By November this area is covered by the northeast trades as the ITC has retreated rapidly south toward its winter position for the northern hemisphere (Fig. 9.18). Swarming populations do *not* move downwind to the southwest, toward the convergence at this period, except occasionally in the horn of east Africa and in western Africa, where a few swarms move to the south but return with the advance of the ITC without having reproduced. Most swarming populations in west Africa move west and remain immature whereas those in eastern Africa mostly move northeast, possibly under the influence of the local Red Sea convergence. Populations from India also move west across the prevailing northeast airflow. In west Africa the westward displacement could be maintained either by moving downwind only on days when the trades veer to the east or by adopting an off-wind heading in the prevailing northeast airflow. (2) From November to March swarms appear in the spring breeding grounds of western North Africa and the Middle East *against* the prevailing northeast airflow. These areas are up to 1000 km from the summer breeding grounds and swarms moving downwind at 50–200 km a day would take 5–10 days of northward movements to reach them. Arrivals of swarms coincide with the eastward passage of temperate depressions over the Mediterranean region when warm, short-lived southerly flows develop across the Sahara and Middle East (Pedgley, 1981; Shaw, 1965) (Fig. 9.18). The northwest airflows in the wake of fronts are generally too cool for flight by mid-winter. The southerly flows persist for only 1–2 days at any location because the depressions and associated fronts move from northwest to northeast Africa in 2–3 days, traveling eastward at 50–100/km. The northward movements are probably accomplished in a series of steps, corresponding to the passage of successive depressions. Rain is sometimes associated with these systems, which also have the potential for moving populations into areas where breeding is likely to be successful. Such rain-bearing systems are also essential for plagues to persist. Swarms rarely invade North Africa east of 20°E, where winter rains are less reliable, because of the earlier westward displacements which remove populations from areas downwind of this region. (3) The bulk of the spring generation is produced in high latitudes from February to June with an occasional smaller contribution from the short rains areas of tropical east Africa. Southward movements toward the ITC from northern Africa and the Middle East can be simply attributed to downwind movement on the prevailing northeast airflow (Fig. 9.18). However, these airflows do not warm up until late spring and if it is argued that low temperatures are inhibiting southward movements in the previous generation, the same restrictions could also apply to the spring generation. Furthermore, temperate depressions also cross the spring breeding areas at the time that the new generation of adults is appearing, but northward movements, similar to those of autumn but extending into Europe, are rarely observed on such warm, southerly prefrontal airflows.

As the ITC advances to its most northerly position many swarms remain to the south of the ITC in the westerly flow of the monsoon (Rainey, 1963) and easterly movements predominate. Rainey (1976) has emphasized the importance of the ITC in concentrating swarming populations of *S. gregaria* in areas where rain is likely to fall. However, it should be remembered that in west Africa, for example, significant rains occur several hundred kilometers to the south of the actual surface convergence and that daily shifts of the convergence are often too rapid and too extensive to be tracked by slow-flying swarms, particularly because winds are often light and variable in its vicinity. Of possibly greater significance to movement and concentration in this species are (1) the westward-moving cyclonic disturbances that form at the interface of the ITC, which have been shown to be important in the nocturnal migrations of acridoids described earlier, and (2) the tropical cyclones (Pedgley, 1972) that typically decay overland into tropical rain-bearing depressions (Venkatesh, 1971) and influence transport and possibly concentration of swarms as well as of any solitarious individuals. The most obvious pattern to the tropical displacements are the westward movements when the ITC is to the south of populations (namely at the start and end of the rains) and eastward movements when it is to the north (middle rains).

Gregarious populations of other species of locust produced in similar latitudinal areas to *S. gregaria* generally move in opposite directions to this species. Swarms of *L. m. migratorioides* and *N. septemfasciata* move toward the equator at the end of the rains before the trade winds have become too cool for flight. At higher latitudes, plagues of a range of species move poleward.

The failure of swarming populations of *S. gregaria* to move south at the end of the monsoon and north in late spring suggests that downwind movements of immature swarms from fledging areas are restricted in some way to preferred wind systems, possibly to disturbed weather at the end of the monsoon season. Although Draper's reanalysis (1980) of Rainey's 1963 data confirmed that the swarms observed in Kenya in January and February 1954 were all migrating downwind in the vicinity of the convergence, it should be emphasized that this region is the only part of the *S. gregaria* invasion area where downwind displacements of swarming populations may follow seasonal shifts of the ITC for a considerable part of the year.

There are sound adaptive reasons why southward movements are avoided by *S. gregaria* in tropical west Africa and India, because swarms would be taken into arid woodland and forest unsuitable for reproduction. In east Africa the semiarid grasslands extend to the south of the equator, where a double rainy season supports "spring" breeding. Similarly, plagues avoid advancing into climatically unfavorable temperate latitudes of the Mediterranean region, where only occasional swarms appear in spring.

In the southern hemisphere, the relationships between swarm movements and wind direction in *S. cancellata* and *S. g. flaviventris* are equally complex. It is likely that southward expansions of plagues in spring and autumn are associated with the approach of eastward-moving depressions and the development of warm but short-lived northerly airflows (Fig. 9.18) (Waloff and Pedgley, 1986). These may transport high-flying swarms to higher latitudes from which there is possibly no return because the return airflows are relatively cool. Such movements are analogous to those recorded in *C. terminifera* in Australia. However, it is difficult to explain eastward expansions of plagues in terms of downwind displacement because of the persistence of easterlies associated with the quasi-stationary oceanic anticyclones (Fig. 9.18), and these may be undertaken by low-flying swarms advancing in a preferred direction. The northward component to movements in summer are associated with the seasonal appearance of southerly airflows caused by the development of semipermanent, subtropical cyclonic centers in both South America and southern Africa (Fig. 9.18).

9.3.7.3 Temperate Systems

In temperate latitudes, intermittent poleward invasions of plagues by the downwind movement of high-flying swarms are well documented in prefrontal airflows (Fig. 9.17) (Farrow, 1975a) and in subtropical airflows on the western side of anticyclones (Parker et al., 1955); these systems are probably responsible for the enhanced northward expansion of plagues observed in Eurasia and North America. Movements in different directions are also observed, but these do not generally concur with the prevailing wind directions, suggesting that in species such as *C. terminifera*, *M. spretus*, and *M. sanguinipes*, the spread of plagues is also influenced by low-level movements in preferred compass directions and by topography. In at least the first two species these movements take swarms into regions of higher rainfall where conditions are likely to be more favorable for gregarious breeding in summer. However in *C. terminifera* this advantage is probably offset by the adverse effect on development and survival caused by declining temperatures as elevations increase in the invaded areas of southeast Australia.

9.3.8 Nightime Flight

Swarms normally descend at dusk and roost overnight. When swarms of *S. gregaria* find themselves over water at dusk they do not appear to alight (Ritchie and Pedgley, 1989), contrary to some speculation (Baron, 1972). Little is known of how swarm structure is maintained in darkness when the range over which visual responses can be maintained is presumably much reduced, and the effectiveness of auditory responses is largely unknown. Short-lived dusk flights by first-generation aggregations of *L. m. migratorioides* in Australia (Symmons, 1979) appear to be a continuation of the milling flights described earlier. There are two genera of locusts in which swarms move predominantly at night. These are the tree locusts in the genus *Anacridium* and the Bombay locust, *Nomadacvis (Patanga) succincta*. The former appear to move in cohesive aggregations at low altitude over short distances under bright moonlight, which presumably permits visual contacts (Popov and Ratcliffe, 1968).

C. terminifera is the only species of acridoid known in which locusts that have been aggregated in swarms readily migrate at night as individuals and afterwards re-form into swarms in a new region. Individual night flights have been reported from first-generation aggregations of *L. m. migratorioides* (Farrow, 1975b) and in *S. gregaria* (Waloff, 1963), but occurred shortly after fledging probably before the start of daytime swarm flight. Dusk takeoff and individual night flights have never been detected in these species once daytime movements have started. Such behavior could, however, explain the sudden appearance of swarms of *S. gregaria* from distant source areas in periods of disturbed weather as well as the rapid displacement of gregarious populations between widely separated summer and winter breeding areas.

9.3.9 Relationships between Invasion and Distribution Areas

Migratory insects have the potential to respond rapidly to changes in the distribution of favorable habitats. The areas in which successful reproduction takes place expand and contract in relation to both short- and long-term changes in habitat favorability, largely as a result of variations in weather and climate, although man-made changes may also be important. In acridoids, population displacements can be the result of movements by both solitarious individuals (particularly during recessions) and gregarious aggregations during plagues. These can make different contributions to the spread of populations. During the exceptionally wet years of 1974–1976 in eastern Australia, solitary populations of *L. m. migratorioides* invaded much of southeast Australia while gregarious populations, with their limited powers of displacement, remained within a well-defined area of central Queensland (Fig. 9.19). This subspecies is widely distributed throughout the tropical Old World, and is present on many remote oceanic islands. Colonization of these islands and of sites in continental areas is more likely a function of migration by solitarious individuals than movement of swarms during plagues.

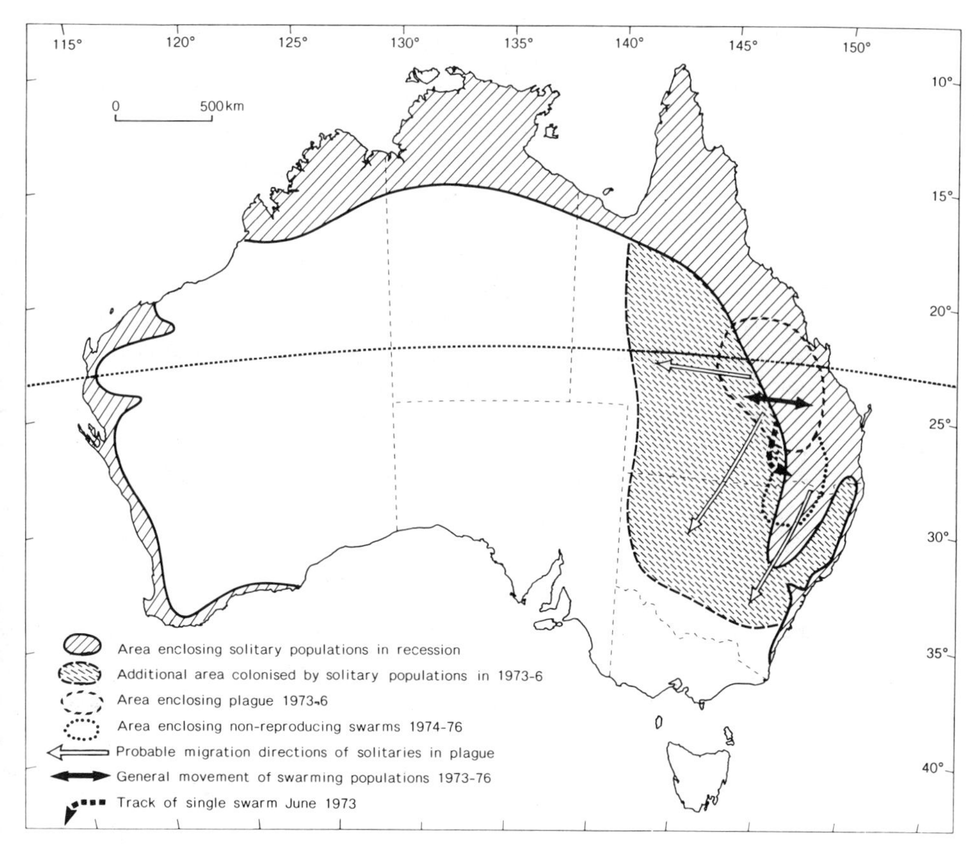

Fig. 9.19. Diagram of the distribution and invasion areas of *L. m. migratorioides* in Australia during the recession and plague periods. (Modified from Farrow, 1979b.)

During the current plague of *S. gregaria* (1987 onward) swarms have invaded places in Europe and the Caribbean where they are currently unlikely to survive. In contrast to *L. m. migratorioides*, movements by swarms outside the normal breeding range of this species appear to be more extensive than those of solitarious individuals and swarms are probably responsible for colonization of islands, remote localities in Africa and the Middle East, and possibly, the New World (Ritchie and Pedgley, 1989). In most other acridoids, migration of solitarious individuals is responsible for colonization of available habitats because aggregations move such short distances.

9.3.10 Functional Significance of Gregarious Flight Behavior

The movement of swarms of *S. gregaria* in the ITC in east Africa during the 1950–1964 plague provided much of the initial evidence for the theory of downwind migration of airborne locusts and other insects into zones of atmospheric convergence and rainfall (Rainey, 1951, 1963). The large-scale displacements achieved by swarms of this species have been used to support the idea that migration is essentially a spatial concept (Uvarov, 1977). Movements of gregarious individuals in swarms of *S. gregaria* were also used by Kennedy (1951, 1961) to develop his behavioral theory of migration in terms of a "straightening out" of flight and a temporary cessation of responses to vegetative stimuli. Phase transformation from the solitarious to the gregarious state is seen as a response to deteriorating environmental conditions to which the more migratory gregarious state is better adapted (Kennedy, 1956). Thus gregarization and the formation of mobile, high-flying swarms was viewed as a process by which large populations of locusts could locate habitats favorable for reproduction, which were often widely separated (Taylor and Taylor, 1983). However, these theories do not provide satisfactory explanations for gregarization and swarming in other species of acridoid nor of the short duration of individual flights in swarms and of the control of direction in low-flying swarms. These theories also tend to underestimate the extent and significance of displacements that result from nocturnal migrations in solitarious acridoids. Finally, less emphasis is now placed on the ability of the ITC to concentrate individual and aggregated night- and day-flying insects in the tropics (Drake and Farrow, 1988).

The phenomenon of gregarization is expressed in acridoids that live in two distinct habitat types: (1) those that tend to concentrate in restricted habitats, particularly where these represent dry season refuges such as flood plains; and (2) those that inhabit those grasslands of the semiarid zone in which variations in weather, particularly rainfall, cause considerable spatial and temporal fluctuations in the abundance of resources. In these environments, as in any other, organisms attempt to maximize their reproductive success. Consequently, substantial population increases by herbivorous insects are inevitable when resources change from being limiting to being abundant, for example, following drought-breaking rains, particularly if density-dependent responses of natural enemies are delayed. Where large populations are concentrated into limited areas suitable for survival and breeding, analogous increases in density occur. In both circumstances a directional foraging strategy, shifting individuals from centers of high density, will reduce the chances of mortality from intraspecific competition for food, compared with a strategy in which insects move at random within a circumscribed, dwindling resource. This kind of behavior is observed in its simplest form in outbreaks of a brachypterous grasshopper, *Phaulacridium vittatum*, where there is a net movement of high-density populations from overgrazed fields to fence lines and hedges along their boundaries (Farrow, 1982b). This activity and the more concerted movements of other acridids differ little from the behavior of those insects that exploit spatially separated resources by extended foraging within a circumscribed home range (Kennedy, 1986) or in a fixed direction (Jones et al., 1980). An aggregative response is probably an inevitable outcome of such concerted movements in order to coordinate the rate of displacement and appears to have evolved independently several times among the insect orders and also within the different subfamilies of Orthoptera. It is proposed here that the bright colors of gregarious individuals serve for intraspecific recognition and maintenance of cohesion. Aggregative behavior is also thought to confuse potential predators and disruptive color patterns are supposed to accentuate this effect (Gillett and Gonta, 1978); however, the numbers of predators are proportionally too low to exert any significant control of aggregated populations in most field situations, suggesting that predation has not played a significant role in the evolution of the gregarious habit.

Gregarious movements are redefined here in terms of *extended foraging behavior*, after Kennedy (1986), and marching by hopper bands is its first manifestation in acridoids. Because densities at hatching are often very high, multidirectional movements away from hatching sites, in relatively straight lines, would confer the highest chances of survival, and could be maintained by gregarious inertia (Kennedy, 1951). The same hypothesis can be extended to the behavior of adults at high densities in rolling streams and swarms where the food source is systematically consumed by advancing aggregations. In most gregarious species, swarms move slowly within their flight boundary layer. Plagues of *L. m. migratorioides* arise in years of above-average rainfall in semiarid and floodplain grasslands, where relative continuity of lush vegetation ensures that limited swarm movements are quite adequate to ensure swarm survival. Plagues of *S. gregaria* are generally initiated by the onset of drought-breaking rainfall produced at the northern limits of the ITC in Africa and India, by cyclones in Arabia, and occasionally by unusually intense temperate depressions at higher latitudes. These systems lead to concentration of solitarious locusts and to successful multiplication and gregarization (Hemming et al., 1979; Roffey and Popov, 1968). However, suitable food sources for gregarious populations are more widely separated in desert and semidesert zones than in the savannas, because rainfall is often effective for vegetation growth only after runoff and accumulation into widely separated drainage systems. Consequently long-range swarm movements by swarming populations have assumed much greater significance in *S. gregaria* than other locusts, and enable it to take advantage of the complementary seasonal breeding areas that are already being exploited by solitarious populations. Far from being able to survive in environments too hostile for individuals in the solitary phase, the survival and persistence of swarms invariably depend on adequate rains and vegetation growth during each breeding period.

In any given synoptic situation, movements by swarming populations during the daytime are less closely associated with prevailing wind trajectories than migrations of individuals at night. This is attributed to the influence of visual signals from the ground on individuals in low-flying swarms moving within their flight boundary layer and to physical obstacles in the path of swarms. Most nocturnal migrants are displaced downwind in the geostrophic flow above their flight boundary layer, where individual headings depart from a downwind direction only at lower wind speeds.

Gregariously behaving individuals are recognizable by a suite of diagnostic morphological and physiological characteristics, including factors such as relative elytron length. Jago (1983) has suggested that sensitivity of these responses is related to the extent to which populations shift spatially between complementary breeding areas and is best developed in a species such as *L. m. migratorioides*, which migrates between rainy season habitats and dry season refuges. It is, however, difficult to reconcile this hypothesis with the fact that (a) many acridoids that show no morphological phase characteristics migrate seasonally over greater distances than *L. m. migratorioides* (e.g., *C. terminifera*); and (b) swarm movements are generally less extensive than those undertaken by solitarious individuals of migratory species.

The main cost of a swarming strategy is the additional energy required to sustain the reiterative movements that last throughout the active life of gregarious individuals. This is probably one cause of their slower rate of development and maturation in the field so that, for example, solitary populations of *L. m. migratorioides* pass through four or five generations a year in Africa but gregarious populations have only three (Farrow and Longstaff, 1986), although laboratory-derived development rates tend to predict an opposite effect (Cheke, 1978). There is no evidence that the reproductive ratio per generation is lower in gregarious individuals. Juvenile survival is usually higher because mortality from natural enemies is much reduced and invariably makes a much greater contribution to reproductive success than fecundity or development rate (Farrow and Longstaff, 1986). Juvenile survival of 20–30% is common in gregarious populations but survival of <0.1–10% is more frequent in solitarious populations. This shift in the balance between natality and mortality is associated with a reduction in fecundity in gregarious females, which lay fewer, larger eggs with greater reserves (Farrow and Longstaff, 1986) so that hatchlings presumably have a longer period in which to locate suitable food in a competitive environment. Although there are elaborate mechanisms to ensure survival of eggs of solitarious individuals, through control of maturation and oviposition in relation to weather and selection of suitable oviposition sites, few such adaptations are found in gregarious units, where maturation and oviposition proceed en route, and are often independent of extrinsic factors. Eggs are laid at synchronized intervals wherever the swarm finds itself, and egg survival depends largely on chance encounters between the swarm and sites favorable for egg development. This strategy is probably a reflection of the "excess" reproductive capacity of swarming populations and suggests that movement in such populations is not primarily aimed at maximizing survival of progeny, as is suggested for solitary populations, but is directed at ensuring adult survival.

9.4 GENERAL CONCLUSIONS

The important role that long-range migration plays in the populations dynamics of many species of grasshopper and solitary-living locust is still often overlooked, and both Rankin and Singer (1984) and Taylor and Taylor (1983) have wrongly equated the solitary state with a sedentary existence. Most acridoids of semiarid environments are adapted to exploit temporally and spatially varying resources and in warmer climates may undertake regular movements at night between widely separated habitats in which the success of recolonization appears to be quite high. By contrast, most aggregations of acridoids persist only where favorable habitats are not widely separated or when conditions are uniformly suitable for extended periods. This usually arises from sustained periods of above-average rainfall in the subtropics and increased insolation in temperate areas. Only two species of locust, *S. gregaria* and *C. terminifera*, appear to have become adapted to regularly reaching widely separated favorable habitats in the gregarious phase, by means of downwind displacement in the upper air, by day and by night, respectively, and few other gregarious species ever displace over long range in the upper air. Only *S. gregaria*, however, moves in a circuit between widely separated, seasonally discrete habitats. Nocturnal movements have the advantage that populations can reach widely separated breeding areas in one step because migrants can take advantage of the relatively strong geostrophic winds. These may eventually carry populations into areas unsuitable for breeding, an inevitable consequence of expansive movements of the kind seen in *C. terminifera*.

Plagues of only two species of locust are known to have been sustained during a long run of fluctuating weather conditions, including seasons of below-average rainfall, namely, those of *L. m. migratorioides* (1928–1941) and *N. septemfasciata* (1930–1944) in Africa. This is possibly related to movements of swarms within relatively well-watered parts of the tropic zone and to avoidance of the extremes of the dry season. The plague of *L. m. migratorioides* in Australia (1973–1976), with its restricted seasonal displacements, was sustained only during the 2 years of above-average and appropriately timed rainfall (Farrow, 1979b).

The view presented here is that movement by aggregated populations of acridoids is primarily an adaptation to exploit a situation in which environmental conditions have become unusually favorable over large areas for a limited period, owing to climatic fluctuations, whereas the movements of solitarious individuals are an adaptation to the more permanent conditions of a patchy environment produced by less favorable, more normal, climatic conditions.

ACKNOWLEDGEMENTS

I would like to thank my colleagues Alistair Drake and Lindsay Barton-Browne of CSIRO and Don Reynolds of ODNRI for their helpful comments on an earlier draft of this manuscript and Chris Hunt of CSIRO for production of the illustrations.

REFERENCES

Ahluwalia, P. J. S., H. L. Sikka, and M. V. Ventakesh. 1976. Behaviour of swarms of *Oedalus senegalensis* Krauss (Orthoptera, Acrididae — subfamily Oedipodinae). *Indian J. Entomol.* **38**, 114–117.

Alexander, G. 1964. Occurrence of grasshoppers as accidentals in the Rocky Mountains of northern Colorado. *Ecology* **45**, 77–86.

Angelo, M. J. and F. Slansky. 1984. Body building by insects, trade-offs with resource allocation with particular reference to migratory species. *Fla. Entomol.* **67**, 22–114.

Anonymous. 1982. *The Locust and Grasshopper Agricultural Manual.* Centre for Overseas Pest Research, London.

Baker, G. L. 1975. *The Locust Outbreak in the Markham and Ramu Valleys, 1973–75*, Bubia Inf. Bull. no. 15. Department of Agriculture, Stock and Fisheries, Lae.

Baker, G. L. and M. Casimir. 1986. Current locust and grasshopper control technology in the Australasian region. *Proc. Pan Am. Acridol. Soc., 4th, 1985*, pp. 191–220.

Baker, P. S., M. Gewecke, and R. J. Cooter. 1984. Flight orientation of swarming *Locusta migratoria. Physiol. Entomol.* **9**, 247–52.

Baker, R. R. 1978. *The Evolutionary Ecology of Animal Migration.* Hodder & Stoughton, London.

Baron, S. 1972. *The Desert Locust.* Methuen, London.

Barry, R. G. and R. J. Chorley. 1987. *Atmosphere, Weather and Climate*, 5th ed. Methuen, London and New York.

Batten, A. 1966. The course of the last major plague of the African Migratory Locust, 1928 to 1941. *FAO Plant Prot. Bull.* **14**, 1–16.

Batten, A. 1967. Seasonal movements of swarms of *Locusta migratoria migratorioides* R. & F. in western Africa in 1928 to 1931. *Bull. Entomol. Res.* **56**, 357–80.

Batten, A. 1969. The Senegalese grasshopper *Oedaleus senegalensis* Krauss. *J. Appl. Ecol.* **6**, 27–45.

Begon, M. 1983. Grasshopper populations and weather, the effects of insolation on *Chorthippus brunneus. Ecol. Entomol.* **8**, 361–370.

Bennet, L. 1976. The development and termination of the 1968 plague of the desert locust, *Schistocerca gregaria* (Forskål) Orthoptera, Acrididae. *Bull. Entomol. Res.* **66**, 511–552.

Bergh, J. -E. 1979. Electromagnetic activity in the VLF range and take-off by locusts. *Int. J. Biometeorol.* **23**, 195–204.

Bergh, J. -E. 1988. Take-off activity in caged desert locusts, *Schistocerca gregaria* (Forskål) (Orthoptera, Acrididae) in relation to meteorological disturbances. *Int. J. Biometeorol.* **32**, 95–102.

Bond, K. M. G. and R. E. Blackith. 1987. A desert locust of the solitary phase in Ireland. *Ir. Nat. J.* **22**, 356–357.

Botha, D. H. and A. Jansen. 1969. Night flying by brown locusts, *Locustana pardalina* (Walker). *Phytophylactica* **1**, 79–92.

Brown, H. D. 1980. *Annual Report for 1980, Red Locust Control Service.* Plant Research Institute, Pretoria.

Browning, K. A. and C. W. Pardoe. 1973. Structure of low-level jet streams ahead of mid-latitude cold fronts. *Q. J. R. Meteorol. Soc.* **99**, 619–638.

Casimir, M. 1958. An experimental campaign with light aircraft against flying locust swarms in New South Wales. *Bull. Entomol. Res.* **49**, 497–508.

Casimir, M. and R. C. Bament 1974. An outbreak of the Australian plague locust, *Chortoicetes terminifera* (Walk.), during 1966–67 and the influence of weather on swarm flight. *Anti-Locust Bull.* no **44**.

Casimir, M. and V. Edge. 1979. The development and impact of a control campaign against *Austracris guttulosa* in New South Wales. PANS **23**, 223–236.

Chapman, R. F. 1959. Observations on the flight activity of the red locust, *Nomadacris septemfasciata* (Serville). *Behaviour* **14**, 300–344.

Chapman, R. F. 1972. The movement of acridoid populations. In C. F. Hemming and T. H. C. Taylor, Eds. *Proceedings of the International Study Conference on the Current and Future Problems of Acridology*. Centre for Overseas Pest Research, London, pp. 239–252.

Chapman, R. F., A. G. Cook, G. A. Mitchell, and W. W. Page. 1978. Wing dimorphism and flight in *Zonocerus variegatus* (L.) (Orthoptera, Acridoidea). *Bull. Entomol. Res.* **68**, 229–242.

Chapman, R. F., W. W. Page, and A. R. McCaffery. 1986. Bionomics of the variegated grasshopper (*Zonocerus variegatus*) in west and central Africa. *Annu. Rev. Entomol.* **31**, 479–505.

Cheke, R. 1978. Theoretical rates of increase of gregarious and solitarious populations of the desert locust. *Oecologia* **35**, 161–171.

Clark, D. P. 1969. Night flights of the Australian plague locust *Chortoicetes terminifera* (Walk.) in relation to storms. *Aust. J. Zool.* **17**, 329–352.

Clark, D. P. 1971. Flights after sunset by the Australian plague locust, *Chortoicetes terminifera* (Walk.) and their significance in dispersal and migration. *Aust. J. Zool.* **19**, 159–176.

Clark, D. P. 1972. The plague dynamics of the Australian plague locust, *Chortoicetes terminifera* (Walk.). In C. F. Hemming and T. H. C. Taylor, Eds. *Proceedings of the International Study Conference on the Current and Future Problems in Acridology*. Centre for Overseas Pest Research, London, pp. 275–87.

Clark, D. P. 1978. The significance of the availability of water in limiting invertebrate numbers. In K. M. W. Howes, Ed., *Proceedings of the Symposium on Studies of the Australian Arid Zone III, Water on Rangelands*. CSIRO, Melbourne, pp. 198–207.

Clark, D. P., C. Ashall, Z. Waloff, and L. Chinnick. 1969. Field studies on the Australian plague locust *Chortoicetes terminifera* (Walk.) in the channel country of Queensland. *Anti-Locust Bull.* **44**.

Davey, J. T. 1953. Possibility of movements of the African migratory locust in the solitary phase and the dynamics of its outbreaks. *Nature (London)* **172**, 720–721.

Davey, J. T. 1959. The African migratory locust (*Locusta migratoria migratorioides* Rch. and Frm., Orth.) in the Central Niger Delta. Part two, The ecology of *Locusta* in the semiarid lands and seasonal movements of populations. *Locusta* no **7**.

De Lépiney, J. 1928. Sur le comportement de *Schistocera gregaria* Forsk, au cours des vols. *Rev. Pathol. Vég. Entomol. Agric. Fr.* **15**, 82–96.

Dempster, J. P. 1963. The population dynamics of grasshoppers and locusts. *Biol. Rev. Cambridge Philos. Soc.* **38**, 490–529.

Den Boer, P. J. 1968. Spreading the risk and stabilisation of animal numbers. *Acta Biotheor.* **18**, 165–194.

Descamps, M. 1965. Acridoides du Mali. Régions de San et Sikasso (Zone soudanaise). *Bull. Inst. Fr. Afr. Noire* **27**, 922–962.

Dingle, H. 1972. Migration strategies of insects. *Science* **175**, 1327–1335.

Dingle, H. 1980. Ecology and evolution of migration. In S. A. Gauthreaux, Ed., *Animal Migration, Orientation and Navigation*. Academic Press, New York, pp. 1–101.

Dingle, H. 1982. Function of migration in the seasonal synchronisation of insects. *Entomol. Exp. Appl.* **31**, 36–48.

Dingle, H. 1985. Migration and life histories. *Contrib. Mar. Sci. Suppl.*, **27**, 27–42.

Drake, V. A. 1982. Insects in the sea breeze front at Canberra, a radar study. *Weather* **37**, 134–143.

Drake, V. A. 1983. Collective orientation by nocturnally migrating Australian plague locusts *Chortocietes terminifera* (Walker) (Orthoptera, Acrididae), a radar study. *Bull. Entomol. Res.* **73**, 679–692.

Drake, V. A. 1985. Solitary wave disturbances of the nocturnal boundary layer revealed by radar observations of migrating insects. *Bound. Layer Meteorol.* **31**, 269–286.

Drake, V. A. and R. A. Farrow 1983. The nocturnal migration of the Australian plague locust, *Chortoicetes terminifera* (Walker) (Orthoptera, Acrididae), quantitative radar observations of a series of northward flights. *Bull. Entomol. Res.* **73**, 567–585.

Drake, V. A. and R. A. Farrow. 1988. The influence of atmospheric structure and motions on insect migration. *Annu. Rev. Entomol.* **33**, 183–210.

Draper, J. 1980. The direction of desert locust migration. *J. Anim. Ecol.* **49**, 959–974.

Duranton, J. F., M. Launois, M. H. Launois-Luong, and M. Lecoq. 1979a. Biologie et écologie de *Catantops haemorroidalis* en Afrique de l'Ouest (Orthopt. Acrididae). *Ann. Soc. Entomol. Fr.* **15**, 391–343.

Duranton, J. F., M. Launois, M. H. Launois-Luong, and M. Lecoq. 1979b. Les voies privilégiées de déplacement du criquet migrateur malgache en phase solitaire. *Bull. Ecol.* **10**, 107–123.

Duranton, J. F., M. Launois, M. H. Launois-Luong, and M. Lecoq. 1983. De l'étude des Criquets à l'écologie opérationnelle. *Pour la Science*, Janvier, pp. 54–56.

Ellis, P. E. 1951. The marching behaviour of hoppers of the African migratory locust (*Locusta migratoria migratorioides* R. and F.) in the laboratory. *Anti-Locust Bull.* no. **7**.

Farrow, R. A. 1974. Comparative plague dynamics of tropical *Locusta* (Orthoptera, Acrididae). *Bull. Entomol. Res.* **64**, 401–411.

Farrow, R. A. 1975a. Offshore migration and the collapse of outbreaks of the Australian plague locust (*Chortoicetes terminifera* Walk.) in southeast Australia. *Aust. J. Zool.* **23**, 569–595.

Farrow, R. A. 1975b. The African migratory locust in its main outbreak area of the Middle Niger: Quantitative studies of solitary populations in relation to environmental factors. *Locusta* no **11**.

Farrow, R. A. 1977a. Origin and decline of the 1973 plague locust outbreak in central western New South Wales. *Aust. J. Zool.* **25**, 455–489.

Farrow, R. A. 1977b. Maturation and fecundity of the spur-throated locust, *Austracris guttulosa* (Walker), in New South Wales during the 1974/75 plague. *J. Aust. Entomol. Soc.* **16**, 27–39.

Farrow, R. A. 1977c. First captures of the migratory locust *Locusta migratoria* L. at light traps and their ecological significance. *J. Aust. Entomol. Soc.* **16**, 59–61.

Farrow, R. A. 1979a. Population dynamics of the Australian plague locust (*Chortoicetes terminifera* (Walker)) in central western New South Wales. I. Reproduction and migration in relation to weather. *Aust. J. Zool.* **27**, 717–745.

Farrow, R. A. 1979b. Changes in the distribution and abundance of the migratory locust (*Locusta migratoria* L.) in Australia between 1973 and 1977 in relation to outbreaks. *CSIRO Aust. Div. Entomol. Rep.* no. **9**.

Farrow, R. A. 1982a. Population dynamics of the Australian plague locust, *Chortoicetes terminifera* (Walker) in central western New South Wales. III. Analysis of population processes. *Aust. J. Zool.* **30**, 569–579.

Farrow, R. A. 1982b. Phenology of the Braidwood grasshopper outbreak. In K. E. Lee, Ed. *Proceedings of the 3rd Australian Conference on Grassland Invertebrate Ecology*. S. A. Govt. Adelaide, pp. 63–66.

Farrow, R. A. 1984. The locust problem in China today. *J. Aust. Inst. Agric. Sci.* 161–166.

Farrow, R. A. 1986. Interactions between synoptic scale and boundary-layer meteorology on micro-insect migration. In W. Danthanarayana, Ed., *Insect Flight, Dispersal and Migration*. Springer-Verlag, Berlin, pp. 185–195.

Farrow, R. A. 1988. The effects of changing land use on outbreaks of tropical migratory locust. *Insect Sci. Appl.* **8**, 969–975.

Farrow, R. A. and J. C. Daly. 1987. Long range movement as an adaptive strategy in the genus *Heliothis* (Lepidoptera, Noctuidae), a review of its occurrence and detection in four pest species. *Aust. J. Zool.* **35**, 1–24.

Farrow, R. A. and B. C. Longstaff. 1986. Comparison of the intrinsic rates of natural increase of locusts in relation to the incidence of plagues. *Oikos* **46**, 207–222.

Farrow, R. A., G. K. Nicolas, and J. E. Dowse. 1982. Migration in the macropterous form of the wingless grasshopper, *Phaulacridium vittatum* (Sjösdedt) during an outbreak. *J. Aust. Entomol. Soc.* **21**, 307–308.

Fishpool, L. D. C. and G. B. Popov. 1984. The grasshopper faunas of the savannas of Mali, Niger, Benin and Togo. *Bull. Inst. Fondam. Afr. Noire, Ser, A* **43**, 275–410.

Gatehouse, A. G. 1987. Migration, a behavioural process with ecological consequences. *Antenna* **11**, 10–12.

Gatehouse, A. G. 1988. Genes environment and insect flight. In G. J. Goldsworthy and G. H. Wheeler, Eds. *Insect Flight*. CRC Press, Boca Raton, Florida, pp. 115–138.

Gillett, J. D. 1979. Out for blood, flight orientation upwind in the absence of visual clues. *Mosq. News* **39**, 221–229.

Gillett, S. D. and E. Gonta. 1978. Locusts as prey, factors affecting their vulnerability to predation. *Anim. Behav.* **26**, 292–298.

Golding, F. D. 1948. The Acrididae (Orthoptera) of Nigeria. *Trans. R. Entomol. Soc. London* **99**, 517–587.

Gould, J. L. 1980. The case for magnetic sensitivity in birds and bees (such as it is). *Am. Sci.* **68**, 256–267.

Grossman, R. L. and G. A. Friehe. 1986. Vertical structure of the south-west monsoon low-level jet over the central and eastern Arabian Sea. *J. Atmos. Sci.* **43**, 3266–3272.

Gurney, A. B. 1953. Grasshopper glacier of Montana and its relation to long distance flights of grasshoppers. *Rep. Smithson. Inst.*, pp. 305–325.

Harrison, R. G. 1980. Dispersal polymorphisms in insects. *Annu. Rev. Ecol. Syst.* **11**, 95–108.

Haskell, P. T. 1957. The influence of flight noise on behaviour in the desert locust, *Schistocerca gregaria*. *J. Insect Physiol.* **1**, 52–75.

Hemming, C. F., G. B. Popov, J. Roffey, and Z. Waloff. 1979. Characteristics of desert locust upsurges. In D. L. Gunn and R. C. Rainey, Eds, *Strategy and Tactics of Control of Migrant Pests*. Royal Society, London, pp. 127–138.

Hughes, R. D. 1979. Movement in Population Dynamics. In R. L. Rabb and G. G. Kennedy, Eds. *Movement of Highly Mobile insects: Concepts and Methodology in Research*. North Carolina State University, Raleigh, pp. 14–34.

Hunter, D. M. 1982. Adult development in the Australian plague locust, *Chortoicetes terminifera* (Walker) (Orthoptera, Acrididae). *Bull. Entomol. Res.* **72**, 589–598.

Hunter, D. M., L. McCulloch, and D. E. Wright, 1981. Lipid accumulation and migratory flight in the Australian plague locust, *Chortoicetes terminifera* (Walker) (Orthoptera, Acrididae). *Bull. Entomol. Res.* **71**, 543–546.

Jago, N. D. 1979. Light trap sampling of the grasshopper *Oedaleus senegalensis* (Krauss, 1877) (Acrididae, Oedipodinae) and other species in West Africa, a critique. *Proc. Pan. Am. Acridol. Soc., 2nd, 1979*, pp. 165–198.

Jago, N. D. 1983. The evolutionary interrelationships of phase attributes in the Acridoidea. *Proc. Trienn. Meet. Pan Am. Acridol. Soc., 3rd, 1981*, pp. 65–91.

Joern, A. 1983. Small-scale displacements of grasshoppers (Orthoptera, Acrididae) within arid grasslands. *J. Kans. Entomol. Soc.* **56**, 131–139.

Johnson, C. G. E. 1969. *Migration and Dispersal of Insects by Flight*. Methuen, London.

Jones, R. E., N. Gilbert, M. Guppy, and V. Nealis. 1980. Long-distance movement of *Pieris rapae*. *J. Anim. Ecol.* **49**, 629–642.

Joyce, R. J. V. 1952. The ecology of grasshoppers in east central Sudan. *Anti- Locust Bull.* no. **11**.

Jungreis, S. A. 1987. Biomagnetism, an orientating mechanism in migrating insects. *Fla. Entomol.* **70**, 277–283.

Kennedy, J. S. 1939. The behaviour of the desert locust *Schistocerca gregaria* Forsk. (Orthopt.) in an outbreak centre. *Trans. R. Entomol. Soc. London* **89**, 385–542.

Kennedy, J. S. 1951. The migration of the desert locust. *Schistocerca gregaria* Forsk. I. The behaviour of swarms. II. A theory of long range migrations. *Philos. Trans. R. Soc. London, Ser. B* **235**, 163–290.

Kennedy, J. S. 1956. Phase transformation in locust biology. *Biol. Rev. Cambridge Philos. Soc.* **31**, 349–370.

Kennedy, J. S. 1961. A turning point in the study of insect migration. *Nature (London)* **189**, 785–791.

Kennedy, J. S. 1986. Migration, behavioural and ecological. In M. A. Rankin, ed. *Migration, mechanisms and adaptive significance. Contrib. Mar. Sci.* **27**, Suppl., 1–20.

Kennedy, J. S. and M. J. Way. 1979. Summing up the Conference. In R. L. Rabb and G. G. Kennedy, Eds., *Movement of Highly Mobile Insects, Concepts in Methodology and Research.* North Carolina State University, Raleigh, pp. 446–455.

Launois, M. 1978. *Modelisation écologique et simulation opérationelle en Acridologie. Application à* Oedaleus senegalensis (*Krauss*). Ministère de la Coopération, Paris.

Lea, A. 1968. Natural regulation and artificial control of Brown Locust numbers. *J. Entomol. Soc. South. Afr.* **31**, 97–112.

Lean, O. B. 1931. The effect of climate on the migrations and breeding of *Locusta migratorioides. Bull. Entomol. Res.* **22**, 551–569.

Lecoq, M. 1978a. Biologie et dynamique d'un peuplement acridien de zone soudanniene en Afrique de l'Ouest. Orthoptera, Acrididae. *Ann. Soc. Entomol. Fr.* **14**, 603–681.

Lecoq, M. 1978b. Les déplacements par vol à grande distance chez les acridiens des zones saheliënne et soudanniënne en Afrique de l'Ouest. *C. R. Hebd. Seances Acad. Sci., Ser. D* **286**, 419–422.

Magor, J. I. 1970. Outbreaks of the Australian Plague Locust (*Chortoicetes terminifera* Walk.) in New South Wales during the period 1939–1962, particularly in relation to rainfall. *Anti-Locust Mem.* no. **11**.

Marsh, D., J. S. Kennedy, and A. R. Ludlow. 1978. An analysis of anemotactic zigzagging flight in male moths stimulated by pheromone. *Physiol. Entomol.* **3**, 221–240.

McAnelly, M. L. 1985. The adaptive significance and control of migratory behaviour in the grasshopper *Melanoplus sanguinipes. Contrib. Mar. Sci.* **27**, Suppl., 687–703.

McAnelly, M. L. and M. A. Rankin. 1986a. Migration in the grasshopper *Melanoplus sanguinipes* (Fab). I. The capacity for flight in non-swarming populations. *Biol. Bull. (Woods Hole, Mass.)* **170**, 368–377.

McAnelly, M. L. and M. A. Rankin. 1986b. Migration in the grasshopper, *Melanoplus sanguinipes* (Fab). II. Interactions between flight and reproduction. *Biol. Bull. (Woods Hole, Mass.)* **170**, 378–392.

McCaffery, A. R. 1975. Food quality and quantity in relation to egg production in *Locusta migratoria migratorioides. J. Insect Physiol.* **21**, 1551–1558.

Mikkola, K. 1986. Insect migration in relation to the wind. In W. Danthanarayana, Ed., *Insect Flight, Dispersal and Migration.* Springer-Verlag, Berlin, pp. 152–171.

Moran, C., P. Wilkinson, and D. D. Shaw. 1980. Allozymic variation across a narrow hybrid zone in the grasshopper *Caledia captiva. Heridity* **44**, 69–81.

Nicolas, G. and R. A. Farrow. 1984. First recorded breeding by the tropical migratory locust (*Locusta migratoria migratorioides* R. and F.) in the Australian Capital Territory. *Aust. Entomol. Mag.* **11**, 34.

Parker, J. R., R. C. Newton, and R. L. Shotwell. 1955. Observations on mass flights and other activities of the Migratory Grasshopper. *U. S. Dep. Agric., Tech. Bull.* no **110**.

Parsons, P. A. 1982. Adaptive strategies of colonising animal species. *Biol. Rev. Cambridge Philos. Mag.* **57**, 117–148.

Pedgley, D. E. 1972. Recent studies on weather systems affecting movements and breeding of the desert locust, *Schistocerca gregaria* (Forsk.) in countries around the Red Sea. In C. F. Hemming and T. H. C. Taylor, Eds., *Proceedings of the International Study Conference on the Current and Future Problems of Acridology*. Centre for Overseas Pest Research, London, pp. 221–228.

Pedgley, D. E. (ed.) 1981. *Desert Locust Forecasting Manual*, Vols. 1 and 2. Centre for Overseas Pest Research, London.

Pedgley, D. E. 1982. *Wind-borne Pest and Diseases, Meteorology of Airborne Organisms*. Ellis Horwood, Chichester, England.

Pedgley, D. E., D. R. Reynolds, J. R. Riley, and M. R. Tucker. 1982. Flying insects reveal small scale wind systems. *Weather* **37**, 295–306.

Pener, M. P. 1985. Hormonal effects on flight and migration. In G. A. Kerkut and L. I. Gilbert, Eds., *Comprehensive Insect Physiology, Biochemistry and Pharmacology*, Vol. 8, Pergamon, Oxford, pp. 491–550.

Popov, G. B. 1965. *Review of the Work of the Desert Locust Ecological Survey June 1958–March 1964 and the Considerations and Conclusions Arising from it*, Prog. Rep. UNSF/DL/ES/8. FAO, Rome.

Popov, G. B. 1988. *Sahelian Grasshoppers*, Bull. No, 5. Overseas Development Natural Resources Institute, Chatham.

Popov, G. B. and M. Ratcliffe. 1968. The sahelian tree locust, *Anacridium melanorhodon. Anti-Locust Mem.* no **9**.

Rabb, R. L. and G. G. Kennedy, (Eds). 1979. *Movement of Highly Mobile Insects: Concepts and Methodology in Research.* North Carolina State University, Raleigh.

Rainey, R. C. 1951. Weather and the movements of locust swarms: A new hypothesis. *Nature (London)* **168**, 1057–1060.

Rainey, R. C. 1963. Meteorology and the migration of desert locusts. Applications of synoptic meteorology in locust control. *Anti-Locust Mem.* no **7**.

Rainey, R. C. 1976. Flight behaviour and features of the atmospheric environment. *Symp. R. Entomol. Soc.* **7**, 75–112.

Rainey, R. C. and Z. Waloff. 1951. Flying locusts and convection currents. *Anti-Locust Bull.* **9**, 51–70.

Rankin, M. A. 1978. Hormonal control of insect migration. In H. Dingle, Ed., *Evolution of Migration and Diapause in Insects*. Springer-Verlag, New York, pp. 5–32.

Rankin, M. A. 1985. Endocrine influences on insect migratory behaviour. *Contrib. Mar. Sci.* **27**, Suppl., 817–841.

Rankin, M. A. and M. C. Singer. 1984. Insect Movement, mechanisms and effects. In C. B. Huffaker and R. L. Rabb, Eds., *Ecological Entomology*. Wiley, New York, pp. 185–216.

Rankin, M. A., M. L. McAnelly, and J. E. Bodenhamer. 1986. The oogenesis flight syndrome revisited. In W. Danthanarayana, Ed., *Insect Flight, Dispersal and Migration*. Springer-Verlag, Berlin, pp. 27–48.

Rao, R. Y. 1960. The desert locust in India. *Monogr. Ind. Counc. Agric. Res.* no **21**.

Reid, D. G., K. G. Wardhaugh, and J. Roffey. 1979. Radar studies of insect flight at Benalla, Victoria in February 1974. *CSIRO Div. Entomol. Rep.* no **16**.

Reynolds, D. R. and J. R. Riley. 1988. A migration of grasshoppers, particularly *Diabolocatantops axillaris* (Thunberg) (Orthoptera, Acrididae), in the West African Sahel. *Bull. Entomol. Res.* **78**, 251–271.

Richards, O. W. and N. Waloff. 1954. Studies on the biology and population dynamics of British grasshoppers. *Anti-Locust Bull.* no 17.

Riley, J. R. 1974. Radar observations of individual desert locusts (*Schistocerca gregaria*) (Forsk.). *Bull. Entomol. Res.* **64**, 19–32.

Riley, J. R. 1975. Collective orientation in night-flying insects. *Nature (London)* **253**, 113–114.

Riley, J. R. and D. R. Reynolds. 1979. Radar-based studies of the migratory flight of grasshoppers in the middle Niger area of Mali. *Proc. R. Soc. London, Ser. B* **204**, 67–82.

Riley, J. R. and D. R. Reynolds. 1983. A long-range migration of grasshoppers observed in the Sahelian zone of Mali by two radars. *J. Anim. Ecol.* **52**, 167–183.

Riley, J. R. and D. R. Reynolds. 1986. Orientation at night by high-flying insects. In W. Danthanarayana, Ed., *Insect Flight, Dispersal and Migration.* Springer-Verlag, Berlin, pp. 71–87.

Riley, J. R., U. Krueger, C. M. Addison, and Gewecke, M. 1988. Visual detection of wind-drift by high-flying insects at night, a laboratory study. *J. Comp. Physiol. A* **162**, 793–798.

Ritchie, M. and D. E. Pedgley. 1989. Desert locusts cross the Atlantic. *Antenna* **13**, 10–12.

Ritchie, M. G., R. K. Butlin, and G. M. Hewitt. 1987. Causation, fitness effects and morphology of macropterism in *Chorthippus parallelus* (Orthoptera, Acrididae). *Ecol. Entomol.* **12**, 209–218.

Roff, D. A. 1977. Dispersal in dipterans, its costs and consequences. *J. Anim. Ecol.* **46**, 443–456.

Roffey, J. 1963. Observations on night flight in the Desert Locust (*Schistocerca gregaria* Forskål). *Anti-Locust Bull.* no **39**, 1–32.

Roffey, J. 1969. Radar studies on the desert locust. Niger, September — October 1968. *Occas. Rep. Anti-Locust Res. Cent.* no **17**.

Roffey, J. 1972. Migration and dispersal in the Australian Plague Locust. *Abstr., Int. Congr. Entomol., 14th, 1972*, p. 155.

Roffey, J. and G. B. Popov. 1968. Environmental and behavioural processes in a desert locust outbreak. *Nature (London)* **219**, 446–450.

Rogers, C. 1983. Pattern and process in large-scale animal movement. In I. R. Swingland and P. J. Greenwood, Eds., *The Ecology of Animal Movement.* Oxford Univ. Press (Clarendon), pp. 160–180.

Rose, A. H., R. H. Silversides, and O. H. Lindquist. 1975. Migration flight by an aphid, *Rhopalosiphum maidis* (Hemiptera, Aphididae), and a noctuid, *Spodoptera frugiperda* (Lepidoptera, Noctuidae). *Can. Entomol.* **107**, 567–576.

Sayer, H. J. 1962. The desert locust and the tropical convergence. *Nature (London)* **194**, 330–336.

Schaefer, G. W. 1969. Radar studies of locust, moth and butterfly migration in the Sahara. *Proc. R. Entomol. Soc. London, Ser. C* **34**, 33, 39–140.

Schaefer, G. W. 1976. Radar observations of insect flight. *Symp. R. Entomol. Soc.* **7**, 157–197.

Shaw, B. 1965. Depressions and associated desert locust swarm movements in the Middle East. *Tech. Notes World Meteorol. Organ.* **69**, 194–198.

Shaw, D. D., C. Moran, and P. Wilkinson. 1980. Chromosomal reorganisation, geographic differentiation, and the mechanism of speciation in the genus *Caledia.* In R. L. Blackman, G. M. Hewitt and M. Ashburner, Eds. *Insect Cytogenetics.* Blackwell, Oxford, pp. 171–194.

Simberloff, D. 1981. What makes a good island colonist? In R. F. Denno and H. Dingle, Eds., *Insect Life History Patterns, Habitat and Geographic Variation.* Springer-Verlag, Berlin, pp. 195–206.

Simpson, J. E. 1987. *Gravity Currents: In the Environment and in the Laboratory.* Ellis Horwood, Chichester, England.

Smit, C. J. B. 1939. Field observations on the brown locust in an outbreak centre. *Dept. Agric. For. Res. Serv.* no **15**.

Solbreck, C. 1978. Migration, diapause and direct development as alternative life histories in a seed bug, *Neacoryphus bicrucis.* In H. Dingle, Ed., *Evolution of Insect Migration and Diapause.* Springer-Verlag, New York, pp. 195–217.

Solbreck, C. 1985. Migration strategies and population dynamics. *Contrib. Mar. Sci.* **27**, Suppl., 642–659.

Sotthibandhu, S. and Baker, R. R. 1979. Celestial orientation by the large yellow underwing moth, *Noctua pronuba* L. *Anim. Behav.* **27**, 786–800.

Southwood, T. R. E. 1962. Migration of terrestrial arthropods in relation to habitat. *Biol. Rev. Cambridge Philos. Soc.* **37**, 171–214.

Southwood, T. R. E. 1977. Habitat, the template for ecological strategies? *J. Anim. Ecol.* **46**, 337–365.

Southwood, T. R. E. 1981. Ecological aspects of insect migration. In D. J. Aidly, Ed., *Animal Migration.* Cambridge Univ. Press, London and New York, pp. 196–208.

Stower, W. J. and D. J. Greathead. 1969. Numerical changes in a population of the desert locust with special reference to the factors responsible for mortality. *J. Appl. Ecol.* **6**, 203–235.

Symmons, P. M. 1979. Flight after sunset by small swarms of *Locusta migratoria* L. *J. Aust. Entomol. Soc.* **18**, 191.

Symmons, P. M. 1986. Locust displacing winds in Eastern Australia. *Int. J. Biometeorol.* **30**, 53–64.

Symmons, P. M. and E. J. Luard. 1981. The simulated distribution of night-flying insects in a wind convergence. *Aust. J. Zool.* **30**, 187–198.

Symmons, P. M. and L. McCulloch 1980. Persistence and migration of *Chortoicetes terminifera* (Walker) (Orthoptera, Acrididae) in Australia. *Bull. Entomol. Res.* **70**, 197–201.

Symmons, P. M. and D. E. Wright. 1981. The origins and course of the 1979 plague of the Australian plague locust *Chortoicetes terminifera* (Walker) (Orthoptera, Acrididae) including the effect of chemical control. *Acrida* **10**, 159–190.

Taylor, L. R. 1974. Insect migration, flight periodicity and the boundary layer. *J. Anim. Ecol.* **3**, 225–238.

Taylor, L. R. 1986. The four kinds of migration. In W. Danthanarayana, Ed., *Insect Flight, Dispersal and Migration.* Springer-Verlag, Berlin, pp. 265–280.

Taylor, L. R. and R. A. J. Taylor. 1977. Aggregation, migration and population dynamics. *Nature (London)* **265**, 415–421.

Taylor, L. R. and R. A. J. Taylor. 1983. Insect migration as a paradigm for survival by movement. In I. R. Swingland and P. J. Greenwood, Eds., *The Ecology of Animal Movement.* pp. 181–214.

Taylor, R. A. J. and D. Reling. 1986. Preferred wind direction of long distance leaf hopper (*Empoasca fabae*) migrants and its relevance to the return migration of small insects. *J. Anim. Ecol.* **55**, 1103–1114.

Têtefort, J. P., P. Dechappe, and J. M. Rakotoharison. 1966. Etude des migrations du criquet migrateur malgache *Locusta migratoria capito* Sauss. dans sa phase solitaire. *Agron. Trop.* **12**, 1389–1397.

Thorsen, J. 1966. Small signal analysis of a visual reflex in the locust. *Kibernetik* **3**, 41–53.

Uvarov, B. P. 1932. Ecological studies on the Moroccan locust in Western Anitolia. *Bull. Entomol. Res.* **23**, 273–287.

Uvarov, B. P. 1957. The aridity factor in the ecology of locusts and grasshoppers of the Old World. *Arid Zone Res.* **8**, 164–198.

Uvarov, B. P. 1969. *Grasshoppers and Locusts*, Vol. 1. Cambridge Univ. Press, London and New York.

Uvarov, B. P. 1977. *Grasshoppers and Locusts*, Vol. 2. Centre for Overseas Pest Research, London.

Venkatesh, M. V. 1971. A case study of the influence of a cyclone on the incursion of desert locusts into Rajasthan, India during June 1964. *Indian J. Entomol.* **33**, 1–16.

Waloff, Z. 1963. Field studies on solitary and *transiens* Desert Locust in the Red Sea Area. *Anti-Locust Bull.* no **40**.

Waloff, Z. 1966. The upsurges and recessions of the desert locust plague, an historical survey. *Anti-Locust Mem.* no **8**.

Waloff, Z. 1972. Orientation of flying locusts (*Schistocerca gregaria* Forsk.) in migrating swarms. *Bull. Entomol. Res.* **62**, 1–72.

Waloff, Z. and D. E. Pedgley. 1986. Comparative biogeography and biology of the South American locust, *Schistocerca cancellata* (Seville), and the South African desert locust, *S. gregaria flaviventris* (Burmeister) (Orthoptera, Acrididae), a review. *Bull. Entomol. Res.* **76**, 1–20.

Waloff, Z. and R. C. Rainey. 1951. Field studies on the factors affecting the displacement of desert locust swarms in eastern Africa. *Anti-Locust Bull.* no **9**.

Wehner, R. 1984. Astronavigation in insects. *Annu Rev. Entomol.* **29**, 277–298.

Wellington, W. G. 1954. Atmospheric circulation processes and insect ecology. *Can. Entomol.* **86**, 313–333.

Williams, C. B. 1958. *Insect Migration.* Collins, London.

Woodrow, D. F. 1965. Observation on the Red Locust (*Nomadacris septemfasciata* Serv.) in the Rukwa Valley, Tanganyika, during its breeding season. *J. Anim. Ecol.* **34**, 187–200.

Wright, D. E. 1988. Analysis of the development of major plagues of the Australian plague locust *Chortoicetes terminifera* (Walker) using a simulation model. *Aust. J. Ecol.* **12**, 423–438.

Young, G. R., F. M. Dori, and K. G. Gorea. 1984. *A plague of Locusta migratoria L. (Orthoptera, Acrididae) in the Markham and Upper Ramu Valley of Papua New Guinea, March 1975 to November 1976*, Res. Bull. no. 3a. Department of Primary Industry, Port Moresby.

10

Territory-Based Mating Systems in Desert Grasshoppers: Effects of Host Plant Distribution and Variation

M. D. GREENFIELD AND

T. E. SHELLY

10.1 INTRODUCTION

The notion of a territorial grasshopper may well be an oxymoron to some readers. Given the high population densities of many species and the tracking of changing weather and food supplies by some of these populations (Uvarov, 1977), the occurrence of site fidelity and aggressive defense of resources would seem unlikely. Nonetheless, a curious group of gomphocerine grasshoppers in which adult males are clearly territorial does occur in the deserts of southwestern North America. Three species, ranging from oligophagous to completely monophagous in feeding habits, exhibit male defense of individual host plants as mating territories. In this chapter we focus on studies of the evolution of territoriality in these unusual grasshoppers. In particular, we describe how these studies have expanded our understanding of territoriality both in herbivores and in species where the quality of resources within a territory as well as the nature of epigamic displays (i.e., signaling) of males on the territory influence the settlement of females.

Otte and Joern (1975) first called attention to the existence of territoriality in certain acridids through a seminal article on species found on creosote (*Larrea*) bushes in North and South American deserts. They described the defense of individual *Larrea tridentata* bushes by *Ligurotettix coquilletti* McNeill males in the Sonoran Desert and the apparent lack of this behavior in *Bootettix argentatus* Bruner, another species found on the same host plant. Additionally, the potential for territoriality in *Ligurotettix planum* (Bruner), a species associated with tarbush (*Flourensia cernua*) in the Chihuahuan Desert, was noted but its actual occurrence was not observed. Since 1975, though, territoriality has been documented in both *B. argentatus* (Schowalter and Whitford, 1979) and *L. planum* (Shelly and Greenfield, 1989). Moreover, territorial behavior and sexual selection in *L. coquilletti* have been studied in depth recently (Greenfield and Shelly, 1985; Shelly and Greenfield, 1985; Greenfield et al., 1987; Shelly et al., 1987; Greenfield et al., 1989b), and many of the findings compiled in this chapter were extracted from these articles.

10.2 NATURAL HISTORY OF *LIGUROTETTIX* AND *BOOTETTIX*: HOST PLANT ASSOCIATIONS AND MATING BEHAVIOR

Ligurotettix coquilletti, the most thoroughly studied member of the group, is an oligophagous species usually found on *Larrea* bushes, and occasionally on *Atriplex* or *Lycium*, in the Sonoran, Mohave, and Colorado Deserts (Rehn, 1923; Otte, 1981). In populations occurring on *Larrea*, both male and female adults and immatures feed on the foliage and perch on the stems of the bushes. Adults are first seen in late May or early June and may survive more than 2 months in the field.

Male *L. coquilletti* are remarkably sedentary and often spend most of their adult life within a single *Larrea* bush. They defend these bushes, passively via stridulation (a "keep-out" signal) and actively via overt aggression, from occupation by other males. Stridulation consists of an incessant clicking produced during morning and midday hours; an additional peak also occurs at dusk on many days. The song serves to attract females to the defended bush as well as to repel males. Once on the same bush as a signaling male, however, females never make the final approach in pair formation. Rather, the males, who are extremely sensitive to movement, orient toward females. Similarly, males often visually detect and approach other nearby males. In the resulting encounters, the intruder normally jumps away and remains silent within the same bush. Only rarely does the interaction escalate to a grapple in which the participants fight with or attack one another with their legs and mandibles.

Courtship and mating occur on the host plant, with copulation lasting approximately 10 min. Mate guarding is not evident, and males usually resume stridulation immediately after mating. Both males and females can and do mate multiply, but the extent and timing of remating are unknown. Females later oviposit in the soil, though usually not at the base of a *Larrea* bush. The nymphs hatch early the following spring and pass through five (males) or six (Females) instars. Adult eclosion is typically protandrous.

Ligurotettix planum is quite similar to *L. coquilletti* in most aspects of behavior and life history, the major difference being that *L. planum* is found on *Flourensia* bushes in the Chihuahuan Desert. This association is more specific than that between *L. coquilletti* and *Larrea*; in our study area in New Mexico, very few *L. planum* have ever been found perched off *Flourensia* [but see Otte and Joern (1975) for different findings in Coahuila, Mexico], and feeding trials conducted in the laboratory showed that *L. planum* would not eat any non-*Flourensia* foliage from its habitat. *L. planum* also exhibits a lower degree of site fidelity than *L. coquilletti*. Although some *L. planum* males remain in and defend the same *Flourensia* bush for several weeks, during any 24-h period 25% of the insects move to bushes more than 5 m away.

Bootettix argentatus, like *L. planum*, is an extreme host plant specialist, feeding and perching exclusively on *Larrea tridentata* bushes in the Sonoran, Mohave, Colorado, and Chihuahuan Deserts. Unlike the two *Ligurotettix* species, individuals of *B. argentatus* rest on the foliage of their host plant and are exceedingly cryptic on this background. Stridulation by male *B. argentatus* is rather sporadic, and territorial defense of host bushes may occur only in low-density populations (Schowalter and Whitford, 1979). Both pre- and postcopulatory mate guarding and associated intermale aggression commonly occur, however, suggesting that male *B. argentatus* typically defend females rather than resources (M. Barzman, unpublished data). In the Colorado Desert, it is likely that two, largely overlapping generations of *B. argentatus* occur annually.

10.3 ON THE FUNCTION OF TERRITORIALITY IN INSECT HERBIVORES

The nature of the resources defended by male *Ligurotettix* grasshoppers provided us with a unique opportunity to investigate territoriality in insect herbivores. Although territorial behavior has been studied in many insect species (Baker, 1983; Thornhill and Alcock, 1983), and its existence is no longer a cause for surprise among ethologists or entomologists, few workers have explored the defense of host plants (but see Whitham, 1979; Mitchell, 1980; Goldsmith, 1987). In this regard, the work of Otte and Joern (1975), which focused on the existence of territoriality in *L. coquilletti* as a consequence of its use of isolated, regularly spaced host plants, served as a landmark study. This factor, in addition to the diurnal activity pattern of the insects and their low population density (200–400 adults/ha), called our attention to the "*Ligurotettix/Larrea* system" as one eminently amenable to field observation and experimentation.

We embarked on a long-term study of the *Ligurotettix/Larrea* system in July 1982 at the Deep Canyon Desert Research Center (hereafter referred to as the Deep Canyon site), near Palm Desert, California. The study plots established at this site are situated in an alluvial plain floristically and climatically typical of the Colorado Desert; *Larrea tridentata* is the dominant perennial in the plots and exists at a density of approximately 190 bushes/ha. Although we have conducted various field manipulations and laboratory experiments on *L. coquilletti*, much of our work has employed systematic observation of unconstrained, individually marked insects within the plots.

Early in our study we recognized two features of the distribution of male *L. coquilletti* that ultimately proved to be central to an understanding of their territoriality. First, despite the generally aggressive behavior of the males, multiples of from two to seven individuals were often found occupying a single bush (Fig. 10,1). Second, each year the same *Larrea* bushes were occupied by territorial males (and by females), whereas others consistently remained vacant (Fig. 10.2). Bushes heavily occupied in a given season were also the earliest sites to be occupied by territorial males during the succeeding year. Curiously, female–male ratios and the numbers of matings per male were higher at these heavily occupied bushes than at bushes harboring only a single male. For example, during systematic observations in 1982 and 1983, 24 of 31 matings occurred at bushes harboring two or more males, whereas aggregated individuals accounted for only 52% of the male population. These findings raised several questions pertinent to site selection by territorial herbivores in general and to how variation in site quality might affect a herbivore's mating system.

10.3.1 Territory Selection by *L. coquilletti*: Host Plant Quality as a Primary Factor

To elucidate the essential characteristics of territorial sites preferred by *L. coquilletti*, we learned the necessity of measuring occupancy of the available sites by an appropriate method (Greenfield et al., 1987). If merely the number of different males arriving at a site throughout the season was considered, then the bushes with higher occupancy values were the larger (in area) ones, possibly as a result of chance or an attraction toward larger silhouettes during settlement. However, our records showed that male *L. coquilletti* frequently

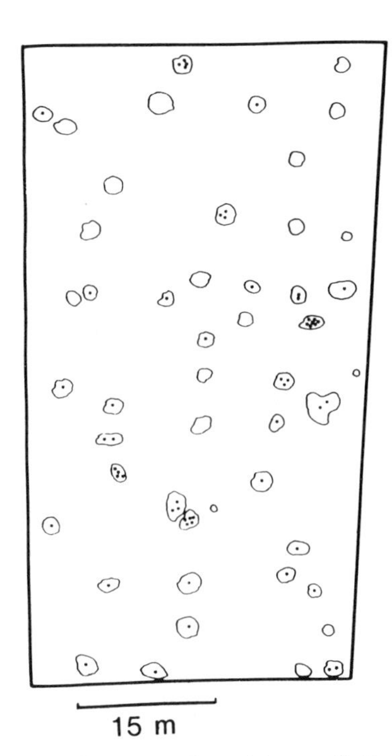

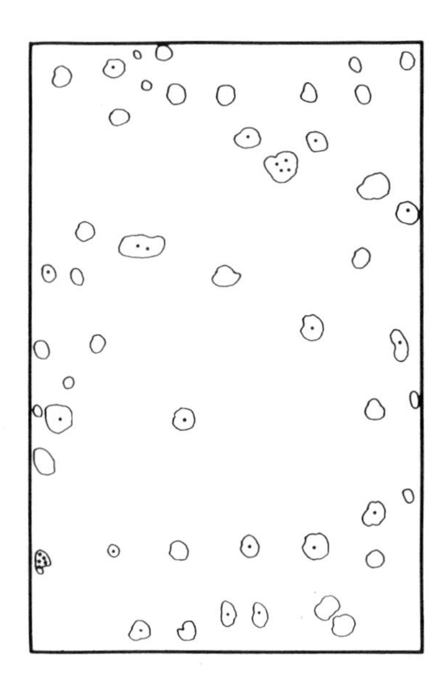

Fig. 10.1. Distribution of *L. coquilletti* males among *Larrea* bushes in two study plots at Deep Canyon, California on 19 and 18 July 1983, respectively. Each solid dot represents a single male individual. (From Greenfield and Shelly, 1985. Copyright 1985 by Baillière Tindall.)

moved among the bushes at the beginning of the season, and >50% of the insects arriving at a bush during this time departed within a few days: obviously, such individuals should not be deemed territorial. By restricting the "territorial occupants" of a bush to long-term occupants, we noted that bushes harboring more of such insects were physiognomically indistinguishable from other bushes. That is, physical attributes, such as height, area, stem density, or various "neighborhood parameters" (e.g., distances to and sizes of neighboring *Larrea* bushes), were not critical features.

We did realize, however, that the *Larrea* bushes bore the primary, if not sole, food resources of the insects, and following the lead of Rhoades (1977a,b), we explored the possibility that prefered and avoided sites differed in food quality. This was tested through both chemical analyses of the foliage of bushes with known histories and short-term feeding trials testing the consumption of foliage taken from the bushes. These two methods demonstrated unequivocally that food quality was a major characteristic distinguishing the perennially used sites.

Measurement of water and nitrogen content, and of hexane- and methanol-extractable compounds, including HPLC analysis of the latter, revealed that

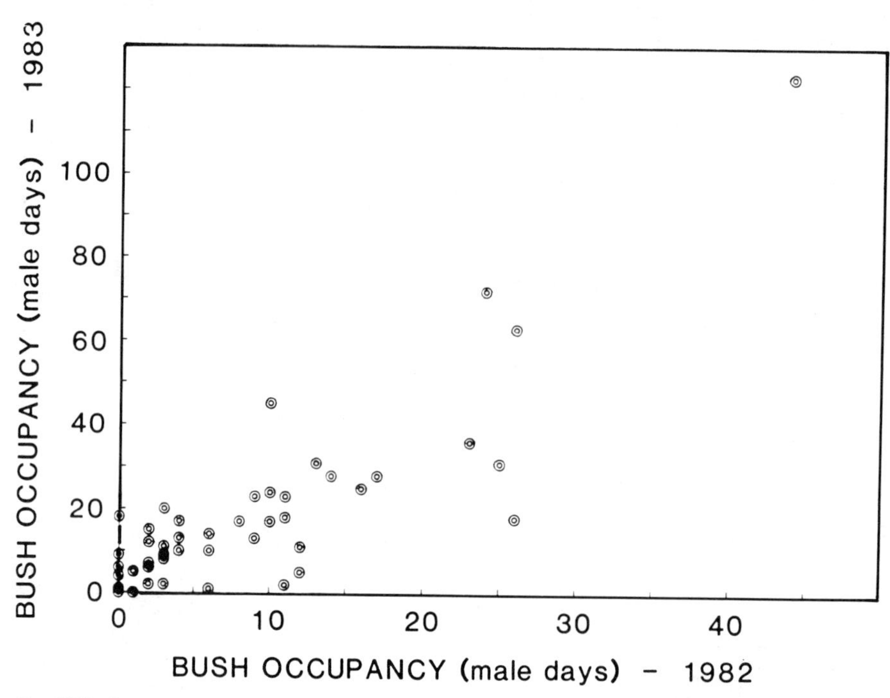

Fig. 10.2. Occupancy by *L. coquilletti* males of *Larrea* bushes in a study plot at Deep Canyon, California, in 1983 versus 1982. The observation of one male on one census day constituted one male day. Censuses were made on 15 and 13 days in 1983 and 1982, respectively.

preferred bushes had low titers of hexane-extractable compounds (mainly wax esters; these ranged from 6 to 10% of leaf dry weight) and of a particular substance [nordihydroguaiaretic acid (NDGA), an antioxidant lignan catechol] in the methanol extract (this extract was mainly extrafoliar phenolic resin and made up approximately 20% of the leaf dry weight; from 10 to 40% of the extract was NDGA). Earlier, Rhoades (1977b) had suggested that NDGA served as a major component of the antiherbivore system of *Larrea* by interfering with protein and starch digestion.

Laboratory feeding trials bolstered our impression that foliage quality was the major issue in site selection. In these 40-min tests both male and female adults fed much less readily on sprigs obtained from bushes with histories of low occupation and high NDGA titers. To substantiate further that the qualitative difference between preferred and avoided bushes was specifically of a chemical nature, resin (methanol extracts) from the bushes was presented on sucrose disks to the insects (Chapman et al., 1988). The results showed that disks impregnated with resin from an avoided bush were significantly more deterrent; that is, the insects nibbled only slightly on them. Additional feeding trials, using either *Larrea* foliage (Greenfield et al., 1987) or sucrose disks (Chapman et al., 1988) as substrates, confirmed that the specific compound

NDGA was deterrent and acted in even low concentrations (<1% of dry weight).

Most certainly, however, other compounds present in *Larrea* foliage also play a role in determining its suitability as food. Although NDGA titer was significantly related to the occupancy of a bush, this compound explained only 10% of the total variance in occupancy. Clearly, much exploratory chemistry will be required to characterize fully the *Larrea* bushes preferred as territories by *L. coquilletti*. Possibly, ages and karyotypes (several polyploid races of *Larrea tridentata* exist in North America) of the bushes influence the essential chemical characteristics. It may also be fruitful to examine the features of high-quality oviposition sites to learn whether these tend to be coupled spatially to the preferred food resources. If not, conflicting demands may be acting upon territory selection.

10.3.2 Territory Selection by *L. coquilletti*: The Consequences of Defending High- versus Low-Quality Sites

Although the characterization of heavily occupied *Larrea* bushes is obviously incomplete, our evidence indicates that males do defend such sites because they include superior food resources. After finding that foliage from occupied bushes was preferred in feeding trials, we then sought the long-term consequences of association with preferred versus avoided bushes (Greenfield et al., 1989b). We gathered late-instar nymphs in the field, held them collectively until the adult molt in screen cages containing *Larrea* foliage, and then transferred the insects to large net bags maintained on bushes with known histories. Insects were checked daily for mortality and weighed weekly. Adult males seldom survived when confined on bushes that were perennially avoided (a 75% mortality rate occurred due to desiccation resulting from starvation), but 90% of the females survived whether kept on preferred or avoided ones. Females confined on the preferred bushes were distinguished, however, by significantly greater relative growth rates (RGR; Waldbauer, 1968) 7–14 days posteclosion. Values on preferred bushes averaged 0.0165 mg/mg/day, as opposed to 0.0046 on avoided bushes.

The preceding information, though further substantiating that unoccupied bushes essentially represented low-quality food resources, failed to differentiate low RGRs due to a mere behavioral aversion to certain (chemical) factors (i.e., the insects simply fed insufficiently on the available leaves; see Bernays and Chapman, 1987) from detrimental physiological effects resulting from feeding. Consequently, we repeated the growth experiment but kept insects in individual screen cages and provided them with sprigs of the test foliage (Greenfield et al., 1989b). Thus we could monitor the amount of food an insect consumed in conjunction with its gain in weight. The former quantity was computed from the weight of the feces (W_{fe}) produced by the insect, which was then converted to the weight of food ingested (W_{fo}) by the following adjustment that considered lignin to be an indigestible "marker":

$$W_{fo} = W_{fe} \cdot \text{feces lignin concn./foliage lignin concn.}$$

Our approach of relying on a marker was necessary because the amount of foliage eaten could not be determined directly: *Larrea* leaves were small and irregular in outline, their weight changed when sprigs were supported in water, and the insects seldom consumed entire leaves.

From the above experiment we learned that adult females consuming a given amount of food gained more weight when that food was obtained from a preferred site; that is, their efficiency of conversion of ingested food (ECI; Waldbauer, 1968) was greater ($\bar{X}$ = 0.0474 on preferred foliage; $\bar{X}$ = 0.0180 on avoided foliage). On the other hand, long-term consumption rates (RCR) of foliage from preferred and avoided bushes were equivalent. These results implied that the behavioral preference observed in short-term feeding trials had an underlying physiological basis.

Further analysis showed that the difference in ECIs between preferred and avoided bushes resulted from higher rates of conversion of digested food (ECD; Waldbauer, 1968), rather than greater digestibility (AD) of the foliage as suggested by Rhoades (1977b). Our interpretation of this is that certain substances (e.g., NDGA) in the foliage of avoided *Larrea* bushes exert a toxic effect on metabolic processes. In the field *L. coquilletti* females normally increase 80% in weight during the 3 weeks following eclosion. Consequently, females feeding on preferred bushes and experiencing higher ECDs would be expected to enjoy more rapid maturation, reduced time and movement associated with feeding, and/or a greater final weight that would probably translate into higher fecundity and fitness (unfortunately, we were unable to obtain data on fecundity, because females did not fully mature in confinement, possibly due to spending excessive durations on sides of the net bags and screen cages).

Findings on ECIs and ECDs similar to those reported above were obtained with males and nymphs of each sex. Because adult males normally increase in weight by 45%, they would also benefit directly from feeding on preferred bushes. For males, greater final weight may result in louder stridulations (hind femur length and intensity of stridulation are positively related), which could be more attractive to females (see Section 10.4.4). Additionally, males that remain on and defend preferred bushes may encounter more females (see Section 10.3.3). This possible supplementary effect may account for our observation that males suffered higher mortality than females when confined to avoided bushes. Normally, males on such locations would depart not only to obtain higher-quality (less toxic) foliage elsewhere but also because of the selection pressure to defend a territory where numerous females would be expected. In confinement, however, this "motivation" may be manifested in frequent attempts to escape, low consumption rates, desiccation, and ultimately death. Males that did survive on avoided bushes, however, did not exhibit lower RCRs, perhaps indicating an intra-population variation in behavioral deterrence.

10.3.3 Territory Selection by *L. coquilletti*: The Basis of a Resource-Defense Polygynous Mating System

Based on the outcomes of the feeding and growth studies and an additional experiment described below, we propose that male *L. coquilletti* defend particular bushes both because of their own food requirements and because of their elevated encounter rates with females at these bushes; that is, the mating system is a case of resource-defense polygyny, dominant males occupying economically defendable sites crucial to females. The first explanation is a straightforward interpretation of male ECDs on different bushes, but the second ultimately depends on a link between differential conversion rates of females and their actual settlement pattern among the *Larrea* bushes. We could only ascertain existence of this link experimentally because males, as well as females, occurred disproportionately on the high-quality bushes under natural conditions. It was therefore uncertain whether females had settled there in large numbers because of resource quality, attractiveness of the signaling males present, or both.

To unravel these possibilities, we conducted a field experiment (Shelly et al., 1987) in which the numbers of males signaling on 12 high- and 12 low-quality bushes within a study plot were maintained, via removal and replacement, at one and only one individual per "test bush"; at the same time, we immediately removed all males found on "nontest bushes." Thus we attempted to equalize male epigamic effects at territories with different resource qualities. By monitoring the settlement of both naturally occurring females and those released into the plot, we found that significantly more (35 versus 14) females remained as long-term occupants on the high-quality bushes.

A problem, not resolved by the above experiment, was the possibility that males placed on high-quality bushes signaled distinctively and thereby retained more females. Recognizing that this possible interaction between resource quality and signaling could not be removed as long as live males were utilized, we repeated the experiment during a following year, replacing the 24 males with a computer-operated sound system that deployed loudspeakers broadcasting identical tape recordings of stridulatory clicks, one loudspeaker per test bush. A similar pattern of female settlement was obtained, reaffirming that a skewed distribution of females could be sustained due to variation in resource quality alone and that males establishing territories at the high-quality bushes early in the season could expect to experience higher encounter rates with maturing and arriving females.

Additional evidence suggests that females evaluate bushes at close range and probably through gustation. Whereas long-term female occupants accumulated primarily on the high-quality bushes, released individuals arrived in approximately equal numbers (24 versus 27) at low- and high-quality sites: a decision to depart or remain was made only after contact. Also, we released into the experimental plot a set of females whose labial and maxillary palps and antennae (the primary chemosensory structures, possibly excepting the tarsi) had been excised, and these insects exhibited no tendency toward remaining on high-quality bushes. Eight and 10 of them, respectively, remained as long-term occupants of the high- and low-quality sites. The settlement pattern of males, in which equal numbers of individuals also arrive initially at low- and high-quality sites, may be partially controlled by similar mechanisms.

In summary, the *Ligurotettix/Larrea* system underscores the relevance of intrapopulation variation in host plant quality to territoriality in herbivores. Such variation may actually lead to the evolution of resource-defense polygyny in some species by restricting high quality resources to a reduced number of sites, each one economically defensible because of the disproportionate number of females expected (see Emlen and Oring, 1977).

10.3.4 Do Insect Herbivores Learn the Locations of High-Quality Host Plants? Homing Orientation in *Ligurotettix*

Territoriality can exist at various levels of development. At a primary level, animals merely defend an economically valuable area in which they reside but fail to distinguish that area from similar, equally valuable sites in the vicinity (Otte and Joern, 1975). Many animals are capable of such discernment, though, and they manifest this ability through homing orientation — they return to their original location following displacement. Among insects, this ability has been recognized in the Hymenoptera and in several other groups relying heavily on visual orientation, but in general it has been discounted (Otte and Joern, 1975). Our recent work on *Ligurotettix*, particularly that on *L. planum*, however, demonstrates that this view may not be correct, and that advanced levels of territory recognition and concomitant homing behavior can evolve when compelling ecological circumstances (e.g., high variation in host plant quality) exist (Greenfield et al., 1989a).

We have been working on *L. planum* at a study plot situated 8 km north of Rodeo, New Mexico (hereafter referred to as the "Rodeo site") since August 1984. The plot is located in a region of typical Chihuahuan Desert vegetation, and *Flourensia cernua* is the dominant perennial. As in *L. coquilletti*, a year-to-year consistency in heavily occupied (by both males and females) bushes is evident, and feeding trials showed that both males and females preferred foliage from perennially occupied bushes. The preferred bushes do not differ from avoided ones in size or leaf water content, and preliminary chemical analyses (hexane and chloroform extracts) have not revealed any distinctions, but this work is continuing and we eventually do expect to find chemical differences.

In the course of our study, we noted that when certain *L. planum* males left a *Flourensia* bush where they had been seen repeatedly, they later reappeared at the original bush. To document that these were not random events or simple movements toward unusually tall or large bushes, we closely monitored the activity of 28 males at the Rodeo site, noting each interbush movement over a 12-day period. The observations verified that males often pursued other males or females to nearby bushes (or were chased there themselves) and then homed, sometimes making a direct flight, despite the presence of many other nearer and larger targets. Some of these homing events extended more than 10 m, several times the mean distance between neighboring territories (Fig. 10.3).

Displacement trials, in which territorial males were removed in a rotated, opaque container and released at points several meters from the "home" bush and equidistant from various large, neighboring bushes, showed that the insects could return without access to information acquired on the outbound journey. These returns were made regardless of wind direction, indicating that olfactory cues were not used. We then altered the profiles of home bushes and their neighbors in six additional displacement trials. Three released insects, which had each previously homed twice, failed to return in the time allotted, but the three others still homed directly. Evidently *L. planum* males learn the local landscape and rely on certain visual features to home, but the precise features learned remain unclear.

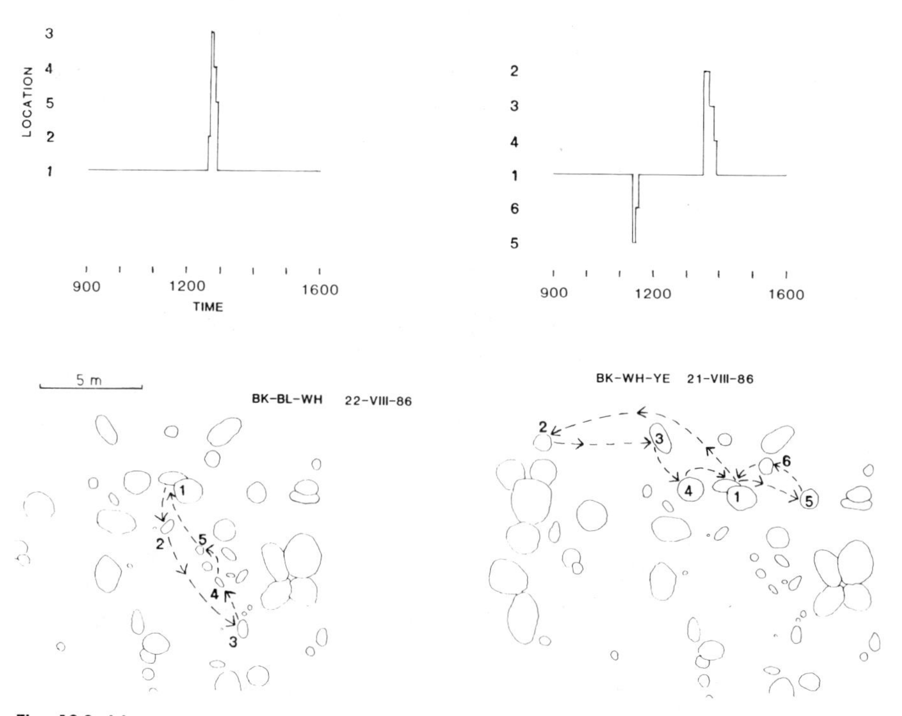

Fig. 10.3. Movement of two different *L. planum* males (BK-BL-WH and BK-WH-YE) at Rodeo, New Mexico during "homing orientation" events on 22 and 21 August 1986, respectively. Outlines of all *Flourensia* bushes in the vicinity of the insects are shown. The upper part of the figure shows the duration of perching at each of the locations shown in the lower part.

Importantly, homing events occurred only at perennially occupied bushes. Extrapolating from our information on the *Ligurotettix/Larrea* system and the *L. planum* feeding trials, it is likely that site recognition learning evolved in *L. planum* because of variation in the quality of its host plant and a skewed distribution of females resulting from this variation. Males departing a high-quality site who could return directly would be at a selective advantage over those having to locate, by trial and error, another territory with comparable resource quality and female numbers.

Since conducting the above work on *L. planum* we have learned that *L. coquilletti* also possess the ability to home. The ability in this species is likewise explained best as an adaptation to variation in host plant quality. Evidence for homing in *L. coquilletti* comes from observations of both subordinate males (see Section 10.4.1) and ovipositing females. By closely monitoring the movements of males, we found that many subordinate individuals repeatedly departed their home bushes for various nonhost plants or low-quality *Larrea* bushes 2–10 m distant, stridulated at the new site for several hours, and then returned, resuming subordinate (nonstridulating, nonterritorial) status (see Section 10.4.2). Bushes that were the focuses of homing movements had relatively low NDGA titers and were perennially occupied. Similarly, we labeled individual females with ^{32}P and tracked their early-morning ground movements with a Geiger counter (the insects were extremely cryptic on the ground and could not be found reliably without this technique). This study revealed that females sometimes oviposited >6 m from their home bush and then returned directly (G.-Y. Wang, unpublished data). Additional findings that some nymphs of each sex make similar, though shorter, movements (G.-Y. Wang, unpublished data) suggest that the learning inherent in homing ability may commence during immature development.

10.4 ON THE CO-OCCURRENCE OF TERRITORIALITY AND AGGREGATION

The distribution of *Ligurotettix* males is notable not only for the selection of certain host plant bushes, but also for the clustering of males at these sites (see Fig. 10.1 for *L. coquilletti*). Because it was difficult for us to reconcile clustering with the expected overdispersed distribution due to territorial behavior of the insects, we investigated whether all males were equally territorial and whether males accrued any particular advantages through clustering. In both investigations, we learned that the nature of male signaling activity and its influence on females were inseparable from the resource quality at the signaling location.

10.4.1 Alternative Mating Tactics in *L. coquilletti*: Inactive (Nonterritorial) versus Actively Signaling Behaviors

By monitoring singing of *L. coquilletti* males in our study plots at the Deep Canyon site, we found that approximately 20% of the male population was not territorial on any given day (Greenfield and Shelly, 1985). These insects, designated as "inactive," either produced stridulatory clicks at a very low rate or produced them only irregularly throughout the day. Inactive behavior did not grade into active signaling. Rather, the two behavioral states were distinct: most males sang at a high rate (>20 clicks/min) and regularly, or they did so only at sporadic intervals. Although some males behaving inactively did so for many consecutive days, transitions between inactive behavior and active signaling, and vice versa, did occur (see Section 10.4.2).

Intensive observation of the insects showed that males behaving inactively were sexually mature and were pursuing a mating tactic alternative to active signaling. Inactive males remained in the same *Larrea* bush as an active signaler, sometimes approached females, and on rare occasions mated (only 1 of 31 matings witnessed during systematic observations involved an inactive male). Usually, however, active signalers chased inactive males if the latter moved (as during courtship). In all such encounters, inactive males retreated, indicating a strong relationship between signaling and dominance in *L. coquilletti*.

Our initial interest in nonterritorial, inactive behavior focused on its maintenance in *L. coquilletti* populations. Theory predicts that alternatives such as inactive (nonterritorial) versus actively signaling (territorial) tactics co-occur either because they are equally rewarding (on a short- or long-term basis) and/or because various conditions (e.g., size, vigor, age) predispose certain individuals to less rewarding tactics, especially when population density is high (Rubenstein, 1980; Waltz, 1982). In *L. coquilletti*, males behaving inactively were found to mate infrequently (relative to their percentage of the population) on a daily basis, but assessing mating success over longer periods (e.g., male lifetimes) was problematic because some males often switched from inactivity to active signaling, and vice versa. That is, on 7% of the occasions that a male was seen actively signaling, he switched to inactivity during the succeeding 3 days; the corresponding value for the reverse switch was 55%. To circumvent this problem, we checked the relationship between adult longevity and the percentage of that longevity spent inactive. We measured longevity by establishing a 15-m "buffer zone" around a 40 x 60 m study plot and monitoring all *L. coquilletti* in this zone daily. Because the probability of a male moving >15 m in a 24-h period was extremely small, we were able to determine with a high degree of confidence whether males no longer seen in the plot had died or migrated. Contrary to expectations based on the energy costs (see Prestwich and Walker, 1981) and risks (vulnerability to natural enemies; see Cade, 1975) of signaling, males in the plot that remained inactive during a high proportion of adulthood did not live longer than consistently active signalers. This implied that the generally inactive males did not achieve a lifetime mating success equivalent to that of active signalers, and we were then obliged to consider inactive behavior as a conditional tactic.

In many animal species, nonterritorial alternatives are adopted primarily by smaller or younger individuals who cannot compete effectively with more vigorous males (Dominey, 1984). Inactive behavior in *L. coquilletti*, however, was not explicable in such simplistic terms, because the incidence of the tactic was unrelated to male size and, other than the silence of teneral adults (0–5 days posteclosion), age. The date of adult eclosion also appeared to have no bearing on the tactic adopted by a male. Through inspection for parasitic microorganisms in the grasshoppers (work performed by Dr. John E. Henry of the U.S.D.A. Rangeland Insect Laboratory in Bozeman, Montana), we then tested the possibility that such agents rendered males inactive (see Hamilton and Zuk, 1982; Zuk, 1987), but no evidence for this was detected either.

We were therefore faced with a dilemma: a substantial fraction of the male population in *L. coquilletti* was employing a (nonterritorial) mating tactic that relegated them to lower reproductive success, yet no condition causing this predisposition was apparent. Although we had experimentally demonstrated that incidence of the nonterritorial tactic increased at higher population density (Greenfield and Shelly, 1985), density in itself could not be construed as a predisposing condition because it did not explain why any given individual resorted to a tactic offering few rewards. It is likely that the dilemma we encountered in studying *L. coquilletti* is far from unique: in other animal species fitnesses conferred by alternative tactics are rarely equal on either a daily or a lifetime basis (Dominey, 1984), and in various cases size and age likewise do not influence signaling activity (e.g., see Cade and Wyatt, 1984).

10.4.2 The Nonterritorial Alternative in *L. coquilletti*: Silence at High-Quality Sites or Singing at Low-Quality Ones

We advanced our understanding of alternative tactics in *L. coquilletti* only after fully recognizing that available territories were not equal in quality. This progress materialized specifically through our analysis of how the males switched between tactics. Transition matrices revealed that males behaving inactively typically became active signalers by moving to a previously vacant (low-quality) bush and singing at the new location (Shelly and Greenfield, 1985). Conversely, the reverse behavioral transition was accomplished most often when an active signaler, usually the sole occupant of a bush (hereafter individuals found in such circumstances are referred to as A^s males), moved to a bush previously occupied (often by several males; hereafter, males singing on bushes that harbor >1 male are referred to as A^m males) and became silent. That is, behavioral transitions in either direction were most often accompanied by a change in location. Few females normally settled on the bushes to which previously inactive males moved when they began singing, and the mating success of these A^s males remained low despite their stridulation. Later, we discovered that many of these bushes contained high NDGA titers, and some were tested in feeding and growth trials and found to be deterrent.

Our conception of the nonterritorial alternative, constructed from the above transitions, is that subordinate males have two options, silence on high-quality bushes defended by dominant (actively signaling) males or singing at low-quality sites, and that many of the (A^s) males categorized initially as active signalers were in fact subordinate and nonterritorial. This was also indicated by a playback experiment demonstrating that many A^s males failed to respond aggressively to broadcasts of taped *L. coquilletti* stridulations, with some individuals actually retreating from the playback (G.-Y. Wang, unpublished data). A^m males tested in other trials, however, typically approached the broadcasting speaker as they would an intruding male.

A parallel to the behavioral transitions between inactive and A^s states existed on a smaller spatial scale and corroborated our above conception (Shelly and Greenfield, 1985). On any given day during systematic observations in 1983, approximately 4% of *L. coquilletti* males sang while perched on non-*Larrea* vegetation adjacent to *Larrea* bushes occupied by territorial males. Encounter rates with females were substitute "low" for negligible off *Larrea*, presumably owing to the inadequacy of the foliage (food). Males who sang on the non-*Larrea* bushes often made regular migrations to the nearby *Larrea* where they would remain silent for several hours until returning to the peripheral location (see Section 10.3.4).

10.4.3 Protandry in *L. coquilletti*: Do Early Eclosing Males Always Gain an Advantage?

After realizing that many of the *L. coquilletti* males who sang regularly were nevertheless nonterritorial, we reevaluated the conditions that could potentially render a male subordinate. This showed that males eclosing later in the season, although not necessarily spending more time inactive (silent), more commonly occurred on bushes where female numbers were low (G.-Y. Wang, unpublished data), probably indicating the low resource quality present. Typically, these males have the opportunity only to remain inactive on high-quality bushes or to signal on low-quality ones; that is, the peak in male eclosion occurs several weeks prior to that in females (differences were 3 and 4 weeks during the 2 years for which we obtained data), and most of the early eclosing males soon settle on and defend the preferred sites. That late-eclosing males seldom vie with the territorial defenders of high-quality bushes over ownership and dominance there remains an enigma. A game theoretical explanation commonly invoked for such situations, the resolving of an ownership dispute via the "uncorrelated asymmetry" of resident (early-eclosing, territorial male) versus intruder (later-eclosing, nonterritorial male) roles to avoid an escalated fight (Maynard Smith and Parker, 1976), does not seem to be applicable in this case. Uncorrelated asymmetries are most likely to evolve when desired territories are plentiful and can easily be obtained by a retreating intruder (Grafen, 1987). This option is not available to nonterritorial *L. coquilletti* males, for high-quality bushes are not superabundant.

A partial solution to the problem posed by nonterritorial males in *L. coquilletti* may be found, however, through scrutiny of the variance in time of eclosion (males typically eclose over a 6-week period; the range in female eclosion is much narrower). In the preceding paragraph we implicitly assumed that early eclosing males defended preferred territories, subsequently encountered the numerous females there, and thereby achieved a higher reproductive success. This advantage of early eclosion would require males to survive at least until peak female eclosion, and probably longer owing to the sexual nonreceptivity of newly eclosed females. During many years this requirement may be met, but unusual climatic conditions during other years might widen the interval between male and female eclosion peaks and leave many males to die prior to the appearance of sexually receptive females. Late-eclosing males would be favored selectively under such circumstances, and the possibility therefore exists that genetic variance in eclosion time remains in *L. coquilletti* populations because of environmental uncertainty. Alternatively, it might be postulated that variance in eclosion time, whether it be due to variance in the hatching date of first-instar nymphs and/or in the period needed for nymphal development, is entirely under environmental control and not subject to selection. Late eclosing males, consequently, would occur generation after generation. However, the mating disadvantage normally experienced by these insects suggests that an absence of stabilizing selection on the time of eclosion would be unlikely unless a hidden benefit, as outlined above, exists.

10.4.4 Preferred Sites of *L. coquilletti*: Do Clusters of Males Represent Resource-Based Leks?

Our treatment of alternative mating tactics in *L. coquilletti* may have inadvertantly created the image that when several males occupy a *Larrea* bush simultaneously, only one remains actively signaling and territorial. Although this scenario often does occur, two or more actively signaling (A^m) males also frequently co-occur on a bush, and the contradiction between male territoriality and aggregation cannot be resolved entirely via focusing on the nonterritorial individuals. To assess this second aspect of male clustering, we tested the hypothesis that some clustering occurs because females preferentially orient toward groups of males per se and that males therefore "actively" aggregate (i.e., they exhibit mutual attraction), forming "resource-based leks" (sensu Alexander, 1975). Previously, this has been argued on theoretical grounds amid claims that females could thereby select an optimal mate by simultaneous, as opposed to sequential, comparison [Bradbury, 1981; note that recent theoretical work by Queller (1987) suggests that such female preference could also evolve in the absence of any initial selective value]. In *L. coquilletti*, A^m males did achieve a significantly higher mating success than A^s males, and we were intrigued by the possibility that females were attracted to A^m males because they preferred clusters of males per se as well as the (high-quality) bushes typically defended by these males.

Our tests of the resource-based lek hypothesis employed the computer-operated sound system, described earlier (see Section 10.3.3), to simulate single versus clustered males in the experimental plot. We selected 12 high-quality *Larrea* bushes and placed one loudspeaker in six of the bushes and three in each of the remaining six. Thus the rate of stridulatory clicks emanating from "multiple-speaker bushes," was triple that from "single-speaker bushes," a not unrealistic difference. Females were released into the plot (which had been cleared of males) at centrally located positions, and their settlement was monitored. Following each replicate of this experiment, we rotated the sites at which single and grouped speakers were placed. We detected no preferences of females for settling or remaining on multiple-speaker bushes after four replicates.

In natural situations, however, clusters of singing males differ from lone singers not only in their overall (collective) click rate but in their sound intensity as well; that is, males elevate their sound intensity by 4–8 dB when singing adjacent to others, possibly because of intrasexual selection. We therefore continued our use of the sound system and selected 16 high-quality *Larrea* bushes for the placement of 16 loudspeakers broadcasting either high-intensity (64 dB at 1 m, measured at 8 kHz, the dominant frequency of *L. coquilletti* clicks; the mean sound intensity of the males is 60 dB) or low-intensity (56 dB) signals. Four replicates of this experiment were conducted, with the eight high- and eight low-intensity signals rotated between the bushes after each. A significant majority (60–80%) of the females released in each replicate moved onto the high-intensity bushes during the initial three days following release. Because signals from the low-intensity speakers as measured at the release points were generally >40 dB, the neurophysiologically determined hearing threshold of *L. coquilletti* (W. Bailey, unpublished data), one interpretation of these results is that females orient toward louder sources of sound, ceteris paribus, because such sources usually harbor clustered males. Even if the accumulation of females on high-intensity bushes was due to "passive" (sensu Parker, 1982), as opposed to "active," attraction (i.e., females simply did not hear the low-intensity signals), selection for louder (typically clustered) males would yet occur. A remaining question, therefore, is why solitary males do not "subvert" such selection pressure by merely increasing the intensity of their songs.

Several variations exist of the above explanation for the attractiveness of clustered males to females and for the initial clustering of the males, none of which can be ruled out at present. Foremost among these is that the range in resource quality between the various sites is sufficiently great that some (A^m) males are neither repelled nor silenced by the aggressive advances of co-occupants of high-quality bushes. If this explanation for male clustering is true, females might orient toward louder stridulations because such sounds are typically associated with high-quality sites and not necessarily because of any advantage accrued through the availability of a large selection of potential mates.

The clustering of males may also be explained following the logical arguments used above. That is, some males may mutually attract one another via stridulation, and some evidence for this does exist in *L. coquilletti*, but such aggregation would not definitely imply that they do so to attract more females. Males may simply orient to the songs of others as a "shortcut" to locating high-quality resources (see Stamps, 1987), as already pointed out for females. Clearly, many potential interactions exist between male clustering and site quality which we are only beginning to explore.

10.5 INFLUENCE OF RESOURCE DISPERSION ON MATING TACTICS

A central theme of this chapter is the influence of resource quality and its variation on territoriality and mating systems. Our work on *L. planum* at the Rodeo site shows how dispersion of resources can be another critical factor in this regard. Specifically, we have found that resource dispersion may constitute the primary influence on the nature of alternative mating tactics adopted by subordinate, nonterritorial males (Shelly and Greenfield, 1989).

As noted earlier, *L. planum* and *L. coquilletti* are quite similar in ecology and behavior. The similarity extends to the existence of alternative (subordinate) mating tactics in both species. A major difference, however, lies in the fact that subordinate *L. planum* males are transient; that is, they repeatedly move between territories while continuing to signal rather than remaining silent and inactive at one site as subordinate *L. coquilletti* males do. Typically, transient males move from one bush to another several times daily. These moves do not merely represent relocation on neighboring *Flourensia* bushes during the course of normal movement among the foliage of contiguous plants. Rather, a third of all moves exceeded the mean nearest-neighbor distance between bushes and traversed a considerable expanse of bare ground. The subordination of these transient *L. planum* males was clearly substantiated by their retreating during aggressive encounters and their low mating success.

Reasons for the different alternative tactics exhibited by *L. coquilletti* and *L. planum* appear unrelated to demography. Population density, sex ratio, and the degree of clustering at particular host plant bushes in both species are all quite similar. The distributions of the host plant bushes of the two species, however, differ greatly. In relative terms, *L. coquilletti* have available a small number of large bushes, whereas *L. planum* encounter numerous small ones (cf. Figs. 10.1 and 10.3). Because this was the only major ecological difference apparent to us, we have argued (Shelly and Greenfield, 1989) that transient behavior is favored by *L. planum* subordinates because of the great number of potential female encounter sites (*Flourensia* bushes); in part, this potential number is elevated because of a relatively high level of female movement. Additionally, the small size of *L. planum* encounter sites may make it easy for territorial males to eject silent subordinates.

10.6 OVERVIEW AND PROGNOSIS

We have attempted to portray *Ligurotettix* as a model genus for the investigation of territoriality in both acoustic insects and herbivores. Because of their conspicuous activity, site fidelity, and clearly demarcated territories, these grasshoppers have provided us with the opportunity to monitor accurately the activity of all individuals within our study plots. Thus we were able to acquire estimates of the relative mating success of males situated at different sites or performing different behaviors. From these relationships we tested various hypotheses concerning alternative mating tactics and the clustering of males, and we learned the significance of resource quality to territory selection. Moreover, we formed the opinion that variation in resource quality, by establishing a mosaic of low- and high-quality sites in the landscape, greatly influences the mating system of these grasshoppers. In particular, subordinate male mating tactics consisting either of signaling at low-quality sites or inactive (silent) behavior at high-quality sites are maintained in *L. coquilletti* populations. On the other hand, subordinate males in *L. planum* are transient, a characteristic probably effected by the dispersion of the host plant resources of this species.

Questions of a broader scope on the territoriality of acridids have proved to be more intractable, however. *Ligurotettix* and *Bootettix* are the only acridid genera in which the territorial phenomenon is known (see Otte and Joern, 1975), and reasons for this restriction, other than a sampling bias due to the paucity of detailed studies, are not obvious. Although monophagy, low population density, and association with small, easily defended bushes would seem to predispose these grasshoppers toward territoriality, the absence of territorial behavior in other species with similar traits (e.g., several other uncommon species associated with *Larrea* and other desert shrubs; see Otte and Joern, 1977) poses a problem. Additionally, territorial mating systems are widespread among other orthopterans (gryllids, tettigoniids) lacking some of these attributes, suggesting that our conception of the factors critical for territorial behavior is inadequate. It is certain, however, that if these factors are to be elucidated, we will need many more thorough studies on a wide range of acridid species, and these studies will have to be conducted in the field, paying particular attention to the distribution of resources and variation in their quality. Only then will it be possible to assess objectively whether *Ligurotettix* and *Bootettix* are indeed unique among the Acrididae, and what the prerequisites for territoriality in this family actually encompass.

ACKNOWLEDGMENTS

This work was supported by grants from the U.S. National Science Foundation, the National Geographic Society, and the U.C.L.A. Academic Senate. The authors wish to thank Theodore Burk, Reginald Chapman, and Anthony Joern for many useful comments.

REFERENCES

Alexander, R. D. 1975. Natural selection and specialized chorusing behavior in acoustical insects. In D. Pimentel, Ed., *Insects, Science, and Society*. Academic Press, New York, pp. 35–77.

Baker, R. R. 1983. Insect territoriality. *Annu. Rev. Entomol.* **28**, 65–89.

Bernays, E. A. and R. F. Chapman. 1987. The evolution of deterrent responses in plant-feeding insects. In R. F. Chapman, E. A. Bernays, and J. G. Stoffolano, Jr., Eds., *Perspectives in Chemoreception and Behavior*. Springer-Verlag, New York, pp. 159–174.

Bradbury, J. W. 1981. The evolution of leks. In R. D. Alexander and D. W. Tinkle, Eds., *Natural Selection and Social Behavior*. Chiron Press, New York, pp. 138–169.

Cade, W. 1975. Acoustically orienting parasitoids: Fly phonotaxis to cricket song. *Science* **190**, 1312–1313.

Cade, W. and D. R. Wyatt. 1984. Factors affecting calling behaviour in field crickets, *Teleogryllus* and *Gryllus* (age, weight, density, and parasites). *Behaviour* **88**, 61–76.

Chapman, R. F., E. A. Bernays, and T. Wyatt. 1988. Chemical aspects of host-plant specificity in three *Larrea*-feeding grasshoppers. *J. Chem. Ecol.* **14**, 561–579.

Dominey, W. J. 1984. Alternative mating tactics and evolutionarily stable strategies. *Am. Zool.* **24**, 385–396.

Emlen, S. T. and L. W. Oring. 1977. Ecology, sexual selection, and the evolution of mating systems. *Science* **197**, 215–223.

Goldsmith, S. K. 1987. The mating system and alternative reproductive behaviors of *Dendrobias mandibularis* (Coleoptera: Cerambycidae). *Behav. Ecol. Sociobiol.* **20**, 111–115.

Grafen, A. 1987. The logic of divisively asymmetric contests: Respect for ownership and the desperado effect. *Anim. Behav.* **35**, 462–467.

Greenfield, M. D., E. Alkaslassy, G.–Y. Wang, and T. E. Shelly. 1989a. Long-term memory in territorial grasshoppers. *Experientia* **45**, 775–777.

Greenfield, M. D. and T. E. Shelly. 1985. Alternative mating strategies in a desert grasshopper: Evidence of density-dependence. *Anim. Behav.* **33**, 1192–1210.

Greenfield, M. D., T. E. Shelly, and K. R. Downum. 1987. Variation in host-plant quality: Implications for territoriality in a desert grasshopper. *Ecology* **68**, 828–838.

Greenfield, M. D., T. E. Shelly, and A. Gonzalez-Coloma. 1989b. Territory selection in a desert grasshopper: The maximization of conversion efficiency on a chemically defended shrub. *J. Anim. Ecol.* **58**, 761–771.

Hamilton, W. D. and M. Zuk. 1982. Heritable true fitness and bright birds: A role for parasites? *Science* **218**, 384–387.

Maynard Smith, J. and G. A. Parker. 1976. The logic of asymmetric contests. *Anim. Behav.* **24**, 159–175.

Mitchell, P. L. 1980. Combat and territorial defense of *Acanthocephala femorata* (Hemiptera: Coreidae). *Ann. Entomol. Soc. Am.* **73**, 404–408.

Otte, D. 1981. *The North American Grasshoppers*. I. *Acrididae, Gomphocerinae and Acridinae*. Harvard Univ. Press, Cambridge, Massachusetts.

Otte, D. and A. Joern. 1975. Insect territoriality and its evolution: Population studies of desert grasshoppers on creosote bushes. *J. Anim. Ecol.* **44**, 29–54.

Otte, D. and A. Joern. 1977. On feeding patterns in desert grasshoppers and the evolution of specialized diets. *Proc. Acad. Nat. Sci. Philadelphia* **128**, 89–126.

Parker, G. A. 1982. Phenotype-limited evolutionarily stable strategies. In King's College Sociobiology Group, Eds., *Current Problems in Sociobiology*. Cambridge Univ. Press, London and New York, pp. 173–201.

Prestwich, K. N. and T. J. Walker. 1981. Energetics of singing in crickets: Effects of temperature in three trilling species (Orthoptera: Gryllidae). *J. Comp. Physiol., B* **143**, 199–212.

Queller, D. C. 1987. The evolution of leks through female choice. *Anim. Behav.* **35**, 1424–1432.

Rehn, J. A. G. 1923. North American Acrididae (Orthoptera). 3. A study of the Ligurotettigi. *Trans. Am. Entomol. Soc.* **49**, 43–92.

Rhoades, D. F. 1977a. Integrated antiherbivore, antidesiccant and ultraviolet screening properties of creosotebush resin. *Biochem. Syst. Ecol.* **5**, 281–290.

Rhoades, D. F. 1977b. The antiherbivore chemistry of *Larrea*. In T. J. Mabry, J. H. Hunziker, and D. R. DiFeo, Jr., Eds., *Creosote Bush: Biology and Chemistry of* Larrea *in New World Deserts*. Dowden, Hutchinson & Ross, Stroudsburg, Pennsylvania, pp. 135–175.

Rubenstein, D. I. 1980. On the evolution of alternative mating strategies. In J. E. R. Staddon, Ed., *Limits to Action: The Allocation of Individual Behavior*. Academic Press, New York, pp. 65–100.

Schowalter, T. D. and W. G. Whitford. 1979. Territorial behavior of *Bootettix argentatus* Bruner (Orthoptera: Acrididae). *Am. Midl. Nat.* **102**, 182–184.

Shelly, T. E. and M. D. Greenfield. 1985. Alternative mating strategies in a desert grasshopper: A transitional analysis. *Anim. Behav.* **33**, 1211–1222.

Shelly, T. E. and M. D. Greenfield. 1989. Satellites and transients: Ecological constraints on alternative mating tactics in male grasshoppers. *Behaviour* **109**, 200–221.

Shelly, T. E., M. D. Greenfield, and K. R. Downum. 1987. Variation in host plant quality: Influences on the mating system of a desert grasshopper. *Anim. Behav.* **35**, 1200–1209.

Stamps, J. A. 1987. Conspecifics as cues to territory quality: A preference of juvenile lizards (*Anolis aeneus*) for previously used territories. *Am. Nat.* **129**, 629–642.

Thornhill, R. and J. Alcock. 1983. *The Evolution of Insect Mating Systems*. Harvard Univ. Press, Cambridge, Massachusetts.

Uvarov, B. P. 1977. *Grasshoppers and Locusts. A Handbook of General Acridology*, Vol. 2. Cambridge Univ. Press, London and New York.

Waldbauer, G.P. 1968. The consumption and utilization of food by insects. *Adv. Insect Physiol.* **5**, 229–288.

Waltz, E. C. 1982. Alternative mating tactics and the law of diminishing returns: The satellite threshold model. *Behav. Ecol. Sociobiol.* **10**, 75–83.

Whitham, T. G. 1979. Territorial behaviour of *Pemphigus* gall aphids. *Nature (London)* **279**, 324–325.

Zuk, M. 1987. Seasonal and individual variation in gregarine parasite levels in the field crickets *Gryllus veletis* and *G. pennsylvanicus*. *Ecol. Entomol.* **12**, 341–348.

11

Pheromones and Phase Transformation in Locusts

W. LOHER

11.1 INTRODUCTION

Locusts are generally distinguished by several features: as polymorphic species, they range from the solitarious phase through a variety of transient forms to the gregarious phase, differing from one another morphologically, such as in their size, shape, and color, as well as in their physiology, ecology, and behavior. In this, locusts are different from the monophasic grasshoppers. The value of a second character once held as important as the first, that gregarious locusts have the ability to swarm, that is, migrate together in more or less dense formations, has been somewhat diminished, for it has been shown that not only solitarious locusts (Roffey and Popov, 1968), but also certain grasshopper species can swarm (Farrow, 1977; Chapman et al., 1986). A more valid third feature of gregarious locusts is the tendency of individuals to group together, whereas the solitarious form displays the opposite effect (Ellis, 1953). Locusts, therefore, are capable of changing from the solitarious to the gregarious phase when a rise in population density brings solitary locust hoppers in close contact. Such aggregation then causes the development of gregarization generated by chemical factors that may accelerate the transformation or delay it.

These chemical factors are called pheromones (Karlson and Butenandt, 1959), which are defined for the purpose of this chapter as substances produced by individuals in glands and transferred to individuals of the same or a related species at close range or by contact. They induce long-term alterations in the physiology of the recipient that can be accompanied by morphometric, behavioral, and genetic changes (primer pheromones), or they act as a signal that is quickly perceived and causes an immediate behavioral response (releaser pheromones).

In locusts, physiological and behavioral research has demonstrated five likely categories of pheromones that are listed according to their functions as follows:

1. A gregarizing pheromone acting in the hopper (larval) stage and participating in the transition from the solitarious to the gregarious phase (Nolte, 1963, 1968, 1969, 1974, 1976, 1977; Gillett, 1968, 1972, 1973, 1975, 1983).
2. A maturation pheromone accelerating sexual maturity in adult locusts (Norris, 1954, 1962, 1963, 1964, 1968, 1970; Loher, 1958, 1959, 1960, 1961; Strong, 1970, 1971; Amerasinghe, 1978).
3. Oviposition pheromones, which either aggregate locust females at certain locations for egg laying or attract them to sites where conspecific females have previously deposited their egg pods (Lauga and Hatté, 1977, 1978).
4. A solitarizing pheromone produced by adult locusts (Gillett and Phillips, 1977; Gillett, 1983).
5. A maturation-retarding pheromone, which delays sexual maturation (Norris, 1962, 1968).

Whereas the first three types of pheromones promote gregarization during various stages of locust development, the other two counteract gregarization and slow down the maturation process. These five functionally different types of pheromones are discussed in relation to the morphological, physiological, biochemical, genetic, and behavioral responses observed during phase transformation.

11.2 THE GREGARIZATION PHEROMONE

Phase transformation in locust hoppers from the solitarious to the gregarious form or vice versa is induced by changes in population density and involves reversible changes in their pigmentation, morphometry, physiology, and behavior. In the species *Schistocerca gregaria*, *Locusta migratoria*, and *Locustana pardalina*, crowded hoppers are characterized by black color patterns consisting of melanin that, according to the species, either turn the body almost completely black (e.g., *S. gregaria*, first instar) or cover more than 50% of the body surface, especially on the thorax and head. After isolation of crowded third-instar hoppers, the quantity of black pigment is significantly reduced, following a molt, and dark areas are replaced by green, buff, or fawn colors. Nolte (1963) observed that when such experiments on color changes were performed in a room normally used for locust rearing, the melanin was only partially reduced, whereas it disappeared in a room devoid of other locusts. Following several translocations between the two environments and subsequent comparisons of color change, Nolte postulated the presence of a pheromone in the air surrounding crowded locust hoppers that was responsible for their melanization. He also tentatively suggested that morphometric characters in the adult, such as the elytron/femur ratio (E/F) and the femur/head width ratio (F/C), changed in the direction of the crowded phase and were thus related to the pheromone. Gillett (1968) experienced similar difficulties when she tried to rear adult *S. gregaria* to become behaviorally solitarious in a mass-rearing room at the Anti-Locust Research Centre (ALRC). Even culturing locusts in isolation for five generations did not totally efface their grouping behavior. In contrast, when locusts were raised singly elsewhere and in a room not previously used for locust rearing, grouping behavior subsided. Morphometric comparisons among three laboratory cultures (crowded or isolated in a mass-rearing room and isolated in a locust-free room) and two field populations that were extremely gregarious and solitarious, respectively, showed that laboratory cultures were intermediate between the two field populations, but the isolated culture raised outside the ALRC exhibited morphological indices closest to those describing the solitary phase. Gillett (1968) proposed an airborne factor that sponsors grouping behavior in both hoppers and adults and shifts adult morphometrics toward the gregarious phase. The persistence of grouping behavior in isolated adults reared in a mass-breeding room was later reinterpreted to mean that it must have been a residual effect from exposure to the pheromone *before* the adult molt, that is, during larval life (1975), which indicated that Nolte and Gillett dealt with the same pheromone.

Nolte et al. (1970) then introduced the term "gregarization pheromone," which in the course of density-induced phase transformation from the solitarious to the gregarious from controls the increase of melanization in hoppers, changes the adult morphometrics of the E/F and F/C quotient in the direction of the gregarious phase, and stimulates grouping behavior. A fourth criterion was added to these three indicators of the gregarious phase: the pheromone causes an increase in the chiasma frequency of chromosomes from spermatocytes during meiosis and thus facilitates genetic recombination. According to Nolte, such rearrangements in the gene pool would be adaptive, because they offer an increased variability in the new environment to which the adult locusts migrate (Nolte, 1974). So far, no field data exist that would support that claim. This criterion has been used by Nolte as the main indicator for transformation toward the gregarious phase. However, chiasma induction has also been the most problematic issue and has aroused fierce controversy, as discussed below.

11.2.1 The Site of Production

To find the production site of the pheromone, air from a locust-breeding room was sucked out and trapped in one of two solvents, risella oil or dimethyl sulfoxide, for a period of 1 month (Nolte, 1968). The following bioassay was performed on solitarious fourth-instar hoppers from two subspecies of *Locusta migratoria* and from *S. gregaria*, which had been isolated during the third instar: a pad of cotton wool was moistened with extract and placed, underneath a false floor of wire mesh, at the bottom of a 1-l jar containing one hopper. Exposure to the volatile substance during the fourth and fifth larval stadia yielded high chiasma frequencies in the three long and five medium chromosomes from cells of testes during the mid-diplotene stage of meiosis, which were counted during the first week of adult life. Individuals also retained the remaining dark coloration during their larval life, but no significant morphometric changes toward the gregarious phase were found in the adult stage (Nolte et al., 1970).

Oil extracts were then made from different body parts of crowded hoppers, such as head, thorax, abdomen, and portions of the alimentary tract, and tested for induction of higher chiasma frequency on solitarious hoppers. The highest values, an increase of chiasma frequency by 30%, were obtained with extracts of the foregut, particularly of the crop. Treatment with crop extracts also caused retention of 65% of the initial third-instar black coloration. Although the morphometric shifts were heterogeneous, the E/F and F/C ratios tended to change toward the gregarious form. With pheromone synthesis probably occurring in the crop, the substance is excreted with the feces, as can be conveniently demostrated by exposing solitarious, single hoppers to fecal material in a jar, separated by a false floor. The feces, which seem to release the volatile pheromone gradually, cause an increase in chiasma frequency, black hopper pigmentation is retained, and morphometric ratios point toward the gregarious phase. It was further found that feces from solitarious hoppers caused the same effects as those from crowded hoppers of either sex, but autostimulation did not occur either because of high thresholds or of low concentrations. The pheromone also acts interspecifically to some extent and is at least active among the three locust species mentioned. In contrast, feces from adults of any of these species have no effect on transformation toward the gregarious phase.

The gregarizing effect of feces from crowded larvae of retaining black pigmentation in solitarized hoppers has been confirmed for second-instar larvae of *S. gregaria*, and it was further shown that the pink/beige background coloration persisted as well (Gillett and Phillips, 1977). In addition, the hoppers exhibited strong social behavior by grouping closely together under the influence of the feces, in contrast to the controls. In the view of Gillett and Phillips (1977), gregarization criteria of black color retention and grouping behavior were linked with the same pheromone.

Dearn (1974a,b) in comparing the chiasma frequency of chromosomes in the testes of solitarious and gregarious locusts of *S. gregaria* and *L. migratoria*, did not find significant differences between the phases and criticized Nolte's technique, the small number of samples employed, and the lack of relevant statistical methods. Nolte (1976), on the other hand, pointed out that Dearn's technique of producing solitarized hoppers was flawed, because by using white paper around the jars holding a hopper to achieve visual isolation, he unintentionally induced homochromy, so that many solitarized hoppers were of pale cream to off-white color. Nolte had expressly initiated such homochromy experiments in 1973 and published results showing that the chiasma frequency of such off-white solitarized locusts was much higher [by up to 29% (Nolte, 1976)] than in green controls. This explains why Dearn (1974a,b) could not find significant differences in chiasma frequency between gregarious and solitarious locusts.

Dearn (1977), working with *S. gregaria*, also took issue with Nolte's claim (Nolte, 1968) that albino mutants of the desert and migratory locusts are solitariform in morphometrics and chiasma frequency. Although he himself found that the morphology of the crowded albino strain suggested a similarity with the *solitaria* phenotype, and that albino individuals have a lower chiasma frequency than normal-colored, wild-type animals, Dearn went back to his earlier conclusions (1974a,b) that there was no difference in chiasma frequency between normal crowded and solitary locusts. He also asserted that the low chiasma frequency of albinos did not in itself reflect equivalence with the solitary phase. This stand, however, becomes difficult to maintain in view of the work on aggregation behavior of crowded albino *S. gregaria*, which showed that they do not group as well as wild-type individuals (Gillett, 1973).

11.2.2 Pheromonal Structure

Nolte's group began to determine the structure of the gregarization pheromone, in cooperation with chemists (Nolte et al., 1973). They started out with 2 kg of feces from gregarious hoppers, steam-distilled the material, and extracted the distillate with pentane. After several steps of purification and reduction, a brownish oil was obtained that after further separation by thin-layer chromatography yielded two major and several minor components, and so did gas chromatography of the pentane extract. Guaiacol and 5-ethylguaiacol were identified by mass spectrometry, and both showed considerable activity in increasing chiasma frequency. The two compounds had, however, different effects on the other three parameters of phase transformation: following guaiacol exposure, only 20% of the black hopper color was retained in comparison to 10% by the controls, whereas 5-ethylguaiacol preserved 80% of the pigmentation. Guaiacol had no effect on the F/C ratio, the most reliable morphometric ratio for phase transformation (Dirsh, 1953), but 5-ethylguaiacol significantly lowered the F/C value toward the gregarious state. Finally, guaiacol had no effect on larval marching behavior, whereas the other compound enhanced it. For these reasons, 5-ethylguaiacol was considered to possess all the properties exhibited by the pheromone from the hopper feces. This pheromone, which gregarized isolated solitarious hoppers and had formerly been called locustone (Nolte et al., 1970), was now renamed locustol (Nolte et al., 1973), apparently because of its phenolic basis.

The synthesis of locustol in the crop from guaiacol, a by-product of lignin degradation in the grass, was indirectly confirmed by applying the antibiotic thypyrameth to protozoan-infested locusts, which killed not only the parasites but also the bacteria responsible for the decomposition process of lignin. After that treatment, both mean chiasma frequency and F/C ratios shifted in the direction of the solitarious phase (Nolte, 1977). The vital role of the bacteria for gregarization was recently confirmed by raising axenic *S. gregaria*, i.e., locusts free from any detectable form of microbial life including bacteria (Charnley et al., 1985). Using the reliable F/C ratio as an indicator for phase transformation, it became evident that germfree gregarious locusts were morphometrically more solitarious than stock-reared, protozoan-infested counterparts, which had also more bacteria than a parasite-free colony. Axenic locusts exhibited a higher F/C ratio, and lost developmental synchrony in the population, both characteristics of solitarization.

In recent work, air from cages containing different developmental stages of *S. gregaria* or *L. migratoria* was captured and analyzed for volatile substances (Fuzeau-Braesch et al., 1988). Using combined gas chromatography–mass spectrometry, four aromatic compounds were found, three of which were identified as guaiacol, phenol, and veratrole. Both species produced all four compounds, but in different ratios depending on whether the air derived from cages containing fifth-instar larvae, immature adults, or copulating or egg-laying locusts. However, in each sample guaiacol was the most abundant compound. When tested singly or as mixtures in an olfactometer on groups of either 10 last-instar hoppers or 5 adult locusts, guaiacol, phenol, and a mixture of all compounds caused aggregation of the locusts, but to a varying degree.

This multicomponent material, called cohesion pheromone, differs from locustol in that it derives from adults and crowded hoppers, and both stages respond by aggregating, whereas locustol is produced only by hoppers and causes gregarizing effects exclusively in solitarious larvae (Nolte et al., 1973). That Fuzeau-Braesch et al. (1988) did not find 5-ethylguaiacol may have been because they extracted contaminated air, whereas Nolte et al. (1973) used locust feces, in which most likely that compound occurs in higher concentrations. Besides, air extracts from adults would not have contained locustol anyway.

For these reasons, it seems unjustifiable to cast doubt on the identity of locustol. Also, the claim that "the volatile substance may come from the animals themselves . . . without feces" (Fuzeau-Braesch et al., 1988) will have to be experimentally tested.

Nolte (1976) extracted 5-ethylguaiacol from hopper feces and compared its effect with that of several part-analogues and other substances on chiasma frequency, hopper color, and F/C ratios. Among the other components of the distillate, only guaiacol caused an increase in chiasma frequency, whereas isoacetovanillone was inactive, as was 5-methylguaiacol. Methyl formate and amyl acetate, tested for comparison, did not affect chiasma frequencies, but the first compound increased melanin formation and the second reduced the F/C ratio towards gregarious values. In contrast, 5-ethylguaiacol changed all those parameters of phase transformation toward their gregarious value.

Nolte (1977) next examined the action of locustol in recipient locust hoppers. He had shown earlier that injection of norepinephrine into fourth- or fifth-instar solitaries increased the chiasma frequency in adult spermatocytes (Nolte, 1969). Separate injections of norepinephrine and locustol had strong effects on chiasma frequency and F/C ratios and also shortened the duration of the fifth larval instar, whereas hopper melanization remained unaffected. Simultaneous injection of both compounds led to only slight gregarization, implying that both substances competed for the same substrate.

Neither DOPA, an important link in both the norepinephrine and the melanin cycles, nor the antagonist aldometh were capable of inducing gregarization, regardless of whether the injections occurred in the presence of aerial locustol or not. However, when both compounds were injected together they provoked a strong increase in chiasma frequency and retention of melanin, possibly a synergistic effect. Nolte (1977) then postulated that locustol, in acting like epinephrine, stimulates the production of cAMP, which causes the gregarization of solitarized hoppers. The evidence is as follows: cAMP, dibutyryl cAMP, or IBMX, with the latter preventing the breakdown of cAMP, duplicated all actions of locustol in gregarizing solitarized hoppers. In addition, cAMP levels in the testes of gregarious hoppers were slightly but significantly higher than in solitarious ones. These results do not yet indicate whether locustol acts like epinephrine and is directly mediated by cAMP. Nolte's hypothesis that cAMP is the second messenger of locustol (Nolte, 1977) would be strengthened if a specific locustol-sensitive adenyl cyclase that produces cAMP was found in target organs of that pheromone, such as the testes.

11.2.3 Pheromone Reception

Although the site of production of the gregarization pheromone in the crop, pheromone release from feces, and the structure of locustol seem to be accepted, the issue of pheromone reception by the affected hoppers is still unsettled. Exposure of antennectomized solitarized mid-fourth-instar hoppers of *L. migratoria* to mineral oil extract did not adversely affect the induction of high chiasma frequencies or black pigment retention, and the F/C ratio of the subsequent adult locusts was close to those of crowded populations (Nolte et al., 1970). Gillett (1983), removed the antennae of early second-instar isolated or crowded *S. gregaria* while exposing them to feces from crowded hoppers. Behavioral tests and color scoring were carried out in the mid-third instar; the controls were deprived of their hind tarsi. Although these operations did not significantly change grouping behavior or coloration of these hoppers, among the antennaless isolated larvae there were more green ones than in the tarsiless group, suggesting the tendency to solitarization. Regardless of whether antennaless isolated hoppers were exposed to feces or not, no gregarizing effect was noted. On the other hand, tarsiless solitarized larvae in the presence of feces became more gregarious in behavior and color than those not exposed to feces and their volatile pheromone. However, removal of the antennae, without any manipulation of the pheromone, has a solitarizing effect as seen by the appearance of green color (Mordue, 1977). This makes attempts to determine the role of the antennae for pheromone reception difficult and the outcome ambiguous, and Gillett wisely considered her results as inconclusive. Nolte, however, discards the antennae as receptor sites for the pheromone and suggests that it enters the hemolymph via the spiracles and proceeds to the site of action in the testes (Nolte et al., 1970; Nolte, 1974). This assumption is not supported by any experimental evidence! Nolte may have been led to that conclusion from earlier results in which he injected hemolymph from crowded into solitarized hoppers and subsequently obtained an increased chiasma frequency later in the adult testes (Nolte, 1968).

The following approach might help to solve the pheromone perception problem: in order to examine the tracheal system as a pathway of pheromone entrance into the hemocoel, carbon labeling of 5-ethylguaiacol is recommended. Solitarized hoppers would then be exposed to the volatile and the labeled molecules could be traced in the hemolymph, if they are present. The presumed target organs of the pheromone, the testes, could then be examined for receptor sites by employing autoradiography.

11.3 THE MATURATION PHEROMONE

The experimental work of Norris (1954, 1962, 1963, 1964, 1968) on sexual maturation in *S. gregaria* suggested that pheromones played an important role in the timing of sexual maturity, as expressed in the onset of sexual behavior, a change in body coloration, and the maturation of the gonads. She found immature males to be pinkish brown, whereas the onset of sexual maturity was indicated by the appearance of yellow on the abdominal tergites, which spread over much of the body as maturation progressed. Both copulatory readiness and color change are influenced by the presence of a yellow male, as the following experiments show. Pairs of males and females of the same age mate only 4 weeks or later after the imaginal molt, whereas young males kept together with a yellow male matured after only 17 days and were well synchronized. Fledgling females exposed to a yellow male also changed color more quickly, although the change was less pronounced than in the male, and they copulated and oviposited earlier. The accelerating effect on maturation is even more distinct in crowded conditions. Similarly, in *Locusta migratoria* males copulated with females of their own age 17–25 days after emergence, but in the presence of a mature male they did so after 13–14 days, and the females also showed signs of rapid maturation.

The stimulus to maturation proved to be chemical in nature and Loher (1960) demonstrated that yellow males of *S. gregaria* produced the substance in the epidermis of the abdominal tergites. In *L. migratoria*, mature males exerted the same accelerating effect on immature conspecific locusts and a pheromone is also suspected because there are glands in the epidermis (Thomas, 1972).

In the absence of mature locusts, young *L. migratoria* react differently from young *S. gregaria*. Isolated pairs of *L. migratoria* of the same age copulated 7 days after the imaginal molt and laid eggs before the fourteenth day, but pairs of *S. gregaria* took about 4 weeks to mate and to oviposit, and maturation was clearly delayed. Norris (1964) explained this discrepancy by assuming that in crowded conditions young *L. migratoria* adults mutually inhibited each other from becoming mature over a period. The inhibitory effect of crowding is removed by isolation. The nature of that inhibition in crowded, immature *L. migratoria* is unknown. For *S. gregaria*, the situation according to Norris is just the opposite. Here, isolation prevented pheromone production and delayed maturation, whereas crowding promoted it. Much more work has to be done to explain this phenomenon satisfactorily.

11.3.1 The Site of Production

Oglobin (1947) was the first to describe vacuolated gland cells in the epidermis of the first to ninth abdominal segments of crowded, mature males of *Schistocerca cancellata*. He correctly associated these glands with the production of an aromatic, sticky substance present on the surface of a male locust at the time of maturation. Oglobin's light microscope descriptions were very detailed and recently confirmed and extended by electron microscopy (Hawkes et al., 1987). These authors also suggest a possible role for that substance in sexual maturation.

Mature, yellow males of *S. gregaria* also have an aromatic smell (Norris, 1954), but attempts to determine the chemical nature of that volatile substance were unsuccessful (Blight, 1969; Blight et al., 1969). Loher (1960, 1961) experimentally demonstrated that the insects produce an oil-soluble substance in the vacuolated epidermis of different parts of the body, especially on the abdominal tergites. Strong (1970, 1971), using a better histological technique, found ducts leaving the vacuolated cells, which coil around and go through the duct nuclei on the way to the epicuticle, where the product of the vacuoles spreads (Fig. 11.1). Loher (1960) could not see these ducts in his thick paraffin sections, although he noted two types of nuclei, which almost certainly belonged to the gland cells and duct cells. Thomas (1972) found these epidermal glands not only in mature males but also in mature females, albeit in smaller numbers, and the vacuoles were less developed. This might explain the very faint odor noticed from mature females, whereas immatures of both sexes are scentless. In addition, she found the same structures in both sexes of *Locusta migratoria* and the red locust, *Nomadacris septemfasciata*, and even in a gomphocerine grasshopper, *Chorthippus parallelus*. The ultrastructure of the epidermal glands from *S. gregaria* was elucidated in a thorough and definitive investigation (Fig. 11.2) that confirmed and elaborated on the elements found in previous light microscope studies (Cassier and Delorme-Joulie, 1976; Cassier, 1977).

Sexual maturation, yellowing, and the production of the pheromone are under the control of the corpora allata. Removal of the corpora allata renders males immature indefinitely, they turn brown with age without any sign of the typical yellow color, the epidermis is devoid of vacuoles, and no pheromone can be collected from the body surface (Loher, 1960). Reimplantation of corpora allata reverses all these processes (Loher, 1960; Pener, 1965), although in older males the yellow color is localized mainly around the site of the implant (Loher, 1959).

Kendall (1972), in a very interesting light and electron microscope study, found the glandular epidermis in parts of the body other than the abdomen, including the dorsal part of the first three tarsal segments of adult male *S. gregaria*. The glands were very similar to the structures found by Strong (1971) and Oglobin (1947). Although they are already present at the time of emergence, the glandular vacuoles develop only at the onset of sexual maturation. The similarity between the tarsal glandular cells and those distributed over the body extends even to the control of their development, which depends on the corpora allata. Allatectomy prevents the tarsal glands from producing vacuoles and the corresponding structures remain immature. It would be exciting to know whether the glandular products of the tarsi are identical with the maturation pheromone.

11.3.2 Pheromone Reception

An oil extract of the maturation pheromone presented on cotton wool 1 cm away from the antennae of immature males of *S. gregaria* causes the vibration reaction, consisting of antennal waving and lateral movements of the hind legs without touching the elytra (Loher, 1958). Although it is a disturbance reaction that can be released by touch and is also seen in mature locusts, the vibration reaction serves as a good indicator of pheromonal presence. That Amerasinghe (1978) obtained only vibrational responses from 7.7% of his immature males

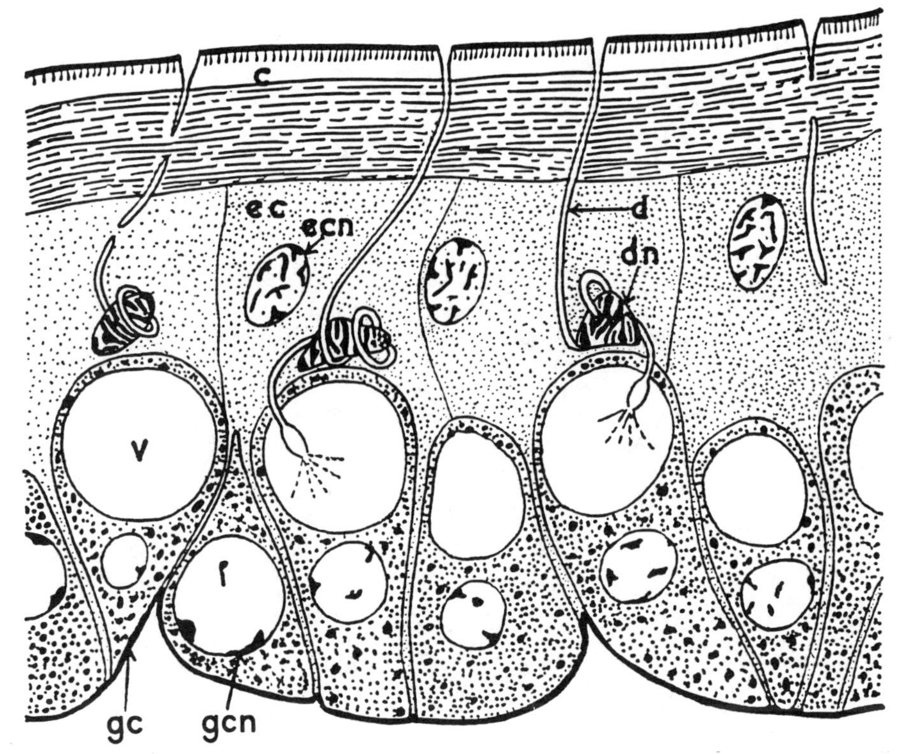

Fig. 11.1. Drawing of a section of integument from mature male of *Schistocerca gregaria.* Glandular cells produce the maturation pheromone that reaches the cuticular surface via individual ducts, which coil around the duct cell nucleus. c, cuticle; d, duct; dn, duct nucleus; ec, epidermal cell; ecn, epidermal cell nucleus; gc, glandular cell; gcn, glandular cell nucleus; v, vacuole. (From Strong, 1971.)

either tested with pheromone extracts or when offered a yellow locust is most likely due to the specific nature of the locust strain employed. In any event, Loher's results indicate that the vapor phase of the pheromone is received by olfactory receptors on the antennae. In further studies, exposure of immature, isolated male locusts to the extract by holding a pad in front of the recipients' antennae daily for 5–10 min and over a period of 3–6 weeks led to sexual maturity and partial yellowing before the control insects, but it never induced full yellow coloration. Norris (1954) had previously observed a similar deficiency. In her experiments, single immature males were put into small double-walled perforated tin cages and exposed in a large cage to 40 yellow, pheromone-producing males, but they did not even then reach the bright yellow coloration. This was also confirmed by Amerasinghe (1978) performing similar experiments.

When, however, the yellow males were able to touch the recipients, the latter matured faster and became bright yellow (Norris, 1954). Because daily

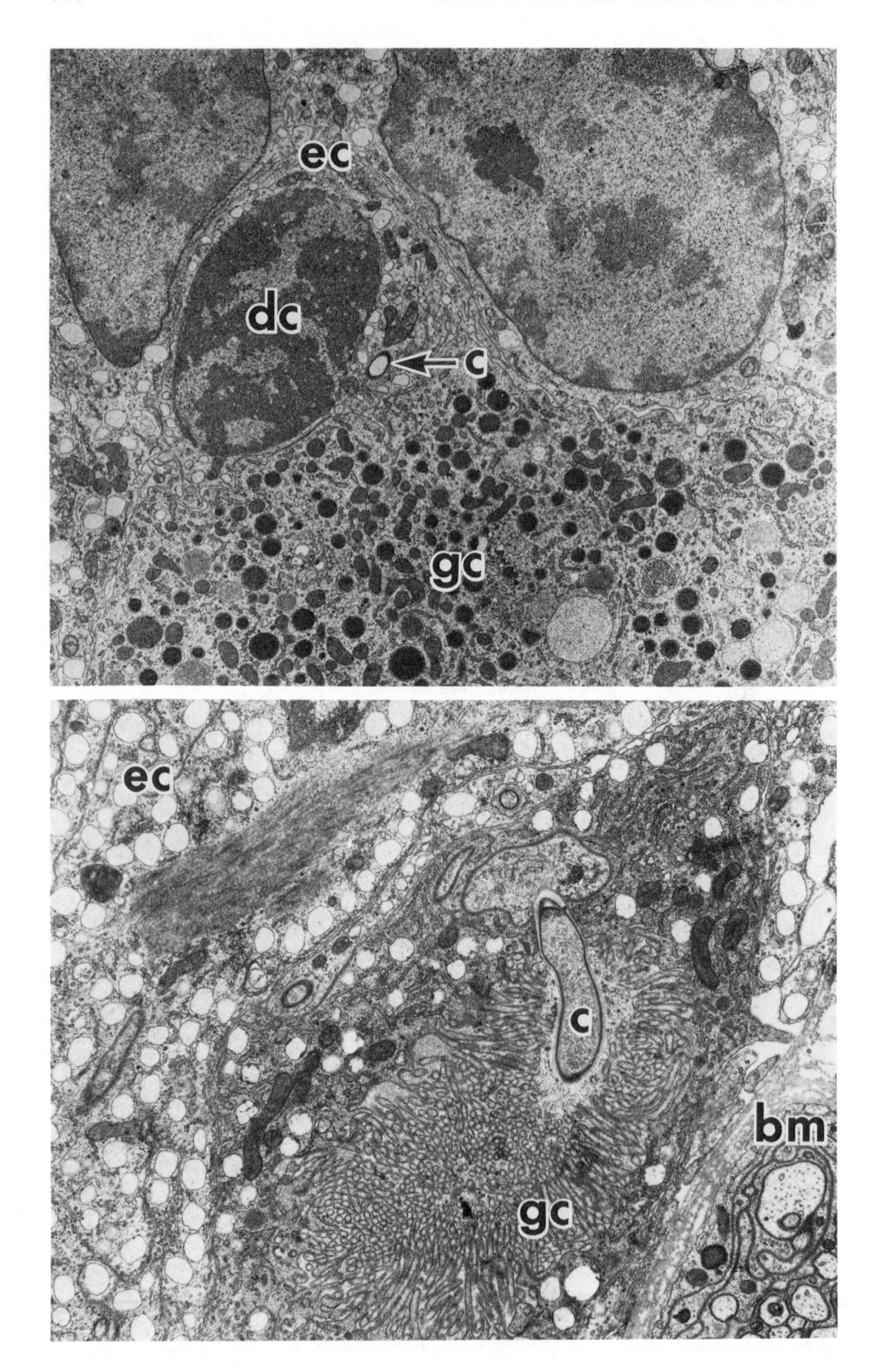
ec
dc
c
gc
ec
c
bm
gc

tactile stimulation with wires or brushes in the absence of the pheromone did not produce that effect, it must be due to the direct transfer of the stimulant to the body of the recipient.

Which sense organs are stimulated when a pheromone-producing locust touches an immature one, or does that substance directly enter the recipient's body? Sensory perception could be mediated via chemoreceptors concentrated on the antennae, but also spread over the body surface; there, the concentration of the pheromone caused by repeated contact should be greater than in the vapor phase. If the glandular products secreted via the tarsi (Kendall, 1972) turn out to be the maturation pheromone, then direct transfer by contact would be assured. Penetration of the substance into the body could occur through the pore canals in the cuticle, facilitated by the lipophilic character of the pheromone. But the possibility of an entry through the spiracles of the tracheal system as postulated for locustol (Nolte et al., 1970) cannot be excluded.

The pheromone seems to act to some extent interspecifically, because mature *S. gregaria* can accelerate sexual maturation of young *L. migratoria*, and mature *L. migratoria* probably have the same effect on immature *S. gregaria* (Norris, 1968). In the face of so many questions, which could be answered if the chemical structure of the maturation pheromone were known, it is disappointing to see that no chemical analysis has been published after all these years since its discovery.

11.4 OVIPOSITION-MEDIATING PHEROMONES

Field observations (Popov, 1958; Uvarov, 1966, 1977) and laboratory experiments have shown that once a *S. gregaria* female has found an oviposition site and lays her eggs, other females tend to form a group and to deposit their egg pods in close proximity. An oviposition site near live decoys tethered to the floor was highly preferred to equally good sites without them (Norris, 1963). The presence of decoys has an overriding effect even over more favorable soil moisture or air temperature conditions and females oviposit near decoys that are tethered on a drier and cooler site. Immature or mature males and last-instar hoppers can be used as decoys, and they are functional even in total darkness. Under night conditions, touching the decoys was necessary for a locust in an egg-laying mood to deposit near them, and the decoys could be alive or dead. The suspicion that a chemical stimulus was responsible for the

Fig. 11.2. Electron micrographs of integument from a mature, 30-day-old *Schistocerca gregaria* male of the gregarious (above) and the solitarious (below) phases. The glandular cell from the gregarious male shows many electron-dense inclusions, whereas the microvilli from the solitarious male are inactive and empty. bm, basement membrane; c, duct cell canal; dc, duct cell nucleus; ec, epidermal cell; gc, glandular cell. (From Cassier and Delorme-Joulie, 1976.)

attraction was confirmed by offering pieces of paper that had been contaminated previously by locusts crawling over them. Contaminated paper functioned as decoys and egg-laying females aggregated nearby.

Antennaless females still found live decoys, but they were much less responsive to dead ones. When live decoys were enclosed in a gauze cage and could no longer be touched, the response was much reduced, which suggests that ordinarily chemotactile stimuli must have been involved. Thus, aggregation can be mediated by three types of stimuli: visual, tactile, and chemical; contact chemoreception seems more important than olfaction. The active substance is believed to be a pheromone spread over the locust's entire body surface. Further work by Norris (1970) has shown that the aggregation-mediating pheromone was present not only in fifth-instar larvae, but also in their freshly shed exuviae. Exuviae from *L. migratoria* were also responded to by *S. gregaria*, but less so than to those from conspecific locusts, suggesting the partially species-specific nature of the substance. Solitarious locusts also produce the pheromone, although to a lesser extent. Contact chemoreceptors all over the body have been assumed to perceive the pheromone, but this has never been demonstrated, and the chemical nature of the pheromone is unknown. Interestingly, sand into which *S. gregaria* had formerly laid many egg pods was not attractive to females ready for egg laying, even when sand had been contaminated with relicts of oviposition ducts, feces, or pieces of the froth plug.

L. migratoria females readily aggregated and oviposited near decoys in the light, but they did very poorly in darkness, suggesting that their chemotactile sensitivity was limited. It is then surprising that *L. migratoria* responded strongly to sand alone, into which gregarious females had repeatedly laid (Lauga and Hatté, 1977). The sand was attractive to females of both phases over a distance of half a meter as shown in Y-tube experiments, where they had to choose between contaminated and pure sand. On arriving at the activated oviposition site, they ingested sand and laid egg pods. The attractive quality of the sand increased with its frequent use for oviposition, suggesting the accumulation of a substance whose activity lasted for at least 6 months. Lauga and Hatté (1978) were then able to correlate the sand contaminated by egg-laying gregarious females with a significant change in the number of eggs per pod from solitary females when they laid into it. It is known that gregarious females lay an increasing number of eggs with each pod, which then levels off after the fifth pod. In solitarious females, however, the number of eggs decreases from one egg pod to the next. When solitary females were given activated sand for several weeks, then their egg number per pod increased as in gregarious females and the overall weight of the egg pod rose, and with it the mean weight of the eggs. This means that egg pods and their eggs laid by solitarious females became progressively determined toward the gregarious phase. The activated sand undoubtedly contained pheromones, perceived via the antennae and possibly also by means of chemoreceptors located in the buccal cavity, on the palps, or on the tarsi; neither origin nor nature of these substances is known.

Because *L. migratoria* is virtually unresponsive to decoys during darkness (Norris, 1970), in contrast to *S. gregaria*, apparently another strategy has evolved to contribute to phase transformation for the next generation. It seems also that *L. migratoria* may have two pheromones: the pheromone attracting a locust to a distant oviposition site acts as a releaser pheromone, the signal is received via the antennae, and a locust responds by approaching the source. The pheromone with the long-term effect that may be perceived when the locust ingests sand could be a primer pheromone, because it alters the developmental physiology of the ovaries and leads to the production of eggs, similar to those laid by gregarious females. Or could one pheromone exercise both functions and act as a releaser and a primer? In the case of the maturation pheromone, which is a primer pheromone, the mediation of the volatile compound via sensory receptors to achieve accelerated maturation is likely (Loher, 1959), but a possible direct action of primers circumventing the sensory pathway should not be dismissed. Such fundamental questions, however, can be addressed only when the chemical structure of these pheromones is known.

11.5 THE SOLITARIZATION PHEROMONE

When the tendency of second-instar hoppers of *S. gregaria* to group together is compared with that of immature adults, the young adults are much less attracted to each other than the larvae (Gillett, 1973). Also, fledglings in the field clearly disperse (Ellis and Ashall, 1957), and this was confirmed in the laboratory (Gillett, 1973). The idea that asocial behavior in adults might be due to yet another pheromone, produced by the latter and excreted with the feces like the gregarization pheromone of hoppers (Nolte et al., 1970), has been tested on second-instar hoppers (Gillett and Phillips, 1977; Gillett, 1983). During the first 4 days of their larval life, the hatchlings were crowded, then they were isolated for another 4 days either in the presence of feces from immature or mature crowded locusts or without feces. Tests for grouping behavior of the 8-day-olds, now second instars, over 90 min showed that hoppers exposed to feces from adults formed significantly fewer groups and were further apart than the controls, and that the feces from mature adults had more noticeable dispersive effects than those from immature ones. No difference in black pigmentation between the differently treated hoppers was noticed, although the pink background coloration was weaker. Interestingly, exposure of two hoppers reared together for about 8 days to feces from mature adults during that period did not reduce the level of grouping, but the background coloration was more gregarious-like. Thus in terms of social behavior, mutual sensory stimulation overcame the effect of the solitarizing pheromone from adults.

The chemical structure of the solitarizing pheromone is unknown. Should it be produced in the foregut by bacterial degradation of lignin, it can be speculated that the solitarizing pheromone might have replaced the gregarizing pheromone, although different in chemical structure. As to the reception of the pheromone, the removal of the antennae has again proved to give ambiguous results and the issue awaits further studies (Gillett, 1983).

11.6 THE MATURATION-RETARDING PHEROMONE

When fledglings of *S. gregaria* or males and females under the age of 1 week are crowded with older, yet still immature males, maturation is retarded and locomotor activity depressed. Also, when they finally reach maturation, the males do not develop yellow coloration (Norris, 1962). The same retarding influence is exerted by an older male that had been previously allatectomized (Norris and Pener, 1965). Although that operation renders males permanently immature (Loher, 1960), the delaying influence of normal fledglings and, by implication, that of allatectomized locusts on maturation may be due to yet another pheromone, although no direct evidence exists (Norris, 1968).

11.7 CONCLUSIONS

Locust gregarization as a consequence of high population density is brought about by favorable climatic and nutritional factors. However, the mediators for achieving such phase transformation are pheromones, of which three types are essential: the gregarization pheromone, the sex maturation pheromone, and the oviposition pheromones. Each of these pheromones sponsors gregarization differently and at a different stage of locust development. Locustol, the gregarizing pheromone produced in both gregarious and solitarious hoppers, may act on the larval genome (Nolte, 1974). By an increase in chiasma frequency more variability in the gene pool could be generated, which should be adaptive if it can be demonstrated that it leads to a change of characters in the adult that are favorable for migration. Locustol initiates changes in the gregarious direction in the larval stage as shown by grouping behavior and color change.

Crowding of adults then provokes the production of the maturation pheromone, which accelerates and synchronizes sexual maturation in both sexes, resulting in earlier copulation, egg production, and oviposition. The presence of mature, yellow males in a population of immatures can bring about maturation within a few days!

The oviposition pheromones act during yet another part of the sexual cycle: in *S. gregaria*, they sponsor aggregation and egg laying in groups, producing crowded conditions at the time of hatching and in the early larval stadia. But at least in *L. migratoria*, a volatile pheromone stored in the sand attracts females to sites where other females have previously laid their egg pods, again creating crowded conditions for the time of hatching. Finally, activated sand gradually transforms the ovaries of solitarious females so that they produce increasingly gregarious-like eggs, which amounts to functional gregarization in the mother.

On the other hand, the solitarization pheromone contained in the feces of adults not only led to the dissolution of grouping behavior in hoppers (Gillett and Phillips, 1977) but could also be the cause for dispersal among adults in the field (Ellis and Ashall, 1957; Gillett, 1973). The maturation-retarding pheromone (Norris, 1962, 1968) cannot prevent swarming, but delays maturation. Both pheromones counter the gregarizing effect and have therefore, applied potential.

As to the pheromones that promote gregarization, can the pathways of their production be disrupted and pheromone release and perception be interfered with, so that swarm formation and migration might be prevented and with it the enormous reproductive potential? In other words, is there a way to turn polyphasic locusts into monophasic grasshoppers by making them permanently solitarious?

REFERENCES

Amerasinghe, F. P. 1978. Pheromonal effects on sexual maturation, yellowing, and the vibration reaction on immature male Desert Locusts (*Schistocerca gregaria*). *J. Insect Physiol.* **24**, 309–314.

Blight, M. M. 1969. Volatile nitrogenous bases emanating from laboratory reared colonies of the Desert Locust, *Schistocerca gregaria*. *J. Insect Physiol.* **15**, 259–272.

Blight, M. M., J. F. Grove, and A. McCormick. 1969. Volatile compounds emanating from laboratory-reared colonies of the Desert Locust, *Schistocerca gregaria*. *J. Insect Physiol.* **15**, 11–24.

Cassier, P. 1977. La différenciation imaginale du tégument chez le cricket pélerin, *Schistocerca gregaria* Forsk. IV. Les étapes de la morphogenèse des unité glandulaires. *Arch. Anat. Microsc. Morphol. Exp.* **66**, 145–161.

Cassier, P. and C. Delorme-Joulie. 1976. La différenciation imaginale du tégument chez le criquet pélerin, *Schistocerca gregaria* Forsk. III. Les différences phasaires et leur déterminisme. *Insectes Soc.* **23**, 179–198.

Chapman, R. F., W. W. Page, and A. R. McCaffery. 1986. Bionomics of the variegated grasshopper (*Zonocercus variegatus*) in West and Central Africa. *Annu. Rev. Entomol.* **31**, 479–505.

Charnley, A. K., J. Hunt, and R. J. Dillon. 1985. Germ-free culture of Desert Locusts *Schistocerca gregaria*. *J. Insect Physiol.* **31**, 477–486.

Dearn, J. M. 1974a. Phase transformation and chiasma frequency variation in locusts. I. *Schistocerca gregaria*. *Chromosoma* **45**, 321–338.

Dearn, J. M. 1974b. Phase transformation and chiasma frequency variation in locusts. II. *Locusta migratoria*. *Chromosoma* **45**, 339–352.

Dearn, J. M. 1977. Pleiotropic effects associated with the albino mutation in the Desert Locust, *Schistocerca gregaria* (Forsk.), and their relationship to phase variation. *Acrida* **6**, 43–53.

Dirsh, V. M. 1953. Morphometrical studies on phases of the Desert Locust (*Schistocerca gregaria* Forskål). *Anti-Locust Bull.* **16**, 1–34.

Ellis, P. E. 1953. Social aggregation and gregarious behaviour in hoppers of *Locusta migratoria migratorioides* (R. & F.). *Behaviour* **5**, 225–260.

Ellis, P. E. and C. Ashall. 1957. Field studies on diurnal behavior, movements and aggregation in the Desert Locust (*Schistocerca gregaria* Forskål). *Anti-Locust Bull.* **25**, 1–94.

Farrow, R. A. 1977. Origin and decline of the 1973 plague locust outbreak in central western New South Wales. *Aust. J. Zool.* **25**, 455–489.

Fuzeau-Braesch, S., E. Genin, R. Jullien, E. Knowles, and R. Papin. 1988. Composition and role of volatile substances in atmosphere surrounding two gregarious locusts, *Locusta migratoria* and *Schistocerca gregaria*. *J. Chem. Ecol.* **14**, 1023–1033.

Gillett, S. D. 1968. Airborne factor affecting the grouping behavior of locusts. *Nature (London)* **218**, 782–783.

Gillett, S. D. 1972. Social aggregation of adult *Schistocerca gregaria* and *Locusta migratoria migratorioides* in relation to the final molt and ageing. *Anim. Behav.* **20**, 526–533.

Gillett, S. D. 1973. Social determinants of aggregation behavior in adults of the desert locust. *Anim. Behav.* **21**, 599–606.

Gillett, S. D. 1975. Changes in the social behavior of the desert locust, *Schistocerca gregaria*, in response to the gregarizing pheromone. *Anim. Behav.* **23**, 494–503.

Gillett, S. D. 1983. Primer pheromones and polymorphism in the desert locust. *Anim. Behav.* **31**, 221–230.

Gillett, S. D. and M. Phillips. 1977. Faeces as a source of a locust gregarization stimulus. Effects on social aggregation and on cuticular color of nymphs of the desert locust, *Schistocerca gregaria* (Forsk.). *Acrida* **6**, 279–286.

Hawkes, F., J. Rzepka, and G. Gontrand. 1987. The scent glands of the male South American locust, *Schistocerca cancellata*, an electron microscope study. *Tissue Cell* **19**, 687–703.

Karlson, P. and A. Butenandt. 1959. Pheromones (ectohormones) in insects. *Annu. Rev. Entomol.* **4**, 39–58.

Kendall, M. D. 1972. Glandular epidermis on the tarsi of the Desert Locust, *Schistocerca gregaria* Forskål. *Acrida* **1**, 121–147.

Lauga, J. and M. Hatté. 1977. Acquisition de propriétés grégarisantes par le sable utilisè à la ponte répetée des femelles grégaires de *Locusta migratoria* L. (Ins. Orthop.). *Acrida* **6**, 307–311.

Lauga, J. and M. Hatté. 1978. L'activité grégarisante du sable de ponte chez *Locusta migratoria* L.: Action sur le comportement et la reproduction des individus. *Ann. Sci. Nat., Zool. Biol. Anim.* [12] **20**, 37–52.

Loher, W. 1958. An olfactory response of immature adults of the desert locust. *Nature (London)* **181**, 1280.

Loher, W. 1959. Studies of chemoreception in the desert locust. Ph.D. Thesis, University of London, Imperial College.

Loher, W. 1960. The chemical acceleration of the maturation process and its hormonal control in the male of the desert locust. *Proc. R. Soc. London, Ser. B.* **153**, 380–397.

Loher, W. 1961. Die Beschleunigung der Reife durch ein Pheromon des Mānnchens der Wūstenheuschrecke und die Funktion der Corpora allata. *Naturwissenschaften* **48**, 657–661.

Mordue, A. J. 1977. Some effects of amputation of the antennae on pigmentation, growth and development in the locust, *Schistocerca gregaria*. *Physiol. Entomol.* **2**, 293–300.

Nolte, D. J. 1963. A pheromone for melanization of locusts. *Nature (London)* **200**, 660–661.

Nolte, D. J. 1968. The chiasma-inducing pheromone of locusts. *Chromosoma* **23**, 346–358.

Nolte, D. J. 1969. Chiasma-induction and tyrosinase metabolism in locusts. *Chromosoma* **26**, 287–297.

Nolte, D. J. 1974. The gregarization of locusts. *Biol. Rev. Cambridge Philos. Soc.* **49**, 1–14.

Nolte, D. J. 1976. Locustol and its analogues. *J. Insect Physiol.* **22**, 833–838.

Nolte, D. J. 1977. The action of locustol. *J. Insect Physiol.* **23**, 899–903.

Nolte, D. J., I. R. May, and B. M. Thomas. 1970. The gregarization of locusts. *Chromosoma* **29**, 462–473.

Nolte, D. J., S. J. Eggers, and I. R. May. 1973. A locust pheromone: Locustol. *J. Insect Physiol.* **19**, 1547–1554.

Norris, M. J. 1954. Sexual maturation in the desert locust (*Schistocerca gregaria* Forskål) with special reference to the effect of grouping. *Anti-Locust Bull.* **7**, 1–44.

Norris, M. J. 1962. Group effects on the activity and behavior of adult males of the Desert Locust (*Schistocerca gregaria* Forsk.) in relation to sexual maturation. *Anim. Behav.* **10**, 275–291.

Norris, M. J. 1963. Laboratory experiments on gregarious behavior in ovipositing females of the Desert Locust (*Schistocerca gregaria* (Forsk.)). *Entomol. Exp. Appl.* **6**, 279–303.

Norris, M. J. 1964. Accelerating and inhibiting effects of crowding on sexual maturation in two species of locusts. *Nature (London)* **203**, 784–785.

Norris, M. J. 1968. Some group effects on reprodution in locusts. *Colloq. Int. C.N.R.S.* **173**, 147–161.

Norris, M. J. 1970. Aggregation response in ovipositing females of the desert locust, with special reference to the chemical factor. *J. Insect Physiol.* **16**, 1493–1515.

Norris, M. J. and M. P. Pener. 1965. An inhibitory effect of allatectomized males and young females on the sexual maturation of young male adults of *Schistocerca gregaria* (Forsk.) (Orthoptera: Acrididae). *Nature (London)* **208**, 1122.

Oglobin, A. 1947. Las glandulas odoriferas de la langosta *Schistocerca cancellata* (Serville) (Orthopt. Acridoidea). *Arthropoda* **1**, 54–77.

Pener, M. P. 1965. On the influence of the corpora allata on maturation and sexual behavior of *Schistocerca gregaria*. *J. Zool.* **147**, 119–136.

Popov, G. B. 1958. Ecological studies on oviposition by swarms of the desert locust (*Schistocerca gregaria* Forskål) in eastern Africa. *Anti-Locust Bull.* **31**, 1–70.

Roffey, J. and G. B. Popov. 1968. Environmental and behavioural processes in a desert locust outbreak. *Nature (London)* **219**, 446–450.

Strong, L. 1970. Epidermis and pheromone production in males of the desert locust. *Nature (London)* **228**, 285–286.

Strong, L. 1971. Intracellular ducts in the epidermis of the male desert locust. *J. Insect Physiol.* **17**, 1823–1831.

Thomas, J. G. 1972. Epidermal glands in the abdomen of acridids. *Acrida* **1**, 223–232.

Uvarov, B. P. 1966. *Grasshoppers and Locusts*, Vol. 1. Cambridge Univ. Press, London and New York.

Uvarov, B. P. 1977. *Grasshoppers and Locusts*, Vol. 2. Centre for Overseas Pest Research, London.

12

Grasshopper Chemical Communication

D. W. WHITMAN

12.1 INTRODUCTION

Chemical communication is probably the foremost mode of communication in animals (Birch, 1974; Shorey, 1976). This is especially true for insects, where a great variety of pheromones and allelochemicals has been discovered (Blum, 1981; Bell and Cardé, 1984; Kerkut and Gilbert, 1985). In grasshoppers, chemical communication is less well known, perhaps because these diurnal insects rely so heavily on vision and sound to communicate with one another and to monitor the events that transpire in their habitats. However, many grasshopper species do employ chemical signals either for intraspecific communication or to warn and deter natural enemies.

This chapter examines chemical communication in grasshoppers. For our purposes, chemical communication is defined broadly as the release of a chemical substance by one individual which results in a behavioral or physiological change in another individual. First, intraspecific communicative chemicals (pheromones) are discussed, then interspecific defensive substances (allomones) will be examined. A more extensive account of locust pheromones is given in Chapter 11.

12.2 INTRASPECIFIC CHEMICAL COMMUNICATION IN GRASSHOPPERS

Since the coining of the term "pheromone" [a substance secreted by an individual to the outside that causes a specific reaction in a receiving individual of the same species (Karlson and Butenandt, 1959; Karlson and Lüscher, 1959)], hundreds of these intraspecific communication substances have been discovered across a wide range of arthropod taxa. It is not surprising, then, that grasshoppers have also been found to produce pheromones. Indeed, one of the earliest examples of pheromonal activity discovered for any insect was the maturation pheromone of the locust *Schistocerca gregaria* (Norris, 1954). Since then, workers have demonstrated that pheromones influence or control numerous grasshopper processes including gregarization, maturation, oviposition, mate identification and stimulation, and perhaps aggregation (Table 12.1).

12.2.1 Gregarization Pheromones

Certain species of acridoids (termed locusts) are polymorphic and polyethic. Locust individuals can exist in either a "solitarious" or a "gregarious" phase, or in any number of intermediate states that can differ in behavior, morphology, ecology, physiology, and biochemistry (Uvarov, 1966, 1977; Génin et al., 1986, 1987; Colgan, 1987). The processes eliciting such changes are complex and not completely understood. Phase transformation is thought to be controlled by the environment acting through its effects on grasshopper density.

TABLE 12.1. Pheromones That Have Been Proposed in Various Species of Acridoidea

Pheromone Effect	Species	Identity	Reference
Gregarization	*Locusta migratoria* *Schistocerca gregaria* *Locustana pardalina*	5-Ethyl-2-methoxyphenol	Nolte (1973) Nolte et al. (1973)
Solitarization	*S. gregaria*	Unknown	Gillett (1983)
Maturation stimulation	*S. gregaria* *L. migratoria*	Δ^1-Pyrroline?	Loher (1960); Norris (1964); Blight (1969)
Maturation inhibition	*S. gregaria* *L. migratoria*	Unknown	Norris (1954, 1964) Norris and Pener (1965)
Increased fecundity (volatile stimulus)	*Schistocerca vaga* *Aiolopus thalassinus*	Unknown	Okelo (1979) Schmidt and Osman (1988)
Increased fecundity (mating factor)	*Melanoplus sanguinipes* *S. gregaria* *L. migratoria*	Protein?	Leahy (1973a,b); Friedel and Gillott (1976); Lange and Loughton (1985)
Oviposition	*S. gregaria* *L. migratoria* *Zonocerus variegatus*	Unknown	Norris (1970); Lauga and Hatté (1977a,b); McCaffery and Page (1982)
Mate identification and stimulation	*Hieroglyphus nigrorepletus* *Taeniopoda eques* *Elaeochlora trilineata* *Chromacris speciosa* *Melanoplus sanguinipes*	Long-chain esters?	Pickford and Gillott (1972) Warthen and Uebel (1980) Siddiqi and Khan (1981) Whitman (1982a); Riede (1987)
Cohesion	*S. gregaria* *L. migratoria*	Phenol and guaiacol	Fuzeau-Braesch et al. (1988)

Locust populations can exist in the solitarious phase for many years until favorable environmental conditions lead to a local outbreak, or until patchy conditions elicit clumping (Uvarov, 1966, 1977; Nolte, 1974). Increased density presumably results in increased visual (Ellis and Pearce, 1962), tactile (Ellis, 1959, 1962), and, most importantly, pheromonal stimulation, which in turn initiates the transformation from the solitarious to the gregarious phase.

The potential role of chemical communication in phase dynamics was first recognized by Nolte (1963). By that time, phase transformation was already known to be reversible: when gregarious phase nymphs were isolated, they reverted to the solitarious phase. Working with *S. gregaria, L. migratoria*, and *L. pardalina*, Nolte discovered that reversal did not occur if the isolated hoppers were kept in the same room with gregarious hoppers; volatile substances from the nearby gregarious hoppers somehow blocked the solitarization process. Building on Nolte's discoveries, Gillett (1968, 1975a,b) found that gregarious *S. gregaria* nymphs produced an airborne factor that retarded the gregarious to solitarious changes in isolated nymphs. Increased melanization, developmental rates, and aggregative tendencies (all gregarious-type traits) were observed in isolated nymphs following exposure to these volatiles.

Working with frass extracts of *L. migratoria*, Nolte isolated and identified this "gregarization pheromone" as 5-ethyl-2-methoxyphenol (Fig. 12.1a), a degradation product of plant lignin thought to be produced in the crop (perhaps by gut microflora) and released from the frass of both solitarious and gregarious nymphs, but not adults (Nolte et al., 1970, 1973; Nolte 1977). Termed locustol, this volatile primer pheromone was hypothesized to be absorbed directly through the spiracles and into the hemolymph, where it elicits the hormonal changes triggering phase transformation (Nolte, 1968, 1976; Nolte et al., 1970, 1973). Other workers suggested that the pheromone was perceived by the antennae (Gillett, 1983).

Whether locustol is the actual gregarization pheromone, a substance that mimics the pheromone, or just one of many similarly acting substances that together initiate phase transformation remains unknown. Also unclear is the relationship of locustol to the effects observed by Gillett (1968, 1975a,b). In *L. migratoria* and *S. gregaria*, locustol influences some of the morphological and physiological traits associated with phase transformation. However, locustol does not appear to affect or be produced by all locust species, nor does it ever elicit complete phase transformation (Nolte et al., 1973). Adding to this confusion is the recent work of Fuzeau-Braesch et al. (1988) who isolated phenol, guaiacol, and veratrole (Fig. 12.1) (but not 5-ethyl-2-methoxyphenol) from the air surrounding gregarious phase *Schistocerca gregaria* and *Locusta migratoria* larvae and adults. In olfactometer tests, locusts were not attracted to the three isolated phenolics. However, phenol and guaiacol elicited significant clumping in both locust species and were therefore termed "cohesion pheromones" (Fuzeau-Braesch et al., 1988).

Similar phenolics (see Fig. 12.5) are secreted by the New World lubber grasshoppers *Romalea guttata* and *Taeniopoda eques* and are thought to serve a

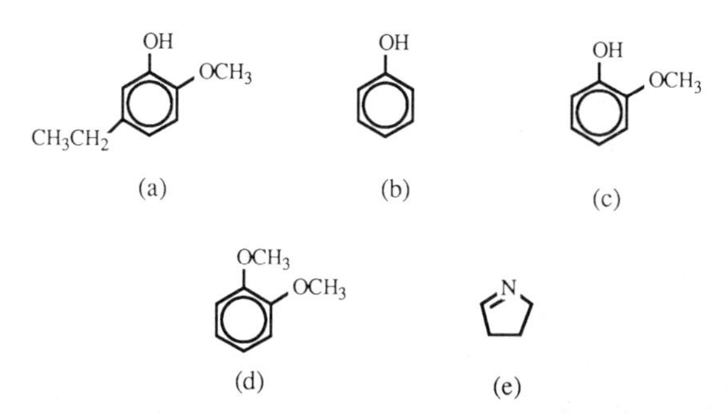

Fig. 12.1. The chemical structures of some proposed grasshopper pheromones. (a) 5-ethyl-2-methoxyphenol, (b) phenol, (c) guaiacol, (d) veratrole, (e) Δ^1-pyrroline.

defensive function (Whitman et al., 1985; Jones et al., 1986, 1988). In *Romalea guttata*, phenolic composition is partially influenced by diet (Jones et al., 1987).

Whether the various phenolics released by locusts and lubbers are derived directly from plants, modified by gut microflora (Charnley et al., 1985), or synthesized by the grasshoppers remains to be discovered. Also unknown is the extent of phenolic production across acridoid taxa; perhaps many grasshopper species produce phenolics. In any case, the role of phenolics in phase transformation is still very confused. A strong possibility exists that other factors or additional pheromones influence phase transformation, and remain to be discovered. Indeed, recent work suggests that a chemical factor in the frass of adult gregarious *S. gregaria* stimulates solitarization in isolated nymphs (Gillett, 1983).

12.2.2 Maturation Pheromones

In some locust species, sexually mature adult males produce pheromones that induce rapid maturation in sexually immature adults of either sex. Such an effect is seen in gregarious phase *Schistocerca gregaria*, where immature adults are pink and brown in color, immature females have undeveloped ovarioles, and immature males demonstrate no interest in mating and possess a thin, nonvacuolated abdominal epidermis. Mature adult *S. gregaria*, on the other hand, are often yellow in color; the females contain developing eggs, and the males readily copulate and possess a thickened, glandular abdominal tergal epidermis (Cassier and Delorme, 1974; Cassier, 1977). These maturational changes normally occur over a period of 48 days for females and 28 days for males when they are kept with other immature adults. However, this period is significantly shortened to 27 days for females and 17 days for males when they are reared with sexually mature adult males (Norris, 1952, 1954; Uvarov, 1966).

The stimulus responsible for this accelerated maturation is a pheromone (Loher, 1960; Amerasinghe, 1978). Apparently, only mature adult males produce the active substance. It is oil soluble, and its effects can be induced either through physical contact between mature males and immatures, or, surprisingly, when immatures are simply exposed to volatiles emanating from mature males. Loher (1960) suggested that the substance was produced and released from glandular abdominal epidermis found only in mature males. He further determined that the antennae were the site of reception, and suggested that neural signals from the antennae stimulated the corpora allata to release a hormone that entered the blood to produce the maturation changes. An alternative hypothesis was proposed by Highnam and co-workers, who found that immature adult females were more active when reared with mature adult males. Noting that electrical stimulation of the central nervous system and enforced activity stimulated neurosecretion and maturation, they suggested that the pheromone acted by simply eliciting increased activity in recipients (Highnam, 1962; Highnam and Lusis, 1962).

Similar maturation effects are observed in *Locusta migratoria* and *Schistocerca cancellata*. In both species, male-released pheromones may be involved (Ali, 1987; Amerasinghe, 1978; Hawkes et al., 1987). In *L. migratoria*, adult males normally copulate 17–25 days after emergence, but in the presence of mature males, copulate in 13–14 days (Norris, 1950, 1964). In male *S. cancellata*, a sticky odorous substance that exudes from dorsal abdominal integumental glands is thought to stimulate faster maturation in conspecifics. The glands are similar to those found in *S. gregaria* (Hawkes et al., 1987).

The chemical identities of the accelerating pheromones have not been established. However, a potential candidate in *S. gregaria* is the volatile amine Δ^1-pyrroline (Fig. 12.1e) found in the air surrounding adult males and also in male accessory sexual glands. This substance is produced in relatively large quantities by mature males, is produced in trace amounts by mature females, and has not been obtained from immatures (Blight, 1969). However, the pheromonal activity of this substance has not been tested, and because Δ^1-pyrroline is a common microbial degradation product of putrescine, which also could originate via microbial action, its origin and role are still in dispute.

In addition to the stimulatory effect of mature adult *Schistocerca gregaria* males on immature adults, there appears to be an inhibitory effect originating from larvae or immature adults, which has been hypothesized to be pheromonally based (Norris, 1954, 1964; Norris and Pener, 1965; Richards and El Mangoury, 1968).

It is unknown if these two pheromonal effects operate in the field. If so, the demographic consequences of simultaneous inhibition and stimulation would be great. In dense grasshopper populations, inhibitory effects might tend to retard the maturation of the oldest individuals until a turning point is reached when the increasing numbers of maturing individuals would begin to stimulate the maturity of the entire population. In this way, maturation synchrony of the entire population would result, an event that could be considerably important in the dynamics of locust phase transformation (Norris, 1962, 1964; Uvarov, 1966; Richards and El Mangoury, 1968).

12.2.3 Male-Produced Oviposition Stimulants

The acceleration of female maturity (discussed above) is a complex process and may be related in part, to another pheromonal phenomenon, increase in female fecundity. In many grasshoppers, mating or simply close proximity of males elicits vitellogenesis and/or oviposition in females (Hamilton, 1936; Husain and Mathur, 1946; Norris, 1954; Highnam, 1962; Riegert, 1965). This phenomenon is thought to be caused by two separate but not mutually exclusive processes. In the first, volatile male pheromones are perceived by females over a distance, and stimulate increased oocyte maturation or oviposition. Copulation or physical or visual contact between the sexes are not necessary for the effect to occur. In the second process, copulation elicits vitellogenesis or oviposition, but olfactory stimulation from the male has little effect.

An example of the first case is seen in *Schistocerca vaga* where virgin females exposed to male odors but prevented from visual and physical contact produced more eggs per egg pod than virgins not exposed to male odors (Okelo, 1979). Indeed, these virgin females produced as many eggs per pod as females allowed both physical contact and copulations. A similar phenomenon occurs in the oedipodine *Aiolopus thalassinus* (Schmidt and Osman, 1988). Virgin females exposed to male odors laid more eggs and egg pods per lifetime, and exhibited shorter pre- and interoviposition periods than virgin females not exposed to male odors. Oviposition rates in the former group were even higher than those exhibited by mated females. As with *S. vaga*, this effect did not require visual or physical contact among the sexes, only that air be allowed to move between the male and female cages. Additional experiments suggested that the antennae were the sites of pheromone reception in females and that males released the pheromone from their abdomens; isolated virgin females exposed to ether extracts of male abdomens produced significantly more eggs and egg pods than similar females not exposed (Schmidt and Osman, 1988).

The second type of pheromonal process is seen in *Melanoplus sanguinipes*, In this species, virgin females oviposit approximately every 4.5 days whereas mated females average one egg pod every 2.9 days (Pickford et al., 1969). This effect is thought to be caused by a chemical factor produced by the male and transferred via the spermatophore to the female. In *M. sanguinipes* and also in *Schistocerca gregaria*, implantation of male accessory gland complexes into virgin females resulted in greater rates of oviposition (Pickford et al., 1969; Leahy, 1973a). The effect was dose dependent in *S. gregaria*; implantation of extra male accessory glands resulted in even greater egg laying (Leahy, 1973a). Implantation of sperm, seminal vesicles, or sperm-filled spermathecae did not significantly increase oviposition, nor did mating with males whose accessory glands had been extirpated (Pickford et al., 1969; Leahy, 1973a,b). These results suggested that the factor responsible for increased oviposition resided not in the sperm, but in one or more of the many accessory gland tubules that make up the male accessory gland complex.

In subsequent experiments, Friedel and Gillott (1976) demonstrated that water extracts of male *M. sanguinipes* accessory glands could induce oviposition when injected into virgin females. Injection of gland extracts induced oviposition in 75% of experimental virgin females within 24 h, whereas all control females retained their eggs (Friedel and Gillott, 1976). Interestingly, gland extracts from allatectomized males had no activity when injected into females, suggesting that production of the male oviposition stimulant was under the control of the corpus allatum. Friedel and Gillott (1976) hypothesized that a proteinaceous male factor was deposited in the female spermatheca and moved across the spermathecal wall into the hemolymph and hence to the brain where it stimulated the female neurosecretory system.

A similar phenomenon occurs in *L. migratoria*, where a proteinaceous factor originating in the opalescent gland (one of 15 glands that make up the male accessory gland complex) is passed to the female during mating and serves to induce oviposition (Lange and Loughton, 1985). Interestingly, extracts from this gland stimulate muscle contractions and adenylate cyclase activity in female *L. migratoria* oviducts (Lafon-Cazal et al., 1987). These authors suggest that the oviposition-stimulating factor may actually consist of multiple substances that act at different sites to produce the complex oviposition effect.

The value of female stimulation for the male is apparent. Males who can induce rapid oviposition in females suffer less sperm competition from other males and lessen the detrimental impact of premature female mortality. Although the chemical identities of such factors in male grasshoppers are yet to be identified, recent studies with crickets have shown that males produce arachidonic acid and prostaglandin synthetase. During mating, these are transferred to the female where they react to produce prostaglandins, which then stimulate oviposition (Loher et al., 1981; Loher, 1984). Significantly, prostaglandin synthetase has been found in the male accessory gland of *L. migratoria* (Lange and Loughton, 1984).

The phenomenon of male-induced oviposition is complicated by yet another factor, the transfer of nutrients during copulation. Although females can store viable sperm for long periods, the females of many grasshopper species mate repeatedly. Other species mate for exceedingly long durations (up to 95 h in the laboratory), during which multiple spermatophores are passed to the females (Uvarov, 1966, 1977; Pickford and Gillott, 1971, 1976; Pickford and Padgham, 1973; Whitman and Loher, 1984). Because a single mating presumably supplies enough sperm to fertilize all the eggs a female will ever produce, researchers have wondered why multiple and long-duration matings occur. One possibility is that females obtain nutrients from males during copulation. Indeed, protein is produced in male accessory glands and is passed in spermatophores to the female (Friedel and Gillott, 1977; Lange and Loughton, 1984).

Radio-label experiments with *M. sanguinipes* and *Chorthippus brunneus* demonstrated that a significant amount of protein passed in spermatophores and deposited into the spermatheca eventually became incorporated into the eggs (Friedel and Gillott, 1977; Butlin et al., 1987). This process can result in real benefits in terms of fecundity. For example, multiple-mated *C. brunneus* produced more eggs than females mated only once (Butlin et al., 1987). The effect was particularly strong in food-restricted females, implying that mating males make an important contribution of protein to the developing oocytes. Friedel and Gillott (1977) also suggested that multiple spermatophore transfer served more of a nutritive than an insemination function, and Wickler and Seibt (1985) considered long copulation in *Zonocerus elegans* as a form of postcopulatory guarding, designed to protect the male's investment of a costly proteinaceous spermatophore delivered to the female.

It is unknown how nutrient transfer relates to the pheromonal processes described previously. Although a nutrient contribution could hasten oocyte development and result in increased oviposition, the short interval which sometimes occurs between mating and oviposition implies the existence of a faster acting stimulus (i.e., a pheromone).

Much is still unknown regarding the effects of maturation-, vitellogenesis-, and oviposition-inducing pheromones, spermatophore-passed stimulants, and male protein contributions. Because all these factors could potentially influence oviposition, teasing out their individual effects is difficult. Indeed, some of the maturation events observed when immature adult females are paired with mature males may not relate to airborne pheromones, but may be the direct result of forced copulation (Norris, 1964). This area needs work and with modern techniques of chemical separation and identification offers much potential.

12.2.4 Pheromones Influencing Group Oviposition

In most grasshoppers, oviposition is an individual event unrelated to the activities of other females. In some species, however, the timing and location of oviposition is strongly influenced by other ovipositing females. Sometimes dense groups of laying females form in relatively uniform habitats, depositing up to 4000 egg pods/m^2 (Uvarov, 1966; Filip'ev, 1926). Although patchy environmental conditions or visual or tactile attraction to conspecifics often trigger such oviposition aggregations, in some cases pheromones may be involved.

In *Schistocerca gregaria*, a chemical factor produced by adults of both sexes serves as a contact pheromone, stimulating grouping by ovipositing females. Paper from the floor of a dense colony of mature adults elicited 200 ovipositions, whereas clean paper elicited only 36 (Norris, 1963). The pheromone was present on the cuticle of adults, but not internally, and was relatively heat stable; heating at 100°C for an hour did not destroy its activity (Norris, 1970).

Evidence suggested that ovipositor sensilla were the receptor organs because blockage of antennae, palps, and tarsi did not appreciably reduce the response. The chemical stimulus was species specific; dead *S. gregaria*, but not dead *Locusta migratoria*, elicited grouping by ovipositing *S. gregaria*.

Additional evidence for pheromonal stimulation of oviposition comes from *Locusta migratoria*. In this species, sand used for oviposition attracted females who preferentially oviposited into it and even ingested it. The activity lasted for up to a year and was destroyed by detergents and acids (Lauga and Hatté, 1977a,b, 1978).

In *L. migratoria* and *S. gregaria*, the chemical releaser appears to be a short-range or contact pheromone. *Zonocerus variegatus*, however, may possess a more volatile attractant, for both sexes moved upwind to join mixed-sex aggregations at oviposition sites (McCaffery and Page, 1982). These oviposition sites remained attractive for several days, even when the ground was devoid of insects as in the early morning.

Group laying could have multiple benefits. Females are vulnerable to predation during the oviposition process; aggregating individuals may suffer a lower mortality rate from predators. Soils containing high densities of egg pods may remain at a higher humidity or contain greater amounts of female-produced antimicrobial substances. Perhaps the greatest benefit of aggregated oviposition may be that the nymphs will be highly aggregated upon hatching (Norris, 1963). On the other hand, group laying seems in some ways maladaptive. Aggregations of ovipositing grasshoppers are more conspicuous to predators and parasitoids, and ovipositing females might dig through previously laid egg pods, killing eggs and eliciting fungal growth which could spread to nearby pods. Clumped egg pods might also foster increased parasitoid attack.

12.2.5 Sexual Pheromones

It is perhaps because grasshoppers are so well known for their visual and acoustic acuities, and for the use of these modalities during pair formation and courtship (Jacobs, 1953; Faber, 1953; Otte, 1970), that so little research has been conducted on grasshopper sexual pheromones. Yet many related orthopteroid groups are known to possess sexual pheromones (Whitman, 1982a), and the ancestral stock for this diverse group is presumed to have relied upon chemical signals for mate recognition (Otte, 1970). This is not to say that acridoid sexual pheromones have not been considered. Indeed, numerous anecdotal references to this possibility exist (Varde, 1934; Hubbell and Cantrall, 1938; Slifer, 1940; Oglobin, 1947; Perdeck, 1958; Laub-Drost, 1959; Gregory, 1965; Otte, 1970). For example, Pickford and Gillott (1972) noted that *Melanoplus sanguinipes* males were able to differentiate between mature and immature females and between virgin and mated females. However, to date, only two grasshopper species have rigorously been shown to possess sexual pheromones.

In *Hieroglyphus nigrorepletus* (Acrididae: Hemiacridinae) males became sexually aroused and were attracted to females in an olfactometer (Siddiqi and Khan, 1981). Males were not attracted to the odor of males, and females were not attracted to either sex. Pheromone production in females began at about the third day of adulthood, peaked at 8–9 days (which coincided with the time of first mating), and continued for up to 29 days. The source of the pheromone was traced to the frass, and then to the gut. Perception in males was via the antennae.

In *Taeniopoda eques* (Romaleidae), blinded males readily mounted and attempted to copulate with dead females, but not dead males (Whitman, 1982a). An active substance could be rubbed from a female's body onto a finger or rubber tube, which, when presented to males, elicited sexual behavior (Fig. 12.2). Interestingly, pheromonal activity was traced to the female defensive secretion: the exudate from females, but not males, elicited sexual behavior in males. Female *T. eques* first became chemically attractive to males about 16–18 days after eclosion, which coincided with the onset of sexual behavior in the males. As in *Hieroglyphus*, the antennae were found to be the site of pheromone detection in males. Because males were not attracted in olfactometer and wind tunnel tests (Whitman, 1982b), the active substance in *Taeniopoda* appears to act as a contact pheromone, in contrast to *Hieroglyphus*, where males were attracted over a short distance.

Fig. 12.2. Male *Taeniopoda eques* responding sexually to presence of female sexual pheromone applied to fingers. The response is most obvious in the lowest, largely hidden, insect.

There is some evidence that two other Romaleidae, *Elaeochlora trilineata* and *Chromacris speciosa*, also possess sexual pheromones. When females of these South American species were placed in perforated semiopaque containers, conspecific males appeared to be attracted. A few males even demonstrated heightened sexual arousal by attempting to copulate with other males (Riede, 1987).

Among the possible sources of grasshopper sexual pheromones are the Comstock–Kellogg glands, which are paired invaginated sac-like structures found at the anterior end of the vagina in many species of Cyrtacanthacridinae, Catanopinae, Romaleinae, and related subfamilies (Slifer and King, 1936; Slifer, 1940). These glands are sometimes everted before and during copulation, leading some authors to speculate that they were the source of sexual pheromones (Varde, 1934; Kyl, 1938; Hubbell and Cantrall, 1938; Laub-Drost, 1959; Thomas, 1965). However, attempts to demonstrate this function in *Melanoplus* and *Taeniopoda* fail (Pickford and Gillott, 1972; Whitman, 1982b). Perhaps the true function for Comstock–Kellogg glands was discovered by Eisner et al. (1966), who found that the glandular secretion in *Romalea guttata* (=*microptera*) served to promote tanning of fresh-laid eggs and egg pod froth.

Another potential source of sex or species identifying pheromones are the lipid soluble components of the cuticle. The chemistries of the cuticular lipids of about 10 grasshopper species have been studied. These analyses show a moderate qualitative and quantitative variability among distantly related taxa (Lockey, 1976), but little variability among congenerics (Soliday et al., 1974; Jackson, 1981). Generally, long-chain normal and methyl-branched alkanes make up 50–80% of the lipid-soluble cuticular fraction. The remaining 20–50% consists of wax esters, fatty acids, aliphatic alcohols, *n*-alkyl ethers, triglycerides, and sterols (Blomquist et al., 1972, 1987; Jackson, 1982; Nelson et al., 1984; Génin et al., 1987). Some of these components (e.g., *n*-alkanes) can be sequestered directly from the diet (Blomquist and Jackson, 1973), whereas others are autogenously synthesized (Blomquist and McCain, 1975; Lockey, 1976).

Some sexual and developmental differences in grasshopper cuticular chemistries are also found (Jackson, 1982; Génin et al., 1987), although these are generally quantitative rather than qualitative in nature. For example, Warthen and Uebel (1980) noted a significantly greater amount of octadecanoates of 11-tricosanol and 12-pentacosanol in male than in female *Melanoplus differentialis*. The authors speculated that these primarily male substances might rub off onto females during mating and thus provide a means for males to differentiate among virgin and mated females. Sexual differences have also been found in cuticular proteins from mature *Locusta migratoria*. However, the function of these proteins is thought to be structural (allowing the female to stretch her intersegmental membranes during oviposition), not pheromonal (Cassier and Papillon, 1983; Andersen and Højrup, 1987). Thus sexual differences in cuticular chemistry are known but are not great, and a pheromonal function for these differences remains to be demonstrated.

It appears that we still know little about sexual pheromones in grasshoppers. In many subfamilies, visual and acoustic signals undoubtedly function in mate location, identification, courtship, and species isolation (Faber, 1953; Jacobs, 1953; Otte, 1970). However, in other grasshopper subfamilies (e.g., Cyrtacanthacridinae and Romaleinae), premating signals are reduced or absent. In these species males are usually aggressive and often stalk, and mount, any small moving object in their habitat (Otte, 1970; Whitman and Orsak, 1985). Thus identification occurs primarily following physical contact. For members of these groups, species and sex specific pheromones may play an important but hitherto unrecognized role. Even for those grasshoppers that employ complex visual and acoustic signals, final close-range sex and species discrimination may be determined chemically.

12.2.6 Pheromones and Postmating receptivity in Females

In some grasshoppers, mating appears to lower the subsequent sexual receptiveness of females (Renner, 1952; Mika, 1959; Haskell, 1960). Although attempts have been made to relate these effects to gonotropic cycles or mechanical stimulation of the spermatheca, no clear cause and effect have been established (Mika, 1959; Haskell, 1960). Male-produced factors that lower sexual receptivity in females are known for a number of insects. A similar process may explain such refractory periods in female grasshoppers, and should be explored.

12.2.7 Aggregation Pheromones

Many grasshopper species aggregate and therefore one might expect at least some members of this diverse group to possess aggregation pheromones. Indeed, the early literature suggested that such pheromones might exist for *Schistocerca gregaria* (deLépiney, 1930; Volkonsky, 1942). However, Haskell et al. (1962), working with this species in a wind tunnel, found little evidence for intraspecific olfactory orientation. Taking a different approach, Gillett et al. (1976) tested not for attraction, but for the ability of odors to hold animals in defined areas. They reported that *S. gregaria* remained for longer periods on filter paper treated with feces or extracts of adult cuticle than in odorfree areas. The volatiles from live conspecifics or from aqueous or chloroform–methanol extracts of frass sometimes elicited reduced locomotion. However, these responses were for the most part barely significant and rather erratic, leading the authors to conclude that aggregation pheromones were of little importance in the biology of *S. gregaria*.

A recent study by Fuzeau-Braesch et al. (1988) partially contradicts the above conclusions. These authors isolated, identified, and tested three volatiles (Fig. 12.1) that were present in the atmosphere surrounding *L. migratoria* and *S. gregaria* colonies. They found that the odors of phenol or guaiacol, but not veratrole, elicited clumping in both species, and that the strongest clumping occurred when all three volatiles were presented simultaneously. Locusts were never attracted to these odors, but responded to their presence by aggregating to one another.

The attraction of females to oviposition sites has already been discussed. For example, in *Zonocerus variegatus*, both males and females move upwind to join mixed-sex aggregations. This behavior is thought to be in response to an olfactory stimulus, but at present its source and identity are unknown (McCaffery and Page, 1982).

12.3 CHEMICAL DEFENSE AGAINST PREDATORS AND PARASITES

An enormous diversity of predators attacks and consumes grasshoppers. Included are representatives of nearly all terrestrial vertebrate and arthropod classes (Bryant, 1912; Martin, et al., 1951; Greathead, 1963, 1966; Uvarov, 1928, 1977; Joern, 1986). In response to these predative pressures, grasshoppers have evolved numerous defensive features, including crypsis, wariness, strong jumping legs, flight, and autotomy. Chemical defenses are also employed in the form of regurgitants, anal discharges, glandular secretions, and internal toxins.

12.3.1 Regurgitation

Perhaps the best known form of grasshopper chemical defense is regurgitation, the voluntary oral discharge of gut contents (Fig. 12.3a). Many, if not most, grasshoppers regurgitate when molested (Rabaud, 1916), a behavior colloquially referred to as "spitting tobacco." This sticky, brown, and odorous fluid, derived from the capacious crop, is a complex mixture of recently eaten plant materials and digestive and salivary secretions. Ranging in pH from 4.9 to 6.4, it contains a variety of enzymes including several proteolytic ones (Uvarov, 1966; Freeman, 1967, 1968; Lymberry and Bailey, 1980).

Although regurgitation in grasshoppers is well known, there have been surprisingly few reports of its defensive efficacy. Earlier studies showed that *Locusta migratoria* regurgitants caused hemorrhaging, paralysis, and/or death when injected into mice, guinea pigs, sheep, and goats (Curasson, 1934; Freeman, 1968). However, these studies were of little ecological relevance because grasshopper regurgitants are seldom introduced into predators subcutaneously. Exceptions might be grasshoppers with particularly powerful mandibles, such as *Romalea guttata* and *Brachystola magna*, which bite and regurgitate simultaneously (Whitman et al., 1985).

More appropriate evidence for the defensive nature of regurgitants comes from oral and topical tests. For example, oral administration of grasshopper regurgitants produced vomiting in guinea pigs, sheep, and goats (Currasson, 1934), but not in mice or jirds (*Meriones shawi*) (Freeman, 1968). Topical application of as little as 2 μl of *L. migratoria* regurgitant in the eyes of jirds

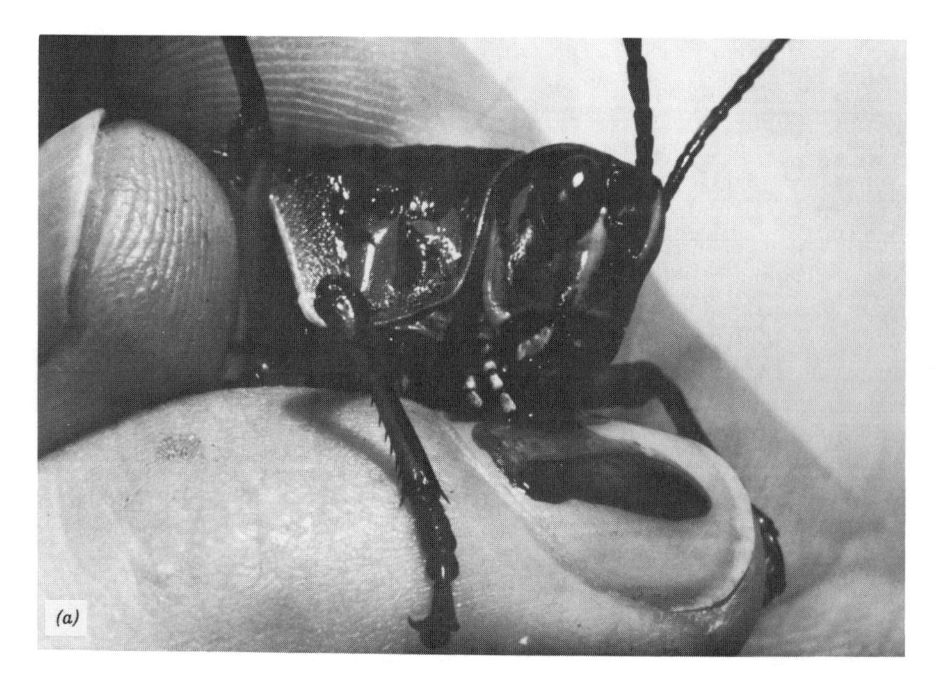

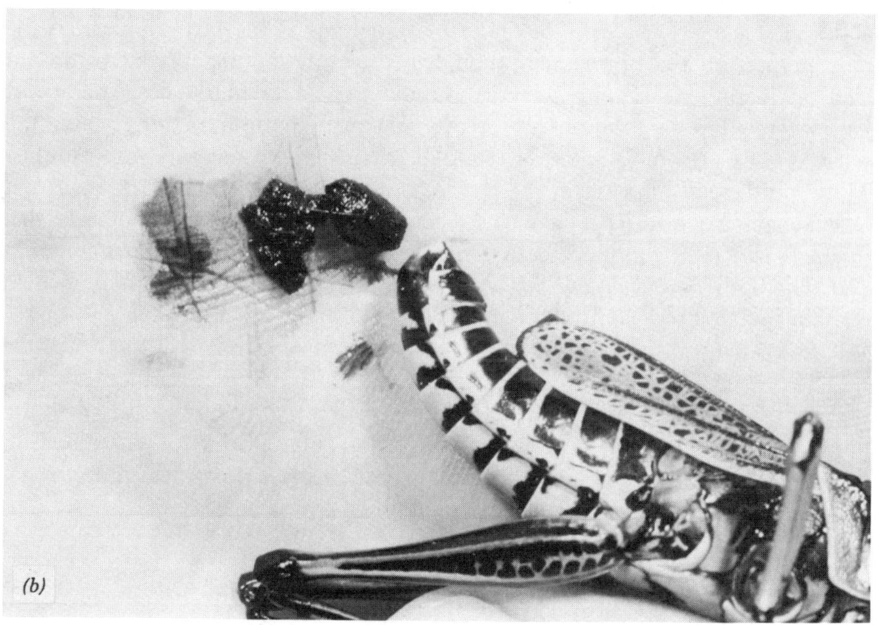

Fig. 12.3. *Romalea guttata* (a) regurgitating and (b) defecating.

elicited immediate distress, including scratching and plowing of the head through the substrate (Freeman, 1968).

Regurgitants also act against invertebrate predators as shown by Eisner (1970) who glued *Romalea guttata* (=*microptera*) and *Brachystola magna* to tethers and placed them near *Pogonomyrmex* ant colonies. The aggressive ants first attacked but withdrew following grasshopper regurgitation. The ants were clearly repelled; those contacting the discharge performed extensive and prolonged cleaning behaviors. In additional tests, grasshopper parts, which were normally eagerly taken by the ants, were rejected if treated with regurgitant.

The Australian acridid *Goniaea* sp. also produces a viscous regurgitant when disturbed (Lymberry and Bailey, 1980). In laboratory trials using a diversity of predators, regurgitating *Goniaea* repelled skinks 33%, lycosid spiders 20%, and ants (*Camponotus*) 80% of the time. These small predators also rejected regurgitant-treated mealworms. Conversely, larger predators were not deterred by the grasshopper's enteric discharge. Marsupial mice (*Antechinus stuartii*), king quail (*Excalfactoria chinensis*), western bullfrogs (*Litoria mourei*), and western bearded dragons (*Amphibolurus minimus*) attacked and consumed all *Goniaea* and regurgitant-treated mealworms that were offered (Lymberry and Bailey, 1980).

Grasshopper regurgitants might derive their repellent properties from several sources including acidic pH, enzymatic activity (particularly protease activity), or ingested plant secondary compounds. This last source may be more important than previously thought. Most grasshopper predators do not feed on plant leaves, and are not adapted to handle plant allelochemicals. Disgorgement of a solution of recently masticated plant material onto the sense organs of some predators might be a shocking and repellent stimulus. The cyanide that appears in the gut of *Zonocerus variegatus* following ingestion of cyanogenic plants (Bernays and Chapman, 1978) might function in this manner. Indeed, it may not be a coincidence that *Goniaea* feeds on *Eucalyptus*, and that both *B. magna* and *R. guttata* consume a variety of plants known to contain toxic or repellent allomones (Capinera and Sechrist, 1982; Whitman, 1988).

The efficacy of grasshopper regurgitants may also derive from their physical properties. Lymberry and Bailey (1980) noted that regurgitants adhered to the mouthparts of ants and skinks that attempted to rub off the viscous, entangling fluid. Most likely, all of the above features of grasshopper regurgitants act synchronously or synergistically to deter predators.

12.3.2 Defecation

Many grasshoppers defecate when molested; however, the potential defensive role of this behavior has been generally ignored. Although grasshopper fecal pellets are often dry, many species produce watery, sticky feces (Fig. 12.3b) that might deter potential predators for many of the same reasons as regurgitants. When considering the defensive efficacy of regurgitation and defecation,

PLATE 1. Aposematic chemically defended pyrgomorphid grasshoppers from Africa. *(a) Phymateus morbillosus* (Pyrgomorphidae) from S. Africa (Photo: Ed Ross). *(b) Taphronota occidentalis* (Pyrgomorphidae) from the Congo. Note defensive froth (Photo: Ed Ross). *(c) Maura regulosa* (Pyrgomorphidae) from South Africa (Photo: M. Tweedie, courtesy of Octopus Books Limited). *(d) Zonocerus* sp. (Pyrgomorphidae) from the Congo (Photo: Ed Ross).

(e) (f) (g) (h)

PLATE 2. Aposematic chemically defended grasshoppers from Australia and North America. *(e) Taeniopoda eques* (Romaleidae) from Arizona performing warning display in response to inquisitive mouse. Note frothy defensive secretion emerging from metathoracic spiracle. *(f) Romalea guttata* (Romaleidae) from Georgia performing warning display. *(g) Dactylotum* sp. (Melanoplinae) from Arizona (Photo: Larry Orsak). *(h)* The Australian pyrgomorphid *Petasida ephippigera* (Photo: Densey Clyne, courtesy of University of Queensland Press).

one must remember that grasshoppers are successfully preyed upon by a great many predator species. For these insectivores, enteric discharges may have little repellent value. However, there are also numerous facultative predators that are more cautious and hesitant and that attack grasshoppers only under certain favorable conditions. For these, a slight increase in prey repellency due to distasteful, sticky, or distracting enteric discharges might be enough to tip the balance away from further attack.

12.3.3 Glandular Defensive Secretions

Although chemical defenses (other than enteric discharges) are relatively rare in grasshoppers, exocrine defensive glands appear to have independently evolved in three separate acridoid taxa. Certain Pyrgomorphidae possess mid-dorsal abdominal glands, two Romaleinae species have metathoracic tracheal glands, and some Oedipodinae exhibit eversible pronotal glands.

12.3.3.1 Pyrgomorphid Abdominal Glands

Many Old World Pyrgomorphidae are defended with repugnant glandular secretions. Included are species of *Phymateus*, *Poekilocerus*, *Zonocerus*, *Pyrgomorpha*, *Colemania*, *Maphyteus*, *Taphronota*, and possibly *Pyrgomorphella*, *Atractomorpha*, *Rubellia* and *Dictyophorus* (Coleman, 1911; Pawlowsky, 1916; Swynnerton, 1919; Hingston, 1927; Grassé, 1937; De Lotto, 1950; Ewer, 1957; Fishelson, 1960; Descamps and Wintrebert, 1966; Uvarov, 1966; Euw et al., 1967; Qureshi and Ahmad, 1970; Abushama, 1972; Chapman et al., 1986; Schmidt et al., 1987).

Many of these species have forsaken the cryptic, wary life style common to other Acridoidea, and have instead acquired a set of attributes characteristic of other chemically defended insects (see Color Plate 1). These chemically protected pyrgomorphids tend to be flightless, sluggish, and large – up to 9 cm long in *Phymateus purpurascens* (Rowell, 1967). Nymphs are often brightly aposematic and highly gregarious, gathering together in aggregations of hundreds or thousands of individuals per square meter. Although adults tend to be less gregarious and aposematic, many possess brightly colored hind wings that are raised in a warning display when the insects are threatened (Hingston, 1927; Carpenter, 1938; Bishop, 1940; Kevan, 1949; Chapman, 1962; Roffey, 1964; Euw et al., 1967; Bernays and Chapman, 1978; Toye, 1982; Chapman et al., 1986).

When molested, both sexes and all instars expel an odorous, milky secretion from a gland opening on the dorsal intersegmental membrane between the first and second abdominal tergites. The method of discharge varies between adults and larvae, and depends upon the degree of stimulation. When grasshoppers are mildly disturbed, the secretion oozes out and collects as a droplet or film on the integument. In contrast, highly agitated larvae expel the secretion with great force, usually as double jets of spray that, in some species, can travel up to 60 cm. Discharging larvae assume an arched stance by curling the head and abdominal tip ventrad, thus exposing the gland orifice and increasing the hemostatic pressure (Fig. 12.4a). During such eruptions, an appreciable amount of hemolymph is sometimes ejected with the gland contents. This hemolymph probably does not originate from the gland itself, but from the lips of the glandular orifice, which becomes turgid with blood during discharge. In some adults, the secretion erupts beneath the wings and flows down the abdomen along lateral grooves to the abdominal spiracles, where excurrents of air produce masses of froth up to 1.5 cm in diameter (see Color Plate). Frothing is sometimes accompanied by a hissing noise that combines with conspicuous chemical and visual stimuli to produce a potent and memorable aposematic display (Ebner, 1914; Carpenter, 1938).

The ejection of this secretion can be elicited by both mechanical and visual stimuli, and individuals can fire up to eight times in short succession. In some species, larvae can aim their secretory volleys anteriorly or posteriorly, depending on the direction of the threat. Exudates vary between species and have been described by different authors as watery, slimy, sticky, pungent, foul smelling, and distasteful (Ebner, 1914; Hingston, 1927; Carpenter, 1938; Fishelson, 1960; Euw et al., 1967; Qureshi and Ahmad, 1970; Abushama, 1972).

A bilobed gland lying just under the first abdominal tergite is the source of the secretion. This cuticle-lined sac is wrinkled when empty, but assumes a more turgid, spherical form when full. In adult *Poekilocerus pictus* it can be as large as 6 by 6 mm (Qureshi and Ahmad, 1970). Compressor muscles are apparently lacking (secretion ejection being effected by hemostatic pressure); however, a variety of muscles inserted near the gland orifice control opening and aiming (De Lotto, 1950; Ewer, 1957; Fishelson, 1960; Qureshi and Ahmad, 1970). Following total discharge, 8–14 days are required for total regeneration of the gland contents in *Poekilocerus bufonius* (Euw et al., 1967).

Species possessing such glands are generally shunned by predators. Indeed, spiders, scorpions, solpugids, ants, mantids, reduviid bugs, toads, skinks, geckoes, cuckoos, jays, chickens, laboratory mice, domestic cats, hedgehogs, mongooses, and monkeys have all been known to reject these pyrgomorphids as prey (Marshal, 1902; Carpenter, 1938, 1946; Bishop, 1940; Rothschild, 1966; Abushama, 1972; Chapman and Page, 1979; Chapman et al., 1986). Fishelson (1960) tested 28 species of potential arthropod, reptile, bird, and mammal predators, and found none that would feed on *Poekilocerus bufonius*. Apparently, *Zonocerus variegatus* are so offensive that some army ants reject them, and spiders will cut them out of their webs (Chapman and Page, 1979). A guinea fowl that ate a *Phymateus* died (Descamps and Wintrebert, 1966). Even humans may be susceptible to pyrgomorphid toxins; Steyn (1962) records the death of a Bantu child following ingestion of one *Phymateus leprosus*. However, not all predators are repelled by all species and stages of these Pyrgomorphidae. Individual spiders, mantids, ants, tettigoniids, robber flies, reduviids, frogs, lizards, African rollers, guinea fowl, kites, grackles, cranes, cattle egrets, skunks, and baboons are known to have consumed at least one grasshopper, although

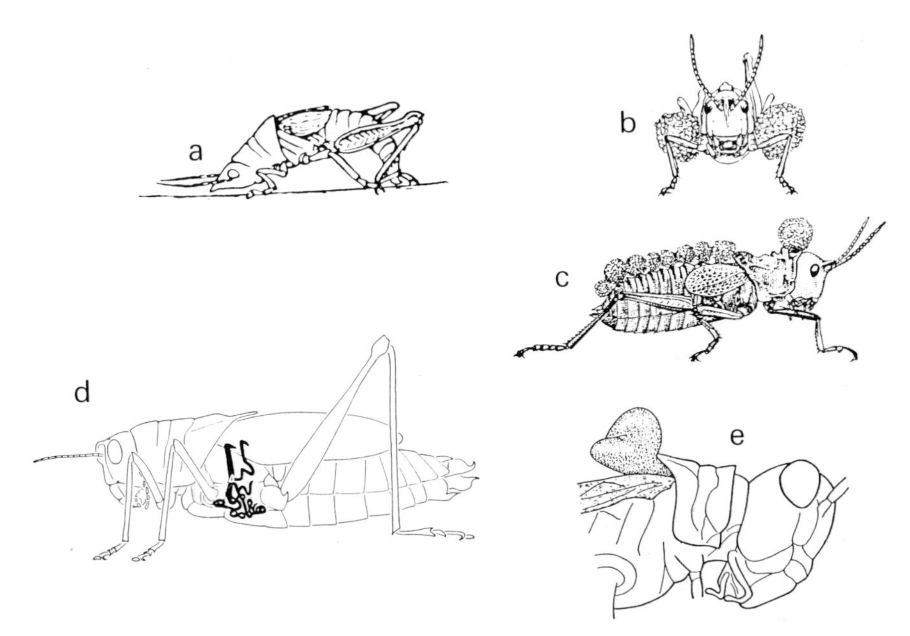

Fig. 12.4. Defensive glands and defensive postures in grasshoppers. (a) Defensive posture of *Poekilocerus bufonius* (Fishelson, 1960). (b,c) *Dictyophorus oberthuri* discharging defensive foam (Grassé, 1938). (d) *Romalea guttata* metathoracic glandular tracheal trunks, the source of the defensive secretion (Alsop, 1970). (e) *Acrotylus patruelis* showing everted prothoracic gland (Schmidt et al., 1987).

often with evident displeasure (Marshal, 1902; Coleman, 1911; Carpenter, 1946; Rothschild, 1966; Chapman, 1962). Rothschild (1966) records that a European hedgehog *Erinaceus europeas* appeared to be insensitive to the toxins of *Poekilocerus bufonius* and ate one to three each day for a week.

Although it is clear that many predator species are repelled by these pyrgomorphids, the exact mechanism of deterrence is unknown. The defensive secretion in some species is certainly highly repugnant, as attested to by the Afrikaans name for *Zonocerus elegans*, "die stinksprinkaan" (Smit, 1964). Predators attacking pyrgomorphids are often wetted about the head by successive secretory salvos of this odorous and distasteful fluid, an assuredly unpleasant event. In addition to these olfactory and gustatory affronts, the secretion of some species appears to act as a general irritant to the mucous membranes of the eyes, noses, and mouths of vertebrates (Fishelson, 1960; Abushama, 1972). Perhaps the most significant function of the defensive exudate is to serve as a potent aversive chemical conditioning agent, reminding predators of the unhappy consequences of previous pyrgomorphid predation. Predators that ignore these blatant chemical signals and proceed to ingest these insects are often poisoned by the grasshopper's internal toxins. Such predators become listless or timid, vomit, and acquire a strong food aversion conditioning against subsequent feeding (Fishelson, 1960; Rothschild and Kellet, 1972). Although some predators are probably deterred by certain specific defense attributes, most likely all of the above repellent features work in unison to deter most predators.

Surprisingly, the defensive chemistry of these pyrgomorphids is only partially known. Most of the defended species are polyphagous, and some, such as *Colemania sphenarioides*, prefer grasses (Coleman, 1911). However, many species appear to specialize on toxic plants in the Asclepiadiceae and Euphorbiaceae (Kevan, 1949; Fishelson, 1960; Smit, 1964; Roffey, 1964; Rowell, 1967; Abushama, 1968, 1972; Qureshi and Ahmad, 1970; Chapman et al., 1986), and it has been suggested that such plants supply the insects with repellent allomones. Indeed, cardenolides have been reported to occur in the bodies or secretions of *Poekilocerus bufonius*, *Poekilocerus hieroglyphicus*, *Poekilocerus pictus*, *Phymateus viridipes*, and *Phymateus baccatus* (Reichstein, 1967; Rothschild, 1972).

Only in *Poekilocerus bufonius* has the defensive chemistry been extensively studied (Fishelson, 1960; Euw et al., 1967; Reichstein, 1967). This species obtains the toxic cardenolides calactin and calotropin (Fig. 12.5a) from its milkweed (Asclepiadaceae) food plant. Secretions from insects reared on *Asclepias* showed 10 times more cardenolide activity than secretions from grasshoppers reared on plants lacking cardenolides. Interestingly, cardenolide concentrations in the blood and secretion were similar, suggesting that both fluids were equally noxious. An average grasshopper contained a dose sufficient to kill one cat. The defensive secretion of *P. bufonius* also contained histamine (approximately 1% by weight) (Fig. 12.5b), which appeared to be synthesized by the insect itself. In contrast to the cardenolides, only low levels of histamine were present in the blood.

A similar scenario probably occurs in some species of *Zonocerus* that are thought to employ plant toxins for their own defense (Boppré et al., 1984). For example, *Z. variegatus* sequesters pyrrolizidine alkaloids from its food plants, and under laboratory conditions will sequester cannabinoids (Fig. 12.5c–e) (Bernays et al., 1977; Rothschild et al., 1977). The congener *Z. elegans* is attracted to the odor of pyrrolizidine alkaloids and attempts to feed whether the source is a withered plant or the pure compound. Because *Zonocerus* spp. feed on numerous toxic plants and may actually seek out certain toxins, they have been referred to as "poison specialists" and "pharmacophagous" (Bernays et al., 1977; Boppré, 1984; Boppré et al., 1984).

Other pyrgomorphids exude fluids or froths when threatened. For example, some *Dictyophorus* spp., which have been reported as lacking the abdominal gland (Uvarov, 1966), expel a liquid described as hemolymph from a variety of pores on the thorax and abdomen (Fig. 12.5b,c) (Grassé, 1937). The distasteful fluid is audibly ejected as a froth, and protects these aposematic, gregarious insects from predators (Grassé, 1937; Carpenter, 1938). Another gregarious, aposematic pyrgomorphid, *Aularches miliaris*, can discharge up to a teaspoon of slimy, bitter tasting froth from five pairs of openings distributed about the thorax (Hingston, 1927; Carpenter, 1938). The secretion smells like

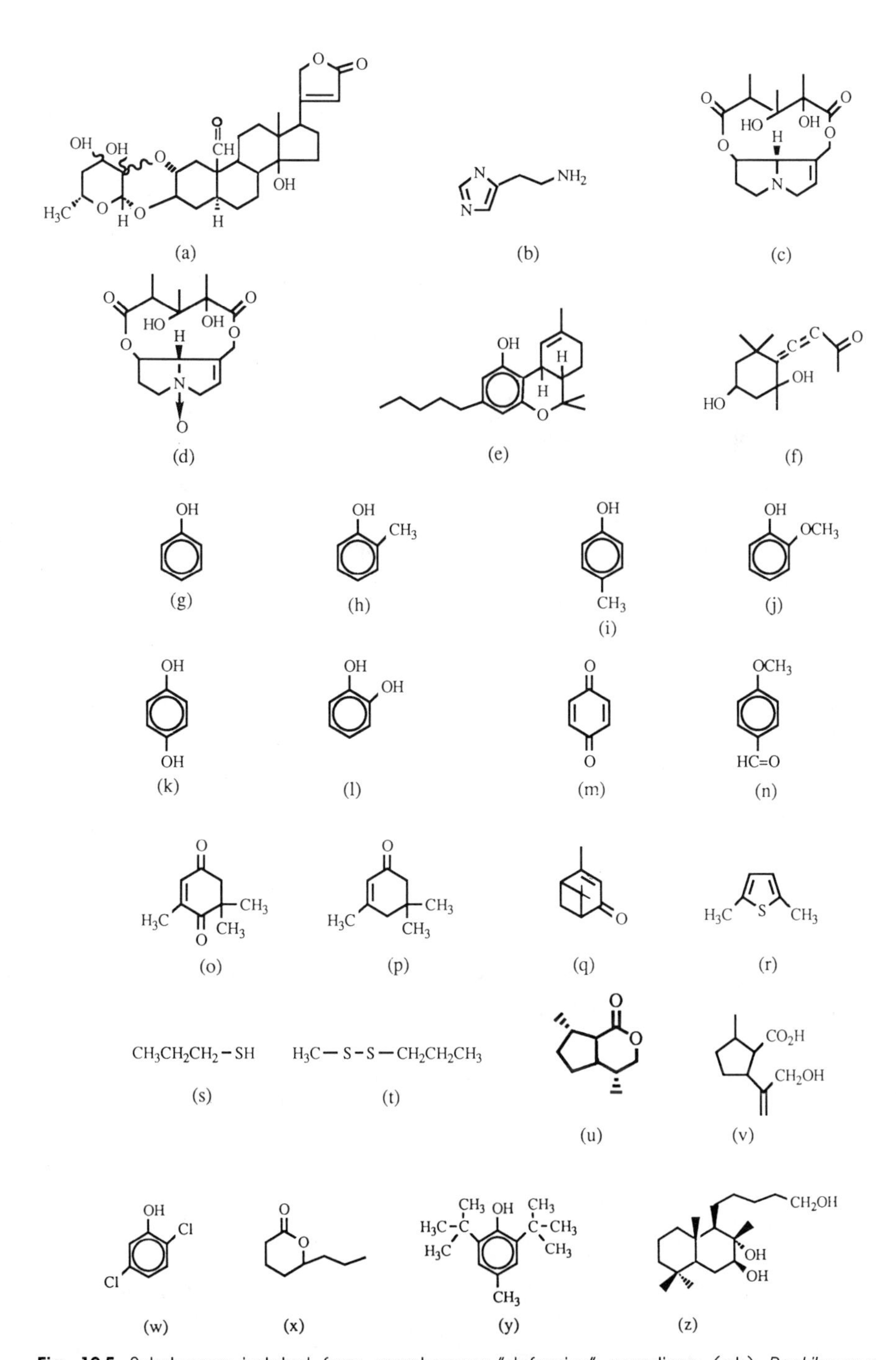

Fig. 12.5. Substances isolated from grasshopper "defensive" secretions, (a,b) *Poekilocerus bufonius*, (c–e) *Zonocerus variegatus*, (f–w) *Romalea guttata*, (x,y) *Acrotylus patruelis*, (z) *Dactylotum variegatum*. (a) Calactin–calotropin, (b) histamine, (c) monocrotaline, (d) monocrotaline *N*-oxide, (e) Δ^1-tetrahydrocannabinol, (f) romallenone, (g) phenol, (h), *o*-cresol, (i) *p*-cresol, (j) guaiacol, (k) hydroquinone, (l) catechol, (m) *p*-benzoquinone, (n) 4-methoxybenzaldehyde, (o) 2,6,6-trimethylcyclohex-2-ene- 1,4-dione, (p) isophorone, (q) verbenone, (r) 2.5-dimethylthiophene, (s) 1-propanethiol, (t) methylpropyl disulfide, (u) dihydronepetalactone, (v) hydroxy acid metabolite of nepetalactone, (w) 2,5-dichlorophenol, (x) δ-octalactone, (y) ional, (z) gymnosper min.

the insect's toxic liliaceous or bignoniaceous food plants (McCann, 1953). As in *Dictyophorus*, the froth is audibly ejected with air. *Aularches punctatus* produces a similar secretion and is rejected by birds (Katiyar, 1955).

At present, it is unclear if the fluids ejected by *Dictyophorus* and *Aularches* represent blood, secretion, or a mixture of both. The fact that other pyrgomorphids produce similar foamy secretions from a single abdominal gland suggests that the descriptions for *Dictyphorus* and *Aularches* may be in error and that the process in these two genera should be critically reexamined.

12.3.3.2 Romaleine Tracheal Glands

Paralleling the Old World Pyrgomorphidae are two species of New World Romaleinae (commonly referred to as lubbers), *Taeniopoda eques* and *Romalea guttata* (=*microptera*). Like their Old World counterparts, these polyphagous grasshoppers ingest a wide variety of toxic plants and are generally shunned by predators. The larvae are aposematic and gregarious and the adults are large, sluggish, flightless, and exhibit an extensive warning display that includes the raising of brightly colored hind wings (see Color Plate 2) (Whitman and Orsak, 1985; Whitman et al., 1985; Jones et al., 1987). Late instars and adults of both species produce an odorous defensive secretion that issues from the metathoracic spiracles with an audible hiss. Adult female *R. guttata* can discharge up to 10 μl of the exudate. The secretion is produced by glands surrounding the metathoracic spiracular tracheae, and is stored behind the closed spiracles in the tracheal lumina until needed (Fig. 12.4d). Thus the metathoracic spiracles are not normally used for respiration. When molested, these insects open the metathoracic spiracles, close all other spiracles, and compress the abdomen. This forces the secretion out first as a fine spray, which can travel up to 15 cm, then as a bubbly adherent froth, which can persist up to 10 s (D. W. Whitman, C. G. Jones, M. S. Blum, J. P. J. Billen and D. Alsop, unpublished).

The chemistry of the secretion has been studied only in *R. guttata*. It is exceedingly complex, containing over 50 compounds. The major volatile constituents are an allenic sesquiterpenoid called romallenone, and a variety of phenols, ketones, and quinones, including phenol, hydroquinone, *p*-benzoquinone, guaiacol, *p*-cresol, catechol, isophorone, and verbenone (Fig. 12.5f–q) (Meinwald et al., 1968; Eisner et al., 1971; Jones et al., 1986). Many of these constituents are synthesized by the insect (Jones et al., 1987), others are apparently sequestered from the diet, and some of them can arise from either source (Blum et al., 1990). Indeed, the tracheal defense gland in *R. guttata* seems to be a repository for a wide diversity of plant secondary substances. For example, insects fed wild onion, a highly preferred host plant, sequester isopropyl sulfide, methylpropyl disulfide, isopropyl disulfide, methyl disulfide, propyl disulfide, propanethiol, methylthiirane, and 2,5-dimethylthiophene (Fig. 12.5r–t) (Jones et al., 1989). These onion compounds are not simply trace contaminants, but give the defense secretion a strong onionlike odor.

When *R. guttata* is reared on catnip (*Nepeta cataria*), the defensive secretion acquires a number of nepetalactone metabolites (Fig. 12.5u,v) (Blum et al., 1987, 1990). The most astounding illustration of *R. guttata* sequestrative abilities is that grasshoppers feeding on plants sprayed with the herbicide 2,4-dichlorophenoxyacetic acid apparently sequester the breakdown product 2,5-dichlorophenol (Fig. 12.5w) in their defensive secretion (Eisner et al., 1971; Berger, 1976).

The consequence of this broad sequestrative ability is that lubbers raised on different plants often exhibit extreme qualitative and quantitative differences in their defensive secretion composition. This variation is in addition to an innate variability shown even by grasshoppers raised on the same diet (Jones et al., 1986, 1987). This defensive variability may be quite important for lubbers because potential predators are faced with and must adapt to not one, but a wide range of secretion types. Indeed, secretions from insects reared on either wild onion or catnip were significantly more repellent to two species of ant predators than secretions from grasshoppers on other diets (Blum et al., 1987; Jones et al., 1989).

The above studies are also significant with regard to the evolution of sequestered chemical defenses in insect herbivores. It is often assumed that a high degree of adaptation or even coevolution between an insect and a toxic plant precedes the sequestration and defensive use of plant allelochemicals. Indeed, most insects known to sequester plant toxins are dietary specialists (Ehrlich and Raven, 1964; Rothschild, 1972). The *R. guttata* studies suggest that the evolution of sequestered chemical defenses could occur through an alternative pathway. *R. guttata* is polyphagous and already chemically defended by virtue of its ability to synthesize deterrent compounds. However, forced monophagy can result in the sequestration of plant allelochemicals and an increase in the deterrency of secretions. This increase does not derive from a highly coevolved relationship, but appears to result from a casual accumulation of plant substances that, because of monophagy, occur in high concentrations in the gut. Thus evolution of sequestered defenses may begin with the fortuitous bioaccumulation of plant allomones following dietary specialization and may not necessarily require specific adaptation to, or coevolution with, a toxic plant (Jones et al., 1988, 1989).

In addition to their tracheal secretions, *R. guttata* and *T. eques* also possess potent internal toxins capable of inducing toxicosis and vomiting in predators (Fig. 12.6) (Whitman and Orsak, 1985). However, as with the Pyrgomorphidae, not all predators are equally affected by lubber toxins. In a comprehensive survey of 88 potential predator species, it was found (D. W. Whitman, C. G. Jones and M. S. Blum, unpublished) that 20 species tended to feed heavily on lubber grasshoppers, and that 50 species were unable or unwilling to feed or, after short- or long-term feeding, demonstrated aversion or vomiting. Although variable responses were observed, broad taxonomic trends were evident. Generally, invertebrates were immune, birds and lizards were strongly affected, and mammals and amphibians were intermediate in their abilities to handle lubber toxins. Birds and lizards that consumed lubber grasshoppers often

Fig. 12.6. *Anolis carolinensis* lizard which has just regurgitated a toxic *Romalea guttata* hopper.

vomited, and most acquired strong conditioned aversions to subsequent feeding. Some individuals died following lubber ingestion.

Chemical defense in romaleines has been described only for *T. eques* and *R. guttata*. However, a strong possibility exists that many of the other 10 species of *Taeniopoda*, as well as species of the related genera *Chromacris* and *Xestotrachelus* possess deterrent allomones. Many of these species are large, gregarious, aposematic, sluggish, and feed on toxic solanaceous plants, traits suggestive of chemical defense (Hebard, 1925; Mathieu, 1970; Gangwere and Ronderos, 1975; Roberts and Carbonell, 1982).

12.3.3.3 Oedipodine Prothoracic Eversible Glands

Species in the Old World Oedipodinae genera *Oedaleus*, *Acrotylus*, *Morphacris*, and possibly *Locusta* possess an eversible pronotal gland that everts through a middorsal transverse slit lying on the pro-mesonotal intersegmental membrane (Fig. 12.4e) (Vosseler, 1902, 1903; Nikolskii, 1925; Hollande, 1926; Miller, 1936; Jannone, 1938, 1939; Uvarov, 1928, 1966).

The gland of *Acrotylus patruelis* is typical for the group. It is found just under the pronotum in all instars and both sexes (Schmidt et al., 1987). Disturbed insects spread their reddish hind wings, lower the head, and raise the posterior edge of the pronotum to expose the gland orifice. Usually the secretion is released only following mechanical stimulation. The small amount of odorous secretion (ca. 0.15 μl/adult) that emerges quickly wets the thorax, so that evaporation is increased. If a grasshopper is squeezed, the entire gland is everted from under the posterior edge of the pronotum. Although compressor muscles are lacking, retractor muscles serve to return the gland to its normal position. The secretion is thought to contain octalactone and ional (Fig. 12.5x,y). However, the function of these substances remains unknown, for the secretion did not repel birds, lizards, mantids, mice, or ants, or prevent the growth of several entomophagous fungi (Schmidt et al., 1987).

12.3.4 Internal Toxins

In addition to the secretion-producing grasshoppers discussed above, numerous other species may be defended with internal toxins. Indeed, scattered throughout the world are grasshoppers that are not known to discharge secretions, yet are aposematic, gregarious, flightless, or stenophagous on noxious plants — traits suggestive of chemical defenses that may be internal. Unfortunately, few of these grasshoppers have been examined in detail.

Such a species is the North American melanopline *Dactylotum variegatum*. This flightless brilliantly colored blue, red, and yellow insect (see Color Plate 2) is reported to be unpalatable to birds (Isely, 1938; Roberts, 1947). *D. variegatum* feeds heavily on the composite *Baccharis glutinosa*, which contains the potentially toxic diterpenic alcohol gymnospermin (Fig. 12.5z) (Miyakado et al., 1974). Similar (but as yet unidentified) substances have been recently isolated from *D. variegatum* (M. Aregullin, personal communication), suggesting that this insect sequesters plant compounds for its protection. Similar sequestrative phenomena probably occur in several other aposematic pyrgomorphids (see Color Plate 2) that feed on toxic or aromatic plants such as the African grasshopper *Maura regulosa* and the Australian pyrgomorphid *Petasida ephippigera* (Key, 1985).

12.3.5 Chemical Defense of Eggs

Little is known about the chemical defense of grasshopper eggs. However, because eggs are attacked by numerous predators, nematodes, and microorganisms (Greathead, 1963, 1966; Uvarov, 1977), and because grasshoppers spend a major part of their existence in this stage (Uvarov, 1966), strong selection pressures undoubtedly exist for egg defense. To date, the eggs of only one species have been shown to possess defensive allomones; the cardenolides calactin and calotropin were found in the eggs of *Poekilocerus bufonius* that were fed on Asclepiadaceae (Euw et al., 1967). Egg pod foam represents another potential source of defensive chemicals. This material is derived from a variety of glands, any of which might produce repellent or antimicrobial chemicals. Such substances, surrounding the eggs, could act as a chemical barrier, deterring or preventing natural enemy attack.

12.4 CONCLUSIONS

It is clear that grasshoppers have evolved a multitude of chemical substances that allows them not only to communicate among themselves, but to thwart the attacks of hostile natural enemies. However, as this chapter has shown, we have only begun to understand these phenomena; many questions remain unanswered concerning the functioning of pheromone systems, the glands and receptor organs involved, and the identities of the chemical participants.

When surveying the organic products of grasshoppers, one is immediately struck by the variety of phenols and quinones produced. Guaiacol, veratrole, phenol, and 5-ethyl-2-methoxyphenol have been implicated as grasshopper pheromones. Phenol, cresol, guaiacol, hydroquinone, catechol, and benzoquinone have been found in grasshopper defensive secretions, and ional occurs in the eversible gland of *Acrotylus*. Is this merely a coincidence, or is this the result of some underlying physiological process common among grasshoppers? One must remember that phenols and quinones are common constituents of grasshopper food plants and that they play an important role in acridid cuticular chemistry. Thus it should not be surprising that these substances are common products of grasshopper cuticular glands. The exact relationship between phenolics in the food, the gut, the cuticle, and in exocrine secretions, and the role of phenolics in intraspecific communication are presently unclear and represent important areas for future acridological research.

Perhaps the greatest problem with regard to grasshopper chemical communication is that, although numerous pheromonal processes have been proposed for the Acridoidea (Table 12.1), not a single grasshopper pheromone has been unambiguously identified. Those substances touted as having pheromone activity produce only partial effects at best. Clearly, the chemical identification of these communicative agents is of the highest priority. Given the state of the art of modern chemical separation and identification, there is no reason why these substances cannot be quickly isolated and characterized, and their relationships revealed.

Acknowledgements

I would like to thank Matt Hameister, Scott Sakaluk, Charles Thompson, and Kathy Smith-Whitman for editorial suggestions, David Alsop, Larry Orsak, and Edward Ross for kindly supplying figures or photographs, and Cheryl Winchester for manuscript preparation.

REFERENCES

Abushama, F. T. 1968. Food-plant selection by *Poecilocerus hieroglyphicus* (Klug) (Acrididae: Pyrgomorphinae) and some of the receptors involved. *Proc. R. Entomol. Soc. London*, A *Ser.* **43**, 96–104.

Abushama, F. T. 1972. The repugnatorial gland of the grasshopper *Poecilocerus hieroglyphicus* (Klug.) *J. Entomol.*, A *Ser. Gen. Entomol.* **47**, 95–100.

Ali, Y. 1987. Pheromone cells in epidermis of *Locusta migratoria. Pak. J. Zool.* **19**, 127–132.

Alsop, D. W. 1970. Defensive glands of arthropods: Comparative morphology of selected types. Ph.D. Dissertation, Cornell University, Ithaca, New York.

Amerasinghe, F. P. 1978. Pheromonal effects on sexual maturation, yellowing, and the vibration reaction in immature male desert locusts (*Schistocerca gregaria*). *J. Insect Physiol.* **24**, 309–314.

Andersen, S. O. and P. Højrup. 1987. Extractable proteins from abdominal cuticle of sexually mature locusts. *Locusta migratoria. Insect Biochem.* **17**, 45–51.

Bell, W. J. and R. T. Cardé (Eds.) 1984. *Chemical Ecology of Insects.* Sinauer, Sunderland, Massachusetts.

Berger, R. S. 1976. 2,5-Dichlorophenol: Not synthesized by eastern lubber grasshopper. *Ann. Entomol. Soc. Am.* **69**, 1–2.

Bernays, E. A. and R. F. Chapman. 1978. Plant chemistry and acridoid feeding behaviour. In J. B. Harborne, Ed., *Biochemical Aspects of Plant and Animal Coevolution* Academic Press, London, pp. 99–141.

Bernays, E. A., J. A. Edgar, and M. Rothschild. 1977. Pyrrolizidine alkaloids sequestered and stored by the aposematic grasshopper, *Zonocerus variegatus. J. Zool.* **182**, 85–87.

Birch, M. C. (Ed.) 1974. *Pheromones.* North-Holland, Amsterdam.

Bishop, H. J. 1940. The bush locust (*Phymateus leprosus*) in the Eastern Cape Province. *Bull. Dept. Agric. For. S. Afr.* **208** 2–10.

Blight, M. M. 1969. Volatile nitrogenous bases emanating from laboratory-reared colonies of the desert locust, *Schistocerca gregaria. J. Insect Physiol.* **15**, 259–272.

Blomquist, G. J. and L. L. Jackson. 1973. Incorporation of labelled dietary *n*-alkanes into cuticular lipids of the grasshopper *Melanoplus sanguinipes. J. Insect Physiol.* **19**, 1639–1647.

Blomquist, G. J. and D. C. McCain. 1975. Incorporation of oxygen-18 into secondary alcohols of grasshopper *Melanoplus sanguinipes. Lipids* **10**, 303–306.

Blomquist, G. J., C. L. Soliday, B. A. Byers, J. W. Brakke, and L. L. Jackson. 1972. Cuticular lipids of insects. V. Cuticular wax esters of secondary alcohols from the grasshoppers *Melanoplus packardii* and *Melanoplus sanguinipes. Lipids* **7**, 356–362.

Blomquist, G. J., D. R. Nelson, and M. de Renobales. 1987. Chemistry, biochemistry, and physiology of insect cuticular lipids. *Arch. Insect Biochem. Physiol.* **6**, 227–265.

Blum, M. S. 1981. *Chemical Defenses of Arthropods.* Academic Press. New York.

Blum, M. S., R. F. Severson, R. F. Arrendale, D. W. Whitman, P. Escoubas, O. Adeyeye, and C. G. Jones. 1990. A generalist herbivore in a specialist mode: Metabolic, sequestrative, and defensive consequences. *J. Chem. Ecol.* **16**, 000–000.

Blum, M. S., D. W. Whitman, R. F. Severson, and R. F. Arrendale. 1987. Herbivores and toxic plants: Evolution of a menu of options for processing allelochemicals. *Insect Sci. Appl.* **8**, 459–463.

Boppré, M. 1984. Redefining "pharmacophagy." *J. Chem. Ecol.* **10**, 1151–1154.

Boppré, M., U. Seibt, and W. Wickler. 1984. Pharmacophagy in grasshoppers? *Zonocerus* attracted to and ingesting pure pyrrolizidine alkaloids. *Entomol. Exp. Appl.* **35**, 115–117.

Bryant, H. C. 1912. Birds in relation to a grasshopper outbreak in California. *Univ. Calif., Berkeley, Publ. Zool.* **11**, 1–20.

Butlin, R. K., C. W. Woodhatch, and G. M. Hewitt. 1987. Male spermatophore investment increases female fecundity in a grasshopper. *Evolution* (*Lawrence Kans.*), **41**, 221–225.

Capinera, J. L. and T. S. Sechrist. 1982. Grasshoppers (Acrididae) of Colorado. *Colo. Agric. Exp. Stn., Bull.* 584S.

Carpenter, G. D. H. 1938. Audible emission of defensive froth by insects. *Proc. Zool. Soc. London, Ser.* (A) **108**, 243–252.

Carpenter, G. D. H. 1946. The relative edibility and behaviour of some aposematic grasshoppers. *Entomol. Mon. Mag.* **82**, 5–10.

Cassier, P. 1977. La différenciation imaginale du tégument chez le criquet pélerin *Schistocerca gregaria* Forsk. IV. Les étapes de la morphogenèse des unités glandulaires. *Arch. Anat. Microsc. Morph. Exp.* **66**, 145–161.

Cassier, P. and C. Delorme. 1974. La sécrétion de la phéromone sexuelle mâle chez le criquet pélerin, *Schistocerca gregaria Forsk. Ann. Zool. Ecol. Anim.* **6**, 174–176.

Cassier, P. and M. Papillon. 1983. Le dimorphisme sexuel des proteines cuticulaires du Criquet migrateur, *Locusta migratoria migratorioides* (Reiche & Farmaire). *Can. J. Zool.* **61**, 2976–2986.

Chapman, R. F. 1962. The ecology and distribution of grasshoppers in Ghana. *Proc. Zool. Soc. London*, **139**, 1–66.

Chapman, R. F. and W. W. Page, 1979. Factors affecting the mortality of the Grasshopper, *Zonocerus variegatus*, in Southern Nigeria. *J. Anim. Ecol.* **48**, 271–288.

Chapman, R. F., W. W. Page, and A. R. McCaffery. 1986. Bionomics of the variegated grasshopper (*Zonocerus variegatus*) in west and central Africa. *Annu. Rev. Entomol.* **31**, 479–505.

Charnley, A. K., J. Hunt, and R. J. Dillon. 1985. The germ-free culture of desert locusts, *Schistocerca gregaria. J. Insect Physiol.* **31**, 477–485.

Coleman, L. C. 1911. The Jola or deccan grasshopper (*Colemania sphenarioides*, Bol.). *Dep. Agric. Mysore Entomol. Bull.* **2**, 1–43.

Colgan, D. J. 1987. Developmental changes of isoenzymes catalysing glycolytic and associated reactions in *Locusta migratoria* in relation to the rearing density of hatchlings. *Insect Biochem.* **17**, 303–308.

Curasson, G. 1934. Sur la toxicité de la sécrétion buccale des Sauterelles. *Bull. Acad. Vét. Fr.* **17**, 377–382.

de Lépiney, J. 1930. Sur le comportement des larves de *Schistocerca gregaria* Forsk. Concentration et dissémination des individus, voyages des bandes larvaires, nutrition. *C. R. Séances Soc. Biol. Ses Fil.* **104**, 256–267.

De Lotto, G. 1950. Sulla presenza di una ghiandola ripugnatoria in due Ortotteri de genere *Phymateus. Boll. Soc. Ital. Med., Sez. Eritrea* **10**, 195–201.

Descamps, M., and D. Wintrebert. 1966. Pyrgomorphidae et Acrididae de Madagascar. Observations biologiques et diagnoses (Orth., Acridoidea). *EOS Rev. Esp. Entomol.* **42**, 41–263.

Ebner, R. 1914. Ein eigentümliches Verteidigungsmittel bei *Poecilocerus hieroglyphicus* (Klug) (Orthoptera). *Anz. Akad. Wiss. Wien, Math.-Naturwiss, Kl.* **51**, 395–397.

Ehrlich, P. R. and P. H. Raven. 1964. Butterflies and plants: a study in coevolution. *Evolution (Lawrence, Kans.)* **18**, 586–608.

Eisner, T. 1970. Chemical defense against predation in Arthropods. In E. Sondheimer and J. B. Simeone, Eds., *Chemical Ecology*. Academic Press, New York, pp. 157–217.

Eisner, T., J. Shepherd, and G. M. Happ. 1966. Tanning of grasshopper eggs by an exocrine secretion. *Science* **152**, 95–97.

Eisner, T., L. B. Hendry, D. B. Peakall, and J. Meinwald. 1971. 2,5-Dichlorophenol (from ingested herbicide?) in defensive secretion of grasshopper. *Science* **172**, 277–278.

Ellis, P. E. 1959. Learning and social aggregation in locust hoppers. *Anim. Behav.* **7**, 91–106.

Ellis, P. E. 1962. The behaviour of locusts in relation to phases and species. *Colloq. Int. C.N.R.S.*, **114**, 123–143.

Ellis, P. E., and A. Pearce. 1962. Innate and learned behaviour patterns that lead to group formation in locust hoppers. *Anim. Behav.* **10**, 305–318.

Euw, J. V., L. Fishelson, J. A. Parsons, T. Reichstein, and M. Rothschild. 1967. Cardenolides (Heart poisons) in a grasshopper feeding on milkweeds. *Nature*, (*London*) **214**, 35–39.

Ewer, D. W. 1957. Notes on acridid anatomy, IV. The anterior abdominal musculature of certain acridids. *J. Entomol. Soc. Southern Africa* **20**, 260–279.

Faber, A. 1953. Laut-und gebärdensprache bei insekten: Orthoptera (Geradflüger). I. *Mitt. Staatl. Mus. Naturk. Stuttgart* **287**, 1–198.

Fishelson, L. 1960. The biology and behaviour of *Poekilocerus bufonius* Klug with a special reference to the repellent gland. *Eos* **36**, 41–62.

Filip'ev, I. N. 1926. Injurious insects and other animals in U.S.S.R. in the years 1921–1924. No. 2. Acridoidea (in Russian with English summary). *Tr. Otd. prikl. Entomol.* **13**, 57–176.

Freeman, M. A. 1967. Proteolytic enzymes of the crop fluid from *Locusta migratoria* L. *Comp. Biochem. Physiol.* **20**, 1013–1015.

Freeman, M. A. 1968. Pharmacological properties of the regurgitated crop fluid of the African migratory locust, *Locusta migratoria* L. *Comp. Biochem. Physiol.* **26**, 1041–1049.

Friedel, T., and C. Gillott. 1976. Male accessory gland substance of *Melanoplus sanguinipes*: an oviposition stimulant under the control of the corpus allatum. *J. Insect Physiol.* **22**, 489–495.

Friedel, T., and C. Gillott. 1977. Contribution of male-produced proteins to vitellogenesis in *Melanoplus sanguinipes*. *J. Insect Physiol.* **23**, 145–151.

Fuzeau-Braesch, S., E. Génin, R. Jullien, E. Knowles, and C. Papin. 1988. Composition and role of volatile substances in atmosphere surrounding two gregarious locusts, *Locusta migratoria* and *Schistocerca gregaria*. *J. Chem. Ecol.* **14**, 1023–1032.

Gangwere, S. K. and R. A. Ronderos. 1975. A synopsis of food selection in Argentine Acridoidea. *Acrida* **4**, 173–194.

Génin, E., R. Jullien, F. Perez, and S. Fuzeau-Braesch. 1986. Cuticular hydrocarbons of gregarious and solitary locusts *Locusta migratoria cinerascens*. *J. Chem. Ecol.* **12**, 1213–1238.

Génin, E., R. Jullien, and S. Fuzeau-Braesch. 1987. New natural aliphatic ethers in cuticular waxes of gregarious and solitary locusts *Locusta migratoria cinerascens* (II). *J. Chem. Ecol.* **13**, 265–282.

Gillett, S. D. 1968. Airborne factor affecting grouping behaviour in locusts. *Nature, London* **218**, 782–783.

Gillett, S. D. 1975a. The action of the gregarisation pheromone on five non-behavioural characters of phase polymorphism of the desert locust *Schistocerca gregaria* (Forskål). *Acrida* **4**, 137–149.

Gillett, S. D. 1975b. Changes in the social behaviour of the desert locust, *Schistocerca gregaria*, in response to the gregarizing pheromone. *Anim. Behav.* **23**, 494–503.

Gillett, S. D. 1983. Primer pheromones and polymorphism in the desert locust. *Anim. Behav.* **31**, 221–230.

Gillett, S. D., J. M. Packham, and S. J. Papworth. 1976. Possible pheromonal effects on aggregation and dispersion in the desert locust, *Schistocerca gregaria* (Forsk.). *Acrida* **5**, 287–297.

Grassé, P. P. 1937. L'hémaphrorrhée, rejet-réflexe de sang et d'air par les acridiens phymatéides. *C. R. Hebd. Séances. Acad. Sci.* **204**, 65–67.

Greathead, D. J. 1963. A review of the insect enemies of Acridoidea (Orthoptera). *Trans. R. Entomol. Soc. London* **114**, 437–517.

Greathead, D. J. 1966. A brief survey of the effects of biotic factors on populations of the desert locust. *J. Appl. Ecol.* **3**, 239–250.

Gregory, G. E. 1965. On the initiation of spermatophore formation in the African migratory locust, *Locusta migratoria migratorioides* Reiche and Fairmaire. *J. Exp. Biol.* **42**, 423–435.

Hamilton, A. G. 1936. The relation of humidity and temperature to the development of three species of African locusts — *Locusta migratoria migratorioides* (R. & F.), *Schistocera gregaria* (Forsk), *Nomadacris septemfasciata* (Serv.) *Trans. R. Entomol. Soc. London* **85**, 1–60.

Haskell, P. T. 1960. Stridulation and associated behaviour in certain Orthoptera. 3. The influence of the gonads. *Anim. Behav.* **8**, 76–81.

Haskell, P. T., M. W. J. Paskin, and J. E. Moorhouse. 1962. Laboratory observations on factors affecting the movements of hoppers of the desert locust. *J. Insect Physiol.* **8**, 53–78.

Hawkes, F., J. Rzepka, and G. Gontrand. 1987. The scent glands of male South American locust *Schistocerca cancellata*, an electron microscope study. *Tissue and Cell* **19**, 687–703.

Hebard, M. 1925. A revision of the genus *Taeniopoda* (Orthoptera, Acrididae, Cyrtacanthacrinae) *Trans. Am. Entomol. Soc.* **50**, 253–274.

Highnam, K. C. 1962. Neurosecretory control of ovarian development in *Schistocera gregaria*. *Q. J. Microsc. Sci.* **103**, 57–72.

Highnam, K. C., and O. Lusis. 1962. The influence of mature males on the neurosecretory control of ovarian development in the desert locust. *Q. J. Microsc. Sci.* **103**, 73–83.

Hingston, R. W. G. 1927. The liquid-squirting habit of oriental grasshoppers. *Trans. Entomol. Soc. London* **75**, 65–68.

Hollande, A. C. 1926. La signification de l'autohémorrhée des insectes. *Arch. Anat. Microsc.*, **22**, 374–412.

Hubbell, T. H. and I. J. Cantrall. 1938. A new species of *Appalachia* from Michigan (Orthoptera, Acrididae, Cyrtacanthacrinae). *Occas. Pap. Mus. Zool. Univ. Mich.* **389**, 1–22.

Husain, M. and C. B. Mathur. 1946. Studies on *Schistocerca gregaria* Forsk. XIII. Sexual life. *Indian J. Entomol.* **7**, 89–101.

Isely, F. B. 1938. Survival value of acridian protective coloration. *Ecology* **19**, 370–389.

Jackson, L. L. 1981. Cuticular lipids of insects — IX. Surface lipids of the grasshoppers *Melanoplus bivittatus, Melanoplus femurrubrum* and *Melanoplus dawsoni*. *Comp. Biochem. Physiol. B* **70B**, 441–445.

Jackson, L. L. 1982. Cuticular lipids of insects. X. Normal and branched alkanes from the surface of the grasshopper *Schistocerca americana*. *Comp. Biochem. Physiol. B* **71B**, 739–742.

Jacobs, W. 1953. Verhaltensbiologische studien an feldheuschrecken. *Z. Tierpsychol.* (*Beih.*) **1**, 1–228.

Jannone, G. 1938. Osservazioni sulla presenza, struttura e funzione d'una vescicola ghiandolare confinata nel protorace della specie mediterranee del gen. *Acrotylus* Fieb., con particolare riguardo all'*A. insubricus* (Scop.) (Orthoptera, Acridioidea). *Boll. Lab. Zool. Portici 31*, 41–62.

Jannone, G. 1939. Sulla diffusione della vescicola ghiandolare protoracica negli Ortotteri della subfam. Oedipodinae. *Boll. Zool. Agrar. Bachi.*, **10**, 105–107.

Joern, A. 1986. Experimental study of avian predation on coexisting grasshopper populations (Orthoptera: Acrididae) in a sandhills grassland. *Oikos* **46**, 243–249.

Jones, C. G., T. A. Hess, D. W. Whitman, P. J. Silk, and M. S. Blum. 1986. Idiosyncratic variation in chemical defenses among individual generalist grasshoppers. *J. Chem. Ecol.* **12**, 749–761.

Jones, C. G., T. A. Hess, D. W. Whitman, P. J. Silk, and M. S. Blum. 1987. Effects of diet breadth on autogenous chemical defense of a generalist grasshopper. *J. Chem. Ecol.* **13**, 283–297.

Jones, C. G., D. W. Whitman, P. J. Silk, and M. S. Blum. 1988. Diet Breadth and Insect Chemical Defenses: A Generalist Grasshopper and General Hypotheses. In K. C. Spencer, Ed., *Chemical Mediation of Coevolution*. Academic Press, San Diego, California, pp. 477–512.

Jones, C. G., D. W. Whitman, S. J. Compton, P. J. Silk, and M. S. Blum. 1989. Reduction in diet breadth results in sequestration of plant chemicals and increases efficacy of chemical defense in a generalist grasshopper. *J. Chem. Ecol.* **15**, 1811–1822.

Karlson, P., and A. Butenandt. 1959. Pheromones (ectohormones) in insects. *Annu. Rev. Entomol.* **4**, 39–58.

Karlson, P. and M. Lüscher. 1959. "Pheromones," a new term for a class of biologically active substances. *Nature (London)*, **183**, 55–56.

Katiyar, K. N. 1955. The life-history and ecology of northern spotted grasshopper. *Aularches punctatus* Drury (Orthoptera: Acrididae). *Agra Univ. J. Res. Sci* **4**, 397–414.

Kerkut, G. A. and L. I. Gilbert (Eds.) 1985. *Comprehensive Insect Physiology, Biochemistry, and Physiology*, Vol. 9. Pergamon, Oxford.

Kevan, D. K. Mc.E. 1949. Notes on east African bush locusts with special reference to *Phymateus aegrotus* (Gerstaecker, 1869) (Orth., Acrid., Pyrgomorphinae). *Bull. Entomol. Res.* **40**, 359–369.

Key, K. H. L. 1985. Monograph of the Monistriini and Petasidini (Orthoptera: Pyrgomorphidae). *Aust. J. Zool. Suppl. Ser.* **107**, 1–213.

Kyl, G. 1938. A study of copulation and the formation of spermatophores in *Melanoplus differentialis*. *Proc. Iowa Acad. Sci.* **45**, 299–308.

Lafon-Cazal, M., D. Gallois, J. Lehouelleur, and J. Bockaert. 1987. Stimulatory effects of male accessory-gland extracts on the myogenicity and the adenylate cyclase activity of the oviduct of *Locusta migratoria*. *J. Insect Physiol.* **33**, 909–915.

Lange, A. B. and B. G. Loughton. 1984. An analysis of the secretions of the male accessory reproductive gland of the African migratory locust. *Int. J. Invertebr. Reprod. Dev.* **7**, 73–81.

Lange, A. B. and B. G. Loughton. 1985. An oviposition-stimulating factor in the male accessory reproductive gland of the locust, *Locusta migratoria*. *Gen. Comp. Endocrinol*, **57**, 208–215.

Laub-Drost, I. 1959. Verhaltensbiologie, besonders Ausdrucksäusserungen (einschliesslich Lautäusserungen) einiger Wanderheuschrecken und anderer Orthopteren (Orthopt., Acrid.: Catantopinae und Oedipodinae). *Stutt. Beitr. Naturk.* **30**, 1–27.

Lauga, J. and M. Hatté. 1977a. Propriétés grégarisantes acquises par le sable dans lequel ont pondu à de nombreuses reprises des femelles grégaires de *Locusta migratoria migratorioides*, R. & F. (Orthoptère, Acrididae). *Acrida* **6**, 307–311.

Lauga, J. and M. Hatté. 1977b. Mise en évidence d'une activité de type phéromone dans le milieu de ponte du criquet migrateur *Locusta migratoria* L. *C. R. Hebd. Séances Acad. Sci.* Sér. D **285**, 57–59.

Lauga, J. and M. Hatté. 1978. L'activité grégarisante du sable de ponte chez *Locusta migratoria* L.: Action sur le comportement et la reproduction des individus. *Annl. Zool. Biol. Anim. Sci. Nat., (Sér. 12E)*, **20**, 37–52.

Leahy, M. G. 1973a. Oviposition of virgin *Schistocerca gregaria* (Forskål) (Orthoptera: Acrididae) after implant of the male accessory gland complex. *J. Entomol., Ser. A: Gen. Entomol.* **48**, 69–78.

Leahy, M. G. 1973b. Oviposition of *Schistocerca gregaria* (Forskål) (Orthoptera: Acrididae) mated with males unable to transfer spermatophores. *J. Entomol., Ser. A: Gen. Entomol.* **48**, 79–84.

Lockey, K. H. 1976. Cuticular hydrocarbons of *Locusta*, *Schistocerca* and *Periplanta*, and their role in waterproofing. *Insect Biochem.* **6**, 457–472.

Loher, W. 1960. The chemical acceleration of the maturation process and its hormonal control in the male of the desert locust. *Proc. R. Soc. London, Ser. B* **153**, 380–397.

Loher, W. 1984. Behavioral and physiological changes in cricket-females after mating. *Adv. Invertebr. Reprod. 3*, 189–201.

Loher, W., I. Ganjian, I. Kubo, D. Stanley-Samuelson, and S. S. Tobe. 1981. Prostaglandins: Their role in egg-laying of the cricket *Teleogryllus commodus*. *Proc. Natl. Acad. Sci. U.S.A.* **78**, 7835–7838.

Lymbery, A. and W. Bailey. 1980. Regurgitation as a possible anti-predator defensive mechanism in the grasshopper *Goniaea* sp. (Acrididae, Orthoptera). *J. Aust. Entomol. Soc.* **19**, 129–130.

Marshall, G. A. K. 1902. Five years' observations and experiments (1896–1901) on the bionomics of South African insects, chiefly directed to the investigation of mimicry and warning colors. *Trans. Entomol. Soc. London* **50**, 287–584.

Martin, A. C., H. S. Zim, and A. L. Nelson. 1951. *American Wildlife and Plants: A Guide to Wildlife Food Habits*. General Publishing Co., Toronto.

Mathieu, J. M. 1970. Biological studies on *Chromacris colorata* (Orthoptera: Romaleinae). *J. Kans. Entomol. Soc.* **43**, 262–269.

McCaffery, A. R. and W. W. Page. 1982. Oviposition behaviour of the grasshopper *Zonocerus variegatus*. *Ecol. Entomol.* **7**, 85–90.

McCann, C. 1953. Aposematic insects and their food plants. *J. Bombay Nat. Hist. Soc.* **51**, 752–754.

Meinwald, J., K. Erickson, M. Hartshorn, Y. C. Meinwald, and T. Eisner. 1968. Defensive mechanisms of arthropods. XXIII. An allenic sesquiterpenoid from the grasshopper *Romalea microptera*. *Tetrahedron Lett.* **25**, 2959–2962.

Mika, G. 1959. Über das Paarungsverhalten der Wanderheuschrecke *Locusta migratoria* R. und F. und deren Abhängigkeit vom Zustand der inneren Geschlechtsorgane. *Zool. Beitr.* **4**, 153–203.

Miller, N. C. E. 1936. A collection of Acrididae made in Southern Rhodesia (Orthoptera). *Proc. R. Entomol. Soc. Ser. B* **5**, 153–161.

Miyakado, M., N. Ohno, H. Yoshioka, T. J. Mabry, and T. Whiffin. 1974. Gymnospermin: A new Ladan Triol from *Gymnosperma glutinosa*. *Phytochemistry* **13**, 189–190.

Nelson, D. R., N. J. Nunn, and L. Jackson. 1984. Re-analysis of the methylalkanes of the grasshoppers, *Melanoplus differentialis*, *M. packardi*, and *M. sanguinipes*. *Insect Biochem.* **14**, 677–683.

Nikolskii, V. V. 1925. The Asiatic locust *Locusta migratoria* L. (in Russian). *Tr. Otd. Prikl. Entomol.* **12**, 1–332.

Nolte, D. J. 1963. A pheromone for melanization of locusts. *Nature (London)* **200**, 660–661.

Nolte, D. J. 1968. The chiasma-inducing pheromone of locusts. *Chromosoma* **23**, 346–358.

Nolte, D. J. 1973. The relationship of locust species to locustol. *Acrida* **2**, 53–60.

Nolte, D. J. 1974. The gregarization of locusts. *Biol. Rev. Cambridge Plilos. Soc.* **49**, 1–14.

Nolte, D. J. 1976. Locustol and its analogues. *J. Insect Physiol.* **22**, 833–838.

Nolte, D. J. 1977. The action of locustol. *J. Insect Physiol.* **23**, 899–903.

Nolte, D. J., I. R. May, and B. M. Thomas. 1970. The gregarisation pheromone of locusts. *Chromosoma* **29**, 462–473.

Nolte, D. J., S. H. Eggers, and I. R. May. 1973. A locust pheromone: Locustol. *J. Insect Physiol.* **19**, 1547–1554.

Norris, M. J. 1950. Reproduction in the African migratory locust (*Locusta migratoria migratorioides* R. & F.) in relation to density and phase. *Anti-Locust Bull.* **6**, 1–48.

Norris, M. J. 1952. Reproduction in the desert locust (*Schistocerca gregaria* Forsk.) in relation to density and phase. *Anti-Locust Bull.* **13**, 1–49.

Norris, M. J. 1954. Sexual maturation in the desert locust (*Schistocerca gregaria* Forskål) with special reference to the effects of grouping. *Anti-Locust Bull.* **18**, 1–44.

Norris, M. J. 1962. Group effects on the activity and behaviour of adult males of the desert locust (*Schistocerca gregaria* Forsk.) in relation to sexual maturation. *Anim. Behav.* **10**, 275–291.

Norris, M. J. 1963. Laboratory experiments on gregarious behavior in ovipositing females of the desert locust, (*Schistocerca gregaria* [Forsk.]). *Entomol. Exp. Appl.* **6**, 279–303.

Norris, M. J. 1964. Accelerating and inhibiting effects of crowding on sexual maturation in two species of locusts. *Nature (London)* **203**, 784–785.

Norris, M. J. 1970. Aggregation response in ovipositing females of the desert locust, with special reference to the chemical factor. *J. Insect Physiol.* **16**, 1493–1515.

Norris M. J. and M. P. Pener. 1965. An inhibitory effect of allatectomized males and females on the sexual maturation of young male adults of *Schistocerca gregaria* (Forsk.) (Orthoptera: Acrididae). *Nature (London)* **208**, 1122.

Ogloblin, A. 1947. Las glandulas odoriferas de la langosta *Schistocerca cancellata* (Serville) (Orthopt., Acridoidea). *Arthropoda* **1**, 54–77.

Okelo, O. 1979. Influence of male presence on clutch size in *Schistocerca vaga* Scudder (Orthoptera: Acrididae). *Int. J. Invertebr. Reprod.* **1**, 317–321.

Otte, D. 1970. A comparative study of communicative behavior in grasshoppers. *Misc. Publ. Mus. Zool. Univ. Mich.* **141**, 1–168.

Pawlowsky, E. 1916. On the anatomy of *Phymateus hildebrandti* Bol. (Orthoptera, Phymateidae) in connection with the pecularities of its dermal secretion. In V. Dogiel and I. Sokolow, Eds., *Scientific Results of the Zoological Expedition to British East Africa and Uganda* (in Russian with English summary). Petrograd, pp. 23–28.

Perdeck, A. C. 1958. The isolating value of specific song patterns in two sibling species of grasshoppers (*Chorthippus brunneus* Thunb. and *C. biguttulus* L.). *Behaviour* **12**, 1–75.

Pickford, R. and C. Gillott. 1971. Insemination in the migratory grasshopper, *Malanoplus sanguinipes* (Fabr.). *Can. J. Zool.* **49**, 1583–1588.

Pickford, R. and C. Gillott. 1972. Courtship behavior of the migratory grasshopper *Melanoplus sanguinipes* (Orthoptera: Acrididae). *Can. Entomol.* **104**, 715–722.

Pickford, R. and C. Gillott. 1976. Effect of varied copulatory periods of *Melanoplus sanguinipes* (Orthoptera: Acrididae) females on egg hatchability and hatchling sex ratios. *Can. Entomol.* **108**, 331–335.

Pickford, R. and D. E. Padgham. 1973. Spermatophore formation and sperm transfer in the desert locust, *Schistocerca gregaria* (Orthoptera: Acrididae) *Can. Entomol.* **105**, 613–618.

Pickford, R., A. B. Ewen, and C. Gillott. 1969. Male accessory gland substance: An egg-laying stimulant in *Melanoplus sanguinipes* (F.) (Orthoptera: Acrididae). *Can. J. Zool.* **47**, 1199–1203.

Qureshi, S. A. and I. Ahmad. 1970. Studies on the functional anatomy and histology of the repellent gland of *Poekilocerus pictus* (F.) (Orthoptera: Pyrgomorphidae). *Proc. R. Entomol. Soc. London, Ser. A* **45**, 149–155.

Rabaud, E. 1916. Le dégorgement réflexe des Acridiens. *Bull. Soc. Zool. Fr.* **40**, 223–238.

Renner, M. 1952. Analyse der Kopulationsbereitschaft des Weibchens der Feldhuschrecke *Euthystria brachyptera* Ocsk. in ihrer Abhangigkeit vom Zustand des Geschlechtsapparates. *Z. Tierpsychol.* **9**, 122–154

Reichstein, T. 1967. Cardenolide (herzwirksame glykoside) als abwehrstoffe bei insekten. *Naturwiss. Rundsch.* **20**, 499–511.

Richards, M. J. and M. A. El Mangoury. 1968. Further experiments on the effects of social factors on the rate of sexual maturation in the desert locust. *Nature (London)* **219**, 865–866.

Riede, K. 1987. A comparative study of mating behaviour in some neotropical grasshoppers (Acridoidea). *Ethology* **76**, 265–296.

Riegert, P. W. 1965. Effects of grouping, pairing, and mating on the bionomics of *Melanoplus bilituratus* (Walker) (Orthoptera: Acrididae). *Can. Entomol.* **97**, 1046–1051.

Roberts, H. R. 1947. Revision of the Mexican Melanoplini (Orthoptera: Acrididae: Cyrtacanthacridinae). *Proc. Acad. Nat. Sci. Philadelphia* **49**, 201–230.

Roberts, H. R. and C. S. Carbonell. 1982. A revision of the grasshopper genera *Chromacris* and *Xestotrachelus* (Orthoptera, Romaleidae, Romaleinae). *Proc. Calif. Acad. Sci.*, **43**, 43–58.

Roffey, J. 1964. Note on gregarious behaviour exhibited by *Phymateus aegrotus* Gerstaecker (Orthoptera: Acrididae). *Proc. R. Entomol. Soc. London, Ser. A* **39**, 47–49.

Rothschild, M. 1966. Experiments with captive predators and the poisonous grasshopper *Poekilocerus bufonius* (Klug). *Proc. R. Entomol. Soc. London, Ser. C* **31**, 32, 40–41.

Rothschild, M. 1972. Secondary plant substances and warning colouration in insects. *Symp. R. Entomol. Soc. London* **6**, 59–83.

Rothschild, M. and D. N. Kellett. 1972. Reactions of various predators to insects storing heart poisons (cardiac glycosides) in their body tissues. *J. Entomol. Ser. A.: Gen. Entomol.* **46**, 103–110.

Rothschild, M., M. G. Rowan, and J. W. Fairbairn. 1977. Storage of cannabinoids by *Arctia caja* and *Zonocerus elegans* fed on chemically distinct strains of *Cannabis sativa*. *Nature (London)* **266**, 650–651.

Rowell, C. H. F. 1967. Experiments on aggregations of *Phymateus purpurascens* (Orthoptera, Acrididae, Pyrgomorphinae). *J. Zool.* **152**, 179–193.

Schmidt, G. H. and K. S. A. Osman. 1988. Male pheromones and egg production in acrididae. In F. Sehnal, A. Zabza and D. L. Denlinger, Eds., *Endocrinological Frontiers in Physiological Insect Ecology*, Wroctaw Technical Univ. Press, Wroctaw, pp. 701–706.

Schmidt, G. H., W. Krempien, and B. Johannes. 1987. Studies on the secretion of the prothoracic epidermal gland in *Acrotylus patruelis* (Insecta: Saltatoria: Acrididae). *Zool. Anz.* **219**, 357–368.

Shorey, H. H. 1976. *Animal Communication by Pheromones.* Academic Press, New York.

Siddiqi, J. I. and M. A. Khan. 1981. The secretion and perception of a sex pheromone in the grasshopper *Hieroglyphus nigrorepletus* Bolivar (Orthoptera: Acrididae). *Acrida* **10**, 233–242.

Slifer, E. H. 1940. The internal genitalia of female Ommexechinae and Cyrtacanthacridinae (Orthoptera, Acrididae). *J. Morphol.* **67**, 199–239.

Slifer, E. H. and R. L. King. 1936. An internal structure in the Cyrtacanthacrinae (Orthoptera, Acrididae) of possible taxonomic value. *J. N.Y. Entomol. Soc.* **44**, 345–348.

Smit, B. 1964. *Insects in Southern Africa; How to Control Them.* Oxford Univ. Press, Capetown.

Soliday, C. L., G. J. Blomquist, and L. L. Jackson. 1974. Cuticular lipids of insects. VI. Cuticular lipids of the grasshoppers *Melanoplus sanguinipes* and *Melanoplus packardii. J. Lipid Res.* **15**, 399–405.

Steyn, D. G. 1962. Grasshopper (*Phymateus leprosus* Fabr.) poisoning in a Bantu child. *S. Afr. Med. J.* **36**, 822–823.

Swynnerton, C. F. M. 1919. Experiments and observations bearing on the explanation of form and colouring, 1908–1913. *J. Linn. Soc. London Zool.* **33**, 203–385.

Thomas, J. G. 1965. The abdomen of the female Desert Locust (*Schistocerca gregaria* Forskål) with special reference to the sense organs. *Anti-Locust Bull.* **42**, 1–20.

Toye, S. A. 1982. Studies on the biology of the grasshopper pest *Zonocerus variegatus* (L.) (Orthoptera: Pyrgomorphidae) in Nigeria: 1911–1981. *Insect Sci. Appl.* **3**, 1–7.

Uvarov, B. P. 1928. *Locust and Grasshoppers. A Handbook for their Study and Control.* Imperial Bureau of Entomology, London.

Uvarov, B. P. 1966. *Grasshoppers and Locusts*, Vol. 1. Cambridge Univ. Press, London and New York.

Uvarov, B. P. 1977. *Grasshoppers and Locusts*, Vol. 2. Centre for Overseas Pest Research, London.

Varde, V. P. 1934. The protrusible vesicles in Cyrtacanthacrinae. Acridinae (Orthoptera). *J. Univ Bombay Biol. Sci.* **2**, 53–57.

Volkonsky, M. 1942. Observations sur le comportement du Criquet Pèlerin (*Schistocerca gregaria* Forsk.) dans le Sahara algéro-nigérien. *Arch. Inst. Pasteur Algér 20*, 236–248.

Vosseler, J. 1902. Beiträge zur Faunistik und biologie der Orthopteren Algeriens und Tunesiens. *Zool. Jahrb. Abt. Syst. (Oekol.) Geogr, Biol.* **16**, 337–404.

Vosseler, J. 1903. Beiträge zur Faunistik und biologie der Orthopteren Algeriens und Tunesiens. Teil 2. *Zool. Jahrb. Abt., Syst.* (*Oekel.*) *Geogr, Biol.* **17**, 1–98.

Warthen, J. D., Jr. and E. C. Uebel. 1980. Differences in the amounts of two major cuticular esters in males, females and nymphs of *Melanoplus differentialis* (Thomas) (Orthoptera: Acrididae). *Acrida* **9**, 101–106.

Whitman, D. W. 1982a. Grasshopper sexual pheromone: A component of the defensive secretion in *Taeniopoda eques. Physiol. Entomol.* **7**, 111–115.

Whitman, D. W. 1982b. The reproductive biology of *Taeniopoda eques* (Burmeister). Ph.D. Dissertation, University of California, Berkeley.

Whitman, D. W. 1988. Allelochemical Interactions Among Plants, Herbivores, and their predators. In P. Barbosa and D. Letourneau, Eds., *Novel Aspects of Insect–Plant Interactions.* Wiley, New York, pp. 11–64.

Whitman, D. W. and W. Loher. 1984. Morphology of male sex organs and insemination in the grasshopper *Taeniopoda eques* (Burmeister). *J. Morphol.* **179**, 1–12.

Whitman, D. W. and L. Orsak. 1985. Biology of *Taeniopoda eques* (Orthoptera: Acrididae) in southeastern Arizona. *Ann. Entomol. Soc. Am.* **78**, 811–825.

Whitman, D. W., M. S. Blum, and C. G. Jones. 1985. Chemical defense in *Taeniopoda eques* (Orthoptera: Acrididae): Role of the metathoracic secretion. *Ann. Entomol. Soc. Am.* **78**, 451–455.

Wickler, W. and U. Seibt. 1985. Reproductive behaviour in *Zonocerus elegans* (Orthoptera: Pyrgomorphidae) with special reference to nuptial gift guarding. *Z. Tierpsychol.* **69**, 203–223.

13

The Endocrine Basis of Locust Phase Polymorphism

J. F. DALE and

S. S. TOBE

13.1 INTRODUCTION

Uvarov has defined locust phase polymorphism as that property of certain grasshopper species that renders them capable of existing as a series of forms differing from each other not only morphologically but also biologically (Uvarov, 1928). In the field, as well as in the laboratory, changes in the density of locust populations are reflected in diverse changes in the appearance and physiology of individual insects. For example, locusts in a population of high density have a different color from locusts in a scattered population. Moreover, they are generally less fecund and are more tolerant of the approach of conspecifics. Thus a relatively simple change in the external environment in some way causes a complex of profound internal changes and for this reason alone phase polymorphism has always been a phenomenon of great interest. However, no small part of its fascination for entomologists has related to the hope that an understanding of phase polymorphism would offer an opportunity to define a key internal control mechanism acting to translate changes in the external environment into morphological and physiological effects. The findings of P. Joly (e.g., 1949, 1952) suggested that this control mechanism might be endocrine. However, unequivocal evidence that this is the case has not yet been forthcoming.

The "phase theory" was initially proposed by Uvarov (1921) as an explanation for the origin and disappearance of locust plagues. In the last century and early in this, locusts of the gregarious and solitarious phases were thought to be separate species (in the case of *Locusta migratoria migratoria*, *L. migratoria* and *L. danica*, respectively). There was considerable puzzlement regarding the apparent disappearance and periodic de novo reappearance of the gregarious "species." Uvarov (1921) proposed that only a single species was involved but that this species was capable of movement between two phenotypes — solitarious (isolated locusts) and gregarious (locusts in swarms). Solitarious and gregarious are now considered to constitute the two extreme locust phases; locusts with intermediate characteristics are usually included in a third phase — *transiens* (Uvarov, 1966).

Phase polymorphism is a complex phenomenon and the differences between phases extend to many aspects of the insects' biology. The differences in coloration between the solitarious and gregarious phases are striking. For example, solitarious larvae of *Schistocerca gregaria* are usually a uniform green, whereas gregarious larvae are dark in the early stadia and orange and black during the fourth and fifth stadia (Pener, 1983). In both *L. migratoria* and *S. gregaria*, the development of a yellow color over all or part of the integument, which is associated with sexual maturation (Pener, 1967), is much more pronounced in crowded adults (which show many of the features of the gregarious phase) than in isolated adults (which resemble those of the solitarious phase) (Norris, 1952, 1954; Pener, 1976). There are also behavioral differences. Crowded *L. migratoria* larvae show a more marked tendency to aggregate than isolated (Ellis, 1953, 1963) and also have a greater tendency to "march" (Ellis, 1951). Not unexpectedly, it has been shown (Michel, 1970) that crowded *S. gregaria* adults will fly on a roundabout for longer than isolated adults. Some of the most remarkable differences relate to the insects' reproductive physiology. The fecundity of crowded *L. migratoria* and *S. gregaria* females is considerably lower than that of isolated females. This is a reflection of the number of egg pods per female, number of eggs per pod, and the total number of eggs per reproductive lifetime being lower in crowded females and also of the proportion of nonfunctional ovarioles being higher (Norris, 1950; Albrecht et al., 1958; Papillon, 1960; Injeyan and Tobe, 1981a). Fewer hatchlings emerge from pods laid by crowded females but these hatchlings are heavier than those from pods of isolated females (Albrecht et al., 1958; Hunter-Jones, 1958; Papillon, 1960). In addition, hatchlings from pods laid by crowded *L. migratoria* (Albrecht et al., 1958) and *S. gregaria* (Papillon, 1960) have fewer ovarioles than hatchlings of isolated females, the number of ovarioles being determined prior to hatching (Albrecht and Verdier, 1956). There are also phase-dependent differences in the rate of sexual maturation of adult locusts. However, in this case, the nature of the differences is species-specific. Sexual maturation is swifter in crowded than in isolated *S. gregaria* females (Norris, 1952) whereas in the case of *L. migratoria*, the first wave of oocytes develops much more swiftly in isolated than in crowded females (Norris, 1950; Dale and Tobe, 1986).

The fact that the differences between gregarious and solitarious locusts have been interpreted as adaptive constitutes a strong argument for considering them as more than mere biological curiosities. The adaptive significance of some individual phase characters is fairly clear. For example, Albrecht (1962) has shown that the heavier, more resilient hatchlings that develop from eggs laid by gregarious *S. gregaria* females are better adapted to survive in the arid conditions of this species' invasion area than the smaller solitarious hatchlings. Probably the greater capacity for survival of the heavier hatchlings is not directly related to their ability to withstand aridity but rather to their enhanced success in intense competition for limited food. The task of proposing an adaptive rationale for the phenomenon of phase polymorphism as a whole is, however, rather more demanding and only rather generalized discussions of this topic are to be found in the literature. Kennedy (1956, 1962) has argued that two considerations are key — (1) a buildup of numbers always precedes a phase change to the gregarious state and (2) in their solitarious phase, all locust species occupy habitats that are, to a greater or lesser extent, unstable and subject to seasonal aridity. Kennedy maintains that the change to the gregarious phase is an adaptation to the fact that the area supporting a large solitarious population must inevitably become incapable of so doing. He proposes that the more resilient and mobile gregarious phase exploits a different habitat (i.e., the whole "invasion area") from the solitarious phase. The gregarious phase is capable of ranging over the entire invasion area and is adapted to surviving in it until, in a different location or season, solitarious habitat becomes available again. Kennedy thus regards the plague cycle as a "kind of ecological opportunism."

13.2 ENDOCRINES AND PHASE POLYMORPHISM

The endocrine basis of phase polymorphism was recently the subject of an extended study by Pener (1983). This chapter surveys, relatively briefly, research in this area completed prior to 1980; work since that time is examined in greater detail.

Many workers in the field of phase polymorphism have proposed that differences in morphology, color, and so forth are manifestations of profound physiological differences between locusts of the two extreme phases (Uvarov, 1928, 1961, 1966; Faure, 1932; Kennedy, 1961; Staal, 1961; Nolte, 1974; Pener, 1983). Moreover, it has also been specifically suggested that humorally acting factors are responsible for the differential expression of phase characters in gregarious and solitarious locusts. Faure (1932) argued that excess metabolic products, resulting from higher activity levels, accumulate in the hemolymph of gregarious locusts and are deposited as black pigment in the cuticle. He called these excretory products "locustine" and further proposed that "locustine" was deposited in the eggs of gregarious locusts, thus accounting for the dark color of gregarious hatchlings. Nolte has claimed that a pheromone, which has been identified as 5-ethyl 2-methoxyphenol (Nolte et al., 1973), accumulates around roosting swarms and is taken up into the hemolymph of gregarious locusts where it sets in train changes in phase characters (Nolte, 1974). It is important to note, however, that direct demonstrations of the existence of factor(s) in the hemolymph of locusts that are capable of modifying the expression of phase characters are extremely rare. Nickerson (1956) has provided one of the few. He showed that injection of hemolymph from gregarious *S. gregaria* larvae into solitarious larvae induced gregarious coloration, whereas injection of Ringer or hemolymph from solitarious larvae did not. Interestingly, he also showed that epidermis grafted from solitarious to gregarious larvae swiftly lost its green color whereas epidermis from gregarious larvae did not lose its yellow color when grafted into solitarious larvae.

Review of the literature also shows that there has been widespread agreement that endocrine differences between locust phases both exist and play a causative role in the expression of phase differences (Kennedy, 1956, 1961; Staal, 1961; Uvarov, 1961, 1966; Nolte, 1974; de Wilde, 1984; but see Pener, 1983).

Studies of the endocrine differences between phases have concentrated upon the role of juvenile hormone (JH) to an extent that suggests an almost a priori assumption of the importance of this hormone. A central role has frequently been attributed to differences in JH titer between gregarious and solitarious locusts (e.g. P. Joly, 1949; P. Joly and L. Joly, 1953; Kennedy, 1956, 1961; Nijhout and Wheeler, 1982). Kennedy in particular has proposed that the "neotonous" features of solitarious locusts (e.g., reduced wing length and the retention of the prothoracic glands during adult life) are correlated with greater activity of the corpora allata (CA, the glands that secrete JH) in this phase (Kennedy, 1956). Also, P. Joly (1972) suggested that there are differences between phases in the timing and duration of JH secretion during the second half of the last larval stadium. Examination of the role and action of JH in the wider context of insect physiology seems to justify, or at least explain, the manifest emphasis upon the role of JH in research concerning phase polymorphism. First, it has long been widely accepted (e.g., Wigglesworth, 1964) that JH, acting in concert with ecdysteroids, controls the developmental polymorphism demonstrated to a greater or lesser extent by all insects as they progress from egg to adult. Wigglesworth (1961) has suggested that there is little difference between the action of JH in controlling developmental polymorphism and its putative role in phase polymorphism. Second, there is increasing evidence that JH can interact directly with the genome to modulate its expression. For example, a JH analogue appears to initiate the production of mRNA coding for vitellogenin in adult *L. migratoria* females (Chinzei et al.,

1982; Wyatt et al., 1984). Third, as described above, the differences in reproductive physiology between locust phases are profound and JH is known to play a crucial role in insect reproduction. It has been shown in many insect species that the biosynthesis and release of JH by the CA is an essential prerequisite for the maturation of oocytes in adult females (see Tobe and Stay, 1985, for review). In locusts, JH is known to stimulate both vitellogenin production by the fat body (Chen et al., 1976, 1979) and uptake of vitellogenin by the developing oocytes (Ferenz et al., 1981).

13.3 JH TITER AND PHASE POLYMORPHISM

The methodology of most of the early investigations into the role of JH in phase polymorphism relied upon the presumed manipulation of JH titers in locusts by CA implantation or allatectomy, followed by monitoring of the effects upon phase characters. These data provided an indirect assessment of the significance of the normal endocrine balance in solitarious and gregarious locusts, because the experiments of necessity involved disturbing the in vivo situation. When considered in totality, the evidence provided by "implantation/extirpation" experiments regarding the role of JH in phase polymorphism is equivocal. Certainly in the cases of integument color and reproduction, the data are fully compatible with the contention that the solitarious phase is characterized by a relatively elevated JH titer. Implantation of CA into gregarious larvae of *L. migratoria* (L. Joly, 1954; P. Joly, 1949, 1952, 1958, 1972; P. Joly and Joly, 1953; Staal, 1961) and *S. gregaria* (Novak and Ellis, 1967) undoubtedly causes them to assume a green color like that of solitarious larvae, although the effectiveness of the implantation varies according to the point in the larval stadium at which it is performed. The color change seems definitely to be due to the action of the implanted CA because in some recipients (L. Joly, 1954), a localized patch of especially dark green has been observed in the integument above the site of the implanted glands. Regarding reproductive physiology, it has been shown (Cassier, 1964, 1965) that implantation of CA into gregarious adult *L. migratoria* females causes them to lay pods from which hatchlings with solitarious characteristics emerge (lighter color, reduced weight, and increased number of ovariole rudiments). These data are certainly persuasive, although in retrospect, interpretation of the experiments that rely on implantation is complicated by the knowledge that JH biosynthesis in isolated CA (i.e., those that have been disconnected from the brain and corpus cardiacum) drops within 6–12 hours to a very low rate both in *S. gregaria* (Tobe et al., 1977) and *L. migratoria* (Couillaud et al., 1984). However, this very low rate of biosynthesis has been shown to be adequate to support development of the first wave of oocytes in crowded *L. migratoria* adults at the same rate as in controls (Couillaud et al., 1984) and must presumably be adequate to account for the effects observed for implanted glands. More disturbing, perhaps, is the fact that a number of widely accepted phase characters cannot be manipulated by CA implantations and extirpations. For example, the shape of the dorsal surface of the pronotum of adult *L. migratoria* varies in a characteristic fashion between the phases (Uvarov, 1921), being arched in the solitarious and flat or even concave in the gregarious phase. P. Joly (1956) found that the shape of the pronotum in this species could not be changed by CA implantations into larvae. Similarly unaffected (P. Joly, 1962) was the femur length/caput width (F/C) ratio, which was recommended by Dirsh (1953) as an excellent indicator of phase status. Staal also failed to observe a clear-cut effect of CA implantation upon this ratio (Staal, 1961). Moreover, implantation and allatectomy studies in the areas of activity levels and general metabolism (e.g., heartbeat frequency and rate of oxygen consumption) have yielded results that are inconclusive or incompatible with the concept of the solitarious phase being characterized by a higher JH titer (see Pener, 1983, for summary). Thus some phase characters have been shown to respond to manipulations of the JH titer but this was not possible for others, suggesting that differential JH titer cannot be the only factor of importance in the physiological control of locust phase polymorphism.

The development of techniques for the bioassay (de Wilde et al., 1968), radioimmunoassay (e.g., Baehr et al., 1976) and gas chromatographic–mass spectrometric (e.g., Bergot et al., 1981a,b) assay of JH titer facilitated the direct measurement of JH titers in vivo in locusts that had not been disturbed by CA implantations and extirpations. This raised the exciting possibility of clarifying the issue of whether there actually were, under normal circumstances, any differences in JH titer between solitary and gregarious locusts. A few such studies have been completed; disappointingly, the results can at best be described as confusing. L. Joly and Joly (1974) and L. Joly et al. (1977) used bioassays to measure JH titer during the fourth- and fifth-larval stadia and the adult stage in *L. migratoria*. These studies appeared to show that there is a difference in hormone titer between the two phases, the titer being higher in the solitarious phase, particularly at the end of the fifth larval stadium and during the first 5–7 days of the adult stage. However, interpretation of these data is rendered difficult by the fact that the statistical significance of the differences between the phases was not assessed. It has also been shown (Dahm et al., 1976) that the bioassay used in the above studies is rather insensitive to JH III, which is the naturally occurring JH of locusts (Trautmann et al., 1974; Bergot et al., 1981b). Subsequently, Fuzeau-Braesch et al. (1982) published data from radioimmunological studies of JH titers in larval and adult gregarious and solitarious *L. migratoria* that failed to confirm the existence of consistent differences between the phases. However, the reliability of radioimmunoassays for JH has been challenged (Tobe and Feyereisen, 1983; Tobe and Stay, 1985) and physicochemical (i.e., mass spectrometric) methods of JH titer measurement are in general (e.g., Gilbert, 1984) considered to provide more accurate data. Using physicochemical methods, Dale and Tobe (1986) carried out a limited investigation of JH titer in the early part of the first gonotrophic cycle in female solitarious and gregarious *L. migratoria*. Duplicate assays showed (Fig. 13.1) that, although there was no difference in JH titer between the two phases on day 1 of adult life, on day 4 JH titer was approximately twice as high in solitarious as in gregarious females. Of particular interest was the fact that the timing of the appearance of a difference in JH titer correlated rather closely with the development of a difference in basal oocyte length (as described above, the rate of sexual maturation, in particular the rate of development of the first wave of oocytes, is recognized as a most important phase character). Oocyte length was significantly greater in females of the solitarious phase from day 3 to day 8 of adult life (solitarious females oviposited for the first time at day 8). However, it must be stressed that the methodology of titer measurement used in this study (which required the

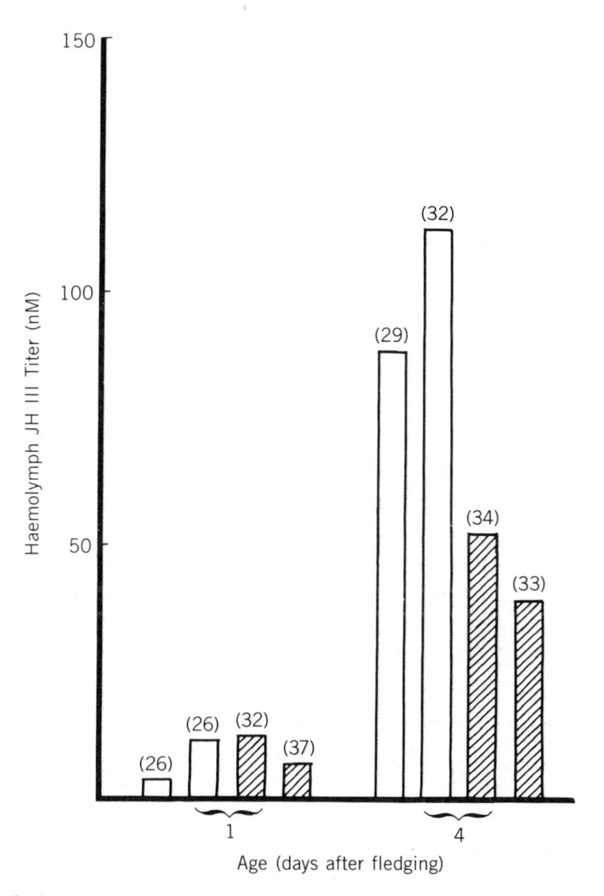

Fig. 13.1. Juvenile hormone III titer of hemolymph samples collected from *Locusta migratoria* during the early part of the first gonotrophic cycle. Nonshaded columns show results for solitarious females; shaded columns show results for gregarious females. Values in parentheses give number of females from which hemolymph was collected for each sample. (Redrawn from Dale and Tobe, 1986.)

pooling of hemolymph samples) precluded the attribution of statistical significance to the difference in JH titer found on day 4. Thus many data remain to be gathered regarding the issue of whether locust phases are characterised by differences in JH titer. Above all, there is a need for studies that provide titer data for individual solitarious and gregarious locusts. Only from such data can the statistical significance of phase-related differences be assessed. As yet, such studies have been performed only for crowded *L. migratoria* (Rembold, 1981).

13.4 JH BIOSYNTHESIS AND PHASE POLYMORPHISM

The rate of JH biosynthesis by the CA has been widely considered (see, for example, Tobe and Stay, 1985) to be one of the most important factors in the regulation of hemolymph JH titer. Thus comparisons of JH biosynthesis in solitarious and gregarious locusts are of considerable interest. Such comparisons have been performed using the radiochemical in vitro assay (RCA) developed by Tobe and Pratt (Pratt and Tobe, 1974; Tobe and Pratt, 1974a; Feyereisen and Tobe, 1981) for both *S. gregaria* and *L. migratoria*. In general, these studies have failed to demonstrate clear-cut differences in rates of JH biosynthesis and release between the phases. Injeyan and Tobe (1981b) investigated fourth- and fifth-instar *S. gregaria* larvae and adults during the first gonotrophic cycle. Although there was limited evidence of qualitative temporal differences in the peaks of JH biosynthesis in the adults (Fig. 13.2) there were no statistically significant quantitative differences between the phases. Dale and Tobe (1986) studied JH biosynthesis in solitarious and gregarious *L. migratoria* females during the first 8 days of adult life but found a statistically significant difference between the two phases on day 8 only. In view of the temporal correlation in this species of the development of differences between the phases in basal oocyte length and in JH titer (see above), it was noteworthy that the difference in the rate of oocyte growth (significant between days 3 and 8 of adult life) was apparently unrelated to differences in the rate of JH biosynthesis (apparent on day 8 of adult life and not before). The relationship in adult female *L. migratoria* of differences in CA activity to the rate of basal oocyte development was further examined (Dale, 1988) using locusts subjected to a novel rearing regime: gregarious females raised in crowded conditions until fledging and then isolated during adult life. Such females mature the first wave of oocytes as quickly as solitarious females (Dale and Tobe, 1988a), although they are of gregarious lineage and are themselves raised in crowded conditions for the majority of their lives. It was clearly of interest to determine whether the activities of their CA were similar to those of gregarious or of solitarious locusts during the early part of adult life. Basal oocyte length and rate of JH release were determined several times after fledging for such females. Although JH release was measured (as opposed to JH biosynthesis in the earlier studies of solitarious and gregarious adults), the results may be compared because in *L. migratoria*, as in *S. gregaria* (Tobe and Pratt, 1974b), JH is

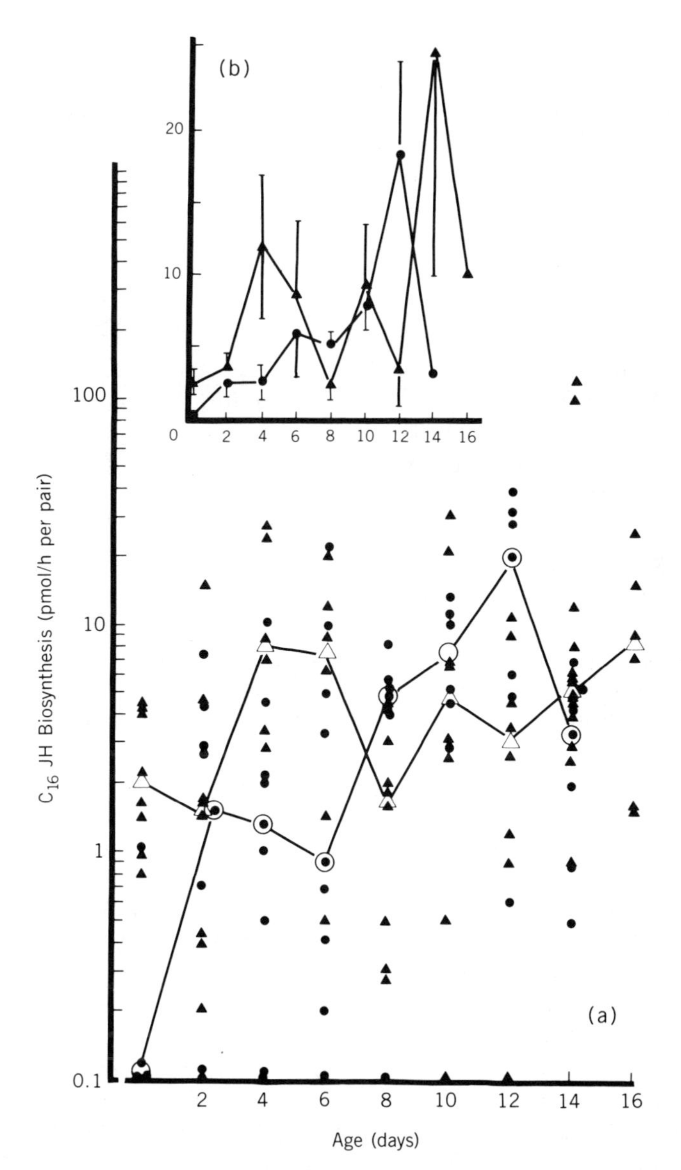

Fig. 13.2 Rate of juvenile hormone III biosynthesis by adult gregarious (●) and solitarious (▲) *Schistocerca gregaria* females. (a) Individual (solid symbols) and median values (open symbols); (b) mean values of the individual determinations shown in a. Bars represent S. E. M. (Redrawn from Injeyan and Tobe, 1981b.)

released from the CA immediately after biosynthesis and is not stored in the glands (Dale and Tobe, 1988b). Thus rates of JH biosynthesis and release approximate closely in this species. Figure 13.3 shows results for JH biosynthesis and release and oocyte length at day 8 of adult life for solitarious females, gregarious females, and gregarious females isolated at fledging. Because rates of JH biosynthesis and release for locusts of the same age and sex are not distributed normally (Injeyan and Tobe, 1981b; Dale, 1988), median values for JH biosynthesis and release are shown in Figure 13.3; the statistical significance of differences between rates for locusts raised under different rearing regimes was determined using the Mann–Whitney test. Means and standard errors are shown for oocyte lengths and the significance of differences were determined using the Student *t*-test. It is apparent that on day 8 of adult life, rates of JH release from CA of gregarious females isolated during adult life, like the rates of JH biosynthesis of CA from gregarious females kept crowded throughout life, were significantly lower than rates of biosynthesis of CA from solitarious locusts (there were no significant differences between the three categories of locusts on days 3 and 5). However, the basal oocytes of gregarious females isolated during adult life are as large by day 8 as those of solitarious females. Thus it seems that the rates of JH biosynthesis and release during the first 8 days of adult life and the length of the basal oocytes at the end of that period can vary independently between stocks of *L. migratoria* females reared in different social environments. The degree of development of the oocytes on day 8 appears not to be, in any obvious way, a function of the relative magnitude of the rates of JH biosynthesis and release in the females of the different stocks. Thus, at least in the context of this phase character, the significance of such differences in JH biosynthesis as can be measured between the phases is uncertain.

There is much evidence for the presence of allatotropin(s) in the brain and corpora cardiaca of *L. migratoria* (Ferenz, 1984; Gadot and Applebaum, 1985; Rembold et al., 1986; Gadot et al., 1987b). It has been suggested (Highnam, 1962) that the timing of the release of neurosecretion from cells in the brain and from the corpora cardiaca may differ between solitarious and gregarious locusts. Once an allatotropin has been characterized, studies of the pattern of its release in the two phases are clearly a possibility, although the fact that clear-cut, general differences in JH biosynthesis have not been detected must render it less likely that differences in allatotropin release could be demonstrated.

The very marked variation in JH biosynthesis (often well over a hundredfold) found between individual locusts of the same age and phase (Girardie et al., 1981; Injeyan and Tobe, 1981b; Couillaud, 1986) has repeatedly underlined the possibility that the temporal pattern of JH biosynthesis in locusts may be complex. There is some evidence that this is so. Couillaud and Girardie (1984) have demonstrated that CA from *L. migratoria* that showed very low rates of JH biosynthesis during short-term in vitro incubations were able to induce, over a period of 2 days, the same degree of green coloration when implanted into gregarious larvae as those showing high rates of JH biosynthesis in the short-term assay. Furthermore, it has been found that data for *L. migratoria* adults are consistent with the occurrence of a pulsatile mechanism of JH biosynthesis (Dale, 1988), as has been suggested previously (e.g. Girardie et al., 1981). Luteinizing hormone (LH) release from the pituitary has been

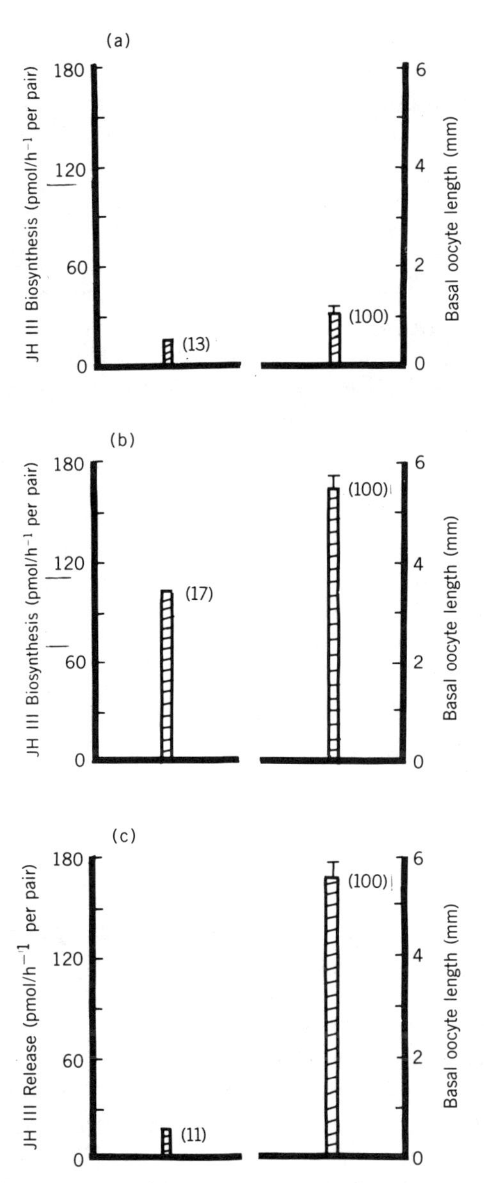

Fig. 13.3 (left) Median rate of juvenile hormone III biosynthesis (a, b) or release (c); (right) mean basal oocyte length for day-8 *Locusta migratoria* raised under the following rearing regimes: (a) gregarious females crowded throughout life; (b) solitarious females; (c) gregarious females raised under crowded conditions until fledging and then isolated during adult life. Numbers in parentheses show number of oocytes (right) or pairs of CA (left) sampled. Bars indicate 1 S. E. M. For JH biosynthesis and release, there are statistically significant (Mann–Whitney *U*-test) differences ($p < .01$) between (a) and (b) and between (b) and (c). For oocyte length, there are statistically significant (Student *t*-test) differences ($p < .01$) between (a) and (b) and between (a) and (c). (Data from Dale, 1988.)

shown to be pulsatile in some vertebrates, and in rats (Gallo, 1981) one of the factors controlling LH titer was found to be modulation of the amplitude of the peaks in the pulsatile pattern — titer increases as a reflection of an increase in the height of the peaks. It can be seen (Table 13.1) that if samples of pairs of CA from solitarious and gregarious *L. migratoria* are compared during the first 8 days after fledging, the maximum value for the rate of JH biosynthesis within the sample is greater (generally 50–100% greater) for the solitarious than for the gregarious females at all data points except for day 1. This finding raises the possibility that, if JH biosynthesis in *L. migratoria* is pulsatile, the amplitude of the peaks in the rate of biosynthesis may be higher in the solitarious than in the gregarious phase. However, such a proposition will remain speculative until a methodology capable of measuring JH biosynthesis over time in vivo becomes available.

Interestingly, it has also been suggested (Gadot and Applebaum, 1986; Gadot et al., 1987a) that the general increase in CA activity during the first cycle of oogenesis in locusts involves two independent processes — short-term activation of the CA by allatotropin(s) and the development of competence of the CA to respond to the latter. Potential differences between locust phases in respect of the timing of the development of responsiveness constitute attractive subjects for research into the endocrine aspects of phase polymorphism that have not as yet been explored.

Thus the present state of knowledge regarding the importance of JH titer and biosynthesis in phase differentiation in locusts is unsatisfactory. Although considerable amounts of data have been accumulated, no totally convincing evidence has been produced that JH biosynthesis or titer varies in any clear and consistent fashion between the phases. Nonetheless, it is not clear whether the failure to date to demonstrate any difference in these parameters is a function of a real absence of a difference or of the fact that the methodologies of measurement employed are not sufficiently sophisticated to accommodate ill-understood complexities in the temporal pattern of JH biosynthesis and, perhaps, titer. However, it is essential to consider even the possibility of as yet undetected differences in JH biosynthesis and titer between locust phases in the context of three important factors.

First, recent evidence has indicated that the relationship between JH biosynthesis and titer may not be as simple as previously thought and may indeed be poorly understood. Couillaud et al. (1984, 1985) have shown that following transection of the nervi corporis allati (NCA I) the rate of JH biosynthesis of adult *L. migratoria* CA falls to low levels, considerably lower than those found in control locusts. Nonetheless, JH titer (measured by gas chromatography–mass spectrometry) was found to be much higher in the former than in the latter. Furthermore, Tobe et al. (1984, 1985) have produced evidence for the existence in *Diploptera punctata* of an extra-hemolymph pool of JH which is thought to be in equilibrium with the JH in the hemolymph and to have a role in preventing rapid changes in the latter. It is not known if such a pool exists in locusts, but if it does, its existence would constitute a complicating factor in the relationship between JH biosynthesis and titer.

TABLE 13.1. Rates (picomoles per hour per pair) of JH III Biosynthesis in Samples of Corpora Allata from Gregarious and Solitarious *Locusta migratoria* Females during the First Eight Days of Adult Life[a]

		Adult Age (days)							
Phase	Rate	0	1	2	3	4	5	6	8
	Minimum	0.17	0.11	0.12	0.61	0.37	0	0.29	3.04
Solitarious	Median	3.78	11.88	0.82	5.24	5.95	2.16	18.46	104.60
		(10)	(15)	(13)	(16)	(12)	(23)	(16)	(17)
	Maximum	42.91	23.85	25.56	38.01	66.85	59.65	229.71	397.35
	Minimum	0.09	0.12	0.19	0.12	0.11	0.18	0.12	0.92
Gregarious	Median	4.06	3.12	6.53	9.38	2.41	6.61	9.50	14.31
		(11)	(13)	(15)	(12)	(13)	(13)	(14)	(13)
	Maximum	6.45	27.92	15.64	29.62	30.32	28.50	75.61	56.51

[a] Figures in parentheses show number of pairs of corpora allata assayed (data from Dale, 1988).

Second, JH titer is clearly influenced by factors other than the rate of JH biosynthesis. The activity of JH esterases may well be important and the presence or absence of JH binding proteins is presumably relevant. In locusts, hemolymph JH esterases (Peter et al., 1983) and binding proteins (Peter et al., 1979; de Bruijn et al., 1986) as well as a peptide ('neuroparsin A') with an antijuvenile hormone effect (Girardie et al., 1987) are known to be present. However, no determination of their relative abundance in the gregarious and solitarious phases has been made. In this context, it is of interest that Zera and Tiebel (1986) have shown that hemolymph JH esterase activity differs considerably in the last larval instar between long- and short-winged morphs of the cricket *Gryllus reubens*.

Third, and perhaps most important, there is good evidence that phase differentiation involves causative factors that are independent of any variations in JH titer that may exist between the phases or that can be induced in locusts. Pener (1976, 1983) has illustrated this very clearly with reference to the color of the integument of adult male *L. migratoria*. He has pointed out that the yellowing of the integument, which is so typical of mature gregarious *L. migratoria*, depends entirely on the presence of the CA (Loher, 1960; Pener, 1965, 1967, 1976; Pener and Lazarovici, 1979). However, multiple implantations of CA into solitarious males fail to cause the integument to become more yellow than that of solitarious controls. On the other hand, solitarious males are capable of developing a yellow color over much of the integument if transferred to crowded cages (Pener, 1976). It is concluded that the intrinsic competence of the integument to respond to JH must be in itself differentially affected by the social environment of the individual.

13.5 PHASE POLYMORPHISM AND ENDOCRINES OTHER THAN JH

The importance of endocrine agents other than JH in the context of phase polymorphism has been studied, but in no cases have the investigations provided conclusive evidence for any of these factors having a causative role. For example, the persistence throughout life of the prothoracic glands in solitarious locusts, in contrast to their disappearance soon after the molt to the adult stage in gregarious locusts (Carlisle and Ellis, 1959), has prompted a number of workers to explore a possible role for differences in ecdysteroid biosynthesis and titer in the control of phase polymorphism. Some investigators, for example, Ellis and Carlisle (1961), claimed to have demonstrated such a role for the prothoracic glands in the color polymorphism associated with phase, but others, for example, Staal (1961), found that implantation or removal of these glands had no appreciable effect on the color of *L. migratoria*. Interest in the role of the prothoracic glands and ecdysteroids has diminished since radioimmunoassay studies showed that there is no quantitative difference in the hemolymph ecdysteroid titers of gregarious and solitarious *L. migratoria* (L. Joly et al., 1977) and *S. gregaria* (Wilson and Morgan, 1978).

There is evidence of differences between locust phases in the abundance of certain neurochemicals in the nervous system and endocrine glands. A number of studies have shown that the amount of octopamine extractable from solitarious *L. migratoria* adults is considerably greater than that extractable from gregarious adults (Fuzeau-Braesch and David, 1978, 1980; Fuzeau-Braesch et al., 1979; Benichou-Redouane and Fuzeau-Braesch, 1982). Conversely, dopamine is more abundant in gregarious than in solitarious *L. migratoria* (Fuzeau-Braesch, 1977a,b). However, at present the functional implications of these data are unclear.

Padgham (1976) has shown that a hormone analogous to bursicon is present in both dark and green hatchlings of *S. gregaria*. Although the hatchlings compared were not of distinct phases, darker hatchlings are present in a higher proportion in the pods of gregarious females and are typical of that phase, and green hatchlings are present in a higher proportion in the pods of solitarious females (Papillon, 1960). The factor was found to promote both sclerotization and melanization in the dark hatchlings, but only sclerotization in the green hatchlings. These findings are of note in that they are consistent with Pener's contention, discussed above, that there are differences between the phases in the competence of comparable tissue to respond to the same hormone.

There are a number of other endocrine factors known to be present in locusts for which no comparative studies between gregarious and solitarious locusts have been attempted. For example, as has been pointed out (Pener, 1985), no attempt has been made to compare the release rate, titer, or mode of action of adipokinetic hormones in solitarious and gregarious locusts. Furthermore, it has been postulated that certain neurosecretions may have a role in the expression of characteristics important in phase polymorphism. For example, Girardie (1967, 1970) has shown, using microcautery, that the "C" cells of the pars intercerebralis of *L. migratoria* secrete a neurohormone that acts directly on the epidermis to promote melanization. Girardie (1970) has further proposed that a gonadotrophic factor is released from the pars intercerebralis of *L. migratoria*, which acts directly on the ovary. Once these neurosecretions have been isolated and identified, comparative studies between locust phases of the dynamics of their release might well yield data of interest.

13.6 DIFFERENCES AT THE MOLECULAR LEVEL BETWEEN PHASES

Two observations have figured repeatedly in studies of the physiology of phase polymorphism.

First, as has been described above, there is clear evidence that there are differences between the phases in the character of the response of comparable tissue to a particular endocrine factor. Presumably this is a reflection of differences at the molecular level of cellular function, and indeed, recent studies have shown that there are differences at a molecular level between locusts of different phases. Génin et al. (1986, 1987) have shown that there are differences between phases in the relative abundance of alkane and ether molecules of different chain lengths in the cuticle of *L. migratoria*. Colgan (1987) has demonstrated, using the same species, that there are differences between locusts raised at high and low densities in the relative abundance of isoenzymes catalyzing certain steps in the glycolytic pathway. It is of interest that treatments of crowded locusts with JH were unsuccessful in eliciting the appearance of the novel isoenzymes typical of locusts raised at low densities.

Second, it is clear that the differential expression of at least some characters important in phase polymorphism is determined at a very early stage in the development of an individual. A number of studies (e.g., Albrecht et al., 1958; Injeyan and Tobe, 1981a) have shown that in both *L. migratoria* and *S. gregaria*, the social environment of the mother (crowded or isolated) can affect organogenesis in the hatchlings developing from her oocytes. Specifically, crowding of mothers from isolated stocks leads to a reduction in the number of ovarioles in their offspring compared with offspring from mothers kept isolated. Conversely, isolation of mothers from crowded stocks leads to an increase in the number of ovarioles in the offspring (Fig. 13.4). Clearly this "maternal effect" must be at least initiated while the offspring are still oocytes in the ovaries of the mother. The nature of the maternal effect is not understood but

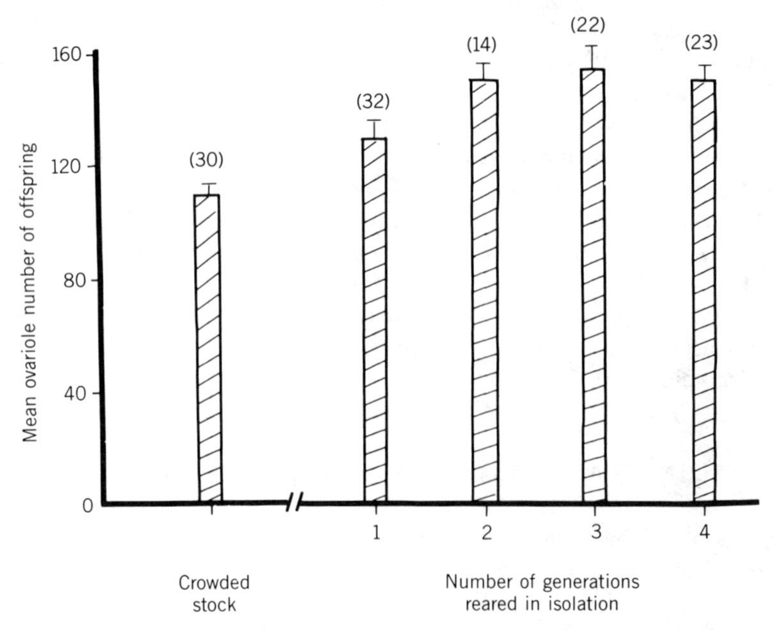

Fig. 13.4 Mean ovariole number in the two ovaries in offspring of *Schistocerca gregaria* females from stocks isolated for one, two, three, and four generations. Ovariole number in offspring of crowded females also shown. Bars indicate 1 S. E. M. Numbers in parentheses show number of locusts sampled. (Data from Injeyan and Tobe, 1981a.)

there is no evidence that it is due to the effect of differential titers of JH within the oocytes (Injeyan et al., 1979). The latter found that application of JH III to eggs of gregarious *S. gregaria* females did not result in any phase change among the larvae that developed from the eggs.

These facts suggest that to some extent phase is a reflection of differences in the pathway of cellular differentiation during embryonic development. On the other hand, other studies demonstrate that change in some phase characters can be achieved in the short term, long after embryonic tissue differentiation has occurred. For example, as shown earlier (Fig. 13.3), isolation of gregarious *L. migratoria* females as late as the day of fledging causes them to mature the first wave of oocytes as fast as true solitarious females and much faster than gregarious females kept crowded during adult life.

In conclusion, no very startling evidence has yet been yielded by the comparative study of the action of endocrine agents in locusts of different phases. However, as has been indicated previously with respect to JH (Pener, 1983), there is no a priori reason for believing that endocrine differences stand at the beginning of the chain of physiological causative factors involved in phase differentiation. It is more likely that they are in the middle. This leaves an understanding of the fundamental nature of phase polymorphism still to be sought.

Data are beginning to appear that hint that there are differences between the phases in the pathways of cellular differentiation in the embryo. However, as is indicated above, these differences cannot constitute the "fundamental cause" of phase polymorphism because other phase differences can be evoked long after hatching by changing the social environment of the locusts. It seems likely, therefore, that a number of different mechanisms must be involved in the physiological control of phase polymorphism. It must unfortunately also be admitted that all of these remain ill-understood and essentially undefined.

REFERENCES

Albrecht, F. O. 1962. Some physiological and ecological aspects of locust phases. *Trans. R. Entomol. Soc. London* **114**, 335–375.

Albrecht, F. O. and M. Verdier. 1956. Le poids et le nombre d'ovarioles chez les larves nouveau-nées de *Locusta migratoria migratorioides* (R. & F.). *C. R. Hebd. Seances Acad. Sci.* **243**, 203–205.

Albrecht, F. O., M. Verdier, and R. E. Blackith. 1958. Détermination de la fertilité par l'effet de groupe chez le criquet migrateur (*Locusta migratoria migratorioides* R. & F.). *Bull. Biol. Fr. Belg.* **92**, 349–427.

Baehr, J. C., P. Pradeles, C. Lebreux, P. Cassier, and F. Dray. 1976. A simple and sensitive radioimmunoassay of insect juvenile hormone using an iodinated tracer. *FEBS Lett.* **69**, 123–128.

Benichou-Redouane, K. and S. Fuzeau-Braesch. 1982. Comparaison des taux d'octopamine chez *Locusta migratoria cinerescens* grégaire, solitaire et solitarisé au gaz carbonique, dans différents organes nerveux. *C. R. Seances Acad. Sci.* ser. III **294**, 385–388.

Bergot, B. J., M. Ratcliff, and D. A. Schooley. 1981a. Method for quantitative determination of the four known juvenile hormones in insect tissue using gas chromatography-mass spectroscopy. *J. Chromatogr.* **204**, 231–244.

Bergot, B. J., D. A. Schooley, and C. A. D. de Kort. 1981b. Identification of JH III as the principal juvenile hormone in *Locusta migratoria. Experientia* **37**, 909–910.

Carlisle, D. B. and P. E. Ellis. 1959. La persistance des glandes ventrales céphaliques chez les criquets solitaires. *C. R. Hebd. Seances Acad. Sci.* **249**, 1059–1060.

Cassier, P. 1964. Effets immediats et transmis des implantations de corps allates sur la fecondité et la descendance des femelles de *Locusta migratoria migratorioides* (R. & F.), phase gregaire (Insecte, Orthopteroide). *C. R. Hebd. Seances Acad. Sci.* **259**, 2706–2708.

Cassier, P. 1965. Determinisme endocrine de quelques caractéristiques phasaires chez *Locusta migratoria migratorioides* (R. & F.) (Insecte orthopteroide, Acrididae). *Insectes Soc.* **12**, 71–79.

Chen, T. T., P. Couble, F. L. Delucca, and G. R. Wyatt. 1976. Juvenile hormone control of vitellogenin synthesis in *Locusta migratoria.* In L. I. Gilbert, Ed., *The Juvenile Hormones.* Plenum, New York, pp. 505–529.

Chen, T. T., P. Couble, R. Abu-Hakima, and G. R. Wyatt. 1979. Juvenile hormone controlled vitellogenin synthesis in *Locusta migratoria* fat body. *Dev. Biol.* **69**, 59–72.

Chinzei, Y., B. N. White, and G. R. Wyatt. 1982. Vitellogenin mRNA in locust fat body: Identification, isolation and quantitative changes induced by juvenile hormone. *Can. J. Biochem.* **60**, 243–251.

Colgan, D. J. 1987. Developmental changes of isoenzymes catalyzing glycolytic and associated reactions in *Locusta migratoria* in relation to the rearing density of hatchlings. *Insect Biochem.* **17**, 303–308.

Couillaud, F. 1986. Influence of sexual organs on corpora allata biosynthetic activity in *Locusta migratoria. Physiol. Entomol.* **11**, 397–403.

Couillaud, F. and A. Girardie. 1984. Ovariectomie et biosynthèse de l'hormone juvenile C16 (JH-3) chez le criquet migrateur. *C. R. Seances Acad. Sci.* ser. III **298**, 157–162.

Couillaud, F., J. Girardie, S. S. Tobe, and A. Girardie. 1984. Activity of disconnected corpora allata in *Locusta migratoria*: Juvenile hormone biosynthesis *in vitro* and physiological effects *in vivo. J. Insect Physiol.* **30**, 551–556.

Couillaud, F., B. Mauchamp, and A. Girardie. 1985. Regulation of juvenile hormone titer in african locust. *Experientia* **41**, 1165–1167.

Dahm, K. H., G. Bhaskaran, M. G. Peter, P. D. Shirk, K. R. Seshan, and H. Roller. 1976. On the identity of the juvenile hormone in insects. In L. I. Gilbert, Ed., *The Juvenile Hormones.* Plenum, New York, pp. 19–47.

Dale, J. F. 1988. Juvenile hormone biosynthesis and phase polymorphism in adult female *Schistocerca gregaria* and *Locusta migratoria.* Ph.D. Thesis, University of Toronto.

Dale, J. F. and S. S. Tobe. 1986. Biosynthesis and titer of juvenile hormone during the first gonotrophic cycle in isolated and crowded *Locusta migratoria* females. *J. Insect Physiol.* **32**, 763–769.

Dale, J. F. and S. S. Tobe. 1988a. Effect of high potassium concentrations on juvenile hormone release *in vitro* from corpora allata of *Locusta migratoria. Physiol. Entomol.* **13**, 21–27.

Dale, J. F. and S. S. Tobe. 1988b. The effect of a calcium ionophore, a calcium channel blocker and calcium-free medium on juvenile hormone release *in vitro* from corpora allata of *Locusta migratoria. J. Insect Physiol.* **34**, 451–456.

de Bruijn, S. M., A. B. Koopmanschap, and C. A. D. de Kort. 1986. High-molecular-weight serum proteins from *Locusta migratoria*: Identification of a protein specifically binding juvenile hormone. III. *Physiol. Entomol.* **11**, 7–16.

de Wilde, J. 1984. Extrinsic control of physiological processes mediated by the insect neuroendocrine system. In A. B. Borkovec and T. J. Kelly, Eds., *Insect Neurochemistry and Neurophysiology*. Plenum, New York, pp. 151–169.

de Wilde, J., G. B. Staal, C. A. D. de Kort, A. de Loof, and G. Baard. 1968. Juvenile hormone titer in the haemolymph as a function of photoperiodic treatment in the adult Colorado beetle (*Leptinotarsa decemlineata* Say). *Proc. K. Ned. Akad. Wet., Ser. C* **71**, 321–326.

Dirsh, V. M. 1953. Morphometric studies on phases of the desert locust (*Schistocerca gregaria* Forskål). *Anti-Locust Bull.* **16**.

Ellis, P. E. 1951. The marching behaviour of hoppers of the african migratory locust (*Locusta migratoria migratorioides* R. & F.) in the laboratory. *Anti-Locust Bull.* **7**.

Ellis, P. E. 1953. Social aggregation and gregarious behaviour in hoppers of *Locusta migratoria migratorioides* (R. & F.). *Behaviour* **5**, 225–260.

Ellis, P. E. 1963. Changes in the social aggregation of locust hoppers with changes in rearing conditions. *Anim. Behav.* **11**, 152–160.

Ellis, P. E. and D. B. Carlisle. 1961. The prothoracic gland and colour change in locusts. *Nature (London)* **190**, 368–369.

Faure, J. C. 1932. The phases of locusts in South Africa. *Bull. Entomol. Res.* **23**, 293–427.

Ferenz, H. J. 1984. Isolation of an allatotropic factor in *Locusta migratoria* and its effect on corpus allatum activity *in vitro*. In J. Hoffman and M. Porchet, Eds., *Biosynthesis, Metabolism and Mode of Action of Invertebrate Hormones*. Springer-Verlag, Berlin and New York, pp. 92–96.

Ferenz, H. J., E. Lubzens, and H. Glass. 1981. Vitellin and vitellogenin incorporation by isolated oocytes of *Locusta migratoria migratorioides* (R. & F.). *J. Insect Physiol.* **27**, 869–875.

Feyereisen, R. and S. S. Tobe. 1981. A rapid partition assay for routine analysis of juvenile hormone release in insect corpora allata. *Anal. Biochem.* **111**, 372–375.

Fuzeau-Braesch, S. 1977a. Comportment et taux de catecholamines: Etude comparative des insectes grégaires, solitaires et traites au gaz carboniques chez *Locusta migratoira* (Insectes, Orthoptères). *C. R. Seances Acad. Sci.* ser. D **284**, 1361–1363.

Fuzeau-Braesch, S. 1977b. Neurophysiological aspect of phase differentiation in the migratory locust: Biogenic amines and membrane permeability. *Bull. Soc. Zool. Fr.* **102**, 327–328.

Fuzeau-Braesch, S. and J. C. David. 1978. Etude du taux d'octopamine chez *Locusta migratoria* (Insecte: Orthoptère): comparaison entre insectes grégaires, solitaires et traités au gaz carbonique. *C. R. Seances Acad. Sci.* ser. D **286**, 697–699.

Fuzeau-Braesch, S. and J. C. David. 1980. Taux d'octopamine, mutation et différentiation phasaire chez *Locusta migratoria*. *C. R. Seances Soc. Biol. Ses Fil.* **174**, 6–9.

Fuzeau-Braesch, S., J. F. Coulon, and J. C. David. 1979. Octopamine levels during the moult cycle and adult development in the migratory locust, *Locusta migratoria*. *Experientia* **35**, 1349–1350.

Fuzeau-Braesch, S., G. Nicolas, J. C. Baehr, and P. Porcheron. 1982. A study of hormonal levels of the locust *Locusta migratoria cinerascens* artificially changed to the solitary state by a chronic CO_2 treatment of 1 minute per day. *Comp. Biochem. Physiol. A* **71A**, 53–58.

Gadot, M. and S. W. Applebaum. 1985. Rapid *in vitro* activation of corpora allata by extracted locust brain allatotropic factor. *Arch. Insect Biochem. Physiol.* **2**, 117–129.

Gadot, M. and S. W. Applebaum. 1986. Farnesoic acid and allatotropin stimulation in relation to locust allatal maturation. *Mol. Cell. Endocrinol.* **48**, 69–76.

Gadot, M., O. Faktor, and S. W. Applebaum. 1987a. Maturation of locust corpora allata during the reproductive cycle: Effect of reserpine on allatotropic activity, juvenile hormone-III biosynthesis and oocyte development. *Arch. Insect Biochem. Physiol.* **4**, 17–28.

Gadot, M., A. Rafaeli, and S. W. Applebaum. 1987b. Partial purification and characterization of locust allatotropin 1. *Arch. Insect Biochem. Physiol.* **4**, 213–223.

Gallo, R. V. 1981. Pulsatile LH release during the ovulatory LH surge on proestrous in the rat. *Biol. Reprod.* **24**, 100–104.

Génin, E., R. Jullien, F. Perez, and S. Fuzeau-Braesch. 1986. Cuticular hydrocarbons of gregarious and solitary locusts *Locusta migratoria cinerascens*. *J. Chem. Ecol.* **12**, 1213–1238.

Génin, E., R. Jullien, and S. Fuzeau-Braesch. 1987. New natural aliphatic ethers in cuticular waxes of gregarious and solitary locusts. *J. Chem. Ecol.* **13**, 265–295.

Gilbert, L. I. 1984. Introduction. In J. Hoffman and M. Porchet, Eds., *Biosynthesis, Metabolism and Mode of Action of Invertebrate Hormones*. Springer-Verlag, Berlin, pp. 350–354.

Girardie, A. 1967. Controle neuro-hormonal de la métamorphose et de la pigmentation chez *Locusta migratoria cinerascens*. *Bull. Biol. Fr. Belg.* **101**, 79–114.

Girardie, A. 1970. Neurosécrétions cérébrales chez les acridiens. *Bull. Soc. Zool. Fr.* **95**, 783–802.

Girardie, J., S. S. Tobe, and A. Girardie. 1981. Biosynthèse de l'hormone juvenile C16 (JH-III) et maturation ovarienne chez le criquet migrateur. *C. R. Seances Acad. Sci.* ser. III **293**, 443–446.

Girardie, J., D. Boureme, F. Couillaud, M. Tamarelle, and A. Girardie. 1987. Anti-juvenile effect of neuroparsin A, a neuropeptide isolated from locust corpora cardiaca. *Insect Biochem.* **17**, 977–983.

Highnam, K. C. 1962. Neurosecretory control of ovarian development in *Schistocerca gregaria*, and its relation to phase differences. *Colloq. Int. C. N. R. S.* **114**, 107–121.

Hunter-Jones, P. 1958. Laboratory studies on the inheritance of phase characters in locusts. *Anti-Locust Bull.* **29**.

Injeyan, H. S. and S. S. Tobe. 1981a. Phase polymorphism in *Schistocerca gregaria:* Reproductive parameters. *J. Insect Physiol.* **27**, 97–102.

Injeyan, H. S. and S. S. Tobe. 1981b. Phase polymorphism in *Schistocerca gregaria:* Assessment of juvenile hormone synthesis in relation to vitellogenesis. *J. Insect Physiol.* **27**, 203–210.

Injeyan, H. S., S. S. Tobe, and E. Rapport. 1979. The effects of exogenous juvenile hormone treatment on embryogenesis in *Schistocerca gregaria. Can. J. Zool.* **57**, 838–845.

Joly, L. 1954. Résultats d'implantations systématiques de corpora allata à des jeunes larves de *Locusta migratoria* L. *C. R. Seances Soc. Biol. Ses Fil.* **148**, 579–585.

Joly, L. and P. Joly. 1974. Comparaison de la phase grégaire et de la phase solitaire de *Locusta migratoria migratorioides* (Orthoptére) du point de vue de la teneur de leur hémolymphe en hormone juvénile. *C. R. Seances Acad. Sci.* **279**, 1007–1009.

Joly, L., J. Hoffman, and P. Joly. 1977. Contrôle humoral de la différenciation phasaire chez *Locusta migratoria migratorioides* (R. & F.) (Orthoptères). *Acrida* **6**, 33–42.

Joly, P. 1949. Le système endocrine retrocérébral chez les acridiens migrateurs. *Ann. Sci. Nat., Zool. Biol. Anim.* [11] **11**, 255–262.

Joly, P. 1952. Production d'adultoides chez *Locusta migratoria* L. (Insecte orthopteroide). *C. R. Hebd. Acad. Sci. Seances* **235**, 1555–1557.

Joly, P. 1956. Croissance et indices de grégarisation chez *Locusta migratoria* (L). *Insectes Soc.* **3**, 17–24.

Joly, P. 1958. Les corrélations humorales chez les acridiens. *Annee Biol.* **62**, 97–118.

Joly, P. 1962. Rôle joué par les corpora allata dans la réalisation du polymorphisme de phase chez *Locusta migratoria* L. *Colloq. int. C. N. R. S.* **114**, 77–88.

Joly, P. 1972. Environmental regulation of endocrine activity of Acridids. *Gen. Comp. Endocrinol., Suppl.* **3**, 459–465.

Joly, P. and Joly, L. 1953. Résultats de greffes de corpora allata chez *Locusta migratoria* L. *Ann. Sci. Nat., Zool. Biol. Anim.* [11] **15**, 331–345.

Kennedy, J. S. 1956. Phase transformation in locust biology. *Biol. Rev. Cambridge Philos. Soc.* **31**, 349–370.

Kennedy, J. S. 1961. Continuous polymorphism in locusts. In J. S. Jennedy, Ed., *Insect Polymorphism*. Royal Entomological Society, London, pp. 80–90.

Kennedy, J. S. 1962. La division du travail entre les phases acridiennes. *Colloq. Int. C. N. R. S.* **114**, 269–297.

Loher, W. 1960. The chemical acceleration of the maturation process and its hormonal control in the male of the desert locust. *Proc. R. Soc. London, Ser. B* **153**, 380–397.

Michel, R. 1970. Etude expérimentalle de l'activité maximum de vol journalière du criquet pélérin (*Schistocerca gregaria* Forsk.) élevé en groupe ou en isolement. *Behaviour* **36**, 286–299.

Nickerson, B. 1956. Pigmentation of hoppers of the desert locust (*Schistocerca gregaria* Forskål) in relation to phase coloration. *Anti-Locust Bull.* **24**.

Nijhout, H. F. and D. E. Wheeler. 1982. Juvenile hormone and the physiological basis of insect polymorphisms. *Q. Rev. Biol.* **57**, 109–133.

Nolte, D. J. 1974. The gregarization of locusts. *Biol. Rev. Cambridge Philos. Soc.* **49**, 1–14.

Nolte, D. J., S. H. Eggers, and I. R. May. 1973. A locust pheromone: locustol. *J. Insect Physiol.* **19**, 1547–1554.

Norris, M. J. 1950. Reproduction in the African migratory locust (*Locusta migratoria migratorioides* R. & F.) in relation to density and phase. *Anti-Locust Bull.* **6**.

Norris, M. J. 1952. Reproduction in the desert locust (*Schistocerca gregaria* Forsk.) in relation to density and phase. *Anti-Locust Bull.* **13**.

Norris, M. J. 1954. Sexual maturation in the desert locust (*Schistocerca gregaria* Forskål) with special reference to the effects of grouping. *Anti-Locust Bull.* **18**.

Novak, V. J. and P. E. Ellis. 1967. The metamorphosis hormones and the phase dimorphism in *Schistocerca gregaria*. II. Implantations of the glands into hoppers reared in crowded conditions. *Gen. Comp. Endocrinol.* **9**, 477–478.

Padgham, D. E. 1976. Bursicon-mediated control of tanning in melanizing and non-melanizing first-instar larvae of *Schistocerca gregaria*. *J. Insect Physiol.* **22**, 1447–1452.

Papillon, M. 1960. Etude préliminaire de la répercussion du groupment des parents sur les larves nouveau-nées de *Schistocerca gregaria* Forsk. *Bull. Biol. Fr. Belg.* **93**, 203–263.

Pener, M. P. 1965. On the influence of corpora allata on maturation and sexual behaviour of *Schistocerca gregaria*. *J. Zool.* **147**, 119–136.

Pener, M. P. 1967. Effects of allatectomy and sectioning of the nerves of the corpora allata on oocyte growth, male sexual behaviour and colour change in adults of *Schistocerca gregaria*. *J. Insect Physiol.* **13**, 665–684.

Pener, M. P. 1976. The differential effect of the corpora allata on yellow colouration in crowded and isolated *Locusta migratoria migratorioides* (R. & F.) males. *Acrida* **5**, 269–285.

Pener, M. P. 1983. Endocrine aspects of phase polymorphism in locusts. In R. G. H. Downer and H. Laufer, Eds., *Endocrinology of Insects*. Liss, New York, pp. 379–394.

Pener, M. P. 1985. Hormonal effects on flight and migration. In G. A. Kerkut and L. I. Gilbert, Eds., *Comprehensive Insect Physiology, Biochemistry and Pharmacology*, Vol. 8. Pergamon, Oxford, pp. 491–550.

Pener, M. P. and P. Lazarovici. 1979. Effect of exogenous juvenile hormones on mating behaviour and yellow colour in allatectomized adult male desert locusts. *Physiol. Entomol.* **4**, 251–261.

Peter, M. G., S. Gunawan, G. Gellissen, and H. Emmerich. 1979. Differences in hydrolysis and binding of homologous juvenile hormones in *Locusta migratoria* haemolymph. *Z. Naturforsch., C: Biosci;* **34C**, 588–589.

Peter, M. G., H. P. Stupp, and K. U. Lentes. 1983. Reversal of the enantioselectivity in the enzymic hydrolysis of juvenile hormone as a consequence of a protein fractionation. *Angew. Chem.* **95**, 794.

Pratt, G. E. and S. S. Tobe. 1974. Juvenile hormones radiobiosynthesised by corpora allata of adult female locusts *in vitro*. *Life Sci.* **14**, 575–586.

Rembold, H. 1981. Modulation of JH III-titer during the gonotrophic cycle of *Locusta migratoria*, measured by gas chromatography — selected ion monitoring mass spectrometry. In G. E. Pratt and G. T. Brooks, Eds., *Juvenile Hormone Biochemistry*. Elsevier, Amsterdam, pp. 11–20.

Rembold, H., B. Schlangintweit, and G. M. Ulrich. 1986. Activation of juvenile hormone synthesis *in vitro* by a corpus cardiacum factor from *Locusta migratoria*. *J. Insect Physiol.* **32**, 91–94.

Staal, G. B. 1961. Studies on the physiology of phase induction in *Locusta migratoria migratorioides* R. & F. *Publ. Fds. Landb. Expt. Bur., 1916–1918* No. 40.

Tobe, S. S. and R. Feyereisen. 1983. Juvenile hormone biosynthesis: Regulation and assay. In R. G. H. Downer and H. Laufer, Eds., *Endocrinology of Insects*. Liss, New York, pp. 161–178.

Tobe, S. S. and G. E. Pratt. 1974a. The influence of substrate concentrations on the rate of insect juvenile hormone biosynthesis by corpora allata of the desert locust. *Biochem. J.* **144**, 107–113.

Tobe, S. S. and G. E. Pratt. 1974b. Dependence of juvenile hormone release from corpus allatum on intraglandular content. *Nature (London)* **252**, 474–476.

Tobe, S. S. and B. Stay. 1985. Structure and regulation of the corpus allatum. *Adv. Insect Physiol.* **18**, 305–432.

Tobe, S. S., C. S. Chapman, and G. E. Pratt. 1977. Decay in juvenile hormone synthesis by insect corpus allatum after nerve transection. *Nature (London)* **268**, 728–730.

Tobe, S. S., B. Stay, F. C. Baker, and D. A. Schooley. 1984. Regulation of juvenile hormone titre in the adult female cockroach *Diploptera punctata*. In J. Hoffman and M. Porchet, Eds., *Biosynthesis, Metabolism and Mode of Action of Invertebrate Hormones*. Springer-Verlag, Berlin and New York, pp. 397–406.

Tobe, S. S., R. P. Ruegg, B. A. Stay, F. C. Baker, C. A. Miller, and D. Schooley. 1985. Juvenile hormone titre and regulation in the cockroach *Diploptera punctata*. *Experientia* **41**, 1028–1034.

Trautmann, K. H., P. Masner, A. Schuler, M. Suchy, and H. K. Wipf. 1974. Evidence of the juvenile hormone methyl (2E.6E)-10, 11-epoxy-3, 7, 11-trimethyl-2, 6-didecadienoate (JH-3) in insects of four orders. *Z. Naturforsch., C: Biosci.* **29C**, 757–759.

Uvarov, B. P. 1921. A revision of the genus *Locusta* L. (= *Pachytylus* Fieb.), with a new theory as to the periodicity and migrations of locusts. *Bull. Entomol. Res. 12*, 135–163.

Uvarov, B. P. 1928. *Locusts and Grasshoppers: A Handbook for Their Study and Control*. Imperial Bureau of Entomology, London.

Uvarov, B. P. 1961. Quantity and quality in insect populations. *Proc. R. Entomol. Soc. London, Ser. C* **25**, 52–59.

Uvarov, B. P. 1966. *Grasshoppers and Locusts*, Vol. 1. Cambridge Univ. Press, London and New York.

Wigglesworth, V. B. 1961. Insect polymorphism — tentative synthesis. In J. S. Kennedy, Ed., *Insect Polymorphism*. Royal Entomological Society, London, pp. 102–113.

Wigglesworth, V. B. 1964. The hormonal regulation of growth and reproduction in insects. *Adv. Insect Physiol.* **2**, 247–336.

Wilson, I. D. and E. D. Morgan. 1978. Variations in ecdysteroid levels in 5th instar larvae of *Schistocerca gregaria* in gregarious and solitary phases. *J. Insect Physiol. 24*, 751–756.

Wyatt, G. R., T. S. Dhadialla, and P. E. Roberts. 1984. Vitellogenin synthesis in locust fat body: Juvenile hormone-stimulated gene expression. In J. Hoffman and M. Porchet, Eds., *Biosynthesis, Metabolism and Mode of Action of Invertebrate Hormones*. Springer-Verlag, Berlin and New York, pp. 475–483.

Zera, A. J. and K. C. Tiebel. 1986. Variations in the developmental profile of JH esterase activity between presumptive longwinged and shortwinged morphs of the cricket *Gryllus reubens*. Poster, Fourth International Symposium on Juvenile Hormones, Niagara-on-the-Lake, Ontario, Canada.

14

Population Dynamics and Regulation in Grasshoppers

A. JOERN and

S. B. GAINES

14.1 INTRODUCTION

Great variation in patterns of population fluctuation clearly exists among grasshopper species. The obvious contrast between the dynamics of typical grasshopper versus locust populations highlights one extreme potential difference. Many species, especially locusts that exhibit phase polymorphism, undergo extreme increases in numbers. In phase polymorphism, very different phenotypes (morphological, behavioral, physiological, and ecological) are observed in individual locusts from high-versus low-density conditions, which are related directly to the population dynamics and the likelihood of dispersal/migration in these species. When coupled with extensive long-distance movements, plagues are formed that may infest large portions of whole continents. In contrast, most other grasshopper species are rare, and seldom if ever build up to densities observed in locust plagues. Nor is the extensive movement often associated with locust migrations seen in most grasshopper species. Even from such a simple comparison, it appears that population processes among acridoid species may be aligned along a continuum (or perhaps a multidimensional space), a notion suggesting that no single, simple mechanism or model of population growth is likely to be an adequate representation of all species. Rather, relative contributions from among a complex sampling of available mechanisms must be assessed.

In this chapter we first summarize proposed theory and mechanisms responsible for driving insect population dynamics, best understood in forest insect pests, and then examine the evidence available from grasshopper populations. The result will be a conceptual blueprint that outlines the possible set of important interactions responsible for changes in grasshopper population size. In this way, the relative roles of deterministic physiological processes, biotic interactions, and external events that are not affected by processes of the local system (e.g., weather) on grasshopper populations can be assessed. In addition, the degree of stochasticity involved in driving population change must be recognized. Such an analysis presents clues regarding whether unexplained variance for a specific population study represents a lack of understanding of the system or whether it is an embedded ingredient of the particular population process.

The extensive diversity of life histories among grasshoppers from tropical and temperate zones severely challenges our ability to construct a synthetic view of population dynamics and regulation in this insect group at this time. An unfortunate heritage of the lack of sufficient detail from population studies representing the range of population responses known to exist hinders construction of an adequate summary of the problem. Many studies have lumped species and presented results of aggregate species responses to various environmental changes, a situation that confounds the interpretation of species-specific responses. Or, the age/stage-specific responses of key population parameters responsible for population change have not been examined with enough care or in enough instances. A proper attack on the problem will center on individual species by comparing responses by different populations or individuals within populations to similar and dissimilar environmental conditions.

This chapter focuses on how a complex network of factors might influence the dynamics of grasshopper populations. It builds upon the excellent previous reviews that have presented extensive details related to many of the topics discussed here (Dempster, 1963; Uvarov, 1961, 1966, 1977; Capinera, 1987). We forge links between population level responses through changes in mortality and fecundity by examining underlying physiological and ecological mechanisms acting at the individual level. However, our understanding of grasshopper population processes will not be complete until both levels of understanding are sufficiently appreciated in the context of specific environmental patterns, especially the degree of spatial and temporal heterogeneity (Roff, 1974a,b; Morrison and Barbosa, 1987) as well as the consequences of individual differences that are also potentially very important but poorly understood (Begon, 1984; Lomnicki, 1988).

14.2 GENERAL PRINCIPLES OF INSECT POPULATION DYNAMICS AND REGULATION

14.2.1 Overview

Conceptually, the basic population growth process is very simple. Populations wax and wane in size according to the processes of birth, death, emigration, and immigration. Regulation consists of those processes that maintain population size within restricted limits. The general life cycle for grasshoppers outlined in Figure 14.1 indicates the types of factors that determine the number of individuals at any given stage. Unfortunately, the dynamic, biological details that drive these "straightforward processes" are complex. Associate with evolved, species-specific differences, interactions among the many forces that affect population change, such as the sampling of topics shown in Figure 14.2, must be delineated and their direct and indirect effects evaluated. These interactions need not be intense, pervasive, consistent, or even important to all populations involved, nor must all of the forces be symmetrically linked to have an important influence (Strong, 1986). Such complexities and asymmetries add to the conditional nature of the problem and the difficulty of dealing with grasshopper population change in a general framework. On the other hand, it is our opinion that a general framework capable of coupling mechanistic and phenomenological models will be much more fruitful in generating the appropriate specific, testable hypotheses.

Discussions of proposed primary processes involved in insect population dynamics often stress the exceptional role of external events such as weather-related forces, especially temperature and precipitation, with some attention paid to density-dependent biotic forces (Andrewartha and Birch, 1954; Milne, 1957; Dempster, 1963; Strong, 1986). As background for the specific details

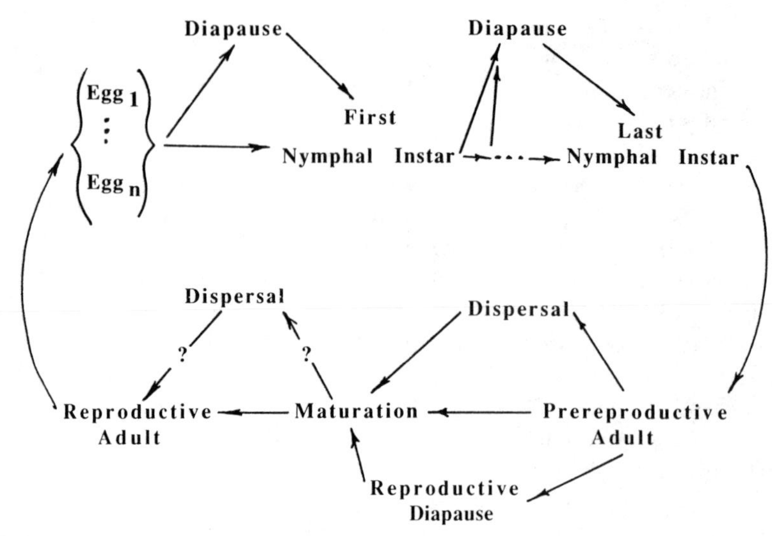

Fig. 14.1. Generalized grasshopper life cycle. Individual species vary greatly in actual paths exhibited. Diapause may occur at a variety of stages although it typically occurs only once for a particular species. Subscripts associated with eggs refer to egg pod number. Egg diapause may take several paths within a generation, even among eggs within a single egg pod.

concerning grasshopper population dynamics, we briefly outline the major conceptual themes that have directed studies of insect herbivores to this point. Past studies were primarily planned and results largely interpreted within the conceptual frameworks favored by the investigators, especially in the context of the acrimonious debate over density independence versus density dependence. Because this framework has recently expanded significantly in scope, past results may need to be evaluated in a new light.

14.2.2 Conceptual View of Insect Herbivore Population Dynamics

Dynamic changes in the densities of grasshopper populations should be consistent with the same general patterns and principles that have been described for other insect herbivores. At present, the most complete understanding of insect population dynamics is based on studies of forest insect herbivores that often exhibit outbreaks (Peterman et al. 1979; Wallner, 1986; Barbosa and Schultz, 1987; Berryman et al., 1987). Significant insights have been gained from intensive studies of these species which should have direct applicability to understanding grasshopper population dynamics.

We distinguish those processes that *regulate* the population from among all factors that may cause grasshopper populations to change. Population regulation implies tight control of population processes in response to increases or decreases in population size around an equilibrium level. Following Berryman et al. (1987), population density can be *regulated* in a strict sense (*sensu* control

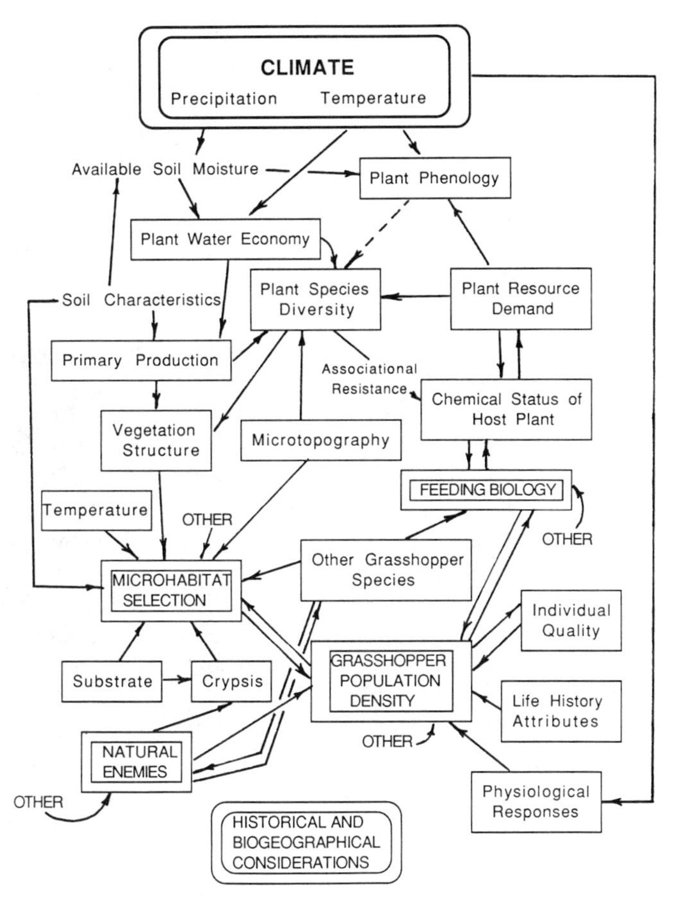

Fig. 14.2. Factors involved in population dynamics of grasshoppers to illustrate the potential complexity of the problem. Indirect interactions are not easily represented in such a diagram but potentially play very significant roles.

theory) only by negative feedback processes. Negative feedback requires that a factor (or factors) increase its negative impact on key population processes, such as survival, reproduction, or movement, in response to increases in population density. This view has several consequences. Only some factors can be involved in strict regulation even though a larger number of elements may influence changes in population density. These other factors may influence the dynamics of population change by setting absolute limits to change or by altering rates of change, but not by directing immediate convergence toward equilibrium. Finally, the notion of negative feedback is important to understanding the regulation of population buildup as well, especially the negative effects when populations exceed equilibrium levels.

Other exogenous factors such as climate or other abiotic components (soil, topographic features) often influence the patterns and rates of population change as well, but not as regulatory processes. These must be integrated with those factors acting in a strictly regulatory fashion if the dynamics of a particular population is to be understood. One plausible interpretation is that exogenous factors may set the framework within which the feedback processes operate to mediate the final level and severity of population regulation. Without doubt, physiological mechanisms that influence population parameters will provide key links between regulating and exogenous factors, for these often define the limits and the corresponding time lags of the response. Yet recognizing such physiological processes is not sufficient by itself to understand population change and regulation. Our challenge is to identify the critical links.

Key density-dependent feedback mechanisms most likely responsible for regulating several forest insect herbivore populations have been identified (Fig. 14.3). Many of these same factors can be shown to be potentially important to grasshopper population change. In one interpretation of the process, density-independent variables set the environmental limits within which the density-dependent feedback processes operate by defining the favorability of the environment for the herbivore population (Berryman et al., 1987). Population regulation may be best understood by relating the population size of the previous generation with the present one while simultaneously examining the effects of the key variables indicated in Figure 14.3 on any potential changes; the processes associated with each mechanism act in different ways and within different time frames. However, the regulation of a population can be described in a general sense using replacement curves (*R* functions; Berryman et al., 1987). For populations with discrete generations, the population replacement rate R_t is

$$R_t = N_{t+1} \,/\, N_t = \mathrm{f}(N_{t-T})$$

where N_t and N_{t+1} are the density in generations t and $t+1$, respectively, $f(N_{t-T})$ is a function that describes the combined feedback effects of all density-dependent variables on the survival and fecundity of the average individual over one generation, and T is the time delay in the feedback process. Stylized examples of *R* functions for forest insect herbivores are shown in Figure 14.3. Many of these responses are reasonable for a variety of grasshopper species.

Two *R* functions are depicted to illustrate potential differences depending on whether the environment is favorable or not to the herbivore population, where favorableness is set by exogenous conditions. The population is unchanging when $R = 1$. For each environment, a single stable equilibrium point is seen corresponding to population changes that fluctuate around either level. Changes in the overall favorableness of the environment may result in a change regarding which equilibrium is tracked, but the population will remain in this new domain until further environmental changes are experienced. Some populations exhibit cycling, largely in response to time delays of the negative

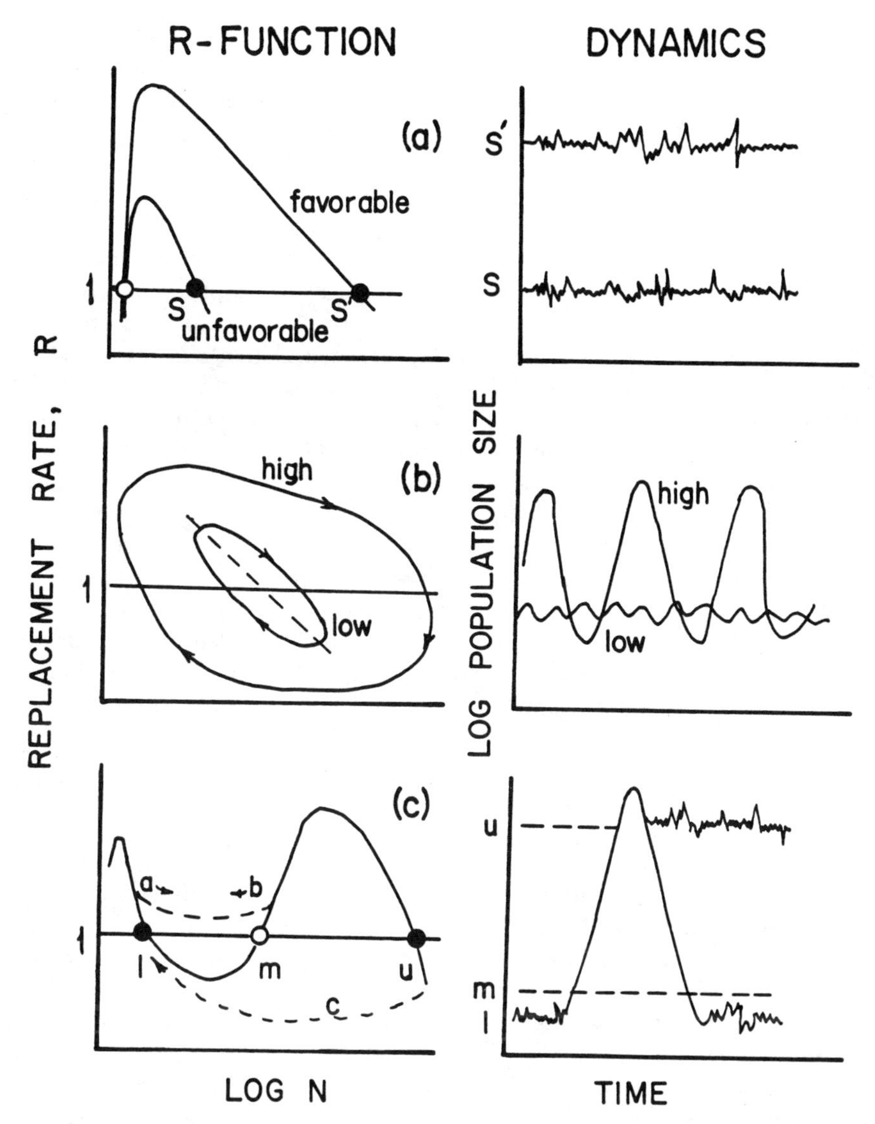

Fig. 14.3 Conceptual model of insect population change with an emphasis on forest insect outbreaks described by Berryman et al. (1987). Details of *R* functions are described in the text. Representative dynamic population change for each *R* function is illustrated alongside it.

feedback in the system (Fig. 14.3b); the longer the time lags, the greater the amplitude and periodicity of the fluctuations. It is widely believed that exogenous conditions are responsible for this response and that more extreme environments are more likely to lead to wildly fluctuating populations. For example, the larch bud moth in Switzerland exhibits larger replacement orbits at higher elevations than at lower ones (Baltensweiler et al., 1977; Berryman et al., 1987) in a manner similar to that indicated in Figure 14.3b. Finally, more complex R functions can be constructed based on the combined effects of several density-dependent factors acting at different densities. In Figure 14.3c, the R function has one unstable (m) and two stable (l, u) equilibrium points. Two domains are depicted in this figure which correspond to the two stable equilibria and are separated by the unstable equilibrium point. Population densities below m tend to fluctuate around the lower equilibrium point (l) whereas populations above m tend to fluctuate around the upper (u) equilibrium.

Transitions between domains, especially if quickly accomplished, correspond to outbreak or irruptive population cycles. Such changes can be fostered in this model by any ecological force that causes population densities to reach levels above those at m, such as through immigration, or through the loss of the density-dependent negative feedback mechanism that had been previously responsible for maintaining this equilibrium point. Vertebrate predation may often be responsible for the lower equilibrium point l in some forest insect populations (Peterman et al., 1979) studied to date. If the environment becomes more favorable through exogenous factors (b or c in Fig. 14.3c), so that the impact of vertebrate predation is ameliorated, the population escapes the domain dominated by this equilibrium point and grows until it reaches the upper equilibrium density. Ultimately, the population density exceeds the available resources and rapidly declines (negative feedback delayed) or continues to fluctuate about the upper equilibrium level (rapid negative feedback). The key feature of the R function described in Figure 14.3c includes the dual (or even multiple) domains and the associated dynamics that can result in very rapid increase and then decrease of population densities. Outbreaks may spread easily if this R function describes the population. At high densities, individuals may emigrate into new areas. If adjacent populations are shifted into the upper domain by the inclusion of migrant individuals, these populations will also erupt owing to the triggering action of emigration from neighboring outbreaks. Spatial advances observed in erupting species can be understood in this context.

Both density-dependent and density-independent factors are incorporated into this model of population dynamics but do not manifest their influence in the simple dichotomous fashion as was previously debated so heatedly. Density-independent responses interact with the R functions to influence the manifestation and stability of the equilibrium points (Berryman et al., 1987). Because density-independent factors are not influenced by changes in population density (by definition), these act as exogenous forces that may influence reproduction or survival either directly or indirectly through their effect on the density-dependent variables. By influencing the effectiveness of particular regulating factors, say, by reducing the impact of vertebrate predators through providing more hiding places, certain equilibrium points may be shifted or even eliminated by density-independent factors (Fig. 14.3a, and c, broken line). Because of such actions, changes in density-independent variables can precipitate further outbreaks under special conditions (Berryman et al., 1987).

In addition, density-independent variables may determine the levels at which populations attain their steady states or have a pronounced effect on herbivore population cycles (Berryman et al., 1987). As suggested in Figure 14.3b, the high amplitude fluctuations at higher elevations are not seen at lower elevations, presumably because differences in the abiotic environment change the regulatory processes. A central but unstudied problem concerns how the regulatory processes are modified by abiotic pressures. In the larch bud moth example, it is possible (though not proved) that the generally cooler conditions at higher elevations introduce longer time lags into population processes through effects on physiological processes (Berryman et al., 1987). Similar situations undoubtedly exist among areas for a particular grasshopper species when different population processes are in effect (also see Section 14.5).

How can density-independent factors determine the suitability of the environment to insect herbivore populations in such a model? Most likely, except for catastrophic events in which large numbers of individuals are suddenly killed by unexpected environmental changes during vulnerable life stages (e.g., severe cold), density-independent characters operate by influencing physiological responses that modify response times of key density-dependent regulating responses and introduce or otherwise influence time lags. For example, developmental rates can easily be modified by changes in either temperature or food quality. Faster developmental rates may result in earlier maturation and more clutches compared with individuals with slower developmental rates. In the case of developmental rates altered strictly in terms of temperature, the effect of this density-independent influence may be truly exogenous to the regulating factors. However, changes in food quality in response to temperature-related stress (Mattson and Haack, 1987a,b) would be indirect in that the density-independent influences are experienced by the herbivore through changes in food plant quality. Clearly, a rich set of interactions can be operating when both density-dependent and density-independent influences are considered.

14.3 PATTERNS OF GRASSHOPPER POPULATION DYNAMICS

What is the nature (pattern) of grasshopper population fluctuations within and among species? This section describes representative patterns of change in population densities. Although extensive documentation exists for population changes and characteristics during outbreaks of locust plagues in Africa and Australia, outbreaks often cover several million square kilometers, making precise estimation of absolute densities and spatial relationships difficult or even impossible to obtain. Similar information for nonoutbreak acridoid species is presently limited as well. Long-term grasshopper density surveys in North America are available for some areas where infestations threaten croplands or rangelands. However, many of these surveys are either estimates of absolute densities of entire grasshopper assemblies or of relative densities of three or four dominant species combined. Examples of available studies are presented in this section.

14.3.1 General Patterns

Periodic outbreaks of many acridoid species share a number of qualitative characteristics. Representative examples are pictured in Figures 14.4–14.6. Initiation often begins with two to eight generations of population growth (and gregarization, if applicable), the duration of this period being strongly influenced by the number of generations produced annually by each species. This is followed by one to several years of sustained high-density populations and a subsequent "crash" when populations return to low densities. In locusts, gregarization begins with local increases in nymphal densities and continues until adulthood, when large-scale migrations are likely to occur, a process that may take several generations though it need not do so. Phase changes and outbreaks are often associated with weather patterns (Uvarov, 1977) whereas declines are sometimes related to predators, parasites, and reductions in egg viability (Farrow, 1977b, 1979). Some data suggest that density-dependent regulation mechanisms may be operating during decline periods (Farrow, 1977a), but not all studies are consistent with this interpretation (Farrow and Longstaff, 1986).

Repeated large-scale locust infestations represent one extreme pattern of population changes. *Schistocerca gregaria* (Forsk.) has long been known to exhibit phase changes and massive population fluctuations (Z. Waloff and Rainey, 1951; Uvarov, 1977; N. Waloff, 1972; Z. Waloff, 1976). The frequency of plagues has been described as irregular but quasi-periodic (Blackith and Albrecht, 1979); recessions between plagues rarely last more than 7 years and are of variable duration (Z. Waloff, 1976) (Fig. 14.5). During infestations covering over 2.9×10^6 km^2 in Africa and southwestern Asia, estimates of gregarious phase hopper bands were as high as 500/m^2 under unusual conditions. These can be compared with solitarious hopper estimates of 0–5/m^2 (Uvarov, 1977; Z. Waloff, 1976). Both initiation and movement of *S. gregaria* swarms are related to weather patterns with precipitation and wind the primary elements useful for forecasting future population densities and plague locations (Rainey, 1972).

Other locust species also exhibit massive outbreaks and phase polymorphism. Detailed studies of large-scale outbreaks have shown that substantial spatial variation in density exists (Z. Waloff and Rainey, 1951; Merton, 1959; Z. Waloff, 1963; Uvarov, 1977); values reported here do not necessarily represent universal values. Maximum adult population densities for *Locusta migratoria* (L.) have been estimated at up to 150/m^2. Nymphs can reach densities of several thousand per square meter in Russia, where outbreaks occur with a 2–3-year periodicity in the south, increasing in duration with latitude (Tsyplenkov, 1972). Adult densities of *Nomadacris septemfasciata* (Serv.) and *Locustana pardalina*

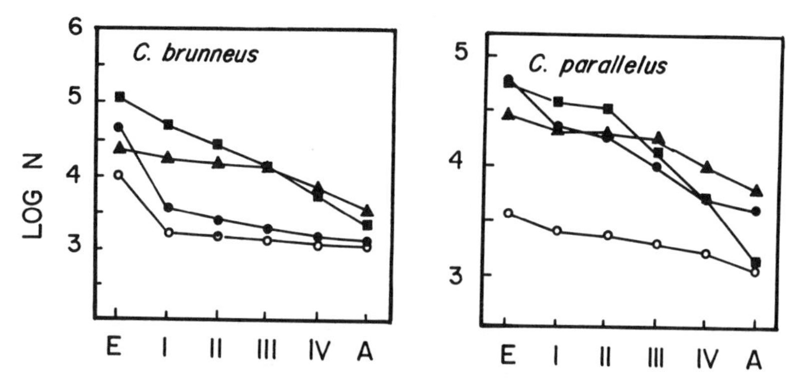

Fig. 14.4 Population flux over four years for *C. brunneus* and *C. parallelus* from Great Britain (data from Richards and Waloff, 1954). Years (eggs laid–adult) are represented as 1947–1948 (●), 1948–1949 (▲), 1949–1950 (■), 1950–1951 (○).

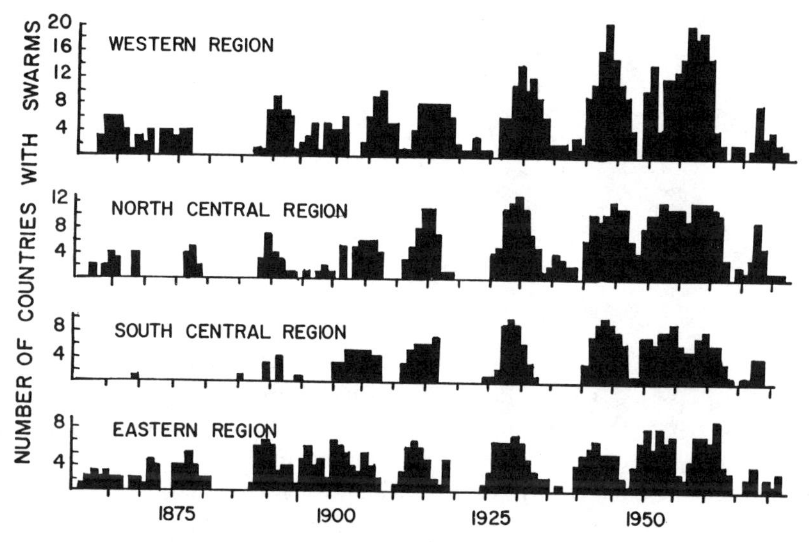

Fig. 14.5 Regional plagues of the desert locust (*S. gregaria*), based on countries reporting presence of swarms (after Z. Waloff, 1976).

(Walk.) can become as high as $100/m^2$ (Uvarov, 1977) and $8-35/m^2$, respectively. Both these species were characterized by long plagues and recessions, up to 30-year periods, in South Africa prior to 1907 when control measures began but have since exhibited shorter plague and recession periods (Lea, 1972). Dempster (1957) estimated adult *Dociostaurus maroccanus* Thunberg population densities at $11-14/m^2$ in a high-density area in Cyprus during swarming.

Chortoicetes terminifera (Walker) is a phase-polymorphic plague species especially common in Queensland and New South Wales, Australia. Outbreaks

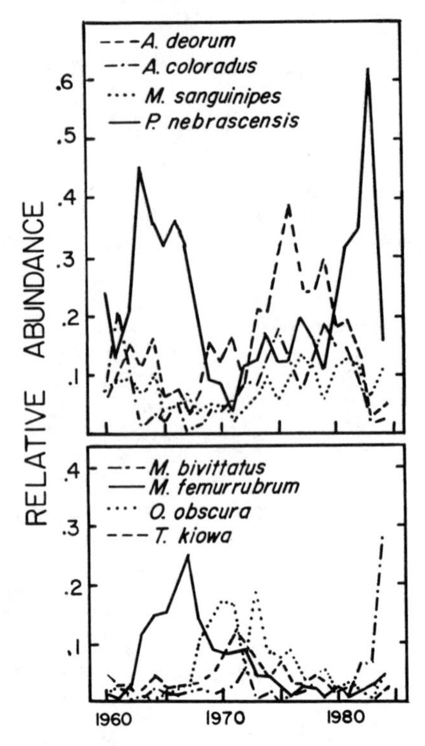

Fig. 14.6 Fluctuations in relative abundance of the eight most common grasshopper species from a Nebraska grassland (Joern and Pruess, 1986).

occur frequently but are short, lasting only 1 year; plagues are less frequent and last 2–4 years (Clark, 1972). *C. terminifera* produces from one to three generations per year depending on rainfall. Dramatic increases in population densities often occur in years favoring production of three generations and population reductions tend to occur during droughts, when one generation is produced per year (Farrow, 1977b, 1982a,b; Clark, 1972). Detailed studies of some populations of this species indicated that emigration was the primary factor affecting population changes (Farrow, 1979, 1982a,b).

Grasshoppers that exhibit high population densities elicit population responses besides phase transformation. Populations of *Zonocerus variegatus* (L.) from southern Nigeria exhibited increased nymphal mortality associated with high densities (Chapman et al., 1979). The largest increase in adult population density (five times the previous season) was followed by a population reduction to one-third during the next year. *Nomadacris (Austracris) guttulosa* (Walker) is a highly fecund, univoltine species whose population fluctuations were related to rainfall patterns that affect availability of plants as well as egg survival (Farrow, 1977b). Population densities of *Parapleurus alliaceus* Germar varied from $0.08/m^2$ to $0.65/m^2$ with 3 years of constant densities ($0.2/m^2$) during 5 years of study (Matsumoto, 1971; Nakamura et al., 1971; 1975). Two other species, *Monglotettix japonicus* and *Chorthippus latipennis*, maintained constant relative densities compared with *P. alliaceus*.

North American grasshoppers also show substantial variation in population trends. Outbreaks occur in many species but are usually of short duration (2–4 years) (Pfadt, 1982). More typical examples include temporal fluctuations of smaller magnitude (Pfadt, 1982; Capinera and Thompson, 1987).

Constancy of the relative abundances of component species of North American grassland grasshopper assemblies was examined by Joern and Pruess (1986). Even though absolute densities fluctuated from year to year, the relative ranks of individual, coexisting species were retained for long periods. Densities of common grass-feeding species tended to fluctuate together, suggesting that common factors may be influencing population dynamics among these species.

Demographic comparisons of plague and nonoutbreak species might suggest that these life histories are discrete, alternative tactics. This may prove to be misleading, however, if these tactics are not considered as component strategies along a continuum. The location of each species along this continuum is determined by a combination of historical and proximate factors. Consideration of a continuous set of population responses by acridoids does not preclude that causal factors also lie on a continuum or are even similar for species with similar population trends. More detailed information concerning (1) the type and magnitude of population changes for species expressing a variety of life-history strategies and (2) characteristics of plague species during recessions will prove to be essential to the construction of more generalized theories of the population dynamics of acridoids.

14.3.2 Life Cycle Diversity

Grasshoppers and locusts that live in highly variable environments must adjust their life cycles in order to survive and reproduce. Migration and diapause are the most likely mechanisms available that allow populations to cope with inhospitable periods (Dingle, 1985; Tauber et al., 1986). In a simplistic sense, migration is an effective response to rapidly changing but aperiodic portions of a variable environment, especially in a spatial context, whereas diapause is more likely if environmental variability is predictable and especially if seasonal. As a result, significant variation is observed in grasshopper life cycles and populations of the same species experiencing different environments may exhibit a different number of generations per year (e.g., univoltine versus bivoltine) or undergo a different number of developmental stages.

Life cycles may differ greatly among grasshopper species although a general picture of possibilities can be pictured (Fig. 14.1). In most species, there is a portion of the year in which the environment is inhospitable, although the feature that makes it so varies greatly among grasshopper species depending on where they live (e.g., cold freezing winters or hot periods with drought).

Life cycles often mold themselves, in an evolutionary sense, around the periods of adversity to maximize the likelihood and amount of time available for recruitment into the next generation. In comparisons among species, diapause has been documented for each of the grasshopper life cycle stages (Tauber et al., 1986).

Many tropical species experience prolonged dry spells that alternate with short but intense wet seasons. The long dry period is a very critical period for survival in most tropical grasshoppers and locusts because suitable green food is scarce, especially preferred hosts. Breeding centers of the spur-throated locust (*Nomadacris (Austracris) guttulosa*) from Australia provide an example (Farrow, 1977a). *N. guttulosa* survives the dry season as adults by feeding on the available leaves of trees and shrubs (e.g., species of *Eucalyptus*, *Acacia*, and *Dodonaea*). Breeding is typically restricted to the rainy season. At its onset, adults rapidly mature, oviposit in moist soil, and the eggs develop directly to hatching without a diapause. Nymphs develop on the fresh growth of ephemeral grasses and forbs in addition to leaves of trees and shrubs and reach adulthood before suitable host plants disappear. Because the onset of the monsoon is abrupt, synchronous maturation and oviposition ensures that adequate moisture for egg development and hatching as well as adequate food exists for developing nymphs (Farrow, 1977a). Although oviposition occurs in dry soil and eggs can withstand lack of moisture for a short period (20–30 days), survival requires dependable moisture at this stage. Females lay repeated clutches during the rainy season (~5 pods in *N. guttulosa* over a 4–5 month period; Farrow, 1977a). A broad-spectrum diet enables the species to survive and mature under extremely adverse conditions. In contrast, droughts have tremendous impact on survival, maturation, and fecundity in the grass-feeding *Chortoicetes terminifera* and *Locusta migratoria* (Farrow, 1975, 1977a).

Diapause. Initiation of diapause in grasshoppers can be keyed to a variety of environmental cues including temperature, photoperiod, or responses to complex interactions among the cues; it may be either obligate or facultative (Tauber et al., 1986). An example of the complexity of diapause induction is illustrated in the Australian plague locust (*C. terminifera*) (Wardhaugh, 1977, 1980a,b, 1986). This multivoltine, highly migratory species is wide ranging throughout Australia. Suitable habitat conditions are largely ephemeral, especially because of the patchiness and unpredictability of rainfall. This species exploits ephemeral habitats by combining migration and diapause with the ability of the eggs to withstand long periods (2–3 months) of drought. During winter, most areas throughout the range of this species are too dry and/or too cool to support normal development of active stages. During this period, a facultative egg diapause occurs (Wardhaugh, 1986).

Complex interactions of environmental cues direct diapause induction in *C. terminifera* (Wardhaugh, 1986), in which both facultative and obligate diapause states exist depending on the suite of conditions encountered by all developmental stages (Fig. 14.7). Photoperiod and temperature interact to induce

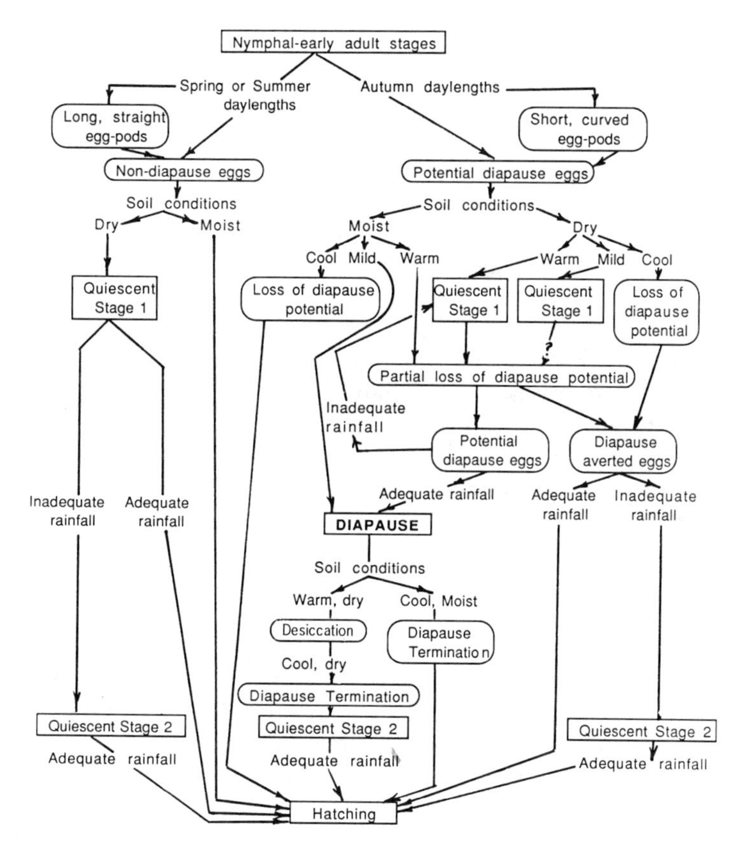

Fig. 14.7. Developmental pathways of eggs of *C. terminifera* illustrating the importance of the key proximate factors of day length, temperature, and rainfall in influencing egg diapause (Wardhaugh, 1986).

embryonic diapause. During long or increasing days (spring and summer periods), primarily nondiapausing eggs were produced (Wardhaugh, 1980a,b). However, a variable fraction of the eggs laid in autumn entered diapause. Short or decreasing photophases during sensitive nymphal and adult stages favored the production of diapause eggs if moisture availability and at least two temperature-mediated processes were appropriate to control entry into diapause (Hunter and Gregg, 1984; Wardhaugh, 1986). The result is 18 different paths of embryonic development employed to track environmental conditions during the life cycle of the laying female and the egg.

14.3.3 Estimation and Variation in Key Population Parameters

Although it is readily acknowledged that population dynamics and life-history evolution can be best understood by obtaining detailed estimates of demographic parameters (Stearns, 1976, 1977), this has seldom been done for grasshoppers. Only a handful of detailed life tables for grasshopper populations exist with even fewer good estimates from naturally occurring populations (e.g., Chapman et al., 1979; Onsager and Hewitt, 1982; Sanchez and Onsager, 1989). In this section we provide an overview of available estimates of demographic parameters and their variation.

Variations in parameters defining different stages of the life cycle result in grasshopper population change. Loss of reproductive potential in the adult population has been attributed as the largest single factor affecting the size of the adult breeding population in the following year for *Zonocerus variegatus* (Chapman et al., 1979). On the other hand, variation in nymphal mortality, where loss of recruits from one reproductive stage to the next may reach 90–99%, could provide a sensitive link for explaining fluctuations in other species (N. Waloff, 1972). Also, qualitative differences among individuals could have important consequences in explaining parameter variation and subsequent population dynamics (Hassell and May, 1985; Smith and Sibly, 1985; Wall and Begon, 1987a,b) as in the processes leading to phase change in locusts among others.

Populations change as age-specific survivorship and fecundity schedules vary in response to environmental changes coupled with underlying genetically determined capacity. Variation at any of the life stages depicted in Figure 14.1 has the potential for altering population growth. Excellent reviews of the detailed responses to environmental pressures by different life cycle stages are provided in Dempster (1963), Uvarov (1966, 1977), and Hewitt (1985). From a population dynamics point of view, the key population attributes are mortality rates in each stage, developmental rates within and between stages, diapause, fecundity, growth rate, sex ratio, and movement (dispersal or migration). In many cases, the timing of events in association with external processes such as food availability or suitable weather conditions is a key issue to understanding population dynamics of particular species as well.

Survivorship. Survivorship in grasshoppers is very similar to that in many other insects in that mortality rates are often quite high at early stages (N. Waloff, 1972). Survivorship curves for two species that were estimated for naturally occurring populations (Fig. 14.8) indicate that much potential variation within and among species is likely to exist. In addition, it is likely that causes of mortality are quite variable in time and space within species populations and very different factors may be most important among species.

Egg mortality can be high and generally depends on moisture and temperature, although egg predation and fungal attack can also be important (Dempster, 1963; Hewitt, 1985). Studies that have estimated a range of possible mortality factors in the field have shown this often to be less important than other stages, however (Chapman and Page, 1979; Farrow, 1982a,b). For example, population changes in both *Zonocerus variegatus* (Chapman and Page, 1979) and *Chortoicetes terminifera* (Farrow, 1982a,b) are much more affected by nymphal

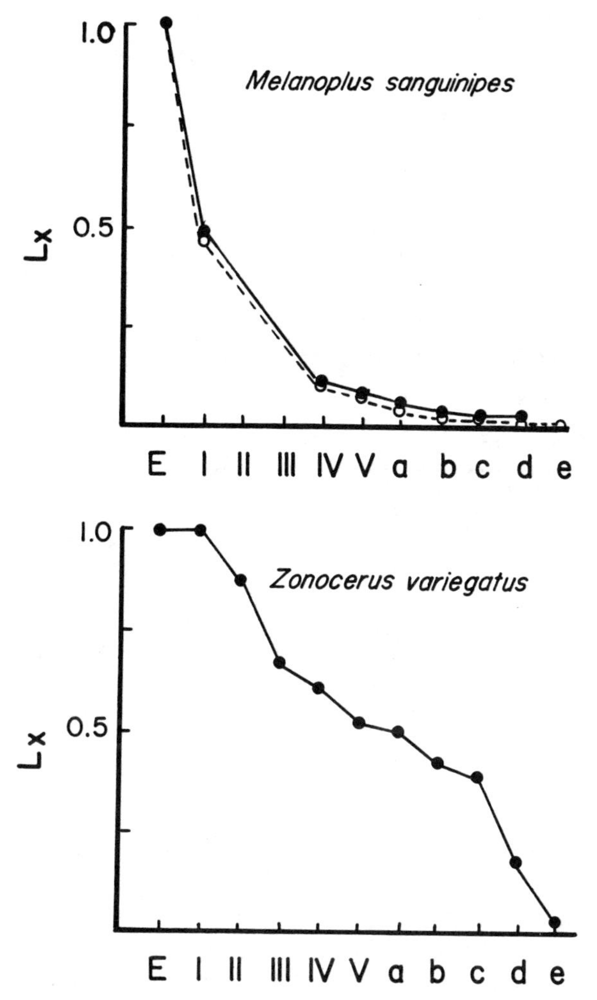

Fig. 14.8. Survivorship curves for two grasshopper species: temperate-region *M. sanguinipes* (data from Sanchez and Onsager, 1989) and tropical *Z. variegatus* (data from Chapman and Page, 1979).

mortality than egg mortality (Fig. 14.9). On the other hand, the egg stage is likely to be important at the edge of the normal range (such as during outbreaks) where unpredictable, extreme conditions are more likely and the life cycle is less likely to have had time to mold itself to the new conditions.

Fecundity. Age-specific reproduction is another key attribute for understanding grasshopper population change. Potential fecundity is determined by the number of ovarioles (the maximum number of eggs that can be laid per oviposition) and the number of egg pods laid. Each is potentially influenced by both genetic and environmental factors.

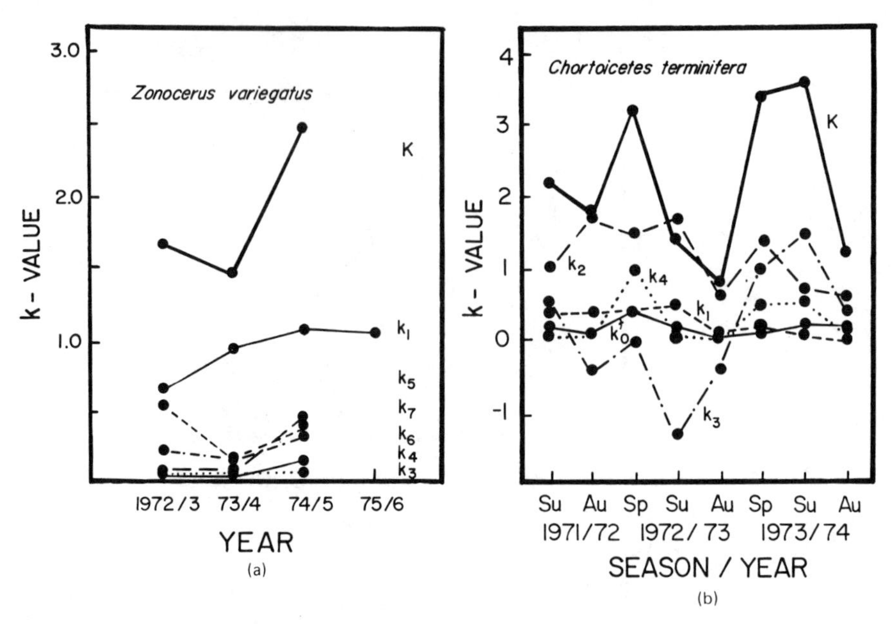

Fig. 14.9. Key factor analysis of population trends in two grasshopper species: (a) *Zonocerus variegatus* (data from Chapman et al., 1979) and (b) *C. terminifera* (after Farrow, 1982b). Key factors for *Z. variegatus* include *K*, generation mortality (log total mortality); $k_1 + k_2$, failure to lay and egg mortality (mostly k_1); k_3, loss of early instars, k_4, loss of instar 4; k_5, loss of instar 5; k_6, loss of instar 6; k_7, loss of immature adults. Key factors for *C. terminifera* include k_0, unrealized natality; k_1, egg mortality; k_2, nymphal mortality; k_3, migration gains or losses; k_4, adult parasitism.

Great variation in the number of ovarioles exists among species (Fig. 14.10). However, few species oviposit more than three or so times under natural conditions and most individuals in these populations do not even live this long (Fig. 14.8). In other words the contribution of late laying females to subsequent population density is low relative to early laying individuals and unrealized natality is the norm. Perhaps more importantly, there was no significant relationship between ovariole number and the ranked density among coexisting species in a Nebraska grassland (A. Joern, unpublished).

Fecundity is greatly affected by environmental factors. Seasonal patterns of fecundity for *M. sanguinipes* studied in field cage experiments are indicated in Figure 14.11. Both the number of eggs per female per day and egg pods per female drop as females age and the season progresses (Pickford, 1960, 1966a, b). Also note that the number of egg pods per female does not drop in a study performed simultaneously in the lab, suggesting that the observed drop is due to deteriorating food resources or an otherwise deteriorating environment. Fecundity in *Phaulacridium vittatum* clearly depends on the availability of food (Fig. 14.12), for example.

Rate of egg production has been examined for several species in both the lab and in field cages. The mean number of eggs per pod is seasonally

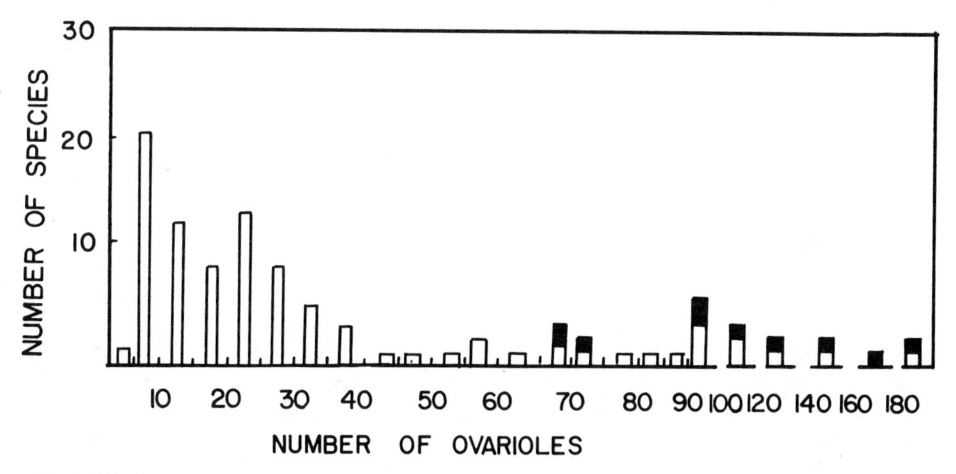

Fig. 14.10. Frequency distribution of ovariole numbers in a representative sample of grasshopper species. Dark bars represent locust species (gregarious phase). Note that the *x* axis changes scale.

adjusted in *Camnula pellucida*: June, 20.9; July, 21.8; August, 23.5; and September, 25.5 eggs/pod (Pickford, 1966a) but the number of eggs per pod dropped in lab-studied *M. sanguinipes*. For *Chorthippus brunneus*, compensation in number of eggs per pod versus the rate of pod production was density dependent (Wall and Begon, 1987a); more eggs were produced per pod when fewer pods were laid under low-density situations but not when densities were high. In another lab study (Fig. 14.13), both *M. sanguinipes* and *M. bivittatus* exhibited constant rates of egg production throughout the female life-span when food was of constant quality (D. S. Smith, 1966). In natural populations of *M. sanguinipes*, actual fecundity did not reach individual potential and varied greatly between seasons (Sanchez et al., 1988). Females that hatched earliest had greatest fecundity owing to shorter preoviposition time, greater longevity, and higher oviposition rates.

Behavioral attributes of individuals may contribute to fecundity as well. Length of copulation and opportunity for multiple copulations can sometimes significantly influence fecundity. Multiple matings in *Chorthippus brunneus* and *Melanoplus sanguinipes* increased fecundity (Pickford and Gillot, 1976; Butlin et al., 1987; but see D. S. Smith, 1968). In *C. brunneus*, nutrient transfer to the female through the spermatophore was demonstrated; radio-labeled amino acids injected into the males were ultimately incorporated into the eggs (Butlin et al., 1987). Spermatophore transfer as male reproductive investment has been previously documented in other Orthoptera and other insects (Bowen et al., 1984; Gwynne, 1984; Gwynne et al., 1984). As a consequence, females may discriminate among males based on their assessment of spermatophore nutrients; this may contribute to variance in male reproductive success (Butlin et al.,

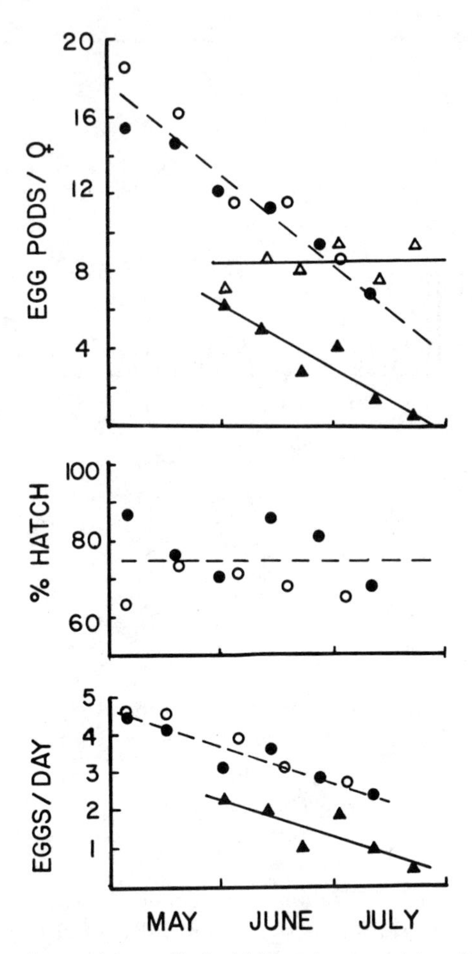

Fig. 14.11. Seasonal patterns of fecundity in *Melanoplus sanguinipes* in field and lab cages (data from Pickford, 1960). Data from two sites including a lab study and sampling from 2 years are presented: interior, British Columbia, field (▲); interior, British Columbia, lab (△); Delisle, Saskatchewan — 1957 (●); Delisle — 1958 (○).

1985). Although difficult to study directly in the field, such individual differences may have important population-level consequences (Section 14.4.4).

Reproductive Capacities. Rates of increase that rely on combined contributions of both age-specific fecundity and survivorship have been calculated for some species. *Melanoplus sanguinipes* may lay 500–600 eggs over an individual female life-span when reared under favorable conditions with an individual record of 861 eggs in one lab-reared female (D. S. Smith, 1966, 1968, 1970). Based on life table analysis, the intrinsic rate of increase (r) when reared under controlled conditions (lab or field cages) was 4.1–4.4 per generation.

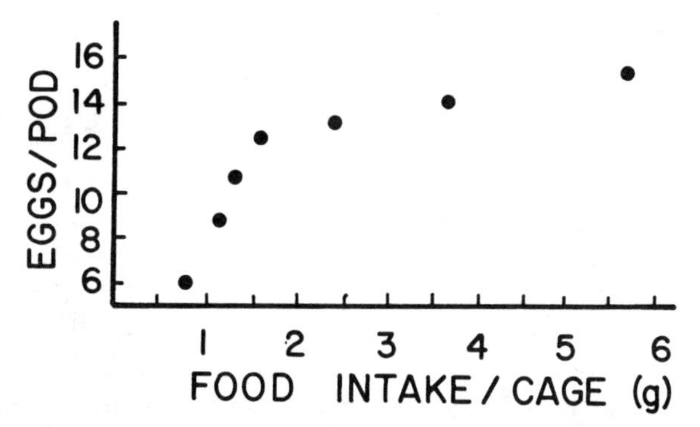

Fig. 14.12. Dependence of egg production rate on food availability in *Phaulacridium vittatum* in a laboratory study. (Data from Clark, 1967.)

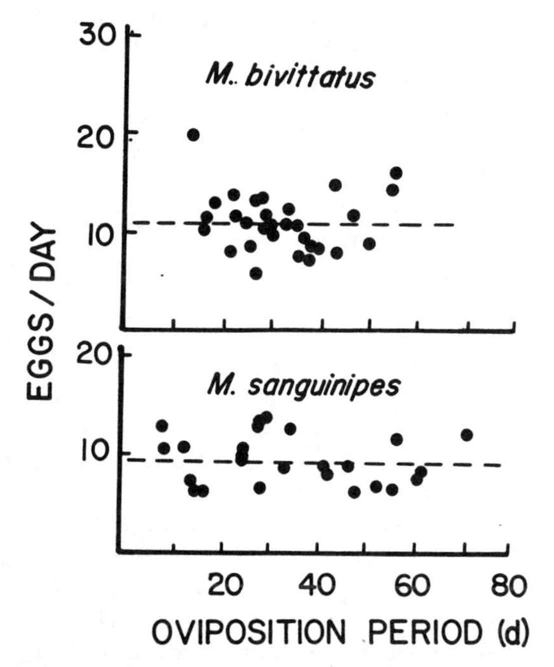

Fig. 14.13. Egg production rate (eggs/female day) throughout reproductive period of *M. bivittatus* and *M. sanguinipes* (data from D. S. Smith, 1966).

Leslie population projection matrices have been used to estimate rates of increase, r, for solitarious versus gregarious forms of *Schistocerca gregaria* (Cheke, 1978). Although the number of eggs per pod and the total number of pods per female is significantly greater in solitarious than gregarious morphs (Uvarov, 1966), the gregarious form actually has a greater reproductive potential. Based on representative phenologies and age-specific survivorship in addition to fecundity estimates for each phase type, the *r* for the gregarious morph (0.0248 individuals/week) is about 10 times greater than that calculated for the solitary morph (0.0023 individuals/week). Gregarious forms appear to achieve this reproductive edge despite lower rates of egg production by being synchronized, developing faster, and being exposed to predators for shorter periods of time (Cheke, 1978).

Rate of Development. Developmental rates at each critical stage in the grasshopper life cycle can have pronounced effects on key population parameters. In particular, developmental variance can exert a significant impact on reproductive capacity, intrinsic rates of increase, and generation times in mathematical models of insect populations (Bellows, 1986a,b). In addition, the same environmental pressures that directly affect fecundity and survivorship also influence developmental rates. Developmental rate can be influenced by abiotic factors (see below), food quality and availability (Haniffa and Periasamy, 1981), or demographic factors such as crowding (D. S. Smith, 1970, 1972).

As is true in most insects, developmental rates are greatly influenced by abiotic factors, especially temperature (Logan et al., 1976; Wagner et al., 1984). Embryonic development, for example, is primarily temperature driven, although soil moisture often plays a key role as well (Dempster, 1963; Mukerji and Randell, 1975; Mukerji and Gage, 1978). As a result, numerous models of grasshopper embryonic and nymphal development have been devised based on accumulated temperature (Gage et al., 1976; Mukerji and Gage, 1978; Hilbert and Logan, 1983; Kemp and Onsager, 1986; Kemp, 1987a; Kemp and Dennis, 1989), using either linear or nonlinear relations.

Although detailed physiological mechanisms are not understood in great detail, grasshopper embryonic developmental rates are typically strongly correlated with accumulated temperature above some critical threshold as long as there is sufficient soil moisture (Dempster, 1963; Uvarov, 1977; Mukerji and Randell, 1975; Gage et al., 1976; Hilbert and Logan, 1983; Kemp, 1987a; Kemp and Dennis, 1989). The effect of this relationship is extended to include temperature accumulation before and after egg diapause (Mukerji and Randell, 1975; Hilbert et al., 1985). Eggs that are laid early and that receive sustained high temperatures achieve maximum prediapause development and hatch first when the postdiapause conditions are again favorable (Pickford, 1960, 1966a,b, 1972, 1976; Mukerji and Randell, 1975; Hilbert et al., 1985). Hatching date and variability in emergence in reference to resource quality and availability can greatly influence subsequent population processes (Mukerji et al., 1977; Logan and Hilbert, 1983).

Embryonic development exhibits much within- and between-species variation. Although *Aulocara elliotti* was observed to hatch before or coincidently with *M. sanguinipes*, laboratory studies indicated that *M. sanguinipes* actually required fewer accumulated degree-days to complete postdiapause development than did *A. elliotti* (at a constant temperature of 23.3°C) (Kemp and Sanchez, 1987). Apparent lack of coincidence between lab and field results can be explained by considering oviposition preferences between these species: *A.*

elliotti oviposits in open areas whereas *M. sanguinipes* prefers ovipositing in sod formed by a short grass species (*Bouteloua gracilis*), and *A. elliotti* lays its eggs at shallower depths (midportion of pod ~1 cm below surface) and horizontally to the surface compared with the pods of *M. sanguinipes*, which were laid almost twice as deep (midpoint of pod ~2 cm below surface) and vertically. *A. elliotti* achieves its elevated heat requirements earlier in the season than does *M. sanguinipes* because the soil temperatures increase more quickly in open microhabitats. In addition, differences in pod orientation explain why hatching in *M. sanguinipes* is spread out over a longer period: deeper eggs do not heat as fast as top eggs so that hatching date among eggs within a vertical pod exhibit higher variance than in a horizontal pod. Similar results were observed among sites for *Chorthippus parallelus* and *C. brunneus* (Monk, 1985). Eggs of both species collected from sites that exhibited the lowest soil temperatures in the spring hatched first in the lab.

Nymphal developmental rates and rates of sexual maturation are temperature dependent in addition to effects of host plants, pathogens, or individual differences. Nymphal development rate of *M. sanguinipes* in the lab increased with temperature up to about 37–38°C, with the duration of the nymphal stage dropping from 79 days at 22°C to 17 days at 37.5°C (Fig. 14.14). In a lab study of *Chorthippus brunneus*, direct insolation from a radiant heat source (light bulb) greatly accelerated nymphal development, rate of adult maturation, and egg production (Begon, 1983). This result has been documented in its general form many times (Pickford, 1972; Gage et al., 1976; Hewitt, 1979; Logan and Hilbert, 1983; Kemp and Onsager, 1986; Logan, 1988; Kemp and Dennis, 1989).

It may be that grasshoppers seek to achieve body temperatures close to that which induces maximal rate of development (Hardman and Mukerji, 1982; Chappell and Whitman, Chapter 6). This seems reasonable because faster developmental time would permit increased exposure to higher-quality plant tissue during the reproductive stage while lessening exposure to predators, parasites, or physical stress during prereproductive and reproductive periods; it may also result in increased size, which leads to a variety of advantages (Section 14.4.4). Also, increased body temperature could facilitate greater efficiency of food use, which would translate into increased growth or egg production rate.

Developmental effects carry into the reproductive stage. Length of oviposition period varies as a function of food quality and abundance, temperature, and date of maturation (Uvarov, 1966, 1977). Reproductive attributes such as interlaying interval and eggs per day also vary according to these same influences (D. S. Smith 1966; Uvarov, 1966, 1977; Pfadt and Smith, 1972).

Sex Ratio. Sex ratios in grasshoppers are generally found to be 1:1 in a wide variety of species (Pfadt, 1949; D. S. Smith and Northcott, 1951; Pickford, 1962; Pfadt and Smith, 1972; Farrow, 1982a). These studies include grasshoppers fed with different food plants (Pfadt, 1949) or host plants in which quality was altered with nitrogen fertilization (D. S. Smith and Northcott,

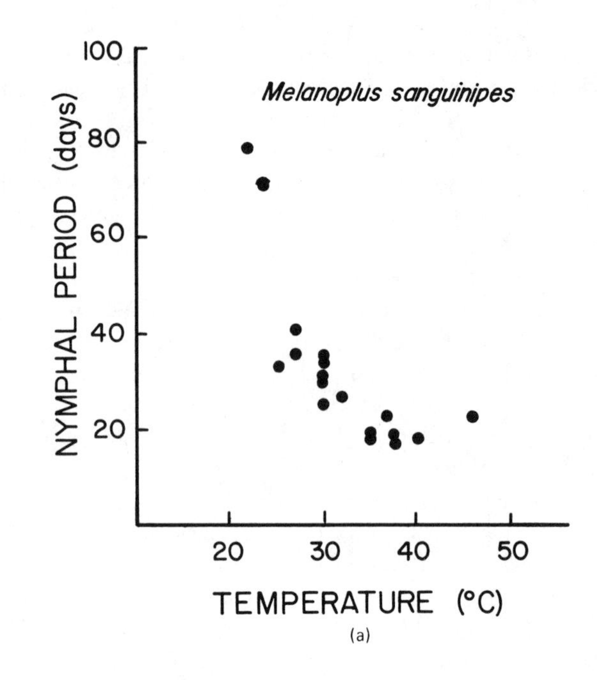

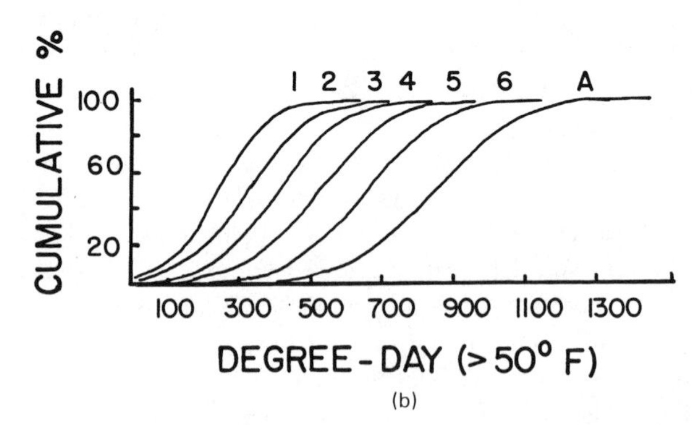

Fig. 14.14. Influence of temperature on grasshopper development. (a) Developmental rate of *M. sanguinipes* in response to temperature in a mixture of laboratory studies (summarized in Hilbert and Logan, 1981). (b) Phenology of developmental stages in response to accumulated temperature (degree-days, > 50°F after April 1) at reference position in the field. 1–6 indicate nymphal instars. (Redrawn from Gage and Mukerji, 1977.)

1951). The sex ratio in a nondiapausing strain of *M. sanguinipes*, developed in the lab, was also 1:1 (Pfadt and Smith, 1972). Sex ratios of individual pods of *M. sanguinipes* ranged from all male to all female but the overall ratio did not significantly differ from 1:1 (Mulkern, 1983). However, the phenologies of the two sexes typically differ, with the males emerging as adults first and then dying first, so that the sex ratio in the field often differs from 1:1 but shifts from male to female biased as the season progresses (Mulkern, 1983).

14.4 KEY MECHANISMS DRIVING GRASSHOPPER POPULATION DYNAMICS

In this section, we examine how specific changes in a variety of attributes influence demographic parameters and potentially drive grasshopper population dynamics. Berryman (1987) has summarized a series of proximate hypotheses to explain outbreaks of forest insect herbivores (Table 14.1) providing a good starting framework for examining grasshopper population dynamics.

14.4.1 Climate and Other Abiotic Factors

Fluctuations in grasshopper abundance are commonly attributed to changes in the weather (Andrewartha and Birch, 1954; Dempster, 1963; Uvarov, 1966, 1977; Capinera, 1987; Kemp, 1987a,b). The hypothesis that weather directly influences population fluctuations is generally examined using correlational analyses where grasshopper population sizes are related to weather variables in present or previous years (McCarthy, 1956; Wakeland, 1958; Edwards, 1960; Gage and Mukerji, 1977; Guseva, 1979; Farrow, 1979). One general mechanism heavily relied upon for explaining significant correlations between population fluxes and weather-related variables is the sensitivity of key physiological or developmental processes to thermal or other abiotic environments at some critical life stage. Possible implications of such individual physiological

TABLE 14.1. Key hypotheses Proposed to Explain Outbreaks of Insect Herbivores

1. Dramatic changes in the physical environment cause outbreaks (Andrewartha and Birch, 1954).
2. Intrinsic shifts in the genetic or physiological state of the individuals in a population cause outbreaks (e.g., Wellington, 1960; Krebs, 1978).
3. Population interactions between trophic levels cause oscillation or periodic escape from control and results in outbreaks (e.g., Hassell, 1978).
4. Qualitative or quantitative increases in host plant quality (nutritional or antibiotic attributes), usually in response to stress, cause outbreaks (e.g., White, 1978; Rhoades, 1983; Mattson and Haack, 1987a,b).
5. Specific life-history strategies (*r* strategists or opportunistic species) are responsible for pest species status and the likelihood that outbreaks will occur (Southwood and Comins, 1976).
6. Escape from the regulating influence of natural enemies results in outbreaks (e.g., Holling, 1965; Southwood and Comins, 1976).
7. Outbreaks occur when the insect herbivore populations cooperatively overwhelm the defensive systems of their hosts (e.g., Berryman, 1982).

Source: After Berryman (1987). References cited with each hypothesis provide examples.

responses translated to population processes are often easily seen (Dempster, 1963; Uvarov, 1966, 1977). The critical link between individual and population responses is seldom forged in an empirically rigorous fashion, however. This section examines the general relationships between population change and climate and assesses possible mechanisms that best explain the observed correlations.

Clear relationships between abiotic factors and important life cycle stages can be identified. Dempster (1963), for example, provides a comprehensive listing concerning the potential impact of abiotic factors on each life cycle stage. The central issue here concerns the likelihood that specific regulating mechanisms are altered by climatic variation in such a way that directly accounts for observed population size fluctuations or regulation. Conclusions based on such correlations will always appear vague unless specific mechanisms are identified, for some of the correlations will undoubtedly reflect complex biotic interactions mediated by weather conditions or indirect interactions (Fig. 14.2). Indirect relationships are largely unaccounted for in the original experimental and statistical designs used to examine the issue up to this point. Another important issue concerns the mode of action of climate. Are observed population responses due to direct manipulation of population parameters by abiotic factors or to combined responses of biotic interactions within the limits of change set by abiotic forces that determine specific population trajectories (Section 14.2.2)?

In general, there appears to be a positive correlation between grasshopper abundance and hot, dry weather in temperate latitudes and high elevations or seasonal rainfall abundance in generally tropical or arid regions (Capinera, 1987; Capinera and Thompson, 1987). High July–September temperatures coupled with abnormally low rainfall were correlated with high grasshopper population densities in western Canada (Edwards, 1960). Although temperature and rainfall in the preceding year was most important, grasshopper abundance in Saskatchewan (Canada) was also significantly affected by conditions during the previous 2 or 3 years (Gage and Mukerji, 1977). Heat units and heat–precipitation ratios were positively correlated with density ratings over the 32-year period of Gage and Mukerji's (1977) analysis. However, the most highly significant correlation with density rating was the grasshopper density rating from the previous year. Using Markov transition analyses, Kemp (1987b) analyzed the likelihood of outbreaks from coincident regions in western U.S. grassland. Certain areas were much more likely to support high grasshopper populations independent of climate. A similar result was observed in three of four locust species compared by Farrow and Longstaff (1986), which exhibited high population growth rates during favorable climatic periods and were capable of initiating outbreaks in 2 or 3 years. This trend was not necessarily observed in other North American grassland species (Mulkern, 1980; Capinera and Thompson, 1987).

Widespread drought or above-average rainfall is correlated with population change in several well-studied tropical locusts (Sayer, 1962; Farrow and

Longstaff, 1986) and desert grasshoppers (Nerney, 1961). This response differs from those populations discussed above and suggests that a major difference exists among species from habitats experiencing different stresses. The same key abiotic factors driving population change are involved but differ in importance to such a degree as to be almost qualitatively different.

Four long-term studies from different North American grasslands have yielded similar results concerning the general influence of weather on grasshopper abundance (Pfadt, 1977; Mulkern, 1980; Joern and Pruess, 1986; Capinera and Thompson, 1987). Each study considered coexisting species separately. In each case, species dominance did not shift radically among years. Favorable environmental factors were probably responsible for increases in grasshopper density because most species responded in a similar fashion in each study. For example, 23 of 25 species from a mixed grassland in Nebraska exhibited positive correlations (14 of 25, statistically significant) with total grasshopper density (Joern and Pruess, 1986). However, not all species-specific grasshopper abundance responses were in close synchrony (Joern and Pruess, 1986; Capinera and Thompson, 1987). Species densities followed in another study varied independently if not erratically; presumably in response to many different factors among species (Mulkern, 1980).

Any number of pressure points in the grasshopper life cycle (Fig. 14.1) are susceptible to alteration from abiotic factors and may be sufficiently critical to influence significantly the species' population dynamics (Table 14.2).

Eggs are especially sensitive to both moisture and temperature, and hatching may be prevented under extreme conditions (Dempster, 1963; Uvarov, 1966; Hewitt, 1985). Timing of hatching is largely keyed to accumulated temperature in many species (Randell and Mukerji, 1974; Mukerji and Randell, 1975; Gage et al., 1976; Kemp and Onsager, 1986; Kemp and Dennis, 1989) and precise timing may be of critical importance if quality resources (especially plants) are available only for brief periods (Kemp, 1987a).

Clear species-specific differences in life cycles exist among coexisting species (Joern, 1979, 1982; Capinera and Thompson, 1987), among individuals within the same population (Slifer and King, 1961; Walker, 1980); among generations within a species, especially in bi- or multivoltine species (Tauber et al., 1986); and certainly among geographic locations (Capinera and Horton, 1989). Delayed or accelerated hatching may have important effects on the resulting population dynamics and ability to track resources (Kemp and Onsager, 1986; Kemp and Dennis, 1989), although the appearance of key plant resources is often closely tied to or at least correlated with the same cues (Kemp, 1987a). Yet different species that are seemingly tracking the same resources may employ different physiological thresholds, as in egg development in *Aulocarca elliotti* and *Melanoplus sanguinipes* (Kemp and Sanchez, 1987). Nymphal development and survivorship exhibit similar sensitivities to abiotic conditions (Uvarov, 1966, 1977; Begon, 1983) and phenological relationships are often temperature driven (Fig. 14.14) (Begon, 1983; Kemp and Onsager, 1986; Kemp and Dennis, 1989) if suitable quality food is available. Starvation has

TABLE 14.2. Selected Examples of Abiotic Factors Influencing Grasshoppers

Life Stage	Factor	Effect	Reference
Egg	Moisture	Desiccation	Hewitt (1985)
		Flooding	Hewitt (1985)
		Diapause: initiation/termination	Tauber et al. (1986)
	Temperature	Developmental rate	Dempster (1963); Hilbert and Logan, 1983
		Lethal limits	Hewitt (1985)
		Hatching, phenology	
	Photoperiod	Hatching, phenology	Wardhaugh (1986)
	Soil type	Interaction with temperature and moisture	Hewitt (1985)
Nymph	Moisture	Plant quality	Lewis and Bernays (1985)
		Fungal attacks	Dempster (1963); Hewitt (1979)
	Temperature	Development rate	Begon (1983); Hilbert and Logan (1983)
		Microhabitat selection	
	Photoperiod	Phenology	Tauber et al. (1986)
Adult	Moisture	Plant quality	Lewis and Bernays (1985); Mattson and Haacke, 1987a,b
		Oviposition sites	Hewitt (1985)
	Temperature	Fecundity	Visscher et al. (1979)
		Activity levels	Kemp (1986)
		Microhabitat selection	Joern (1982); Gillis and Possai (1983)
	Photoperiod	Phenology	Tauber et al. (1986)
	Wind patterns	Orientation and displacement	Rainey (1976)
	Soil type	Oviposition sites	Dempster (1963); Hewitt (1985)

been recorded in nymphs that have experienced several days of cold, wet weather during which nymphal feeding is inhibited (Hewitt, 1979). Weather also contributes to the rate at which adults become sexually mature and, for females, the rate of oocyte development between clutches (Uvarov, 1966, 1977). Interestingly, the rate of embryonic development in *M. sanguinipes* is explained by a combination of adult maturity date in addition to current temperature (Mukerji and Randell, 1975), illustrating the interplay among life cycle stages.

In dry environments, adults often experience reproductive diapause and exist in a nonreproductive state for as many as several months until quality food becomes available after rains (Ellis et al., 1965; Tauber et al., 1986). Breeding is often synchronized and rapid in such species and egg diapause is typically absent (Tauber et al., 1986). Day length and temperature may also trigger the reversal of reproductive diapause in males as well as females (Middlekauf, 1964; Orshan and Pener, 1979; Pener and Orshan, 1980).

Specific spatial and temporal mosaics of key weather variables, simultaneously considered, may be more important than temporal patterns alone. This is especially true in locusts, where movement from one site to another is a key feature of the population process. Inherent fluctuations in the numbers and distribution of the Australian plague locust (*Chortoicetes terminifera*), for example, are largely caused by spatial and temporal patterns of suitable weather conditions (Chapter 9). Individuals move in space while tracking suitable locations, often in response to rainfall. There was no evidence of negative feedback for either likelihood of emigration or immigration in response to density. The most reasonable explanation is that population increases resulting from favorable weather in a defined breeding area make it more likely that the proportion of emigrating individuals that colonize new, climatically suitable breeding areas increases (Farrow, 1979, 1982b). Similarly, long-distance dispersal in several locust species is strongly influenced by weather patterns, especially wind currents that arise as a result of major weather systems associated with rainfall (Uvarov, 1977; Pedgley, 1979; Farrow, Chapter 9). Widespread winds can easily move flying locusts about and either promote crowding of individuals from widespread areas via convergent currents or drive individuals apart (Pedgley, 1979). There is little doubt that grasshopper population processes are greatly affected by weather acting in this fashion.

Correlations between key climatic parameters and population changes are often weak and nonexistent (Riegert, 1972). Sometimes significant relationships with key weather variables typically explain less than half of the variation in population change (Gage and Mukerji, 1977). When such a pattern exists, it suggests that the relationship between climatic changes and the key processes that drive population change may be exceedingly convoluted or unimportant relative to other overriding population processes.

14.4.2 Biotic Factors

Interactions among species within and between trophic levels may have important implications for grasshopper population dynamics. Generally, these interactions are the least studied aspect of the problem, especially in the field, because of the more extensive survey and experimental effort required. Yet, as this section illustrates, the potential importance of these interactions to understanding grasshopper population dynamics is great.

14.4.2.1 Influence of Host Plants on Population Dynamics

Host plants may contribute much to grasshopper dynamics through their influence on key population parameters. Specific characteristics responsible for the acceptance or rejection of a host plant are relevant (reviewed in Chapters 2, 4, and 5) because these determine whether a species will be present in an area and define its overall population status. Perhaps more important, however, are temporal and spatial shifts in host plant quality and the associated grasshopper responses. Quality is defined in the present context to include both nutritive and defensive properties of the host plant. Central to this view is the importance of the variation in quality of food plants available to a grasshopper population and the response of population sizes to this change because of the significant effect of the host plant on survivorship, growth rate, fecundity, and dispersal.

Host plant quality often varies in response to environmental stress. Plant stress, such as that due to drought, can lead to altered proximate nutritional quality of the hosts (White, 1978; 1984; Rhoades, 1983; Mattson and Haack, 1987a,b). In addition, production of active chemical defenses by plants may fluctuate in response to environmental stresses (Mattson and Haack, 1987a,b); this fluctuation in turn alters the acceptability and quality of the host plants to grasshoppers. The individual and population responses may result from enhanced nutritional quality of stressed plants alone but more probably represents an imbalance between nutritional quality and intrinsic plant defenses (Rhoades, 1983).

Naturally occurring plants may be generally poor in quality. A native sod mixture composed of grasses and sedges in Canada, for example, was consistently unfavorable to *Camnula pellucida* in terms of developmental rate, nymphal and adult survival, and fecundity when compared with spring and summer wheat (Pickford, 1963).

Insect herbivores feeding on high-quality plants often exhibit increased survivorship and fecundity (White, 1976, 1984; Rhoades, 1979, 1983; Mattson and Haack, 1987a,b). Dynamic shifts in these life-history attributes in response to food plant quality may greatly contribute to population fluctuations and partly explain the correlations between extreme weather and grasshopper outbreaks. The influence of weather in these situations is indirect, with effects mediated through changes in the host plant.

Plant defensive chemistry coupled with nutritional characteristics of leaf tissue is central to understanding host plant choice, especially for grasshoppers that feed on nongrasses (Bernays and Chapman, 1978; Bernays, Chapter 5; Bernays and Simpson, Chapter 4; Chapman, Chapter 2). Although many plant secondary chemicals are toxic, some grasshoppers (especially forb feeders) may encounter a wide variety of host plants and associated defenses during their lifetime (Chapter 2). The presence of defensive chemicals may have several effects on grasshopper feeding as it relates to population dynamics. Such toxic chemicals may cause grasshoppers to shift from eating one plant species to another. Such shifts may result in no obvious effect to the individuals, or the new plants eaten may be otherwise less nutritious even while less toxic. Or, there may be less food available that is easily consumed and digested. In any of these cases, the capacity for population growth may be decreased.

Different grasshopper species are not equally susceptible to plant defenses, however. *Schistocerca gregaria* is not negatively affected by condensed tannins whereas the grass-feeding *Locusta migratoria* is (Bernays et al., 1980, 1981). If grass-feeding species are in general less resistant to condensed tannins or other chemical defenses, the impact on population dynamics could be dramatic. Because C_3 grass species tend to have higher tannin levels than C_4 grass species (Capinera et al., 1983), this additional factor could play a role in many grasslands that contain both grass types.

Variation in host plant quality and quantity alter a variety of key life-history traits in grasshoppers, including survivorship, developmental rate, rate of egg production, total fecundity, and oviposition rates. Although the key studies that directly link large-scale population processes to changes in host plant quality have yet to be performed, a variety of carefully crafted experiments indicate the likelihood of their importance. There is typically a direct relationship between the species of host plant taken and grasshopper survival, growth, and reproductive output (Mulkern, 1967). In general, approximate digestibility is correlated with preference rankings (Bailey and Mukerji, 1976a) and different species of host plants have been repeatedly shown to influence fecundity (Pickford, 1962, 1963; Putnam, 1962; Pfadt, 1949; Bailey and Mukerji, 1976b).

The quality and quantity of specific food plants often have marked influences as well. Numerous accounts suggest that poor-quality food plants associated with extreme drought results in poor reproductive performance (Pickford, 1963; Chapman et al., 1979). More graded responses are also observed. Oocyte development is not initiated in *Locusta migratoria* females when fed low-protein *Agropyron repens*. When females are switched from high- to low-protein *A. repens*, egg production drops and terminal oocytes are resorbed (McCaffery, 1975). Few eggs were laid when the mixed feeding *Melanoplus sanguinipes* was fed on wheat seedlings with low nitrogen content (Smith and Northcott, 1951; Krishna and Thorsteinson, 1972).

Physiological responses such as oocyte production in response to quality and quantity of protein in the food plants and the hormonal processes mediating the responses illustrate a close tie. Protein composition of ingested food affects the synthesis of proteins, qualitatively as well as quantitatively, in the fat body (*S. gregaria*; Hill, 1965). Ovarian development in *Melanoplus sanguinipes* was greatly influenced by availability and quality of water-soluble proteins in host plants (Krishna and Thorsteinson, 1972). Female-specific proteins are incorporated into the yolk of developing oocytes after synthesis in the fat body and transport via the hemolymph. Larger primary oocytes contain more vitellinogenic protein (Ferenz et al., 1981). The hormonal feedbacks are very complicated and beyond the scope of this chapter, but the existence of processes that depend on protein availability in the host plant indicates the direct route that can be mapped between host plants and egg production.

Temperatures during growth of western wheatgrass, *Agropyron smithii*, resulted in biochemical changes in the plant which altered both fecundity and survival of *Aulocara elliotti*. When fed grass grown at cool temperatures *A. elliotti* exhibited increased egg production and decreased average longevity compared with individuals fed grass grown at warm temperatures (Visscher et al., 1979). The mechanism responsible for such changes is not known.

Reproductive cycles in grasshoppers can be modified by changes in food plant quality mediated by plant growth hormones. Adult *Schistocerca gregaria* do not mature sexually during the long, dry periods when they feed on senescent vegetation. When new green foliage is available after rains, the locusts mature very quickly. This response has been attributed to high concentrations of the plant growth hormone giberellin A_3 (GA_3) found in the leaves of new growth but only in small amounts in senescent foliage (Ellis et al., 1965). Other responses to plant hormones have been observed in other grasshoppers where GA_3 and abscissic acid both had negative effects on fecundity (Visscher, 1980). Ethylene had both positive and negative effects on *Melanoplus sanguinipes* life histories (Chrominski et al., 1982). At present, it is unclear whether these plant hormones directly influence the physiology of the grasshopper or act secondarily through other unmeasured physiological changes in the food plant (Rhoades, 1983).

Detailed studies have demonstrated the importance of particular chemical constituents in host plants to grasshopper survival, including both defensive secondary chemicals (Harley and Thorsteinson, 1967; Mulkern and Toczek, 1972; Bernays et al., 1974, 1980) and nutrients (McGinnis and Kasting, 1966; Nayar, 1964; D. S. Smith, 1959; D. S. Smith and Northcott, 1951). For example, nitrogen content was very important to nymphal growth of *Melanoplus bivittatus* (McGinnis and Kasting, 1966). *Melanoplus sanguinipes* fed wheat with low phosphorus content developed faster, with increased survival and fecundity, compared with individuals fed plants with higher phosphorus content. In the same context, Lewis (1979, 1984) showed that *M. differentialis* preferred wilted sunflowers to turgid individuals, possibly because important nutrients other than water were altered.

Absolute availability of food may be important and populations may fluctuate in response to changing food supplies (Bernays and Chapman, 1973; Dempster and Pollard, 1981; Belovsky, 1986a,b). It is possible that polyphagous species

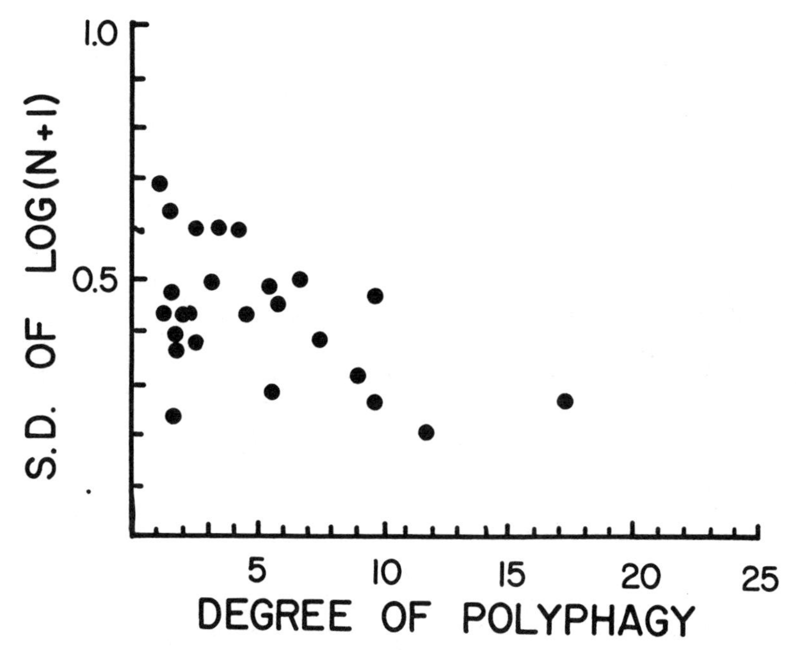

Fig. 14.15. Relationship between polyphagy and variability in population size. Population variability was calculated from 25 years of sampling at North Platte, Nebraska (Joern and Pruess, 1986) and diets of grasshoppers from this site were reported in Mulkern et al. (1969). Degree of polyphagy is calculated as diet breadth using exp (H'), where $H' = \Sigma_i P_i \ln P_i$ (see Joern, 1979).

experience a wider range of food plants during harsh times and the population does not fluctuate as greatly as in more host-specific feeders. Dry conditions had little impact in the survival of the forb- and tree-feeding spur-throated locust, *Nomadacris (Austracris) guttulosa*, during the maturation period. This should be contrasted with the drastic effects of drought on survival, maturation, and fecundity in grass-feeding *Locusta migratoria* and *Chortoicetes terminifera* (Farrow, 1977b).

Population variability may be negatively correlated with the degree of polyphagy in insect herbivores (Redfearn and Pimm, 1988), a prediction that suggests that population sizes vary less when the food resource base is more predictable. Coexisting grasshopper species from Nebraska mixed grassland exhibit a negative relationship between the degree of polyphagy and population variability (Fig. 14.15). No relationship existed between mean densities and degree of polyphagy. This result is consistent with the hypothesis that polyphagous grasshopper populations are more buffered from variation in food resources, partially independent of shifts in availability.

14.4.2.2 Competition

Competition among coexisting individuals occurs when critical resources are in short supply relative to population size. The potential consequences of either intraspecific or interspecific competition for food resources to grasshopper population dynamics include reduced survivorship, lowered fecundity, altered growth and development rates, potentially increased dispersal rates, and stabilized or destabilized population processes. In addition, interspecific competition can lead to host shifts among coexisting species, decreased population sizes of competing species, and possible local extinction of the poorer competitor. How significant are competitive interactions to grasshopper population dynamics? Unfortunately, very few studies have rigorously investigated the importance of either intra- or interspecific competition. Yet sufficient examples and counter examples exist to keep the problem an open one.

Intraspecific competition is regularly noted in terms of density-dependent responses to key demographic parameters (see Section 14.3.3), although immediate ties to resulting population dynamics are seldom made. Recent inquiries into this problem in the context of individual differences are outlined in Section 14.4.4.

Interspecific competition and the consequences among grasshoppers are less well studied. Descriptive studies of host plant use patterns are consistent with the idea that interspecific competition can occur, but alternative explanations for the patterns are just as compelling (Joern and Lawlor, 1980; Joern, 1986b). Evans (1989) manipulated densities of *Phoetaliotes nebrascensis* in otherwise undisturbed plots of tallgrass prairie in Kansas. However, no density responses by other coexisting grasshopper species were detected, even when *P. nebrascensis* densities were increased up to four times normal levels.

On the other hand, population cage studies performed in the field have demonstrated that populations of four species (*Melanoplus femurrubrum*, *M. sanguinipes*, *Arphia pseudonietana*, and *Dissosteira carolina*) from Montana Palousse grassland drop when grown with another species (Belovsky, 1986a,b, unpublished). Equilibrium densities among competing species were readily observed and competitive isoclines derived (Fig. 14.16). In these studies, Belovsky documented good correspondence between resulting equilibrium densities in cages (scaled by available food supply) and grasshopper densities under natural conditions. In addition, both daily mortality and egg production rate were density dependent (G. E. Belovsky, unpublished). Mechanistic interpretations of the process were made by comparing results to expectations from foraging theory using a linear programming model that incorporated physiological constraints on grasshopper feeding (Belovsky, 1986a,b). Results closely fit predictions that assumed energy maximization as opposed to time minimization.

It appears from these results that density-dependent interactions among species may play a role in setting both upper limits to population size as well as greatly influencing the survivorship trajectories and reproductive output throughout the growing season. However, other studies, including experimental ones, do not support the notion that interspecific competition has much of a role in determining population sizes relative to other environmental forces. Future studies must further examine the likelihood and consequences of interspecific competition as well as the conditions under which it is most likely to

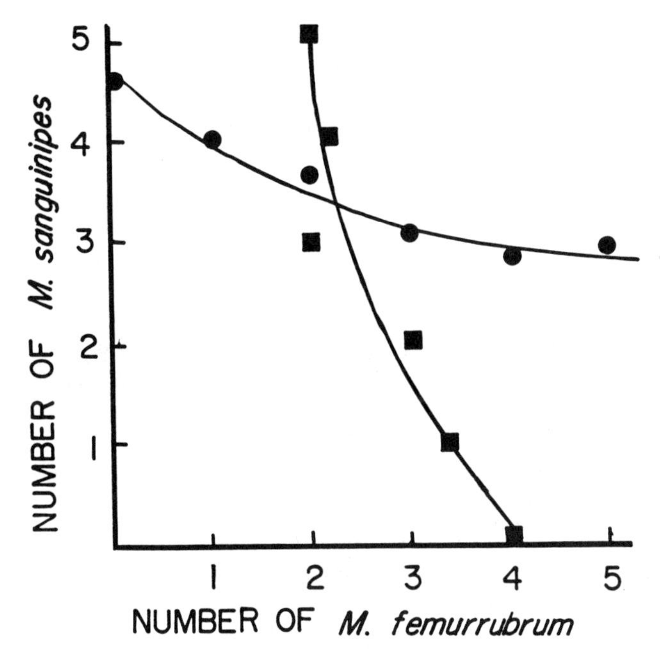

Fig. 14.16. Interspecific competition isoclines at equilibrium in field-cage experiments between *M. sanguinipes* and *M. femurrubrum*. One species (*M. femurrubrum* ■, *M. sanguinipes* ●) is allowed to vary while the other is kept at constant density. (Belovsky, 1986a).

occur. Studies that include both mechanistic and phenomenological responses are needed.

14.4.2.3 Natural Enemies

A wide range of natural enemies are known to attack grasshoppers including microbial and fungal pathogens, parasites and parasitoids, and predators (Dempster, 1963; Greathead, 1963; Rees, 1973; Streett and McGuire, Chapter 15). All life cycle stages are attacked (Greathead, 1963, 1966). In some cases, especially for fungal pathogens under moist conditions, it is very clear that grasshopper populations can be severely affected rather quickly, often resulting in high mortality (Greathead, 1966; Hewitt, 1979; Chapman et al., 1979; Chapman and Page, 1979). Although population sizes can be readily altered, it is unclear whether the impact on grasshopper populations from natural enemies is great enough to be an important regulating factor. Some argue that the numbers of natural enemies are insufficient relative to the number of grasshoppers or else the response is too slow (Dempster, 1963).

Relative to climatic influences, Dempster (1963) argued that interactions with natural enemies are seldom sufficiently density-dependent to do more than dampen the peaks of population fluctuations. However, in a population of the desert locust (*S. gregaria*), Stower and Greathead (1969) estimated that about 50% of mortality resulted from natural enemies, and Chapman and Page (1979) document even higher pressure from biotic factors in *Zonocerus variegatus* from southern Nigeria. The potential importance of natural enemies in grasshopper population dynamics and regulation is illustrated by examining survivorship curves significantly altered due to their impact (Fig. 14.17). Greathead (1966) concludes that natural enemies are probably able to hasten the decline of locust populations but do not materially affect large populations under optimal conditions at the height of a plague. This claim again does not address the important issue of whether natural enemies keep populations at relatively low levels under many situations (i.e., actually regulate populations). Quantitative ecological studies on the impact of natural enemies on grasshopper populations and the conditions that favor regulation by these agents are generally lacking and clearly needed.

Microbial and Fungal Pathogens. Fungal and microbial pathogens clearly have the potential to exert great influence on grasshopper populations (Chapter 15). Few studies have documented the impact of pathogens relative to other mortality factors and examined the impact of such mortality on population

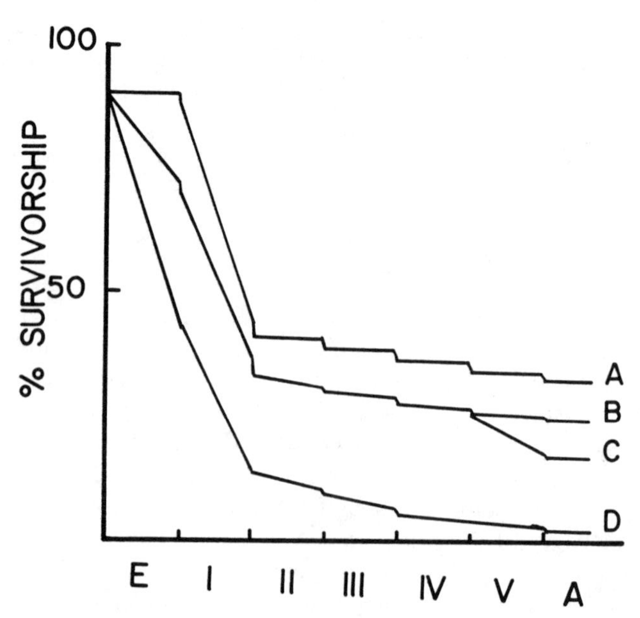

Fig. 14.17. Effect of natural enemies on survivorship curves of the desert locust, *S. gregaria* (Greathead, 1966). Life stages include eggs (E), five nymphal instars (I–V), and adults (A). Curves: A, hypothetical survivorship curve with no natural enemies; B, survivorship with the addition of the impact of observed predation rates (especially birds); C, added effect of a second natural enemy, the parasitoid *Symmictus* (Nemestrinidae); D, hypothetical survivorship curve illustrating the required impact of natural enemies for the locust population to just replace itself.

level changes, however. The fungus *Entomophthora grylli* can be devastating (Pickford and Riegert, 1964; Hewitt, 1979). Extensive mortality of late-instar *Zonocerus variegatus* nymphs and immature adults was common especially after heavy rains (Chapman and Page, 1979; Chapman et al., 1979). On some occasions, 10% of the population died in one night. Life table analysis indicated that *E. grylli* was responsible for significant mortality of *Z. variegatus* over several years (Chapman and Page, 1979), for example.

The importance of these pathogens may often go unrecognized, in part because the effects may be subtle and indirect. For example, multiple pathogens may interact and make an individual grasshopper either more or less susceptible to their action (Chapter 15). Equally important, grasshopper populations may share pathogens, where the limiting attribute in long-term effect of the pathogen depends on the reservoir effect, that is, how pathogens are retained in the system. The protozoan *Nosema locustae* can kill its hosts (Onsager et al., 1981; Street and McGuire, Chapter 15). Equally important, feeding rates of infected grasshoppers may be decreased by as much as 50–75% resulting in morbidity from lack of food but generated because of the presence of *Nosema* (Oma and Hewitt, 1984; Johnson, 1986; D. Johnson and Paulikova, 1986). Infection by *Nosema* may also make the individual grasshopper more susceptible to other environmental stresses such as temperature or ultraviolet light (Wilson, 1974) [e.g., temperature-dependent responses to insecticide toxicity (Hinks, 1985)]. Such indirect effects are poorly understood, especially in natural populations. Also, egg production may be reduced by as much as 60–80% in *M. packardi* or *M. sanguinipes*, respectively (Ewen and Mukerji, 1980). Lowered average egg weight and significantly reduced hatchability also result from *Nosema* infections; little or no hatchability was observed in light eggs (Erlandson et al., 1986).

Parasitoids and Predators. All stages of the grasshopper life cycle are known to be attacked by parasitoids or predators. Many insects (e.g., Bombyliidae, Calliphoridae, Meloidae, Trogidae) spend a portion of their life cycle underground and feed on grasshopper eggs, resulting in egg predation rates that are sometimes great, especially when grasshopper population sizes are large (Greathead, 1962). Once hatched, a great series of parasitoids and predators, including those of early instars, are often present (Greathead, 1962; Rees, 1973).

Most studies consist of mere tabulations of prey parasitized or taken by predators. However, detailed studies illustrate the need for additional research on the dynamics of this interaction. Parasitoids can have a significant impact in some years, primarily on adult stages (Chapman and Page, 1979). Several field experiments have also demonstrated that predators such as spiders (Kajak et al., 1968), robber flies (Joern and Rudd, 1982; Rees and Onsager, 1982), and birds (Joern, 1986a, 1988) can significantly decrease grasshopper population densities. These experimental studies coupled with the very strong inferential evidence that most grasshoppers are cryptic against naturally chosen backgrounds (Joern, 1988; Bernays and Hamai, 1987) suggest that ecological and evolutionary pressure from visually orienting natural enemies can be significant.

Circumstantial evidence suggests that parasitoids have important but intermittent effects on grasshopper population dynamics. In most years, parasitoid prevalence levels for a particular grasshopper population may be very low but occasionally reach high densities. The calliphorid (Diptera) *Blaesoxipha filipjevi* is a principal parasitoid of *Zonocerus variegatus* adults in southern Nigeria (Chapman and Page, 1979). Parasitism rates of greater than 40% were recorded there for several years running in the early 1970s (Chapman and Page, 1979) and at other sites in this region of Africa (Sierra Leone, 1934; Ivory Coast, 1950–1951) (Chapman et al., 1986). However, at nearby sites 10 years earlier, parasitism rates on the order of 2–3% were recorded. The discrepancy is presently unexplained but may reflect temporal variation in infection levels. At present, it is difficult to determine the precise timing of parasitoid infestations or the consequences for actual population dynamics. Parasitoids and predators built up rapidly during upsurges of the Australian plague locust (*C. terminifera*) (Farrow, 1979). Although these might have been expected to exert a delayed density-dependent effect on locust numbers, they appeared to become effective only following population declines that resulted from emigration (not the impact of the parasitoid) and did not directly influence variations in grasshopper density (Farrow, 1977b, 1982b). It was not determined, however, whether the parasitoids kept the population in check during environmental conditions that were less favorable for outbreaks.

Robber flies (Diptera; Asilidae) are very conspicuous insect predators in many arid grassland systems, and several species take primarily grasshoppers (Rees, 1973; Dennis and Lavigne, 1975). For example, diets of 6 of 33 species (18%) from shortgrass regions of Wyoming and Colorado contained a large proportion of orthopterans (primarily acridids) and another six species preyed on grasshoppers to a lesser degree (Dennis and Lavigne, 1975; Lavigne and Holland, 1969; Rogers and Lavigne, 1972). The largest proportion of grasshoppers in their diets was observed in late July into September, the period with the greatest species diversity and density of adult grasshoppers. Two studies have carefully examined the impact of robber flies on grasshopper populations but with interesting differences (Joern and Rudd, 1982; Rees and Onsager, 1982).

Proctacanthus milbertii preys primarily on grasshoppers in North American grasslands; 94% of the observed diet in one year was grasshoppers (Joern and Rudd, 1982). This large robber fly is a generalist (grasshoppers were largely taken in proportion to their abundance), sit-and-wait predator. Grasshoppers are taken after movement. Based on individual predation estimates of 0.5–2 grasshopper prey/day, the *P. milbertii* population took from 0.5 to 2% of the composite grasshopper assemblage per day in years when the robber fly population was quite large. This predation rate occurred just as grasshoppers at this site were at their reproductive peak (July–August) and a total of ~25% of the grasshoppers were taken during this period. In most years, however, robber fly populations were smaller, so these estimates must be considered an upper limit.

Although Joern and Rudd (1982) were able to document significant predation pressure from robber flies in some years, this is not always the case (Rees and Onsager, 1982, 1985). In Montana grasslands, dominant robber fly species (*Efferia bicaudata*, *E. staminea* and *Mallophorina guildiana*) fed primarily on Diptera, including dipterous parasitoids of grasshoppers. Presence of these robber fly species increased the survival rate of grasshoppers living in high-density patches (30–40 individuals/m^2 at the beginning) (Rees and Onsager, 1982), presumably by removing important parasitoids. Reduction of robber fly densities by approximately 40% resulted in a doubling of rate of parasitism (from 2–3% to 5.5–6.5%) from several Dipteran families (Anthomyiidae, Nemestrinidae, Tachinidae, and especially Sarcophagidae) (Rees and Onsager, 1985). When robber flies were nearly absent in a natural experiment, parasitism rate was about fourfold greater. A significant 12-day reduction in average longevity of grasshoppers (40 to 28 days) resulted from parasitoids. In this case, then, complex interactions among parasitoids, predators, and prey resulted in significant effects on predation rates of grasshopper populations but the effect of dominant robber flies was to negate the impact of dipteran parasitoids.

Losses from avian predators can be quite large (Greathead, 1966). In terms of the importance of birds as regulating agents several key questions remain: is avian predation significant on a regular basis or only infrequently? Are birds typically important as regulators of grasshopper populations? Do avian predators respond to grasshopper populations in density- or frequency-dependent fashion? Does switching among alternative prey occur and to what degree? Is there compensation by other mortality factors when avian predation pressure changes? Answers to each of these questions are generally not known and require quantitative studies, preferably experiments under field conditions. Given the potential importance of frequency and density-dependent predation as a dynamic regulator of insect herbivore populations, especially at low densities, this area deserves more intense study.

Birds regularly remove approximately 25% of the adult grasshoppers over a 4–5 week period in July and August from a North American sandhills grassland site, a period when the maximum number of adult grasshoppers is present and at their reproductive peak (Joern, 1986a, unpublished). Both densities and species diversity drop significantly in the presence of avian predators. Field experiments were performed during years when grasshopper population densities were average to below average in size, so it is not known what happens at high densities. Because of other compensating factors responding to losses to predation, the influence of avian predation on actual dynamics and regulation of grasshopper populations is still open. The dominant grasshopper sparrow, for example, was shown to switch among different grasshoppers in a frequency- and density-dependent fashion (Joern, 1988, unpublished). Synoptic studies that include predation among the many factors affecting grasshopper population dynamics are clearly needed.

14.4.3 Dispersal

Dispersal plays a major role in acridoid population dynamics. A wide range of behaviors, from large-scale seasonal migrations (Uvarov, 1977) to sedentary territoriality (Chapter 10), demonstrates the diversity of movement strategies. Dispersal serves as a means of escape from unfavorable conditions such as drought or reduced food availability, immigration to more favorable conditions, or opportunity to use more than one habitat, as when seasonal weather patterns lead to geographic variations in food abundance or access to mates. Adaptations for variations in displacement capabilities are associated with temporal and spatial stability of residence sites (Southwood, 1988). An understanding of dispersal patterns is essential to investigations of population structure and rates of gene flow within and among populations. Additionally, many populations exhibit dispersal polymorphisms, ranging from complete physiological and morphological phase changes, to wing-length polymorphisms or variations in flight capabilities within monomorphic populations (Harrison, 1980). Analysis of the extent of and variation in dispersal capabilities demonstrates the disparity of strategic responses to the abiotic and biotic factors affecting movement.

Circuit migrations, large-scale seasonal migrations involving movement from one breeding area to another along a more or less repeated route, are characteristic of many locusts (Chapman, 1972; Uvarov, 1977). High densities of prereproductive adults are drawn together in areas where air masses converge, the direction of movement being influenced by regular seasonal reversals of winds such as those found in convergence zones (Rainey, 1962; 1963; C. G. Johnson, 1969). These converging air masses can also lead to precipitation so that movement is often directed toward areas of future lush, green vegetation.

Although geographic variation in breeding seasons and migration routes occurs (Davies, 1952; Donnelly, 1947; Fortescue-Foulkes, 1953; Z. Waloff, 1946), many characteristics are similar among widespread populations. Large locust hopper bands have been recorded marching at rates ranging from 80 yd/day for first-instar larvae to 650 yd/day for fifth-instar larvae, with rates being strongly affected by both the size of the band and the speed of individual movements (Ellis and Ashall, 1957). Swarms of adults moving at speeds between 9 and 23 km/h (Rainey, 1963) are usually day flying and composed of both flying and intermittently resting individuals (Z. Waloff, 1966; Z. Waloff and Rainey, 1951). On the other hand, solitary- and *transiens*-phase locusts exhibit daily displacement averages of only 1.5 m (Z. Waloff, 1963), with night flights commonly occurring among individuals from low-density populations (Roffey, 1963).

An interesting example of intraspecific variation in dispersal capability exists in *Schistocerca cancellata* (Drury). This locust follows north–south migration routes between breeding sites during outbreaks of gregarious individuals in Argentina, but remains in the solitarious, sedentary phase in Chile (Lieberman, 1972). Additionally, swarms have been reported moving into Paraguay, Bolivia,

Brazil, and Uruguay but did not return to their original breeding grounds in Argentina. *Locusta migratoria migratoria* (R. & F.) swarms followed north–south migration routes in western Africa but swarms disperse away from breeding areas in other parts of its distribution (Batten, 1972). Although *Nomadacris septemfasciata* (Serv.) migrants never return to previous breeding areas, large-scale movements of dispersing swarms repeatedly followed the same route during the 1933–1943 plague in Africa (Chapman, 1972).

Expansive migration is the second type of large-scale dispersal characteristic of Acridoidea. In this case movement of adult swarms, for one to many generations, is directed into new areas without subsequent return to previously exploited breeding areas (Chapman, 1972; Uvarov, 1977). As discussed, even species that are clearly characterized by circuit migrations also have some populations that disperse into new habitats. Regular seasonal reversals in wind direction may not occur in certain parts of these species' ranges, thereby greatly reducing the likelihood of return after dispersing.

Populations of *Chortoicetes terminifera* periodically increase to outbreak levels in central western New South Wales as a result of increased natality and immigration-coupled limited dispersal (Farrow, 1979). Invasions of new areas have been primarily influenced by nocturnal migrations during wheather disturbances leading to massive population translocations (Symmons and Wright, 1982). Typical movements of *Melanoplus sanguinipes* during an outbreak are indicated in Figure 14.18. Similarly, movement of outbreak populations of the now extinct *Melanoplus spretus* from the permanent breeding grounds of the Colorado plains across Kansas occurred throughout the late 1800s through extensive dispersal across the Colorado plains (R. C. Smith, 1954).

Mark and recapture methods have been utilized in a number of studies to estimate small-scale movement rates. Average daily movement rates of most populations are very low compared with those of outbreak swarms. Many of these movements are associated with dispersal to new habitats. Early instars of *Chorthippus bruneus* and *C. parallelus* were found more frequently in bare ground areas, with later instars moving into tall grass habitats (Dempster, 1955; Richards and Waloff, 1954). Additionally, adult females returned to the same bare areas for oviposition. *Anacridium melanorhodon* (Walker) characteristically disperses widely while feeding before aggregating in breeding areas (Popov and Ratcliffe, 1968).

Another interesting result of these studies is the range of displacement distances exhibited by individuals. Comparisons of mean, minimum, and maximum values demonstrate that in all populations certain individuals have a stronger tendency for dispersal than others. McAnelly and Rankin (1986a,b) found that at least some but not all *Melanoplus sanguinipes* individuals, in all populations examined, exhibited dispersal flight behavior when tested in tethering experiments. The proportion of migrants was higher among individuals from high-density populations for both field-collected and lab-reared grasshoppers. In addition, evidence suggests that these differences among populations and among individuals within populations have a genetic basis. In contrast,

Fig. 14.18. Dispersal of *M. sanguinipes* in North America during outbreak periods (Chapman, 1972). Main oviposition areas in different years are represented as solid area, 1937; stippled areas, 1938; and hatched areas, 1939. General direction of flights is indicated by heavy arrows, 1938, and open arrows, 1939.

several studies suggest that territorial behavior by some species can greatly limit dispersal under certain conditions (Aikman and Hewitt, 1972; Clark, 1962; Greenfield and Shelly, 1985; Otte and Joern, 1975; Shelly and Greenfield, 1985).

It is clear that a wide range of dispersal capabilities exists in grasshoppers. Density-dependent factors, acting either directly or indirectly, are often primary influences on movement for many species (Riegert et al., 1954; Uvarov, 1966, 1977). Habitat quality as it relates specifically to food quality and quantity, availability of suitable oviposition sites, and availability of suitable resting sites also has a strong influence on dispersal patterns and rates (Baldwin et al., 1958; Joern, 1983; Gaines, 1989). Specific attention paid to the variation among individuals within populations will greatly contribute to our understanding of population structure, including more precise estimates of gene flow and neighborhood sizes that are integral to generalizations concerning grasshopper population dynamics.

14.4.4 Qualitative Differences among Individuals

Individuals in a population are not uniform, and the variation that exists in traits among individuals strongly influences survival or reproductive abilities. How central are these individual differences among grasshoppers to the resulting population dynamics? Although they are generally ignored in population level thinking, evidence is accumulating that individual differences can be quite significant, for these individual differences can fundamentally alter the population growth trajectory as well as expose important mechanistic underpinnings of population growth (Begon, 1984; Hassell and May, 1985; R. H. Smith and Sibly, 1985; Begon and Wall, 1987). Qualitative changes (genotypic or phenotypic) in the composition of individuals within a population may accompany or even cause outbreaks (Uvarov, 1961; Krebs, 1978; Barbosa and Baltensweiler, 1987; Lomnicki, 1988). Uvarov (1961) held stronger views: "It is the physiological variability and the adaptability of a temporary, reversible kind closely linked with equally temporary environmental changes, which are important in population dynamics."

Locust outbreaks and the associated phase polymorphism syndrome provide an excellent though extreme example (Uvarov, 1966; Dale and Tobe, Chapter 13). Phase polymorphism is an example of environmentally driven phenotypic plasticity that facilitates survival and reproduction of some individuals (or the progeny of certain females) in spite of the unstable conditions that may exist in the normal environment (Barbosa and Baltensweiler, 1987). Clear phenotypic, largely physiologically driven differences exist among individual locusts from high- versus low-density phases. In other cases, genetic shifts among individuals may be contributing elements to population change in response to changing selective pressures in varying environments (including the demographic environment). A recursive relationship exists: populations change in response to the qualitative differences in population composition but changes in the population dynamics can alter the types of individuals making up the population. The contribution of "qualitative differences" due to both genetic and nongenetic control to grasshopper population dynamics in general, although possibly extremely important, is poorly understood.

Examples of striking intrapopulational differences among individual grasshoppers within habitats that have strong ties to survival and reproduction are accumulating. In this category are color polymorphisms that vary in response to the predominant habitat or background color of that season (Rowell, 1971; Gill, 1979; Dearn, chapter 16) and presumably diminish detection from visually orienting predators; wing-length polymorphisms that influence dispersal and fecundity (Chapman et al., 1978; Dearn, 1978; McCaffery and Page, 1978; Ritchie et al., 1987); phase polymorphism in locusts resulting in morphological, physiological, behavioral, and ecological changes in response to density (Uvarov, 1966, 1977); and environmentally determined, size-related differences in survival or fecundity (Wall and Begon, 1986, 1987a,b).

Diapause variation in the Australian plague grasshopper (*C. terminifera*) is accompanied by differences in egg pods and oviposition depth (Wardhaugh, 1977, 1986). Diapause state and the nature of the egg pod are influenced by changing temperature and photoperiod. Egg pods laid in the autumn are curved and confined to the upper 5 cm of soil whereas those laid in the spring and summer are straight and reach depths of 10 cm. Long or increasing days coupled with high or increasing temperatures resulted in long pods with nondiapausing eggs, whereas short or intermediate daylengths coupled with low or decreasing temperatures resulted in curved pods with diapausing eggs.

The qualitative differences in oviposition depth have putative adaptive value by regulating temperature and moisture conditions experienced by the eggs and thus directly affect rates of development, survival, and the proportion of eggs entering diapause (Wardhaugh, 1986). *Chortoicetes terminifera* is multivoltine where eggs complete development without diapause during portions of the year. Eggs laid during the spring and summer encounter periods of hot, dry weather. Even though eggs of this grasshopper are desiccation resistant for 2–3 months, eggs laid at increased depths are better protected from desiccation and high soil temperatures. In addition, embryonic development of eggs laid at lower depths is less likely to be cued by light rains, which are unable to promote sufficient plant growth required for survival. On the other hand, eggs laid during the cooler autumn experience different environments. Available rainfall during this period is sufficient to support plant growth and eggs near the surface are better placed to respond to smaller increments of precipitation. In addition, the higher soil temperatures at shallower depths, no longer extreme, encourage rapid egg development in eggs conditioned to hatch.

Chitty Self-Regulation Hypothesis. Locust outbreaks and phase polymorphism may be manifestations of the Chitty hypothesis (Krebs, 1978), a view that invokes the contributions of individual qualitative differences to population change, especially with an underlying genetic basis. At the extreme, this hypothesis assumes that density-dependent self-regulation of a population is maintained in a changing environment because of intense natural selection for qualitatively different phenotypes in ecological time. Excessive population increase is ultimately prevented by changes in the quality of individuals in the population (as physiological/behavioral, genotypic, or phenotypic responses) (Krebs, 1978), a result clearly observed by Wellington (1960, 1964).

Whether locust populations are self-regulated and the degree to which responses responsible for self-regulation are under strict genetic control are unknown (Krebs, 1978; Garland, 1982). Pheromones appear to play major roles in the phase transformation process (Loher, Chapter 11). The bacterial metabolite locustol seems to initiate gregarization (Nolte et al., 1973). Other important pheromones in this process that influence the degree and rate of phase transformation include a male maturation pheromone that accelerates and synchronizes the process of adult maturation (Amerasinghe, 1978) and an unidentified pheromone produced by ovipositing gregarious females that contributes to increased aggregation (Lauga and Hatté, 1977). Presence of these pheromonally induced responses prompted Garland (1982) to reject the Chitty hypothesis in the strict sense as an explanation for the evolutionary maintenance of phase polymorphisms in locusts (i.e., the process is not under strict genetic control). On the other hand, important aspects of phase polymorphism (possibly the threshold for response to pheromones?) are clearly under genetic control, as indicated in laboratory selection experiments for morphological differences in *L. migratoria* (Hunter Jones, cited in Chapman, 1976) and color patterns in *Locustana pardalina* and *Locusta migratoria* (Nel, 1967).

Individual Ecological Responses and Effects. Variance among individual grasshoppers within a population is widely recognized but seldom examined for any consequences at the population level. Yet such variation may determine which individuals contribute to the next generation and by how much. Asymmetric abilities to obtain key resources among individuals may greatly magnify the impact of intraspecific "scramble" competition and influence population dynamics in grasshoppers (Begon, 1984). This is generally poorly studied.

The effect of population density on mortality, development, and fecundity has been examined in *Chorthippus brunneus* in the context of the individual phenotypic character, size. Phenotypic differences that arise early in the lifespan can result from combinations of genetic differences that influence differential individual growth and development rates, from maternal influences such as the initial size of individuals at hatching, or from stochastic variation in the manner in which different individual nymphs find and use food of sufficient quality (Wall and Begon, 1987a,b). In field cage experiments, Pickford (1960) demonstrated that grasshoppers that hatched earliest laid the greatest numbers of eggs because of both greater adult longevity and a higher rate of oviposition. Size-related effects were not reported but could have easily been involved. Clear-cut differences in developmental rate are related to size, for small adults also took longer to develop (Wall and Begon, 1987b; Fig. 14.19a). Mortality pressure tended to remove disproportionately smaller individuals at higher population densities or at later instars (Wall and Begon, 1986). These differences become increasingly apparent throughout the remainder of the lifespan if density-dependent pressures exist. Similarily, reproductive output can be size related under some circumstances, a consequence of individual differences accumulating during nymphal development, especially at high densities (Wall and Begon, 1987a). At low densities, small females compensated for smaller clutches by producing clutches at a faster rate. At higher densities with limited food, a small proportion of the females, the larger ones, produced the majority of the eggs (Fig. 14.19b). However, individual phenotypic differences do not explain the differences in egg production between the low- and high-density treatments for there was no size-related difference in reproductive output in the low-density treatment.

14.5 EVOLUTION OF GRASSHOPPER LIFE HISTORIES

Life histories record the statistical responses of individuals and indirectly reflect responses of populations to local environments. Probability distributions of environmental conditions that directly influence mortality and fecundity interact to influence the evolution of key life-history traits such as numbers

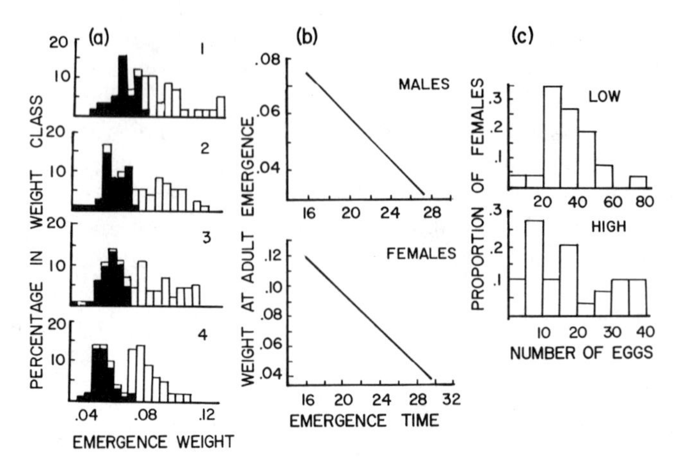

Fig. 14.19. Individual differences in key fitness traits. (a) Percentage of total number of individuals that emerged at four increasing densities (1–4). Males are represented by solid bars. Female bars (open) are placed on top of male bars. (b) Time to adult emergence as a function of weight at adult emergence (redrawn from Wall and Begon, 1987b). (c) Distribution of reproductive output of *C. brunneus* in low- and high-density treatments. Two additional points are not shown but important: only a small proportion of individuals contribute eggs at high densities and a positive relationship exists between size and number of eggs at laid, suggesting that at high densities, only large individuals contribute substantially to the eggs for the next generation (Wall and Begon, 1987a).

and timing of egg pods produced, the number and size of eggs per clutch, the phenology of individual species populations, generations per growing season (voltinism), the nature of diapause (if any), patterns and degree of dispersal, and the thresholds and mechanisms responsible for phase polymorphism in locusts. The summation of these life-history traits provides the statistical basis of population growth in response to local environmental conditions. When expression of these traits has a genetic basis and phenotypic variation exists, it becomes subject to the action of natural selection.

To understand population growth of any particular species requires that we have some understanding of such adaptive responses to a changing environment. It is not possible to be the best in all possible biological worlds and even adaptation to local conditions requires compromise (Stearns, 1976; Reznick, 1985). Understanding the evolution of life-history traits in acridoids would allow us to develop a framework for predicting types of population responses expected in grasshoppers with different life-history traits when environmental conditions vary. At present, very little effort has been directed at disentangling forces responsible for the evolution of life histories in grasshoppers.

14.5.1 Potential Fecundity

Potential fecundity varies dramatically among species. The number of ovarioles provides the upper limit for the number of eggs that can be laid in each clutch.

This varies among species from the minimum of 2 to well over 100 (Fig. 14.10). Although individuals seldom produce the maximum number of eggs possible with each clutch and the number of clutches produced per lifetime may vary depending on external conditions (e.g., food quality or temperature), the number of ovarioles represents an anatomical (evolutionary) commitment to reproduction on the part of the individual. The number of ovarioles may represent an evolved life-history trait that corresponds to the likelihood of reproducing in a given environment or correlates negatively with the degree of environmental permanence. How much of the variation in ovariole number among species is an adaptive evolutionary response to local environmental conditions and how much merely reflects phylogenetic history? In other words, does potential fecundity vary in a predictable way among different environments within species? Is the total number of ovarioles predicted based on attributes of an individual such as size in an allometric fashion, or does it vary in other ways, such as a response to environmental patterns? Intraspecific phenotypic variation (with an underlying genetic basis) in specific life-history traits along environmental gradients is expected but unstudied.

Phylogenetic constraints may direct some of the variation in key life-history traits that is observed (Ballinger, 1983). Ovariole number varies directly in relation to body size in some grasshoppers (Bellinger and Pienkowski, 1985). Interspecific variation in ovariole number for species in the genus *Melanoplus* correlates very highly with femur length (Fig. 14.20). For the 10 species that were originally used to draw attention to this relationship, body size correlated with elevation and length of the growing season (accumulated temperature) (Bellinger and Pienkowski, 1985). Note also, however, that other *Melanoplus* species from distant sites also fit the relationship nicely. Such a result suggests that traits such as potential fecundity as measured by ovariole number are not strictly adaptive in the context of present-day environments. Life-history parameters have not necessarily evolved to meet present conditions encountered by individuals at the site but rather reflect past evolutionary relationships that are encoded in the phylogenetic lineage and expressed as taxonomic differences. However, the relationship is not predictive for all grasshoppers, as evidenced by the lack of fit when data for other taxa are superimposed on the regression line obtained for the *Melanoplus* species that have been studied (A. Joern and S. B. Gaines, unpublished). In addition, there was no reliable relationship between hind femur length and ovariole number within populations for any of three Australian grasshoppers studied along an elevational gradient (Dearn, 1977).

Within- versus among-species comparisons sometimes give different trends. For example, among-population (within-species) comparisons in Australian grasshoppers documented that the number of ovarioles decreased with increased elevation while the hind femora simultaneously increased in length. However, there appears to be a positive correlation between hind femur length and ovariole number when comparisons are made among the three species, a result consistent with the results of Bellinger and Pienkowski (1985). Clearly, very

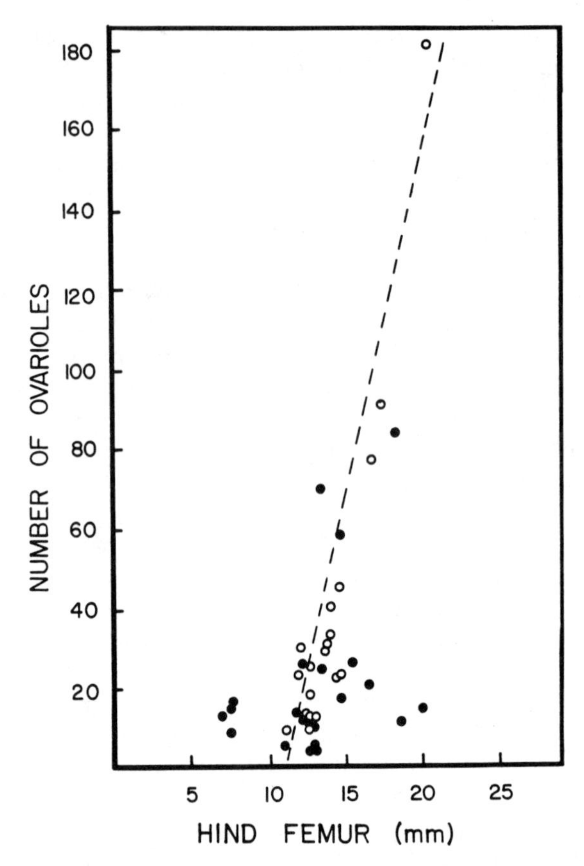

Fig. 14.20. Relationship between ovariole number and body size using hind femur as an index. Ovariole numbers from *Melanoplus* species are indicated by open circles (from Bellinger and Pienkowski (1985). Other grasshopper species are represented by closed circles (from A. Joern, unpublished). The dashed regression line represents the original regression relationship between ovariole number and hind femur length based on nine *Melanoplus* species presented in Bellinger and Pienkowski (1985). Additional *Melanoplus* species fall along the original regression line.

contradictory conclusions can be drawn depending on the focus of the question. These results do highlight the interesting variation observed in critical grasshopper life-history traits such as potential fecundity and the possibilities for testing life-history theory using grasshoppers. Perhaps equally important is that these results demonstrate that evolutionary responses must be carefully evaluated if we are to understand the parameters that drive population changes. Only after reasonable understanding of underlying variations in key life-history traits will we understand why some populations/species undergo wildly fluctuating changes in density whereas other species are relatively constant in numbers.

Intraspecific variation in key life-history traits along environmental gradients has also been documented in several instances (Phipps, 1959, 1962; Blackith and Blackith, 1969). Is this variation typical and does it represent some adaptive response to environmental variation?

Life-history traits of three brachypterous Australian grasshopper species (*Kosciuscola cognatus*, *Kosciuscola usiatus*, and *Praxibulus* sp.) were examined along altitudinal gradients (Dearn, 1977). Ovariole number increased with altitude in all species. Females at higher altitudes laid fewer clutches over the season than females at lower altitudes and overall fecundity (total number of eggs) decreased significantly with increased altitude (Dearn, 1977). In this study, few ovarioles failed to contribute eggs to a specific clutch so that potential and actual fecundity were nearly coincident in each clutch. Total number of clutches over the season also decreased with increased elevation probably because of a shorter growing season. Clearly, rearing individuals from all populations in common environments is required to assess whether the difference is genetic.

14.5.2 Allocation and Tradeoffs

Egg Size and Number. A central assumption in the development of theory of the evolution of life histories is the allocation principle (Stearns, 1976; Reznick, 1985). Organisms have a finite pool of energy or other required resources to be allocated among reproduction, somatic tissue maintenance, growth, and storage needs (Calow, 1983). Such allocations should reflect differences in life histories among organisms in response to differences in environmental conditions and degree of variation. If the allocation of required resources to reproduction is finite, an additional tradeoff must be made: an individual can produce a few large eggs or many small eggs. Potentially more offspring can be produced when many small eggs are produced. This results in increased fitness to the parent only if hatching nymphs survive! Nymphs hatching from larger eggs tend to be larger and have an increased probability of survival and a faster developmental rate (Capinera, 1979). There should be strong selective pressures directing allocation of resources available for reproduction along the continuum bracketed by few large or many small eggs depending on the predictability and harshness of the environment. In seasonal environments with some uncertainty in timing and duration, an individual's fitness can be enhanced by laying the maximum number of eggs that survive and develop quickly enough to reach reproductive maturity during the "reproductive window" of that year. If individuals develop slowly, they would most likely leave fewer clutches than individuals that develop more quickly and mature just as conditions are suitable for reproduction. Individuals from populations in different habitats should exhibit differences in egg number and/or size in response to the timing and harshness of the environment.

Egg sizes and consequent hatchling weights in two grasshopper species from southern England, *Chorthippus brunneus* and *Chorthippus parallelus*, both exhibited considerable variation among different sites located in different habitats (dry grassland, wet meadow, and heathland) (Monk, 1985). Eggs hatched first in dry grassland and last in heathland, developed most quickly in heathland and slowest in dry grassland, and exhibited no differences in reproductive maturation of adult females among the sites, results that suggest local adaptation. The increased developmental rate associated with heavier hatchlings resulted in approximately equal timing of first oviposition by females in each habitat. Females from dry grassland sites could lay more eggs because they had to invest less per offspring to support a developmental rate synchronizing reproductive maturation with the appearance of suitable conditions. If individuals developed slowly and missed the "reproductive window," they would lay fewer clutches and have reduced fitness compared with those that matured just as the window appeared (Sibly and Monk, 1987).

A theoretical treatment of the evolution of grasshopper life cycles for univoltine grasshoppers that overwinter as eggs developed predictions relating to egg size versus adult size (Sibly and Monk, 1987). Maximum fecundity is expected for those individuals that develop at a rate that is synchronized with the "reproductive window," for they can lay the most eggs. Positive effects of egg size, discounted by negative effects of egg size, on nymphal survival were modeled. These theoretical predictions were tested with the data of Monk (1985) for the two *Chorthippus* species. Egg size of *C. brunneus* equaled optimal expectations. *Chorthippus parallelus* would have done better to produce smaller eggs at all sites. Clearly the model has captured important elements of the life cycle for *Chorthippus* based on first principles of the evolution of life histories. In the case of *C. parallelus*, either the structure of the model is incorrect or the parameters are imprecisely estimated. Although sensitivity analysis indicated that predictions depended strongly on juvenile mortality, estimates of this parameter had large confidence intervals, suggesting that the test must be considered provisional. An interesting physiological wrinkle suggesting local adaptation is that individuals from the heath population that hatched later than the other populations in the field actually hatched first under controlled laboratory conditions. The nature of the experiment highlights previous claims in this chapter that there is an underlying genetic basis for key physiological responses and expression thresholds of traits directly influencing life histories. Perhaps more importantly, the results suggest the presence of complicated relationships underlying developmental rate in the field and the potential problems of investigating responses of natural populations using only laboratory studies.

Life-Span and Reproduction. Life-history theory has repeatedly assumed a cost to reproduction (Stearns, 1976, 1977; Calow, 1983). This is the basis of the allocation principle underlying most models. The commonly assumed hypothesis that increased reproduction results in decreased longevity was not upheld in laboratory studies of *Melanoplus sanguinipes* (Dean, 1981). Longer-lived females did not differ in the number of eggs laid per clutch or interclutch interval. However, virgin females from a parthenogenic line of this species produced fewer eggs and lived longer than mated females. Although no virgins had died by day 20, over 50% of the mated females had died (Dean, 1981); the number of eggs produced per day of adult life or the interclutch interval did not differ between mated and unmated females. Understanding such tradeoffs may require knowledge of the covariation of these traits with other life-history atributes and laboratory studies must be carefully coupled with field studies to understand the implications of the relationship between life-span and survivorship. Finally, unless genetic tradeoffs associated with physiological relationships can be defined, it will be difficult to interpret such tradeoffs in an evolutionary context (Reznick, 1985).

Tradeoffs are difficult to document, especially in a rigorous genetic context. A phenotypic reproductive cost to survival has been detected in *Chorthippus brunneus* (de Souza Santos and Begon, 1987). Results suggest that survival is sensitive to somatic investment and reproductive effort drops proportionally when food is limited. A negative correlation was observed between survival and reproductive effort when individuals given the same ration schedule were compared. In addition, results of this study suggest that females modulate their reproductive output in accordance with food availability, the result being almost constant survivorship between ration treatments. A reproductive cost argument would explain this result. However, expression of such costs is complex and includes an interaction between body size and food availability.

Heritability of Life-History Traits. Genetic variability must accompany phenotypic variation if natural selection is to operate. Heritability underlying life-history traits has been demonstrated in several instances, although the selective pressures responsible for the observed natural patterns have seldom been rigorously analyzed.

Egg diapause in grasshoppers clearly has a genetic basis (Slifer and King, 1961; Pickford and Randell, 1969). Artificial selection has resulted in a non-diapausing strain of *Melanoplus sanguinipes*, a species that typically exhibits egg diapause (Pickford and Randell, 1969).

Additional evidence that natural differences in life-history traits have a genetic basis comes from "common garden experiments," where individuals from different populations are raised in common environments. Differences that remain are genetically based, especially if they persist for two generations to remove effects of maternal and grand-maternal effects. Significant differences in egg size and clutch size were observed in several species when examined in this fashion. Differences among *Myrmeleotettix maculatus* populations for both egg size and clutch size were observed in lab-reared stocks from different sites (Atkinson and Begon, 1987a,b). Selective pressures responsible for the among-population differences in these traits were not readily explained but may represent selective pressures on nymphs (as a consequence of size). Different selective pressures or very complex interactions among alternate selective pressures may have been operating at different sites.

14.6 CONCLUSIONS

Key issues remain to be resolved. Past reviews have concluded that biotic factors are unlikely to be very important and that abiotic factors are of overwhelming impact (Dempster, 1963; Uvarov, 1977). At best, they argue, biotic influences only hasten the decline of populations after they have peaked (Dempster, 1963). We disagree. Based on recent theoretical developments such as that summarized by Berryman et al. (1987) and new empirical results (reviewed above), we believe that the set of important factors affecting grasshopper populations is much more complex than previously presented and that interactions among biotic and abiotic agents will prove central. Both types of interactions will play key roles with the population dynamics largely determined by the conditional nature of the interactions. Depending on the state the population is in, a change in any one important force may result in a different population response. Future research must focus on defining these conditional relationships.

A clear picture of grasshopper population dynamics and regulation will result from increased attention to the following elements: individual or other qualitative differences in addition to more easily studied quantitative changes; overall stability of population trajectories through both long-term descriptive studies and insightful field experiments; synoptic views in which multiple factors are simultaneously investigated and conditional responses recognized for what they are; the evolution of life-history traits in the context of environmental variation; and the context within which other coexisting species play a role in the dynamics of single populations.

At present, no global overviews have been presented that adequately and explicitly describe grasshopper population dynamics and regulation. The assumption that weather is responsible for the ebb and flow of grasshopper densities (Gyllenberg, 1974; Rodell, 1977) strikes us as too simplistic. In the same vein, we do not deny the great impact that abiotic factors have on key links in a grasshopper life cycle or that it is possible to attribute much of the among-year variation in population densities to density-independent, causal mechanisms. Because many physiological processes that affect key life-history parameters are clearly temperature dependent, potential mechanistic links exist between population change and abiotic forces. But it is misleading to attempt to explain population processes only in these terms. We primarily argue that grasshopper population responses are complex, conditional, and indirect in most cases, a situation that requires a more sophisticated ecological analysis.

The synthetic model (Section 14.2.2) proposed by Berryman et al. (1987) is *one* concise statement (among others) of the possible interplay between abiotic and biotic factors which appears to have great utility. Though devised for forest insect herbivore pests, this viewpoint appears to capture the important elements involved in grasshopper population change and provides a rich global framework for attacking the population dynamics–regulation problem. It also provides a satisfying, through unproved, resolution of the manner in which equilibrium and nonequilibrium forces can interact while also accommodating the different types of biotic versus abiotic interactions. Although some data relative to grasshopper populations are consistent with this particular outlook (reviewed above), additional specific hypotheses must be devised and subsequently rigorously tested.

What important features of the Berryman et al. (1987) model are relevant and why is the model useful to understanding grasshopper populations? Admitting that the model is very general in its application, three features make this model compelling with several additional wrinkles also providing potential important insights. The interplay between abiotic and biotic factors (Fig. 14.3a,b) leading to both wide and narrow bands of population fluctuations could prove very important to grasshopper population dynamics. This is best illustrated in Figure 14.3b and c. As reviewed above, abiotic factors can easily alter many important life-history parameters, the effect being to introduce time lags into the response of the population to traditionally envisioned (in a population model sense) density-dependent pressures.

The model also suggests an interpretation for explaining among-year or consistent between-site differences in population size. Sites in which the plants are regularly stressed (e.g., due to edaphic or topographic conditions) may contain higher-quality food on a more regular basis (Mattson and Haack, 1987a, b). This might lead to regular differences in population size even though the superficial aspect of the sites is similar; Figure 14.3a represents this situation.

The third central aspect of the model concerns the explicit existence of density-dependent domains, each with its own equilibrium and maintained by different biotic factors. Although two domains are presented in the Berryman et al. (1987) model, more could easily be incorporated although multiple domains, let alone stable states, may be difficult to detect rigorously (Connell and Sousa, 1983). This prediction is untested except in its most general form (see also Southwood and Comins, 1976; Peterman et al., 1979) but could provide very important understanding to the population dynamics and regulation processes if true.

The model at least provides rich fodder for hypothesis generation and testing. As presented, the equilibrium points defining the domains in the model result from density-dependent biotic interactions, such as frequency- and density-dependent predation by natural enemies defining the domain at 1, with intra- and interspecific competition defining the equilibrium point at the upper domain. Circumstantial evidence reviewed above at least notes the existence of these mechanisms in some systems and Belovsky (1986a) has documented the existence of a stable equilibrium point resulting from interspecific competitive interactions (Fig. 14.16) in cage experiments performed in the field. It is as likely, however, that the upper equilibrium may be due to intraspecific competition. Although we believe that the real truth will probably fall far from the simple relationships diagramed in Figure 14.3, we also believe that the framework provided by this model will go far in organizing a global attack on the problem and spawn more interesting and insightful hypotheses than those often asked to this point.

In addition to the somewhat deterministic view offered by the Berryman et al. (1987) model, three additional proposals also require careful consideration. Environmental heterogeneity (especially spatial) (Reddingius and den Boer 1970; Roff, 1974a,b; Chesson, 1978; Strong, 1986; Morrison and Barbosa, 1987), density-imperfect (Milne, 1957) or density-vague population fluctuations (Strong, 1984, 1986), and the impact of strange attractors with impending chaos embedded in the dynamics (May, 1976; Schaeffer, 1985; Schaeffer and Kot, 1986) may each provide interesting and possibly central insights.

Dynamics of local dispersal against a spatially heterogeneous environment provides a template against which complex interactions can easily modify population dynamics. In simplified theoretical models, dispersal among subpopulations may result in increased persistence for both single-species (Roff, 1974a,b) and multiple-species (Slatkin, 1974) populations. Roff (1974a) also suggests that dispersal in a heterogeneous arena reduces the variance in total population size caused by environmental fluctuations. For grasshoppers, the dynamics of interactions between coexisting grasshoppers and their natural enemies are greatly modified by frequency- and density-dependent responses from avian predators, resulting in complex interspecific interactions among grasshoppers (Joern, 1988, unpublished) and the dispersion patterns among available patches. Additionally, quality of food plants is quite patchy and grasshoppers tend to move about and accumulate on high-quality patches (Joern, 1987; Heidorn and Joern, 1987). Finally, patterns of microhabitat use differ among grasshopper species resulting in a regular mosaic of species coexistence and density distribution in space for each species (Joern, 1986b, 1987). It is clear that local heterogeneity required for such models to be applicable exists in natural grasshopper systems although population-level consequences have not yet been carefully examined.

Stochastic elements from a variety of sources may greatly contribute to grasshopper population dynamics. These will be particularly vexing to study because simple cause–effect pathways and associated sources of population flux are difficult to trace. For example, the "open," stochastic nature of weather with no negative feedback and the pervasive effects felt from this force on so many physiological and ecological attributes in grasshoppers illustrate potential problems to be faced.

Density-vague or stochastically bounded population dynamics (Chesson, 1978; Strong, 1984, 1986) are patterns that fall into this category. In such models, extreme upper and lower limits of population size exist (although they may be "soft"), but little intense density-dependent regulation operates for mid-range population sizes. As a result, although density-dependent regulating forces may be present at intermediate densities, they are either not intense or else they are repeatedly thrown "off-course" by other forces resulting in high variance, making detection difficult. The consequence is fuzzy population regulation, which will be difficult to disentangle analytically, in accord with other empirical studies (Stiling, 1988).

Chaos and strange attractors have emerged in some basic but important population models (May, 1976; Schaeffer, 1985; Schaeffer and Kot, 1986). Periodicity and seeming randomness (chaos) in model populations appear as parameters describing the rate of growth are increased. Although not definitive, patterns of locust outbreaks superficially resemble output from simple population models with relatively high parameter values (Blackith and Albrecht, 1979). Theoretical studies and examination of long population time series indicate the existence of additional and meaningful "biological signals" embedded in variance about a mean that would otherwise be considered random (Schaeffer, 1985; Schaeffer and Kot, 1986). It is too soon to know what such an analysis holds for understanding long-term population dynamics or regulation in grasshoppers. Yet the approach holds clear intellectual promise with the possibility of predicting real, presently unrecognized dynamical mechanisms. At the very least, it may indicate phenomenological patterns that, when recognized, suggest straightforward (though possibly complex) mechanistic interpretations.

Grasshopper population dynamics and regulation are poorly understood in the sense that no study has adequately followed the multiple factors that are undoubtedly involved in order to assess relative and conditional contributions. There has been too much reliance on old aphorisms regarding how grasshopper populations wax and wane with insufficient attempts to dissect the key underlying mechanisms. In part, this reflects the relatively small number of investigators who have attempted to solve this vexing problem, rather that the quality of available work, and the lack of sustained, long-term studies. In addition, modeling efforts, which have been so effective in understanding dynamics of other insect populations, are poorly developed for grasshopper populations. There is much room for modeling studies based on sophisticated ecological reasoning which is now available for exploitation. Gaming with such models could easily suggest important avenues for clear-cut empirical testing. Finally, it seems likely that grasshopper populations are not closed systems and population processes at a site will be understood only in the context of external events. This presents its own set of unique problems in that the spatial scale of study becomes almost unmanageable, especially for critically needed field manipulations. Fortunately, this seems to be the case for a number of ecological systems, and very interesting theoretical work is in progress which treats such problems (Underwood et al., 1983; Roughgarden, 1989). Such insights should be applied to grasshopper population systems.

Although we have presented the problem of grasshopper population dynamics as a complex problem of conditional relationships, we see room for immediate and rapid progress rather than despair at ever knowing what key interactions are involved. The problem is certainly a difficult one, but many guideposts are now available.

ACKNOWLEDGMENTS

R. F. Chapman, L. S. Vescio, and Y. Yang kindly commented on earlier versions of the manuscript and greatly improved the readability and content. Cedar Point Biological Station provided excellent logistical support for much of our research on questions related to this topic and for a portion for the time used to write this chapter. Research grants from NSF (BSR−840897) and USDA Competitive Grants Program (86-CRCR-1−1974) are gratefully acknowledged.

REFERENCES

Aikman, D. and G. Hewitt. 1972. An experimental investigation of the rate and form of dispersal in grasshoppers. *J. Appl. Ecol.* **9**, 807−817.

Amerasinghe, F. P. 1978. Pheromonal effects on sexual maturation, yellowing, and the vibrational reaction in immature male desert locusts (*Schistocerca gregaria*). *J. Insect Physiol.* **24**, 309−314.

Andrewartha, H. G. and L. C. Birch. 1954. *The Distribution and Abundance of Animals.* Univ. of Chicago Press, Chicago, Illinois.

Atkinson, D. and M. Begon. 1987a. Reproductive variation and adult size in two co-occurring grasshopper species. *Ecol. Entomol.* **12**, 119−127.

Atkinson, D. and M. Begon. 1987b. Ecological correlates and heritability of reproductive variation in two co-occurring grasshopper species. *Ecol. Entomol.* **12**, 129−138.

Bailey, C. G. and M. K. Mukerji. 1976a. Feeding habits and food preferences of *Melanoplus bivittatus* and *Melanoplus femurrubrum* (Orthoptera: Arcidadae). *Can. Entomol.* **108**, 1207−1212.

Bailey, C. G. and M. K. Mukerji. 1976b. Consumption and utilization of various host plants by *Melanoplus bivittatus* (Say) and *Melanoplus femurrubrum* (DeGeer) (Orthoptera: Acrididae). *Can. J. Zool.* **54**, 1044−1050.

Baldwin, W. F., D. F. Riordan, and R. W. Smith. 1958. Notes on dispersal of radio-active grasshoppers. *Can. Entomol.* **90**, 374−376.

Ballinger, R. E. 1983. Life-history variation In R. Huey, E. R. Pianka, and T. W. Schoener, Eds., *Lizard Ecology: Studies of a Model Organism.* Harvard Univ. Press, Cambridge, Mass. pp. 241−260.

Baltensweiler, W., G. Benz, P. Bovey, and V. Delucchi. 1977. Dynamics of larch bud moth populations. *Annu. Rev. Entomol.* **22**, 79−100.

Barbosa, P. and W. Baltensweiler. 1987. Phenotypic plasticity and herbivore outbreaks. In P. Barbosa and J. C. Schultz, Eds., *Insect Outbreaks.* Academic Press, San Diego, California, pp. 469−503.

Barbosa, P. and J. C. Schultz. 1987. *Insect Outbreaks.* Academic Press, San Diego, California.

Batten, A. 1972. Early stages of the 1928−1941 plague of the African migratory locust, *Locusta migratoria migratorioides* (R. & F.). *Proc. Int. Study Conf. Curr. Future Probl. Acridology, 1970*, pp. 331−334.

Begon, M. 1983. Grasshopper populations and weather: The effects of insolation on *Chorthippus brunneus*. *Ecol. Entomol.* **8**, 361−370.

Begon, M. 1984. Density and individual fitness: Asymmetric competition. In B. Shorrocks, Ed., *Evolutionary Ecology*. Blackwell, Oxford, pp. 175–194.

Begon, M. and R. Wall. 1987. Individual variation and competitor coexistence: A model. *Functional Ecol.* **1**, 237–241.

Bellinger, R. G. and R. L. Pienkowski. 1985. Non-random resorption of oocytes in grasshoppers (Orthoptera: Acrididae). *Can. Entomol.* **117**, 1067–1069.

Bellows, T. S. Jr. 1986a. Impact of developmental variance on behavior of models for insect populations. I. Models for populations with unrestricted growth. *Res. Popul. Ecol.* **28**, 41–49.

Bellows, T. S., 1986b. Impact of developmental variance on behavior of models for insect populations. II. Models for populations with density dependent restrictions on growth. *Res. Popul. Ecol. (Kyoto)* **28**, 63–67.

Belovsky, G. E. 1986a. Generalist herbivore foraging and its role in competitive interactions. *Am. Zool.* **26**, 51–69.

Belovsky, G. E. 1986b. Optimal foraging and community structure: Implications for a guild of generalist grassland herbivores. *Oecologia* **70**, 35–52.

Bernays, E. A. and R. F. Chapman. 1973. The role of food plants in the survival and development of *Chortoicetes terminifera* (Walker) under drought conditions. *Aust. J. Zool.* **21**, 575–592.

Bernays, E. A. and R. F. Chapman. 1978. Plant chemistry and acridoid feeding behaviour. In J. B. Harborne, Ed., *Biochemical Aspects of Plant and Animal Coevolution*. Academic Press, London, pp. 99–141.

Bernays, E. A. and J. Hamai. 1987. Head size and shape in relation to grass feeding in Acrididae (Orthoptera). *Int. J. Insect Morphol. Embryol.* **16**, 323–330.

Bernays, E. A., J. Horsey, and E. M. Leather. 1974. The inhibitory effect of seedling grasses on feeding and survival of acridids. *Bull. Entomol. Res.* **64**, 413–420.

Bernays, E. A., D. J. Chamberlain, and P. McCarthy. 1980. The differential effects of ingested tannic acid on different species of Acridoidea. *Entomol. Exp. Appl.* **28**, 158–166.

Bernays, E. A., D. J. Chamberlain, and E. M. Leather. 1981. The tolerance of acridids to condensed tannin. *J. Chem. Ecol.* **7**, 247–256.

Berryman, A. A. 1982. Biological control, thresholds, and pest outbreaks. *Environ. Entomol.* **11**, 544–549.

Berryman, A. A. 1987. The theory and classification of outbreaks. In P. Barbosa and J. C. Schultz, Eds., *Insect Outbreaks*. Academic Press, San Diego, California, pp. 3–30.

Berryman, A. A., N. C. Stenseth, and A. S. Isaev. 1987. Natural regulation of herbivorous forest insect populations. *Oecologia* **71**, 174–184.

Blackith, R. E. and F. O. Albrecht. 1979. Locust plagues: The interplay of endogenous and exogenous control. *Acrida* **8**, 83–94.

Blackith, R. E. and R. M. Blackith. 1969. Observations on the biology of some morabine grasshoppers. *Austr. J. Zool.* **17**, 1–12.

Bowen, B. J., C. G. Codd, and D. T. Gwynne. 1984. The katydid spermatophore (Orthoptera: Tettigoniidae): Male nutritional investment and its fate in the mated female. *Aust. J. Zool.* **32**, 361–363.

Butlin, R. K., G. M. Hewitt, and S. F. Webb. 1985. Sexual selection for intermediate optimum in *Chorthippus brunneus* (Orthoptera: Acrididae). *Anim. Behav.* **33**, 1281–1292.

Butlin, R. K., C. W. Woodhatch, and G. M. Hewitt. 1987. Male spermatophore investment increases female fecundity in a grasshopper. *Evolution (Lawrence, Kans.)* **41**, 221–225.

Calow, P. C. 1983. Energetics of reproduction and its evolutionary implications. *Biol. J. Linn. Soc.* **20**, 153–165.

Capinera, J. L. 1979. Qualitative variation in plants and insects: Effects of propagule size on ecological plasticity. *Am. Nat.* **114**, 350–361.

Capinera, J. L. 1987. Population ecology of rangeland grasshoppers. In J. L. Capinera, Ed., *Integrated Pest Management on Rangeland: A Shortgrass Prairie Perspective*. Westview Press, Boulder, Colorado, pp. 162–182.

Capinera, J. L. and D. R. Horton. 1989. Geographic variation in effects of weather on grasshopper populations. *Environ. Entomol.* **18**, 8–114.

Capinera, J. L. and D. C. Thompson. 1987. Dynamics and structure of grasshopper assemblages in shortgrass prairie. *Can. Entomol.* **119**, 567–575.

Capinera, J. L., A. R. Renaud and N. E. Roehrig. 1983. The chemical basis for host selectivity by *Hemileuca oiliviae*: The role of tannins in preference of C_4 grasses. *J. Chem. Ecol.* **9**, 1425–1437.

Chapman, R. F. 1972. The movements of Acridoid populations. *Proc. Int. Study Conf. Curr. Future Probl. Acridology, 1970*, pp. 239–252.

Chapman, R. F. 1976. *A Biology of Locusts*. Arnold, London.

Chapman, R. F. and W. W. Page. 1979. Factors affecting the mortality of the grasshopper, *Zonocerus variegatus*, in southern Nigeria. *J. Anim. Ecol.* **48**, 271–288.

Chapman, R. F., A. G. Cook, G. A. Mitchell, and W. W. Page. 1978. Wing dimorphism and flight in *Zonocerus variegatus* (L.) (Orthoptera: Pyrgomorphidae). *Bull. Entomol. Res.* **68**, 229–242.

Chapman, R. F., W. W. Page, and A. G. Cook. 1979. A study of population changes in the grasshopper, *Zonocerus variegatus*, in Southern Nigeria. *J. Anim. Ecol.* **48**, 247–270.

Chapman, R. F., W. W. Page, and A. R. McCaffery. 1986. Bionomics of the variegated grasshopper (*Zonocerus vareigatus*) in west and central Africa. *Annu. Rev. Entomol.* **31**, 479–506.

Cheke, R. A. 1978. Theoretical rates of increase of gregarious and solitary populations of the desert locust. *Oecologia* **35**, 161–171.

Chesson, P. 1978. Predator-prey theory and variability. *Annu. Rev. Ecol. Syst.* **9**, 323–347.

Chrominski, A. S., S. Neumann Visscher, and R. Jurenka. 1982. Exposure to ethylene changes nymphal growth rate and female longevity in the grasshopper *Melanoplus sanguinipes*. *Naturwissenschaften* **69**, 145–46.

Clark, D. P. 1962. An analysis of dispersal and movement in *Phaulacridium vittatum* (Sjöst) (Acrididae). *Aust. J. Zool.* **10**, 382–399.

Clark, D. P. 1967. A population study of *Phaulacridium vittatum* Sjöst. (Acrididae). *Aust. J. Zool.* **15**, 799–872.

Clark, D. P. 1972. The plague dynamics of the Australian plague locust, *Chortoicetes terminifera* (Walk.). *Proc. Int. Study Conf. Curr. Future Probl. Acridology, 1970*, pp. 257–287.

Connell, J. H. and W. P. Sousa. 1983. On the evidence needed to judge ecological stability or persistence. *Am. Nat.* **122**, 661–696.

Davies, D. E. 1952. Seasonal breeding and migrations of the desert locust (*Schistocerca gregaria* F.) in western and north-western Africa and the Middle East. *Anti-Locust Mem.* **4**, 1–51.

Dean, J. M. 1981. The relationship between lifespan and reproduction in the grasshopper *Melanoplus*. *Oecologia* **48**, 385–388.

Dearn, J. M. 1977. Variable life history characteristics along an altitudinal gradient in three species of Australian grasshopper. *Oecologia* **28**, 67–85.

Dearn, J. M. 1978. Polymorphisms for wing length and colour pattern in the grasshopper *Phaulacridium vittatum* (Sjöst.). *J. Aust. Entomol. Soc.* **17**, 135–137.

Dempster, J. P. 1955. Factors affecting small-scale movements of some British grasshoppers. *Proc. R. Entomol. Soc. London A, Ser.* **30**, 145–150.

Dempster, J. P. 1957. The population dynamics of the Moroccan locust (*Dociostaurus maroccanus* Thunb.) in Cyprus. *Anti-Locust Bull.* **27**, 1–60.

Dempster, J. P. 1963. The population dynamics of grasshoppers and locusts. *Biol. Rev. Cambridge Philos. Soc.* **38**, 490–529.

Dempster, J. P. and E. Pollard. 1981. Fluctuation in resource availability and insect populations. *Oecologia* **50**, 412–416.

Dennis, D. W. and R. J. Lavigne. 1975. Comparative behavior of Wyoming robber flies (Diptera: Asilidae). II. *Wyo., Agric. Exp. Stn., Sci. Monog,* **30**.

de Souza Santos, P., Jr. and M. Begon. 1987. Survival costs of reproduction in grasshoppers. *Funct. Ecol.* **1**, 215–221.

Dingle, H. 1985. Migration. In G. A. Kerkut and L. I. Gilbert, Eds., *Comprehensive Insect Physiology, Biochemistry and Pharmacology*, Vol. 9. *Pergamon* Press, *Oxford*, pp. 375–415.

Donnelly, U. 1947. Seasonal breeding and migrations of the desert locust (*Schistocerca gregaria* F.) in western and north-western Africa. *Anti-Locust Mem.* **3**, 1–43.

Edwards, R. L. 1960. Relationship between grasshopper abundance and weather conditions in Saskatchewan, 1930–1958. *Can. Entomol.* **92**, 619–624.

Ellis, P. E. and C. Ashall. 1957. Field studies on diurnal behaviour, movement and aggregation in the desert locust. *Anti-Locust Bull.* **25**, 1–94.

Ellis, P. E., D. B. Carlisle, and D. J. Osborne. 1965. Desert locusts: Sexual maturation delayed by feeding on senescent vegetation. *Science* **149**, 546–547.

Erlandson, M. A., A. B. Ewen, M. K. Mukerji, and C. Gillott. 1986. Susceptability of immature stages of *Melanoplus sanguinipes* (Fab.) (Orthoptera: Acrididae) to *Nosema cunneatum* Henry (Microsporida: Nosematidae) and its effect on host fecundity. *Can. Entomol.* **118**, 29–35.

Evans, E. W. 1989. Interspecific interactions among phytophagous insects of tallgrass prairie: An experimental test. *Ecology* **70**, 435–444.

Ewen, A. B. and M. K. Mukerji. 1980. Evaluation of *Nosema locustae* (Microsporidia) as a control agent of grasshopper populations in Saskatchewan. *J. Invertebr. Pathol.* **35**, 295–303.

Farrow, R. A. 1975. Quantitative studies of solitary populations of the African migratory locust in the Middle Niger outbreak area. *Locusta* **11**.

Farrow, R. A. 1977a. Maturation and fecundity of the spur-throated locust *Austracris guttulosa* (Walker), in New South Wales during the 1974/1975 plague. *J. Aust. Entomol. Soc.* **16**, 27–39.

Farrow, R. A. 1977b. Origin and decline of the 1973 plague locust outbreak in central western New South Wales. *Aust. J. Zool.* **25**, 455–489.

Farrow, R. A. 1979. Population dynamics of the Australian plague locust, *Chortoicetes terminifera* (Walker), in central western New South Wales. I. Reproduction and migration in relation to weather. *Aust. J. Zool.* **27**, 717–745.

Farrow, R. A. 1982a. Population dynamics of the Australian plague locust, *Chortoicetes terminifera* (Walker), in Central New South Wales. II. Factors influencing natality and survival. *Aust. J. Zool.* **30**, 199–222.

Farrow, R. A. 1982b. Population dynamics of the Australian plague locust, *Chortoicetes terminifera* (Walker), in Central New South Wales. III. Analysis of population processes. *Aust. J. Zool.* **30**, 569–579.

Farrow, R. A. and B. C. Longstaff. 1986. Comparison of the annual rates of increase of locusts in relation to the incidence of plagues. *Oikos* **46**, 207–222.

Ferenz, H. -J., E. Lubzens, and H. Glass. 1981. Vitellin and vitellogenin incorporation by isolated oocytes of *Locusta migratoria migratorioides* (R. & F.). *J. Insect Physiol.* **27**, 869–875.

Fortescue-Foulkes, J. 1953. Seasonal breeding and migration of the desert locust (*Schistocerca gregaria* F.) in south-western Asia. *Anti-Locust Mem.* **5**, 1–36.

Gage, S. H. and M. K. Mukerji. 1977. A perspective of grasshopper population distributions in Saskatchewan and interrelationships with weather. *Environ. Entomol.* **6**, 469–479.

Gage, S. H., M. K. Mukerji, and R. L. Randell. 1976. A predictive model for seasonal occurrence of three grasshopper species in Saskatchewan (Orthoptera: Acrididae). *Can. Entomol.* **108**, 245–253.

Gaines, S. B. 1989. Experimental analysis of costs and benefits of wing length polymorphism in grasshoppers. PhD Dissertation. University of Nebraska-Lincoln. 280 p.

Garland, J. A. 1982. An addition to Krebs' restatement of the Chitty Hypothesis from an entomological perspective. *Can. J. Zool.* **60**, 833–837.

Gill, P. D. 1979. Colour-pattern variation in relation to habitat in the grasshopper *Chorthippus brunneus* (Thunberg). *Ecol. Entomol.* **4**, 249–257.

Gillis, J. E. and K. W. Possai. 1983. Thermal niche partitioning in the grasshoppers *Arphia conspersa* and *Trimerotropis suffusa* from a montane habitat in central Colorado. *Ecol. Entomol.* **8**, 155–161.

Greathead, D. J. 1962. The biology of *Stomorhina lunata* (Fabricius) (Diptera: Calliphoridae), predator of the eggs of Acrididae. *Proc. Zool. Soc. London* **139**, 139–180.

Greathead, D. J. 1963. A review of the insect enemies of Acridoidea (Orthoptera). *Trans. R. Entomol. Soc. London* **114**, 437–517.

Greathead, D. J. 1966. A brief study of the effects of biotic factors on populations of the desert locust. *J. Appl. Ecol.* **3**, 239–250.

Greenfield, M. D. and T. E. Shelly. 1985. Alternative mating strategies in a desert grasshopper: evidence of density-dependence. *Anim. Behav.* **33**, 1192–1210.

Guseva, V. 1979. The biotic potential and its realization as exemplified by three species of non-gregarious grasshoppers (Orthoptera: Acrididae). *Entomol. Rev. (Engl. Transl.)* **58**, 28–44.

Gwynne, D. T. 1984. Male mating effort, confidence of paternity and insect sperm competition. In R. L. Smith, Ed., *Sperm Competition and the Evolution of Animal Mating Systems*. Academic Press, Orlando, Florida, pp. 117–149.

Gwynne, D. T., B. J. Bowen, and C. G. Codd. 1984. The function of the katydid spermatophore and its role in fecundity and insemination (Orthoptera: Tettigoniidae). *Aust. J. Zool.* **32**, 15–22.

Gyllenberg, G. 1974. A simulation model for testing the dynamics of a grasshopper population. *Ecology* **56**, 645–650.

Haniffa, M. A. and K. Periasamy. 1981. Effect of ration level on nymphal development and food utilization in *Acrotylus insubricus* (Scopali) (Orthoptera: Acrididae). *Acrida* **10**, 91–103.

Hardman, J. M. and M. K. Mukerji. 1982. A model simulating the population dynamics of the grasshoppers (Acrididae) *Melanoplus sanguinipes* (Fabr.), *M. packardii* Scudder and *Camnula pellucida* (Scudder). *Res. Popul. Ecol.* **24**, 276–301.

Harley, K. L. S. and A. J. Thorsteinson. 1967. The influence of plant chemicals on the feeding behavior, development, and survival of the two-striped grasshopper, *Melanoplus bivittatus* (Say), Acrididae: Orthoptera. *Can. J. Zool.* **45**, 302–319.

Harrison, R. G. 1980. Dispersal polymorphisms in insects. *Annu. Rev. Ecol. Syst.* **11**, 95–118.

Hassell, M. P. 1978. *The Dynamics of Arthropod Predator-Prey Systems*. Princeton Univ. Press, Princeton, New Jersey.

Hassell, M. P. and R. M. May. 1985. From individual behaviour to population dynamics. In R. M. Sibly and R. H. Smith, Eds., *Behavioural Ecology*. Blackwell, Oxford, pp. 3–32.

Heidorn, T. J. and A. Joern. 1987. Feeding preference and spatial distribution of grasshoppers (Acrididae) in response to nitrogen fertilization of *Calamovilfa longifolia*. *Funct. Ecol.* **1**, 369–375.

Hewitt, G. B. 1979. Hatching and development of rangeland grasshoppers in relation to forage growth, temperature and precipitation. *Environ. Entomol.* **8**, 24–29.

Hewitt, G. B. 1985. Review of factors affecting fecundity, oviposition and egg survival of grasshoppers in North America. *U.S., Agric. Res. Serv., ARS* **ARS–36**.

Hilbert, D. W. and J. A. Logan. 1981. A review of the population biology of the migratory grasshopper, *Melanoplus sanguinipes*. *Colo., Agric. Exp. Stn., Bull.* **577S**, 1–10.

Hilbert, D. W. and J. A. Logan. 1983. Empirical model of nymphal development for the Migratory Grasshopper, *Melanoplus sanguinipes* (Orthoptera: Acrididae). *Environ. Entomol.* **12**, 1–5.

Hilbert, D. W., J. A. Logan, and D. M. Swift. 1985. A unifying hypothesis of temperature effects on egg development and diapause of the Migratory Grasshopper, *Melanoplus sanguinipes* (Orthoptera: Acrididae). *J. Theor. Biol.* **112**, 827–838.

Hill, L. 1965. The incorporation of C^{14} glycine into the proteins of the fat body of the desert locust during ovarian development. *J. Insect Physiol.* **11**, 1605–1615.

Hinks, C. F. 1985. The influence of temperature on the efficacy of three pyrethroid insecticides against the grasshopper *Melanoplus sanguinipes* (Fab.) (Orthoptera: Acrididae) under laboratory conditions. *Can. Entomol.* **117**, 1007–1012.

Holling, C. S. 1965. The functional response of predators to prey density and its role in mimicry and population regulation. *Mem Entomol. Soc. Can.* **45**, 1–60.

Hunter, D. M. and P. C. Gregg. 1984. Variation in diapause potential and strength in eggs of the Australian Plague Locust, *Chortoicetes terminifera* (Walker) (Orthoptera: Acrididae). *J. Insect Physiol.* **30**, 867–870.

Joern, A. 1979. Resource utilization and community structure in assemblages of arid grassland grasshoppers (Orthoptera: Acrididae). *Trans. Am. Entomol. Soc.* **150**, 253–300.

Joern, A. 1982. Distribution, densities, and relative abundances of grasshoppers (Orthoptera: Acrididae) in a Nebraska sandhills grassland. *Prairie Nat.* **14**, 37–45.

Joern, A. 1983. Small-scale displacements of grasshoppers (Orthoptera: Acrididae) within arid grasslands. *J. Kans. Entomol. Soc.* **56**, 131–139.

Joern, A. 1986a. Experimental study of avian predation on coexisting grasshopper populations (Orthoptera: Acrididae) in a sandhills grassland. *Oikos* **46**, 243–249.

Joern, A. 1986b. Resource utilization by a grasshopper assemblage from grassland communities. *Proc. 4th Trienn. Meet. Pan Am. Acridol. Soc.* 75–100.

Joern, A. 1987. Behavioral responses underlying ecological patterns of resource use in rangeland grasshoppers. In J. L. Capinera, Ed., *Integrated Pest Management on Rangeland: A Shortgrass Perspective.* Westview Press, Boulder, Colorado, pp. 137–161.

Joern, A. 1988. Foraging behavior and switching by the grasshopper sparrow *Ammodramus savannarum* searching for multiple prey in a heterogeneous environment. *Am. Midl. Nat.* **119**, 225–234.

Joern, A. and L. R. Lawlor. 1980. Arid grassland grasshopper community structure: comparisons with neutral models. *Ecology* **61**, 591–599.

Joern, A. and K. P. Pruess. 1986. Temporal constancy in grasshopper assemblies (Orthoptera: Acrididae). *Ecol. Entomol.* **11**, 379–385.

Joern, A. and N. T. Rudd. 1982. Impact of predation by the robber fly *Proctacanthus milbertii* (Diptera: Asilidae) on grasshopper (Orthoptera: Acrididae) populations. *Oecologia* **55**, 42–46.

Johnson, C. G. 1969. *Migration and Dispersal of Insects by Flight.* Methuen, London.

Johnson, D. L. 1986. Reduction in the rate of feeding by grasshoppers infected with *Nosema locustae. Proc. 4th Trienn. Meet. Pan Am. Acridol. Soc.* p. 158.

Johnson, D. L. and E. Paulikova. 1986. Reduction of consumption by grasshoppers (Orthoptera: Acrididae) infected with *Nosema locustae* Canning (Microsporida: Nosematidae). *J. Invertebr. Pathol.* **48**, 232–238.

Johnson, D. L., B. D. Hill, C. F. Hinks, and G. B. Schaalje. 1986. Aerial application of the pyrethroid deltamethrin for grasshopper (Orthoptera: Acrididae) control. *J. Econ. Entomol.* **79**, 181–188.

Kajak, A., L. Andrzejewska, and Z. Wojcik. 1968. The role of spiders in the decrease of damage caused by Acridoidea on meadows — experimental investigations. *Ekol. Pol. Ser., A* **16**, 1–10.

Kemp, W. P. 1986. Thermoregulation in three rangeland grasshopper species. *Can. Entomol.* **118**, 335–343.

Kemp, W. P. 1987a. Predictive phenology modeling in rangeland pest management. In J. L. Capinera, Ed., *Integrated Pest Management on Rangeland: A Shortgrass Prairie Perspective.* Westview Press, Boulder, Colorado, pp. 351–368.

Kemp, W. P. 1987b. Probability of outbreak for rangeland grasshoppers (Orthoptera: Acrididae) in Montana: Application of Markovian principles. *J. Econ. Entomol.* **80**, 1100–1105.

Kemp, W. P. and B. Dennis. 1989. Development of two rangeland grasshoppers at constant temperatures: Development thresholds revisited. *Can. Entomol.* **121**, 363–371.

Kemp, W. P. and J. A. Onsager. 1986. Rangeland grasshoppers (Orthoptera: Acrididae): Modeling phenology of natural populations of six species. *Environ. Entomol.* **15**, 924–930.

Kemp, W. P. and N. E. Sanchez. 1987. Differences in post-diapause thermal requirements for eggs of two rangeland grasshoppers. *Can. Entomol.* **119**, 653–661.

Krebs, C. J. 1978. A review of the Chitty hypothesis of population regulation. *Can. J. Zool.* **56**, 2464–2480.

Krishna, S. S. and A. J. Thorsteinson. 1972. Ovarian development of *Melanoplus sanguinipes* (Fab.) (Acrididae: Orthoptera) in relation to utilization of water-soluble food proteins. *Can. J. Zool.* **50**, 1319–1324.

Lauga, J. and M. Hatté. 1977. Propriétés grégarisantes aquisés par le sable dans lequel ont pondu à de nombreuses reprises des femelles grégaires de *Locusta migratoria migratorioides* R & F. (Orthoptère, Acrididae). *Acrida* **6**, 307–311.

Lavigne, R. J. and F. R. Holland. 1969. Comparative behavior of Wyoming robber flies (Diptera: Asilidae). *Wy., Agric. Exp. Stn., Sci. Monogr.* **18**.

Lea, A. 1972. The plague dynamics of the brown locust, *Locustana pardalina* (Walk.). *Proc. Int. Study Conf. Curr. Future Probl. Acridology, 1970*, pp. 289–298.

Lewis, A. C. 1979. Feeding preference for diseased and wilted sunflower in the grasshopper *Melanoplus differentialis*. *Entomol. Exp. Appl.* **26**, 202–207.

Lewis, A. C. 1984. Plant quality and grasshopper feeding: Effects of sunflower condition on preference and performance in *Melanoplus differentialis*. *Ecology* **65**, 836–843.

Lewis, A. C. and E. A. Bernays. 1985. Feeding behavior: selection of both wet and dry food for optimal growth by *Schistocerca gregaria* nymphs. Entomol. Exp. Appl. **37**, 105–112.

Lieberman, J. 1972. The current state of the locust and grasshopper problem in Argentina. *Proc. Int. Study Conf. Curr. Future Probl. Acridology, 1970*, pp. 191–198.

Logan, J. A. 1988. Toward an expert system for development of pest simulation models. *Environ. Entomol.* **17**, 359–376.

Logan, J. A. and D. W. Hilbert. 1983. Modeling the effects of temperature on arthropod population systems. In W. K. Lauenroth, G. V. Skogerboe, and M. Flug, Eds., *Analysis of Ecological Systems: State-Of-The-Art in Ecological Modeling*. Elsevier, Amsterdam, pp. 113–122.

Logan, J. A., D. H. Wolkind, S. C. Hoyt, and L. K. Tanigoshi. 1976. An analytic model for description of temperature dependent rate phenomenon in arthropods. *Environ. Entomol.* **5**, 1133–1140.

Lomnicki, A. 1988. *Population Ecology of Individuals*. Princeton Univ. Press, Princeton, New Jersey.

Matsumoto, T. 1971. Estimation of population productivity of *Parapleurus alliaceus* Germar (Orthoptera: Acrididae) on a *Miscanthus sinensis* Anders. grassland. II. Population productivity in terms of dry weight *Oecologia* **7**, 16–25.

Mattson, W. J. and R. A. Haack. 1987a. The role of drought in outbreaks of plant eating insects. *BioScience* **37**, 110–118.

Mattson, W. J. and R. A. Haack. 1987b. The role of drought stress in provoking outbreaks of phytophagous insects. In P. Barbosa and J. C. Schultz, Eds., *Insect Outbreaks*. Academic Press, San Diego, California, pp. 365–407.

May, R. M. 1976. Simple Mathematical models with very complicated dynamics. *Nature (London)* **261**, 459–467.

McAnelly, M. L. and M. A. Rankin. 1986a. Migration in the grasshopper *Melanoplus sanguinipes* (Fab.). I. The capacity for flight in non-swarming populations. *Biol. Bull. (Woods Hole, Mass.)* **170**, 368–377.

McAnelly, M. L. and M. A. Rankin. 1986b. Migration in the grasshopper (*Melanoplus sanguinipes* (Fab.) II. Interactions between flight and reproduction. *Biol. Bull. (Woods Hole, Mass.)* **170**, 378–392.

McCaffery, A. R. 1975. Food quality and quantity in relation to egg production in *Locusta migratoria migratorioides*. *J. Insect Physiol.* **21**, 1551–1558.

McCaffery, A. R. and W. W. Page. 1978. Factors influencing the production of long-winged *Zonocerus variegatus*. *J. Insect Physiol.* **24**, 465–472.

McCarthy, H. R. 1956. A ten-year study of the climatology of *Melanoplus mexicanus mexicanus* (Sauss.) (Orthoptera: Acrididae) in Saskatchewan. *Can. J. Agric. Sci.* **36**, 445–462.

McGinnis, A. J. and R. Kasting. 1966. Comparison of tissues from solid- and hollow-stemmed spring wheats during growth. IV. Apparent dry matter utilization and nitrogen balance in the two-striped grasshopper, *Melanoplus bivittatus* (Say). *J. Insect Physiol.* **12**, 671–678.

Merton, L. F. H. 1959. Studies on the ecology of the Moroccan Locust (*Dociostaurus maroccanus* Thunberg) in Cyprus. *Anti-Locust Bull.* **34**, 1–123.

Middlekauf, W. W. 1964. Effects of photoperiod upon oogenesis of *Melanoplus devastator* Scudder. *J. Kans. Entomol. Soc.* **37**, 163–168.

Milne, A. 1957. The natural control of insect populations. *Can. Entomol.* **89**, 193–213.

Monk, K. A. 1985. Effect of habitat on the life history strategies of some British grasshoppers. *J. Anim. Ecol.* **54**, 163–177.

Morrison, G. and P. Barbosa. 1987. Spatial heterogeneity, population "regulation", and local extinction in simulated host-parasitoid interactions. *Oecologia* **73**, 609–614.

Mukerji, M. K. and S. H. Gage. 1978. A model for estimating hatch and mortality of grasshopper egg populations based on soil moisture and heat. *Ann. Entomol. Soc. Am.* **71**, 183–190.

Mukerji, M. K. and R. L. Randell. 1975. Estimation of embryonic development in populations of *Melanoplus sanguinipes* (Fabr.) (Orthoptera: Acrididae) in the fall. *Acrida* **4**, 9–18.

Mukerji, M. K., S. H. Gage, and R. L. Randell. 1977. Influence of embryonic development and heat on population trend of three grasshopper species in Saskatchewan (Orthoptera: Acrididae). *Can. Entomol.* **109**, 229–236.

Mulkern, G. B. 1967. Food selection by grasshoppers, *Annu. Rev. Entomol.* **12**, 59–78.

Mulkern, G. B. 1980. Population fluctuations and competitive relationships of grasshopper species (Orthoptera: Acrididae). *Trans. Am. Entomol. Soc.* **106**, 1–41.

Mulkern, G. B. 1983. Sex ratios of *Melanoplus sanguinipes* and other Acrididae (Orthoptera). *J. Kans. Entomol. Soc.* **56**, 457–465.

Mulkern, G. B. and D. R. Toczek. 1972. Effect of plant extracts on survival and development of *Melanoplus differentialis* and *Melanoplus sanguinipes* (Orthoptera: Acrididae). *Ann. Entomol. Soc. Am.* **65**, 662–671.

Mulkern, G. B., K. P. Pruess, H. Knutson, A. F. Hagen, J. B. Campbell, and J. D. Lambley. 1969. Food habits and preferences of grassland grasshoppers of the North Central Great Plains. *Agric. Exp. Sta. N.D. Bull.* **481**.

Nakamura, K., Y. Ito, M. Nakamura, T. Matsumoto, and K. Hayakawa. 1971. Estimation of population productivity of *Parapleurus alliaceus* Germar (Orthoptera: Acrididae) on a *Miscanthus sinensis* Anders. grassland. I. Estimation of population parameters. *Oecologia* **7**, 1–15.

Nakamura, K., M. Nakamura, T. Matsumoto, and Y. Ito. 1975. Fluctuations of grasshopper populations on a *Miscanthus sisnensis* grassland. *Res. Popul. Ecol.* **16**, 198–206.

Nayar, J. K. 1964. The nutritional requirements of grasshoppers. II. The effects of plant phospholipids and extracts of bran on growth, development, and survival of *Melanoplus bivittatus* (Say) and *Camnula pellucida* (Scudder). *Can. J. Zool.* **42**, 23–48.

Nel, M. D. 1967. Selection for coloration in *Locustana*. *S. Afr. J. Agric. Sci.* **10**, 823–830.

Nerney, N. J. 1961. Effects of seasonal rainfall on range condition and grasshopper population, San Carlos Apache Indian reservation, Arizona. *J. Econ. Entomol.* **54**, 382–385.

Nolte, D. J., S. H. Eggers, and I. R. May. 1973. A locust pheromone: Locustol. *J. Insect Physiol.* **19**, 1547–1554.

Oma, E. A. and G. B. Hewitt. 1984. Effect of *Nosema locustae* (Microsporida: Nosematidae) on food consumption in the differential grasshopper (Orthoptera: Acrididae). *J. Econ. Entomol.* **77**, 500–501.

Onsager, J. A. and G. B. Hewitt. 1982. Rangeland grasshoppers: Average longevity and daily rate of mortality among six species in nature. *Environ. Entomol.* **11**, 127–133.

Onsager, J. A., N. E. Rees, J. E. Henry, and R. N. Foster. 1981. Integration of bait formulations of *Nosema locustae* and carbaryl for control of rangeland grasshoppers. *J. Econ. Entomol.* **74**, 183–187.

Orshan, L. and M. P. Pener. 1979. Repeated reversal of the reproductive diapause by photoperiod and temperature in males of the grasshopper, *Oedoipoda miniata*. *Entomol. Exp. Appl.* **25**, 219–226.

Otte, D. and A. Joern. 1975. Insect territoriality and its evolution: Population studies of desert grasshoppers on creosote bushes. *J. Anim. Ecol.* **44**, 29–54.

Pedgley, D. E. 1979. Weather during Desert Locust plague surges. *Philos. Trans. R. Soc. London, Ser. B* **287**, 387–391.

Pener, M. P. and L. Orshan. 1980. Reversible reproductive diapause and intermediate states between diapause activity in male *Oedipoda miniata* grasshoppers. *Physiol. Entomol.* **5**, 417–426.

Peterman, R., W. C. Clark, and C. S. Holling. 1979. The dynamics of resilience: Shifting stability domains in fish and insect systems. In R. M. Anderson, B. D. Turner and L. R. Taylor, Eds., *Population Dynamics*. Blackwell, Oxford, pp. 321–342.

Pfadt, R. E. 1949. Food plants as factors in the ecology of the lesser migratory grasshopper (*Melanoplus mexicanus*). *Bull. — Wyo., Agric. Exp. Stn.*, **290**.

Pfadt, R. E. 1977. Some aspects of the ecology of grasshopper populations inhabiting the shortgrass plains. *Minn., Agr. Stn., Tech. Bull.* **310**, 1–107.

Pfadt, R. E. 1982. Density and diversity of grasshoppers (Orthoptera: Acrididae) in an outbreak on Arizona rangeland. *Environ. Entomol.* **11**, 690–694.

Pfadt, R. E. and D. S. Smith. 1972. Net reproductive rate and the capacity for increase in the migratory grasshopper, *Melanoplus sanguinipes sanguinipes* (F.). *Acrida* **1**, 149–165.

Phipps, J. 1959. Studies on east African Acridoidea (Orthoptera), with special reference to egg-production, habitats and seasonal cycles. *Trans. R. Entomol. Soc. London* **111**, 27–56.

Phipps, J. 1962. The ovaries of some Sierra Leone Acridoidea (Orthoptera) with some comparisons between eastern and western forms. *Proc. R. Entomol. Soc. London* A **37**, 13–21.

Pickford, R. 1960. Survival, fecundity and population growth of *Melanoplus bilituratus* (Wlk.) (Orthoptera: Acrididae) in relation to date of hatching. *Can. Entomol.* **92**, 1–10.

Pickford, R. 1962. Development, survival and reproduction of *Melanoplus bilituratus* (Wlk.) (Orthoptera: Acrididae) reared on various food plants. *Can. Entomol.* **94**, 859–869.

Pickford, R. 1963. Wheat crops and native prairie in relation to the nutritional ecology of *Camnula pellucida* (Scudder) (Orthoptera: Acrididae) in Saskatchewan. *Can. Entomol.* **95**, 764–770.

Pickford, R. 1966a. Development, survival and reproduction of *Camnula pellucida* (Scudder) (Orthoptera: Acrididae) in relation to climatic conditions. *Can. Entomol.* **98**, 158–169.

Pickford, R. 1966b. The influence of date of oviposition and climatic conditions on the hatching of *Camnula pellucida* (Orthoptera: Acrididae). *Can. Entomol.* **98**, 1145–1159.

Pickford, R. 1972. The effects of climatic factors on egg survival and fecundity in grasshoppers. *Proc. Int. Study Conf. Curr. Future Probl. Acridology, 1970*, pp. 257–260.

Pickford, R. 1976. Embryonic growth and hatchability of eggs of the two-striped grasshopper, *Melanoplus bivittatus* (Orthoptera: Acrididae) in relation to date of oviposition and weather. *Can. Entomol.* **108**, 621–626.

Pickford, R. and C. Gillott. 1976. Effects of varied copulatory periods of *Melanoplus sanguinipes* (Orthoptera: Acrididae) females on egg hatchability and hatchling sex ratios. *Can. Entomol.* **108**, 331–335.

Pickford, R. and R. L. Randell. 1969. A nondiapause strain of the migratory grasshopper, *Melanoplus sanguinipes* (Orthoptera: Acrididae). *Can. Entomol.* **108**, 331–335.

Pickford, R. and P. W. Riegert. 1964. The fungus disease caused by *Entomphthora grylli* Fres. and its effects on grasshopper populations in Saskatchewan in 1963. *Can. Entomol.* **96**, 1158–1166.

Popov, G. and M. Ratcliffe. 1968. The Sahelian tree locust *Anacridium melanorhodon* (Walker). *Anti-Locust Mem.* **9**, 1–45.

Putnam, L. G. 1962. Experiments with some native and introduced plants as foods for *Camnula pellucida* (Scudder) (Orthoptera: Acrididae) in western Canada. *Can. J. Plant Sci.* **42**, 589–595.

Rabb, R. L. and G. G. Kennedy (Eds.). 1979. *Movement of Highly Mobile Insects: Concepts and Methodology in Research.* North Carolina State University, Raleigh.

Rainey, R. 1962. Some effects of environmental factors on movements and phase change of locust populations in the field. *Colloq. Int. C.N.R.S.* **114**, 175–199.

Rainey, R. 1963. Meteorology and the migration of desert locusts. Applications of synoptic meteorology in locust control. *Anti-Locust Mem.* **7**, 1–115.

Rainey, R. 1972. Wind and distribution of the desert locust, *Schistocerca gregaria* (Forsk.). *Proc. Int. Study Conf. Curr. Future Probl. Acridology, 1970*, pp. 229–238.

Rainey, R. 1974. Biometeorology and insect flight: Some aspects of energy exchange. *Annu. Rev. Entomol.* **19**, 407–441.

Rainey, R. C. (Ed.). 1976. *Insect Flight. Symp. R. Entomol. Soc. London.* **7**.

Randell, R. L. and M. K. Mukerji. 1974. A technique for estimating hatching of natural egg populations of *Melanoplus sanguinipes* (Orthoptera: Acrididae). *Can. Entomol.* **106**, 801–812.

Reddingius, J. and P. J. den Boer. 1970. Simulation experiments illustrating stabilization of animal numbers by spreading of risk. *Oecologia* **5**, 240–284.

Redfearn, A. and S. L. Pimm. 1988. Population variability and polyphagy in herbivorous insect communities. *Ecol. Monogr.* **58**, 39–55.

Rees, N. E. 1973. Arthropod and nematode parasites, parasitoids and predators of Acrididae in America north of Mexico. *U.S., Dep. Agric., Tech. Bull.* **460**.

Rees, N. E. and J. A. Onsager. 1982. Influence of predators on the efficiency of the *Blaesoxipha* spp. parasites of the migratory grasshopper. *Environ. Entomol.* **11**, 426–428.

Rees, N. E. and J. A. Onsager. 1985. Parasitism and survival among rangeland grasshoppers in response to suppression of robber fly (Diptera: Asilidae) predators. *Environ. Entomol.* **14**, 20–23.

Reznick, D. 1985. Cost of reproduction. An evaluation of the empirical evidence. *Oikos* **44**, 257–267.

Rhoades, D. F. 1979. Evolution of plant chemical defenses against herbivores. In G. A. Rosenthal and D. H. Janzen, Eds., *Herbivores: Their Interaction with Secondary Plant Metabolites.* Academic Press, New York, pp. 3–54.

Rhoades, D. F. 1983. Herbivore population dynamics and plant chemistry. In R. F. Denno and M. S. McClure, Eds., *Variable Plants and Herbivores in Natural and Managed Systems.* Academic Press, New York, pp. 155–220.

Richards, O. W. and N. Waloff. 1954. Studies on the biology and population dynamics of British grasshoppers. *Anti-Locust Bull.* **17**, 1–182.

Riegert, P. W. 1972. Surveys of grasshopper abundance and forecasts of outbreaks. *Proc. Int. Study Conf. Curr. Future Probl. Acridology, 1970*, pp. 367–374.

Riegert, P. W., R. A. Fuller, and L. G. Putnam. 1954. Studies on dispersal in grasshoppers (Acrididae) tagged with Phosphorus-32. *Can. Entomol.* **86**, 223–232.

Ritchie, M. G., R. K. Butlin, and G. M. Hewitt. 1987. Causation, fitness effects and morphology of macropterism in *Chorthippus parallelus* (Orthoptera: Acrididae). *Ecol. Entomol.* **12**, 209–218.

Rodell, C. F. 1977. A grasshopper model for a grassland ecosystem. *Ecology* **58**, 227–245.

Roff, D. A. 1974a. Spatial heterogeneity and the persistence of populations. *Oecologia* **15**, 245–258.

Roff, D. A. 1974b. The analysis of a population model demonstrating the importance of dispersal in a heterogeneous environment. *Oecologia* **15**, 259–275.

Roffey, J. 1963. Observations on night flight in the desert locust (*Schistocerca gregaria* F.). *Anti-Locust Bull.* **39**, 1–32.

Rogers, L. E. and R. J. Lavigne. 1972. Asilidae of the Pawnee National Grasslands, in northern Colorado. *Wyo., Agric. Exp. Stn., Sci. Monogr.* **25**, 1–35.

Roughgarden, J. A. 1989. The structure and assembly of communities. In J. A. Roughgarden, R. M. May, and S. A. Levin, Eds., *Perspectives in Ecological Theory*. Princeton Univ. Press, Princeton, New Jersey, pp. 203–226.

Rowell, C. H. F. 1971. The variable coloration of Acridoid grasshoppers. *Adv. Insect Physiol.* **8**, 145–198.

Sanchez, N. E. and G. G. Liljesthrom. 1986. Population dynamics of *Laplatracris dispar* (Orthoptera: Acrididae). *Environ. Entomol.* **15**, 775–778.

Sanchez, N. E. and J. A. Onsager. 1989. Life history parameters in *Melanoplus sanguinipes* (F.) under natural field conditions. *Can. Entomol.* (in press).

Sanchez, N. E., J. A. Onsager, and W. P. Kemp. 1988. Fecundity of *Melanoplus sanguinipes* (F.) under natural conditions. *Can. Entomol.* **10**, 29–37.

Sayer, H. J. 1962. The desert locust and tropical convergence. *Nature (London)* **194**, 330–336.

Schaeffer, W. M. 1985. Order and chaos in ecological systems. *Ecology* **66**, 93–106.

Schaeffer, W. M. and M. Kot. 1986. Chaos in ecological systems: The coals that Newcastle forgot. *Trends Ecol. Evol.* **1** 58–67.

Shelly, T. E. and M. D. Greenfield. 1985. Alternative mating strategies in a desert grasshopper: A transitional analysis. *Anim. Behav.* **33**, 1211–1222.

Sibly, R. and K. Monk. 1987. A theory of grasshopper life cycles. *Oikos* **48**, 186–194.

Slatkin, M. 1974. Competition and regional coexistence. *Ecology* **55**, 128–134.

Slifer, E. H. and R. L. King. 1961. The inheritance of diapause in grasshopper eggs. *J. Hered.* **52**, 39–44.

Smith, D. S. 1959. Utilization of food plants by the migratory grasshopper, *Melanoplus bilituratus* (Walker) (Orthoptera: Acrididae) with some observations on the nutritional value of the plants. *Ann. Entomol. Soc. Am.* **52**, 674–680.

Smith, D. S. 1966. Fecundity and oviposition in the grasshoppers *Melanoplus sanguinipes* (F.) and *Melanoplus bivittatus* (Say). *Can. Entomol.* **98**, 617–621.

Smith, D. S. 1968. Oviposition and fertility and their relation to copulation in *Melanoplus sanguinipes* (F.). *Bull. Entomol. Res.* **57**, 559–565.

Smith, D. S. 1970. Crowding in grasshoppers. I. Effects of crowding within one generation on *Melanoplus sanguinipes. Ann. Entomol. Soc. Am.* **63** 1775–1776.

Smith, D. S. 1972. Crowding in grasshoppers. II. Continuing effects of crowding on subsequent generations of *Melanoplus sanguinipes* (Orthoptera: Acrididae). *Environ. Entomol.* **1**, 314–317.

Smith, D. S. and F. E. Northcott. 1951. The effects on the grasshopper, *Melanoplus mexicanus mexicanus* (Sauss.) (Orthoptera: Acrididae) of varying nitrogen content in its food plant. *Can. J. Zool.* **29**, 297–304.

Smith, R. C. 1954. An analysis of 100 years of grasshopper populations in Kansas. *Trans. Kans. Acad. Sci.* **57**, 397–433.

Smith, R. H. and R. Sibly. 1985. Behavioral ecology and population dynamics: Towards a synthesis. In R. M. Sibly and R. H. Smith, Eds., *Behavioural Ecology*. Blackwell, Oxford, pp. 577–591.

Southwood, T. R. E. 1978. Escape in space and time — Concluding remarks. In H. Dingle, Ed., *Evolution of Insect Migration and Diapause*. Springer-Verlag, Berlin, pp. 277–279.

Southwood, T. R. E. 1988. Tactics, strategies and templets. *Oikos* **52**, 3–18.

Southwood, T. R. E. and H. N. Comins. 1976. A synoptic population model. *J. Anim. Ecol.* **45**, 949–965.

Stearns, S. C. 1976. Life-history tactics: A review of the ideas. *Q. Rev. Biol.* **51**, 3–47.

Stearns, S. C. 1977. The evolution of life history traits: A critique of theory and a review of the data. *Annu. Rev. Ecol. Syst.* **8**, 145–171.

Stiling, P. 1988. Density-dependent processes and key factors in insect populations. *J. Anim. Ecol.* **57**, 581–593.

Stower, W. J. and D. J. Greathead. 1969. Numerical changes in a population of the desert locust, with special reference to the factors responsible for mortality. *J. Appl. Ecol.* **6**, 203–235.

Strong, D. R. 1984. Density vague ecology and liberal population regulation in insects. In P. W. Price, C. N. Slobodchikoff, and W. S. Gaud, Eds., *A New Ecology: Novel Approaches to Interactive Systems*. Wiley, New York, pp. 313–327.

Strong, D. R. 1986. Population theory and understanding pest outbreaks. In M. Kogan, Ed., *Ecological Theory and Integrated Pest Management Practice*. Wiley, New York, pp. 37–58.

Strong, D. R., J. H. Lawton, and T. R. E. Southwood. 1984. *Insects on Plants. Community Patterns and Mechanisms*. Harvard Univ. Press, Cambridge, Massachusetts.

Symmons, P. M. 1959. The effect of climate and weather on the numbers of the Red Locust, *Nomadacris septemfasciata* (Serv.), in the Rukwa Valley outbreak area. *Bull. Entomol. Res.* **50**, 507–521.

Symmons, P. M. and D. E. Wright. 1982. The origins and course of the 1979 plague of the Australian plague locust, *Chortoicetes terminifera* (Walker) (Orthoptera: Acrididae), including the effect of chemical control. *Acrida* **10**, 159–190.

Tauber, M. J., C. A. Tauber, and S. Masaki. 1986. *Seasonal Adaptations of Insects*. Oxford Univ. Press, London and New York.

Tsyplenkov, E. P. 1972. *Locusta migratoria* (L.) in the U.S.S.R. and the dynamics of its numbers. *Proc. Int. Study Conf. Curr. Future Probl. Acridology, 1970*, pp. 361–366.

Underwood, A., E. Denley, and M. Moran. 1983. Experimental analyses of the structure and dynamics of mid-shore rocky intertidal communities in New South Wales. *Oecologia* **56**, 202–219.

Uvarov, B. P. 1961. Quantity and quality in insect populations. *Proc. R. Entomol. Soc. London, (C) Ser.* **25**, 52–59.

Uvarov, B. P. 1966. *Grasshoppers and Locusts: A Handbook of General Acridology*, Vol. 1. Cambridge, Univ. Press, London and New York.

Uvarov, B. P. 1977. *Grasshoppers and Locusts*, Vol. 2. Centre for Overseas Pest Research, London.

Visscher, S. N. 1980. Regulation of grasshopper fecundity, longevity and egg viability by plant growth hormones. *Experientia*, **36**, 130–131.

Visscher, S. N., R. Lund, and W. Whitmore. 1979. Host plant growth temperatures and insect rearing temperatures influence reproduction and longevity in the grasshopper *Aulocara ellioti*. *Environ. Entomol.* **8**, 253–258.

Wagner, T. L., H. Wu, P. J. H. Sharpe, R. M. Schoolfield, and R. N. Coulson. 1984. Modeling insect development rates: A literature review and application of a biophysical model. *Ann. Entomol. Soc. Am.* **77**, 208–225.

Wakeland, C. 1958. The high plains grasshopper. *U.S. Dep. Agric., Tech. Bull.* **1167**.

Walker, T. J. 1980. Mixed oviposition in individual females of *Gryllus firmus*: Graded proportions of fast developing and diapause eggs. *Oecologia* **47**, 291–298.

Wall, R. and M. Begon. 1986. Population density, phenotype, and mortality in the grasshopper *Chorthippus brunneus*. *Ecol. Entomol.* **11**, 445–456.

Wall, R. and M. Begon. 1987a. Population density, phenotype, and reproductive output in the grasshopper *Chorthippus brunneus*. *Ecol. Entomol.* **12**, 331–339.

Wall, R. and M. Begon. 1987b. Individual variation and the effects of population density in the grasshopper *Chorthippus brunneus*. *Oikos* **49**, 15–27.

Wallner, W. E. 1986. Factors affecting insect population dynamics: Differences between outbreak and non-outbreak species. *Annu. Rev. Entomol.* **32**, 317–340.

Waloff, N. 1972. Some thoughts on the studies of population dynamics of Acridoids. *Proc. Int. Study Conf. Curr. Future Probl. Acridology, 1970*, pp. 355–358.

Waloff, Z. 1946. Seasonal breeding and migrations of the desert locust (*Schistocerca gregaria* F.) in eastern Africa. *Anti-Locust Mem.* **1**, 1–74.

Waloff, Z. 1963. Field studies on solitary and *transiens* desert locusts in the Red Sea area. *Anti-Locust Bull.* **140**, 1–93.

Waloff, Z. 1966. The upsurges and recessions of the desert locust plague: An historical survey. *Anti-Locust Memoir* **8**, 1–111.

Waloff, Z. 1976. Some temporal characteristics of desert locust plagues. *Anti-Locust Mem.* **13**.

Waloff, Z. and R. Rainey. 1951. Flying locusts and convection currents. *Anti-Locust Bull.* **9**, 1–72.

Wardhaugh, K. G. 1977. The effects of temperature and photoperiod on the morphology of the egg pod of the Australian Plague locust (*Chortoicetes terminifera* Walker, Orthoptera: Acrididae). *Aust. J. Ecol.* **2**, 81–88.

Wardhaugh, K. G. 1980a. The effects of temperature and moisture on the inception of diapause in eggs of the Australian Plague locust, *Chortoicetes terminifera* Walker (Orthoptera: Acrididae). *Aust. J. Ecol.* **5**, 187–191.

Wardhaugh, K. G. 1980b. The effects of temperature and photoperiod on the induction of diapause in eggs of the Australian Plague Locust, *Chortoicetes terminifera* (Walker) (Orthoptera: Acrididae). *Bull. Entomol. Res.* **70**, 635–647.

Wardhaugh, K. G. 1986. Diapause strategies in the Australian plague locust (*Chortoicetes terminifera* Walker) In F. Taylor and R. Karban, Eds., *The Evolution of Insect Life Cycles*. Springer-Verlag, New York, pp. 89–104.

Wellington, W. G. 1960. Qualitative changes in natural populations during changes in abundance. *Can. J. Zool.* **38**, 289–314.

Wellington, W. G. 1964. Qualitative changes in populations in unstable environments. *Can. Entomol.* **96**, 435–451.

White, T. C. R. 1976. Weather, food, and plagues of locusts. *Oecologia* **22**, 119–134.

White, T. C. R. 1978. The relative importance of food shortage in animal ecology. *Oecologia* **336**, 71–86.

White, T. C. R. 1984. The availability of invertebrate herbivores in relation to the availability of nitrogen in stressed food plants. *Oecologia* **63**, 90–105.

Wilson, G. G. 1974. The effects of temperature and ultraviolet radiation on the infection of *Choristoneura fumiferana* and *Malacosoma pluviale* by a microsporidian parasite, *Nosema* (*Perezia*) *fumiferanae* (Thom.). *Can. J. Zool.* **52**, 59–63.

15

Pathogenic Diseases of Grasshoppers

D. A. STREETT and

M. R. MCGUIRE

15.1 INTRODUCTION

There has been a flood of relevant information on grasshopper pathogens since the last comprehensive review on the biology of grasshoppers (Uvarov, 1966, 1977). Research for the most part has emphasized the use of pathogens for controlling grasshopper populations. Accordingly, reviews on grasshopper pathogens have with had an applications perspective and dealt mainly with suppression of grasshopper populations (Henry, 1977; Streett, 1986). In this chapter we will attempt to provide a thorough coverage of the current literature on grasshopper diseases and to address future research trends in this area.

Grasshopper diseases are usually considered in terms of the etiological agent involved, and thus this chapter is divided into sections on protozoa, viruses, fungi, bacteria, rickettsiae, and nematodes. Each section offers a brief introduction and a generalized life cycle; the remainder of the section discusses our present knowledge and offers some insights and challenges in the field.

15.2 PROTOZOA

Protozoa are essentially unicellular eukaryotic organisms considered a subkingdom in the kingdom Protista. Transmission of the protozoa infecting grasshoppers normally occurs following ingestion of infective stages (e.g., spores or cysts). The principal pathogenic protozoa of grasshoppers are found in the groups amoebae (phylum Sarcomastigophora), gregarines (phylum Apicomplexa), and microsporidia (phylum Microspora) (Fig. 15.1). The most extensively studied group is the microsporidia, primarily because of their potential as biological control agents.

15.2.1 Microsporidian Diseases

Microsporidia are obligate intracellular parasites that lack mitochondria and possess some form of a polar tube. Spores upon ingestion by a suitable host extrude a polar tube that must penetrate the host cell for infection to occur. After cell penetration, the infective stage, a sporoplasm, enters the host cell by passing through the polar tube. The sporoplasm will usually undergo some form of fission called "schizogony", forming meronts that continue to undergo fission until the formation of sporonts. A sporont will then undergoes a fission process called "sporogony" and produces sporoblasts, which develop directly into spores.

Four species of microsporidia are described from grasshoppers. Three are from the genus *Nosema*; they are *N. locustae*, *N. acridophagus*, and *N. cuneatum* (Canning, 1953; Henry, 1967, 1971). A fourth species, *Perezia dichroplusae*, has been described from the grasshopper *Dichroplus elongatus* (Lange, 1987). Most of the studies on microsporidian diseases of grasshoppers are centered on *N. locustae* and its potential for suppression of grasshopper populations. A few

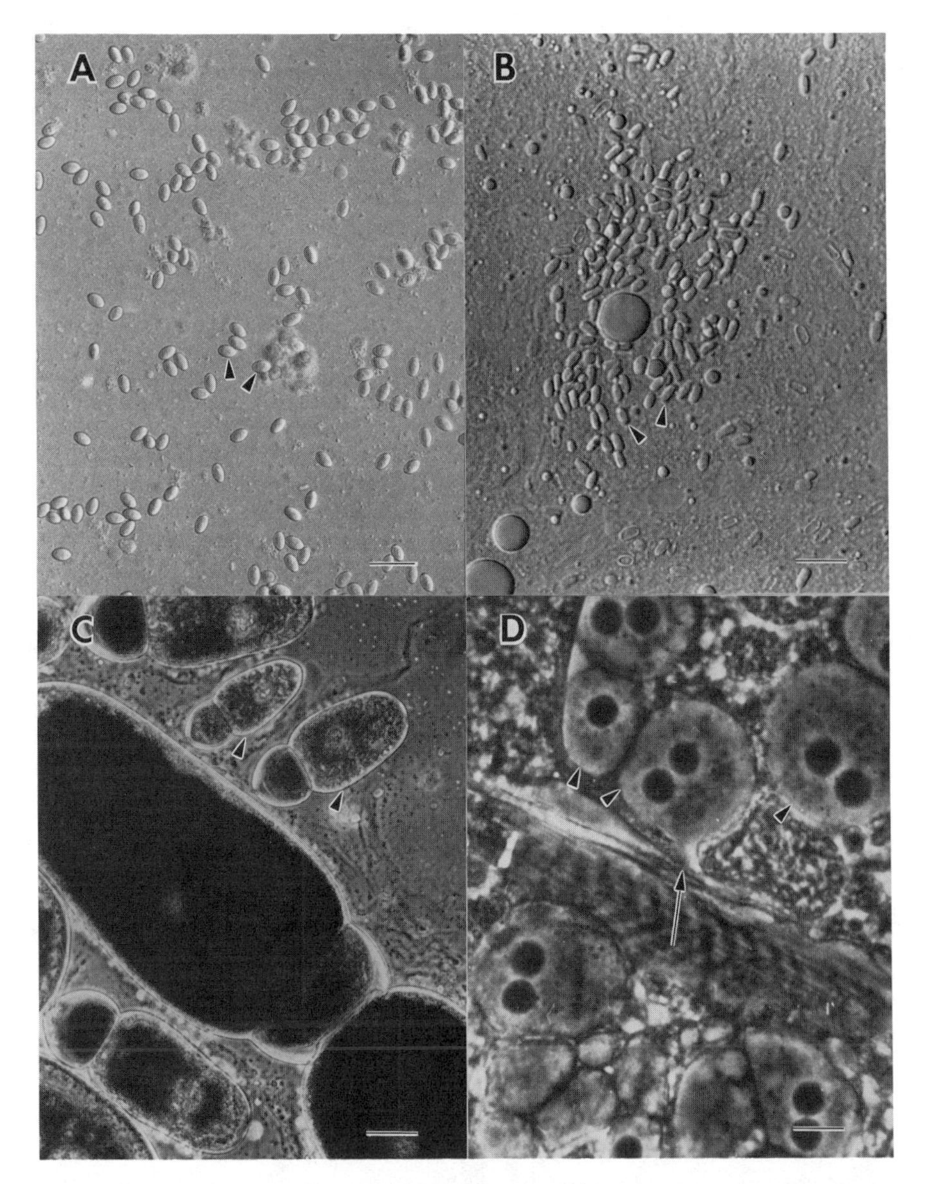

Fig. 15.1. Protozoa commonly found infecting grasshoppers. (A) Fresh preparation of *Malameba locustae* cysts (arrowheads) isolated from the Malpighian tubules of *Melanoplus differentialis*; bar = 20 μm. Courtesy of D. the alimentary tract of A. Street. (B) Differential interference contrast photomicrograph showing fresh spores of an undescribed microsporidium species from *Chorthippus curtipennis*; bar = 12 μm. Courtesy of D. A. Streett. (C) Bright field photomicrograph showing fresh preparation of cephaline eugregarine sporonts (arrowheads) from lumen of grasshopper midgut; bar = 50μm. Courtesy of J. E. Henry. (D) Phase contrast photomicrograph of a fresh preparation showing uninucleate and binucleate schizonts (arrowheads) of a grasshopper neogregarine. A schizont holdfast (arrow) apparently attached to a muscle fiber; bar = 30 μm. Courtesy of J. E. Henry.

of these studies, though, consider the more basic aspects of host–parasite interactions.

Reports on the natural occurrence of microsporidia in grasshopper populations illustrate the complexity of this interaction. Henry's (1972b) study on the epizootiology of *N. locustae* found a generally low prevalence of the disease in grasshopper populations. In contrast, Streett and Henry (1984) found a microsporidium infecting *Aulocara elliotti* with a prevalence of infection reaching 100% and no apparent suppression of the grasshopper population. Many of the more interesting effects of microsporidia on grasshoppers are sublethal or indirect in nature, and include changes in behavior and reductions in feeding and host fecundity.

Significant reductions in feeding by grasshoppers infected with *N. locustae* have been demonstrated for *M. differentialis* females (Oma and Hewitt, 1984). Subsequent studies by Johnson and Pavlikova (1986) show even more pronounced effects on food consumption, with about a 40% reduction in feeding observed for both sexes of infected *M. sanguinipes*.

Grasshoppers infected with microsporidia may also exhibit reduced fecundity. A field trial by Ewen and Mukerji (1980) with *N. locustae* reported a 64–81% reduction in egg deposition by several *Melanoplus* species. Likewise, Erlandson et al. (1986) found that grasshoppers infected with *N. cuneatum* as second or third-instars did not oviposit when they become adult, whereas those infected as fifth-instar nymphs produced almost as many eggs per female as uninfected insects. Furthermore, this study demonstrated the common phenomenon of reduced susceptibility to infection in late-instar grasshoppers. Substantially more *N. cuneatum* spores were required to infect fifth-instar nymphs than second- and third-instar *M. sanguinipes* (Fig. 15.2).

Evidence reported by Boorstein and Ewald (1987) of a behavioral fever in *M. sanguinipes* infected with *N. acridophagus* (Fig. 15.3) and their hypothesis that such a fever was a host defense suggests some intriguing studies at the organism level. Presumably, if disease causes a change in host behavior it may influence many lab and field studies (e.g., such as ambient air temperature preference).

15.2.2 Amoebic Diseases

The genus *Malameba* King and Taylor contains only one species, *M. locustae*, that is responsible for the amoebic disease of grasshoppers. King and Taylor (1936) originally described *M. locustae* from the Malpighian tubules of several *Melanoplus* species. *M. locustae* has been found to have a relatively wide host range and geographical distribution (Ernst and Baker, 1982; Henry et al., 1985). Malpighian tubules of heavily infected grasshoppers are often swollen and lustrous white in appearance (Fig. 15.4). When examined with phase microscopy these tubules are found to be packed with refractile cysts (Fig. 15.5). Diseased grasshoppers are lethargic, consume less food, and have reduced fecundity (King and Taylor, 1936). Most of the studies with *M. locustae*

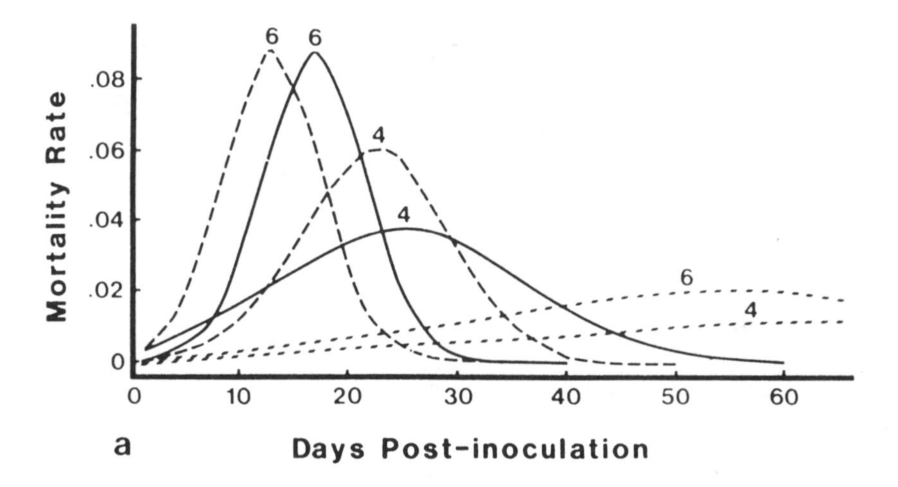

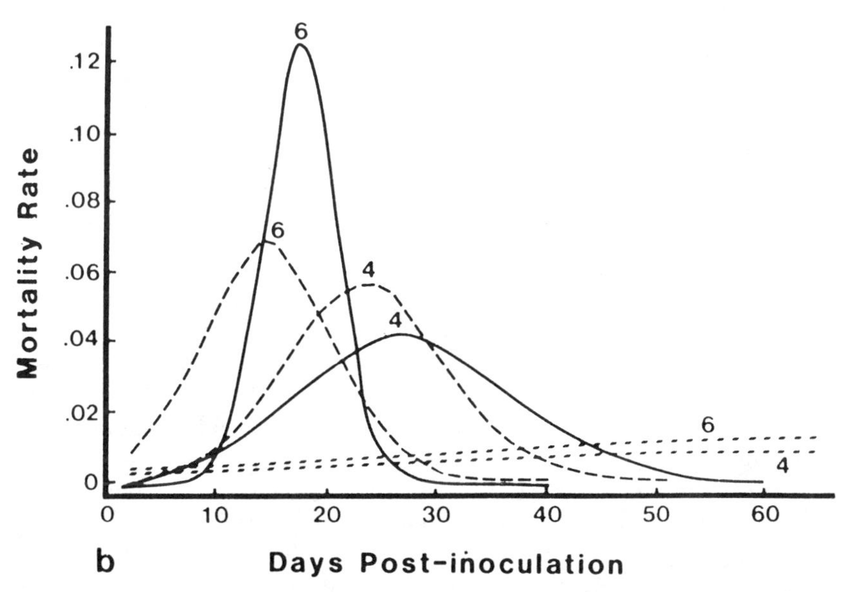

Fig. 15.2. Grasshopper susceptibility to microsporidia. Daily mortality rate curves derived from probit equation for time–mortality responses of second- (— — —), third- (———), and fifth-instar (---) *Melanoplus sanguinipes* inoculated with either 10^4 (4 in figure) or 10^6 (6 in figure) *Nosema cuneatum* spores and reared either individually (a) or in a group (b). Reprinted from Erlandson et al. (1986) by courtesy of Mr. A. Erlandson.

typically deal with host range, life cycle, and natural occurrence; few have considered either the host–parasite relationship or epizootiology.

Horizontal transmission by oral ingestion is the only reported route of

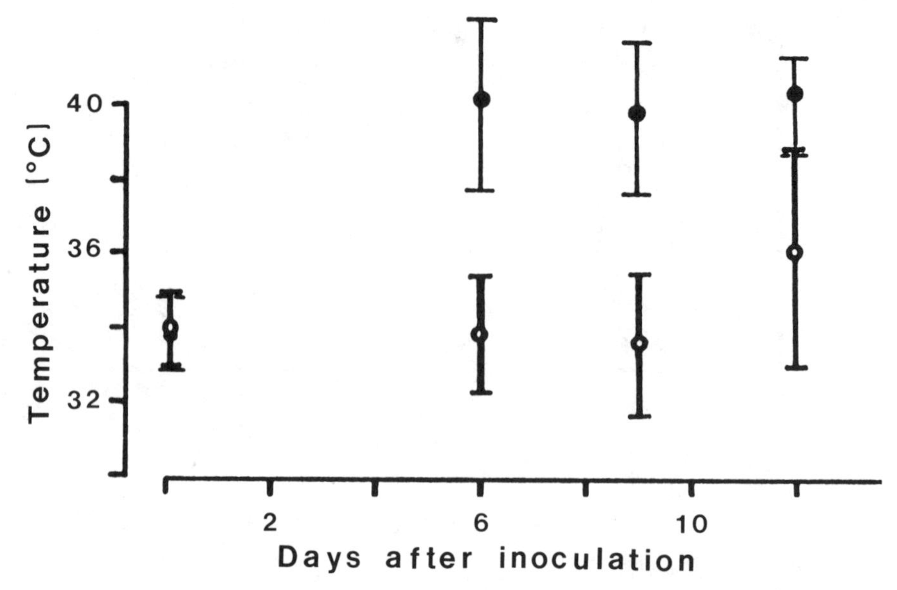

Fig. 15.3. Temperature preference of grasshoppers infected with microsporidia. Selected air temperature of *Melanoplus sanguinipes* either infected with *Nosema acridophagus* (closed circles) or uninfected (open circles) at 6, 9, and 12 days after ingestion of spores. Vertical bars represent ± 1 S.D. Reprinted from Boorstein and Ewald (1987) by courtesy of P. Ewald.

infection, although some controversy exists regarding the course of infection in grasshoppers. In an early report, Evans and Elias (1970) observed primary trophozoites entering midgut epithelial cells. Secondary trophozoites, formed from dividing primary trophozoites, then migrated from the midgut epithelial cells to infect the Malpighian tubules. Harry and Finlayson (1976) later proposed a more direct route with infected degenerating midgut epithelial cells sloughed during a molt entering Malpighian tubules at the midgut junction. The recent findings of Braun et al. (1988) support the direct route hypothesis: trophozoites do not infect the Malpighian tubules via the midgut epithelial cells, but enter the tubules directly from the alimentary tract. They also contend that the trophozoites occasionally found in midgut epithelial cells do not undergo division but appear to degenerate.

15.2.3 Gregarine Diseases

The taxonomy of gregarines has changed significantly since the establishment of the genus *Gregarina* by Dufour (1828). At present, there are over 200 genera and about 1600 species in the class Conoidasida (Levine, 1988). Earlier classification schemes divided the gregarines into the schizogregarines and eugregarines based on the presence or absence of merogony, respectively. The

Fig. 15.4. The effect of amoebae on grasshoppers. Dissected grasshoppers infected with *Malameba locustae* showing the swollen lustrous white malpighian tubules (arrows). Courtesy of J. E. Henry.

schizogregarines have since been split into the more primitive archigregarines and the advanced neogregarines.

Ingestion of spores by the host initiates the infection process. Sporozoites are released from the spores into the alimentary tract and subsequently enter a cell. Development ensues intracellularly for the early trophozoites whereas the later stages are usually intercellular or extracellular. Few studies describe the intracellular development but instead have focused on the more easily detected intercellular or extracellular stages.

Eugregarines are found mainly in the alimentary tract of the host. The group is commonly divided into the acephaline gregarines (body not differentiated into regions) or cephaline gregarines (body differentiated into two main regions).

Life cycles vary almost as much as eugregarines vary in their morphology. A general life cycle, though, will have the spores germinate when ingested by a suitable host species. Sporozoites then emerge and enter the midgut epithelial cells. After some intracellular development the trophozoites reenter the midgut lumen and attach to epithelial cells. Gamogony occurs with the encystment of paired sporadins or sporadins in syzygy to form gametocysts. The gametocysts

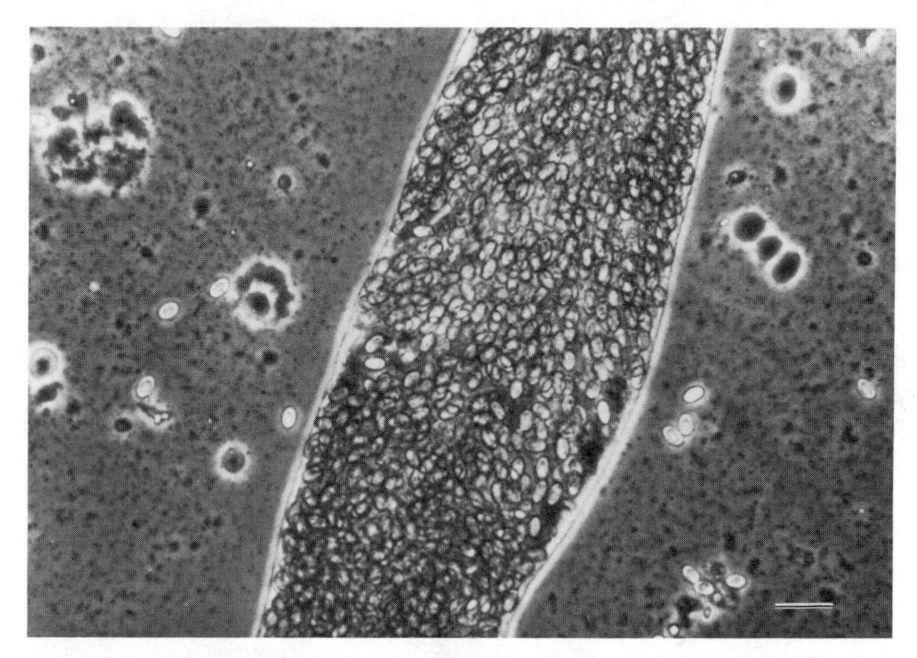

Fig. 15.5. Phase contrast photomicrograph of Malpighian tubule of *Melanoplus sanguinipes* packed with *Malameba locustae* cysts; bar = 37.5μm. Courtesy of J. E. Henry.

are normally passed out in the feces to complete development, eventually forming gametes. Zygote formation initiates sporogony and each zygote forms a spore filled with sporozoites.

In neogregrines, spores germinate after ingestion and release sporozoites, which penetrate into the hemocoel and either attach to or enter host cells. Sporozoites continue to increase in size and form trophozoites. The trophozoites undergo schizogony to form large multinucleate plasmodia. Gametocytes are uninucleate stages formed after division of the large multinucleate plasmodium. Gamogony occurs with the union of gametocytes and encystment within a gametocyst. A zygote is formed from the isogamous union of two gametes. The zygote undergoes sporogony to form spores with eight sporozoites. When sporogony does not occur the zygote becomes a single spore with the gametocyst containing numerous oocysts.

The limited scope of reports on gregarine diseases of grasshoppers certainly provides a fertile area for study. Most of the studies have been descriptive or reports on geographical distribution (Kundu and Haldar, 1983; Harry, 1965). Notable exceptions are the reports by Harry (1970, 1971) on the effect of *Gregarina garnhami* on the growth of the desert locust, *Schistocerca gregaria*, and an infection and reinfection study on *S. gregaria* with the same parasite. Gregarines for the most part, though, have not been thoroughly investigated in the areas of host–parasite relationships and epizootiology.

15.3 VIRUSES

Viruses are one of the most extensively studied groups of invertebrate pathogens and infect virtually every order of insects. Surprisingly, viruses from only three viral groups have been originally isolated from grasshoppers. Henry and Jutila (1966) and Henry et al. (1969) isolated an entomopoxvirus and Jutila et al. (1970) isolated a crystalline array virus thought to be closely related to the picornaviruses. Both viruses were isolated from *Melanoplus* species. A third virus recently identified in *Caledia captiva* (F.) by Colgan (1986) is apparently a cytoplasmic polyhedrosis virus (CPV) from the reovirus group. Significant increases in mortality were observed in *L. migratoria* fed the CPV polyhedral preparation. Apparently these deaths, although associated with the viral treatment, were attributed to a strain of *Enterobacter cloacae*. The initial discovery of a reovirus infecting *C. captiva* was followed by the isolation of a unique reovirus genome from mitochondrial preparations of the same grasshopper, *C. captiva*, by Colgan and Christian (1988).

Recent findings of Bensiom et al. (1987) represent the first report of cross-infection of a lepidopteran nuclear polyhedrosis virus (NPV) in an orthopteran host. Oral transmission of the *Spodoptera littoralis* NPV to *L. m. migratoriodes* and *S. gregaria* was proved with restriction endonuclease analysis. Symptoms of the NPV disease in grasshoppers include the development of a darkened condition on the head and thorax, hence the name "dark cheeks" disease.

15.3.1 Entomopoxviruses

Since the initial discovery in *M. sanguinipes*, entomopoxviruses have been found in nine different species of grasshoppers from around the world (Table 15.1). Six of these viruses have been characterized and shown to be distinct from each other by restriction endonuclease analysis (Streett et al., 1986). Additionally, no relatedness has been demonstrated to vertebrate poxviruses in DNA homology studies.

The genome of grasshopper entomopoxviruses is a linear double-stranded DNA molecule ranging in size from 126 to 158 megadaltons (unpublished data). The grasshopper entomopoxvirus DNA shares with other entomopoxvirus DNAs the unique feature of having a relatively low GC content (Langridge et al., 1977; Langridge and Henry, 1981). The significance of such a low GC content on the virus genome is open to speculation and requires further study.

Grasshopper entomopoxvirus exists in the environment with 100–500 virus particles occluded in a proteinaceous crystalline matrix called a spheroid (Fig. 15.10). Upon ingestion, spheroids break down in the gut, probably due to proteases, and the virions are released. Although the life cycle of grasshopper entomopoxviruses is not entirely worked out, it probably is similar to that of other entomopoxviruses (Granados, 1981). Briefly, virions fuse with midgut epithelial cells and eventually enter the hemolymph. It is not known if replication takes place in the epithelial cells. Virions circulate and are reported

TABLE 15.1. Source and Host Range of Grasshopper Entomopoxviruses

Entomopoxvirus[a]	Known Host Range of Virus	Geographical Location of Original Isolate	Reference[b]
Melanoplus sanguinipes	*M. sanguinipes* *M. packardii* *M. differentialis* *M. brunneri* *M. infantalis* *Camnula pellucida*	Arizona, U.S.A.	Henry and Jutila (1966)
Phoetaliotes nebrascensis	*P. nebrascensis*	Montana, U.S.A.	Oma and Henry (1986)
Arphia conspersa	*A. conspersa* *Camnula pellucida* *Psoloessa delicatula* *Dissosteira carolina* *Spharagemon collare* *Xanthippus corallipes*	Montana, U.S.A.	Oma and Henry (1985)
Aeropedellus clavatus	*A. clavatus* *Chorthippus curtipennis* *Dichromorpha viridis* *Ageneotettix deorum*	Saskatoon, Canada	Oma and Henry (1986)

Oedaleus senegalensis	*Oedaleus senegalensis* *Oedaleus nigeriensis* *Kraussaria angulifera* *Heteracris annulosus* *Camnula pellucida* *Dissosteria carolina*	Senegal, Africa Mauritania, Africa	Henry et al. (1985)
Cataloipus cymbiferus	*Cataloipus cymbiferus* *Cataloipus fuscocoerulipes* *Heteracris annulosus* *Dissosteira carolina*	Senegal, Africa	Henry et al. (1985)
Locusta migratoria	*L. migratoria*	Tanzania, Africa	Purrini et al. (1988)
Shistocerca gregaria	*S. gregaria*	Yemen Arab Republic	Purrini and Rhode (1988)
Chortipes sp.	*Chortipes* sp.	Yemen Arab Republic	Purrini and Rhode (1988)

[a] Virus assumes original host species name.
[b] References included in bibliography refer to original isolation of virus.

to infect only fat body tissue. Replication takes place in the cytoplasm of cells. As the infection cycle continues, virions become occluded in spheroids. Other than somewhat reduced adult and nymphal size, there are no apparent external signs of infection.

Entomopoxvirus spheroids are observable under a compound microscope approximately 2–3 weeks after infection (Fig. 15.7A). They can be distinguished from similarly shaped objects in a homogenate, such as adipose cells or gut contents, by the addition of 1 N KOH. Spheroids swell and small dark masses can be seen within (Fig. 15.7B).

The epizootiology of these viruses is almost completely unstudied. Likewise, studies examining virus survival in the environment and the effect of viruses on fecundity, feeding behavior, reproductive behavior, and other physiological processes are but a few of the many possible effects of entomopoxviruses that require investigation.

15.3.2 Crystalline Array Virus (CAV)

No studies have been conducted on this virus since the initial work by Henry (1972a) and Henry and Oma (1973). The virus was first isolated from *M. bivittatus* and was found also to infect *M. sanguinipes*, *M. differentialis*, *Schistocerca vaga vaga*, and *Schistocerca americana*. CAV replicated in the cytoplasm of pericardial, muscle, and tracheal matrix cells. General symptoms of infection include loss of vigor and coordination. Diagnosis of disease was by microscopic examination for the presence of rod-shaped crystals from 2 to 10 μm in length (Fig. 15.8). Subsequent studies of CAV indicated indirectly that the virus could replicate in vitro in dorsal vessels of *S. americana*. Homogenates of dorsal vessels exposed to CAV in vitro expressed increased virus activity as compared to the original inoculating media (Henry, 1972a). Jutila et al. (1970) reported that the RNA virus could cause up to 91% mortality 8 days following per os infection. Ultrastructural studies demonstrated that CAV replicated very similarly to that of picornaviruses from both invertebrates and vertebrates (Fig. 15.9). This close morphological relationship suggested that the CAV may be very closely related to the vertebrate-infecting picornaviruses (Henry and Oma, 1973) and research was discontinued. Molecular techniques could be utilized to determine if, indeed, CAV was related to the picornaviruses. CAV appeared to be a promising candidate for development before the apparent relationship to vertebrate-infecting viruses was discovered.

Fig. 15.6. Grasshopper entomopoxvirus spheroids and virus particles. (A) Transmission electron micrograph of *Melanoplus sanguinipes* entomopoxvirus spheroids (OB) in fat body tissue with numerous virus particles (V) occluded in each spheroid (×10,000). Courtesy of D. A. Streett. (B) Transmission electron micrograph of *Oedaleus senegalensis* entomopoxvirus spheroid (OB) at high magnification showing viral core (VC), lateral body (LB), and viral envelope (VE) of virus particles (×95,000). Courtesy of D. A. Streett.

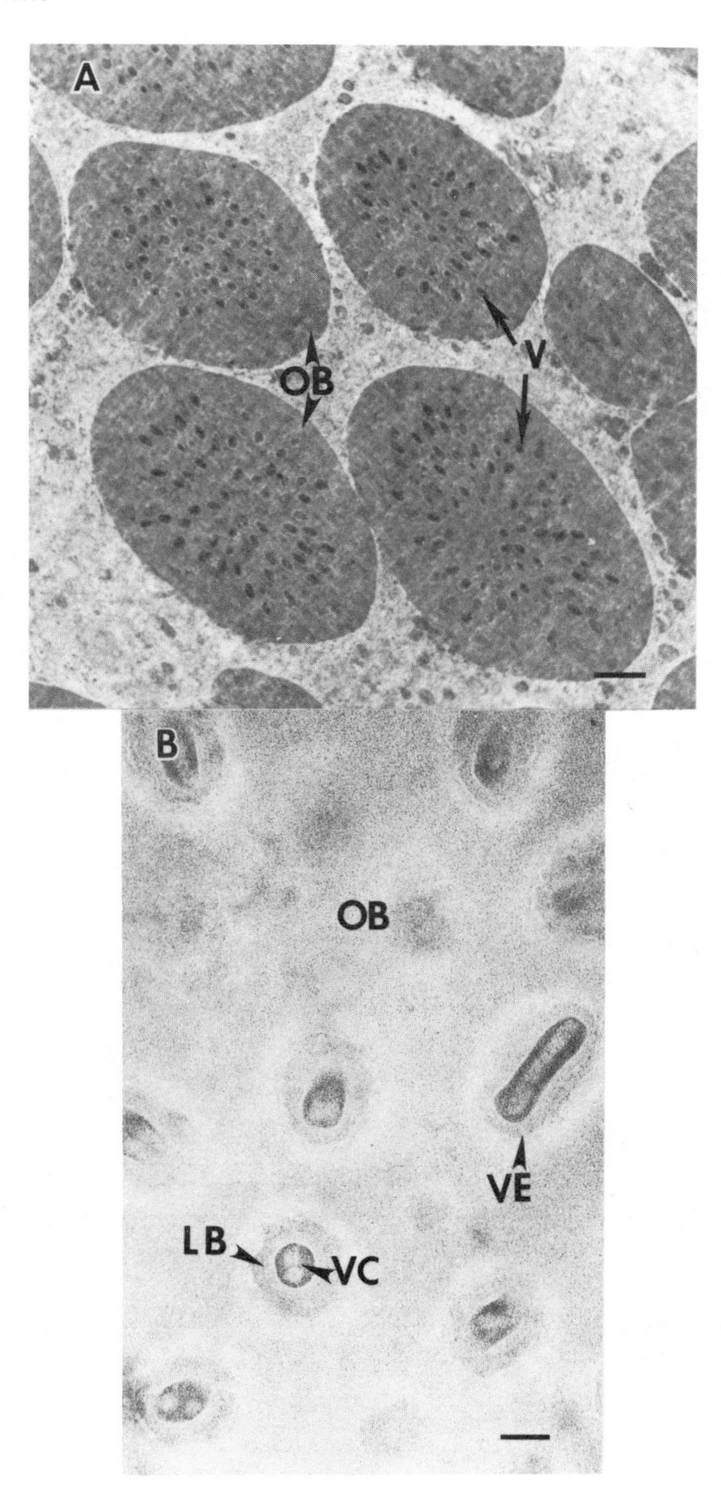
A
OB
V
B
OB
VE
LB
VC

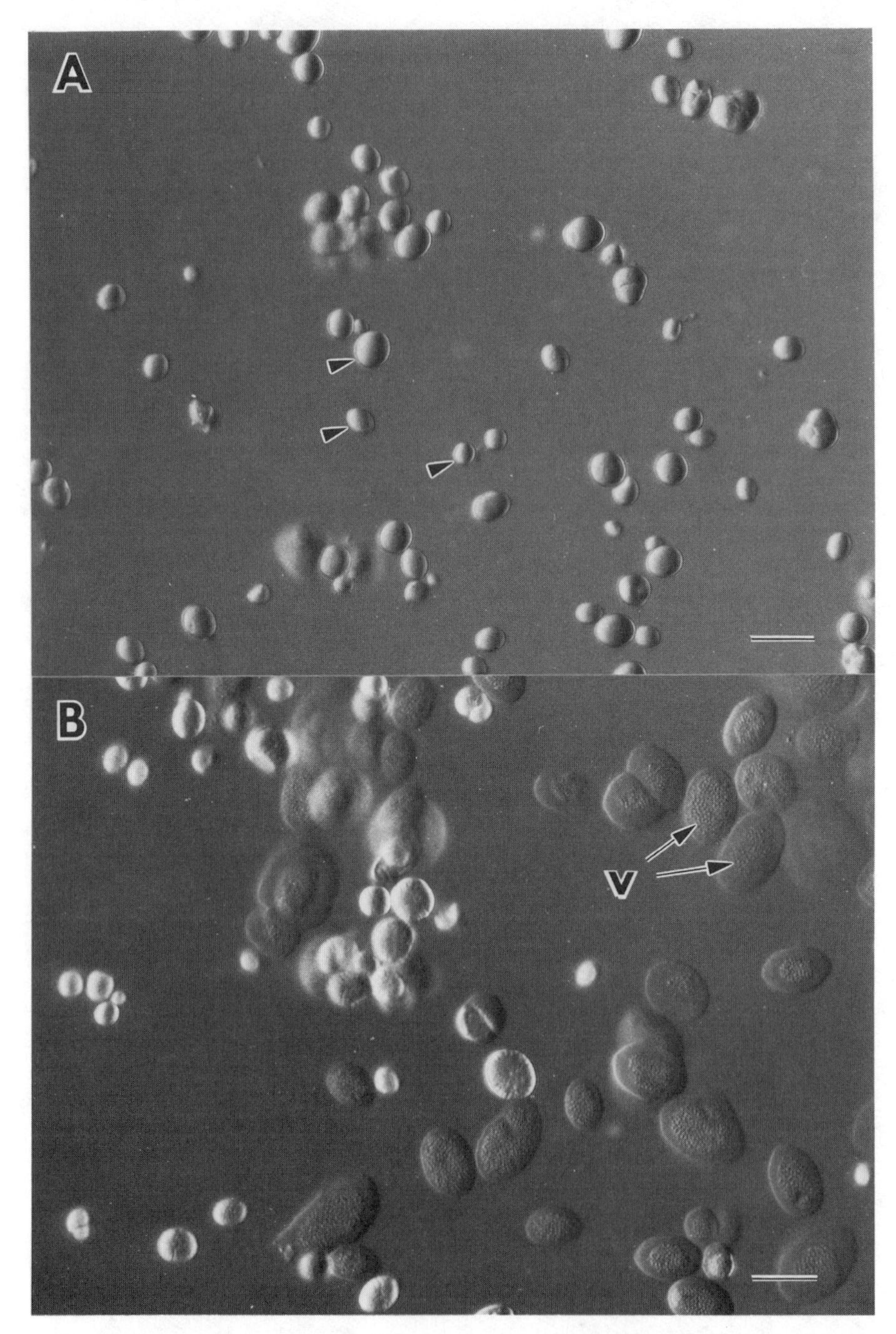

Fig. 15.7. Differential interference contrast photomicrograph of *Melanoplus sanguinipes* entomopoxviruses spheroids. (A) Purified spheroid preparation (arrowheads); bar = 18 μm. Courtesy of D. A. Streett. (B) Purified spheroid preparations treated with 1 N KOH. Note the beadlike structures, presumably virus particles (V), evident in the expanding spheroids; bar = 18 μm. Courtesy of D. A. Street.

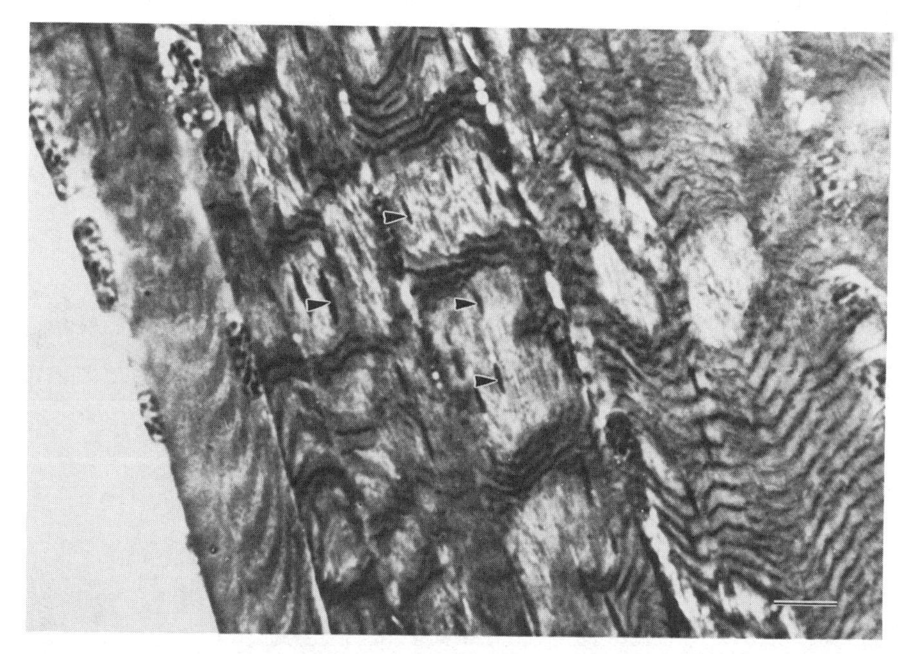

Fig. 15.8. Photomicrograph of thin section from *Melanoplus bivittatus* showing the crystal-line array virus crystal (arrowheads) embedded in muscle tissue (×1000). Reprinted from Jutila et al. (1970) by courtesy of J. W. Jutila.

15.3.3 Detection and Diagnosis

Diagnosis of diseases in grasshoppers has been primarily by microscopic examination. Development of sensitive and specific detection techniques will be of major importance in laboratory and field diagnostic studies of grasshopper pathogens. New technological developments in enzyme immunoassay (Fig. 15.10) and nucleic acid detection techniques (Fig. 15.11) have provided rapid and specific diagnosis of grasshopper diseases (Streett and McGuire, 1988). Earlier studies have demonstrated that these techniques are more sensitive than conventional microscopic examination for virus detection (McGuire and Henry, 1989; Streett et al., 1989).

15.4 FUNGI

For the past century, entomopathogenic fungi have been known to cause drastic declines among grasshopper and locust populations. Consequently, most of the research has described fungal epizootics or attempted to utilize fungi as biological control agents. Detailed studies have only recently been initiated to separate the various environmental and physiological factors responsible for the host–pathogen relationship such that these factors can be studied in a concise, repeatable, and controlled manner.

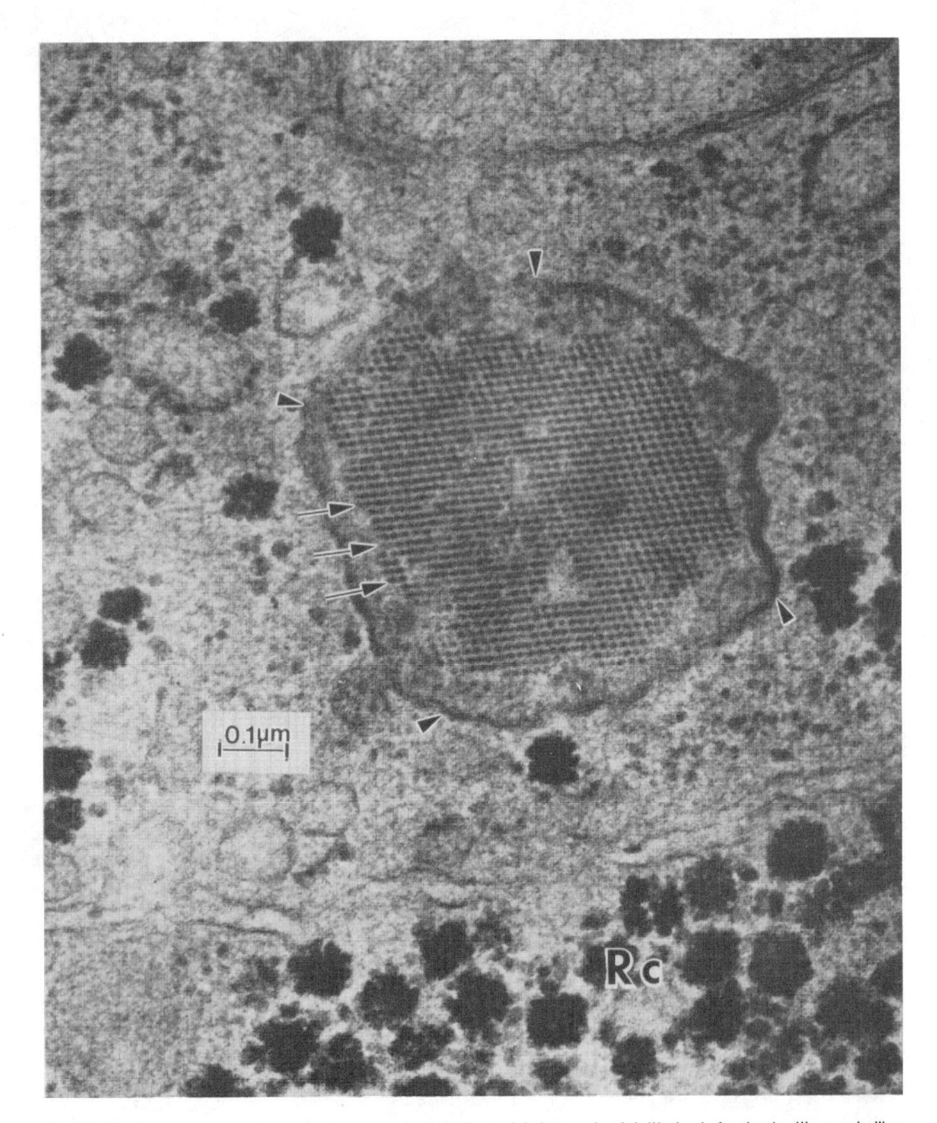

Fig. 15.9. Electron micrograph of a muscle cell from *Melanoplus bivittatus* infected with crystalline array virus showing virus particles (arrowhead) in crystalline array. A cross-section of the microcrystal is shown in an area rich in ribosomal clusters (RC). Reprinted from Jutila et al. (1970) by courtesy of J. W. Jutila.

There are two major groups of fungi that infect grasshoppers: the Entomophthorales (e.g., *Entomophaga grylli*) and the Deuteromycetes (e.g., *Metarrhizium anisopliae*, *Beauveria bassiana*, and *Verticillium lecanii*). The life cycles are somewhat similar and are treated together. In grasshoppers that have been fed a normal diet, infection through the gut has not been observed

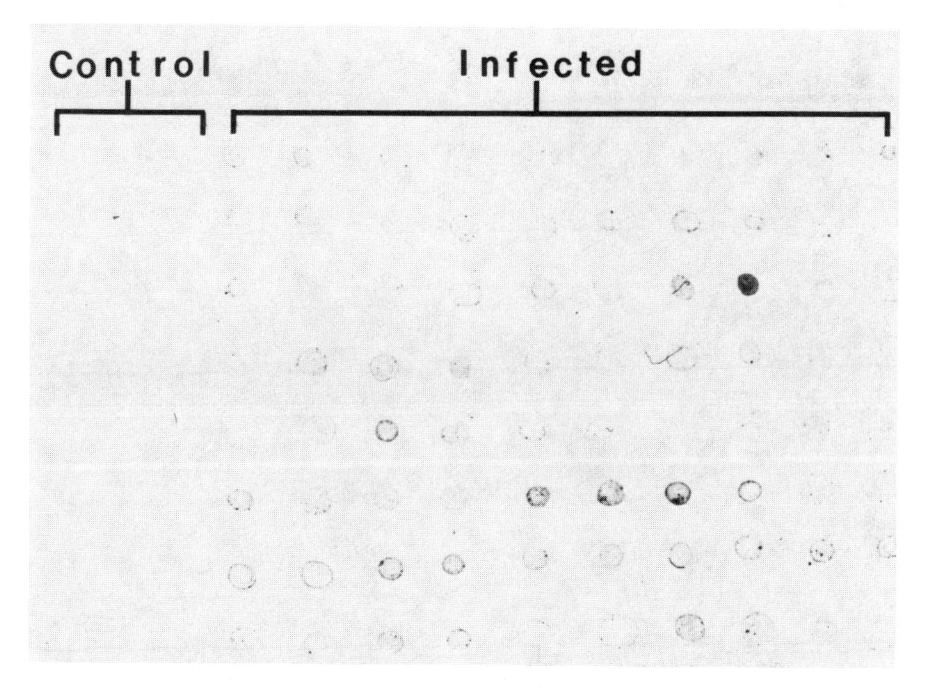

Fig. 15.10. Enzyme-linked immunosorbent assay (ELISA) for diagnosis of entomopoxvirus in grasshoppers. Detection of *Melanoplus sanguinipes* entomopoxvirus in grasshopper hemolymph 13 days after injection of virus particles. Control = uninjected grasshoppers. Infected = virus-injected grasshoppers. Note that the dark spots represent a positive sample. Courtesy of M. R. McGuire.

(Dillon and Charnley, 1986b). Infection is via attachment of a conidium to the cuticle, and integumental penetration by a germ tube. The germ tube grows into the hemocoel where hyphae or protoplasts begin to proliferate and eventually fill the body cavity. No specific tissues are attacked and mycosis continues until host nutrients are depleted. Possibly fungal toxins produced by some of the Deuteromycetes may accelerate death of the host (Roberts, 1981). Depending on the fungal species, either conidiophores erupt through the integument to form conidia (both groups) or resting spores are formed (Entomophthorales only). This cycle takes approximately 13 days for *E. grylli* (Pickford and Riegert, 1964) or 8 days for *B. bassiana* (Marcandier and Khachatourians, 1987).

15.4.1 Deuteromycetes

M. anisopliae, *B. bassiana*, and *V. lecanii* have extremely wide host ranges throughout the class Insecta, and it is not surprising that they infect at least a few grasshopper species. Although *M. anisopliae* has been reported to cause

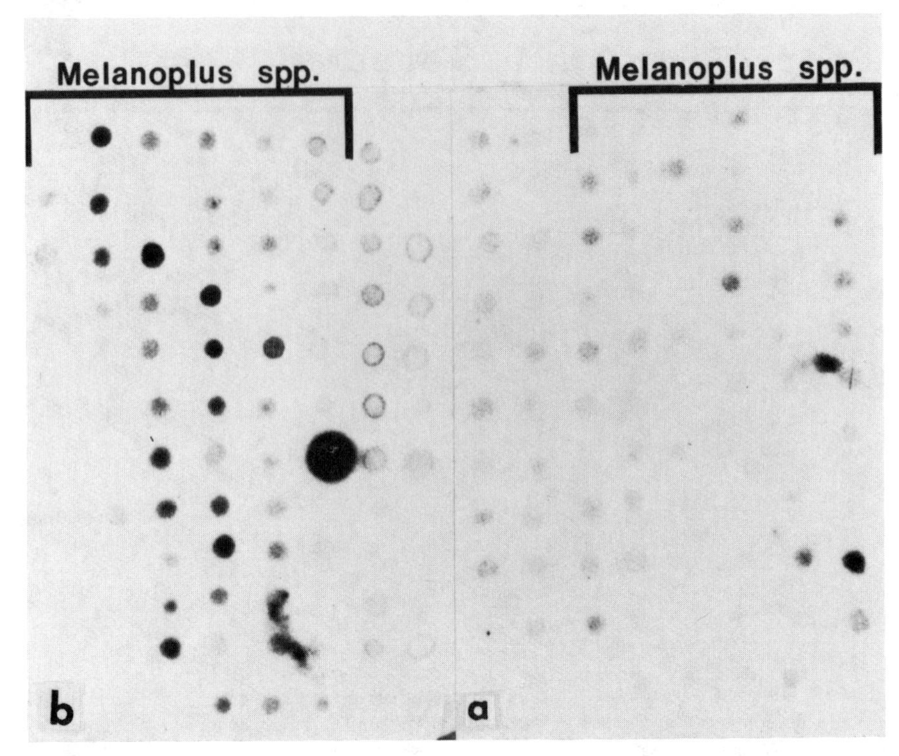

Fig. 15.11. Detection of grasshopper entomopoxvirus with nucleic acid probe. Grasshopper dot blots from field trial collection 3 weeks post-application with *Melanoplus sanguinipes* entomopoxvirus. Autoradiogram of homogenates from field-collected grasshopper applied to nitrocellulose. (a) Untreated control plot. (b) Virus-treated plot. Courtesy of D. A. Streett.

reductions in a swarm of *S. gregaria* (Balfour-Brown, 1960), *B. bassiana* is noted only in anecdotal reports (Marcandier and Khachatourians, 1987) and *V. lecanii* has not been observed infecting grasshoppers in field situations. *M. anisopliae* was observed in South Africa under warm, humid conditions, and it was suggested by Balfour-Brown (1960) that epizootics require such conditions. Balfour-Brown concluded that *M. anisopliae* probably would not work as a mycoinsecticide because of the environmental requirements necessary for establishment and maintenance of *M. anisopliae* populations.

Although *B. bassiana* also requires high humidity for horizontal transmission, Marcandier and Khachatourians (1987) demonstrated that conidia can germinate on *M. sanguinipes* cuticle and initiate infection during times of low ambient humidity. Apparently microclimatic conditions surrounding the cuticle are adequate for fungal germination. However, it was noted that sporulation did not occur in low humidity, thus preventing horizontal transmission. Because both *B. bassiana* and *M. anisopliae* can be produced on a commercial basis, this type of study becomes very important when evaluating these fungi as potential mycoinsecticides.

Only recently has a Canadian isolate of *V. lecanii* been experimentally inoculated into grasshoppers. Harper and Huang (1986) observed high rates of mortality after adult *M. sanguinipes* were dipped in *V. lecanii* spore suspensions. Subsequently, Johnson et al. (1988) determined that infection of three grasshopper species occurred under a variety of inoculation techniques, but horizontal transmission was not observed and efficacy in caged field tests was low. These authors concluded that *V. lecanii* may be used in irrigated alfalfa fields as a control tool. However, serious questions remain to be answered with regard to horizontal transmission, host range, and host colonization before widespread application of the fungus can occur.

Grasshopper cuticle is the first defense system a potential fungal pathogen must overcome. Recently, Gunnarson (1988) examined the cellular reactions that occur on and immediately surrounding the site of *M. anisopliae* penetration of *S. gregaria* cuticle. Hemocytes form mutlilayered melanized aggregates on the basement membrane below the site of penetration. Further experiments using mechanically wounded cuticle demonstrate a similar response, indicating an undescribed form of communication between external layers of cuticle and circulating hemocytes. Despite these cellular reactions, *M. anisopliae* hyphae penetrated through the aggregate into the hemocoel where defense mechanisms were not sufficient to prevent mycosis.

Along similar lines, Dillon and Charnley (1986a,b,c) found that *M. anisopliae* would not initiate mycosis within the *S. gregaria* gut unless individuals were axenic and starved after inoculation. Only then were hyphae observed to penetrate hindgut intima. Infections occurred per os routinely by germination of conidia on mouthparts. Further, they demonstrated that a toxin produced by bacteria residing in the gut inhibited conidial germination and was possibly toxic to *M. anisopliae* and several other fungi.

15.4.2 Entomophthorales

Entomophaga grylli was first described by Fresenius (1858) and has since been extensively studied. The *E. grylli* complex is cosmopolitan and has a wide host range among grasshoppers (Table 15.2). Pickford and Riegert (1964) originally suggested that different strains of *E. grylli* may exist. Their attempts to cross-infect *M. bivittatus* under artificial conditions were unsuccessful even though diseased individuals were found in the field. Subsequently Soper et al. (1983), through isozyme, host range, and life cycle studies proposed that sufficient differences existed to separate *E. grylli* into groups called pathotypes I and II. The taxonomy of the *E. grylli* complex currently is being resolved. The type species representative of *E. grylli* was isolated in Europe and appears to be different from the North American pathotypes. Additionally, several other species may be redescribed from the *E. grylli* complex (R. A. Humber, personal communication).

Grasshoppers undergo an interesting behavioral modification after infection with *E. grylli* and just prior to death (Fig. 15.12). They become sluggish, climb a stalk of grass or other support, clasp their legs around the stem, and die.

TABLE 15.2. Grasshopper Species Known to be Susceptible to Infection by the *Entomophaga grylli* Complex

Subfamily/Species	Country	Source
Pyrgomorphidae		
Zonocerus variegatus	Nigeria	Chapman & Page (1979)
Hemiacridinae		
Hieroglyphus banian	Thailand	Roffey (1968)
Coptacridinae		
Eucopotacra sp.	Nigeria	Chapman & Page (1979)
Calliptaminae		
Calliptamus italicus	Europe	Uvarov (1928)
Cyrtacanthacridinae		
Chondracris rosea	Thailand	Roffey (1968)
Cyrtacanthacris tatarica	Thailand	Roffey (1968)
Nomadacris septemfasciata	South Africa	Skaife (1925)
Nomadacris (Patanga) succincta	Thailand	Roffey (1968)
Ornithacris cyanea	Canada	Macleod & Mueller-Kogler (1973)
Catantopinae		
Anacatantops sp.	Nigeria	Chapman & Page (1979)
Bermius brachycerus	Australia	Milner (1978)
Catantops pingus	Thailand	Roffey (1968)
Catantops sp.	Nigeria	Chapman & Page (1979)
Hesperotettix speciosus	U.S.A	McDaniel (1986)
H. viridis	U.S.A	McDaniel (1986)
Praxibulus sp.	Australia	Milner (1978)
Stenocatantops splendens	Thailand	Roffey (1968)
Melanoplinae		
Melanoplus bivittatus	U.S.A.	Macleod & Mueller-Kogler (1973)
M. confusus	U.S.A.	McDaniel & Bohls (1984)
M. cuneatus	U.S.A.	Soper et al. (1983)
M. differentialis	U.S.A.	McDaniel & Bohls (1984)
M. femurrubrum	U.S.A.	McDaniel & Bohls (1984)
M. flavidus	U.S.A.	Soper et al. (1983)
M. lakinus	U.S.A.	McDaniel & Bohls (1984)
M. packardi	U.S.A.	Soper et al. (1983)
M. sanguinipes	U.S.A.	McDaniel & Bohls (1984)
Phoetaliotes nebrascensis	U.S.A.	McDaniel & Bohls (1984)
Acridinae		
Acrida conica	Australia	Milner (1978)
A. willemsei	Thailand	Roffey (1968)
Oedipodinae		
Camnula pellucida	Canada	Pickford & Reigert (1964)
Dissosteira carolina	Canada	Macleod & Mueller-Kogler (1973)
Trilophidia annulata	Thailand	Roffey (1968)
Gomphocerinae		
Chorthippus parallelus	Finland	Gyllenberg (1974)

Fig. 15.12. Behavioral modification of grasshopper infected with fungus. *Camnula pellucida* infected with *Entomophaga grylli* exhibits characteristic behavior and climbs to top of stalk prior to death; bar = 0.5 cm. Courtesy of B. Gillespie.

Conidial sporulation then occurs (pathotype I) or resting spores are formed (pathotype I or II). The added height gained by the grasshopper presumably aids in the dispersal of the spores across a larger area.

Epizootics of the *E. grylli* complex have been noted many times (Skaife, 1925; Uvarov, 1928; Smith, 1933; Pickford and Riegert, 1964; Chapman and Page, 1979). All of these authors indicated that although epizootics occurred and populations crashed, the fungus probably could not be utilized as a biological control agent. These statements were made in light of environmental factors, such as high humidity, that must coincide with dense grasshopper populations for *E. grylli* to become widespread. Carruthers et al. (1988) and Carruthers and Soper (1987) have begun an elaborate set of field and laboratory experiments designed to identify some of these environmental parameters important to the survival of the fungus. They suggest that a number of ecological factors operate together to regulate disease cycles of 3–5 years in isolated mountain populations (Carruthers et al., 1988). More specifically, Carruthers et al. (1988) demonstrated that sunlight is extremely detrimental to conidial survival. Germination of conidia was 0% following exposure to sunlight for just a few hours.

One of the major stumbling blocks to working with the *E. grylli* complex is the lack of an in vitro production system. Although pure cultures of protoplasts have been initiated (MacLeod et al. 1980), to date, nobody has succeeded in culturing *E. grylli* all the way through its life cycle on artificial media.

Epizootic prediction is another area of fungal research that deserves added research attention. Although epizootics apparently go through fairly predictable multiyear cycles in isolated mountain populations (Carruthers and Soper, 1987), the variable habitat of rangeland grasshoppers virtually precludes general forecasting. However, new strains (or species) of *Entomophaga* are being discovered. The current technology of protoplast fusion coupled with a drought-tolerant *E. grylli* makes possible the introduction of a strain more suited to rangeland habitat. Standard epizootiological techniques could then be employed to investigate *E. grylli*–grasshopper population dynamics thoroughly.

15.5 BACTERIA

Bacteria are ubiquitous microorganisms that, upon entering the hemocoel, may cause a general septicemia. Bacterial insect pathogens can be divided into two groups: spore formers and nonspore formers. Spore-forming bacteria are best illustrated by *Bacillus thuringiensis* (Bt). Although many species of insects are infected by this pathogen, no strain of Bt has been reported to infect Orthoptera.

Likewise, there are only three known obligate bacterial pathogens, and none infect grasshoppers. However, *Serratia marcescens*, a facultative pathogen (one that can cause damage to host tissue and is seldom isolated from the guts of healthy grasshoppers), has been found to be pathogenic in laboratory colonies of grasshoppers and locusts (Bucher, 1959; Stevenson, 1959). Pathogenicity is observed upon injection of 10,000 or fewer bacteria directly into the hemocoel (Bucher, 1959) but, upon feeding, the oral LD_{50} is 300,000–500,000 bacteria. Stevenson (1959) observed epizootics in caged populations of *Schistocerca gregaria* where 72% died 16 days after the addition of two *S. marcescens*-injected locusts (Fig. 15.13). *S. marcescens* was isolated from every cadaver, whereas dead insects from control cages had no *S. marcescens*.

15.5.1 Defense Responses

Grasshoppers are used as models for many studies of host defense mechanisms in response to foreign particles invading the hemocoel. Generally, three types of defense mechanisms are recognized: phagocytosis (unicellular), encapsulation (multicellular), and serum-mediated responses.

Phagocytosis involves the engulfment of microscopic particles, generally smaller than 10 μm, by individual cells. Bt cells disappeared within 4 h following injection into a *Locusta migratoria* hemocoel (Hoffmann et al., 1974). Microscopic examination revealed that the bacteria were taken up by reticular cells of the hemocytopoietic tissue and not by circulating hemocytes (Hoffmann et al., 1974). Eventually, tissue necrosis occurred and capsules formed, isolating the necrotic zones from the hemolymph. Similar reactions were not detected

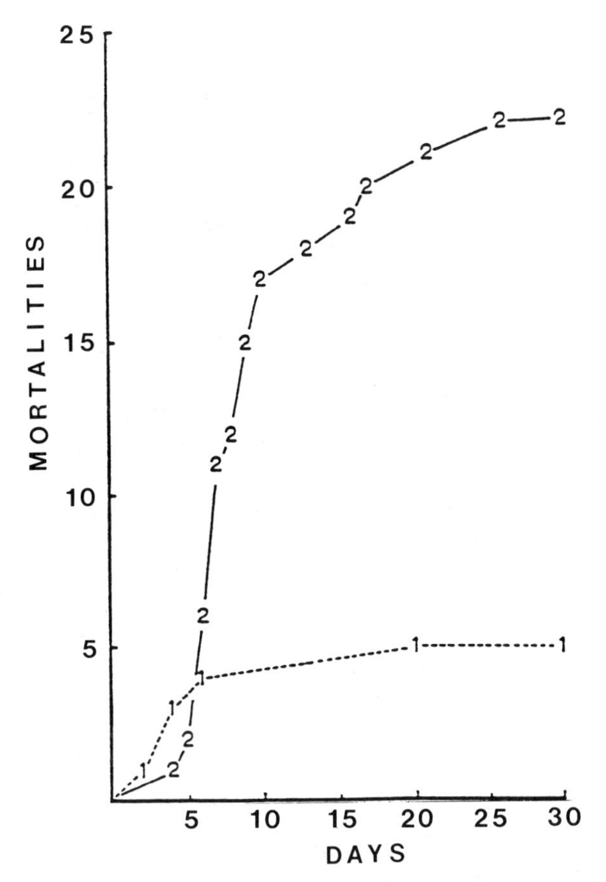

Fig. 15.13. Daily mortality of grasshoppers infected with bacteria. Daily mortality of *Schistocerca gregaria* in caged populations after introduction of locusts infected with *Serratia marcescens.* 1, uninfected cage; 2, infected cage. Reprinted from Stevenson (1959) by courtesy of Academic Press.

following injection of heat-killed bacteria, B-exotoxin, or bacterial culture supernatant.

Encapsulation involves the formation of a multilayered capsule around a foreign particle. Injected substances ranging from inert cellophane (Brehelin et al., 1975) to insect tissues (Lackie, 1981) have been encapsulated to varying degrees. Brehelin et al. (1975) observed capsule formation of sterilized cellophane, cat gut, cotton, and beef muscle. In all cases the capsule formation followed a similar pattern: granular hemocytes attached to the particle forming a multilayered capsule, and outer layers formed fibrous material within the hemocyte cytoplasm and eventually formed thick bundles extending into the elongated cells. Capsules were tracheated in all cases observed and remained in the hemocoel throughout the life of the insect. The factors that affect encapsulation and recognition of self versus nonself are not clearly understood.

Lackie (1981) and more recently Ratner and Vinson (1983) have reviewed the literature pertinent to this subject.

Serum-mediated responses are due to chemicals in the hemolymph that may aid in the antibiotic removal of organisms or in the recognition of foreign substances such that encapsulation can occur. Hoffmann (1980) demonstrated antibacterial activity in the hemolymph of *Locusta migratoria* following injection of nonlethal doses of Bt. Although the factor responsible for this activity has not been isolated or characterized, several responses were noted. The antibacterial activity, as measured by Bt growth in the presence of "immune" hemolymph serum, appeared approximately 4 h after injection, increased dramatically for the next 4–12 h and declined 6–8 days later. Subsequent "immunizing" doses lead to quicker, but not stronger, antibacterial responses. Injections of other bacteria species, bovine IgG, ovalbumin, or iron saccharate also prompted antibacterial activity, but this activity was much weaker than that prompted by Bt injections.

Hemagglutinins are proteins with carbohydrate binding properties found in insect tissue that have been implicated in insect immune responses. Jurenka et al. (1982) demonstrated the presence of agglutinin activity in a number of grasshopper species (Table 15.3). Subsequently, Stebbins and Hapner (1985) demonstrated that a single hemagglutinin protein is responsible for all the agglutinating activity observed in *M. sanguinipes* and *M. bivitattus*. Immunochemical labeling has demonstrated the association of agglutinin with granular hemocytes (Fig. 15.14) (Bradley, 1987), but it is not clear if the agglutinins are synthesized in these cells or stored there. Additionally, Bradley (1987) showed that this membrane-bound agglutinin is not involved in in vitro agglutination activity. Because the work with hemagglutinins has been conducted in vitro, conclusions cannot be drawn as to the absolute function of the proteins in vivo.

Research on bacteria infecting grasshoppers is currently limited to experiments on physiological responses of the hosts. Work on hemocytic aggregation, function and origin of hemagglutinins, and any other antimicrobial activity should continue to play important roles in grasshopper physiological research. Mechanisms affecting grasshopper defense systems continue to be difficult to examine. Methods of studying these mechanisms in vivo must be worked out such that in vitro observations can be compared to processes occurring within a grasshopper. Once these experiments are conducted, the interrelatedness of the three types of defense mechanisms can be determined, for it is likely that they do not occur independently.

15.6 RICKETTSIAE

Rickettsiae were first reported in grasshoppers by Martoja (1964), who experimentally infected *Schistocerca gregaria* and *Locusta migratoria* with a rickettsial microorganism described by Vago and Martoja (1963) as *Rickettsiella grylli*. Vago and Meynadier (1965) later described *Rickettsiella schistocercae* from the

TABLE 15.3. Grasshopper Species with Hemolymph Agglutinin

Subfamily	Species
Gomphocerinae	*Acrolophitus hirtipes* *Aulocara elliotti* *Psoloessa delicatula*
Oedipodinae	*Arphia conspersa* *A. pseudonietana* *Chortophaga viridifasciata* *Circotettix rabula* *Dissosteira carolina* *Spharagemon collare* *S. equale*
Cyrtacanthacridinae	*Schistocerca shoshone*
Melanoplinae	*Hesperotettix viridis* *Melanoplus bivitattus* *M. differentialis* *M. sanguinipes*

Source: After Jurenka et al. (1982).

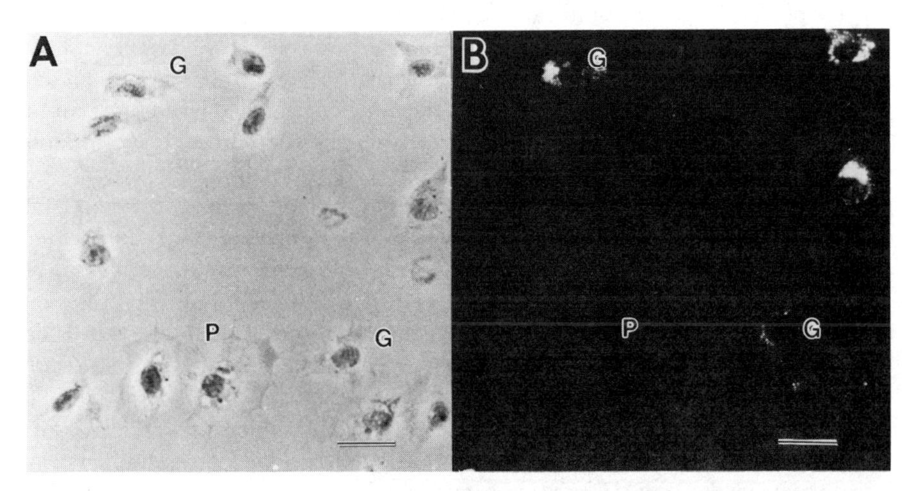

Fig. 15.14. Phase contrast (A) and fluorescence (B) photomicrographs of fixed monolayer hemocytes from *Melanoplus differentialis* subjected to indirect immunofluorescence stain for grasshopper hemagglutinin. Hemocytes were incubated in agglutin-specific monoclonal antibody followed by FITC-conjugated secondary antibody. (G = graunlocyte; P = plasmatocyte) bar = 25 μm Courtesy of R. S. Bradley.

desert locust, *S. gregaria*, and experimentally transmitted the microorganism to *L. migratoria*. When no major differences were found, Weiss and colleagues (1984) placed *R. schistocercae* in synonymy with *R. grylli*. Henry et al. (1986)

also tentatively identified a strain of *R. grylli* infecting the variegated grasshopper, *Zonocerus variegatus*.

Rickettsiae found in grasshoppers are Gram-negative bacilliform or disk-shaped infectious particles. The microorganisms infect grasshoppers orally and usually replicate by binary fission in the cytoplasm of fat body cells. Infectious forms enter the host cell by phagocytosis and grow within a vacuole (Fig. 15.15). Heavily infected cells are closely packed with vacuoles and the cells eventually rupture, releasing infectious particles into the hemocoel to begin another cycle. Signs of disease are evident at 25–28 days after inoculation. Infected grasshoppers appear lethargic and turbid hemolymph is observed in moribund grasshoppers.

Louis et al. (1986) reported that crickets infected with *R. grylli* survived better at febrile temperatures than at the preferred body temperature of uninfected animals. This would likely hold true for grasshoppers and would be a fascinating area of study both at the organismal and ecological level.

15.7 NEMATODES

The two most common nematodes infesting grasshoppers are both mermithids, *Mermis nigrescens* Dujardin and *Agamermis decaudata* Cobb. The first report of *M. nigrescens* dates back to 1844 (Dujardin, 1942). We only briefly cover a few aspects of the life history and host responses of these two species for a major bulk of this research has been recently reviewed (Webster and Thong, 1984).

Parasitism by *Mermis nigrescens* begins when the host ingests embryonated eggs that were deposited on foliage. The second-stage larvae (L_2) hatch within the proventriculus or midgut (Craig and Webster, 1978) and penetrate through the gut wall. The parasitic phase usually lasts from 4 to 10 weeks and the nematodes complete two molts (Fig. 15.16). L_4 larvae exit through the cuticle, thus killing the host as well as any remaining nematodes. Larvae enter the soil, overwinter, and molt into adults the following spring or summer. Adult nematodes mate and overwinter again. During the following spring or early summer, females emerge from the ground following a thunderstorm or very heavy dew, and deposit eggs (Christie, 1937). Eggs may be fertilized or may develop parthenogenetically (Christie, 1929).

The life cycle for *Agamermis decaudata* differs considerably from *M. nigrescens* in that the L_2 larvae hatch from eggs laid in the soil, then infest grasshoppers via integument penetration. Infestation occurs either below ground or, more likely, on foliage above ground as the L_2 disperse. There is no molt within the host during the 1- to 3-month parasitic phase. Larvae emerge through the cuticle, thus killing the host. Larvae enter the soil, overwinter, and molt into adults the following summer. The adults then begin laying eggs (Christie, 1936).

True epizootiological studies are lacking. Several studies (Mongkolkiti and Hosford, 1971) have reported infection levels of 70% or higher, followed by declines of the grasshopper population. These studies are anecdotal and do not discount the effect of other possible mortality factors. Studies of this nature

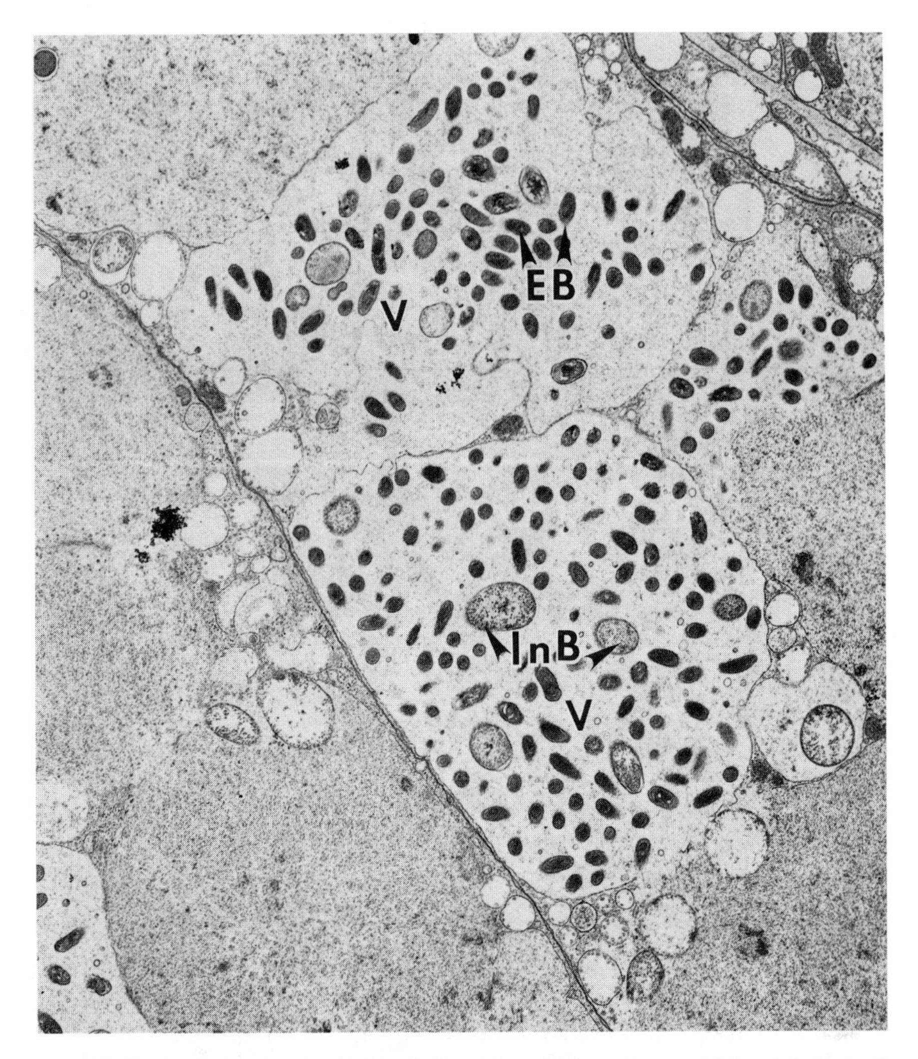

Fig. 15.15. Electron micrograph of fat body tissue from *Melanoplus sanguinipes* infected with *Rickettsiella grylli.* Large intracellular vesicles (V) containing elementary bodies (EB) and intermediate bodies (InB) are evident in the host cell cytoplasm (×10,000). Reprinted from Henry et al. (1986) by courtesy of D. A. Streett.

are difficult to conduct because epizootics are nearly impossible to predict and nematode identification is very difficult. Sampling must begin early each year and grasshoppers must be dissected individually to confirm infestation. At the same time, careful observations must be made concerning other factors such as predators, parasitoids, and pathogens that may affect population sizes of both grasshoppers and nematodes.

Mermithid parasites apparently circulate in the hemolymph at random and do not infest specific tissues (Denner, 1976). Defense reactions elicited by *M. nigrescens* or *A. decaudata* have not been reported. However, Ibrahim et al.

Fig. 15.16. Dissected grasshopper showing larval stage of nematodes (arrowheads); bar = 1.6 mm. Courtesy of J. E. Henry.

(1986) suggested that mermithids may alter the defense system of *Schistocerca gregaria*. When *M. nigrescens*-infested locusts were injected with trypanosomatids, agglutinin titers dramatically decreased and the locusts died. Injected locusts not infested with nematodes survived and agglutinin titers did not decrease.

Another nematode, however, does elicit a defense response from grasshoppers. *Diplotriaena tricuspis*, a diplotriaenoid parasite of corvid birds, spends part of its life cycle in intermediate orthopteran hosts (Cawthorn and Anderson, 1980). The life cycle has been reviewed by Webster and Thong (1984). An interesting cellular response is elicited by the parasite upon infestation. Nematode larvae become encapsulated in fat body globules almost immediately following gut wall penetration. Capsules are fibrous and contain numerous adipocytes. Apparently, larvae feed on the adipocytes and are protected from any further host defense response (Cawthorn, 1980).

Because of their life-history characteristics, mermithids require moist or wet areas to complete development. Arid or semiarid regions of much of the western United States, therefore may not support mermithid populations. Rees (1973) reported that the distribution does not extend west of North Dakota, but *A. decaudata* has occasionally been observed in Montana, Wyoming, and Colorado.

Before fieldwork can be conducted with mermithids, an easy way to distinguish between species must be developed. Currently, adults are sent to nematode taxonomists for identification. Clearly this is not an ideal situation if many nematodes must be identified, for instance, during an epizootic. Possible solutions to this problem are immunodetection systems utilizing monoclonal antibodies, electrophoresis and identification of unique proteins, and biotinylated DNA probes.

Very little research has been conducted with *Agamermis* since Christie's (1936) life-history report. Defense, host reactions, and physiological response are virtually unstudied in this species.

15.8 CONCLUSIONS

Research on grasshopper diseases has basically focused on an applied approach toward the use of pathogenic microorganisms as biological control agents. Unfortunately, the more basic elements of research on grasshopper diseases has been rather seriously neglected. Similar observations have been made to some extent in other general reviews on insect pathology (Steinhaus, 1949, 1963).

A more concerted effort is required in the epizootiology of grasshopper diseases to examine the dynamics of natural infections and the factors governing disease development in grasshopper populations. Moreover, past studies have largely ignored host–pathogen relationships at the intracorpeal or intracellular level. Certainly the major limitation to some of the more basic research questions on host–pathogen relationships for any intracellular pathogen has been the lack of in vitro techniques. Many of the biological processes of intracellular pathogens are obscured by their intimate association with the host. Development of a grasshopper cell culture system susceptible to intracellular pathogens (e.g., viruses, protozoa) would facilitate many of these studies.

REFERENCES

Balfour-Browne, F. L. 1960. The green muscardine disease of insects, with special reference to an epidemic in a swarm of locusts in Eritrea. *Proc. R. Entomol. Soc. London*, ser. A **35**, 65–74.

Bensimon, A., S. Zinger, E. Gerassi, A. Hauschner, I. Harpaz, and I. Sela. 1987. "Dark Cheeks", a lethal disease of locusts provoked by a lepidopterous baculovirus. *J. Invertebr. Pathol.* **50**, 254–260.

Boorstein, S. M. and P. Ewald. 1987. Costs and benefits of behavioral fever in *Melanoplus sanguinipes* infected by *Nosema acridophagus*. *Physiol. Zool.* **60**, 586–595.

Bradley, R. S. 1987. Grasshopper hemagglutinin: Immunochemical localization in hemocytes and confirmation of non-opsonic properties. M.S. Thesis, Montana State University, Bozeman.

Braun, L., A. B. Ewen, and C. Gillott. 1988. The life cycle and ultrastructure of *Malameba locustae* (King & Taylor) (Amoebidae) in the migratory grasshopper *Melanoplus sanguinipes*. *Can. Entomol.* **120**, 759–772.

Brehelin, M., J. A. Hoffmann, G. Matz, and A. Porte. 1975. Encapsulation of implanted foreign bodies by hemocytes in *Locusta migratoria* and *Melolontha melolontha*. *Cell Tissue Res.* **160**, 283–289.

Bucher, G. E. 1959. Bacteria of grasshoppers of western Canada. III. Frequency of occurrence, pathogenicity. *J. Insect Pathol.* **1**, 391–405.

Canning, E. U. 1953. A new microsporidian, *Nosema locustae* n. sp., from the fat body of the African migratory locust, *Locusta migratoria migratorioides* R. and F. *Parasitology* **43**, 287–290.

Carruthers, R. I. and R. S. Soper. 1987. Fungal diseases. In J. R. Fuxa and Y. Tanada, Eds., *Epizootiology of Insect Diseases.* Wiley, New York, pp. 357–416.

Carruthers, R. I., Z. Feng, M. E. Ramos, and R. S. Soper. 1988. The effect of solar radiation on the survival of *Entomophaga grylli* (Entomophthorales: Entomophthoraceae) conidia. *J. Invertebr. Pathol.* **52**, 154–162.

Cawthorn, R. J. 1980. The cellular responses of migratory grasshoppers (*Melanoplus sanguinipes* F.) and African desert locusts (*Schistocerca gregaria* L.) to *Diplotriaena tricuspis* (Nematoda: Diplotriaenoidea). *Can. J. Zool.* **58**, 109–113.

Cawthorn, R. J. and R. C. Anderson. 1980. Development of *Diplotriaena tricuspis* (Nematoda: Diplotriaenoidea), a parasite of Corvidae, in intermediate and definitive hosts. *Can. J. Zool.* **58**, 94–108.

Chapman, R. F. and W. W. Page. 1979. Factors affecting the mortality of the grasshopper, *Zonocerus variegatus*, in southern Nigeria. *J. Anim. Ecol.* **48**, 271–288.

Christie, J. R. 1929. Some observations on sex in the mermithidae. *J. Exp. Zool.* **53**, 59–76.

Christie, J. R. 1936. Life history of *Agamermis decaudata*, a nematode parasite of grasshoppers and other insects. *J. Agric. Res. (Washington, D.C.)* **52**, 161–198.

Christie, J. R. 1937. *Mermis subnigrescens*, a nematode parasite of grasshoppers. *J. Agric. Res. (Washington, D.C.)* **55**, 353–364.

Colgan, D. J. 1986. Studies of the mortality of *Locusta migratoria* (L.) treated with a polyhedrosis virus from the grasshopper *Caledia captiva* (F.) (Orthoptera: Acrididae). *Bull. Entomol. Res.* **76**, 539–544.

Colgan, D. J. and P. D. Christian. 1988. A putative reoviral genome contaminating mitochondrial preparations from the grasshopper, *Caledia captiva*. *J. Invertebr. Pathol.* **52**, 474–476.

Craig, S. M. and J. M. Webster. 1978. Viability and hatching of *Mermis nigrescens* eggs and subsequent larval penetration of the desert locust *Schistocerca gregaria*. *Nematologica* **24**, 472–474.

Denner, M. W. 1976. Preliminary studies on the pathology caused by *Mermis nigrescens* in Orthoptera. *Proc. Indiana Acad. Sci.* **85**, 258–261.

Dillon, R. J. and A. K. Charnley. 1986a. Properties of the antifungal toxin produced by the gut bacterial flora of the desert locust, *Schistocerca gregaria*. In R. A. Samson, J. M. Vlak, and D. Peters, Eds., *Fundamental and Applied Aspects on Invertebrate Pathology*, p. 260.

Dillon, R. J. and A. K. Charnley. 1986b. Inhibition of *Metarhizium anisopliae* by the gut bacterial flora of the desert locust, *Schistocerca gregaria*: Evidence for an antifungal toxin. *J. Invertebr. Pathol.* **47**, 350–360.

Dillon, R. J. and A. K. Charnley. 1986c. Invasion of the pathogenic fungus *Metarhizium anisopliae* through the guts of germ-free desert locusts, *Schistocerca gregaria*. *Mycopathologia* **96**, 59–66.

Dufour, J. M. L. 1828. Note sur la grégarine, nouveau genre de ver qui vit en troupeau dans les intestines de divers insectes. *Ann. Sci. Nat., Zool. Biol. Anim.* **13**, 366–388.

Dujardin, F. 1942. Mémoire sur la structure antomique des *Gordius* et d'un autre helminthe, le Mermis, qu'on a confundu avec eux. *Ann. Sci. Nat., Zool. Biol. Anim.* [10] **18**, 129–151.

Erlandson, M. A., A. B. Ewen, M. K. Mukerji, and C. Gillott. 1986. Susceptibility of immature stages of *Melanoplus sanguinipes* (Fab.) (Orthoptera: Acrididae) to *Nosema cuneatum* Henry (Microsporida: Nosematidae) and its effect on host fecundity. *Can. Entomol.* **118**, 29–35.

Ernst, H. P. and G. L. Baker. 1982. *Malameba locustae* (King and Taylor) (Protozoa: Amoebidae) in field populations of Orthoptera in Australia. *J. Aust. Entomol. Soc.* **21**, 295–296.

Evans, W. A. and R. G. Elias. 1970. The life cycle of *Malamoeba locustae* (King et Taylor) in *Locusta migratoria migratorioides* (R. et F.). *Acta Protozool.* **7**, 229–241.

Ewen, A. B. and M. K. Mukerji. 1980. Evaluation of *Nosema locustae* (Microsporida) as a control agent of grasshopper populations in Saskatchewan. *J. Invertebr. Pathol.* **35**, 295–303.

Fresenius, G. 1858. Notiz, insekten — pilze betreffend. *Bot. Ztg.* **14**, 882–883.

Granados, R. R. 1981. Entomopoxvirus infections in insects. In E. W. Davidson, Ed., *Pathogenesis of Invertebrate Microbial Diseases.* Allanheld, Osmun, New Jersey, pp. 101–126.

Gunnarson, S. G. S. 1988. Infection of *Schistocerca gregaria* by the fungus *Metarhizium anisopliae*: Cellular reactions in the integument studied by scanning electron and light microscopy. *J. Invertebr. Pathol.* **52**, 9–17.

Gyllenberg, G. 1974. A simulation model for testing the dynamics of a grasshopper population. *Ecology* **55**, 645–650.

Harper, A. M. and H. C. Huang. 1986. Evaluation of the entomophagous fungus *Verticillium lecanii* (Moniliales: Moniliaceae) as a control agent for insects. *Environ. Entomol.* **15**, 281–284.

Harry, O. G. 1965. Studies on the early development of the eugregarine *Gregarina garnhami. J. Protozool.* **12**, 296–305.

Harry, O. G. 1970. Gregarines: Their effect on the growth of the desert locust (*Schistocerca gregaria*). *Nature, (London)* **225**, 964–966.

Harry, O. G. 1971. Studies on infection and reinfection by eugregarines. *Parasitology* **63**, 213–223.

Harry, O. G. and L. H. Finlayson. 1976. The life-cycle, ultrastructure and mode of feeding of the locust amoeba *Malpighamoeba locustae. Parasitology* **72**, 127–134.

Henry, J. E. 1967. *Nosema acridophagus* sp. n., a microsporidian isolated from grasshoppers. *J. Invertebr. Pathol.* **9**, 331–341.

Henry, J. E. 1971. *Nosema cuneatum* sp. n. (Microsporida: Nosematidae) in grasshoppers (Orthoptera: Acrididae). *J. Invertebr. Pathol.* **17**, 164–171.

Henry, J. E. 1972a. Development of the crystalline-array virus (CAV) in cultures of dorsal vessels from *Schistocerca americana* (Orthoptera: Acrididae). *J. Invertebr. Pathol.* **19**, 325–330.

Henry, J. E. 1972b. Epizootiology of infections by *Nosema locustae* Canning (Microsporida, Nosematidae) in grasshoppers. *Acrida* **1**, 111–120.

Henry, J. E. 1977. Development of microbial agents for the control of Acrididae. *Rev. Soc. Entomol. Argent.* **36**, 125–134.

Henry, J. E. and J. W. Jutila. 1966. The isolation of a polyhedrosis virus from a grasshopper. *J. Invertebr. Pathol.* **8**, 417–418.

Henry, J. E. and E. A. Oma. 1973. Ultrastructure of the replication of the grasshopper crystalline array virus in *Schistocerca americana* compared with other picornaviruses. *J. Invertebr. Pathol.* **21**, 273–281.

Henry, J. E., B. P. Nelson, and J. W. Jutila. 1969. Pathology and development of the grasshopper inclusion body virus in *Melanoplus sanguinipes. J. Virol.* **3**, 605–610.

Henry, J. E., M. C. Wilson, E. A. Oma, and J. L. Fowler. 1985. Pathogenic micro-organisms isolated from West African grasshoppers (Orthoptera: Acrididae). *Trop. Pest Manage.* **31**, 192–195.

Henry, J. E., D. A. Streett, E. A. Oma, and R. H. Goodwin. 1986. Ultrastructure of an isolate of *Rickettsiella* from the African grasshopper *Zonocerus variegatus. J. Invertebr. Pathol.* **47**, 203–213.

Hoffmann, D. 1980. Induction of antibacterial activity in the blood of the migratory locust *Locusta migratoria* L. *J. Insect Physiol.* **26**, 539–549.

Hoffmann, D., M. Brehelin, and J. A. Hoffmann. 1974. Modifications of the hemogram and of the hemocytopoietic tissue of male adults of *Locusta migratoria* (Orthoptera) after injection of *Bacillus thuringiensis. J. Invertebr. Pathol.* **24**, 238–247.

Ibrahim, E. A., D. H. Molyneux, and G. A. Ingram. 1986. Suppression of immune responses of locusts to trypanosomatid flagellates by concomitant mermithid infection. *J. Invertebr. Pathol.* **48**, 252–253.

Johnson, D. L. and E. Pavlikova. 1986. Reduction of consumption by grasshoppers (Orthoptera: Acrididae) infected by *Nosema locustae* Canning (Microsporida: Nosematidae). *J. Invertebr. Pathol.* **48**, 232–238.

Johnson, D. L., H. C. Huang, and A. M. Harper. 1988. Mortality of grasshoppers (Orthoptera: Acrididae) inoculated with a Canadian isolate of the fungus *Verticillium lecanii*. *J. Invertebr. Pathol.* **52**, 335–342.

Jurenka, R., K. Manfredi, and K. D. Hapner. 1982. Haemagglutinin activity in Acrididae (grasshopper) haemolymph. *J. Insect Physiol.* **28**, 177–181.

Jutila, J. W., J. E. Henry, R. L. Anacker, and W. R. Brown. 1970. Some properties of a crystalline-array virus (CAV) isolated from the grasshopper *Melanoplus bivittatus* (Say) (Orthoptera: Acrididae). *J. Invertebr. Pathol.* **15**, 225–231.

King, R. L. and A. B. Taylor. 1936. *Malpighamoeba locustae* n. sp. (Amoebidae), a protozoan parasite in the malpighian tubes of grasshoppers. *Trans. Am. Microsc. Soc.* **55**, 6–10.

Kundu, T. K. and D. P. Haldar. 1983. Observation on a new species of cephaline gregarine of the genus *Quadruspinospora* Sarkar and Chakravarty, 1969 from a grasshopper *Spathosternum prasiniferum*. *Arch. Protistenkd.* **127**, 97–102.

Lackie, A. M. 1981. Immune recognition in insects. *Dev. Comp. Immunol.* **5**, 191–204.

Lange, C. E. 1987. A new species of *Perezia* (Microsporida: Pereziidae) from the Argentine grasshopper *Dichroplus elongatus* (Orthoptera: Acrididae). *J. Protozool.* **34**, 34–39.

Langridge, W. H. R. and J. E. Henry. 1981. Molecular weight and base composition of DNA isolated from *Melanoplus sanguinipes* entomopoxvirus. *J. Invertebr. Pathol.* **37**, 34–37.

Langridge, W. H. R., R. F. Bozarth, and D. W. Roberts. 1977. The base composition of entomopoxvirus DNA. *Virology* **76**, 616–620.

Levine, N. D. 1988. *The Protozoan Phylum Apicomplexa*, Vol. 1. CRC Press, Boca Raton, Florida.

Louis, C., M. Jourdan, and M. Cabanac. 1986. Behavioral fever and therapy in a rickettsia-infected Orthoptera. *Am. J. Physiol.* **250**, R991–R995.

MacLeod, D. M. and E. Muller-Kogler. 1973. Entomogenous fungi: *Entomophthora* species with pear-shaped to almost spherical conidia (Entomophthorales: Entomophthoraceae). *Mycologia* **65**, 823–893.

MacLeod, D. M., D. Tyrrell, and M. A. Welton. 1980. Isolation and growth of the grasshopper pathogen, *Entomophthora grylli*. *J. Invertebr. Pathol.* **36**, 85–89.

Marcandier S. and G. G. Khachatourians. 1987. Susceptibility of the migratory grasshopper, *Melanoplus sanguinipes* (Fab.) (Orthoptera: Acrididae), to *Beauveria bassiana* (Bals.) Vuillemin (Hyphomycete): Influence of relative humidity. *Can. Entomol.* **119**, 901–907.

Martoja, R. 1964. Sur l'infection expérimentale de quelques insectes Orthopteres par *Rickettsiella grylli*, agent de la rickettsiose de gryllides. *C. R. Hebd. Seances Acad. Sci.* **258**, 1318–1321.

McDaniel, B. 1986. Fungus infection of two species of grasshoppers in western South Dakota. *Southwest. Nat.* **31**, 269–270.

McDaniel, B. and R. A. Bohls. 1984. The distribution and host range of *Entomophaga grylli* (Fresenius), a fungal parasite of grasshoppers in South Dakota. *Proc. Entomol. Soc. Wash.* **86**, 864–868.

McGuire, M. R. and J. E. Henry. 1989. Production and partial characterization of monoclonal antibodies for detection of entomopoxvirus from *Melanoplus sanguinipes*. *Entomol. Exp. Appl.* **51**, 21–28.

Milner, R. J. 1978. On the occurrence of *Entomophaga grylli*, a fungal pathogen of grasshoppers in Australia. *J. Aust. Entomol. Soc.* **17**, 293–296.

Mongkolkiti, S. and R. M. Hosford, Jr. 1971. Biological control of the grasshopper *Hesperotettix viridis pratensis* by the nematode *Mermis nigrescens*. *J. Nematol.* **3**, 356–363.

Oma, E. A. and J. E. Henry. 1986. Host relationships of entomopoxviruses isolated from grasshoppers. *Grasshopper Symp. Proc.*, Bismark, N. D., pp. 48–49.

Oma, E. A. and G. B. Hewitt. 1984. Effect of *Nosema locustae* (Microsporida: Nosematidae) on food consumption in the differential grasshopper (Orthoptera: Acrididae). *J. Econ. Entomol.* **77**, 500–501.

Pickford, R. and P. W. Riegert. 1964. The fungous disease caused by *Entomophthora grylli* Fres., and its effects on grasshopper populations in Saskatchewan in 1963. *Can. Entomol.* **96**, 1158–1166.

Purrini, K. and M. Rhode. 1988. Light and electron microscope studies on two new diseases in natural populations of the desert locust, *Schistocerca gregaria*, and the grassland locust, *Chortipes* sp. caused by two entomopoxviruses. *J. Invertebr. Pathol.* **51**, 281–283.

Purrini, K., G. W. Kohring, and Z. Seguni. 1988. Studies on a new disease in a natural population of migratory locusts, *Locusta migratoria*, caused by an entomopoxvirus. *J. Invertebr. Pathol.* **51**, 284–286.

Ratner, S. and S. B. Vinson. 1983. Phagocytosis and encapsulation: cellular immune responses in arthropoda. *Am. Zool.* **23**, 185–194.

Rees, N. E. 1973. Arthropod and nematode parasites, parasitoids, and predators of Acrididae in America North of Mexico. *U.S., Dep. Agric., Tech. Bull.* **1460**.

Roberts, D. W. 1981. Toxins of entomopathogenic fungi. In H. D. Burges, Ed., *Microbial Control of Pests and Plant Diseases 1970–1980.* Academic Press, London, pp. 441–464.

Roffey, J. 1968. The occurrence of the fungus *Entomophthora grylli* Fresenius on locusts and grasshoppers in Thailand. *J. Invertebr. Pathol.* **11**, 237–241.

Skaife, S. H. 1925. The locust fungus, *Empusa grylli*, and its effects on its host. *S. Afr. J. Sci.* **22**, 298–308.

Smith, R. C. 1933. Fungous and bacterial diseases in the control of grasshoppers and chinch bugs. *Bienn. Rep. Kans. Board Agric.* **28**, 44–61.

Soper, R. S., B. May, and B. Martinell. 1983. *Entomophaga grylli* enzyme polymorphism as a technique for pathotype identification. *Environ. Entomol.* **12**, 720–723.

Stebbins, M. R. and K. D. Hapner. 1985. Preparation and properties of haemagglutinin from haemolymph of Acrididae (grasshoppers). *Insect Biochem.* **15**, 451–462.

Steinhaus, E. A. 1949. *Principles of Insect Pathology.* McGraw–Hill, New York.

Steinhaus, E. A. 1963. Introduction. In E. A. Steinhaus, Ed., *Insect Pathology: An Advanced Treatise*, Vol. I. Academic Press, New York, pp. 1–27.

Stevenson, J. P. 1959. Epizootiology of a disease of the desert locust, *Schistocerca gregaria* (Forskål), caused by nonchromogenic strains of *Serratia marcescens* Bizio. *J. Insect Pathol.* **1**, 232–244.

Streett, D. A. 1986. Future prospects for mcrobial control of grasshoppers. In J. L. Capinera, Ed., *Integrated Pest Management on Rangeland: A Shortgrass Prairie Perspective.* Westview Press, Boulder, Colorado, pp. 205–218.

Streett, D. A. and J. E. Henry. 1984. Epizootiology of a microsporidium in field populations of *Aulocara elliotti* and *Psoloessa delicatula* (Insecta: Orthoptera). *Can. Entomol.* **116**, 1439–1440.

Streett, D. A. and M. R. McGuire. 1988. Microbial control of rangeland grasshoppers: New techniques for the detection of entomopathogens. *Mon. Agric. Res.* **5**, 1–5.

Streett, D. A., E. A. Oma, and J. E. Henry. 1986. Characterization of the DNA from Orthopteran entomopoxviruses. In R. A. Samson, J. M. Vlak, and D. Peters, Eds., *Fundamental and Applied Aspects on Invertebrate Pathology*, p. 408.

Streett, D. A., E. A. Oma, and J. E. Henry. 1989. DNA probe for detection of *Melanoplus sanguinipes* entomopoxvirus in grasshoppers. *Appl. Environ. Microsc.* (to be published).

Uvarov, B. P. 1928. *Locusts and Grasshoppers: A Handbook for Their Study and Control.* Imperial Bureau of Entomology, London.

Uvarov, B. P. 1966. *Grasshoppers and Locusts*, Vol. 1. Cambridge Univ. Press, London and New York.

Uvarov, B. P. 1977. *Grasshoppers and Locusts*, Vol. 2. Centre for Overseas Pest Research, London.

Vago, C. and R. Martoja. 1963. Une rickettsiose chez le Gryllidae (Orthoptera). *C. R. Hebd. Seances Acad. Sci., Ser. D* **256**, 1045–1047.

Vago, C. and G. Meynadier, 1965. Une rickettsiose chez le criquet pèlerin (*Schistocerca gregaria* Forsk.). *Entomophaga* **10**, 307–310.

Webster, J. M. and C. H. S. Thong. 1984. Nematode parasites of orthopterans. In W. R. Nickle, Ed., *Plant and Insect Nematodes*. Dekker, New York, pp. 697–726.

Weiss, E., G. A. Dasch, and K. Chang. 1984. Genus VIII *Rickettsiella* Phillip 1965. In N. R. Krieg and J. G. Holt, Eds., *Bergey's Manual of Systematic Bacteriology*, 9th ed., Vol. 1. Williams & Wilkins, Baltimore, Maryland, p. 713.

16

Color Pattern Polymorphism

J. M. DEARN

16.1 INTRODUCTION

Body coloration can play a major role in the biology of animals, especially small invertebrates such as grasshoppers. Color patterns affect how individuals are perceived, by both predators and other individuals in the same population, as well as influencing internal body temperature and overall metabolism. Compared with some other well-studied groups of insects, grasshoppers are not noted for their bright colors and have a preponderance of species exhibiting various shades of brown and green coloration. However, many species of acridid grasshopper are characterized by color pattern polymorphism, unrelated to sex, age, or state of maturation, where individuals develop into one of two or more color patterns morphs cued by a genetic or environmental developmental switch. It is this aspect of grasshopper coloration that is considered in this chapter.

Although there is still debate over the adaptive significance of polymorphisms observed at the DNA and protein levels, there is abundant evidence that "visible" polymorphisms, such as those involving color pattern variation, are adaptive and can result from selection operating on phenotypic variation in a heterogeneous environment (Ford, 1975). The evolution of polymorphism is only one possible adaptive response to environmental heterogeneity (Levins, 1968; Southwood, 1977) but is a particular feature of the biology of insects (Richards, 1961). However, although the study of color pattern polymorphism in some animal groups, in particular moths, butterflies, and snails, has been central to the development of population and evolutionary biology, grasshoppers have failed to attract the same attention. This is unfortunate because grasshoppers exhibit a wealth of material for ecological and evolutionary research and offer many advantages for experimental studies.

This chapter examines the role of color pattern polymorphisms in the biology of acridid grasshoppers with particular emphasis on the forms of selection that might operate on this variation. General accounts of color pattern variation in acridid grasshoppers can be found in Uvarov (1966, 1977) and Rowell (1971), and reviews of the pigments involved in insect coloration are given in Fuzeau-Braesch (1972), Needham (1978), and Kayser (1985).

16.2 COLOR PATTERN VARIATION IN ACRIDID GRASSHOPPERS

Although coloration can affect many aspects of the biology of organisms, observations on grasshoppers and their habitats demonstrate the importance of crypsis. Many species exhibit a striking resemblance to their background. Some grasshopper species, for example, those resembling desert stones or leaves, provide some of the most spectacular examples of adaptation (Cott, 1940; Fogden and Fogden, 1974), though there is some argument as to whether such protective resemblance should be regarded as crypsis or mimicry (Vane-Wright, 1976, 1980; Robinson, 1981; Cloudsley-Thompson, 1981;

Edmunds, 1981). The importance of crypsis as a factor determining coloration in grasshoppers is supported by the observation that species with variable coloration are associated with habitats, such as grasslands, that exhibit either seasonal change or local spatial variation (Rowell, 1971). Furthermore, when visual displays do occur in species exhibiting cryptic coloration, the bright coloration is confined to regions of the body that are hidden when the grasshoppers are at rest. Not all grasshoppers exhibit cryptic coloration, however, and the coloration of some species is highly conspicuous. Such coloration presumably either serves some function in intraspecific communication or is involved in predator avoidance through flash coloration, warning coloration, or mimetic resemblance (Cott, 1940; Otte, 1977; Uvarov, 1977), though detailed studies in acridid grasshoppers are lacking (but see Chapter 12). Such bright coloration is usually uniform within populations, however, and although of great interest, is not considered here.

There have been a number of different attempts to develop a scheme for describing the polymorphic color pattern variation observed in acridid grasshoppers. The first of these was by Vorontzovsky (1928), who examined 48 species of acridid grasshopper from the southern Urals. He recognized and described 32 distinct morphs, differing mainly in their distribution of green and brown pigments, which he termed varieties and to which he gave descriptive Latinized names, for example, var. *hyalolateralis*, var. *polychloros*, and var. *robusculus*. He considered similar morphs in different species to be homologous, resulting from phylogenetic affinities. In some cases the same name was given to the homologous variety in different species, but this practice was not consistent. For example, the variety that had the face, sides of the head and thorax, the lateral lobes of the pronotum, and the femora of the hind legs bright green while having the frons, vertex, occiput, and dorsal part of the pronotum between the lateral keels ruddy or rust brown was termed var. *hyalolateralis* in *Chorthippus albomarginatus* but var. *prasinolateralis* in *Chorthippus parallelus*. The system was rationalized by Rubtzov (1935), who used the same name for similar varieties in different species. He recognized six morphs among 26 Siberian species (see his Figs 1–25) and although he used Latinized names originally proposed by Vorontzovsky (1928), he prefixed them with the term form, for example, f. *viridis* and f. *rubiginosa*.

An entirely different approach to describing color pattern polymorphisms was adopted by E. J. Clark (1943), who explicitly criticized the approach used by Vorontzovsky and Rubtzov. Clark's approach attempted to address the complexity of the color pattern variation observed in some species as well as avoiding the assumption that similar color patterns observed in different species are indeed homologous. The system uses a coloration formula consisting of a string of symbols describing the individual color of different parts of the body of a resting grasshopper, that is, as it might appear to a predator. As an example, the total color pattern description for a particular individual grasshopper can take the form

T *v b* E *v v b* m F *b b* m C li *b* fpd *s* + [C zl (o) P fdl *n* cl *a* zl (o)]

where individual terms refer to regions of the body, parts of those regions, or the color of those parts. For example, T refers to total coloration of the dorsal surface of the body, E to the fore wings, and C to the head. Color abbreviations are italicized and include *a* for white, *f* for yellow, and *n* for black. Such a system is obviously cumbersome but does have the advantage of being empirically derived at the level of the individual and clearly represents a fuller description of color pattern variation than a limited number of named morphs could possibly do. An account of the Clark system with some modifications may be found in Ragge (1965), which contains splendid color plates illustrating the main color pattern morphs observed in British acridid grasshoppers.

The Clark system of notation, with some modification to take into account the coloration of nymphs, was used with success by Richards and Waloff (1954) in their study of the population biology of five species of British acridid grasshoppers. However, they also found it convenient in some species, such as *Chorthippus parallelus* and *Stenobothrus viridulus*, to recognize a number of distinct color pattern varieties such as var. "dorsal stripe" and var "green brown legs," each of which includes a number of different coloration formulas. Richards and Waloff divided the British acridids into two groups; group I contained the species exhibiting a limited range of variation and group II contained the species that were considered so variable that they did not warrant full classification. To illustrate this latter problem they described 32 of the more easily defined color pattern morphs in *Chorthippus brunneus*, a group II species. Despite its obvious complexity, the genetic basis of the color pattern variation in *C. brunneus* was investigated by Gill (1981d), who described the variation in terms of the action of individual genes each of which appears to determine color and pattern elements in different areas of the body. For example, six color variants of the dorsal pronotum were recognized (brown, black, red–brown, purple, green, and white), each corresponding to the action of different alleles, and three color variants of the upper hind femur (brown, purple, and green), again determined by allelic variation.

In his study of color pattern polymorphism in nine species of acridid grasshoppers belonging to the Australian oedipodine genera *Chortoicetes* and *Austroicetes*, Key (1954) recognized a total of 15 color pattern morphs and described them using the system of Vorontzovsky and Rubtzov. Indeed, he considered five of the morphs homologous to morphs described by Vorontzovsky (1928) and gave them the same names while giving names to 10 new morphs. However, it is clear there is a major difference between the polymorphic systems. In acridids from the southern Urals, Siberia, and Britain the color pattern polymorphism involves the distribution of green and brown coloration, which, as discussed below, is under genetic control. By contrast, the genetically determined variation in *Chortoicetes* and *Austroicetes* involves only the distribution of light and dark areas, with the light areas varying in color from white to medium gray or brown and the dark areas from medium gray or brown to black (see Fig. 16.1). The green coloration is under separate environmental control and super-imposed on the genetically controlled color pattern polymorphism.

In summary, there have been three approaches to describing systematically the color pattern variation in acridid grasshoppers. First, distinct morphs can be given descriptive names that can be used across different species. Second, the phenotype of individual grasshoppers can be described empirically using a coloration formula. Last, phenotypic descriptions can be based on the known action of individual genetic loci. Although this last approach would seem preferable, because it offers the potential for examining how natural selection acts on color pattern variation, it has limited applicability at present because so few species have been studied genetically.

16.3 THE DETERMINATION OF COLOR PATTERN POLYMORPHISMS

Different grasshopper color pattern morphs can be determined by either a genetic or an environmental switch, or indeed by a switch incorporating both genetic and environmental factors. These switches operate to determine the production and deposition of pigments, often through the hormonal system (Rowell, 1967; de Wilde, 1975; Nijhout and Wheeler, 1982), but the biochemical and physiological aspects of grasshopper coloration are not considered here.

Populations are able to track environmental variation as a consequence of selection acting on genetically determined polymorphic variation. Such a system imposes a considerable genetic load on the population (Wallace, 1970), however, and there is a permanent lag between the state of the environment and the adaptive state of the population. The existence of suitable environmental cues that provide information on future environments permits the evolution of polymorphic systems that are determined by an environmental switch and do not involve the time lags and genetic loads that characterize genetic polymorphisms. In populations where there are no suitable environmental cues to the state of future environments, or where individuals live in discrete habitats where there is a high correlation between the environments of parents and progeny, there is no opportunity for environmentally cued polymorphisms to evolve. Moreover, unpredictable environmental variation, which all populations experience, has to be tracked by natural selection acting on genetic variation; high levels of genetic variation seem to characterize natural populations, including those of grasshoppers (Butlin and Hewitt, 1987; Gill, 1981a,b,c; Daly et al., 1981; Halliday et al., 1984; Chapco and Bidochka, 1986).

The factors determining whether selection will favor a genetically determined polymorphism or an environmentally cued polymorphism in a spatially variable environment have been examined by Lively (1986), who concluded that the outcome of selection depends on the frequencies of the patches, the reliability of the environmental cues, the intra- and intermorph interactions, the relative costs of the alternative phenotypes, and the relative survival of the morphs in the different patches. In making this distinction between genetic and environmental switches, however, it must be remembered that developmental systems that respond to environmental cues are themselves the product of specific genotypes that have been favored by natural selection. Moreover, individuals within populations can differ genetically with respect to their sensitivity to environmental cues. Thus the distinction between genetic and environmental switches is not as empirically clear as it might first appear.

16.3.1 Genetic Switches

Studies on both interspecific and long-term temporal variation in acridid grasshoppers suggest the genetic determination of some color pattern morphs, but there have been surprisingly few studies that have directly examined modes of inheritance. Inheritance studies have been carried out on a total of five species: two gomphocerine species, *Chorthippus brunneus* (the common field grasshopper), *C. parallelus* (the meadow grasshopper), one oedipodine species, *Chortoicetes terminifera* (the Australian plague locust), and two catantopines, *Melanoplus sanguinipes* (the migratory grasshopper) and *Phaulacridium vittatum* (the wingless grasshopper).

C. terminifera, *M. sanguinipes*, and *P. vittatum* all exhibit simple, well-defined color pattern morphs, and as an example, the nine phenotypic classes in *C. terminifera* are illustrated in Figure 16.1. In all three species the variation has been shown to be controlled by allelic variation at individual loci (Table 16.1). This does not preclude the existence of supergenes (coadapted closely linked genes) at any of these loci, and there is some evidence for this in *C. terminifera* (Hawke, 1974). Further evidence for the existence of supergenes, or at least genes in linkage disequilibrium, comes from the numerous reports of associations between color pattern morphs and other phenotypic characteristics, which are discussed in Section 16.5. Interestingly, the three loci involved in determining the color pattern variation of the hind tibia, hind femur, and pronotum in *M. sanguinipes* show linkage between themselves, with map distances of about 9 and 20 map units (Chapco, 1980b).

Studies on *Chorthippus* species have revealed a more complex situation. These species exhibit quite variable coloration and the number of loci that have been identified as being involved in determining this variation is correspondingly higher. Sansome and La Cour (1935) identified 14 loci in *C. parallelus*, of which four appeared to be epistatic, that is, they exhibited interactions with the other loci, though there was little evidence of linkage. More detailed studies have been carried out by Gill (1981d) on *C. brunneus*. A total of 20 loci were identified with up to six alleles at a locus, many of the loci exhibiting linkage with each other, and as was the case with *C. parallelus*, some loci exhibited epistatic interactions. This genetic system clearly has the capacity to produce an enormous variety of color pattern phenotypes, but particular

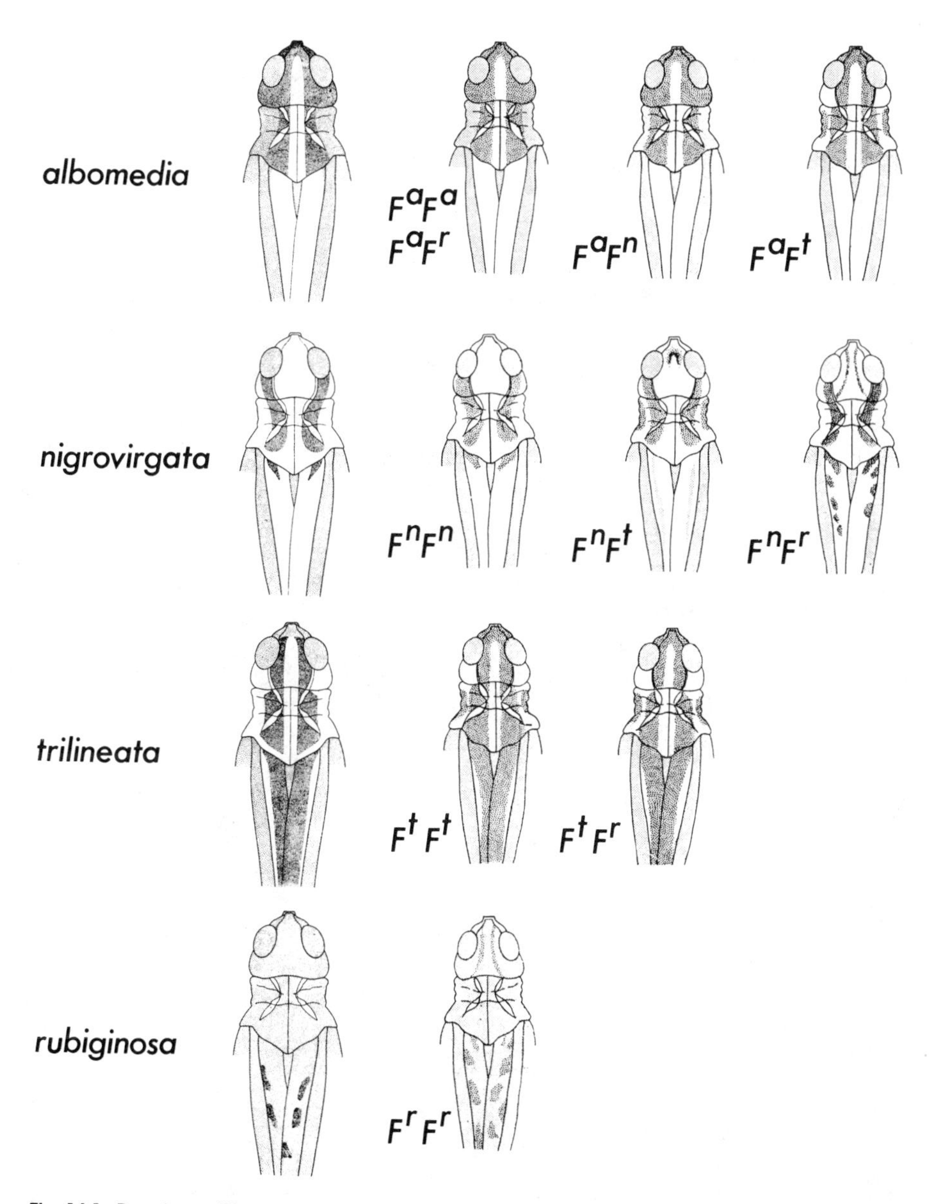

Fig. 16.1. Drawings of the color pattern morphs in *Chortoicetes terminifera*, the Australian plague locust (from Key, 1954), together with the phenotypes of the 10 constituent genotypic classes (from Byrne, 1967a).

allelic combinations producing the overall color pattern phenotypes striped, mottled, and plain, appeared to be common.

With so few studies for comparison it is hard to make any generalizations

TABLE 16.1. The Genetic Basis of Color Pattern Variation in Three Species of Acridid Grasshoppers

Species	Character Affected	Genotype	Phenotype	Reference
Phaulacridium vittatum (wingless grasshopper)	Dorsal surface of the body	C^SC^S	striped	Dearn (1983)
		C^SC^I		
		C^SC^P		
		C^IC^I	incomplete-striped	
		C^IC^P		
		C^PC^P	plain	
Chortoicetes terminifera (Australian plague locust)	Dorsal and lateral surface	F^aF^a/F^aF^r	albomedia	Key (1954); Byrne (1967a); Dearn and Davies (1983)
		F^aF^n		
		F^aF^t		
		F^nF^n	nigrovirgata	
		F^nF^t		
		F^nF^r		
		F^tF^t	trilineata	
		F^tF^r		
		F^rF^r	rubiginosa	
Melanoplus sanguinipes (migratory grasshopper)	Hind tibia	T^RT^R	red	Chapco (1980a,b)
		T^RT^B		
		T^BT^B	blue	
	Pronotum	Pro^RPro^R	red	
		Pro^Rpro		
		$pro\ pro$	normal	
	Hind femur	$Ost\ Ost$	orange stripe	
		$Ost\ Ost$		
		$ost\ ost$	stripeless	

about the genetic systems determining color pattern variation in acridid grasshoppers. Nevertheless, two features are worth noting. First, in many of the systems the dominant alleles do not appear to be found in high frequencies in natural populations (Gill, 1979, 1981d; Dearn, 1984; Dearn and Davies, 1983; Chapco and Bidochka, 1986). A possible explanation for this could be low viability of individual homozygous for dominant color pattern genes; evidence for such a phenomenon is given in Section 16.5. Second, as mentioned above, the loci involved in determining the color pattern variation seem to be either linked or part of a supergene complex. This situation is also observed in tetrigids (Tettrigidae), where *Acridium arenosum*, *Apotettix eurycephalus*, and *Paratettix texanus* were observed to exhibit progressively tighter linkage between the color pattern genes (Nabours, 1929; Nabours et al., 1933).

16.3.2 Environmental Switches

Although many animal species have genetically determined color pattern polymorphisms, grasshoppers are remarkable for the degree to which attributes of the phenotype, including coloration, are affected by environmental factors. This need not be the result of poorly canalized developmental systems but instead the reflection of particular genotypes that determine highly complex developmental and physiological systems that are responsive to environmental cues. Over recent years, environmentally cued polymorphisms have received far less attention than genetically determined ones though the phenomenon is well established in plants (Bradshaw, 1965; Lloyd, 1984; Schlichting, 1986) and there are many examples known in insects (Wigglesworth, 1961; Shapiro, 1976; Harrison, 1980; Tauber et al., 1984).

Environmentally cued polymorphism is similar to the phenomenon of developmental conversion proposed by Smith-Gill (1983), but the criteria she established for developmental conversion appear to be too restrictive for the grasshopper color pattern polymorphisms considered here, and therefore the term environmentally cued polymorphism (W. C. Clark, 1976) will be used in preference. In the case of environmentally cued color pattern polymorphisms in acridid grasshoppers, four environmental factors have been shown to be important determinants: humidity and the moisture content of food, the color of the background, temperature, and population density. Each of these factors has different effects and is discussed separately below.

16.3.2.1 Green–Brown Dimorphism

Many grasshoppers exhibit an environmentally determined green–brown dimorphism, where individuals in a population are either green or brown as nymphs, or as both nymphs and adults. There is a great deal of experimental evidence that the most important environmental factors determining the dimorphism are humidity and the level of moisture in the diet (Rowell, 1971.) In the cyrtacanthacridine grasshopper *Schistocerca vaga*, laboratory experiments by Rowell and Cannis (1972) showed that low humidity results in a high proportion of the green hoppers turning brown in later instars. High humidity resulted in most hoppers retaining their green coloration but only under conditions of low density. Both crowding and low humidity were equally effective and appeared to act additively in promoting the change from green to brown coloration. In the oedipodine grasshopper *Gastrimargus africanus*, the hatchlings are brown and can develop green coloration under conditions of high humidity (Rowell, 1970). Crowding reduces the probability of hoppers developing green coloration, but unlike *S. vaga*, *G. africanus* exhibits a homochromic response (see below) that can inhibit the development of the green morph. The importance of the moisture content of food in determining green coloration has been demonstrated in other species by Okay (1953, 1956), Otte and Williams (1972), and Ibrahim (1974).

The humidity and the moisture content of vegetation are directly correlated with its color. It is clear, then, why natural selection would result in them acting as an environmental switch in response to selection for crypsis but it should be remembered that in some species green–brown variation is only under genetic control and cannot be altered by humidity (Gill, 1981d). Furthermore, not only would green and brown coloration provide appropriate camouflage on green and brown vegetation, but there is evidence that the green and brown morphs differ in their tolerance to different humidity levels (see Section 16.5).

16.3.2.2 Homochromy

Many species of grasshopper have been shown to be able to change their color to match that of their background (Rowell, 1971). Perhaps the most spectacular example of this is the ability of the adults of some grasshopper species to turn dark following fire, their coloration matching that of the blackened vegetation (Poulton, 1926; Burtt, 1951). Homochromic responses have been demonstrated to white, gray, black, yellow, orange, and brown backgrounds but are inhibited in both crowded individuals exhibiting the phase *gregaria* phenotype (see below) and in individuals exhibiting the green phenotype of the green–brown dimorphism. The homochromic response is mediated by the eyes in response to light and seems to involve only the black, yellow, and orange pigments (Rowell, 1971). The black pigment is involved to different degrees on all backgrounds (the albedo response), whereas the orange and yellow pigments are involved only when the backgrounds themselves contain orange, brown, or yellow (Hertz and Imms, 1937; Rowell, 1970).

The homochromic response appears at first sight to be an ideal solution to the problem of matching body coloration to that of the background in cryptic species, and it is perhaps surprising that it is not more common. It must be remembered, however, that the evolution of such a system will occur only in a population that experiences temporal variation in the coloration of the background and, moreover, where suitable environmental cues exist. In addition, a homochromic response has not been observed to green backgrounds (Rowell, 1971) and green–brown background color variation is common for many grasshoppers. It is possible that many genetically determined color pattern polymorphisms are maintained by frequency-dependent selection exerted by predators, resulting in rare, even conspicuous, morphs being protected from predation (see Section 16.6.5). Under such a selection regime a homochromic response, where all individuals adopt the color of their background, might be a disadvantage.

16.3.2.3 Temperature

Temperature has been shown to affect coloration in grasshoppers and individuals reared at low temperatures tend to be darker (Rowell, 1971). *Chortoicetes terminifera* collected at different times over the year show variation in dark coloration related to variation in temperature (Key, 1954), though the ecological significance of this type of variation is unknown. One case where the role of temperature-determined variation in color pattern is clear is the color change seen in the Australian alpine catantopine grasshopper *Kosciuscola tristis* (Key and Day, 1954a,b). Above 25°C individuals are a bright blue but below 15°C are a dull near-black. This response takes only a few hours and has a direct thermoregulatory role. The black coloration appears to enhance the capacity of individuals to utilize solar radiation and maintain a higher internal body temperature when temperatures are low in the morning and in late afternoon, whereas the color change to blue assists in avoiding overheating during the day when insolation levels can be very high. This type of change in coloration is, however, very unusual among acridid grasshoppers.

16.3.2.4 Phase Variation

Phase variation is the name given to phenotypic variation that is determined by population density; it can involve many aspects of the phenotype including behavior, morphology, and physiology, as well as coloration. The phenomenon is exhibited to an extreme degree by a small number of acridid species termed locusts, in which it has been the subject of a great deal of research (Kennedy, 1956, 1961; Albrecht, 1962; Uvarov, 1966). The distinction between locusts and grasshoppers is made on the basis of the phase response and consequently phase variation is not dealt with in detail here. However, many acridid species do exhibit some phenotypic response to changes in density (Uvarov, 1966, 1977), and there is no clear distinction between locusts and grasshoppers. For example, Rubtzov (1935) observed in *Chorthippus albomarginatus* that crowded conditions in the field produced darker coloration analogous to the dark coloration of crowded locusts. Accordingly, he assigned individuals to one of three phase types, *gregaria, transiens,* and *solitaria* on the basis of coloration and demonstrated a correlation between population density and phase status (Table 16.2). This relationship was also demonstrated in the laboratory. Furthermore, Rubtzov showed that the degree of response to population density

TABLE 16.2. Frequency of the Three Phase Types Based on Color Pattern in Relation to Population Density in *Chorthippus albomarginatus*

Number of Individuals per Square Meter	% Phase type		
	gregaria	*transiens*	*solitaria*
< 1	14.2	39.0	46.8
10–20	37.3	42.4	20.3
15–150	57.1	36.4	6.5

Source: From Rubtzov (1935).

varied among the color pattern morphs. The green f. *viridis* individuals showed the least degree of phase variation and the brown f. *rubiginosa* individuals showed the greatest response to density. Similar responses to density analogous to phase variation in locusts were observed for morphology and behavior.

Despite the immense amount of work that has been carried out on phase variation in locusts, the significance of the color patterns observed at different population densities remains unclear. The gregarious nymphs of some species exhibit striking color patterns involving black and orange or yellow markings that are in marked contrast to the uniform green coloration of solitary nymphs. These color patterns are not aposematic and do not appear to play a direct role in grouping (Gillett, 1973). It is possible that they may be involved in temperature regulation, in maintaining the cohesion of mobile groups, or in confusing predators (Gillett and Gonta, 1978), for predation efficiency does appear to be lower when prey are faced with an aggregated group of locusts (Gillett et al., 1979). Certainly cryptic color patterns would be of little use in highly mobile locust hopper bands or swarms, and, in any case, the homochromic response is suppressed in crowded nymphs. Much remains to be learned about the phenomenon of phase variation in acridids, and the color pattern variation associated with phase variation represents an interesting experimental system for studying the role of coloration in grasshoppers.

16.3.3 Combined Genetic and Environmental Switches

Polymorphic systems can be under the simultaneous control of both genetic and environmental switches, and there are examples of this in the case of acridid color pattern polymorphisms. For example, in *Chortoicetes terminifera* the color pattern locus does not directly control the presence of green pigmentation. It has been shown, however, to contain genes that determine sensitivity to the environmental factors that determine the green–brown dimorphism, and the three common color pattern morphs in *C. terminifera* develop green coloration to different extents under the same environmental conditions (Key, 1954). This means that the factors determining one polymorphic system must interact with the factors determining the other system and that the overall pattern of variation could be a compromise between conflicting selection pressures.

The existence of variation for the sensitivity of coloration to density has been demonstrated in the brown locust, *Locustana pardalina*, and the African migratory locust, *Locusta migratoria migratorioides* (Nel, 1967, 1968), though because of the system of cytoplasmic inheritance affecting phase variation in locusts it is not known to what extent this variation in environmental sensitivity is genetic. The extent to which there is intra- and interpopulation genetic variation for sensitivity to environmental cues determining color pattern polymorphisms in grasshoppers remains a fascinating area for future studies.

16.4 PATTERNS OF POLYMORPHIC COLOR PATTERN VARIATION

Spatial and temporal patterns of variation for polymorphic systems have long been used to provide evidence for the existence of selection and the correlation between phenotypic frequencies and environmental variables can suggest a possible adaptive role for particular characters. Although such associations are only correlations and do not necessarily represent causal relationships, they do propose possible selective agents that can be examined experimentally.

16.4.1 Spatial Patterns

Many studies have shown a relationship between the distribution of different color pattern morphs and habitat characteristics, and these support the contention that selective predation is a major factor determining grasshopper color patterns. Gill (1981e) examined color pattern variation in British acridids in relation to habitat and noted that the predominantly brown species were usually found in drier habitats with less vegetation and more bare ground whereas the predominantly green species were found in denser vegetation. The same relationship has been seen within species between genetically determined green and brown morphs. For example, Rubztov (1935) observed that *Chorthippus albomarginatus* individuals of f. *viridis*, the greenest of the four morphs in this species, were more abundant in "humid and stable mesophytic habitats" whereas f. *rubiginosa*, the brownest morph, were more abundant in "more xerophytic and less stable habitats" such as pastures (Table 16.3). Similar observations were made in *Omocestus viridulus*, the common green grasshopper, by Petersen and Treherne (1949) and Ragge (1965).

Associations between habitat and color pattern were observed by Cox and Cox (1974) in the oedipodine grasshopper *Circotettix rabula* in populations on either red sandstone or gray shale substrates. These populations exhibited four different color pattern morphs, gray, yellow—brown, brown, and red, which were observed at different frequencies on the two contrasting substrates (Table 16.4). The role of predation in producing this substrate color matching was examined and is considered in Section 16.6.1 below. Gill (1979) examined color pattern variation in *Chorthippus brunneus*, the common field grasshopper,

TABLE 16.3. Frequency of the f. *rubiginosa* (Brown) and f. *viridis* (Green) Color Pattern Morphs in *Chorthippus albomarginatus* in Two Contrasting habitats in Siberia[a]

Habitat	% Color Pattern Morph		Total
	f. *rubiginosa*	f. *viridis*	
Margin of forest; humid; dense vegetation	37.3	14.2	198
20 km from forests; the driest habitat	71.4	2.1	850

[a] The other morphs (f. *hyalosuperficies* and f. *hyalolateralis*) are intermediate in their distribution of green and brown color.
Source: From Rubtzov (1935).

TABLE 16.4. Frequency of Different Color Pattern Morphs of Nymphs of *Circotettix rabula* Collected from Gray Shale and Red Sandstone Substrates

Substrate Color	% Color Pattern Morph				Total
	Gray	Yellow–Brown	Light–Dark Brown	Red	
Gray	81	2	4	13	48
Red	0	3	36	61	70

Source: From Cox and Cox (1974).

with habitats classified into three types varying from vegetated with little bare ground (type 1) to habitats containing large areas of bare ground with only patches of vegetation (type 3). The analysis showed type-1 habitats had a higher frequency of green morphs and morphs with striped wings, whereas there was a higher frequency of mottled-wing grasshoppers in type-3 habitats.

The catantopine grasshopper *Phaulacridium vittatum*, known as the wingless grasshopper, exhibits a simple color pattern polymorphism under genetic control with a plain brown morph and a morph with two light stripes on the dorsal surface, similar to the classic disruptive pattern discussed by Cott (1940). The pattern of spatial variation for this system analyzed by Dearn (1984) can be interpreted in terms of crypsis because populations in pasture habitats containing a mosaic of bare earth and dried grass have a higher frequency of striped individuals than populations from forest habitats that have a more uniform darker background color pattern (Table 16.5). The marked latitudinal cline in pasture populations, which was first reported for this polymorphic system (Dearn, 1981), could therefore be due to the spatial relationships and gene flow among pasture and forest sites. The spatial distribution of morphs does, however, present some interesting features. First, all

TABLE 16.5. Frequency of the Striped Color Pattern Morph in *Phaulacridium vittatum* in Pooled Samples from Two Habitats at Two Locations in Southeast Australia

Location	Habitat	% Striped Color Pattern Morph	Total
Stratford	Pasture	18.3	21478
	Forest	11.3	902
Mansfield	Pasture	24.6	4239
	Forest	11.9	73

Source: From Dearn (1984).

populations are polymorphic, that is, striped individuals are always found in forest sites even when these sites are remote from pasture sites. Thus either the heterogeneity of these sites is such as to maintain microhabitats conferring differential protection or there are selective forces other than predation maintaining the polymorphism. The second feature of this system is that relatively few populations have been observed with a frequency of the striped individuals greater than 30%.

Spatial patterns have also been recorded for the environmentally determined green–brown dimorphism. A higher proportion of green morphs was observed by Otte and Williams (1972) in areas containing green vegetation compared with adjacent dry vegetation in *Syrbula admirabilis* and *Chortophaga viridifasciata* (Table 16.6). In *Chortoicetes terminifera*, the Australian plague locust, the green–brown dimorphism is virtually restricted to females, and Key (1954) showed a higher frequency of green females in regions of eastern Australia with a higher precipitation–saturation deficit ratio (Table 16.7).

The relationship between color pattern and habitat in grasshoppers shown in these examples could be due to four different causes: to selective predation based on the degree of crypsis, to pleiotropy with selection acting on aspects of the phenotype other than color pattern per se, to individuals directly developing a particular color pattern that matches the color pattern of the background, or to individuals seeking out a habitat with a color pattern that matches their own coloration. All four phenomena have been observed in grasshoppers and are discussed in detail below.

In marked contrast to the examples of spatial variation discussed above, the nine genetically controlled color pattern morphs in *Chortoicetes terminifera* show little geographic variation despite a distributional range covering 95% of Australia (Key, 1954; Dearn and Davies, 1983). Thus it appears that selection regimes within, rather than among, populations are responsible for maintaining this polymorphism and indeed the variation may not be related directly to crypsis. Although the different morphs are distinct (Fig. 16.1), they do not appear to confer any obvious differential protection against different backgrounds.

TABLE 16.6. Distribution of Green and Brown Morphs in Adjacent Green and Dry habitats in *Chortophaga viridifasciata* and *Syrbula admirabilis*

		% Color Pattern Morph		
Species	Habitat	Green	Brown	Total
C. viridifasciata	Green	79	21	39
	Dry	7	93	14
S. admirabilis	Green	68	32	37
	Dry	11	89	35

Source: From Otte and Williams (1972).

TABLE 16.7. Frequency of the Green Color Pattern Morph in Male and Female *Chortoicetes terminifera* Sampled East of Longitude 142°E in Australia from Three Climatic Zones

	% Green Individuals[b]	
Climatic Zone[a]	Males	Females
Zone 2 (P/s.d. 50–70)	0 (63)	11.4 (35)
Zone 3 (P/s.d. 70–150)	0.8 (265)	26.2 (260)
Zone 4 (P/s.d. >150)	0.5 (216)	43.3 (127)

[a] Based on mean annual precipitation/saturation deficit (Meyer) ratio.
[b] Sample size in parentheses.
Source: From Key (1954).

When the spatial distribution of other color pattern polymorphisms not obviously related to crypsis is examined, problems of interpretation arise. For example, geographic variation was examined in *Melanoplus sanguinipes*, the migratory grasshopper, for the hind tibia color and hind femur stripe polymorphisms in the southern Saskatchewan area of Canada (Chapco and Bidochka, 1986). Both the red tibia and the orange femur stripe alleles showed considerable geographic variation. The red tibia morph was positively associated with latitude and elevation and negatively associated with longitude, with minimum annual temperature accounting for 40% of this variation. By contrast, no patterns of variation were observed for the orange femur stripe morph and no correlations were observed with climatic factors. If these polymorphisms are maintained by some form of balancing selection, the selection pressures must be complex and variable. Chapco and Bidochka (1986) discussed several different possibilities to explain the variation for the tibia color morph, including linkage disequilibrium with other genes related to the climatic variables and selection related to soil or vegetation characteristics that are correlated to the climatic variables. Whatever the reasons for the maintenance of both polymorphisms it appears that crypsis per se is unlikely to be responsible.

As a final example of geographic variation, Schennum and Willey (1979) studied the wing color and tegmen stripe polymorphisms in the oedipodine grasshopper *Arphia conspersa*, the speckled rangeland grasshopper, in the southern Rocky Mountains. Both polymorphisms showed geographic variation but no explanation for the maintenance of the polymorphisms was apparent and the pattern of geographic variation was interpreted in terms of previous allopatric differentiation of populations. These studies serve to emphasize the complexity of the selective forces acting on grasshopper color pattern variation and show that not all polymorphisms can be readily interpreted in terms of either predator selection or climatic selection.

16.4.2 Temporal Patterns

Because environmentally cued polymorphisms evolve as a response to predictable patterns of environmental variation, it is not surprising that the most successful attempts to observe temporal variation in the frequency of color pattern morphs have been for environmentally cued polymorphisms. The remarkable diurnal variation in the coloration of *Kosciuscola tristis* in response to temperature has already been mentioned. In *Chortoicetes terminifera* the frequencies of the green and brown morphs have been shown to change over time in relation to rainfall levels (Key, 1954), as might be expected from what is known of the environmental determination of this dimorphism. Similarly, the frequency of brown morphs in *Syrbula admirabilis* were observed to increase during a period of drought and in *Dichromorpha viridis* the frequency of green morphs increased following heavy rain and the growth of vegetation (Otte and Williams, 1972).

Of more interest are observations on temporal variation in genetically determined morphs because it is the maintenance of these different forms, not the environmentally cued morphs, that requires explanation. Unfortunately there are few relevant data available. In an analysis of changes in color pattern morphs in *Chortoicetes terminifera* over the nymphal and early adult stages in two separate populations, Byrne (1967b) showed consistent patterns. For example, the proportion of F^rF^r individuals increased while individuals containing the F^a allele were selected against. During this time each habitat became drier and the increase in the frequency of F^rF^r (*rubiginosa*) individuals is consistent with the associations between brown morphs and habitat type previously discussed in Section 16.4.1. However, the selective disadvantage of the F^a allele must be counterbalanced by selection favoring the F^a acting at other stages of the life cycle or for other fitness components such as fecundity. Byrne continued his examination of the two populations into a second generation when the vegetation was greener because of increased rainfall. Significantly, in this generation no changes in the frequency of F^rF^r individuals were observed. Although data on differential predation on the color pattern morphs are lacking, Byrne was of the opinion that what he called "physiological selection" was more important in determining the morph frequencies. Richards and Waloff (1954) examined changes in the frequencies of the color pattern morphs in *Chorthippus parallelus*, *Omocestus viridulus*, and *Stenobothrus lineatus* at one particular site over five successive years. Although they noted that green forms were somewhat rarer in the year following a very dry year, their overall conclusion was that the frequencies of the different color pattern morphs were extremely constant.

These observations on temporal variation in grasshopper color polymorphisms highlight the general problem of measuring the action of natural selection on balanced polymorphic systems in wild populations. This is difficult to do under any circumstances and requires experiments specifically designed to detect the small variable selection coefficients often found in natural populations (B. Clarke, 1975). There is a clear need, however, to begin such studies in grasshoppers if the nature of the selection pressures acting on color pattern variation is to be determined.

16.5 ASSOCIATIONS WITH COLOR PATTERN POLYMORPHISMS

Color pattern variation has been shown to be associated with aspects of the phenotype of grasshoppers apparently unrelated to the direct or indirect effects of color per se. This could result from allelic association at different loci arising either by chance or from the action of natural selection favoring particular gene complexes (Hedrick et al., 1978).

Rubtzov (1935) showed that four different color morphs in *Chorthippus albomarginatus* differing in their relative distribution of green and brown pigments also differed in their morphology (hind femur length, elytron length, and hind femur length/elytron length ratio) in a similar way to the differences associated with the phenomenon of phase variation. The relationship between color pattern and phase variation is further supported by the observations by Rubtzov (1935) that the f. *rubiginosa* brown morph shows greater mobility than the green f. *viridis* morph and the brown morphs, unlike the green ones, were seen to form aggregations. Thus it appears that the genes associated with color pattern variation either have multiple pleiotropic effects affecting morphology and behavior or, as is more likely, are linked with other genes.

In contrast to the results of Rubtzov, Richards and Waloff (1954) did not observe any morphological differences between the genetically determined color pattern morphs in British grasshoppers. Presumably, either the British grasshoppers, unlike the Siberian populations, do not exhibit phase variation because they do not reach the high densities associated with phase variation or they lack the genetic capacity to exhibit a phase response. It is known, for example, that some populations of locusts have lost the capacity to exhibit phase variation (Uvarov, 1966). In *Chortoicetes terminifera*, the green–brown dimorphism is under environmental control and is superimposed on the genetically controlled color pattern morphs only in low-density populations. Individuals from swarming populations never exhibit green coloration. Key (1954) showed that green and nongreen individuals did indeed exhibit the morphological differences one would expect to be associated with phase variation but the different color morphs within the brown individuals did not differ among themselves.

Differences have been observed between green and brown morphs in locusts related to phase variation. For example, Pick and Lea (1970) observed behavioral differences between the green and brown morphs in hoppers of *Locustana pardalina*, the brown locust, with brown hoppers showing more movement and green hoppers spending more time on green grass. Of great interest is the observation by Albrecht (1965) that in the African migratory locust, *Locusta migratoria migratorioides*, brown hoppers are more resistant to starvation at low humidity than green hoppers whereas green hoppers are more resistant to starvation at high humidities than brown hoppers.

An interesting example of an association with a genetically determined color pattern polymorphism was demonstrated by Nankivell (1974) in the Australian oedipodine grasshopper *Austroicetes interioris*. *A. interioris* shares the same polymorphic system as that seen in the Australian plague locust, *Chortoicetes terminifera*, discussed above. Nankivell grouped individuals into two color pattern groups, the Lineosa group (f. *lineosa*, f. *trilineata* and f. *nigrovirgata*) and the Rubiginosa group (f. *rubiginosa* and f. *transmaculata*). He showed an association between a chromosome inversion polymorphism on the fourth autosome, heat stress tolerance, and color pattern morph, and the same chromosome polymorphism was shown to be associated with variation in body weight. These relationships with the color pattern polymorphisms were discovered only accidentally following high mortality in a sample of grasshoppers collected in the field; it would be interesting to look for similar associations in other systems.

One of the most curious observations to emerge from studies on color pattern polymorphisms is that concerning the viability estimates of different color pattern genotypes. In an analysis of the data collected by Nabours on color pattern polymorphism in two species of tetrigids, *Apotettix eurycephalus* and *Paratettix cucullatus* (formerly *P. texanus*), Fisher (1939) showed that individuals heterozygous for a single dominant color pattern gene and the recessive gene had an inherent selective advantage of about 10%. In addition, an analysis of field samples of *Paratettix cucullatus* revealed that at least 40% of individuals heterozygous for two dominant genes appeared to be eliminated.

These results have some historical significance because they were among the first to demonstrate the existence of large selection coefficients in field populations. They might be dismissed as some sort of curiosity if it were not for similar observations made in other systems. In *Chortoicetes terminifera*, Dearn and Davies (1983) showed that large viability differences appear to exist between the color pattern genotypes with individuals heterozygous for two dominant genes present in frequencies much lower than those expected if there were random mating and equal viability of all color pattern genotypes. Low viability has also been noted for the dominant red back gene (Pro^R) in the migratory grasshopper, *Melanoplus sanguinipes* (Chapco, 1980a). Such low viabilities would certainly put constraints on the frequencies of dominant color pattern genes in addition to any effects associated directly with crypsis. Significantly, the overall low frequency of dominant color pattern genes in field populations of grasshoppers is a feature of some genetically determined polymorphisms, as discussed in Section 16.3.1. The disassortative mating observed with respect to the color pattern variation in *Phaulacridium vittatum* (Dearn, 1979) would clearly serve to maintain the dominant striped pattern gene in the presence of any associated low viability. Exactly this situation has been proposed to explain the maintenance of the *medionigra* color pattern morph in the scarlet tiger moth, *Panaxia dominula* (Ford, 1975).

The significance of these viability differentials are unclear, and much more work is needed in acridid grasshoppers. They could arise from some metabolic cost associated with the development of particular color pattern phenotypes. Alternatively, they could be due to specific low viability genes that have evolved to maintain the integrity of coadapted gene complexes containing color pattern genes. Whatever the reason for the viability differences, their existence will impose limits on how color pattern systems evolve and respond to selection.

The observations discussed in this section indicate that any study of color pattern polymorphism in grasshoppers must take into account the possibility of other effects associated with the color variation and indeed, these effects, however they arise, may be more important than color per se in terms of how selection will act.

16.6 SELECTION AND COLOR PATTERN POLYMORPHISMS

The study of polymorphisms and their developmental switches is of great interest in itself because it focuses attention directly on the ways organisms perceive and respond to their environment. In addition, however, polymorphisms provide an excellent experimental approach to studying the adaptive role of particular characters, and this is particularly true of color pattern variation, where direct comparisons can be made between different morphs within the same population. Color pattern variation can have many effects on the biology of grasshoppers and consequently color pattern variation is subject to different, and in many cases conflicting, selection pressures. It is for this reason that studies on color pattern polymorphisms, in marked contrast to polymorphisms at the DNA or protein levels, have produced a plethora of alternative explanations, a situation well illustrated by the extensive studies on the color pattern polymorphism in the land snail, *Cepaea nemoralis* (Jones et al., 1977).

The observed associations between habitat type and particular cryptic color pattern morphs could be due to the consequences of selective predation but could equally well be due to selection acting on aspects of the phenotype other than color pattern or indeed simply to individuals selecting different habitats. This section will examine two major forms of selection that are known to operate on grasshopper color pattern variation, predation and climatic selection, together with a consideration of mating preferences and habitat choice, both of which will affect the evolution of polymorphic systems. It should be noted that, in addition, color pattern variation may be subject to changes due to the action of selection on linked loci, as discussed in Section 16.5, and the observation of viability differences associated with different color patterns suggests the possibility of other forms of selection.

16.6.1 Predator Selection

Despite the widespread assumption of the role of grasshopper coloration in crypsis, there has been surprisingly little experimental work on predation with respect to color pattern variation. Direct evidence for the protective role of grasshopper coloration was obtained by Isely (1938) using experimental plots containing a mixture of squares of black soil, white soil, red soil, and green grass with a variety of grasshopper species with either overall white, reddish brown, black, or green coloration. Grasshoppers were either tethered or placed anesthetized on different backgrounds and designated either protected or nonprotected on the basis of their color against the color of the background. The grasshoppers were then subject to predation by either domestic or wild birds, the results clearly demonstrated the survival value of cryptic coloration (Table 16.8).

Similar experiments were carried out by Cox and Cox (1974) using lizard predation on the grasshopper *Circotettix rabula*, which exhibits different color pattern morphs that match the heterogeneous coloration of the substrate. The results showed that the color matching provided effective defense against predators (Table 16.9), and significantly, one of the two lizard species used in the experiment was also seen to eat *C. rabula* in the field. The adaptive role of cryptic coloration was further supported by the results of release and recapture experiments in the field, which showed a rapid decline in the ratio of mismatched to matched grasshoppers with time. Unfortunately, nothing is known about the inheritance of these color pattern morphs. This is critical because there must be an underlying genetic basis to any phenotypic variation for it to have any evolutionary significance.

TABLE 16.8. The Protective Value of Grasshopper Coloration

Predator	Coloration status	% Eaten	Total
Bantams	Protected	44	185
	Nonprotected	85	185
Turkeys	Protected	46	80
	Nonprotected	91	80
Wild birds	Protected	34	114
	Nonprotected	84	114

Source: Data from Isely (1983) and based on a table in Cott (1940).

TABLE 16.9. Results of Experiments Involving Predation by the Lizards *Sceloporus undulatus* and *S. graciosus* on Nymphs of *Circotettix rabula* on Substrates of Different Color

Color of Substrate	Color Pattern Morph	% Eaten	Total
Gray	Red	82	50
	Gray	40	50
Red	Red	49	94
	Gray	66	93

Source: (From Cox and Cox 1974).

16.6.2 Climatic Selection

One consequence of variation in color pattern in grasshoppers is a direct effect on the energy budget of individuals. For a grasshopper resting in sunlight, solar radiation is the major form of heat input (May, 1979) and the degree of temperature excess (the amount by which the internal body temperature exceeds the ambient air temperature) varies with wind velocity, body temperature, body size, and the strength and absorptivity of radiation, the latter being a function of the color pattern of the grasshopper (Digby, 1955; Hamilton, 1975; and see Chapter 6). Differences in body temperature between different color pattern morphs in grasshoppers have been observed. For example, the internal temperature of the black color pattern morph of *Calliptamus coelesyriensis* exposed to sunshine is higher than that of the green morph (Buxton 1924), and the black color pattern morph of *Melanoplus differentialis* heats up more rapidly than the yellow morph (Pepper and Hastings, 1952). Similarly, the dark gregarious morph in *Locusta migratoria* has a higher internal body temperature than the green solitarious morphs (Hill and Taylor, 1933). The effect of body color on internal temperature was investigated by Joern (1981) by comparing the body temperatures in sunlight of dead, normal, light brown individuals of the grasshopper *Brachystola magna* with dead individuals painted either black or white. As expected, the black individuals had a higher and the white individuals a lower body temperature than the unpainted individuals.

The environmentally cued color pattern polymorphism in *Kosciuscola tristis* discussed above appears to be directly related to the different heating properties of the blue and black morphs. It is unlikely, however, that many color pattern polymorphic systems have evolved only because of the effect of color on internal body temperature. Whatever the role of the color pattern variation in any population, though, different color pattern morphs will have different heating properties and this must play a role, however small, in the evolution and maintenance of all color pattern polymorphisms. Climatic selection related to coloration has been shown to be an important factor determining the color pattern morph frequencies in a variety of insects such as spittlebugs (Berry and Willmer, 1986), leafhoppers (Stewart, 1986), butterflies (Watt, 1968), and ladybugs (Brakefield and Willmer, 1985).

Grasshoppers are able to regulate their body temperature by behavioral means either by selecting microhabitats or in the way they orient themselves to the sun (Anderson et al., 1979; Willmer, 1982; Kemp, 1986; and see Chapter 6) and different color pattern morphs would be able to maintain the same internal body temperature by behavioral means. This has been demonstrated experimentally by Joern (1981) in the grasshopper *Brachystola magna*. He showed that live grasshoppers, unlike dead ones, were able to maintain a body temperature close to the ambient air temperature by changes in posture and the selection of different microhabitats during the day.

Two questions of interest arise with respect to color pattern polymorphisms and thermoregulation in grasshoppers. The first concerns the costs of regulating body temperature and compensating for color pattern variation. Even though different color pattern morphs do not necessarily have different body temperatures in the field, the behaviors required to maintain body temperature in particular cases may involve the expenditure of different amounts of energy or the exposure of different morphs to different degrees or types of predation. Thus a particular morph may have evolved for crypsis in response to selection pressure exerted by a predator, but this has entailed some cost related to thermoregulation. The net outcome of selection will be determined by the costs and benefits associated with each morph, and the costs associated with thermoregulation may put limits on selection for crypsis. (The same argument applies equally well to the evolution of color pattern in a monomorphic species.)

A second aspect of the heating properties of different color pattern morphs also arises from the behavioral responses required to regulate body temperature. Heat can be regarded as an ecological resource (Magnuson et al., 1979) and different coexisting grasshopper species can be regarded as exhibiting thermal niche partitioning (Gillis and Possai, 1983). Exactly the same analysis can be applied to different color pattern morphs within the same populations; they could be viewed as occupying different thermal subniches. This could mean, for example, that populations polymorphic for color pattern are better able to exploit their environment and maintain higher populations densities than monomorphic populations.

16.6.3 Mating Preferences

One of the advantages offered by grasshoppers for experimental work is that in many species mating pairs remain *in copula* for long periods in the field. This behavior enables information to be gained on the phenotypes or genotypes of mating pairs and, in particular, the extent to which mating preferences with respect to particular phenotypes or genotypes are nonrandom. Models in population genetics often assume random mating with respect to genotype, although this is usually for reasons of simplification rather than because random mating has been demonstrated empirically. A study of mating patterns in field populations of the wingless grasshopper, *Phaulacridium vittatum*, with respect to the color pattern morphs showed an excess of heterogametic matings (Dearn, 1979). Striped individuals were mating with plain individuals more frequently than would be expected by chance; that is, there was disassortative (or negative assortative) mating with respect to this polymorphism.

The disassortative mating observed in *P. vittatum* is the first case of nonrandom mating with respect to a color pattern polymorphism recorded in grasshoppers. The result is significant because such a system of mating can lead, in itself, to the maintenance of a genetic polymorphism (Workman, 1964). However, it would seem likely that such a system of mating would evolve only if it resulted in some underlying selective advantage. Females mating with males of an unlike color pattern genotype may leave progeny of a different morph type, and in a heterogeneous environment such females may leave more descendants than females mating with males of the same genotype. Alternatively, the progeny of females mating with males with unlike color pattern genotype may be more heterozygous for linked genes. In this case the role of the color pattern variation is simply to indicate genotypic status and is not important in itself. The existence of nonrandom patterns of mating with respect to color pattern polymorphisms is of great interest and warrants further investigation.

16.6.4 Habitat Choice

Habitat choice is of great significance with respect to the problem of maintaining polymorphisms. It is still unclear how genetic polymorphisms are maintained in natural populations but the conditions for maintaining a stable polymorphism are far less stringent if there is some linkage between alleles determining preference for a particular niche and alleles conferring some advantage within that niche (Garcia-Dorado, 1986). The associations between habitat and color pattern morph discussed above could be due simply to the results of selection within niches favoring certain morphs, to the differential niche choice of different morphs, or to both factors. If different color pattern morphs are fitter in different microhabitats such as bare soil versus vegetated areas, as seems to be the case, then natural selection will favor an association between the color pattern genes and genes conferring habitat choice.

Evidence for the existence of habitat choice comes from elegant experiments carried out by Gillis (1982) on the oedipodine grasshopper *Circotettix rabula rabula*. A population of *C. rabula* in north central Colorado inhabits a coarse-grained mosaic of red granite and green vegetation and exhibits red and green color pattern morphs. Each morph appears inconspicuous against the appropriate background. In a series of laboratory experiments, grasshoppers of both morphs were given the choice of resting on backgrounds painted to look like the two types of habitat found in the natural habitat. Grasshoppers were painted around the eye with either red or green paint the same as that used to paint the test chamber. The results (Table 16.10) show clearly that grasshoppers choose to rest on a background with a color that matches the color they can see on their own bodies. Control experiments showed that painting did not interfere with their normal color preferences and furthermore that red grasshoppers painted with the novel color blue preferred blue to red backgrounds when given the choice. Clearly more experiments of this type are needed to explore how common this phenomenon is in acridid grasshoppers.

16.6.5 The Maintenance of Color Pattern Polymorphisms

Although the evidence is largely circumstantial, it is clear that selection does act on color pattern variation in grasshoppers. However, it is quite another problem to explain how color pattern polymorphisms are maintained in natural populations. That these polymorphisms are indeed stable can be inferred from the existence of similar color pattern morphs in different species as well as the observed constancy in morph frequencies in particular populations over time. Such balanced polymorphisms arise only from particular selection regimes that result in different alleles being maintained at the color pattern loci. Three mechanisms are commonly proposed to explain the maintenance of genetic

TABLE 16.10. Substrate Color Preferences in Two Color Pattern Morphs of *Circotettix rabula rabula* when unpainted or Painted Around Eye

Subject		Color Preference (%)		
Morph	Eye Paint	Red	Green	Number
Red	None	80	20	30
Red	Red	86	14	29
Red	Green	14	86	28
Greenish gray	None	17	83	29
Greenish gray	Green	23	77	30
Greenish gray	Red	53	47	30

Source: From Gillis 1982.

polymorphisms in populations: a balance between gene flow and directional selection among habitat patches, heterozygote advantage, and frequency-dependent selection. These mechanisms are not mutually exclusive and can operate simultaneously on the same polymorphic system.

There is a great deal of evidence to show that environmental variation promotes genetic variation but the conditions for maintaining polymorphisms through variable selection in heterogeneous environments are very restrictive (Hedrick et al., 1976; Hedrick, 1986). However, as discussed in Section 16.6.4, the existence of habitat choice, which has been observed for one grasshopper color pattern polymorphism, would greatly enhance the probability of a polymorphism being maintained and the possibility of habitat choice for color pattern morphs should be examined in other grasshopper species. Heterozygote advantage has been widely dismissed as a general method for maintaining polymorphisms both on theoretical grounds related to its associated genetic load and because it has been rarely demonstrated for individual loci. However, it should not be discounted in the case of grasshopper color pattern polymorphisms in view of the results on apparent viability differences between color pattern genotypes revealed by Fisher (1939) and Dearn and Davies (1983) (see Section 16.5). It may be inappropriate to regard the color pattern loci as single genes; it is likely that they form part of coadapted gene complexes. It may be more appropriate to compare the color pattern loci to the coadapted gene complexes found in the chromosome inversions in *Drosophila* species that have been studied in so much detail and for which there is clear evidence of heterozygote advantage (Dobzhansky, 1970).

The general failure of heterozygote advantage to explain the extensive level of genetic variation observed in natural populations has prompted a search for alternative forms of selection. Selection that is frequency-dependent and exerted by predators, parasites, and competitors has received considerable attention and has been proposed as a general explanation for genetic variation within natural populations (B. C. Clark, 1979; and see Chapter 14). Of particular relevance to the maintenance of color pattern polymorphisms in grasshoppers is the form of frequency-dependent selection called apostatic selection, which is exerted by predators using specific search images. If it is costly for a predator to search simultaneously for two different color pattern morphs using different cues, it may be more efficient to search only for an abundant prey morph (Greenwood, 1984). This type of selection is frequency dependent and said to be proapostatic because the fitness of color pattern morphs is inversely related to their frequency in the population and leads to the maintenance of different color pattern morphs. Although likely predators of grasshoppers such as birds have been shown to possess the potential to exert apostatic selection (B. C. Clarke, 1979), there have been no studies in grasshoppers demonstrating the action of this type of selection or indeed any quantitative study on the process of predation in the field. Moreover, it would have to be shown that any observed apostatic selection is in fact sufficient to explain the maintenance of a particular polymorphic system.

Despite the lack of evidence for apostatic selection acting on grasshopper color pattern morphs, its potential importance warrants further comment. First, apostatic selection does not in itself require that the prey morphs match their background. However, Cooper (1984) has shown that apostatic selection in an experimental system using wild birds and artificial prey was stronger when the prey matched their background, as is the case for most grasshopper color pattern polymorphisms. This could be due to the less frequent encounters of predators with prey when the latter are matched to their background, and predators may then concentrate on distinguishing the more common morph from the background. Second, apostatic selection at high prey density could actually result in a higher proportion of the rare morphs being taken by predators, the reverse of the situation discussed above (Greenwood, 1984). Such antiapostatic selection would result from the rare morphs being more conspicuous because of the problems encountered by predators hunting prey morphs that are at high density. It has been shown in a locust that predation rates do indeed drop when prey densities increase (Gillett et al., 1979); we need to examine the relation between population density and color pattern variation in grasshoppers.

16.7 CONCLUSIONS

Despite the limited work on grasshopper color pattern polymorphisms, it is clear from the evidence reviewed in this chapter that variation in coloration plays an important and central role in the biology of grasshoppers. It has been widely assumed that the primary selective agent is predation acting on the visual phenotype, though it must be remembered that cryptic coloration is only one means of predator avoidance and other methods of escaping predation in grasshoppers include flight, posturing, surprise behaviour, mimicry, warning coloration, and the regurgitation of noxious fluids (see Chapter 12). Indeed, changes in the mechanisms of predator avoidance can be seen over the lifetime of individual grasshoppers (Schultz, 1981). Furthermore, the significance of color pattern variation cannot be understood without reference to the color pattern of the background and the behavior of predators. The work of Endler (1984) indicates the direction future studies on predation and color pattern variation in grasshoppers must take.

Coloration must, however, be seen as only one component of an integrated coadapted phenotype determining overall fitness; to look at only one aspect of the phenotype such as coloration in isolation can be misleading (Kingsolver, 1987). Although there is strong evidence for visual selection, color patterns in grasshoppers may have functions other than those related to crypsis, or even color per se, which will affect the response to selection. This could mean that effects associated with color pattern loci might evolve over time and that the visual phenotype is no longer the main one affecting the variation or even that polymorphisms might be maintained in the absence of the selective agents responsible for their origin. Goodhart (1987) has argued that despite all the evidence demonstrating visual selection by predators in the highly polymorphic land snail *Cepaea*, the color polymorphism is maintained by effects other than those related to the color pattern phenotype.

It is clear that further understanding of the role of color pattern polymorphism in grasshoppers will depend on detailed studies of the biology of individuals in the field, where color patterns are viewed in the context of the overall biology of the individuals. In particular, the relationship between color pattern morphs and fitness components, such as development rate and egg production, must be examined. Recent developments in ecological studies on grasshoppers (e.g., Atkinson and Begon, 1987; Wall and Begon, 1986, 1987a,b) indicate the type of ecological information that must be collected if progress is to be made in determining the significance of color pattern variation in natural populations. Although there is considerable evidence that selection does act on genetically determined color pattern variation in grasshoppers, how these polymorphisms are in fact maintained is not known. Until this is discovered our understanding of the role of color pattern variation in grasshoppers will remain incomplete.

ACKNOWLEDGMENTS

I am grateful to Dr. John McKenzie for his comments on an earlier version of the manuscript.

REFERENCES

Albrecht, F. O. 1962. Some physiological and ecological aspects of locust phases. *Trans. R. Entomol. Soc. London* **114**, 335–375.

Albrecht, F. O. 1965. Influence du groupment, de l'état hygrometrique et de la photopériode sur la resistance au jeune de *Locusta migratoria migratorioides* (R. & F.). *Bull. Biol. Fr. Belg.* **99**, 287–339.

Anderson, R. V., C. R. Tracy, and Z. Abramsky. 1979. Habitat selection in two species of short-horned grasshoppers. The role of thermal and hydric stresses. *Oecologia* **38**, 359–374.

Atkinson, D. and M. Begon. 1987. Ecological correlates and heritability of reproductive variation in two co-occurring grasshopper species. *Ecol. Entomol.* **12**, 129–138.

Berry, A. J. and P. G. Willmer. 1986. Temperature and the colour polymorphism of *Philaenus spumarius* (Homoptera: Aphrophoridae). *Ecol. Entomol.* **11**, 251–259.

Bradshaw, A. D. 1965. Evolutionary significance of phenotypic plasticity in plants. *Adv. Genet.* **13**, 115–155.

Brakefield, P. M. and P. G. Willmer. 1985. The basis of thermal melanism in the ladybird *Adalia bipunctata*: Differences in reflectance and thermal properties between the morphs. *Heredity* **54**, 9–14.

Burtt, E. 1951. The ability of adult grasshoppers to change colour on burnt ground. *Proc. R. Entomol. Soc. London, Ser. A.* **26**, 45–48.

Butlin, R. K. and G. M. Hewitt. 1987. Genetic divergence in the *Chorthippus parallelus* species group (Orthoptera: Acrididae). *Biol. J. Linn. Soc.* **31**, 301–310.

Buxton, P. A. 1924. Heat, moisture, and animal life in deserts. *Proc. R. Soc. London, Ser. B* **96**, 123–131.

Byrne, O. W. 1967a. Polymorphism in the Australian Acrididae. I. Inheritance of colour patterns in the plague locust, *Chortoicetes terminifera. Heredity* **22**, 561–568.

Byrne, O. W. 1967b. Polymorphism in the Australian Acrididae. II. Changes in colour pattern gene frequencies in the plague locust, *Chortoicetes terminifera. Heredity* **22**, 569–589.

Chapco, W. 1980a. Inheritance of an unusual color variant in the grasshopper *Melanoplus sanguinipes. Can. J. Genet. Cytol* **22**, 315–318.

Chapco, W. 1980b. Genetics of the migratory grasshopper, *Melanoplus sanguinipes*: Orange stripe and its association with tibia color and red-black genes. *Ann. Entomol. Soc. Am.* **73**, 319–322.

Chapco, W. 1983. The genes T^R, *Ost*, Pro^R and *L* in natural populations of *Melanoplus sanguinipes. Can. J. Genet. Cytol.* **25**, 88–92.

Chapco, W. and M. J. Bidochka. 1986. Genetic variation in prairie populations of *Melanoplus sanguinipes*, the migratory grasshopper. *Heredity* **56**, 397–408.

Clark, E. J. 1943. Colour variation in British Acrididae (Orthopt.) *Entomol. Mon. Mag.* **79**, 91–104.

Clark, W. C. 1976. The environment and the genotype in polymorphism. *J. Linn. Soc. London, Zool.* **58**, 255–262.

Clarke, B. 1975. The contribution of ecological genetics to evolutionary theory: Detecting the direct effects of natural selection on particular polymorphic loci. *Genetics* **79**, 101–113.

Clarke, B. C. 1979. The evolution of genetic diversity. *Proc. R. Soc. London, Ser. B* **205**, 453–474.

Cloudsley-Thompson, J. L. 1981. Comments on the nature of deception. *Biol. J. Linn. Soc.* **16**, 11–14.

Cooper, J. M. 1984. Apostatic selection on prey that match the background. *Biol. J. Linn. Soc.* **23**, 221–228.

Cott, H. B. 1940. *Adaptive Coloration in Animals*. Methuen, London.

Cox, G. W. and D. G. Cox. 1974. Substrate color matching in the grasshopper *Circotettix rabula* (Orthoptera: Acrididae). *Great Basin Nat.* **34**, 60–70.

Daly, J. C., P. Wilkinson, and D. D. Shaw. 1981. Reproductive isolation in relation to allozymic and chromosomal differentiation in the grasshopper *Caledia captiva. Evolution (Lawrence, Kans.)* **35**, 1164–1179.

Dearn, J. M. 1979. Evidence of non-random mating for the colour pattern polymorphism in field populations of the grasshopper *Phaulacridium vittatum. J. Aust. Entomol. Soc.* **18**, 241–243.

Dearn, J. M. 1981. Latitudinal cline in a colour pattern polymorphism in the Australian grasshopper *Phaulacridium vittatum. Heredity* **47**, 111–119.

Dearn, J. M. 1983. Inheritance of the colour pattern polymorphism in the wingless grasshopper *Phaulacridium vittatum* Sjöstedt (Orthoptera: Acrididae). *J. Aust. Entomol. Soc.* **22**, 217–218.

Dearn, J. M. 1984. Colour pattern polymorphism in the grasshopper *Phaulacridium vittatum.* I. Geographic variation in Victoria and evidence of habitat association. *Aust. J. Zool.* **32**, 239–249.

Dearn, J. M. and R. A. H. Davies. 1983. Natural selection and the maintenance of colour pattern polymorphism in the Australian plague locust, *Chortoicetes terminifera. Aust. J. Biol. Sci.* **36**, 387–401.

de Wilde, J. 1975. An endocrine view of metamorphosis, polymorphism and diapause in insects. In E. J. W. Barrington, Ed., *Trends in Comparative Endocrinology*. Thomas J. Griffiths Sons, New York, pp. 13–28.

Digby, P. S. B. 1955. Factors affecting the temperature excess of insects in sunshine. *J. Exp. Biol.* **32**, 279–298.

Dobzhansky, T. 1970. *Genetics of the Evolutionary Process.* Columbia Univ. Press, New York.

Edmunds, M. 1981. On defining 'mimicry.' *Biol. J. Linn. Soc.* **16**, 9–10.

Endler, J. A. 1984. Progressive background matching in moths, and a quantitative measure of crypsis. *Biol. J. Linn. Soc.* **22**, 187–231.

Fisher, R. A. 1939. Selective forces in wild populations of *Paratettix texanus. Ann. Eugen. (London)* **9**, 109–122.

Fogden, M. and P. Fogden. 1974. *Animals and their Colors.* Crown, New York.

Ford, E. B. 1975. *Ecological Genetics*, 4th ed. Chapman & Hall, London.

Fuzeau-Braesch, S. 1972. Pigments and color changes. *Annu. Rev. Entomol.* **17**, 403–424.

Garcia-Dorado, A. 1986. The effect of niche preference on polymorphism protection in a heterogenous environment. *Evolution (Lawrence, Kans.)* **40**, 936–945.

Gill, P. D. 1979. Colour-pattern variation in relation to habitat in the grasshopper *Chorthippus brunneus* (Thunberg). *Ecol. Entomol.* **4**, 249–257.

Gill, P. D. 1981a. Heterozygosity estimates in the grasshopper *Chorthippus brunneus* (Thunberg). *Heredity* **46**, 269–272.

Gill, P. D. 1981b. Enzyme variation in the grasshopper *Chorthippus brunneus* (Thunberg). *Biol. J. Linn. Soc.* **15**, 247–258.

Gill, P. D. 1981c. Allozyme variation in sympatric populations of British grasshoppers — evidence of natural selection. *Biol. J. Linn. Soc.* **16**, 83–91.

Gill, P. D. 1981d. The genetics of colour-patterns in the grasshopper *Chorthippus brunneus. Biol. J. Linn. Soc.* **16**, 243–259.

Gill, P. D. 1981e. Colour patterns and ecology of British grasshoppers. *Acrida* **10**, 145–158.

Gillett, S. D. 1973. The role of integumental colour pattern in locust grouping. *Anim. Behav.* **21**, 153–156.

Gillett, S. D. and E. Gonta. 1978. Locust as prey: Factors affecting their vulnerability to predation. *Anim. Behav.* **26**, 282–289.

Gillett, S. D., P. J. Hogarth, and F. E. J. Noble. 1979. The response of predators to varying densities of *gregaria* locust nymphs. *Anim. Behav.* **27**, 592–596.

Gillis, J. E. 1982. Substrate colour-matching cues in the cryptic grasshopper *Circotettix rabula rabula* (Rehn and Hebard). *Anim. Behav.* **30**, 113–116.

Gillis, J. E. and K. W. Possai. 1983. Thermal niche partitioning in the grasshoppers *Arphia conspersa* and *Trimerotropis suffusa* from a montane habitat in central Colorado. *Ecol. Entomol.* **8**, 155–161.

Goodhart, C. B. 1987. Why are some snails visibly polymorphic, and other not? *Biol. J. Linn. Soc.* **31**, 35–58.

Greenwood, J. J. D. 1984. The functional basis of frequency-dependent food selection. *Biol. J. Linn. Soc.* **23**, 177–199.

Halliday, R. B., S. F. Webb, and G. M. Hewitt. 1984. Genetic and chromosomal polymorphism in hybridizing populations of the grasshopper *Podisma pedestris. Biol. J. Linn. Soc.* **21**, 299–305.

Hamilton, W. J. 1975. Coloration and its thermal consequences for diurnal desert insects. In N. F. Hadley, Ed., *Environmental Physiology of Desert Organisms.* Halsted, Stroudsburg, pp. 67–89.

Harrison, R. G. 1980. Dispersal polymorphisms in insects. *Annu. Rev. Ecol. Syst.* **11**, 95–118.

Hawke, A. D. 1974. Genetic studies of polymorphism in laboratory and natural populations of the Australian plague locust, *Chortoicetes terminifera.* Ph.D. Thesis, Australian National University, Canberra.

Hedrick, P. W. 1986. Genetic polymorphism in heterogenous environments: A decade later. *Annu. Rev. Ecol. Syst.* **17**, 535–566.

Hedrick, P. W., M. E. Ginevan, and E. P. Ewing. 1976. Genetic polymorphism in heterogeneous environments. *Annu. Rev. Ecol. Syst.* **7**, 1–32.

Hedrick, P. W., S. Jain, and L. Holden, 1978. Multilocus systems in evolution. *Evol. Biol.* **11**, 101–182.

Hertz, M. and A. D. Imms. 1937. On the responses of the African migratory locust to different types of background. *Proc. R. Soc. London, Ser. B* **122**, 281–297.

Hill, L. and H. J. Taylor. 1933. Locusts in sunlight. *Nature (London)* **132**, 276.

Ibrahim, M. M. 1974. Environmental effects on colour variation in *Acrida pellucida* Klug. *Z. Angew. Entomol.* **77**, 133–136.

Isely, F. B. 1938. Survival value of acridian protective coloration. *Ecology* **19**, 370–389.

Joern, A. 1981. Importance of behavior and coloration in the control of body temperature by *Brachystola magna* Girard (Orthoptera: Acrididae). *Acrida* **10**, 117–130.

Jones, J. S., B. H. Leith, and P. Rawlings. 1977. Polymorphism in *Cepaea*: A problem with too many solutions? *Annu. Rev. Ecol. Syst.* **8**, 109–143.

Kayser, H. 1985. Pigments. In G. Kerkut and L. I. Gilbert, Eds., *Comprehensive Insect Physiology, Biochemistry and Pharmacology* Vol. 10. Pergamon, Oxford, pp. 367–415.

Kemp, W. P. 1986. Thermoregulation in three rangeland grasshopper species. *Can. Entomol.* **118**, 335–343.

Kennedy, J. S. 1956. Phase transformation in locust biology. *Biol. Rev. Cambridge Philos. Soc.* **31**, 349–370.

Kennedy, J. S. 1961. Continuous polymorphism in locusts. *Symp. R. Entomol. Soc. London* **1**, 80–90.

Key, K. H. L. 1954. *The Taxonomy, Phases, and Distribution of the Genera Chortoicetes Brunn. and Austroicetes Uv. (Orthoptera: Acrididae).* C.S.I.R.O., Canberra.

Key, K. H. L. and M. F. Day. 1954a. A temperature controlled physiological colour response in the grasshopper *Kosciuscola tristis* Sjöst. (Orthoptera: Acrididae). *Aust. J. Zool.* **2**, 309–339.

Key, K. H. L. and M. F. Day. 1954b. The physiological mechanism of colour change in the grasshopper *Kosciuscola tristis* Sjöst. (Orthoptera: Acrididae). *Aust. J. Zool.* **2**, 340–363.

King, R. L. and E. H. Slifer. 1955. The inheritance of red and blue hind tibiae in the lesser migratory grasshopper, *Melanoplus mexicanus mexicanus* (Saussure). *J. Hered.* **46**, 302–304.

Kingsolver, J. G. 1987. Evolution and coadaptation of thermoregulatory behavior and wing pigmentation pattern in pierid butterflies. *Evolution (Lawrence, Kans.)* **41**, 472–490.

Levins, R. 1968. *Evolution in Changing Environments.* Princeton Univ. Press, Princeton, New Jersey.

Lively, C. M. 1986. Canalization versus developmental conversion in a spatially variable environment. *Am. Nat.* **128**, 561–572.

Lloyd, D. G. 1984. Variation strategies of plants in heterogenous environments. *Biol. J. Linn. Soc.* **21**, 357–385.

Magnuson, J. J., L. B. Crowder, and P. A. Meduick. 1979. Temperature as an ecological resource. *Am. Zool.* **19**, 331–343.

May, M. L. 1979. Insect thermoregulation. *Annu. Rev. Entomol.* **24**, 313–349.

Nabours, R. K. 1929. The genetics of the Tettigidae (grouse locusts). *Bibliogr. Genet.* **5**, 27–104.

Nabours, R. K., I. Larson, and N. Hartwig. 1933. Inheritance of color patterns in the grouse locust *Acrydium arenosum* Burmeister (Tettigidae). *Genetics* **18**, 159–171.

Nankivell, R. N. 1974. Interactions between inversion polymorphisms and the colour pattern polymorphism in the grasshopper *Austroicetes interioris* (White and Key). *Acrida* **3**, 93–111.

Needham, A. E. 1978. Insect biochromes: Their chemistry and role. In M. Rockstein. Ed., *Biochemistry of Insects.* Academic Press, New York, pp. 237–305.

Nel, M. D. 1967. Selection of phase types based on hopper coloration in *Locusta migratoria migratorioides* (Reiche & Fairmaire). *S. Afr. J. Agric. Sci.* **10**, 461–470.

Nel, M. D. 1968. Selection at a high humidity for green and brown *solitaria* hopper coloration in *Locustana pardalina* (Walker). *S. Afr. J. Agric. Sci.* **11**, 163–172.

Nijhout, H. F. and D. E. Wheeler. 1982. Juvenile hormone and the physiological basis of insect polymorphisms. *Q. Rev. Biol.* **57**, 109–133.

Okay, S. 1953. Formation of green pigment and colour changes in Orthoptera. *Bull. Entomol. Res.* **44**, 299–315.

Okay, S. 1956. The effect of temperature and humidity on the formation of green pigment in *Acrida bicolor* (Thunb.). *Arch. Int. Physiol. Biochim.* **64**, 80–91.

Otte, D. 1977. Communication in Orthoptera. In T. A. Sebeok, Ed., *How Animals Communicate.* Indiana Univ. Press, Bloomington, pp. 334–361.

Otte, D. and K. Williams. 1972. Environmentally induced color dimorphisms in grasshoppers. *Syrbula admirabilis, Dichromorpha viridis,* and *Chortophaga viridifasciata. Ann. Entomol. Soc. Am.* **65**, 1154–1161.

Pepper, J. H. and E. Hastings. 1952. The effects of solar radiation on grasshopper temperatures and activities. *Ecology* **33**, 96–103.

Petersen, B. and J. E. Treherne. 1949. On the distribution of colour forms in Scandinavian *Omocestus viridulus* L. *Oikos* **1**, 175–183.

Pick, F. E. and A. Lea. 1970. Field observations on spontaneous movements of solitarious hoppers of the brown locust, *Locustana pardalina* (Walker), and behavioural differences between various colour forms. *Phytophylactica* **2**, 203–210.

Poulton, E. B. 1926. Protective resemblance borne by certain African insects to the blackened areas caused by grass fires. *Int. Congr. Entomol., 3rd, 1925,* Vol. 2, pp. 433–451.

Ragge, D. R. 1965. *Grasshoppers, Crickets and Cockroaches of the British Isles.* Warne, London.

Richards, O. W. 1961. An introduction to the study of polymorphism in insects. *Symp. R. Entomol. Soc. London* **1**, 2–10.

Richards, O. W. and Waloff, Z. 1954. Studies on the biology and population dynamics of British grasshoppers. *Anti-Locust Bull.* **17**.

Robinson, M. H. 1981. A stick is a stick and not worth eating: On the definition of mimicry. *Biol. J. Linn. Soc.* **16**, 15–20.

Rowell, C. H. F. 1967. Corpus allatum implantation and green/brown polymorphism in three African grasshoppers. *J. Insect Physiol.* **13**, 1401–1412.

Rowell, C. H. F. 1970. Environmental control of coloration in an acridid, *Gastrimargus africanus* (Saussure). *Anti-Locust Bull.* **47**.

Rowell, C. H. F. 1971. The variable coloration of the acridoid grasshoppers. *Adv. Insect Physiol.* **8**, 145–198.

Rowell, C. H. F. and T. L. Cannis. 1972. Environmental factors affecting the green/brown polymorphism in the cyrtacanthacridine grasshopper *Schistocerca vaga* (Scudder). *Acrida* **1**, 69–71.

Rubtzov, I. A. 1935. Phase variation in non-swarming grasshoppers. *Bull. Entomol. Res.* **26**, 499–574.

Sansome, F. W. and L. La Cour. 1935. The genetics of grasshoppers: *Chorthippus parallelus. J. Genet.* **30**, 415–422.

Schennum, W. E. and R. B. Willey. 1979. A geographical analysis of quantitative morphological variation in the grasshopper *Arphia conspersa. Evolution (Lawrence, Kans.)* **33**, 64–84.

Schlichting, C. D. 1986. The evolution of phenotypic plasticity in plants. *Annu. Rev. Ecol. Syst.* **17**, 667–693.

Schultz, J. C. 1981. Adaptive changes in antipredator behavior of a grasshopper during development. *Evolution (Lawrence, Kans.)* **35**, 175–179.

Shapiro, A. M. 1976. Seasonal polyphenism. *Evol. Biol.* **9**, 259–333.

Smith-Gill, S. J. 1983. Developmental plasticity: Developmental conversion *versus* phenotypic modulation. *Am. Zool.* **23**, 47–55.

Southwood, T. R. E. 1977. Habitat, the templet for ecological strategies? *J. Anim. Ecol.* **46**, 337–365.

Stewart, A. J. A. 1986. Nymphal colour/pattern polymorphism in the leafhoppers *Eupteryx urticae* (F.) and *E. cyclops* Matsumura (Hemiptera: Auchenorrhyncha): spatial and temporal variation in morph frequencies. *Biol. J. Linn. Soc.* **27**, 79–101.

Tauber, M. J., C. A. Tauber, and S. Masaki. 1984. Adaptations to hazardous seasonal conditions: dormancy, migration and polyphenism. In C. B. Huffacker and R. L. Rabb, Eds., *Ecological Entomology*. Wiley, New York, pp. 149–183.

Uvarov, B. P. 1966. *Grasshopper and Locusts*, Vol. 1. Cambridge Univ. Press, London and New York.

Uvarov, B. P. 1977. *Grasshoppers and Locusts*, Vol. 2. Centre for Overseas Pest Research, London.

Vane-Wright, R. I. 1976. A unified classification of mimetic resemblances. *Biol. J. Linn. Soc.* **8**, 25–56.

Vane-Wright, R. L. 1980. On the definition of mimicry. *Biol. J. Linn. Soc.* **13**, 1–6.

Vorontzovsky, P. A. 1928. On the question of homologous series of colour variation in Acrididae (in Russian). *Bull. Oren. Plant Prot. Stn.* **1**, 27–39.

Wall, R. and M. Begon. 1986. Population density, phenotype and mortality in the grasshopper *Chorthippus brunneus. Ecol. Entomol.* **11**, 445–456.

Wall, R. and M. Begon. 1987a. Individual variation and the effects of population density in the grasshopper *Chorthippus brunneus. Oikos* **49**, 15–27.

Wall, R. and M. Begon. 1987b. Population density, phenotype and reproductive output in the grasshopper *Chorthippus brunneus. Ecol. Entomol.* **12**, 331–339.

Wallace B. 1970. *Genetic Load: Its Biological and Conceptual Aspects*. Prentice-Hall, Englewood Cliffs, New Jersey.

Watt, W. B. 1968. Adaptive significance of pigment polymorphisms in *Colias* butterflies. I. Variation in melanin pigment in relation to thermoregulation. *Evolution (Lawrence, Kans.)* **22**, 437–458.

Wigglesworth, V. B. 1961. Insect polymorphism — a tentative synthesis. *Symp. R. Entomol. Soc. London* **1**, 103–113.

Willmer, P. G. 1982. Microclimate and the environmental physiology of insects. *Adv. Insect. Physiol.* **16**, 1–57.

Workman, P. L. 1964. The maintenance of heterozygosity by partial negative assortative mating. *Genetics* **50**, 1369–1382.

Subject Index

This index includes the names of plants and animals other than grasshoppers. **Bold** indicates a figure, table, or major entry.

Species Index

Bold indicates a figure or table.